Calculus

Second Edition

Calculus

Second Edition

Lynn Loomis
Harvard University

Addison-Wesley Publishing Company

Reading, Massachusetts
Menlo Park, California
London · Amsterdam · Don Mills, Ontario · Sydney

Frontispiece courtesy Professor Elmer R. Pearson, Institute of Design, I.I.T.

Chapter opening photographs: Christopher S. Johnson, Chapter 1. Stock, Boston: Franklin Wing, Chapters 2 and 9; Clif Garboden, Chapter 11; Frank Siteman, Chapter 15; John Urban, Chapter 18; and Daniel S. Brody, Chapter 19. DeWys, Inc., New York: David W. Hamilton, Chapters 3, 12, and 16. Bill Finch, Chapters 4 and 5. Manolo Guevara, Jr., Chapter 6. Grant Heilman, Lititz, Pennsylvania, Chapters 7, 8, and 10. James Leisy, Chapter 13. Erik Hanson, Chapter 14. Hedrich-Blessing, Chicago: Chapter 17.

ISBN 0-201-04326-2
ABCDEFGHIJ-HA-7987

Preface

The aims of this edition are those of the first. The book is intended as an intuitive, but not superficial, treatment of the standard calculus sequence. Variable and "y is a function of x" are used freely but carefully; Leibniz notation and function notation receive about equal time.

On place of rigor, there is some emphasis on approximation and computation (though this material can be considered to be optional). For the most part this appears as estimation, directed to the question "How good an answer do I have?" However, Appendix 3 goes on to the natural follow-up, "What must I do to get the accuracy I want?", and in this computational context there is a bit of ε, δ reasoning.

Most of the traditional topics from analytic geometry are covered in Chapter 1 and Appendix 2. Chapter 1 treats lines, circles, translation of axes, and completing the square to simplify quadratic equations, while Appendix 2 develops the conic sections from their standard locus definitions. Some related material involving polar coordinates and parametric equations will be found in Chapter 10. There are also a few clusters of problems in other chapters that are relevant. For example, in Section 14.2 the optical property of the ellipse is obtained directly from its locus definition by vector differentiation.

The material is organized into chapters in such a way as to allow considerable syllabus flexibility. Possible realignments can be read off from the flow chart (on page vi), which shows the major dependencies between chapters. Any reordering of material that is consistent with the chart should involve only minor problems of accommodation. For example, Chapter 4 (The Antiderivative) can be postponed and taken up just before Chapter 8, uniting the integration chapters, while Chapter 15 (Functions of Two Variables) can be taken up immediately after Chapter 6, should an early introduction to this topic be desired. Similarly, Chapter 19 on differential equations presupposes only Chapter 11 (and could come even earlier if series solutions were omitted).

Answers (or hints) for most of the odd-numbered problems are given at the back of the book.

Major changes from the first edition are as follows:

1. The expositions of the definite integral, inverse functions, and vectors have been completely rewritten.

2. Several subject areas have been reorganized, to make the level of difficulty more nearly monotone and to improve cohesiveness. For example, the material on curve sketching and estimation in the old Chapter 3 has been moved into Chapter 6, which now contains all the standard applica-

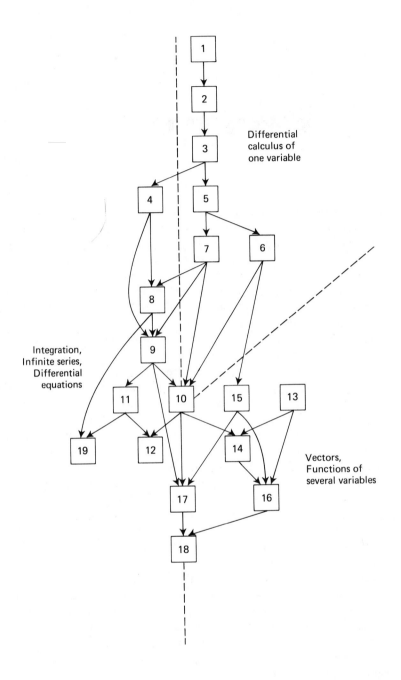

Differential
calculus of
one variable

Integration,
Infinite series,
Differential
equations

Vectors,
Functions of
several variables

tions of the derivative. The applications to estimation have been shifted to the end of the chapter and can be viewed as optional topics, all or in part. The multidimensional material in old Chapters 14 through 17 has been reorganized along algebra vs. calculus lines instead of the earlier two vs. three dimensions. The introduction to polar coordinates is now in the chapter on parametric equations (Chapter 10) along with the polar-coordinate sections on area, arc length and curvature.

3. The antiderivative chapter has been shortened, simplified, and relocated (as Chapter 4), to provide an early introduction to integration should that be desired. The definite integral notation $\int_a^b f$ is no longer used in this chapter.

4. There is a new short chapter on Green's theorem in the plane.

5. Other new material includes a section on the intermediate-value principle in Chapter 2, some theoretical sections in Appendix 4, and expanded discussions at several key points. For example, the treatment of the exponential function in Chapter 5 has been fleshed out, as have been the discussions of graphs in 3-space (Chapter 15) and the vector product (Chapters 13 and 14).

6. The problem sets have been reorganized and extended. The general chapter scheme now is to have 10 to 25 more or less routine problems at the end of each section and to gather together the more difficult problems along with further routine problems in a single collection at the end of the chapter.

7. Old Chapter 20 has been dropped, although some of its material has been repositioned in Appendixes 3 and 4. Also dropped were occasional sections here and there that were judged to be expendable.

I would like to express my sincere appreciation and gratitude to Nancy Larson for her expert typing and to my colleagues at Addison-Wesley for their constant help and encouragement.

January 1977   L. H. L.

Contents

Introduction

NEWTON, LEIBNIZ, AND THE CALCULUS

The invention of calculus is attributed to two geniuses of the 17th century, Isaac Newton in England, and, independently, Gottfried Leibniz in Germany. Earlier mathematicians had uncovered bits and pieces of the subject. Newton and Leibniz discovered its *pattern*, and thereby created an algorithmic discipline of enormous power, applicable to all sorts of fundamental questions about the nature of the world.

Although the new calculus obviously worked, Newton and Leibniz did not have a clear idea of *why* it worked. They tried to explain its successes by geometric reasoning, since at that time all mathematical phenomena were viewed in terms of geometry, but their explanations were unsatisfactory. In fact, the logical foundations of calculus remained a mystery for another century and a half. Some fragmentary progress occurred, and a new point of view gradually formed, based on numbers, variables, and functional relationships between variables. Then, around 1820, the French mathematician, Augustin Cauchy, settled the matter by showing that calculus rests on the properties of the limit operation. This was still not what we today call rigor, and it took another fifty years of deeper probing to reach the bedrock of ϵ, δ reasoning and the completeness of the real-number system.

The chronological development of calculus was thus marked at several points by leaps in precision and sophistication. Now, three hundred years after Newton and Leibniz, we can start our study of the subject at practically any level we wish. Since there seems to be little point in repeating the confusions of the first one hundred fifty years, we shall approach calculus at about the level of Cauchy, which is still very intuitive. We can then increase our precision in a natural manner as the subject unfolds.

1
Graphs

Calculus is about functions, and it is important when approaching calculus to have a reasonably good understanding of what a function is and what its graph can be like. In this preliminary chapter we shall go over the beginnings of analytic geometry, and thus review the graphs of simple equations. Then in Chapter 2 we shall take up the fundamental notion of a *function*.

1. COORDINATES

If a unit of distance is given in the plane or in space, then the distance between any two points is determined. It is the length of the line segment joining the two points, in terms of the given unit. Sometimes it is convenient to use different units for different measurements. For example, we might measure a pipeline and find that it is fifty miles long and one foot in diameter. The possibility of making such measurements of length leads to a systematic scheme for representing points by numbers, called a coordinate system.

The fundamental step is putting coordinates on a geometric line l, and thus making l into a "number line," as pictured below.

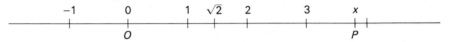

In order to do this, we make l into a *directed line* by choosing one of the two directions along l as the *positive direction*. Then we choose a *unit of distance* (generally by taking some particular line segment as the unit of length). On the line shown below, we have taken the positive direction to the right and have marked off a few points one unit of distance apart.

DEFINITION   *The signed distance from P to Q is the distance between P and Q if the direction from P to Q is positive. It is the negative of the distance between P and Q if the direction from P to Q is negative.*

In the figure above, the signed distance from A to B is 2, and the signed distance from C to B is -3. From C to A it is -5, and from B to D it is 4.

We shall represent the signed distance from P to Q by $\overline{PQ}$. Note then that $\overline{QP} = -\overline{PQ}$. If you try various combinations of three points in the figure above you will see that the following **Addition Law** is true.

ADDITION   *If P, Q, and R are any three points on a directed line, then*
LAW
$$\overline{PQ} + \overline{QR} = \overline{PR}.$$

For example,

$$\overline{BC} + \overline{CA} = 3 + (-5) = -2 = \overline{BA},$$
$$\overline{AD} + \overline{DC} = 6 + (-1) = 5 = \overline{AC},$$
$$\overline{AE} + \overline{BC} = 2 + 3 = 5 = \overline{AC}, \text{ etc.}$$

Ordinary (unsigned) distances satisfy this identity only when Q is between P and R.

Finally we choose an *origin* point O on l and make the following definition:

DEFINITION *The coordinate of any point P on l is the signed distance from O to P.*

In this way we assign to each point P on l a uniquely determined numerical coordinate. Conversely, if we are given any real number x, then we obtain a unique point P having x as its coordinate by measuring off the signed distance x from the origin O. Thus, each point P determines a unique number x, and each number x determines a unique point P. This is what we mean when we call the coordinate assignment a *one-to-one correspondence* between the points of l and the real numbers.

Because of the coordinate correspondence, we can think of the real number system $\mathbb{R}$ in its entirety as though it were a geometric line, and we often refer to numbers like 3, π, and $-\sqrt{2}$ as being "points" on the "number line."

Coordinates are signed distances from O. Other signed distances are determined from coordinates as follows:

THEOREM 1 *The signed distance from the point x to the point t is $t - x$.*

Proof. This is a special case of the Addition Law. Suppose that x is the coordinate of P and that t is the coordinate of Q. That is, $\overline{OP} = x$ and $\overline{OQ} = t$. The identity

$$\overline{OP} + \overline{PQ} = \overline{OQ}$$

then becomes

$$x + \overline{PQ} = t,$$

or

$$\overline{PQ} = t - x. \quad \blacksquare$$

COROLLARY *If a new origin O' is chosen at the point with coordinate a, then the point having the old coordinate t has the new coordinate $t - a$.*

Proof. This is just a restatement of the theorem, in view of the definition of coordinates. See the figure below.

The coordinate plane involves two number lines. We choose a pair of perpendicular lines intersecting in a point O, and on each line set up a coordinate system with O as the zero point. We call O the *origin of coordinates*, or simply the *origin*; the intersecting number lines are the *coordinate axes*. We choose one of the axes as the "horizontal" axis, and picture it horizontal, with the *positive direction to the right*. The positive direction of the vertical axis is pictured *upward*.

Given a coordinate system as described above, then each point in the plane determines, and is determined by, an ordered pair of coordinate numbers, as indicated below.

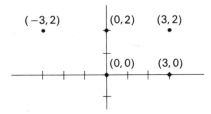

We shall feel free to call (3, 2) a point in the coordinate plane, just as we call 3 a point on the number line.

Notice that the point (0, 2) is different from the point (2, 0). It is crucial that the coordinates form an *ordered* pair. The *first* coordinate is obtained by dropping a perpendicular to the *horizontal* axis, and is the coordinate on the horizontal axis of the foot of this perpendicular. The second coordinate is found similarly on the vertical axis.

The axes divide the plane into four parts called *quadrants*, which are conventionally numbered in the counterclockwise direction, starting from the quadrant that is above the positive half of the horizontal axis.

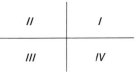

PROBLEMS FOR SECTION 1

For each of the following problems, mark off equally spaced points on a directed line, and then label three of them A, B, and C, in such a way that the addition law $\overline{AB} + \overline{BC} = \overline{AC}$ illustrates the given numerical identity.

1. $3 - 2 = 1$ 2. $-1 + 4 = 3$
3. $1 - 4 = -3$ 4. $-1 - 2 = -3$
5. $5 - 3 = 2$ 6. $5 - 6 = -1$

Mark off the letters A through J, in order and equally spaced, on a line l. Suppose that the positive direction is from A to J and that the letters are a unit of distance apart. Verify each of the following instances of the addition law by determining the three signed distances involved.

7. $\overline{AG} + \overline{GB} = \overline{AB}$ 8. $\overline{FB} + \overline{BJ} = \overline{FJ}$
9. $\overline{IG} + \overline{GD} = \overline{ID}$ 10. $\overline{HC} + \overline{CF} = \overline{HF}$
11. $\overline{DF} + \overline{FC} = \overline{DC}$ 12. $\overline{AJ} + \overline{JE} = \overline{AE}$
13. $\overline{DG} + \overline{GA} = \overline{DA}$

14. Again let the points A through J be in order and equally spaced on the line l, but now suppose that $\overline{AC} = -3$. What does this say about the direction of l and the spacing of the letters? Work out Problems 7, 8, and 9 in this new situation.

For each of the following problems, mark two points P and Q on a line and then determine where the origin of coordinates is if P and Q have, respectively, the given coordinates.

15. (-1), (1) 16. (3), (4)

17. (2), (-3) 18. (-1), (-2)

19. Prove that a coordinate system on a line is entirely determined if we know the coordinates of two points.

20. Suppose that $a < b$ and consider the segment ab on the number line. Show that its midpoint is $(a + b)/2$. [*Hint:* if m is the midpoint, then the signed distance from a to m should equal the signed distance from m to b.]

21. Find the point (number) that is two-thirds of the way from a to b. We say that this point divides the segment ab in the ratio two to one.

22. Find the point (number) that divides the segment ab in the ratio r to s, where r and s are any positive numbers.

23. Draw a plane coordinate system and plot the points $(-2, 1)$, $(2, -1)$, $(-1, 0)$, $(4, 1)$, $(4, -1)$.

24. Draw the collection of all points whose first coordinate is 2 (in a plane coordinate system). Describe this collection geometrically.

25. Draw and describe the collection whose second coordinate is -3.

26. Draw the collection of all points (x, y) for which $y/x = 2$. Is this the same as the collection for which $y = 2x$?

27. Draw the collection of all points (x, y) for which $y = -x$.

28. a) Given the point $P = (3, 4)$, find a point $Q_1 = (x, y)$ such that the perpendicular bisector of the segment PQ_1 is the y-axis.

    b) Now find Q_2 such that the perpendicular bisector of PQ_2 is the x-axis.

    c) Finally, find Q_3 such that the midpoint of the segment PQ_3 is the origin.

    d) What kind of geometric figure is the quadrilateral $PQ_1Q_3Q_2$?

29. Answer the same four questions for the point $P = (-2, 1)$.

30. The same for $P = (-1, -3)$.

2. GRAPHS OF EQUATIONS

In order to graph an equation in two variables, say

$$3x + 4y = 6,$$

it is necessary first to label the axes. That is, we decide which variable is going to have its values measured along the horizontal axis. If the variables are x and y, it is conventional to take the horizontal axis as the x-axis. In fact, the horizontal axis is sometimes regarded as being permanently labelled x. But how, then, do we graph the equation

$$3u + 4v = 6?$$

Anyway, with the axes labelled—and we shall generally use the standard x, y labelling—we can proceed.

A *solution* of an equation in two variables x and y is a pair of values (x, y) that "satisfies" the equation, i.e., that makes it true. For example, $(x, y) = (3, 2)$ is a solution of

$$x^2 + 4y^2 = 25,$$

because it is true that $3^2 + 4(2^2) = 25$. Another solution is $(5, 0)$. The pair $(x, y) = (2, 3)$ is not a solution because

$$2^2 + 4(3^2) = 4 + 36$$
$$= 40 \neq 25.$$

The collection of all solution pairs is called the *solution set* of the equation.

The fact that the ordered pair $(3, 2)$ *is* a solution while $(2, 3)$ is *not* a solution depends on our stipulation that the first number in an ordered pair is a value of x and the second number a value of y. This goes back to the basic convention that the first coordinate of a point in the plane is the one measured horizontally, and the assumption that the horizontal axis is the x-axis.

The *graph* of an equation is just the geometric representation of its solution set. That is, the graph of an equation is the curve or other configuration in the plane consisting of the points whose coordinates satisfy the equation. The graph of the above equation looks like this:

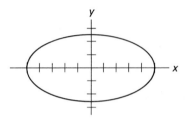

Two equations having the same solution set (i.e., the same graph) are said to be *equivalent*. Here are some other equations equivalent to the equation $x^2 + 4y^2 = 25$:

$$\frac{x^2}{25} + \frac{y^2}{(25/4)} = 1, \qquad \sqrt{x^2 + 4y^2} = 5, \qquad x^2 + 4y^2 - 25 = 0.$$

On the other hand, the two equations

$$x = y \text{ and } x^2 = y^2$$

are *not* equivalent. Why?

If the curve C is the graph of the equation E, then we call E *an* equation of C, because, as we saw above, *other* equations can have the same graph. However, if E has been singled out from various equivalent equations because it has a *standard form*, then we often call it *the* equation of C. For example,

$$\frac{x^2}{a^2} + \frac{y^2}{b^2} = 1$$

is called *the* equation of an ellipse with center at the origin, meaning that this is the standard form for the equation. Sometimes there is more than one standard form. For example, the equations

$$4x - 2y - 8 = 0, \qquad y = 2x - 4, \qquad \frac{x}{2} - \frac{y}{4} = 1$$

are equivalent, and are all standard forms for the equation of a certain straight line.

The *intercepts* of a graph are the points where the graph intersects the axes. An intercept on the x-axis is a point of the form $(a, 0)$, but since the second coordinate of an x-intercept is automatically zero, it is customary to say that the x-intercept is a. Any intercept $(a, 0)$ of

$$x^2 + 4y^2 = 25$$

must satisfy the equation

$$a^2 + 4(0^2) = 25,$$

giving $a^2 = 25$ and $a = \pm 5$. Thus we obtain the x-intercepts of an equation by setting $y = 0$ and solving for x. Similarly, we get the y-intercepts by setting $x = 0$ and solving for y. The above equation has the y-intercepts $b = \pm 5/2$.

Intercepts can be important "positioning" aids. For example, the equation $3x - 4y - 5 = 0$ has the x-intercept $a = 5/3$ and the y-intercept $b = -5/4$, and once we know that the graph of this equation is a straight line, we can just draw it through these two points.

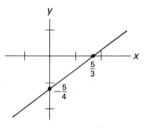

The points whose coordinates satisfy two equations simultaneously, say

$$y = x^2 \qquad \text{and} \qquad y = 2x - 1,$$

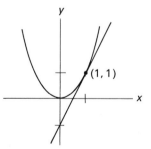

are the points lying on both graphs, i.e., the points of intersection of the graphs. We find these points by the usual techniques for solving simultaneous equations. Here we can eliminate y simply by subtracting the second equation from the first, getting

$$0 = x^2 - 2x + 1 = (x - 1)^2.$$

This gives the solution $x = 1$. The corresponding y value is then obtained from either of the original equations, say $y = x^2 = (1)^2 = 1$, so $(1, 1)$ is the only point common to the two curves.

But now suppose that we want to consider how the two graphs are related to each other over all. Suppose, for instance, that we want to compare the points on the two graphs that have the same x-coordinates. To do this we have to use two different y variables, such as in

$$y = x^2 \quad \text{and} \quad \bar{y} = 2x - 1,$$

since the two equations will generally determine different y values for a given value of x. Then we can conclude from

$$y - \bar{y} = x^2 - (2x - 1) = (x - 1)^2 \geq 0$$

that the first graph lies everywhere above the second graph, with only the point $(1, 1)$ in common.

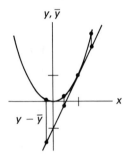

PROBLEMS FOR SECTION 2

The graph of an equation of the form

$$ax + by + c = 0$$

is always a straight line, provided that a and b are not both zero. The graph of

$$x^2 + y^2 + Ax + By + C = 0$$

is always a circle (or nothing). Assuming these facts, sketch the graphs of the equations in Problem 1 through 9 from their intercepts.

1. $x + y = 1$ 2. $x - y = 2$ 3. $2x + y = 2$
4. $3p + 2q = 6$ (with the p-axis horizontal)
5. $3p + 2q = 6$ (q-axis horizontal)
6. $x^2 + y^2 = 2$
7. $x^2 + y^2 + 2x - 2y = 0$
8. $x^2 + y^2 - 4x = 0$ (assume the center is on the x-axis)
9. $x^2 + y^2 + 2x - 3 = 0$
10. a) Draw the graphs of Problems 1 and 2 above on the same coordinate system, and estimate their point of intersection.
    b) Compute the point of intersection by solving the equations simultaneously.
11. Same for (1) and (3).   12. Same for (2) and (3).
13. Same for (2) and (6).   14. Same for (1) and (6).
15. Same for (8) and (9).   16. Same for (7) and (8).
17. Graph the circle $x^2 + y^2 = 4$ and the line $x + y = 3$. Do these graphs appear to intersect? Show by algebra that they do not.
18. Graph the equation $y = x^2$. Here we don't know the shape of the graph, and the only way we can start is to plot a few points on the graph. Make up a little table of values of $y = x^2$, when x has the values -2, $-\frac{3}{2}$, -1, $-\frac{1}{2}$, 0, $\frac{1}{2}$, 1, $\frac{3}{2}$, 2. Then plot these points (x, y) and draw a smooth curve through them.

19. Graph the equation $y = x^2 - 2$, proceeding as in the above problem.

20. Graph the equation $y = x^2 - 2x - 2$, using the x-values of Problem 18.

21. Graph the equation $y = 2 - x^2$.

22. Graph the equation $y^2 = x + 2$. (The easiest way to do this is to give y some simple values and determine the corresponding x values from the equation.)

23. Graph the equation $y^2 = -x$.

24. a) Graph the equations $y = x^2$ and $y = x + 2$ on the same axes. Estimate their points of intersection.

    b) Compute the points of intersection algebraically by solving the equations simultaneously.

25. The graph of the equation $x^2 + 4y^2 = 4$ is a smooth oval-shaped curve called an *ellipse*. It will be longest in the direction of the x-axis and shortest in the direction of the y-axis. Use this information to sketch the graph, starting with only its intercepts.

26. The graph of $9x^2 + 4y^2 = 36$ is also an ellipse, but this time it is longest in the y direction. Plot its intercepts and sketch and graph.

27. Graph $y = x^2/2$, using integer values of x from $x = -3$ to $x = 3$.

28. Graph $y = x^2/4$, using integer values of x from $x = -5$ to $x = 5$.

3. THE STRAIGHT LINE

Horizontal and vertical lines have the simplest equations. Consider the horizontal line 3 units above the x-axis. A point (x, y) lies on this line if and only if its y-coordinate is 3, so this line is the graph of the equation $y = 3$.

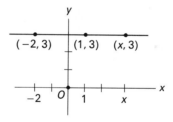

In general, the graph of

$$y = b$$

is the horizontal line with y-intercept b. If b is negative, say $b = -4$, then the line lies 4 units *below* the x-axis.

   Similarly,

$$x = a$$

is the equation of the vertical line with x-intercept a.

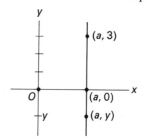

Note that although x is missing from an equation like $y = 3$, we can think of it as being there with a zero coefficient:

$$0x + y = 3.$$

Consider next a nonvertical straight line l through the origin. Suppose, for the sake of definiteness, that l contains the point $(1, 2)$. If a point (x, y) lies on l in the first quadrant, then the similar right triangles shown at the left in the next figure have proportional corresponding sides, so

$$\frac{y}{x} = \frac{2}{1}.$$

If (x, y) lies on l in the third quadrant, then the triangle side lengths are $-x$ and $-y$, so then $(-y)/(-x) = 2/1$. Each of these equations is equivalent to

$$y = 2x,$$

which thus holds for the coordinates of any point (x, y) on l. On the other hand, the reverse of the above line of reasoning shows that if $y = 2x$, then (x, y) lies on l. Therefore, l is the graph of the equation $y = 2x$.

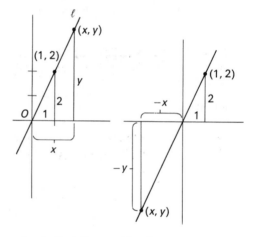

Similarly, we can check that the graph of

$$y = -2x$$

is the straight line through the origin and $(1, -2)$. Here the two ways of rewriting the equation as statements about similar triangles are:

$$\frac{-y}{x} = \frac{2}{1} \quad \text{and} \quad \frac{y}{-x} = \frac{2}{1}.$$

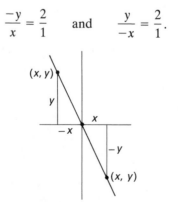

If we use an arbitrary constant m instead of the constants 2 and -2, the above reasoning shows that:

The graph of the equation

$$y = mx$$

is the straight line through the origin and the point $(1, m)$.

A different value of m gives a different line, and every nonvertical line through the origin is obtained in this way, because each such line intersects the vertical line $x = 1$ in a point $(1, m)$. (Strictly speaking, this geometric reasoning applies only when $m \neq 0$, so that the figure involves genuine triangles. But when $m = 0$, the equation $y = mx$ reduces to $y = 0$, which we already know to be the equation of the x-axis. Since the x-axis is the line through the origin and $(1, 0)$, our general conclusion remains correct.)

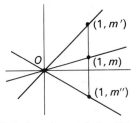

The constant m is called the *slope* of the line. Notice that $m > 0$ for a line *rising* to the right and that $m < 0$ for a line *falling* to the right. A line with large positive slope rises steeply and a line with small positive slope rises gently.

Now consider how the graph of

$$y = 2x - 3$$

is related to the straight line $y = 2x$. In order to consider these two graphs simultaneously, we use a different y variable for the second equation, say

$$\bar{y} = 2x.$$

Then for each value of x we can compare the points (x, y) and $(x, \bar{y})$ on the two graphs, as we did at the end of Section 2. Subtracting the second equation from the first gives

$$y - \bar{y} = -3.$$

That is, for each x the point (x, y) on the graph of $y = 2x - 3$ lies three vertical units below the point $(x, \bar{y})$ on the line $\bar{y} = 2x$. The points (x, y)

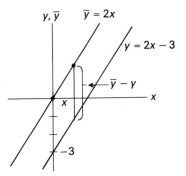

satisfying

$$y = 2x - 3$$

thus make up the parallel line lying three vertical units below the line $\bar{y} = 2x$, i.e., the parallel line that intersects the y-axis in the point $(0, -3)$.

If we carry through this argument using b and m instead of -3 and 2, we have the following general conclusion.

THEOREM 2   *The graph of the equation*

$$y = mx + b$$

is the straight line that is parallel to the line $y = mx$ and has y-intercept b.

If l is any nonvertical line, then l is parallel to a unique nonvertical line through the origin, and l has a unique y-intercept b. Thus each nonvertical line l has a unique equation of the above form. As before, m is called the *slope* of l.

Also, since two lines are parallel to each other if and only if they are parallel to the same line through the origin, we have:

COROLLARY   *Two nonvertical lines are parallel if and only if they have the same slope.*

In summary: the slope m gives the direction of the line l, and the y-intercept b gives the point where l crosses the y-axis. The two numbers m and b together determine l and are determined by l.

EXAMPLE   Identify the graph of the equation $2x + 3y - 6 = 0$.

Solution. Solving for y we obtain the equivalent equation

$$y = -\frac{2}{3}x + 2.$$

This is of the form

$$y = mx + b,$$

with $m = -2/3$ and $b = 2$. The graph is therefore a straight line, with slope $-2/3$ and y-intercept 2.

The equation $y = mx + b$ is called the *slope-intercept* form of the equation for l.

A line is determined if we know its slope m and one point (x_1, y_1) on it, and we can calculate its equation from this data. For example, a line with slope 2 has the equation

$$y = 2x + b,$$

with b still to be determined. If we know that $(2, 1)$ is on the line then

$$1 = 2 \cdot 2 + b,$$

so $b = -3$, and the equation of the line is

$$y = 2x - 3.$$

When this calculation is repeated in general terms, it yields a new standard form for the equation of a line. Thus, if (x_1, y_1) is on the line

$$y = mx + b,$$

then it is true that

$$y_1 = mx_1 + b,$$

and if we subtract the second equation from the first then the b's cancel and we get the *point–slope* form of the equation:

$$y - y_1 = m(x - x_1).$$

We can write this form down directly whenever we know the slope of a line and the coordinates of a point on it. For example, the line through $(2, 1)$ with slope 2 has the point–slope equation

$$y - 1 = 2(x - 2).$$

Solving this for y, we get back to the slope–intercept form

$$y = 2x - 4 + 1 = 2x - 3.$$

The earlier qualitative remarks about the significance of the slope m can be sharpened to an important quantitative fact:

The slope m is exactly the change in y divided by the change in x between any two points on the line.

That is,

THEOREM 3   *If (x_1, y_1) and (x_2, y_2) are any two distinct points on a line with slope m, then*

$$m = \frac{y_2 - y_1}{x_2 - x_1}.$$

Proof. Since each pair of numbers satisfies the equation $y = mx + b$, we have

$$y_2 = mx_2 + b, \qquad y_1 = mx_1 + b;$$

and so, subtracting,

$$y_2 - y_1 = m(x_2 - x_1).$$

Since the line is not vertical, the two x-coordinates must be different and we can divide by the nonzero number $x_2 - x_1$, getting

$$m = \frac{y_2 - y_1}{x_2 - x_1}. \quad \blacksquare$$

We can also obtain this formula by starting from the equation of the line in point–slope form.

The slope formula can be interpreted as saying that the slope m is the constant rate of change of y with respect to x along the line. If $m = 3$, then y increases 3 units per unit increase in x, and if $m = -\frac{1}{2}$, then y *decreases* one-half unit per unit *increase* in x.

Two distinct points (x_1, y_1) and (x_2, y_2) determine a line, and if the line is not vertical then the above formula gives its slope m. Thus the line through $(1, 2)$ and $(3, -1)$ has slope

$$m = \frac{y_2 - y_1}{x_2 - x_1} = \frac{(-1) - 2}{(3 - 1)} = -\frac{3}{2}.$$

For the purposes of geometry it is useful to have a single form for the equation of a line that covers all lines, vertical or nonvertical. The equation

$$Ax + By + C = 0$$

does this (it being assumed that at least one of the coefficients A and B is not zero).

In order to see that the graph of the above equation is always a line, we consider two cases. If $B \neq 0$ we can solve the equation for y and get an equivalent equation of the form $y = mx + b$. In this case the graph is a nonvertical line. For example, the equation

$$x + 2y - 6 = 0$$

can be solved for y, turning into the equivalent equation

$$y = \left(-\frac{1}{2}\right)x + 3.$$

Its graph is thus the nonvertical line with slope $-1/2$ and y-intercept 3.

If B is 0, then A cannot be 0 and we can solve for x, getting an equation of the form $x = a$. In this case the graph is the vertical line with x-intercept a. Thus $3x + 4 = 0$ becomes $x = -\frac{4}{3}$, the equation of the vertical line $\frac{4}{3}$ units to the left of the y-axis.

Conversely, the standard forms

$$y = mx + b \qquad \text{and} \qquad x = a$$

can be rewritten

$$mx - y + b = 0 \qquad \text{and} \qquad x - a = 0,$$

respectively, both of which are of the form $Ax + By + C = 0$.

PROBLEMS FOR SECTION 3

1. Write the equation of the line with y-intercept -6 and slope 3. Find its x-intercept. Find the point where it intersects the vertical line $x = 1$. Same for the horizontal line $y = 1$.

Find the slope m and the y-intercept b of each of the following lines.

2. $x + y = 0$ 3. $x + 2y = 4$

4. $3x - 4y = 4$ 5. $2x + 3y - 6 = 0$

6. $2y - 3x - 6 = 0$ 7. $x - y + 1 = 0$

8. Show that the lines $4x + 2y = 1$ and $y = -2x + 3$ are parallel:

    a) by drawing both graphs;

    b) by proving algebraically that there is no point of intersection.

Draw the graphs of the following equations.

9. $x + y + 1 = 0$ 10. $y + 2x = 0$

11. $3y = x$ 12. $y = x - 1$

13. $2x - 3y = 6$ 14. $x + y = 1$

15. $x + 2y = 3$

16. Determine the slope and y-intercept for the line whose equation is $Ax + By + C = 0$ where A, B, and C are arbitrary constants, and $B \neq 0$.

Determine the slope of the line through the given pair of points, and then write the equation of the line.

17. $(0, 0)$, $(2, 1)$ 18. $(2, -1)$, $(-4, 0)$

19. $(-2, 1)$, $(-2, 4)$ 20. $(3, 1)$, $(-2, 6)$

21. $(2, 1)$, $(-3, 1)$ 22. $(1, 2)$, $(-2, -1)$

Find the (equation of the) line:

23. Through $(1, 2)$, with slope $-(1/2)$.
24. Through the points $(1, 2)$ and $(2, -1)$.
25. Through the points $(-2, -1)$ and $(1, 1)$.
26. Through the points $(1, 2)$ and $(1, 4)$.
27. Through $(2, 3)$, with x-intercept -1.

Find the point of intersection of the two lines:

28. $x + y = 1$, and $y = x$.   29. $x + y = 1$, and $x + 2y = 4$.
30. $3x + y = 2$, and $x = -2$.
31. Find the line through the point $(-1, 2)$ parallel to the line through $(1, 1)$ and $(0, 2)$.
32. Which of the following pairs of lines are parallel?

     a) $2x + y - 2 = 0$, $4x + 2y + 18 = 0$
     b) $x + 2y - 3 = 0$, $2x - y + 3 = 0$
     c) $x + y - 1 = 0$, $y - x + 3 = 0$
     d) $x + 3y + 2 = 0$, $x + 3y - 2 = 0$
     e) $y + x = 1$, $2x + 2y + 5 = 0$

33. Determine the equation of the line passing through the point $(2, 1)$ and parallel to the line through points $(4, 2)$ and $(-2, 3)$.

34. Determine the equation of the line passing through the point $(-2, 5)$ and parallel to the line $2x + y + 6 = 0$.

4. DISTANCE IN THE COORDINATE PLANE

We regard the numbers x and $-x$ as having the same *magnitude*. They are at the same distance from 0 on the number line, and this distance is x or $-x$, whichever one is positive. We call this distance the *absolute value* of x, and designate it $|x|$. Thus,

$$|x| = |-x|,$$

$$|x| = \begin{cases} x & \text{if } x \text{ is positive or zero} \\ -x & \text{if } -x \text{ is positive, i.e., if } x \text{ is negative.} \end{cases}$$

It follows that

$$|x| = \text{the larger of } x \text{ and } -x.$$

Also,

$$|x| = \sqrt{x^2},$$

since by definition $\sqrt{a}$ is the positive square root of a. Theorem 1 says, in particular, that:

THEOREM 4 *The distance between any two numbers x and t, considered as points on the number line, is*

$$|x - t|.$$

EXAMPLE   Graph the equation $y = |x|$.

Solution. By the definition of absolute value, the equation $y = |x|$ says that

$$y = x \qquad \text{if } x \geq 0,$$
$$y = -x \qquad \text{if } x \leq 0.$$

Thus when $x \geq 0$ the graph runs along the straight line $y = x$, and when $x \leq 0$ it runs along the line $y = -x$, as shown below.

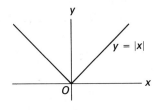

The laws relating absolute value to addition and multiplication are:

$$|x + y| \leq |x| + |y|,$$
$$|xy| = |x| \cdot |y|.$$

These properties of $|x|$ are important for numerical computations and for the rigorous treatment of the foundations of calculus. They will be used off and on throughout the book.

We now turn to the coordinate plane. Up until now we have not needed the same units of distance on the two axes, and it is frequently convenient to use different units, especially when we are graphing the relationship between a pair of unlike quantities. For example, if we wish to show graphically how the average daily temperature varies through the months of the year we would probably use scales somewhat as in the figure below.

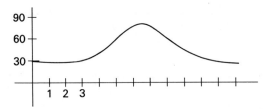

However, in order to obtain a simple formula for distance in the plane, the same unit of distance must be used on the two axes, and from now on a common unit will generally be assumed. Such a coordinate system is said to be *Cartesian.*

The Pythagorean theorem says that a triangle is a right triangle if and only if the square (of the length) of one side is equal to the sum of the squares (of the lengths) of the other two sides. Thus the triangle in the figure below is a right triangle with hypotenuse length h if and only if

$$h^2 = a^2 + b^2.$$

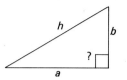

This numerical statement of the theorem depends on a unit of length common to the whole plane, and is the basic reason for using a Cartesian coordinate system. When we combine the Pythagorean theorem with Theorem 4, as shown in the figure below, we obtain a formula for the distance d between any two points (x_1, y_1) and (x_2, y_2) in the plane:

$$a = |x_2 - x_1|$$

$$b = |y_2 - y_1|$$

$$d^2 = a^2 + b^2$$

$$= (x_2 - x_1)^2 + (y_2 - y_1)^2.$$

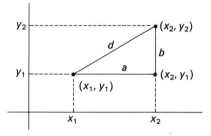

THEOREM 5   *The distance between any two points (x_1, y_1) and (x_2, y_2) in the Cartesian plane is given by the formula*

$$d = \sqrt{(x_1 - x_2)^2 + (y_1 - y_2)^2}.$$

Strictly speaking, the Pythagorean theorem applies only when there is a genuine right triangle. However, the distance formula is correct in all cases. For example, if the two points lie on a horizontal line, then $y_1 - y_2 = 0$ and the formula reduces to

$$d = \sqrt{(x_1 - x_2)^2} = |x_1 - x_2|,$$

which is the correct one-dimensional distance formula.

The slope condition for perpendicular lines follows from these Pythagorean considerations.

THEOREM 6   *Two nonvertical lines are perpendicular if and only if their slopes m and n satisfy*

$$mn = -1.$$

Proof. Consider first the special case of two nonvertical perpendicular lines through the origin, with equations $y = mx$ and $y = nx$. These lines intersect the vertical line $x = 1$ in the points $(1, m)$ and $(1, n)$, respectively, and will be perpendicular if and only if the triangle thus formed is a right triangle. By the Pythagorean theorem, this will be the case if and only if

$$c^2 = a^2 + b^2,$$

or

$$(m - n)^2 = (1 + m^2) + (1 + n^2).$$

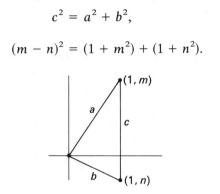

If we expand the square on the left, cancel the terms m^2 and n^2 and finally divide by -2, we end up with

$$mn = -1.$$

Conversely, if this equation holds, then we can work backward through the above steps and conclude that $c^2 = a^2 + b^2$ so that the a, b, c triangle is a right triangle. This proves the theorem for two lines through the origin. In the general case we replace two given lines by the two parallel lines through the origin. Since this replacement changes neither the slopes nor the angle between the lines, the general result follows from the special case. ∎

PROBLEMS FOR SECTION 4

Determine the distance between each of the following pairs of points:

1. $(2, 1)$, $(3, 3)$ 2. $(-1, 2)$, $(2, -3)$
3. $(0, 0)$, $(0, 4)$ 4. $(2, -5)$, $(2, 1)$
5. $(8, 3)$, $(-2, 1)$ 6. $(3, 4)$, $(-6, -1)$
7. $(0, 1)$, $(-2, 0)$ 8. $(5, 2)$, $(3, -1)$
9. $(1, 1)$, $(-1, -1)$ 10. $(2, -1)$, $(-1, 3)$

Find the lengths of the sides of the triangles with the given points as vertices.

11. $A(4, 1)$, $B(2, -1)$, $C(-1, 5)$ 12. $A(1, 2)$, $B(3, 1)$, $C(4, 2)$
13. $A(3, -4)$, $B(2, 1)$, $C(6, -2)$ 14. $A(0, 0)$, $B(2, 1)$, $C(1, 2)$

15. Show that the triangle with vertices at $P_1(1, -2)$, $P_2(-4, 2)$, and $P_3(1, 6)$ is isosceles.

16. Prove that the triangle with vertices at $(2, 1)$, $(1, 3)$, and $(8, 4)$ is a right triangle, by checking the Pythagorean theorem.

17. Prove that the above triangle is a right triangle by computing slopes.

18. Find (the equation of) the line through the point $(1, 1)$ perpendicular to the line $x + 2y = 0$.

19. Find the line through $(2, 0)$ perpendicular to the line through $(2, 0)$ and $(-1, 1)$.

20. Find the equation of the line passing through point $(-2, 3)$ and perpendicular to the line $2x - 3y + 6 = 0$.

21. The midpoint of the line segment joining the points (a_1, b_1) and (a_2, b_2) is the point
$$\left(\frac{a_1 + a_2}{2}, \frac{b_1 + b_2}{2}\right).$$

Verify this analytically in two steps:

a) Show that the three points are collinear.

b) Show that the segment endpoints are equidistant from the claimed midpoint.

22. Using the formula in Problem 21, write the equation of the perpendicular bisector of the segment between $(7, 4)$ and $(-1, -2)$.

Given the vertices of a triangle $P_1(7, 9)$, $P_2(-5, -7)$, and $P_3(12, -3)$, find:

23. The equation of side P_1P_2.

24. The equation of the median through P_1.

25. The equation of the altitude through P_2.

26. The equation of the perpendicular bisector of the side P_1P_2.

27. The points $P_1(3, -2)$, $P_2(4, 1)$, and $P_3(-3, 5)$ are the vertices of a triangle. Show that the line through the midpoints of the sides P_1P_2 and P_1P_3 is parallel to the base P_2P_3 of the triangle.

28. Prove the same result for the triangle with vertices (x_1, y_1), (x_2, y_2), (x_3, y_3).

29. Show that the points equidistant from two given points form a straight line (by equating two distances and simplifying algebraically).

30. Show that if two medians of a triangle are equal, then the triangle is isoceles. (Let the vertices be $(a, 0)$, $(b, 0)$, and $(0, c)$, and equate the lengths of the medians through the vertices $(a, 0)$ and $(b, 0)$.)

5. THE CIRCLE

A point (x, y) lies on the circle about (a, b) of radius r if and only if the distance between (x, y) and (a, b) is r;

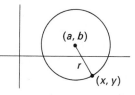

that is, if and only if
$$\sqrt{(x - a)^2 + (y - b)^2} = r.$$

So this is an equation for the circle. We usually eliminate the radical by squaring, to get the equivalent equation

$$(x - a)^2 + (y - b)^2 = r^2. \qquad (*)$$

In particular, the equation for the circle of radius r about the origin is

$$x^2 + y^2 = r^2.$$

Remember that in each case we are saying that the graph of the equation is the circle.

EXAMPLES   1. The circle about the origin with radius $r = 3$ has the equation

$$x^2 + y^2 = 9.$$

2. The equation

$$x^2 + y^2 = 3$$

has the form $x^2 + y^2 = r^2$ with $r = \sqrt{3}$. Therefore its graph is the circle about the origin of radius $r = \sqrt{3}$.

3. The circle about $(1, -2)$ with radius 5 has the equation

$$(x - 1)^2 + (y + 2)^2 = 25.$$

The equation $(x - a)^2 + (y - b)^2 = r^2$ can be directly interpreted as saying that a certain distance is r. But this form is artificial from the point of view of a general polynomial equation. We would more likely find the squares expanded and the constants collected. These steps transform the equation in Example 3 into

$$x^2 + y^2 - 2x + 4y - 20 = 0.$$

This is of the general form

$$x^2 + y^2 + Ax + By + C = 0, \qquad\qquad (**)$$

and leads us to wonder if every equation of this form has a circle as its graph. Simple examples show us that this is too much to expect. Thus

$$x^2 + y^2 = 0$$

is satisfied only for $(x, y) = (0, 0)$, so that the graph of this equation consists of just one point, while

$$x^2 + y^2 + 1 = 0$$

is *never* satisfied and so has an empty graph.

So there is a problem about the graph of an equation having the general form (**). Is it a circle or not? And what circle, if so? The procedure is to try to work back from the general form (**) to the special form (*). We gather together the x terms and make them into a perfect square by adding a suitable constant, and do the same with the y terms. Then the only question is whether the resulting constant on the right side of the equation is positive, zero, or negative.

EXAMPLE 4   If we try to recapture Example 3 from its expanded form by following this prescription, we start with

$$x^2 + y^2 - 2x + 4y - 20 = 0,$$

and first rewrite it as

$$(x^2 - 2x + K) + (y^2 + 4y + L) = 20 + K + L.$$

where we have to discover what values for the constants K and L will make the parentheses into perfect squares. Notice that we have to add these constants to *both sides* of the equation in order for the new equation to be equivalent to the old. In order to find K, we note that $(x^2 - 2x + K)$ must turn out to be $(x - a)^2 = x^2 - 2ax + a^2$. Thus the two expressions

$$x^2 - 2x + K \quad \text{and} \quad x^2 - 2ax + a^2$$

must be the same, so that $2 = 2a$ and $K = a^2$. This gives $a = 1$ and $K = 1^2 = 1$.

Similarly, the two expressions

$$y^2 + 4y + L \quad \text{and} \quad (y - b)^2 = y^2 - 2by + b^2$$

must be the same, so that $4 = -2b$ and $L = b^2$, giving $b = -2$, and $L = (-2)^2 = 4$.

This process is known as *completing the square* (a procedure usually encountered in high-school algebra.) The end result is given by the rule: *to get the constant which must be added, divide the coefficient of x by 2 and then square.*

At this point we have

$$(x - 1)^2 + (y + 2)^2 = 20 + 1 + 4 = 25,$$

which is of the form (*) with $(a, b) = (1, -2)$ and $r = 5$. Now we can identify the graph as the circle with center at $(1, -2)$ and radius 5.

EXAMPLE 5 Find the graph of $x^2 + y^2 - 6x + 2y + 10 = 0$.

Solution. Following the above procedure, we write the equation in the form

$$(x^2 - 6x + K) + (y^2 + 2y + L) = -10 + K + L$$

and see that $K = 9$ and $L = 1$, so that the equation becomes

$$(x - 3)^2 + (y + 1)^2 = 0.$$

Since a sum of squares is zero only if each square is zero, the only solution of this equation is the point $(x, y) = (3, -1)$.

If 10 were replaced by 15 in this example, we would end up with -5 on the right, and since a sum of squares can never be negative there is no graph.

PROBLEMS FOR SECTION 5

Write the equation of each of the following circles in *standard* form. Then expand and collect coefficients to get the general form. Sketch each circle.

1. $C(2, 1)$ and $r = 3$
2. $C(-1, 0)$ and $r = \sqrt{2}$
3. $C(2, -3)$ and $r = 1$
4. $C(-2, -3)$ and $r = 4$
5. $C(-2, 4)$ and $r = 5$
6. $C(2, -1)$ and $r = 0$
7. $C(0, 0)$ and $r = |-4|$
8. $C(2, 5)$ and $r = \sqrt{3}$

By completing the square, convert each of the following equations into the standard form for the equation of a circle. Specify the coordinates of the *center* and the *radius* of the circle.

9. $x^2 + y^2 + 6x - 8y = 0$ 10. $x^2 + y^2 - 4x + 2y + 5 = 0$

11. $x^2 + y^2 + 3x - 5y - \frac{1}{2} = 0$ 12. $2x^2 + 2y^2 + 3x + 5y + 2 = 0$

Identify the graph of each of the following equations. That is, determine whether the graph is a circle, and if so, what circle.

13. $x^2 + y^2 + 2x = 0$ 14. $x^2 + y^2 + 4x - 2y = 4$

15. $x^2 + y^2 + y = 1$ 16. $x^2 + y^2 + 2y + 1 = 0$

17. $x^2 + y^2 + 2x + 2y + 4 = 0$

18. Determine the value of k such that $x^2 + y^2 - 8x + 10y + k = 0$ is the equation of a circle with radius 5.

19. Derive the equation of the circle whose center is at $(5, -2)$ and which passes through the point $(-1, 5)$.

20. Find the equation of the circle having as a diameter the line segment joining $P_1(-4, -3)$ and $P_2(2, 5)$.

21. Find the equation of the circle passing through the origin and the point $(4, 2)$ and having its center on the x-axis.

22. Find the equation of and identify the circle passing through the three points $(1, 0)$, $(-1, 0)$, and $(0, 2)$.

23. Same question for the points $(0, 0)$, $(1, 0)$, and $(2, 1)$.

24. Identify the locus of a point $P = (x, y)$ moving in such a way that the sum of the squares of its distances from $(-2, 0)$ and $(2, 0)$ is a constant k greater than 8.

25. Find the locus of a point $P = (x, y)$ whose distance from the origin is twice its distance from $(3, 0)$.

6. SYMMETRY

Frequently we want to graph an equation in a situation where we don't know ahead of time what kind of curve to expect. The crudest approach is to calculate a few solution points, draw as smooth a curve as we can through them, and hope that we have a reasonable approximation to the graph. Unfortunately we can make gross errors this way, because the few points we choose may fail to indicate some essential feature of the graph which we therefore miss completely. We shall see later on how calculus helps us find the correct general shape. Here we shall look at symmetry properties that can be read off from the algebraic form of the equation.

If a point P is not on a line l, then the *symmetric image* of P in l is the "mirror image" of P in l. It is the unique point P' such that l is the perpendicular bisector of the segment PP'. We obtain P' by dropping the perpendicular from P to l and then continuing an equal distance across l.

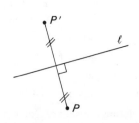

Similarly, a geometric figure F has a mirror image F' consisting of all the symmetric images in l of the points of F. We sometimes call F' the *reflection* of F in the line l.

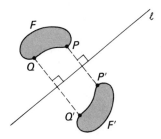

Also, F is the mirror image of F', so F and F' are mirror images of each other in l. A geometric figure F has l as a *line of symmetry* if F is its own mirror image in l, i.e., if F contains the symmetric image in l of each of its points P. Then the 180° rotation of the plane over the axis l just interchanges each symmetric pair of points and carries the figure F exactly into itself.

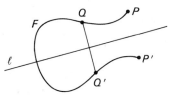

An equilateral triangle has three lines of symmetry; an isosceles triangle that is not equilateral has one line of symmetry; and a nonisosceles (scalene) triangle has no line of symmetry. A square has four lines of symmetry, and a rectangle that is not square has two lines of symmetry. A circle has an *infinite number* of lines of symmetry.

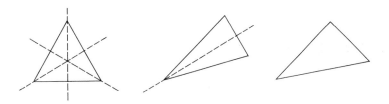

The symmetric (mirror) image of the point (a, b) in the y-axis is the point $(-a, b)$. The following example shows how this carries over to equations.

EXAMPLE 1   Show that the graphs of the equations $xy = 1$ and $-xy = 1$ are mirror images in the y-axis.

Solution. The point $(x, y) = (a, b)$ satisfies the equation $xy = 1$ if and only if the point $(x, y) = (-a, b)$ satisfies the equation $-xy = 1$. That is, the points on the graph of $-xy = 1$ are exactly the mirror images in the y-axis of the points on the graph of $xy = 1$.

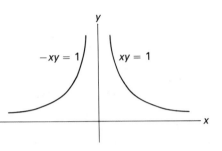

The same reasoning shows in general that:

If the equation E' is obtained from the equation E by replacing x by $-x$, then the graphs of E and E' are mirror images in the y-axis.

In particular:

An equation graph is symmetric in the y-axis (i.e., is its own mirror image) if and only if replacing x by $-x$ in the equation results in an equivalent equation.

Similarly, a graph is symmetric about the x-axis if replacing y by $-y$ in its equation yields an equivalent equation.

EXAMPLE 2   Since $y = x^2$ is equivalent to $y = (-x)^2$, the graph of $y = x^2$ is symmetric about the y-axis.

EXAMPLE 3   The graph of

$$\frac{x^2}{a^2} + \frac{y^2}{b^2} = 1$$

is symmetric about each axis since replacing either x or y by its negative leaves the equation unchanged.

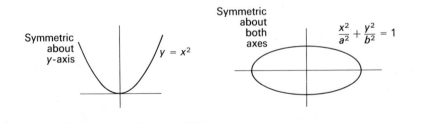

Reflections in the line $y = x$ are also easy to characterize, since the mirror image of the point (a, b) is the point (b, a). Thus, if the equation E' is obtained from the equation E by interchanging x and y, then the graphs of E and E' are mirror images in the line $y = x$. It follows that if the equations E and E' are equivalent, then their common graph is symmetric about the line $y = x$.

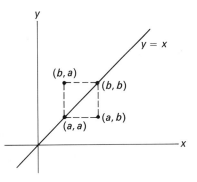

EXAMPLE 4 The graphs of $y = x^2$ and $x = y^2$ are mirror images in the line $y = x$, as shown below.

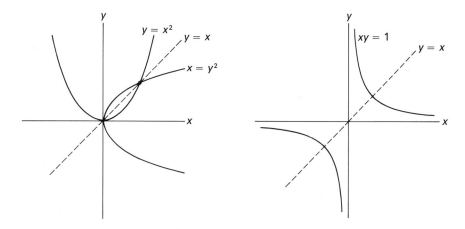

EXAMPLE 5 Since $yx = 1$ is equivalent to $xy = 1$, the graph of $xy = 1$ is symmetric about the line $y = x$.

Another type of symmetry centers about a point Q. Two points P and P' are symmetric *in Q*, or *with respect to Q*, if Q is the midpoint of the segment PP'. Then P' is the symmetric image of P in Q, and vice versa. A figure F has Q as a *point of symmetry*, if F contains the symmetric image in Q of each of its points P.

A parallelogram has a point of symmetry. A triangle does not.

If a figure F has a pair of perpendicular lines of symmetry, then their intersection is necessarily a point of symmetry for F.

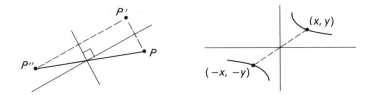

When we introduce coordinates, we see that symmetry with respect to the origin can be stated as follows: if (x, y) is on the graph then so is $(-x, -y)$. That is, replacing (x, y) by $(-x, -y)$ yields an equivalent equation.

EXAMPLE 6   Since $y = 1/x$ is equivalent to $-y = 1/(-x)$, its graph is symmetric in the origin.

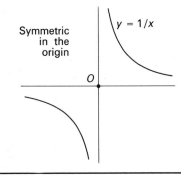

As the above example shows, a curve symmetric in the *origin* need not be symmetric in *either axis*. However, a curve that is symmetric about *both* axes must of necessity also be symmetric in the origin.

PROBLEMS FOR SECTION 6

State what symmetry, if any, the following equations have with respect to the axes, the line $y = x$, and the origin.

1. $9x^2 + 16y^2 = 144$
2. $x^2 + 2xy + y^3 = 0$
3. $2x^3 + 3y + 2y^3 = 0$
4. $2y = x$
5. $2x^2 + y = 0$
6. $y^2 + 2x + 2x^2 + 5 = 0$
7. $x^2 - y^2 = 0$
8. $xy = 1$
9. $x^2 + xy + y^2 = 1$

10. a) Draw all the lines of symmetry of a square.
    b) Draw all the lines of symmetry of a rectangle that is not a square.
    c) Draw all the lines of symmetry of a rhombus that is not a square. (A rhombus is a quadrilateral with all sides equal.)

11. Prove that a triangle has a line of symmetry if and only if it is isosceles.

12. Show that a parallelogram has a point of symmetry.

13. Show that if a figure has a pair of perpendicular lines of symmetry, then it has their point of intersection as a point of symmetry.

7. SECOND-DEGREE EQUATIONS

A treatment of the conic sections is given in Appendix 2. The present section contains a few preliminary facts that can be useful before that more systematic development is likely to be taken up. The standard equations are:

$$y = kx^2 \qquad \text{Parabola}$$

$$\frac{x^2}{a^2} + \frac{y^2}{b^2} = 1 \quad (a > b > 0) \qquad \text{Ellipse}$$

$$\frac{x^2}{a^2} - \frac{y^2}{b^2} = 1 \qquad \text{Hyperbola}$$

The general shapes of the graphs are shown below. The ellipse is symmetric in both axes, and has a center of symmetry at the origin. It has a diameter of maximum length $2a$ along the x-axis and a diameter of minimum length $2b$ along the y-axis. The maximum diameter, and the axis along which it lies, are each referred to as the *major axis* of the ellipse. The minimum diameter and its axis are each called the *minor axis*. The intercepts a and b are thus the lengths of the semimajor and semiminor axes, respectively.

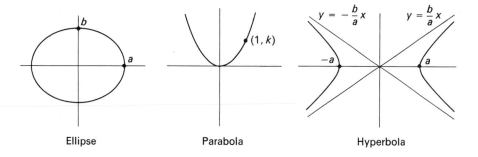

Ellipse   Parabola   Hyperbola

To graph an ellipse, we can calculate a few points in the first quadrant and then use symmetry to obtain the corresponding points in the other quadrants. But for a quick sketch we just plot the intercepts and draw a smooth, symmetric oval through them.

EXAMPLE 1   Quickly sketch the graph of

$$9x^2 + 16y^2 = 144.$$

Solution. Dividing by 144 yields the standard form

$$\frac{x^2}{16} + \frac{y^2}{9} = 1,$$

so we have an ellipse with $a^2 = 16$, $b^2 = 9$. The intercepts are thus

$$\pm a = \pm 4 \quad \text{and} \quad \pm b = \pm 3,$$

and the graph looks roughly like this:

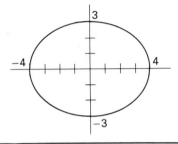

The parabola has one line of symmetry, here the y-axis, and one intersection with its line of symmetry, called its *vertex*, here the origin. Starting at the vertex it opens up along its line of symmetry in one direction, here the positive y-axis.

The hyperbola has two axes of symmetry, but it intersects only one of them and this is called the axis of the hyperbola. To sketch the hyperbola

$$\frac{x^2}{a^2} - \frac{y^2}{b^2} = 1$$

quickly, we use the x-intercepts $\pm a$ and the straight lines

$$y = \pm \frac{b}{a} x.$$

The hyperbola is *asymptotic* to these lines, in the sense that it hugs them more and more closely as $|x|$ gets large. (This is a limit statement that we will take up in the next chapter.)

EXAMPLE 2   Sketch the hyperbola

$$\frac{x^2}{9} - \frac{y^2}{4} = 1.$$

Solution. Here $a^2 = 9$ and the x-intercepts are ± 3. The asymptotes are

$$y = \pm \frac{2}{3} x.$$

We plot the intercepts and asymptotes and then sketch the hyperbola.

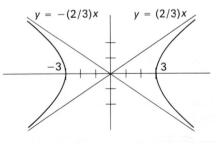

If we interchange x and y in the standard equations we just interchange the roles of the coordinate axes. For example, $x = y^2$ is a parabola opening up along the positive x-axis; $x^2 + y^2/4 = 1$ is an ellipse with major axis along the y-axis; and $y^2 - x^2 = 1$ is a hyperbola with its intercepts on the y-axis. These interchanged forms are equally simple equations, and can also be considered to be standard equations.

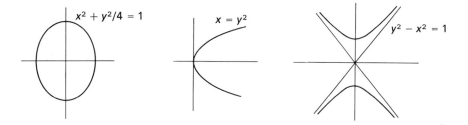

The rest of this section is purely descriptive and is intended only for "general education." Like the circle, each of the above curves can be given various direct geometric definitions, independent of coordinate systems and equations. In Appendix 2 they are defined by locus conditions in the plane. They are also the curves that we get when we intersect a cone with a plane in various ways, and this is why they are called the *conic sections*.

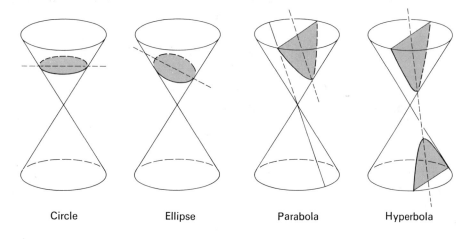

Circle   Ellipse   Parabola   Hyperbola

We should be more explicit. A curve in a geometric plane doesn't have an equation. We have to choose a coordinate system for the plane in order to describe a curve by an equation. Moreover, if we use a different coordinate system, then the same curve will acquire a different equation. Suppose, then, that we are given a plane intersecting a cone. We suppose also that we are given a fixed unit of length, so that we know, for example, whether a circle is large or small. However, we are allowed to place our axes wherever we wish in the plane. Then what we are saying is that the axis system can be so placed that the conic section will have one of the above standard equations.

The proof that these standard equations actually represent conic sections involves a lot of work, and we shall not go into it. However, the equation graphs sketched earlier ought to look the way you visualize the conic sections from the above figure.

If we move each of the intersection planes in the figure parallel to itself until it passes through the vertex of the cone then we get three corresponding "degenerate" conic sections, namely, a single point, a full straight line, and a pair of intersecting straight lines.

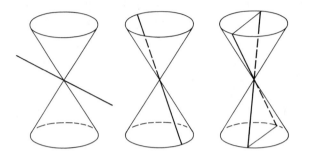

Their corresponding degenerate second-degree equations, in standard form, are

$$x^2 + y^2 = 0,$$
$$(x - y)^2 = 0,$$
$$a^2 x^2 - y^2 = 0.$$

It can be proved that *every* conic section, in standard position or not, has a second-degree equation of the general form

$$Ax^2 + Bxy + Cy^2 + Dx + Ey + F = 0.$$

Conversely, it can be proved that the graph of *any* such second-degree equation is a conic section or a degenerate configuration. This means in particular that a new and better placed axis system can be found with respect to which the graph acquires one of the standard equations. However, if the xy term is present in the equation, then the new axes will be *tilted* with respect to the original system.

EXAMPLE   The graph of

$$x^2 + 2xy + y^2 + x - y = 0$$

is the parabola shown in the next figure.

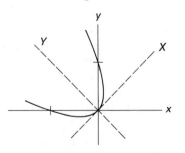

It can be proved that if the axes are rotated $45°$ counterclockwise, this parabola acquires the standard equation

$$Y = \sqrt{2}\, X^2.$$

PROBLEMS FOR SECTION 7

Use intercepts, symmetry and (possibly) asymptotes to make a quick sketch of the graph of each of the following equations.

1. $x^2 + 4y^2 = 1$ 2. $x^2 - 4y^2 = 1$ 3. $x^2 - 4y^2 = -1$
4. $x^2 - 4y = 0$ 5. $4x^2 + 9y^2 = 36$ 6. $2x^2 - y^2 = 4$
7. $x + 4y^2 = 0$ 8. $x^2 + 4y^2 = 0$ 9. $y + x^2 = 0$
10. $y^2 + x^2 = 4$ 11. $x^2 + 9y^2 = 9$ 12. $9x^2 + 4y^2 = 36$
13. $3x^2 + 2y^2 = 6$ 14. $3x^2 - 2y^2 = 6$ 15. $4x^2 + y^2 = 4$
16. $x^2 - 4y^2 = 4$ 17. $4x^2 - y^2 = 4$ 18. $x^2 + 4y = 0$
19. $y^2 + x^2 = 0$ 20. $9x - y^2 = 0$ 21. $4x^2 - y^2 = 0$

8. TRANSLATION OF AXES†

In the last section it was stated that every conic section in the coordinate plane has an equation of the second degree, and that the equation of a tilted conic section contains an xy term. None of this was proved. However, it is relatively easy to see that we will always get a conic section that is *not* tilted as the graph of a second-degree equation *without* xy term. We shall do this here, and shall find that the graph of such an equation can be determined by completing squares and shifting to a new parallel-axis system.

We consider a second set of axes having the same unit of distance and the same directions as the original axes, but a new origin O' at the point having old coordinates (h, k). These new XY-axes can be viewed as obtained by *translating* the old axes, i.e., by sliding them parallel to themselves to a new position.

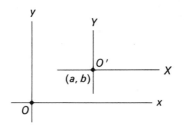

Each point in the plane will now have two sets of coordinates. For example, the new origin O' has old coordinates (h, k) and new coordinates $(0, 0)$. Each curve in the plane will now have two equations. For example, the circle of radius r about the new origin has old and new equations

$$(x - h)^2 + (y - k)^2 = r^2,$$
$$X^2 + Y^2 = r^2.$$

This was shown in Section 5. As this example suggests, if the new coordinate system is in some way more favorably located with respect to the curve than

---

† This section can be postponed; it will not be mentioned until the third section of Chapter 5, at which point there is a reference to Theorem 7 below.

the old, then the new equation for the curve may be simpler and more revealing of its nature than the old. The question is: How do we obtain the coordinates of a point and the equation of a curve in one coordinate system from those in another? The theorem below gives the coordinate change, and the change of equation will then follow.

THEOREM 7   *If a translated XY-axis system is chosen with its origin O' at the point having old coordinates (h, k), then the new coordinates (X, Y) of a point P are obtained from its old coordinates (x, y) by the change-of-coordinate equations*

$$X = x - h,$$
$$Y = y - k.$$

Proof. We look at the first coordinates for both coordinate systems on the horizontal line through $O' = (h, k)$. Considering only the one-dimensional coordinate systems on this line, we know that when the new origin is put at h, then the point whose old coordinate is x acquires the new coordinate $X = x - h$ (by the corollary of Theorem 1). Since the first coordinates x and X remain constant along every vertical line in the plane, the identity $X = x - h$ holds throughout the plane. Similarly, measuring second coordinates along the vertical line through $O' = (h, k)$, we see that a new origin at k replaces the old coordinate y of any point by the new coordinate $Y = y - k$, and this also holds for all points in the plane. We have thus proved the theorem.   ∎

Now consider the equation

$$y = 2x^2 - 4x.$$

We wonder whether it can be simplified by using a translated coordinate system. Our experience with circles suggests that completing the square may lead to a simplifying new origin, and following this lead we get the equivalent equation

$$y + 2 = 2(x^2 - 2x + 1)$$
$$= 2(x - 1)^2.$$

This equation becomes

$$Y = 2X^2$$

under the change of coordinates

$$X = x - 1,$$
$$Y = y + 2.$$

That is, a point has new coordinates (X, Y) which satisfy the equation

$$Y = 2X^2$$

if and only if its old coordinates (x, y) satisfy

$$y = 2x^2 - 4x.$$

The graphs of the two equations are the same curve in the plane, and the common graph is now seen to be a parabola, located in the standard way with respect to the *new axes*, with new origin O' at $(1, -2)$.

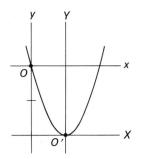

Instead of simplifying by completing the square, we can try a translation change of coordinates and see what happens to the equations. Starting from

$$y = 2x^2 - 4x,$$

the equation in the new coordinates X and Y is obtained by substituting from the change-of-coordinate equations

$$X = x - h,$$
$$Y = y - k.$$

Thus we set

$$x = X + h,$$
$$y = Y + k$$

in the above equation, and get

$$Y + k = 2(X + h)^2 - 4(X + h),$$

or

$$Y = 2X^2 + (4h - 4)X + (2h^2 - 4h - k).$$

This simplifies to

$$Y = 2X^2$$

if

$$4h - 4 = 0,$$
$$2h^2 - 4h - k = 0,$$

and this occurs when $h = 1$ and $k = 2h^2 - 4h = 2 - 4 = -2$, as before.

EXAMPLE Identify the graph of the equation

$$x^2 + 4y^2 + 4x = 0.$$

Solution. We gather together the x terms and complete the square, just as we did for the parabola above and for circles in Section 5. Thus,

$$(x^2 + 4x + 4) + 4y^2 = 4,$$
$$(x + 2)^2 + 4y^2 = 4,$$
$$\frac{(x + 2)^2}{4} + y^2 = 1.$$

The equation is of the form

$$\frac{X^2}{4} + \frac{Y^2}{1} = 1,$$

and it is thus the standard equation of an ellipse in the (X, Y) coordinate system, with semimajor axis 2 and semiminor axis 1. Since

$$X = x + 2 = x - h,$$
$$Y = y = y - k,$$

we see that

$$(h, k) = (-2, 0).$$

These are the (x, y) coordinates of the new origin, and we now have enough information to draw the graph.

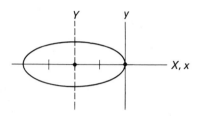

Proceeding just as we did in the above examples, completing squares and then setting $X = x - h$ and $Y = y - k$, we can reduce any second-degree equation in x and y without an xy term to one of the standard forms in X and Y (possibly with variables interchanged), or to a degenerate form. In particular we can identify the graph of any equation

$$Ax^2 + By^2 + Cx + Dy + E = 0$$

as a conic section or a degenerate conic section.

PROBLEMS FOR SECTION 8

1. a) Draw an xy-coordinate system and then draw new XY axes with a new origin at the point $(1, 4)$. From your sketch determine the new coordinates of the point P which has old coordinates $(3, 3)$.
   b) Write down the change-of-coordinate equations and compute the new coordinates of $(3, 3)$ from them.

Follow the above instructions in each of the following situations.

2. New origin at $(-2, 3)$; $P = (-3, 3)$.
3. New origin at $(2, -4)$; $P = (5, 5)$.
4. New origin at $(-5, -2)$; $P = (-6, -3)$.
5. New origin at $(3, 2)$; $P = (1, 1)$.
6. New origin at $(6, -1)$; $P = (6, 0)$.
7. New origin at $(-3, -2)$; $P = (-1, -1)$.
8. New origin at $(-3, 2)$; $P = (2, 1)$.

Reduce each of the following equations to a standard form by a translation of axes. Identify and sketch the graph.

9. $x^2 + y^2 + 4x - 6y = 5$ 10. $25x^2 - 16y^2 = 400$

11. $x^2 + 9y^2 = 9$ 12. $y^2 + 2y - 8x - 3 = 0$

13. $4x^2 + 9y^2 + 16x - 18y - 11 = 0$ 14. $x^2 + y^2 - 2x + 3y + 3 = 0$

15. $4x^2 - y^2 - 8x + 2y + 7 = 0$ 16. $x^2 + 2x + 4y - 7 = 0$

17. $x^2 + 4y^2 - 2x - 16y + 13 = 0$ 18. $9x^2 - 16y^2 + 36x + 96y - 144 = 0$

Find the new equation for the graph of each of the following equations when a new axis is chosen with new origin O' at the given point.

19. $x^2 + y^2 = 4$; $O' = (2, -1)$

20. $xy = 1$; $O' = (1, 3)$

21. $xy - 3x + 2y = 6$; $O' = (-2, 3)$

22. $x^2y - y^3 = 0$; $O' = (1, -1)$

9. INEQUALITIES

There are several important applications of calculus to computations, and whenever such a topic comes up one must be able to use the elementary properties of inequalities in a routine and easy manner. There is a summary of the basic inequality laws in Appendix 1, and this material could be reviewed here if desired.

EXTRA PROBLEMS FOR CHAPTER 1

In the following six problems we consider two coordinate systems on a line l, and we use x and y for the two coordinates of a varying point P.

1. Suppose that the two origins are the same and that $y = 2$ when $x = 1$. Show that $y = 2x$ for every point P.

2. Suppose that the two origins are the same and that $y = -1$ when $x = 1$. Show that $y = -x$ for every point P.

3. Suppose that the two origins are the same and that $y = a$ when $x = 1$. Show that $y = ax$ for every point P.

4. Suppose that $y = b$ when $x = 0$ and that $y = b + 1$ when $x = 1$. Show that $y = x + b$ for every point P.

5. Suppose that $y = b$ when $x = 0$ and that $y = b + a$ when $x = 1$. Show that $y = ax + b$ for every point P.

6. Show that in every case there are constants a and b such that

$$y = ax + b$$

for every point P.

Write the following equations for lines in the form $y = mx + b$. Find the x- and y-intercepts.

7. $y - 1 = (-1/4)(x + 8)$ 8. $3(y + 1/3) = x - 2$

Write an equation for each line specified below.

9. Through the point $(-1/2, 2)$ and parallel to $3x + 4y = 0$.

10. Through the point $(\sqrt{2}, 1)$ and parallel to $y = 4$.

11. Prove algebraically that two nonvertical lines are parallel if and only if their slopes are equal. (The lines l_1 and l_2 are parallel if they never intersect, or if $l_1 = l_2$.)

12. A point P moves along a line with slope $\frac{1}{5}$. What is the change in the y-coordinate of P when its x-coordinate increases by 10? Answer the same question for a line with slope -2.

In Problems 13 through 16, determine algebraically whether or not the three given points lie on a straight line.

13. $(0, 0)$, $(5, 4)$ $(-10, -8)$ 14. $(0, -1)$, $(1, 0)$, $(3, 2)$

15. $(6, 5)$, $(3, 3)$, $(1, 2)$ 16. $(7, 5)$, $(3, 3)$, $(-1, 1)$

17. Find the fourth vertex of the parallelogram with vertices at $(-1, 0)$, $(0, -1)$, $(3, 0)$ in that order.

18. A parallelogram has two sides lying on the lines $y = 2x$ and $3y = x$, and a vertex at $(3, 3)$. Find the equations of the lines containing the other two sides, and find the remaining vertices.

19. If a line has nonzero intercepts a and b on the x and y axes, respectively, show that its equation can be written in the form

$$\frac{x}{a} + \frac{y}{b} = 1.$$

20. Show that the points $(1, 1)$ and $(2, 4)$ are on opposite sides of the line $x - y + 1 = 0$, by reasoning as follows: First, check the sign of $x_0 - y_0 + 1$ for each of the above two points (x_0, y_0). What, then, must happen to the sign of $x - y - 1$ as (x, y) moves along the line segment joining $(1, 1)$ to $(2, 4)$?

21. Show, by reasoning in a manner suggested by the above problem, that the points $(-1, 1)$ and $(2, 4)$ are on the same side of the line $x - y + 1 = 0$.

22. Suppose that the line $Ax + By + C$ does not go through the origin. Show that a point (x_0, y_0) and the origin are on the same side (opposite sides) of this line if $Ax_0 + By_0 + C$ and C have the same sign (opposite signs).

23. Graph the equation $y = x + |x|$.   24. Graph the equation $y = x - |x|$.

25. Graph the equation $|x| + |y| = 1$.

Write an equation for each line specified below.

26. Through $(1/2, 1)$ and perpendicular to $4x - 7y = 8$.

27. Perpendicular bisector of the segment AB where $A(4, -2)$, $B(-1, -8)$.

28. Prove algebraically that the three medians of a triangle are concurrent. (We can simplify the algebra by taking the y-axis through one vertex and the x-axis along the opposite side. The vertices are then $(a, 0)$, $(b, 0)$, $(0, c)$.)

29. Prove that the three perpendicular bisectors of the sides of a triangle are concurrent.

30. Find the distance from the point $(2, 3)$ to the line $x + 2y - 1 = 0$. (Find the point of intersection of the given line and the line perpendicular to it through the given point.)

31. Find the distance from $(1, -1)$ to the line $2x - y + 1 = 0$. (As above, first find the foot of the perpendicular dropped from the point to the l.ne.)

32. Prove that the distance d from the point (x_0, y_0) to the line $Ax + By + C = 0$ is given by

$$d = \frac{|Ax_0 + By_0 + C|}{\sqrt{A^2 + B^2}}.$$

Identify the graph of each of the following. (Specify center and radius of each circle.)

33. $2x^2 + x + 2y^2 + 2y - 32 = 0$ 34. $x^2 + x + y^2 + 4y = 1$

35. Let Q_1 and Q_2 be fixed points and let k be a positive constant. Prove that the locus of a point P, which is moving in such a way that its distance from Q_1 is k times its distance from Q_2, is either a circle or a straight line.

36. Prove algebraically that the line through the origin and the point (a, b) intersects the unit circle $x^2 + y^2 = 1$ in the point $(a/r, b/r)$, where $r = \sqrt{a^2 + b^2}$.

37. Show that the point $(1, -3)$ lies inside the circle $x^2 + y^2 + 6x + 8y = 0$.

38. Show that $x_0^2 + y_0^2 + Ax_0 + By_0 + C < 0$ if the point (x_0, y_0) lies inside the circle

$$x^2 + y^2 + Ax + By + C = 0.$$

39. Show that if $x_0^2 + y_0^2 + Ax_0 + By_0 + C < 0$, then the point (x_0, y_0) lies inside the circle

$$x^2 + y^2 + Ax + By + C = 0.$$

(Reason as follows: If x_1 and y_1 are very large, then

$$x_1^2 + y_1^2 + Ax_1 + By_1 + C$$

is positive. What, therefore, must happen to the sign of

$$x^2 + y^2 + Ax + By + C$$

as the point (x, y) moves along the line segment from (x_0, y_0) to (x_1, y_1)?)

40. Discuss the symmetry of a regular hexagon.

41. Discuss the symmetry of a regular pentagon.

42. If a figure has exactly n lines of symmetry, all passing through a common point 0, show that the angle between adjacent lines of symmetry is $180°/n$.

43. Prove that if a quadrilateral has a line of symmetry through a vertex, then its diagonals must be perpendicular.

44. Prove that if a quadrilateral has a point of symmetry then it is a parallelogram.

45. Show that a quadrilateral with two lines of symmetry is a rhombus or a rectangle.

46. Deduce another symmetry characterization of a rhombus from Problems 43 and 44.

47. Show that if a configuration is bounded (i.e., lies entirely inside some circle), then it cannot have more than one point of symmetry.

48. Assuming the result in Problem 47, show that if a configuration has both a line of symmetry l and a point of symmetry P, and if the configuration is bounded (i.e., lies wholly inside some circle), then P lies on l.

Reduce each equation to standard form by a translation of axes. Identify and sketch the graph.

49. $x^2 + 4y^2 - 4x + 12y + 13 = 0$ 50. $x - y^2 + 2y = 0$

2

Functions and Limits

The operations of calculus apply not to numbers, the way the arithmetic operations do, but to the more complicated things called *functions*.

In this chapter we shall review the notion of function, and introduce the basic new calculation that we make on functions, called the calculation of a limit.

1. FUNCTIONS

Many different operations come up in mathematics. Some apply to pairs of numbers, like the operation of addition, or multiplication, or forming the sum of two squares. However, we shall be concerned here with operations that apply to single numbers, like the operation of squaring, or the operation of taking the positive square root, or the operation of multiplying by 3 and then adding 7. The numbers to which such an operation can be applied make up its *domain*. For example, a number has a square root if and only if it is nonnegative, so the domain of the square root operation is the set of all nonnegative numbers. We can think of an operation as being applied to "input" numbers and determining corresponding "output" numbers. Its domain is then its pool of permissible input numbers.

These operations are all examples of mathematical functions. Moreover, any function can be thought of this way, and this viewpoint toward functions provides a good framework for getting started.

DEFINITION   *A function is an* operation *that determines a unique "output" number y when applied to a given "input" number x. The collection of numbers to which the operation can be applied make up its* domain.

EXAMPLE 1   The operation of taking the reciprocal of a number, i.e., the operation of going from x to $1/x$, is a function defined for all numbers except 0. Its domain is thus the collection of all nonzero numbers.

We use letters such as f and g to represent functions. If f is a function and x is a number in its domain, then we designate by $f(x)$ the unique number obtained by applying the operation f to x. The symbol $f(x)$ is read "f of x," although "f applied to x" would really be more appropriate.

EXAMPLE 2   If f is the reciprocal function considered above, then $f(x) = 1/x$. Therefore

$$f(3) = \frac{1}{3}, \quad f(-1) = -1, \quad f(1 + t) = \frac{1}{(1 + t)}, \quad \text{and} \quad f\left(\frac{1}{a}\right) = a.$$

Also, $1/(f(x)) = x$ and

$$f(x) + \frac{1}{f(x)} = \frac{1}{x} + x = \frac{1 + x^2}{x}.$$

EXAMPLE 3   Let g be the square-root operation, $g(x) = \sqrt{x}$, where $\sqrt{x}$ is always the positive square root of x, as in

$$\sqrt{(-3)^2} = \sqrt{9} = +3.$$

Thus

$$g(4) = 2, \qquad g(x^8) = x^4, \qquad g(y^2) = |y|.$$

The domain of g is the collection of all numbers having square roots, i.e., the collection of all nonnegative numbers.

EXAMPLE 4   Functions frequently arise from relations among variables. Consider the equation

$$y^3 = x.$$

It has the property that *for each value of x there is a unique value of y making the equation true.* We then have the operation of going from x to the unique y that it determines. And we say that *the equation determines y as a function of x.*

We can identify the above function. The unique y such that $y^3 = x$ is the cube root of x, by definition. Thus the equation determines y as a function of x, and the function f is the cube-root function,

$$y = f(x) = x^{1/3}.$$

EXAMPLE 5   Consider next the equation

$$xy = 1.$$

If $x = 0$ there is no value of y satisfying the equation, but for every other value of x there is a unique value of y which makes the equation true. (We could say that for every value of x there is *at most* one corresponding value of y.) The equation thus determines y as a function of x, the domain of the function being the set of all nonzero numbers.

Here, again, we can identify the function. The unique y such that $xy = 1$ is the reciprocal of x, by definition. The equation determines y as a function of x, and the function f is the reciprocal function $y = f(x) = 1/x$, with domain the set of all nonzero numbers.

EXAMPLE 6   The equation

$$y = x^2$$

is already in the form $y = f(x)$, with

$$f(x) = x^2.$$

Here f is the squaring operation.

EXAMPLE 7   On the other hand, the equation

$$y^2 = x,$$

does *not* determine y as a function of x. For each positive value of x there are *two* values of y making the equation true: $y = \pm\sqrt{x}$; so y is not *uniquely* determined by x.

In general, whenever two variable quantities x and y are so related that the value of y depends on and is uniquely determined by the value of x, we say that y *is a function of* x. The function f in question is the operation of going from x to the unique y it determines. We call x the *independent variable*, because its value is chosen arbitrarily, and we call y the *dependent variable*, because its value depends on the value of x.

The phrase "y is a function of x" is very old, and antedates the notion of a function as a separate and distinct object. Thus for many years mathematicians were talking in terms of functionally related variables before they understood functions. This spawned some murky reasoning, and eventually led to such a backlash against the "y is a function of x" language that many authors avoid it entirely. When used *along with* the function concept, however, the old language is perfectly respectable. Moreover, it is an idiom that fits reality, for in practice functions generally arise as relationships among variables.

EXAMPLE 8   In a manufacturing process the cost C of producing x items is a function of x. In a simple situation there might be a fixed initial cost I to purchase a machine, and then an additional cost per item of amount k, for material, labor, etc. The total cost of producing x items is then

$$C = I + kx.$$

Thus, $C = f(x)$, where $f(x) = I + kx$. The average cost, or cost per item, is given by

$$c = \frac{C}{x},$$

and this also is a function of x:

$$c = g(x) = \frac{I}{x} + k.$$

Note that since x is the number of items manufactured, x is a positive integer. The domain of the cost function $f(x) = I + kx$ is thus a set of positive integers. However, if we were to graph the cost equation $C = I + kx$ in order to visualize the relationship between C and x, we would generally draw the smooth curve for all positive x.

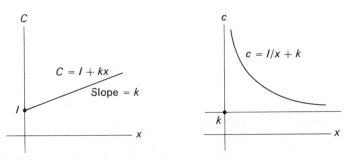

EXAMPLE 9 If a flexible container of gas is kept at a constant temperature, then it is observed that the volume V occupied by the gas is inversely proportional to the pressure p applied to it. That is,

$$pV = \text{constant}.$$

This relationship determines V as a function of p, and also determines p as a function of V. Here, again, the function domains are the *positive* real numbers, because of the physical interpretations of the variables.

The world is full of varying quantities, and the fact the two or more of them satisfy a dependency relationship of this sort amounts to a "law of nature" that we can study mathematically. In particular, we can view it as a cause-and-effect relationship that can be interpreted as a function, and we can analyze the behavior of this function by calculus.

EXAMPLE 10 An object allowed to fall from rest is observed to fall a distance

$$s = 16t^2$$

feet in t seconds. This even applies to a feather, provided it is allowed to fall in a vacuum. In other words, this equation is the universal law for all falling bodies, "neglecting air resistance." It is an observed law of nature expressing s as a function of t, and when this function is examined by calculus it reveals information about how the force of gravity acts.

The notion of a function f is not tied to any particular scheme for obtaining $f(x)$ from x.

EXAMPLE 11 Consider the area of the variable trapezoid drawn below at the left. The area A depends on the position $x = a$ of the righthand base, so A is a function of a, $A = f(a)$. In fact,

$$A = \frac{1}{2}a(1 + (a + 1)),$$

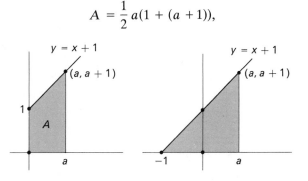

by the formula for the area of a trapezoid. But also

$$A = \frac{1}{2}[a - (-1)](a + 1) - \frac{1}{2}(1 \cdot 1),$$

where the right side is the difference of the two triangle areas in the right figure. We thus have two different explicit ways for expressing the operation f of *going from a to A*.

EXAMPLE 12   If f and g are the functions defined by the equations

$$f(x) = x^2 - 1 \quad \text{and} \quad g(x) = (x - 1)(x + 1),$$

then $f(x) = g(x)$ for all x, so f and g are the same operation. The two formulas present two different ways for carrying out this operation. The inner workings of a particular scheme for obtaining $f(x)$ from x can be thought of as indicating a method for computing the function value $f(x)$, and a given function f can generally be computed in many different ways.

Function domains are generally described in terms of *intervals*. An interval is simply a segment on the number line. If its endpoints are the numbers a and b, then the interval consists of the numbers lying between a and b. Supposing that a is to the left of b, these are the points which are simultaneously to the right of a and to the left of b. However, we may or may not want to include the endpoints themselves in the interval. The *closed interval* from a to b, designated $[a, b]$, includes the endpoints, and so contains exactly those numbers x such that $a \leq x \leq b$. Thus $[1, 2]$ consists of the real numbers from 1 to 2, *inclusive*.

We use a parenthesis to indicate an *excluded* endpoint. Therefore the interval $[a, b)$ contains a but not b, and so contains exactly those numbers x such that $a \leq x < b$. The interval (a, b), with both endpoints excluded, is called the *open* interval from a to b. There is ambiguity here since the same symbol (a, b) is used for a point in the coordinate plane, but the context will always make it clear which meaning is intended.

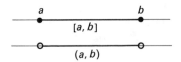

A half-line is considered to be an interval extending to infinity. We use the infinity symbol, ∞, to indicate the unterminated direction of such an interval. Thus $[a, \infty)$ consists of those points x such that $x \geq a$. And $(-\infty, \infty)$ is the whole real-number system. There is, of course, no question of having $[a, \infty]$: ∞ is not a number, and we can't include a number that doesn't exist.

The domains of the functions that we meet in calculus are almost always made up out of one or more intervals. For example:

the domain of $f(x) = \sqrt{x}$ is $[0, \infty)$;

the domain of $f(x) = \sqrt{1 - x^2}$ is $[-1, 1]$;

the domain of $f(x) = 1/x$ is the union of $(-\infty, 0)$ and $(0, \infty)$.

We frequently refer to intervals by letters such as I and J. We do this when we want to talk about an interval but don't need to refer specifically to its endpoints; and we may not care particularly whether the interval is open or closed.

Any point x in an interval I that is not an endpoint of I is called an *interior point* of I. We also say that x is *in the interior of* I. An interval extends *across* an interior point but lies entirely *on one side of* an endpoint. An open interval consists only of interior points.

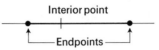

PROBLEMS FOR SECTION 1

Determine the domain of each of the following functions and express it as an interval or a union of intervals:

1. $f(x) = \sqrt{x}$

2. $f(x) = \sqrt[3]{x}$

3. $f(x) = \dfrac{1}{x-2}$

4. $f(x) = x^2 + 2x + 1$

5. $f(x) = \dfrac{x+2}{x^2-4}$

6. $f(x) = \sqrt{1-x^2}$

For each of the following functions find $f(z)$ when z is the given value:

7. $f(t) = t^2 + 2t - 9;\ z = 2$

8. $f(x) = x^3 + 2x^2 - \sqrt{x} + 3;\ z = \pi$

9. $f(u) = u^3 + 3u^2 + 2u;\ z = a + 1$

10. $f(x) = \sqrt{2x + 11};\ z = 2.5$

11. $f(x) = \dfrac{x^2 - 1}{x + 1};\ z = u - 1$

If $H(t) = (1 - t)/(1 + t)$, find

12. $H(-x)$

13. $H(1/t)$

14. $H(1/(1 + x))$

15. $H(H(t))$

In each of the following cases, decide whether or not the equation determines y as a function f of x. If not, show why not. If so, find a formula for f and give its domain.

16. $x^2 + 4y^2 = 1$

17. $x^2 + 4y = 1$

18. $x + 4y^2 = 1$

19. $xy + x - y = 2$

20. $\sqrt{x - y} = 1$

21. $x^2 + y^3 = 0$

22. $x^3 + y^2 = 0$

23. $x^3 + y^3 = 0$

24. $\dfrac{x + y}{x - y} = x$

25. Refer to Example 10 in the text. If an object is dropped from a height h, then the time t it takes to fall to the ground is a function of h. Find a formula for this function $f(h)$.

26. The area A of an equilateral triangle is a function of the side length s, $A = f(s)$. Find this function f.

2. THE GRAPH OF A FUNCTION

As x runs across the domain of a function f, the point $(x, f(x))$ traces a curve in the xy-coordinate plane called the *graph* of f. The graph is thus made up of the points (x, y) satisfying the equation $y = f(x)$, so:

The graph of the function f in the xy-plane is the graph of the equation $y = f(x)$.

EXAMPLE 1   If $f(x) = x^3$, then the graph of f is the set of points (x, x^3), i.e., the set of points (x, y) such that $y = x^3$, i.e., the graph of the equation $y = x^3$.

If we wanted to graph f in the uv-plane, we would graph the equation $v = f(u)$.

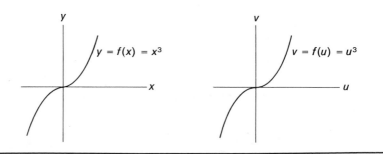

The convention is that:

The independent variable is normally plotted horizontally.

Now suppose we are given an equation in x an y that determines x as a function of y, and we wish to graph *that* function on the standard xy-plane. Then we must interchange variables, as in the following example.

EXAMPLE 2   The equation $y^2 - 2y = x$ expresses x as a function of y,

$$x = f(y) = y^2 - 2y.$$

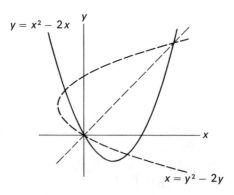

But in the standard xy-plane, the graph of f is the graph of the equation $y = f(x)$, i.e., the graph of the equation

$$y = x^2 - 2x,$$

which is obtained from the given equation by interchanging the variables. The two graphs are mirror images of each other in the line $y = x$. (See Chapter 1, Section 6.)

In summary:

If an equation in x and y determines x as a function of y, x = f(y), then in order to graph f on standard axes (the graph of the equation y = f(x)) we interchange variables in the given equation and graph the resulting equation (which is equivalent to y = f(x)).

It may happen that a graph and its "flipped over" graph are *both* function graphs. This will occur whenever an equation in x and y determines *each* variable as a function of the other. Two functions related this way are said to be *inverse* to each other. This is an important notion and we shall devote a short chapter to it later on (Chapter 7).

EXAMPLE 3 Consider again the cubing function $f(x) = x^3$ of Example 1, and its graph in the xy-plane, the graph of the equation $y = x^3$. This equation also determines x as a function of y, the cube root function $x = y^{1/3} = g(y)$. In order to graph g on standard axes, we interchange variables:

$$y = g(x) = x^{1/3}.$$

The graphs of f and g are then mirror images in the line $y = x$, as shown below.

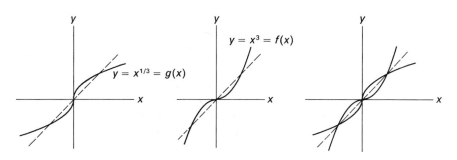

The cubing function and the cube root function are inverse to each other.

Our definition of a function as an operation leads us to view its action one number at a time. A function graph, on the other hand, spreads the whole input–output relationship out before us as a geometric curve. This gives us a panoramic view of the whole function as a global, but passive, entity, and is quite different in nature from the operator-waiting-to-act notion. The two points of view reinforce each other, and are part of our total notion of a function.

It is possible to take the passive function as the basic notion. We note that a curve in the coordinate plane determines y as a function of x if each vertical line cuts it in only one point (or not at all for x not in the domain). For this just says geometrically that only one y value goes with a given x value. We can then define a function by this kind of configuration in the coordinate plane.

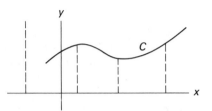

DEFINITION   *A function is given by a configuration C in the coordinate plane (a collection C of ordered pairs of numbers) having the property that each first number x occurring in the points of C is paired with only one second number y.*

This set-of-ordered-pairs definition of a function has become popular recently as part of the emphasis on sets, but by itself it doesn't fully convey the notion of a function.

PROBLEMS FOR SECTION 2

1. Graph the squaring function $f(x) = x^2$ by plotting the points over $x = 0$, $\pm\frac{1}{2}$, ±1, $\pm\frac{3}{2}$, ±2. Name this graph from the list of standard equations in Section 7 of Chapter 1.

2. The graph of $f(x) = \sqrt{x}$ is half of a parabola. How do we know this? Sketch the graph.

3. Show that the graph of $f(x) = \sqrt{1 - x^2}$ is a semicircle about the origin. Then plot one point on the graph and sketch it.

4. Show that the graph of $f(x) = \frac{1}{2}\sqrt{x^2 + 4}$ is half of a hyperbola (by comparing with a nearly equivalent equation in Section 7 of Chapter 1). Plot the intercept and asymptotes and quickly sketch the graph.

5. The equation $y = \sqrt{x}$ determines x as a function f of y, $x = f(y)$. What are $f(0)$ and $f(1)$? Using these two values only, sketch the graph of f, being sure that it has the right general shape. (The equation $y = \sqrt{x}$ is equivalent to $x = y^2$ and $y \geq 0$.)

6. The equation

$$2x - y + 1 = 0$$

determines y as a function f of x, and also determines x as a function g of y. Draw the graphs of these two functions f and g.

7. For positive x and y the equation

$$yx^2 = 1$$

determines each of the variables x and y as a function of the other. Draw the graphs of these two functions.

8. A function f is said to be *even* if

$$f(-x) = f(x)$$

for every number x in its domain. Note that if x is in the domain of f, then $-x$ must also be there if this condition is to make sense. Show that f is even if and only if its graph is symmetric about the y-axis.

9. A function f is *odd* if

$$f(-x) = -f(x)$$

for every x in its domain. Again it is understood that $-x$ must be in the domain for each x in the domain. Show that f is odd if and only if its graph is symmetric with respect to the origin.

3. CONTINUITY

This section and the next will discuss certain ways in which a function can behave near a point, and introduce the language we use to describe this behavior. We shall not give precise definitions, nor is anything going to be proved; if something seems plausible we shall assume it. The idea is to get a *qualitative* feeling for what is happening without going into the quantitative calculations that ultimately back up these intuitions.

The number x^3 varies "continuously" with the number x, in the sense that a small change in x produces only a small change in x^3. Moreover, we can keep the change in x^3 as small as we wish by holding the change in x to within a sufficiently small range.

We can "see" that x^3 behaves this way by looking at the very smooth graph of $y = x^3$ and visualizing the effect of a small change in the values of x. Or we can experiment numerically or algebraically to see what specific small changes in x do to the values of x^3. Since such experimentation gives a very concrete feel for the situation, we include an example.

EXAMPLE 1 We wish to compute how close x^3 is to $2^3 = 8$ when $x = 2.1$. However, instead of just cubing 2.1, we shall write $(2.1)^3$ as $(2 + .1)^3$ and use the expansion

$$(a + h)^3 = a^3 + 3a^2h + 3ah^2 + h^3.$$

The difference $(2.1)^3 - 2^3$ that we want to calculate is thus of the form

$$(a + h)^3 - a^3 = h(3a^2 + 3ah + h^2),$$

with $a = 2$ and $h = .1$, so

$$(2.1)^3 - 2^3 = .1(12 + .6 + .01)$$

$$= .1(12.61)$$

$$= 1.261.$$

For $x = 2.01$ we have

$$(2.01)^3 - 2^3 = .01(12 + .06 + .0001)$$
$$= .01(12.0601)$$
$$= .120601.$$

Such numerical examples make it especially clear that the reason for $(a + h)^3 - a^3$ becoming small when h becomes small is that it contains h as a factor.

In Example 1 we could have explicitly calculated *how* small to hold the change in x in order to keep the change in x^3 within prescribed limits. We shall do this in Appendix 3. Such computations form the core of the technical study of continuity. For now, though, we shall rely on intuitions about continuous variation. Here, then, is our provisional and intuitive definition of continuity.

The function f is continuous if a small change in x produces only a small change in f(x), and if we can keep the change in f(x) as small as we wish by holding the change in x sufficiently small.

Practically every function that we shall want to study in the differential calculus will be continuous in the above sense. The reason is that in building up functions we customarily use operations, like addition and multiplication, that themselves react to small input changes with only small changes in output. Some basic continuity principles are given below. They will probably seem correct on intuitive grounds, and we shall assume them for now.

I. $x^{1/n}$ *varies continuously with x, for any fixed positive integer n. Any constant function is trivially continuous (since it doesn't vary at all).*

II. *If $f(x)$ and $g(x)$ vary continuously with x, then so do $f(x) + g(x)$ and $f(x)g(x)$. So does $1/f(x)$, provided x is kept away from any value for which $f(x) = 0$.*

III. *If $g(y)$ varies continuously with y and if $y = f(x)$ varies continuously with x, then $g(f(x))$ varies continuously with x.*

For example, in order to convince ourselves that $\sqrt{1 - x^2}$ varies continuously with x, we visualize a small change in x, producing only a small change in x^2 and hence in $1 - x^2$, and this small change in $1 - x^2$ then produces only a small change in $\sqrt{1 - x^2}$. This intuitive analysis uses all three of the above principles. Starting off with the continuity of 1, x, and $\sqrt{y}$ (Principle I), we get the continuity of x^2 and then $1 - x^2$ from (II), and the continuity of $\sqrt{y} = \sqrt{1 - x^2}$ from (III).

In the same way, our intuitive feeling that a polynomial like

$$p(x) = x^4 + 2x^3 - x^2 + 5x + 1$$

varies continuously with x can be traced back to (I) and repeated applications of (II). A rational function $f(x)$ is a quotient of two polynomials,

$$r(x) = \frac{p(x)}{q(x)};$$

and, since every polynomial is continuous, we can conclude that $r(x)$ varies continuously so long as we stay away from any value of x for which $q(x) = 0$ (by (II) again).

However, not all functions that we meet in everyday life and in mathematics are continuous.

EXAMPLE 2 When we mail a letter we have to use one stamp of a certain denomination for anything up to and including one ounce in weight, then two stamps up to and including two ounces, etc. Thus, if $p(x)$ is the number of stamps required to mail a letter weighing x ounces, then the graph of p looks like this:

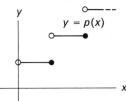

The empty dot ∘ shows a point *not* on the graph, whereas solid dots show points on the graph. The function $p(x)$ is called the postage-stamp function. It is defined for $x > 0$ and has a *jump discontinuity* at each of the points $x = 1, 2, 3, \ldots$. For example, the change in $p(x)$ *cannot* be made as small as we wish when x changes from slightly below 1 to slightly above 1, so $p(x)$ is *not* continuous as x moves through the value 1.

EXAMPLE 3 The greatest integer function $[x]$ is an analogue of the postage-stamp function in pure mathematics: $[x]$ is defined as the largest integer that is $\leq x$. Thus $[2.5] = 2$, $[2] = 2$, and $[1.9] = 1$. Although the domain of $[x]$ might naturally be taken to be $[0, \infty)$, we shall define $[x]$ for negative x in the same way. The integers less than -2.5 are all negative but among them there is a greatest one, namely -3. Thus $[-2.5] = -3$, $[-3] = -3$, etc. The graph of $[x]$ looks like this:

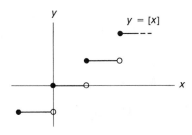

EXAMPLE 4 Here is a glimpse of the graph of $x - [x]$.

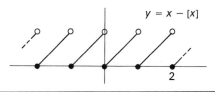

These examples show that even quite simple functions may have isolated points of discontinuity. This focuses our attention on the notion of continuity *at a point x = a*. The idea is the same, except that we restrict the variation of *x* to small changes away from *x = a*.

DEFINITION *The function f is continuous at the point x = a in its domain if a small change in x away from x = a produces only a small change in f(x). More precisely, we require that we can keep the change f(x) − f(a) as small as we wish in magnitude by restricting x to a suitably small interval about x = a.*

Then:

DEFINITION *A function is continuous on an interval I in its domain if it is continuous at each point a in I.*

PROBLEMS FOR SECTION 3

1. a) Compute how close $f(x) = 2x - x^2$ is to $f(3) = -3$ when x is near 3. Do this first algebraically by writing $x - 3 = h$ and factoring h from $f(x) - f(3) = f(3 + h) - f(3)$. Then give $x - 3 = h$ the values .1, .01, .001, and compute the difference $f(x) - f(3)$.

    b) Now compute how close $f(x)$ is to $f(2) = 0$, following the same program as above.

2. Compute how close $f(x) = 1/x$ is to $f(2) = 1/2$ when x is near 2. Follow the same steps as in the above problem.

Verify the continuity of each of the following functions by applying the continuity principles I through III.

3. $3x^2 + 5x$ 4. $x/(1 + x^2)$

5. $1/\sqrt{x}$ 6. $\sqrt{x^2 + 4}$

7. $\sqrt{x^2}$ 8. $|x|$

9. $|x^{1/3}|$ (Sketch this graph.)

10. Draw the graph of $f(x) = x - [x/2]$. What are its points of discontinuity?

11. Define a function (by drawing its graph) that is constant over each of the intervals $(-\infty, -1)$, $(-1, 1)$, $(1, \infty)$, and has discontinuities at $x = -1$ and $x = 1$.

4. LIMITS

The new operation on which calculus rests is the calculation of a limit. In the beginning this calculation will be very easy to make in terms of continuity, and the rules of algebra for limits are obvious. Later on, we shall have to make more difficult limit calculations, and this will force us to examine the nature of the limit operation more critically.

We start by describing the continuity of x^3 at a point in different language. It is reasonable to say that x^3 *approaches* 8 *as x approaches* 2. The arrow → is the symbol for the word *approaches*, so in symbols this statement is:

$$x^3 \to 8 \qquad \text{as } x \to 2.$$

In general,

$$x^3 \to a^3 \qquad \text{as } x \to a.$$

The slightly different connotation of "approaches" is that here we imagine x moving *toward* a, whereas before we imagined x moving slightly *away* from a. But we mean the same thing—we can see to it that x^3 is as close to a^3 as we wish merely by restricting x to a sufficiently small interval about a. So we are just restating the continuity of x^3 at the point $x = a$.

More generally, the continuity of a function f at the point $x = a$ can be described by saying that $f(x)$ *approaches* $f(a)$ *as* x *approaches* a:

$$f(x) \to f(a) \qquad \text{as } x \to a.$$

The value that $f(x)$ approaches is called its *limit*, and another form of the above statement is:

$$f(a) \qquad \text{is the limit of } f(x) \qquad \text{as } x \text{ approaches } a.$$

In symbols this form is written

$$f(a) = \lim_{x \to a} f(x).$$

We thus have a new way of viewing the continuity of f at a. The value $f(a)$ is like a target that $f(x)$ "zeroes in on" as x approaches a. It turns out that this new point of view expresses continuity in terms of a more general phenomenon, for we shall see that a function can exhibit such target-seeking behavior as x approaches a without the target having to be $f(a)$. That is, a function can have a limit as x approaches a without being continuous at a.

EXAMPLE 1   The function

$$f(x) = \frac{1 - \sqrt{x}}{1 - x}$$

is not defined at $x = 1$, so it cannot be continuous there. Yet $f(x)$ approaches the limit $\frac{1}{2}$ as x approaches 1, as we shall see a little later. That is, the difference $f(x) - \frac{1}{2}$ can be made as small as we wish by restricting x to a sufficiently small interval about 1, *but excluding the value $x = 1$*, since $f(1)$ is not defined.

This example suggests how the general description of a limit should be worded.

DEFINITION   *We say that $f(x)$ has the limit l as x approaches a, and write*

$$\lim_{x \to a} f(x) = l,$$

if we can make $f(x)$ as close to l as we wish by restricting x to a sufficiently small interval about a, excluding the value $x = a$.

Continuity at a is then a special case:

A function f is continuous at x = a if and only if a is in the domain of f and

$$\lim_{x \to a} f(x) = f(a).$$

Although the limit property is more general than continuity, it is not *much* more general: there are only two possible ways for a function to have a limit at $x = a$ without being continuous there. First $f(a)$ may not be defined. We noted above that the function $f(x) = (1 - \sqrt{x})/(1 - x)$ is like this at $x = 1$, although we haven't established the limit yet.

The second possibility is that $f(a)$ exists but is different from the limit $\lim_{x \to a} f(x)$.

EXAMPLE 2   An example of a function exhibiting this bizarre behavior is

$$f(x) = p(x) - [x],$$

the difference between the postage-stamp function and the greatest-integer function. If you will reexamine the definitions of these functions carefully, especially around the integer values of x, you will see that their difference has the constant value 1, *except at the integers*, where its value is 0. In particular,

$$\lim_{x \to 2} f(x) = 1,$$

but

$$f(2) = 0,$$

so

$$\lim_{x \to 2} f(x) \neq f(2).$$

The same thing happens at every integer value of x.

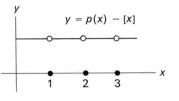

If f has a limit at a without being continuous there, then we can modify f so that it becomes continuous at a simply by changing what happens at the one point a. We just define (or redefine) $f(a)$ to be the limit l, and then have

$$\lim_{x \to a} f(x) = f(a).$$

Thus, if $f(x)$ has a limit as x approaches a, then the limit l is the number that f *ought* to have as its value *at a*, whether or not it does.

Functions don't *have* to have limits, and this anomolous behavior can occur in various ways. We have already met one example: The greatest-integer function $[x]$ has a limit *from the left* at $x = 2$ and also a limit *from the right* at 2, but the two "one-sided" limits are different, so there is no one number that $[x]$ approaches as x approaches 2. However, our principal concern here is evaluating limits that do exist.

Limit evaluations are trivial for continuous functions. (We assume the properties of continuity from the last section.)

EXAMPLE 3 Any polynomial $p(x)$ is everywhere continuous, so $\lim_{x \to a} p(x) = p(a)$ for any a. For example,

$$\lim_{x \to 4} (x^3 - 8x) = 4^3 - 8 \cdot 4 = 64 - 32 = 32.$$

EXAMPLE 4 A rational function is a quotient

$$r(x) = \frac{p(x)}{q(x)}$$

of two polynomials. It follows from our general principles for continuous functions that $r(x)$ is continuous everywhere except at points where the denominator is zero. That is, $\lim_{x \to a} r(x) = r(a)$, provided $g(a) \neq 0$. For example, if

$$r(x) = \frac{x - 1}{x + 1},$$

then

$$\lim_{x \to 3} r(x) = r(3) = \frac{3 - 1}{3 + 1} = \frac{1}{2}.$$

EXAMPLE 5
$$\lim_{x \to 2} \sqrt{x} = \sqrt{2}.$$

EXAMPLE 6 We return now to Example 1, which is a nontrivial limit of the sort we will meet in our early calculus computations. We wish to show that if

$$f(x) = \frac{1 - \sqrt{x}}{1 - x},$$

then

$$\lim_{x \to 1} f(x) = \frac{1}{2}.$$

Solution. If we try substituting $x = 1$, we get 0/0, which is meaningless; so f is not defined at $x = 1$ and we do not have a trivial limit. We can nevertheless calculate the limit as follows. Factoring $1 - x$ as a difference of squares, $1 - x = (1 - \sqrt{x})(1 + \sqrt{x})$, we see that

$$\frac{1 - \sqrt{x}}{1 - x} = \frac{1 - \sqrt{x}}{(1 - \sqrt{x})(1 + \sqrt{x})} = \frac{1}{1 + \sqrt{x}}$$

when $x \neq 1$. Now

$$g(x) = \frac{1}{1 + \sqrt{x}}$$

is continuous on $[0, \infty)$, and its limit at 1 is its value there:

$$\lim_{x \to 1} g(x) = g(1) = \frac{1}{1 + \sqrt{1}} = \frac{1}{2}.$$

Since $f(x) = g(x)$ for $x \neq 1$, we have

$$\lim_{x \to 1} f(x) = \lim_{x \to 1} g(x) = g(1) = \frac{1}{2}.$$

In brief,

$$\lim_{x \to 1} \frac{1 - \sqrt{x}}{1 - x} = \lim_{x \to 1} \frac{1}{1 + \sqrt{x}} = \frac{1}{1 + \sqrt{1}} = \frac{1}{2}.$$

Thus, although f is not defined at $x = 1$, it is equal *everywhere else* to a function g that is continuous at 1, and we use this fact to evaluate the limit of f at 1.

Although we can frequently use this trick to evaluate a limit, it isn't always available. Sometimes, instead of finding $f(x)$ to be *equal* to a known continuous function, we find it *squeezed between two* known continuous functions that come together at the limit point.

EXAMPLE 7   We shall need to evaluate the limit

$$\lim_{t \to 0} \frac{\sin t}{t}.$$

You may not know what the functions $\sin t$ and $\cos t$ are, but read on anyway.

Setting $t = 0$ again gives 0/0 so the function $f(t) = (\sin t)/t$ is not defined at $t = 0$. This time it can be shown that

$$\cos t < \frac{\sin t}{t} < 1$$

on an interval about 0. Now $\cos t$ is continuous everywhere, and in particular

$$\lim_{t \to 0} \cos t = \cos 0 = 1.$$

Thus $(\sin t)/t$ is squeezed between two continuous functions each having the value 1 at $t = 0$. We conclude that therefore $(\sin t)/t$ has the limit 1 at $t = 0$.

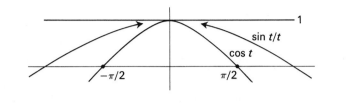

The above examples exhibit the methods by which we shall make most of our explicit limit calculations. In each case, we exploited the connection between limits and continuity, as follows:

a) $\lim\limits_{x \to a} f(x) = f(a)$ if f is continuous at a;

b) $\lim\limits_{x \to a} f(x) = g(a)$ if g is continuous at a and $f = g$ for all $x \neq a$ (in some interval about a);

c) $\lim\limits_{x \to a} f(x) = g(a) = h(a)$ if g and h are continuous at a, $g(a) = h(a)$, and $g(x) \leq f(x) \leq h(x)$ for all $x \neq a$ on some interval about a).

These devices are listed in order of increasing generality, and the squeeze device (c) logically includes both of the others.

PROBLEMS FOR SECTION 4

Evaluate each of the following limits, citing the relevant reasons in each case.

1. $\lim\limits_{x \to 2} \sqrt{x}$

2. $\lim\limits_{x \to 2} \dfrac{x + 1}{x^2 - 1}$

3. $\lim\limits_{x \to 8} x^{1/3}$

4. $\lim\limits_{x \to -1} 5x^3 + 4x^2 + 1$

5. $\lim\limits_{x \to 0} \dfrac{1 - \sqrt{x}}{1 + \sqrt{x}}$

Use the trick of Example 6 to evaluate each of the following limits.

6. $\lim\limits_{x \to 1} \dfrac{x - 1}{x^2 - 1}$

7. $\lim\limits_{x \to 2} \dfrac{x^3 - 8}{x - 2}$

8. $\lim\limits_{x \to 4} \dfrac{\sqrt{x} - 2}{x - 4}$

9. $\lim\limits_{x \to 4} \dfrac{x - \sqrt{x} - 2}{x - 4}$

10. $\lim\limits_{x \to 1} \dfrac{x^3 - 1}{x + \sqrt{x} - 2}$

11. $\lim\limits_{x \to 0} \dfrac{\sqrt{1 + x} - 1}{x}$

Evaluate the following limits (if possible).

12. $\lim\limits_{x \to -1} \dfrac{2x^2 - x - 3}{x + 1}$

13. $\lim\limits_{x \to 0} \dfrac{2x^2 - x - 3}{x + 1}$

14. $\lim\limits_{x \to 4} \dfrac{x - 4}{3(\sqrt{x} - 2)}$

15. $\lim\limits_{x \to 4} \dfrac{x - 4}{x^2 - x - 12}$

16. $\lim\limits_{x \to -3} \dfrac{x - 4}{x^2 - x - 12}$

17. $\lim\limits_{x \to 4} \dfrac{\sqrt{x^2 + 9} - 5}{x^2 - 4x}$

18. $\lim\limits_{x \to 2} \dfrac{\sqrt{x^2 + 9} - 5}{x^2 - 4x}$

19. $\lim\limits_{x \to 0} \dfrac{\sqrt{x^2 + 9} - 5}{x^2 - 4x}$

20. $\lim\limits_{x \to 2} \left(\dfrac{x^4 - 16}{x^3 - 8} \right)^{1/2}$

21. $\lim\limits_{x \to -4} \dfrac{x^2 + x - 6}{x^2 + 7x + 12}$

22. $\lim\limits_{x \to -3} \dfrac{x^2 + x - 6}{x^2 + 7x + 12}$

23. $\lim\limits_{x \to -2} \dfrac{x^2 + x - 6}{x^2 + 7x + 12}$

24. Is it possible for a function f to satisfy the inequality $2x \le f(x) \le x^2 + 1$ on an interval containing $x = 1$? Supposing that this inequality does hold near $x = 1$, prove that $f(x) \to 2$ as $x \to 1$.

25. Evaluate $\lim\limits_{x \to 0} f(x)$ if $\sqrt{1 + x^2} \le f(x) \le 1 + |x|$.

26. It was remarked in the text that although the greatest integer function is discontinuous at $x = 2$, nevertheless it does have the limit 1 as x approaches 2 *from the left*, and also the limit 2 as x approaches 2 *from the right*. Suppose that we designate these one-sided limits by using a minus superscript to indicate approach from below, and a plus superscript to indicate approach from above. Thus

$$\lim_{x \to 2^-} [x] = 1, \qquad \lim_{x \to 2^+} [x] = 2.$$

In this terminology, what are the values of the following one-sided limits?

a) $\lim\limits_{x \to 4^-} [x]$

b) $\lim\limits_{x \to 4^+} [x]$

c) $\lim\limits_{x \to 4^-} p(x)$

d) $\lim\limits_{x \to 4^+} p(x)$

e) $\lim\limits_{x \to 4^-} [x] + p(x)$

f) $\lim\limits_{x \to 4^+} [x] + p(x)$

27. Graph the function $f(x) = |x|/x$. What are its one-sided limits at $x = 0$?

Determine each of the one-sided limits below.

28. $\lim\limits_{x \to 1^+} \dfrac{|x - 1|}{x - 1}$

29. $\lim\limits_{x \to 1^-} \dfrac{|x - 1|}{x - 1}$

30. $\lim\limits_{x \to 2^-} \dfrac{|x - 2|}{x - 2}$

31. $\lim\limits_{x \to 3^+} \dfrac{x - 3}{|x - 3|}$

32. $\lim\limits_{x \to 4^-} \dfrac{x^2 - 16}{|x - 4|}$

33. $\lim\limits_{x \to 1^+} \dfrac{|1 - x^2|}{1 - x}$

34. If $\lim\limits_{x \to 4^-} f(x) = f(a)$, we say that f is continuous from the left at $x = a$. Write down the definition for continuity at the right at $x = a$.

35. Examine the greatest integer function for one-sided continuity and state your conclusions.

36. The same for the postage stamp function $p(x)$.

5. THE LIMIT LAWS

When we reduce a limit calculation to the trivial limit of a continuous function, as in the last section, we don't need to make any explicit use of laws of algebra for limits. Algebraic combinations of continuous functions are continuous, according to the principles in Section 3, and this fact automatically accounts for the limiting behavior of such algebraic combinations. However, there are situations in which the reduction of a limit calculation to the evaluation of a continuous function doesn't occur quite so naturally and easily, and it is useful and customary to replace the continuity laws by their more general limit formulations. We do this below. At present we shall regard these new versions as *axioms* about the continuous and limiting behavior of functions. They can play much the same role as the axioms of geometry. That is, we can start with them as properties that seem obviously true, and then use them to justify steps leading to other conclusions that are less obvious. The kind of reasoning needed to prove the limit laws themselves will be discussed in Appendix 3.

In these laws there are implicit assumptions about domains. Generally it is understood that all the functions in question are defined throughout some interval about the limit point a except possibly at a itself. Occasionally we want to consider a function defined only on *one* side of a. For example, the domain of f might be an interval $[a, b]$ or (a, b). Then the other functions in question will be defined on the *same* side of a, so that the combinations $f(x) + g(x)$, $f(x)g(x)$, etc., can still be formed.

Continuity and Limit Laws

L1. *A function f is continuous at the point $x = a$ if and only if $f(a)$ is defined and $f(x) \to f(a)$ as $x \to a$.*

L2. *For each positive integer n, the function $f(x) = x^{1/n}$ is continuous at each point of its domain, as is the constant function $f(x) = c$.*

L3. *If $f(x) \to l$ and $g(x) \to m$ as $x \to a$, then*

   a) $f(x) + g(x) \to l + m$,

   b) $f(x)g(x) \to lm$,

   c) $\dfrac{1}{f(x)} \to \dfrac{1}{l}$ (provided $l \neq 0$)

as $x \to a$.

L4. *If $f(x) \to l$ as $x \to a$, and if $g(y)$ is continuous at $y = l$, then*

$$g(f(x)) \to g(l) \qquad as \; x \to a.$$

So far this is simply our list from Section 5 partially restated in terms of limits. We now add the squeeze principle.

L5. *If* $f(x)$ *and* $h(x)$ *have the same limit* l *as* $x \rightarrow a$, *and if* $f(x) \leq g(x) \leq h(x)$ *on some interval about* a *(except, possibly, at* $x = a$*), then* $g(x) \rightarrow l$ *as* $x \rightarrow a$. *In particular, if* $f(x) = g(x)$ *except possibly at* $x = a$, *and if* $f(x) \rightarrow l$ *as* $x \rightarrow a$, *then* $g(x) \rightarrow l$ *as* $x \rightarrow a$.

An obvious property that we haven't mentioned yet is what might be called the *contagiousness of positivity*: If f is continuous at a and $f(a) > 0$ then f is necessarily positive on some interval about a.

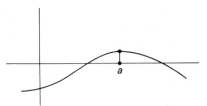

Here is its statement as a limit law:

L6. *If* $f(x) \rightarrow l$ *as* $x \rightarrow a$, *and if* $l > 0$, *then* $f(x) > 0$ *on some interval about* a *(possibly excluding* $x = a$*).*

A corollary of **L6** is worthwhile stating separately:

L7. *If* $f(x) \geq 0$ *on some interval about* a *(possibly excluding* a *itself), and if* $f(x) \rightarrow l$ *as* $x \rightarrow a$, *then* $l \geq 0$.

Our informal computation of the limit of $(1 - \sqrt{x})/(1 - x)$ as x approaches 1 could now be recast into a formal proof in which each step is justified by one of the limit laws. The same will be true of the many limit calculations to be found throughout the book; they will generally be presented informally, but will have formal versions that can be filled in by anyone interested.

Note that **L1** and **L3** imply:

L3′. *If* f *and* g *are continuous at* a, *then so are* $f + g$, fg, *and* $1/f$ *(provided* $f(a) \neq 0$*).*

By repeatedly applying this principle, starting from the continuity of x and the constant functions, we establish the continuity of any function that can be built up from x and constants by using the two operations of addition and multiplication, i.e., any *polynomial* Examples are

$$p(x) = x^2 + 2x + 5,$$
$$q(x) = 4x^{13} - x^9 + 2x^5 + 3x.$$

Knowing that every polynomial is continuous, we can then conclude, from **L3′** again, that the quotient

$$r(x) = \frac{p(x)}{q(x)}$$

of any two polynomials is continuous, except where it is undefined (i.e., where $q(x) = 0$). A quotient of two polynomials is called a *rational function*.

We thus have the theorem:

THEOREM 1   *A polynomial function $p(x)$ is everywhere continuous. A rational function*

$$r(x) = \frac{p(x)}{q(x)}$$

where p and q are polynomials, is continuous except where undefined (i.e., where $q(x) = 0$).

More generally, by using also the continuity of $x^{1/n}$ and **L4**, it can be shown that any function that can be built up from x and the constants by using the operations of addition, multiplication, division, and root extraction, is continuous wherever it is defined.

Our discussion of continuity and limits has been intuitive and qualitative. We have not considered *how* close to a we have to take x in order to ensure that $f(x)$ is closer to l than some preassigned tolerance. It is not too difficult to make these quantitative calculations, and carrying them out for various combinations of functions amounts to quantitative proofs for the above laws. See Appendix 3.

We conclude with three remarks.

A.   To be completely formal, the proof of Theorem 1 has to use a principle called *mathematical induction* that the reader may or may not know about. We shall not give any formal proofs requiring induction in this book.

B.   A word about polynomial notation. A particular polynomial, like the ones displayed above, can always be written out explicitly. This becomes a chore when the degree is large—a polynomial of degree 18 would (normally) have 19 terms, for example—and in many situations there is no particular need to see each individual term. Then we can present the polynomial "in outline" by giving its first and last terms, and indicating the middle terms by three dots, ..., that we read "and so on." Thus,

$$p(x) = a_{18}x^{18} + a_{17}x^{17} + \cdots + a_0$$

is the polynomial $a_{18}x^{18} + a_{17}x^{17} +$ and so on, down to a_0. Of course this notation is not needed for low-degree polynomials.

C.   A first-degree polynomial $ax + b$ is often called a linear function because its graph is a straight line. This use of the word *linear* is related to, but different from, its use in linear algebra.

PROBLEMS FOR SECTION 5

1. List all the steps required to show that the polynomial $p(x) = x^3 - 4x + 2$ is continuous, justifying each step by referring to part of **L3′**.

2. List all the steps required to show that the rational function $r(x) = (x^2 - 1)/(x^2 + 1)$ is continuous, justifying each step by referring to part of **L3′**.

Justify the continuity of the following functions from **L2** and **L3′**, in the manner of the two problems above.

3. $\sqrt{x} + \dfrac{1}{\sqrt{x}}$

4. $\dfrac{1}{x^{1/3} + 1}$

5. $\dfrac{\sqrt{x}}{1 + x^2}$

6. $x^{1/2} + x^{1/3} + x^{1/4}$

7. Write out the limit law **L3** for limits from the left, using the notation:

$$\lim_{x \to a^-} f(x).$$

(See the remarks in Problem 26 in the last section.)

8. Show that $y^2 - 4y$ is continuous at $x = 2$ if $y = f(x)$ and

$$\lim_{x \to 2} f(x) = 1, \qquad f(2) = 3.$$

9. Show that $y^3 - 4y + 1$ is continuous at $x = a$ if $y = f(x)$ and

$$\lim_{x \to a^-} f(x) = -2, \qquad f(a) = 0, \qquad \lim_{x \to a^+} f(x) = +2.$$

10. Show that **L5** can be proved from the following special case **L5'** and the other limit laws.

    L5'. If $0 \le g(x) \le f(x)$ on some interval about a (except, possibly, at $x = a$), and if $f(x) \to 0$ as $x \to a$, then $g(x) \to 0$ as $x \to a$.

11. Show from **L6** and other limit laws that:

    If $f(x) \to l$ as $x \to a$ and if $l > k$, then $f(x) > k$ on some interval about a (possibly excluding $x = a$).

12. State and prove a similar variant of **L6** when $l < 0$.

13. State and prove a similar variant of **L6** when $l < k$.

14. Prove **L7** as a corollary of Problem 12.

15. Prove **L7** as a corollary of **L5**.

6. INFINITE LIMITS AND LIMITS AT INFINITY

Although the symbol $+\infty$ does not represent a number, it is nevertheless used in limit statements with a special convention about what is being said.

EXAMPLE 1   The equation

$$\lim_{x \to 0} \frac{1}{x^2} = +\infty,$$

read literally, says that $1/x^2$ approaches and comes arbitrarily close to $+\infty$ as x approaches 0. We interpret this as meaning that $1/x^2$ *grows without bound* as x approaches 0, in the sense that it becomes and remains larger than any given number when x is taken suitably small. This is true. For example, if we want

$$\frac{1}{x^2} > 10,000,$$

then we only have to take $|x| < 1/100$.

Similarly,

$$\lim_{x \to 0} -\frac{1}{x^2} = -\infty,$$

meaning that $-1/x^2$ becomes and remains *smaller* than any given negative number when $|x|$ is taken suitably small. For example,

$$-\frac{1}{x^2} < -10{,}000$$

when $|x| < 1/100$.

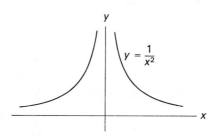

The "approaching symbolism seems especially appropriate for infinite limits. Thus,

$$\frac{1}{x^2} \to +\infty \qquad \text{as } x \to 0.$$

EXAMPLE 2   The function $f(x) = 1/x$ is more complicated in that what it does as x approaches 0 depends on which side of zero x is on. We use 0^- and 0^+ to indicate approach from below and from above respectively. Thus

$$\lim_{x \to 0^-} \frac{1}{x} = -\infty, \qquad \lim_{x \to 0^+} \frac{1}{x} = +\infty.$$

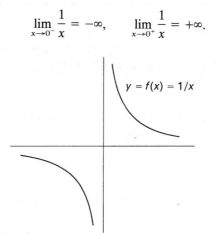

In the same way, the postage-stamp function $p(x)$ has unequal one-sided limits at each of its points of discontinuity. For example,

$$\lim_{x \to 1^-} p(x) = 1, \qquad \lim_{x \to 1^+} p(x) = 2,$$

as you will see upon reviewing the definition and graph of $p(x)$ on page 53. See Problem 26 and the following problems in Section 4.

EXAMPLE 3   Consider next the statement

$$\lim_{x \to +\infty} \frac{1}{x} = 0.$$

Here the limit is a number but the independent variable x is tending to ∞, i.e., growing without bound. The statement means that $1/x$ can be made as close to zero as desired by taking x sufficiently *large*. For example, to get

$$\frac{1}{x} < \frac{1}{1000}$$

we take $x > 1000$.

EXAMPLE 4 Similarly

$$\lim_{x \to +\infty} \frac{1}{\sqrt{x}} = 0.$$

But $1/\sqrt{x}$ gets small much more slowly than $1/x$. In order to ensure that

$$\frac{1}{\sqrt{x}} < \frac{1}{1000},$$

we have to take $x > (1000)^2 = 1,000,000$.

It is customary to combine the two statements

$$\lim_{x \to +\infty} \frac{1}{x} = 0 \quad \text{and} \quad \lim_{x \to -\infty} \frac{1}{x} = 0$$

into the single statement

$$\lim_{x \to \infty} \frac{1}{x} = 0 \quad \text{or} \quad \frac{1}{x} \to 0 \quad \text{as } x \to \infty.$$

Limits at infinity obey the same limit laws (**L3** through **L7**) as for limits at a, but sometimes a preliminary revision is needed before a computation can be made.

EXAMPLE 5 In order to verify that

$$\lim_{x \to \infty} \frac{3 + x^2}{3x^2 + 1} = \frac{1}{3},$$

we first rewrite the expression by dividing top and bottom by x^2, getting

$$\frac{\dfrac{3}{x^2} + 1}{3 + \dfrac{1}{x^2}}$$

(for $x \neq 0$). Now $3/x^2$ and $1/x^2$ both approach 0 as x tends to ∞. Thus

$$\lim_{x \to \infty} \frac{3 + x^2}{3x^2 + 1} = \lim_{x \to \infty} \frac{\dfrac{3}{x^2} + 1}{3 + \dfrac{1}{x^2}}$$

$$= \frac{0 + 1}{3 + 0} = \frac{1}{3}.$$

EXAMPLE 6 Show that the hyperbola

$$y = \frac{b}{a}\sqrt{x^2 - a^2}$$

is asymptotic to the line

$$y' = \frac{b}{a}x$$

in the strong sense that

$$y' - y \to 0 \qquad \text{as } x \to +\infty.$$

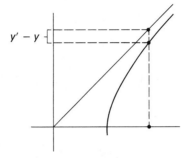

Solution. We rationalize a numerator:

$$y' - y = \frac{b}{a}[x - \sqrt{x^2 - a^2}]$$

$$= \frac{b}{a}[x - \sqrt{x^2 - a^2}]\frac{(x + \sqrt{x^2 - a^2})}{(x + \sqrt{x^2 - a^2})}$$

$$= \frac{ab}{x + \sqrt{x^2 - a^2}}.$$

In particular,

$$0 < y' - y < \frac{ab}{x},$$

and since $1/x \to 0$ as $x \to +\infty$, it follows that

$$y' - y \to 0$$

by the squeeze limit law.

___

If

$$\lim_{x \to \infty} \frac{f(x)}{g(x)} = 1,$$

then f and g must behave in essentially the same way "at ∞." For example, if g has a limit at infinity, then f must have the same limit:

If $g(x) \to l$, then

$$f(x) = \frac{f(x)}{g(x)} \cdot g(x) \to 1 \cdot l = l$$

as $x \to \infty$. This holds for the limits $+\infty$ and $-\infty$ as well.

A polynomial
$$p(x) = a_n x^n + a_{n-1} x^{n-1} + \cdots + a_0$$
behaves like its highest-degree term $a_n x^n$ at ∞. For (supposing $a_n \neq 0$, of course)

$$\frac{p(x)}{a_n x^n} = 1 + \frac{a_{n-1}}{a_n} \cdot \frac{1}{x} + \cdots + \frac{a_0}{a_n x^n}$$

$$\to 1 + 0 + \cdots + 0$$

$$= 1$$

as $x \to \infty$.

In the same way we can show that any rational function

$$r(x) = \frac{p(x)}{q(x)}$$

$$= \frac{a_n x^n + a_{n-1} x^{n-1} + \cdots + a_0}{b_m x^m + b_{m-1} x^{m-1} + \cdots + b_0}$$

behaves like $(a_n / b_m) x^{n-m}$ at ∞.

PROBLEMS FOR SECTION 6

Evaluate the following limits.

1. $\lim\limits_{x \to +\infty} \dfrac{7x + 2}{3 - x}$

2. $\lim\limits_{x \to +\infty} (\sqrt{x^2 + 4} - x)$

3. $\lim\limits_{x \to -\infty} \left(1 + \dfrac{2}{x}\right)$

4. $\lim\limits_{x \to +\infty} \sqrt{x^2 + x + 1} - x$

5. $\lim\limits_{x \to \infty} \dfrac{2x^2 + 1}{6 + x - 3x^2}$

6. $\lim\limits_{x \to \infty} \dfrac{x}{x^2 + 5}$

7. $\lim\limits_{x \to +\infty} \dfrac{3^x - 3^{-x}}{3^x + 3^{-x}}$

8. $\lim\limits_{x \to -\infty} \dfrac{3^x - 3^{-x}}{3^x + 3^{-x}}$

9. $\lim\limits_{x \to \infty} \dfrac{\sqrt{x + 1}}{\sqrt{4x - 1}}$

10. $\lim\limits_{x \to \infty} \left(\dfrac{1}{x} + 1\right)\left(\dfrac{5x^2 - 1}{x^2}\right)$

11. $\lim\limits_{x \to \infty} \dfrac{1 + 99x + 2x^2}{3x^2 + 4}$

Evaluate the following limits.

12. $\lim\limits_{x \to -1^-} \dfrac{3x^2 - 4x - 7}{x^2 + 2x + 1}$

13. $\lim\limits_{x \to 5^+} \dfrac{5x^2 + x^3}{25 - x^2}$

14. $\lim\limits_{x \to 1^+} \dfrac{x^2}{1 - x^2}$

15. $\lim\limits_{x \to 2^+} \dfrac{\sqrt{x^2 - 4}}{x - 2}$

16. $\lim\limits_{x \to 0} \dfrac{x + 1}{|x|}$

17. Evaluate

$$\lim_{x \to 3} \frac{2x^2 - 5x - 3}{x^2 - x - 6}$$

and

$$\lim_{x \to \infty} \frac{2x^2 - 5x - 3}{x^2 - x - 6}.$$

Justify both calculations.

7. THE RANGE OF A CONTINUOUS FUNCTION

The domain of a continuous function f is usually known from inspecting a formula for f, or is given as part of the definition of f. However the range of f (its set of values) may be more elusive. The following theorem is the best general result we have.

THEOREM 2   **(The Range Principle).** *If f is continuous on an interval I, then the range of f over I is also an interval*

This will be self-evident for any continuous graph that we draw; the proof requires an excursion into foundational material and will be mostly put off until Appendix 4.

By virtue of the theorem, the range problem is reduced to finding the *endpoints* of the range interval and determining whether or not they are included. This can be easy, or it can still pose a hard problem.

Find the range of $f(x) = x^4 + 1$.

EXAMPLE 1   ***Solution.*** Note that $f(0) = 1$ and that $f(x) \geq 1$ for all x (because $x^4 \geq 0$). Also that $f(x) \to +\infty$ as $x \to \pm\infty$. The range therefore extends from 1 to $+\infty$. Since the domain of f is an interval (namely $(-\infty, \infty)$), the range must be the interval $[1, +\infty)$, by the theorem.

Find the range of $f(x) = x^5 - x^2$.

EXAMPLE 2   ***Solution.*** Again the domain is the interval $(-\infty, \infty)$, so the range must be an interval, by the theorem. Here $f(x) \to -\infty$ as $x \to -\infty$, and $f(x) \to +\infty$ as $x \to +\infty$, so the range interval is $(-\infty, \infty)$.

Note that this conclusion is equivalent to the assertion that the polynomial equation

$$x^5 - x^2 = c$$

must have at least one solution for each value of the constant c.

EXAMPLE 3   The domain of $f(x) = 1/x$ is not a single interval, but is the union of the two intervals $(-\infty, 0)$ and $(0, +\infty)$.

On $(0, +\infty)$, $f(x) \to +\infty$ as $x \to 0$ and $f(x) \to 0$ as $x \to +\infty$. So the range of f over this interval is $(0, +\infty)$. Similarly, the range over $(-\infty, 0)$ is $(-\infty, 0)$. So here the range equals the domain, a union of two intervals.

Here the theorem was inefficient and distracting. Since the equation $y = 1/x$ is equivalent to $x = 1/y$, all we are saying is that a number has a reciprocal if and only if it isn't zero.

EXAMPLE 4   The function

$$f(x) = \frac{1 + x}{\sqrt{x}} = \frac{1}{\sqrt{x}} + \sqrt{x}$$

is defined and continuous on the interval $(0, \infty)$, and $f(x) \to +\infty$ as x approaches either 0 or ∞. The graph must therefore be something like this.

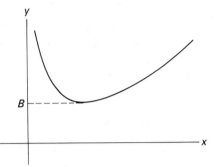

It appears that f must have a positive minimum value B, in which case the range interval is $[B, \infty)$. But finding B is a nontrivial problem that we shall solve in Chapter 6 as an application of calculus. (This particular problem can also be solved by algebra.)

Underlying Theorem 2 is a more basic property of a continuous function called the *intermediate-value principle*. It is related to an intuitive description of continuity that is sometimes used as a definition, namely, that if a function is continuous on an interval I then its graph over I "can be drawn without lifting the pencil." That is, the graph does not fall into two or more separate pieces that would require lifting the pencil in going from one to another; it is a single connected strand. This description is not useful in practice because it is difficult to pin down precisely, but there is a special case that is very important. This is the fact that the graph of a continuous function cannot "jump over" a horizontal line $y = k$.

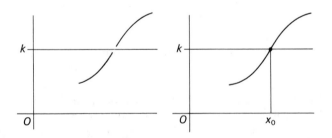

If the graph crosses from a value $y_1 = f(x_1)$ below the line to a value $y_2 = f(x_2)$ above it, then in the process the graph must intersect the line. (Otherwise, the part above $y = k$ would be disconnected from the part below, and the pencil would have to be lifted in going from one part to the other in order to avoid the line.) Here is the formal statement.

Intermediate-Value Principle. *If f continuous on the closed interval $[a, b]$, and if k is a number between $f(a)$ and $f(b)$, then there is a number X in $[a, b]$ such that $f(X) = k$.*

Note that this is an *existence* principle. It says that a number having a certain property exists, but doesn't specify where it is, or how to calculate it. We shall take up the proof in Appendix 4.

We turn now to the relationship between these two principles. Suppose, first, that the range of f on $[a, b]$ is known to be an interval I. Since I contains $f(a)$ and $f(b)$, it also contains any number k lying between $f(a)$ and $f(b)$. That is, any such k is of the form $k = f(X)$, for some X in $[a, b]$. The intermediate-value principle is thus an easy corollary of Theorem 2.

The proof in the other direction is harder, and we shall consider only the special cases that are important for calculus.

If the graph of f steadily rises as x increases, we say that f is an *increasing function*. If the graph moves only downward, f is *decreasing*.

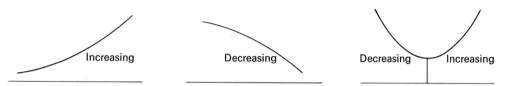

Increasing   Decreasing   Decreasing   Increasing

Generally a function will increase over one interval and decrease over another, and we phrase the definition accordingly.

DEFINITION   *The function f is* increasing *on an interval I in its domain if $f(x_1) < f(x_2)$ whenever $x_1 < x_2$ in I. Similarly, f is* decreasing *on I if $f(x_1) > f(x_2)$ whenever $x_1 < x_2$ in I. Finally, f is* monotone *on I if either f is increasing on I or f is decreasing on I.*

EXAMPLE   The function $f(x) = x^n$ is increasing on $[0, \infty)$ for any positive integer n. So is $f(x) = x^{1/n}$. Over the interval $(-\infty, 0)$, $f(x) = x^n$ is increasing if n is odd and decreasing if n is even. See Problems 70 through 81 at the close of this chapter.

The definition of an increasing function can be interpreted graphically as the fact that the slope of any chord is positive:

$$\frac{f(x_2) - f(x_1)}{x_2 - x_1} > 0$$

for any pair of distinct points x_1 and x_2. This is also equivalent to:

The differences $x_2 - x_1$ and $f(x_2) - f(x_1)$ always have the same sign.

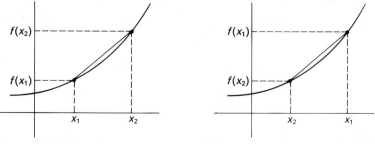

In particular,

$$\text{if } f(x_1) < f(x_2) \qquad \text{then } x_1 < x_2.$$

That is: *an increasing function is also reverse increasing.* Of course, this reverse property can be derived directly from the original definition: If f is increasing then $f(x_1) \geq f(x_2)$ whenever $x_1 \geq x_2$. Therefore we can have the inequality $f(x_1) < f(x_2)$ only if $x_1 < x_2$. In other words,

$$\text{if } f(x_1) < f(x_2) \quad \text{then } x_1 < x_2.$$

Now suppose that f is an increasing function defined on the closed interval $[a, b]$. The range of f then lies in the interval $[f(a), f(b)]$.

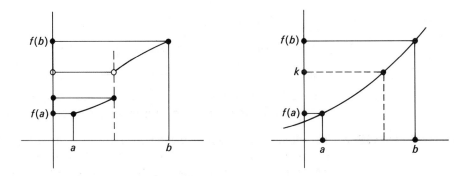

It needn't be the whole interval, as the left figure above shows. However, if f is continuous, then the intermediate-value principle says exactly that the range of f *is* the whole interval, since it says that *every* number k between $f(a)$ and $f(b)$ is a value of f. Thus for such a function Theorem 2 is an immediate corollary of the intermediate-value principle. Moreover, since f is increasing, each such number k is assumed as a value only once.

For example, $f(x) = x^3$ is continuous and increasing on the interval $[1, 2]$, and its range over this interval is therefore the interval $[1^3, 2^3] = [1, 8]$. In particular, since the range interval contains the number 2, we know there is exactly one number X between 1 and 2 such that $X^3 = 2$. So we know that the cube root of 2 exists and is uniquely determined.

In the same way, we can establish the existence of a unique nth root of any positive number k.

On an open interval the situation is more complicated, but if we know the limits of f at the two end points, then we can proceed in the same two steps.

THEOREM 3   *If f is increasing on the open interval (a, b), and if f has the limits A and B as x approaches a and b respectively, then the range of f is included in the open interval (A, B).*

Proof. Let $k = f(x_0)$ be any point of the range. Then $k < f(x_1)$ for any fixed point x_1 between x_0 and b, and $f(x_1) \leq \lim_{x \to b} f(x) = B$, by **L7**, so $k < B$. Similarly, $k > A$, *and k is in (A, B).* The range of f is thus included in (A, B). ▮

THEOREM 4   *If f is continuous and satisfies the hypotheses of Theorem 3, then the range of f is exactly the open interval (A, B). (Here a and/or A can be $-\infty$. Also b and/or B can be $+\infty$.)*

Proof. Let k be any number in (A, B),

$$A < k < B.$$

Since $f(x) \to A$ as $x \to a$, $f(x)$ eventually becomes less than k, and we can choose a point x_1 such that $f(x_1) < k$. Similarly, $f(x)$ eventually becomes greater than k as $x \to b$, so we can choose x_2 with $f(x_2) > k$. Then k is a value of f on the closed interval $[x_1, x_2]$ by the intermediate-value principle, thus proving that the interval (A, B) is included in the range of f. Theorem 3 gave the opposite inclusion, so we are done.   ∎

Note that this procedure of first locating points x_1 and x_2 in (a, b) for which $f(x_1) < k < f(x_2)$ is exactly how we showed earlier that $2^{1/3}$ exists.

As a corollary of Theorem 3, we have the following curious but obvious result.

THEOREM 5   *If f is continuous on the closed interval $[a, b]$ and increasing on the open interval (a, b) then f is increasing on $[a, b]$.*

Proof. The range of f over the open interval (a, b) is included in the open interval $(f(a), f(b))$, by Theorem 3. That is, if $a < x < b$ then $f(a) < f(x) < f(b)$. But this is exactly the extra conclusion asserted by the present theorem.   ∎

Finally, we show for increasing functions that continuity really *can* be characterized as the condition that "the graph can be drawn without lifting the pencil," in the sense that the range fills out an interval.

THEOREM 6   *If f is increasing on an interval I, and if its range over I is an interval, then f is necessarily continuous on I.*

Proof. Suppose we want to prove that f is continuous at an interior point x_0 in the interval I. (At an endpoint the argument would be essentially the same, though differing slightly in detail.) We have to show that $f(x)$ can be kept as close as we wish to $f(x_0)$ simply by restricting x to a suitably small interval J about x_0. So suppose we want $f(x)$ to be closer to $f(x_0)$ than the distance d:

$$f(x_0) - d < f(x) < f(x_0) + d.$$

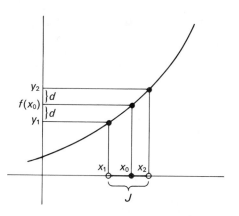

Now the numbers $y_1 = f(x_0) - d$ and $y_2 = f(x_0) + d$ are values of f. (We start with d small enough so that both numbers lie in the range interval.) Thus there are uniquely determined points x_1 and x_2 in I such that $y_1 = f(x_1)$ and $y_2 = f(x_2)$. And since $y_1 < f(x_0) < y_2$, we know that $x_1 < x_0 < x_2$, by the reverse-increasing property of f. Thus the open interval $J = (x_1, x_2)$ contains x_0. Moreover, if x is restricted to J, i.e., if

$$x_1 < x < x_2,$$

then

$$y_1 < f(x) < y_2,$$

or

$$f(x_0) - d < f(x) < f(x_0) + d,$$

because f is increasing. Thus J does what it was supposed to do, and we have shown f to be continuous. ∎

PROBLEMS FOR SECTION 7

Use Theorem 2 to determine the ranges of the following functions.

1. $f(x) = x^6$
2. $f(x) = x^7$
3. $f(x) = x^{1/6}$
4. $f(x) = x^{1/7}$
5. $f(x) = \dfrac{1}{9 - x^2}$
6. $f(x) = \dfrac{1}{x^2 + 2x}$
7. $f(x) = \sqrt{36 - x^2}$
8. $f(x) = \sqrt{x^2 - 16}$
9. $f(x) = (8 - x^2)^{1/3}$
10. $f(x) = (16 - x^2)^{1/4}$
11. $f(x) = x + 1/x$ (Solve $y = x + 1/x$ for x in terms of y to see what values of y are possible.)

Use the intermediate value principle to show that each of the following equations has a solution on the given interval.

12. $x^3 - x^2 + x = 2$, $[1, 2]$
13. $x^5 - x^4 - x^2 = 0$, $[1, 2]$
14. $x^4 - x^3 - x^2 - x = -1$, $[0, 1]$
15. $x^{1/3} + x^{1/2} = 2$, $[0, 4]$
16. $x^{1/2} - x^{1/3} = 3$, $[32, 64]$

EXTRA PROBLEMS FOR CHAPTER 2

Each of the following equations is of the form $y = f(x)$. Determine the domain of f in each case, and express it as an interval or a union of intervals.

1. $y = 8x$
2. $y = \dfrac{1}{x + 3}$
3. $y = \dfrac{x + 4}{(x - 2)(x + 3)}$
4. $y = \dfrac{1}{2x}$
5. $y = \sqrt{16 - x^2}$
6. $y = \dfrac{1}{9 - x^2}$
7. $y = \sqrt{x^2 - 16}$

Show that $x - y$ is a factor of $g(x) - g(y)$ for each of the following functions.

8. $g(x) = x^2$

9. $g(x) = 1/x$

10. $g(x) = x^3$

11. $g(x) = 1/(x^2 + 1)$

12. Let f be the reciprocal function, $f(x) = 1/x$. Show that $f(f(x)) = x$ for all x different from 0.

13. Let g be the function $g(x) = \sqrt{1 - x^2}$. Determine the domain D of g and show that $g(g(x)) = x$ for every x in "one-half" of D.

14. If f is the square-root function and g is the squaring function, show that

$$g(f(x)) = x \qquad \text{for every } x \text{ in domain } f;$$
$$f(g(x)) = |x| \qquad \text{for every } x.$$

15. If f is the cube-root function. $f(x) = x^{1/3}$, what is the function g such that $g(f(x)) = x$ for all x?

16. Find a function f such that

$$\frac{f(x) + 1}{f(x) - 1} = x.$$

(Solve the equation for $f(x)$.) What is the domain of f?

17. Suppose that

$$f = g + h$$

where g is even and h is odd. Prove that

$$g(x) = \frac{f(x) + f(-x)}{2}, \qquad \text{and} \qquad h(x) = \frac{f(x) - f(-x)}{2}.$$

Now show that an *arbitrary* function f, defined on an interval of the form $[-a, a]$, can be written in a *unique* way as a sum of an even function g and an odd function h. We shall call g the *even component* of f, and h the *odd component* of f.

18. Find the even and odd components of the polynomial $1 + x + x^2$.

19. Find the even and odd components of the polynomial $p(x) = a_0 + a_1 x + \cdots + a_n x^n$. When is $p(x)$ itself even? odd? Your answer here should suggest where the words *even* and *odd* come from.

20. Show that the sum of two odd functions is odd, and that the product of two odd functions is even. Write down the rest of the similar statements about even and odd functions.

21. Show how the even and odd components of the product of two functions can be expressed in terms of the even and odd components of the factors. Use the notation

$$f = g + h, \qquad f_1 = g_1 + h_1, \qquad f_2 = g_2 + h_2, \qquad f = f_1 f_2.$$

22. Graph $f(x) = x/|x|$. What symmetry does the graph have?

23. Graph $f(x) = |x| + x$. Find and graph the even and odd components of f.

24. Sketch the graphs of $f(x) = x^2$ and $g(x) = x^4$ on the same axis system, using the function values at $x = 0, 1/2, 1, 3/2$, and symmetry. What configuration in the coordinate plane do the graphs of the even powers x^{2n} approach as n gets very large?

25. Answer the same question for $f(x) = x^3$, $g(x) = x^5$, and the family of all odd powers x^{2n+1}.

26. A polynomial of degree three,
$$p(x) = x^3 + a_2 x^2 + a_1 x + a_0,$$
behaves like x^3 when $|x|$ is very large. Moreover, for *any* polynomial p, $x - a$ is a factor of $p(x)$ if and only if $p(a) = 0$. (This is a theorem of polynomial algebra.) Classify the possible graphs of cubic polynomials p by the ways in which they can cross or touch the x-axis, and draw a sketch of each possibility. (Show that there are four possibilities, as follows:
    a) cross 3 times;
    b) cross once and touch once from above;
    c) cross once and touch once from below;
    d) cross once.)

The following polynomials exhibit the above four possibilities. Determine which is which and sketch very roughly. [*Hint.* Factor.]

27. $x^3 + x$

28. $x^3 - x$

29. $x^3 + 2x^2 + x$

30. $x^3 - 2x^2 + x$

31. Make the same analysis of the possible graphs of a fourth-degree polynomial
$$p(x) = x^4 + a_3 x^3 + a_2 x^2 + a_1 x + a_0.$$

Sketch the following quadratics and cubics.

32. $y = 3x - x^2$

33. $y = (x + 1)^3$

34. $y = x^3 + 2x^2 - x - 2$

35. A rubber strand occupying the segment $[a, b]$ is stretched to the new position $[c, d]$. If y is the new position in $[c, d]$ of the point originally at the position x in $[a, b]$, then y is a function of x. Derive a formula for this function. Draw its graph.

36. Suppose a rubber line segment is held under tension between endpoints a and b. We attach clips to the segment at points $x_1, x_2, \ldots, x_n$, and then move these clips to the new positions (along the same line) $y_1, y_2, \ldots, y_n$. If y is the new position of the point originally at x, then y is a function of x. Draw the graph of this function.

37. Show that when x is within a distance d of 2, then x^2 is within a distance $d(d + 4)$ of 4. [*Hint:* Write $x = 2 + e$, where $|e| < d$, and then expand x^2.]

38. Suppose that x is within a distance d of the point 2, where $d < 1$. Show that then x^2 is within a distance $5d$ of 4. [*Hint:* Show first that $1 < x < 3$. Then factor $x^2 - 4$.)

39. Suppose that x is within a distance d of 2, where $d < 1$. Show that then x^3 is within a distance $19d$ of 8.

40. Strictly speaking, the function $f(x) = (x^2 - x)/(x - 1)$ is not defined at $x = 1$. Draw its graph, indicating the missing domain point. Show that f becomes continuous if $f(1)$ is suitably defined.

41. Suppose that $f(x) = (x^2 + x - 6)/(x - 2)$ for $x \neq 2$. Define $f(2)$ so that f will be everywhere continuous.

42. The same problem for
$$f(x) = \frac{x^2 - 1}{x^3 - 1},$$
and the point $x = 1$.

43. The same problem for
$$f(x) = \frac{\sqrt{x} - 2}{x - 4}$$
at the point $x = 4$.

44. Show that $[x] + p(x)$ is not one-sidedly continuous at $x = 4$.

45. Suppose that

$$\lim_{x \to a^-} f(x), \qquad f(a), \qquad \lim_{x \to a^+} f(x)$$

all exist. List the possibilities for equality and/or inequality among these three values.

46. Show that the following function is continuous everywhere. The crucial point is $t = -1$, where you have to show that the left- and right-hand limits are both equal to $h(-1)$. Graph the function h.

$$h(t) = \begin{cases} 3t + 5 & \text{if } t < -1, \\ t^2 + 1 & \text{if } t \geq -1. \end{cases}$$

47. Determine the constant c so as to make the function

$$f(x) = \begin{cases} x^2 & \text{if } x < 1/2, \\ c - x^2 & \text{if } x \geq 1/2 \end{cases}$$

continuous at $1/2$. Sketch its graph.

Evaluate the following limits (if possible).

48. $\lim\limits_{x \to 1/2} (2x - 1)/(x^2 - 1/4)$

49. $\lim\limits_{x \to 8} (\sqrt{x - 4} - 2)/(x - 8)$

Evaluate the following one-sided limits.

50. $\lim\limits_{x \to 2^-} \sqrt{2 - x}/(\sqrt{x^2 - 3})$

51. $\lim\limits_{x \to 2^-} \sqrt{4 - x^2}/\sqrt{2 - x}$

52. Show by an example that the following "one-sided" version of **L4** is incorrect:

If $f(x) \to l$ as $x \to a^-$ and if $g(y)$ is continuous from the left at $y = l$, then

$$g(f(x)) \to g(l) \qquad \text{as } x \to a^-.$$

53. State a correct version of **L4** involving $x \to a^-$.

54. Prove that

$$\lim_{x \to 0^-} [1 - x^2] = 0$$

from the limit laws and the known behavior of $[x]$. Graph the function $f(x) = [1 - x^2]$.

Evaluate the following limits.

55. $\lim\limits_{x \to \infty} \sqrt{x - 1} - x$

56. $\lim\limits_{x \to 1^-} (x - 1/x)/(x - 1)^2$

57. $\lim\limits_{x \to -2^-} (x + 1)/|x + 2|$

58. $\lim\limits_{x \to 0^+} |x| x/x^3$

59. Prove the assertion made at the end of Section 6 about $r(x)$ behaving like

$$(a_n/b_m)x^{n-m} \qquad \text{at } \infty.$$

60. How large must a positive number x be taken in order to ensure that $1/x^3$ is less than .001?

61. How large must x be taken to ensure that $x^{-1/3}$ is less than .1? .01? .001?

62. Let d be a small positive number. Determine how large a positive number x must be taken in order to ensure that $1/x^3 < d$.

63. Determine how large x must be taken to ensure that $x^{-1/3} < d$.

64. Show that

$$\frac{1}{\sqrt{x}} > 1000$$

for all x smaller than a certain number d. (Find an explicit value of d that will work.)

65. If M is any positive number, show that

$$\frac{1}{\sqrt{x}} > M$$

for all positive x smaller than a certain number d. (Find d in terms of M.)

66. Given a positive number d, determine a positive number M such that

$$\frac{1}{x^2} < d$$

for all x larger than M.

67. Given a positive number d, determine a positive number M such that

$$\frac{1}{x^{1/3}} < d$$

for all x larger than M.

Find the domain and range of each of the following functions.

68. $y = 1/(\sqrt{x} - 1)$ 69. $y = \sqrt{4x^2 - 1}$

The following problems involve working with inequalities. See Appendix 1 for a review of the inequality laws. Later we will find it easier to prove that particular functions are increasing by showing in each case that the derivative is positive.

For Problems 70 through 80 let f and g be functions that are increasing on an interval I. Show that:

70. $f + g$ is increasing.

71. fg is increasing if f and g are both positive.

72. $1/f$ is decreasing if f is positive.

73. $-f$ is decreasing.

74. $f(-x)$ is decreasing on $-I$.

75. Use Problem 71 to show that the functions x, x^2, $x^3, \ldots$ are all increasing on $I = [0, \infty)$.

76. Use Problems 73 through 75 to show that x^n is decreasing on $I = (-\infty, 0]$ if n is even; x^n is increasing on $I = (-\infty, 0]$ if n is odd.

77. Show that $f(g(x))$ is increasing (if the domain of f includes the range of g).

78. Suppose that $f(g(x)) = x$ and that f is increasing. Show that g is increasing. [*Hint:* Use the "reverse-increasing" property of f.]

79. Show from Problems 75 and 78 that $x^{1/n}$ is increasing on $[0, \infty)$.

80. Show that f^n and $f^{1/n}$ are increasing, if f is positive.

81. List the answers to problems 70 through 74 if f and g are decreasing on I.

3

The Derivative

We shall rely heavily on geometric reasoning and intuition in our development of the calculus. This does not mean we have a particular interest in geometry, although it is one of the subjects to which calculus can be applied. (For example, we shall be able to compute areas and volumes that are otherwise inaccessible to us.) Rather, the reason for the special role of geometry is that geometric intuition helps us understand calculus itself, in much the way that the graph of a function f lets us see in a very vivid manner how $f(x)$ varies with x. Furthermore, geometric reasoning will help us to apply calculus to *other* disciplines. Here, for example, is a problem of a type that we shall take up in Chapter 6.

The efficiency E of a certain machine varies with the viscosity x of the lubricating oil that is used. The machine runs badly if the oil is either too thin or too thick, i.e., if x is either too small or too large. Our theoretical calculations (or perhaps our empirical observations) show us that the relationship between E and x is given by

$$E = x - x^3.$$

The problem is to determine the viscosity x that gives the maximum efficiency E. When we plot E against x we find that the graph looks like this in the first quadrant.

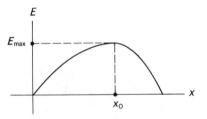

Now we see that our problem can be rephrased geometrically as that of finding the *highest point* on this graph.

This example shows how a nongeometric problem can be cast into geometric terms, but it doesn't suggest the role of calculus. We shall see, however, that calculus will let us compute the slope m of the tangent line as a function of the point of tangency x. At the highest point on the above graph, the tangent line appears to be horizontal, so we want the value of x for which $m = 0$. Therefore, in order to solve our maximum problem, we first compute the tangent slope function $m(x)$ by calculus, and then solve the equation $m(x) = 0$ to find the value $x = x_0$ giving maximum efficiency.

Thus, when we approach the derivative through the tangent-line problem, we will not be solving *only* this geometric problem. We will also be creating a geometric interpretation of the derivative that will help us in calculus itself and in the applications of calculus to other disciplines.

Later in this chapter we shall take up the other major interpretation of the derivative, as a rate of change. Otherwise, the chapter treats the computation of derivatives from scratch, the beginning algorithms of the systematic theory, and a few elementary applications.

1. THE SLOPE OF A TANGENT LINE

The tangent-line problem was frustrating for classical geometers who knew only synthetic geometry. There just is no universal geometric characterization of a tangent line that generalizes the classical circle construction. We can visualize the tangent line as having the same *direction* that the curve has at the point of tangency, and we can find the tangent if we can find this common direction. For the circle and a few other special curves there are geometric characterizations of the tangent line, but there is no purely geometric way to obtain the directions of general curves.

The solution to the tangent problem had to wait for the development of analytic geometry. In the coordinate plane a direction is specified by its *slope*, and the tangent problem is solved by the surprising discovery that we can in fact compute the slope of a smooth graph at any point.

By way of illustration we shall compute the slope of the graph of $f(x) = x^3$ at the point $(x_0, y_0) = (1, 1)$. We start by choosing a second nearby point on the graph of $y = x^3$, say $(x_1, y_1) = (r, r^3)$, and we compute the slope of the *secant* line through these two points on the graph:

$$m_{\text{sec}} = \frac{r^3 - 1}{r - 1}.$$

This is just the slope formula $(y_1 - y_0)/(x_1 - x_0)$ for the two points $(x_0, y_0) = (1, 1)$ and $(x_1, y_1) = (r, r^3)$. We are going to let the second point vary, and we use r instead of x_1 to suggest this "running point," and also to avoid subscripts. We don't replace x_1 by x because we want to use x for another purpose.

Now let r approach 1 and imagine the point (r, r^3) sliding along the graph toward the fixed point $(1, 1)$. As this happens the secant line rotates about $(1, 1)$, and we visualize it approaching the tangent line as its limiting position. Our intuition thus tells us that the slope m_{sec} of the secant line approaches the slope m of the tangent line as its limiting value.

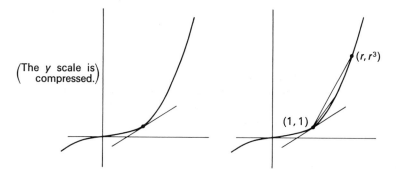

We can't compute this limit by simply setting $r = 1$ in the formula $m_{\text{sec}} = (r^3 - 1)/(r - 1)$ because this would just give the meaningless $0/0$. In fact, it isn't at all obvious what (if anything) m_{sec} approaches as r approaches 1. We have the quotient of two quantities each of which becomes arbitrarily small, and we can't guess how the quotient behaves. However, the reason that the numerator becomes small is that it contains the denominator $r - 1$ as a factor,

$$r^3 - 1 = (r - 1)(r^2 + r + 1).$$

After this common factor is cancelled from top and bottom we *can* see what goes on:

$$m_{\text{sec}} = \frac{r^3 - 1}{r - 1} = r^2 + r + 1 \qquad \text{for } r \neq 1,$$

and $r^2 + r + 1$ approaches $1 + 1 + 1 = 3$ as its limit as r approaches 1. We have thus found that the slope of the secant line approaches 3 as r approaches 1, so 3 is the slope m of the tangent line at the point $(1, 1)$. The equation for the tangent line is then

$$y - 1 = m(x - 1),$$

with $m = 3$, and we end up with

$$y = 3x - 2.$$

As we noted above, the slope of the tangent line is the slope that should be assigned to the graph itself at the point of tangency. Thus, the graph of $y = x^3$ has the slope 3 at the point $(1, 1)$.

In limit notation, the above computation of m is written

$$m = \lim_{r \to 1} m_{\text{sec}} = \lim_{r \to 1} \frac{r^3 - 1}{r - 1} = \lim_{r \to 1}(r^2 + r + 1) = 1 + 1 + 1 = 3.$$

The limit evaluation

$$\lim_{r \to 1}(r^2 + r + 1) = 1 + 1 + 1$$

is the equation

$$\lim_{r \to 1} q(r) = q(1)$$

for the function $q(r) = r^2 + r + 1$; it simply expresses the continuity of q at $r = 1$. We can "see" the value of the limit because the continuity of the polynomial $r^2 + r + 1$ is intuitively obvious.

Now let us rerun the above tangent–slope computation, but this time find the slope m of the tangent line at the general point $(x_0, y_0) = (x_0, x_0^3)$ on the graph of $y = x^3$. The secant slope is now

$$m_{\text{sec}} = \frac{r^3 - x_0^3}{r - x_0},$$

and we have agreed on intuitive grounds that the tangent slope m is the limit of m_{sec} as r approaches x_0. If we can keep in mind that r is the variable in this limit calculation, we can drop the subscript on x_0 and write the secant slope more simply as

$$m_{\text{sec}} = \frac{r^3 - x^3}{r - x}.$$

As before, the numerator has $r - x$ as a factor

$$r^3 - x^3 = (r - x)(r^2 + rx + x^2),$$

so

$$m = \lim_{r \to x} \frac{r^3 - x^3}{r - x} = \lim_{r \to x}(r^2 + rx + x^2) = x^2 + x^2 + x^2 = 3x^2.$$

Again, this limit is obvious because it's obvious that $r^2 + rx + x^2$ is a continuous function of r (x being held fixed). We have now obtained the formula

$$m = 3x^2$$

for the slope m of the graph and its tangent line at the point (x, x^3), This tells us, for example, that at $x = 2$ the curve has slope

$$m = 3 \cdot 2^2 = 12.$$

and that at $x = -1/2$ its slope is

$$m = 3\left(-\frac{1}{2}\right)^2 = \frac{3}{4}.$$

The varying slope $m = 3x^2$ is a new function of x which is called the *derivative* of the original function x^3.

The derivative calculation that we have just made for the function $f(x) = x^3$ can be described for an arbitrary function f. We first compute the slope of the secant line through the two points $(x, f(x))$ and $(r, f(r))$ on the graph of f,

$$m_{\text{sec}} = \frac{f(r) - f(x)}{r - x}.$$

Then we calculate the limit of m_{sec} as r approaches x, obtaining a number that we interpret geometrically as the slope of the graph of f and its tangent line at the point $(x, f(x))$. This limit depends on x and is therefore a new function of x. The new function is called the *derivative* of the function f, and is designated f'. In this notation we calculated above that $f'(x) = 3x^2$ when $f(x) = x^3$.

PROBLEMS FOR SECTION 1

1. Compute the slope of the tangent line to the graph of $f(x) = x^2$ at the point $(1, 1)$. (Imitate the text discussion of $f(x) = x^3$.)

2. Compute the slope of the tangent line to the graph of $y = x^2$ at the point over $x_0 = 2$. Write down the equation of the tangent line.

In each of Problems 3 through 9, compute the slope of the tangent line to the given graph at the point having the given x-coordinate. Then write down the equation of the tangent line.

3. $y = x^2$, $x_0 = -1$

4. $y = x^2$, $x_0 = a$. (For what value of a will the tangent line have slope 1?)

5. $y = 1/x$, $x_0 = 1$

6. $y = 1/x$, $x_0 = 2$

7. $y = x - x^2$, $x_0 = 0$

8. $y = x - x^2$, $x_0 = 2$

9. $y = x - x^2$, $x_0 = a$. (For what value of a will the tangent be horizontal?)

10. Presumably a straight line $y = mx + b$ is its own tangent line at any point. Verify this by the slope-determining procedure, applied to the function $f(x) = mx + b$ at $x_0 = a$.

2. THE DERIVATIVE

Omitting the geometric motivation, we have the following formal definition.

DEFINITION *Given any function f, its derivative f' is the function whose value at x is defined by:*

$$f'(x) = \lim_{r \to x} \frac{f(r) - f(x)}{r - x}.$$

In making the limit calculation, x stays fixed while r varies. The domain of f' then consists of those numbers x for which this r limit exists.

The formal definition makes no reference to geometry. The fact that $(f(r) - f(x))/(r - x)$ and $f'(x)$ can be viewed as slopes constitutes a geometric *interpretation* of what we are doing. There are also, however, other fundamental interpretations related to the notion of velocity (see Sections 7 and 9). It is therefore important to understand that the above definition is a purely *analytic* definition, independent of interpretations. Given a function f, we derive a new function f' from f by making a certain limit calculation that we set up according to the instructions given in the above formula. In this neutral frame of mind we call

$$\frac{f(r) - f(x)}{r - x}$$

the *difference quotient* of the function f.

In applying this definition we always have the problem that the difference quotient is not in a form allowing the limit to be evaluated, because of the 0/0 difficulty that we experienced above. Thus we always have to manipulate the difference quotient into a new form that we can handle. (This may be simple or complicated. It is usually the heart of the problem; the subsequent limit evaluation is usually easy.) So here is our scheme:

a) Write down the difference quotient $(f(r) - f(x))/(r - x)$ for the particular function in question;

b) Manipulate the difference quotient into a form allowing the limit to be evaluated (in these first examples this will always mean cancelling out $(r - x)$ from top and bottom);

c) Evaluate the limit as $r \to x$.

EXAMPLE 1 We compute $f'(x)$ by this procedure when $f(x) = 1/x$.

a) $\dfrac{f(r) - f(x)}{r - x} = \dfrac{\dfrac{1}{r} - \dfrac{1}{x}}{r - x}$

b) $\qquad = \dfrac{\dfrac{x - r}{rx}}{r - x} = -\dfrac{(r - x)}{rx} \cdot \dfrac{1}{r - x} = -\dfrac{1}{rx}.$

c) $f'(x) = \lim_{r \to x} \dfrac{f(r) - f(x)}{r - x} = \lim_{r \to x}\left(-\dfrac{1}{rx}\right) = -\dfrac{1}{xx} = -\dfrac{1}{x^2}.$

Thus $f'(x) = -1/x^2$ when $f(x) = 1/x$.

EXAMPLE 2 Find $f'(x)$ for $f(x) = \sqrt{x}$.

Solution

a) $\dfrac{f(r) - f(x)}{r - x} = \dfrac{\sqrt{r} - \sqrt{x}}{r - x}$.

b) Here we shall "rationalize the numerator" by multiplying top and bottom by $\sqrt{r} + \sqrt{x}$, giving

$$\frac{\sqrt{r} - \sqrt{x}}{r - x} \cdot \frac{\sqrt{r} + \sqrt{x}}{\sqrt{r} + \sqrt{x}} = \frac{r - x}{(r - x)(\sqrt{r} + \sqrt{x})} = \frac{1}{\sqrt{r} + \sqrt{x}}.$$

c) Since

$$\frac{1}{\sqrt{r} + \sqrt{x}} \rightarrow \frac{1}{\sqrt{x} + \sqrt{x}} = \frac{1}{2\sqrt{x}}$$

as $r \rightarrow x$, we see that $f'(x) = 1/(2\sqrt{x})$ when $f(x) = \sqrt{x}$.

EXAMPLE 3 Find $f'(x)$ when $f(x) = x + x^2$.

Solution

$$\frac{f(r) - f(x)}{r - x} = \frac{(r + r^2) - (x + x^2)}{r - x} = \frac{(r - x) + (r^2 - x^2)}{r - x}$$

$$= \frac{(r - x) + (r - x)(r + x)}{(r - x)}$$

$$= 1 + r + x,$$

which approaches $1 + 2x$ as its limit as r approaches x. Thus $f'(x) = 1 + 2x$.

EXAMPLE 4 Find $f'(x)$ when $f(x) = x^{1/3}$.

Solution. Here we break our rule about factoring out $r - x$, but simplify in a similar way. In the difference quotient

$$\frac{f(r) - f(x)}{r - x} = \frac{r^{1/3} - x^{1/3}}{r - x},$$

we may treat $(r - x)$ as a *difference of cubes* and factor it:

$$\frac{r^{1/3} - x^{1/3}}{r - x} = \frac{r^{1/3} - x^{1/3}}{(r^{1/3} - x^{1/3})(r^{2/3} + r^{1/3}x^{1/3} + x^{2/3})}$$

$$= \frac{1}{r^{2/3} + r^{1/3}x^{1/3} + x^{2/3}}.$$

Now $r^{1/3} \rightarrow x^{1/3}$ as $r \rightarrow x$, since $r^{1/3}$ is a continuous function of r, and therefore

$$f'(x) = \lim_{r \rightarrow x} \frac{f(r) - f(x)}{r - x} = \lim_{r \rightarrow x} \frac{1}{r^{2/3} + r^{1/3}x^{1/3} + x^{2/3}} = \frac{1}{3x^{2/3}},$$

provided $x \neq 0$. Thus $f'(x) = \frac{1}{3}x^{-2/3}$ when $f(x) = x^{1/3}$.

EXAMPLE 5   The algebra in Example 4 can be simplified by setting $s = r^{1/3}$ and $y = x^{1/3}$. Then $r = s^3$, $x = y^3$ and

$$\frac{f(r) - f(x)}{r - x} = \frac{r^{1/3} - x^{1/3}}{r - x} = \frac{s - y}{s^3 - y^3}.$$

We can then proceed exactly as in the x^3 example, except that we factor out $(s - y)$ and the formulas are upside down.

In each of these examples the limit computation becomes completely formal if we justify each step by one of the limit laws (pp. 61, 62). For example,

$$\lim_{r \to x} \frac{\frac{1}{r} - \frac{1}{x}}{r - x} = \lim_{r \to x} -\frac{1}{rx} \qquad \textbf{(L5)}$$

$$= -\frac{1}{x^2}. \qquad \textbf{(L1 and Theorem 1)}$$

Instead of relying on Theorem 1, we could work out the particular case occurring here step by step from the earlier principles:

$$\lim_{r \to x}\left(-\frac{1}{rx}\right) = -\frac{1}{\lim_{r \to x}(rx)} = -\frac{1}{x\left(\lim_{r \to x} r\right)}$$

$$= -\frac{1}{x^2}.$$

Such formal arguments do not really make the conclusions any more convincing, because the limits in question are just as obvious as the limit laws from which they are deduced. However, all of the limit calculations occurring in the early chapters *can* be formalized this way in order to emphasize logical structure.

PROBLEMS FOR SECTION 2

Establish the following derivative formulas, using the above scheme in each case to evaluate the limit of the difference quotient.

1. If $f(x) = x^2$, then $f'(x) = 2x$.   2. If $f(x) = x^4$, then $f'(x) = 4x^3$.

3. If $f(x) = 1/x^2$, then $f'(x) = -2/x^3$.

4. If $f(x) = 1/x^{1/2}$, then $f'(x) = -1/2x^{3/2}$.

5. If $f(t) = t^{3/2}$, then $f'(t) = \frac{3}{2}t^{1/2}$.   6. If $f(u) = u^{1/4}$, then $f'(u) = \frac{1}{4}u^{-3/4}$.

7. If $f(s) = -3s$, then $f'(s) = -3$.

8. If $f(x) = 2x - x^2$, then $f'(x) = 2(1 - x)$.

9. If $f(x) = mx + b$, then $f'(x) = m$.

10. If $f(x) = ax^2 + bx + c$, then $f'(x) = 2ax + b$.

11. Each of the first six functions above is of the form $f(x) = x^a$. Examine the form of $f'(x)$ in each of these cases, and then conjecture what the general rule is for the derivative of a power of x: If $f(x) = x^a$, then $f'(x) = $____.

Calculate the derivatives of the following functions.

12. $q(s) = \sqrt{s}/(1 + s)$. 13. $p(t) = t^{-1/3}$.

14. $f(x) = x^{-2/3}$.

15. Calculate the derivative of $f(x) = x^{1/4}$ by the method of Example 5 in the text.

16. Justify the limit calculation in Example 2, appealing to appropriate laws about continuous functions and/or limits from Section 5 in Chapter 2.

17. Calculate the derivative of $f(x) = x^{2/3}$ by setting $s = r^{1/3}$, $y = x^{1/3}$ in the difference quotient and then simplifying.

18. Find the derivative of $f(x) = x^{3/4}$, using substitutions similar to those suggested in the above problem.

19. Use the formula in Problem 10 to show that if $g(x)$ and $h(x)$ are quadratic polynomials and if $f(x) = g(x) + h(x)$, then $f'(x) = g'(x) + h'(x)$.

20. Using the same formula, show that if g and h are linear functions,

$$g(x) = mx + b, \qquad h(x) = nx + c,$$

and if f is the quadratic polynomial $f(x) = g(x)h(x)$, then

$$f'(x) = g(x)h'(x) + h(x)g'(x).$$

3. LEIBNIZ'S NOTATION

In situations where y is a function of x, Leibniz designated the derivative

$$\frac{dy}{dx},$$

which is read "the derivative of y with respect to x." This notation allows us to write down the derivatives of particular functions of x without using the function symbols f and f'. For example,

$$\text{if } y = x^3, \qquad \text{then } \frac{dy}{dx} = 3x^2.$$

Even more simply, we can write

$$\frac{d}{dx}x^3 = 3x^2.$$

In using this notation, however, we must remember that only a *function* can have a derivative. The symbol dy/dx implies that y is a certain function of x, and dy/dx is the derivative of that function. Thus:

$$\text{if } y = f(x), \qquad \text{then } \frac{dy}{dx} = f'(x).$$

Although the form of the Leibniz notation suggests a quotient, such an interpretation is not available until separate meanings are given to dx and *dy*. This will come about eventually, but for the time being dy/dx is just a rather complicated (and indivisible) single symbol for the derivative. The letters x and y in dy/dx are not to be thought of as the variables x and y, but are there to suggest that dy/dx is related to these variables.

Because dy/dx doesn't display the variable x in the way that $f'(x)$ does, substitutions are harder to indicate, and we have to use some such awkward notation as

$$\left.\frac{dy}{dx}\right|_{x=5}$$

for the value of dy/dx when $x = 5$. Thus, if $dy/dx = 3x^2$, then $dy/dx|_{x=5} = 75$. In general,

$$\text{if } y = f(x), \qquad \text{then } \left.\frac{dy}{dx}\right|_{x=a} = f'(a).$$

This awkwardness in showing substitutions is the particular weakness of the Leibniz notation.

From now on we shall use the notations dy/dx and $f'(x)$ freely and interchangeably (as the situation warrants).

4. SOME GENERAL RESULTS ABOUT DERIVATIVES

A function f is *differentiable* at a point x_0 if its derivative exists at x_0. If f is differentiable at every point of its domain we say simply that f is differentiable.

From the several special cases that have been worked out above (and in the problems) you may have conjectured that x^k is differentiable and

$$\frac{d}{dx}x^k = kx^{k-1}$$

for any constant k. We now prove this for *any positive integer n*. The proof is just like the case x^3 that we began with.

THEOREM 1   If $f(x) = x^n$, where n is a positive integer, then f is differentiable and $f'(x) = nx^{n-1}$.

Proof. We use the factorization

$$r^n - x^n = (r - x)(r^{n-1} + r^{n-2}x + \cdots + x^{n-1}).$$

Then

$$\frac{f(r) - f(x)}{r - x} = \frac{r^n - x^n}{r - x} = r^{n-1} + r^{n-2}x + \cdots + x^{n-1}.$$

As r approaches x, this polynomial in r has the limit

$$\underbrace{x^{n-1} + x^{n-1} + \cdots + x^{n-1}}_{n \text{ terms}} = nx^{n-1}.$$

That is,

$$\lim_{r \to x} \frac{r^n - x^n}{r - x} = nx^{n-1};$$

as claimed. ∎

Another thing apparent in the examples and problems is that the derivative of the *sum* of two functions seems always to be the *sum* of the two derivatives. We now check that this always happens.

THEOREM 2   *If the functions g and h are differentiable at x, then so is their sum f = g + h, and*
$$f'(x) = g'(x) + h'(x).$$

Proof
$$\frac{f(r) - f(x)}{r - x} = \frac{[g(r) + h(r)] - [g(x) + h(x)]}{r - x}$$
$$= \frac{g(r) - g(x)}{r - x} + \frac{h(r) - h(x)}{r - x},$$

As $r \to x$ the two terms of this sum approach $g'(x)$ and $h'(x)$ respectively, so the sum approaches $g'(x) + h'(x)$. Thus

$$f'(x) = \lim_{r \to x} \frac{f(r) - f(x)}{r - x} = g'(x) + h'(x). \quad \blacksquare$$

In this limit computation, the difference quotient of f rearranges itself into the sum of the difference quotients of g and h, and this turns into $f'(x) = g'(x) + h'(x)$ upon taking limits. In a similar way, we can show that the derivative of a constant times f is that constant times the derivative of f:

$$(cf)' = cf'.$$

THEOREM 3   *If the function f is differentiable at x, then so is g = cf, and*
$$g'(x) = cf'(x).$$

For example,

$$\frac{d}{dx} 4x^5 = 4\frac{d}{dx} x^5 = 4 \cdot 5x^4 = 20x^4.$$

From the above theorems we can write down the derivative of any polynomial. For example,

$$\frac{d}{dx}(5x^3 + 3x^2) = \frac{d}{dx}(5x^3) + \frac{d}{dx}(3x^2)$$
$$= 5\frac{d}{dx}x^3 + 3\frac{d}{dx}x^2$$
$$= 5 \cdot 3x^2 + 3 \cdot 2x$$
$$= 15x^2 + 6x.$$

After a little practice one can omit the middle steps in this type of computation and just write the answer down.

EXAMPLE   Find the equation of the line tangent to the graph of $y = x^3 - 7x$ at the point given by $x_0 = 2$.

Solution. When $x_0 = 2$,

$$y_0 = x_0^3 - 7x_0 = 8 - 14 = -6.$$

Thus the point of tangency is $(x_0, y_0) = (2, -6)$. The slope of the graph at that point is

$$m = \frac{dy}{dx}\bigg|_{x=2} = 3x^2 - 7\bigg|_{x=2} = 12 - 7 = 5.$$

The tangent line $y - y_0 = m(x - x_0)$ is thus

$$y - (-6) = 5(x - 2), \qquad \text{or} \qquad y = 5x - 16.$$

The use of Theorems 1 through 3 to differentiate polynomials is an example of the process of *systematic differentiation*, which frees us from always having to go back to the basic difference–quotient computation. We work out once and for all the derivatives of various special functions, like x^n, and then we develop formulas for computing the derivatives of *combinations* of functions in terms of the derivatives of their constituents, such as the formula $(f + g)' = f' + g'$. We can then write down the derivative of practically any function, using only these basic formulas and combining rules. This program is carried out systematically in Chapters 5 and 7.

We continue with the power rule for the case $k = 1/n$. This nth root situation generalizes the $x^{1/3}$ example.

THEOREM 4   *If $f(x) = x^k$, where $k = 1/n$ and n is a positive integer, then f is differentiable and*

$$f'(x) = kx^{k-1}.$$

Proof. We proceed exactly as in Example 4 in Section 2. If we write $r - x$ as a difference of nth powers,

$$r - x = (r^{1/n})^n - (x^{1/n})^n,$$

then $r^{1/n} - x^{1/n}$ factors out and cancels against the numerator in the difference quotient, so

$$\frac{f(r) - f(x)}{r - x} = \frac{r^{1/n} - x^{1/n}}{r - x}$$

$$= \frac{1}{(r^{1/n})^{n-1} + (r^{1/n})^{n-2}x^{1/n} + \cdots + (x^{1/n})^{n-1}}.$$

The resulting expression is a continuous function of r, so its limit as r approaches x is its value at $r = x$ (supposing $x \neq 0$). Thus

$$f'(x) = \lim_{r \to x} \frac{r^{1/n} - x^{1/n}}{r - x} = \frac{1}{nx^{1-(1/n)}}$$

$$= kx^{k-1},$$

where $k = 1/n$.   ∎

We conclude our discussion with a general fact that we shall have to use constantly, namely, that a *differentiable function is necessarily continuous*. This is intuitively clear: If a graph has a tangent line at a point, then it *cannot* be discontinuous there. The analytic proof is just a limit evaluation.

THEOREM 5   *If f is differentiable at x, then f is continuous at x.*

Proof

$$\lim_{r \to x}[f(r) - f(x)] = \lim_{r \to x}\left[\frac{f(r) - f(x)}{r - x}\right](r - x)$$

$$= \lim_{r \to x}\left[\frac{f(r) - f(x)}{r - x}\right] \cdot \lim_{r \to x}(r - x)$$

$$= f'(x) \cdot 0 = 0.$$

Therefore

$$\lim_{r \to x} f(r) = f(x),$$

so *f* is continuous at *x*.   ∎

PROBLEMS FOR SECTION 4

Find *dy/dx* for each of the following functions:

1. $y = x^7 + x^{1/7}$ 2. $y = 5x^{120}$

3. $y = x^3 + 3x^2$ 4. $y = x^3 + x - 1$

5. $y = mx + b$ 6. $y = ax^2 + bx + c$

7. $y = 4x^4 - x^2 + 2$ 8. $y = \dfrac{x^5}{5} + \dfrac{x^4}{4} + \dfrac{x^3}{3} + \dfrac{x^2}{2} + x + 1$

9. $y = x(x - 1)$ 10. $y = (x - 2)(x + 3)$

11. $y = 4x^{1/2} + 2x^2 - 3x + 6$

12. Prove that if *m* is a positive integer then

$$\frac{d}{dx}x^{-m} = (-m)x^{(-m)-1}$$

(After a first step simplifying the difference quotient, the proof more or less copies that of Theorem 1.)

Using this new formula, differentiate the following functions:

13. $f(x) = x + \dfrac{1}{x}$ 14. $f(x) = x^3 + x^{-3}$

15. $f(x) = \dfrac{x^3 + 1}{x^2}$ 16. $f(x) = \dfrac{x + 1}{x^{10}}$

17. If $f(x) = x - (1/x)$, prove that $f(x^2) = xf(x)f'(x)$.

18. Find the points on the graph of $y = x^3 + 3x^2 - 9x$ at which the tangent line is horizontal.

19. Sketch the graph of $y = x + (1/x)$ in the first quadrant. Find its lowest point. (Your sketch should suggest that the tangent line must be horizontal at the lowest point.)

20. The parabola $y = ax^2 + bx - 2$ is tangent to the line $y = 4x + 7$ at the point $(-1, 3)$. Find *a* and *b*.

21. Write an equation for the line normal to the curve $y = (x^2 + 1)/x^2$ at the point where $x = 1/2$.

22. Find the points on the curve $y = x^{1/3} + x^{2/3}$ where there are horizontal tangents. (Assume that the power rule is always valid.)

5. POWERS OF DIFFERENTIABLE FUNCTIONS

We have proved several special cases of the power rule

$$\frac{d}{dx} x^k = k x^{k-1},$$

in Sections 2 and 4. If we replace x^k by $[g(x)]^k$, each of these calculations proves the more general formula

$$\frac{d}{dx} [g(x)]^k = k[g(x)]^{k-1} \cdot g'(x).$$

We shall check this below. Note the extra factor $g'(x)$. For example,

$$\frac{d}{dx} (1 + x^3)^2 = 2(1 + x^3) \cdot 3x^2$$

and

$$\frac{d}{dx} (1 + x^3)^{1/2} = \frac{1}{2}(1 + x^3)^{-1/2} \cdot 3x^2.$$

EXAMPLE 1   Find the line perpendicular to the curve

$$y = \sqrt{25 - x^2}$$

at the point $(x_0, y_0) = (3, 4)$.

Solution. The general power formula for $k = 1/2$ tells us that

$$\frac{dy}{dx} = \frac{1}{2}(25 - x^2)^{-1/2}(-2x) = -\frac{x}{\sqrt{25 - x^2}}.$$

The tangent line at $(3, 4)$ therefore has the slope

$$m = -\frac{3}{\sqrt{25 - 9}} = -\frac{3}{4},$$

and the perpendicular line has slope

$$-\frac{1}{m} = \frac{4}{3}.$$

The perpendicular line thus has the point–slope equation

$$y - 4 = \frac{4}{3}(x - 3) \qquad \text{or} \qquad y = \frac{4}{3}x.$$

The general power rule will be subsumed under the chain rule in the next chapter, so in a sense this section is superfluous. However, one may feel more comfortable with the chain rule if some of its special cases have already been obtained independently.

First, a matter of notation. It is conventional to let g^k designate the kth power of the function g; its value at x is of course

$$g^k(x) = [g(x)]^k.$$

For example, g^2 is the square of g, so $g^2(x) = [g(x)]^2$. In this notation the general power formula is:

$$\frac{d}{dx} g^k(x) = kg^{k-1}(x) \cdot g'(x).$$

Setting $u = g(x)$ converts this to the simpler form:

$$\frac{d}{dx} u^k = ku^{k-1} \frac{du}{dx}.$$

Here is the general form of one of the earlier power rule calculations.

EXAMPLE 2 Prove the general power formula for the case $k = 3$:

$$\frac{d}{dx} u^3 = 3u^2 \frac{du}{dx},$$

or

$$\frac{d}{dx} g^3(x) = 3g^2(x) \cdot g'(x).$$

Solution. We can proceed just as in Section 2, factoring a difference of cubes in order to simplify. If $f(x) = g^3(x)$, then

$$\frac{f(r) - f(x)}{r - x} = \frac{g^3(r) - g^3(x)}{r - x}$$

$$= \frac{g(r) - g(x)}{r - x} [g^2(r) + g(r)g(x) + g^2(x)].$$

We know that a differentiable function is always continuous, so $g(r) \to g(x)$ as $r \to x$; and of course the difference quotient $(g(r) - g(x))/(r - x)$ approaches the derivative $g'(x)$. So the right side above has the limit

$$g'(x) \cdot 3g^2(x),$$

and this is $f'(x)$ by definition of the derivative.

Note that if $u = x$, then $du/dx = 1$ and we recapture the earlier formula for x^3.

The other special cases treated earlier generalize in exactly the same way. Thus, Examples 1 and 2 in Section 2 turn into the general power rule for

$$k = -1 \quad \text{and} \quad k = \frac{1}{2},$$

while Theorems 1 and 4 in Section 4 are the cases

$$k = \text{an arbitrary positive integer } m,$$

and

$$k = \text{the reciprocal of a positive integer,} \qquad k = \frac{1}{n}.$$

A different procedure will be given in Section 8.

EXAMPLE 3   For the case $k = -1$ we set

$$f(x) = g^{-1}(x) = \frac{1}{g(x)}$$

and see that

$$\frac{f(r) - f(x)}{r - x} = \frac{\dfrac{1}{g(r)} - \dfrac{1}{g(x)}}{r - x}$$

$$= -\frac{g(r) - g(x)}{r - x} \cdot \frac{1}{g(r)g(x)},$$

which approaches

$$-g'(x) \cdot \frac{1}{g^2(x)}$$

as $r \to x$. Therefore $f(x) = g^{-1}(x)$ is differentiable and its derivative is

$$-g^{-2}(x) \cdot g'(x).$$

We can now do something new that will turn out in Chapter 5 to be a typical chain-rule procedure. We can apply the various power rules already in hand *to each other*, and end up with the general power rule for any rational exponent k, i.e., any exponent k of the form $k = m/n$, where m and n are integers.

THEOREM 6   *If u is a differentiable function of x and k is any rational number, then u^k is differentiable (wherever it is defined) and*

$$\frac{d}{dx} u^k = ku^{k-1} \frac{du}{dx}.$$

Proof. Suppose first that $r = m/n$, where m and n are positive integers. Then

$$\frac{d}{dx} u^r = \frac{d}{dx} (u^m)^{1/n}$$

$$= \frac{1}{n} (u^m)^{(1/n)-1} \frac{d}{dx} u^m \qquad\qquad \text{(By the special case } k = 1/n)$$

$$= \frac{1}{n} u^{(m/n)-m} \cdot mu^{m-1} \frac{du}{dx}$$

$$= \frac{m}{n} u^{(m/n)-1} = ru^{r-1} \frac{du}{dx}.$$

This takes care of all *positive rational* exponents. If k is a *negative* rational number, then $k = -r$, where r is positive, and

$$\frac{d}{dx} u^k = \frac{d}{dx}\left(\frac{1}{u^r}\right) = -\frac{1}{(u^r)^2} \frac{d}{dx}(u^r) \qquad \text{(By Example 3)}$$

$$= -\frac{1}{u^{2r}} \cdot ru^{r-1}\frac{du}{dx} = -ru^{-r-1}\frac{du}{dx}$$

$$= ku^{k-1}\frac{du}{dx}.$$

This finishes the proof of the theorem. ∎

PROBLEMS FOR SECTION 5

Compute dy/dx when:

1. $y = \sqrt{x^2 + 4x}$

2. $y = \sqrt{x^2 + 4x + 4}$

3. $y = 1/(x + 2)$

4. $y = (1 - x)^{2/3}$

5. $y = 1/(x^2 + 1)$

6. $y = 1/\sqrt{x^2 + 1}$

7. $y = \sqrt{1 - x^2}$

8. $y = (x^2 + 1)^{1/3}$

9. $y = (2x + 1)^{4/5}$

10. $y = (3x^2 + 1)^{4/5}$

11. $y = 1/(1 + x^3)^{1/3}$

12. $y = \sqrt{x + (1/x)}$

13. $y = (x^{1/4} - 4x^3)^{-1/5}$

14. $y = (a^2 - x^2)^{1/3}$ (a constant)

15. $y = (a^3 - 3a^2x + 3ax^2 - x^3)^{2/3}$ (a constant)

16. Prove from scratch that if $f(x) = \sqrt{g(x)}$, then

$$f'(x) = \frac{g'(x)}{2\sqrt{g(x)}}.$$

17. The half-hyperbola $y = \sqrt{4x^2 + 1}$ is asymptotic to the line $y = 2x$ in the first quadrant. Prove that its tangent line approaches the asymptote $y = 2x$ as its "limiting position," in the sense that its slope approaches 2 and its intercepts approach 0.

18. Prove that if n is a positive integer, then

$$\frac{d}{dx} g^n(x) = ng^{n-1}(x) \cdot g'(x)$$

by imitating the proof of Theorem 1 in the last section.

19. Prove that if $a = 1/m$, where m is a positive integer, then

$$\frac{d}{dx} g^a(x) = ag^{a-1}(x)g'(x),$$

by imitating the proof of Theorem 4 in the last section.

20. Find the smallest value of the function

$$f(x) = 16x^{2/3} + x^{-2/3}.$$

[*Hint:* The lowest point on the graph of f should have a horizontal tangent.]

21. Prove that a tangent to a circle is perpendicular to the radius drawn to the point of tangency.

6. INCREMENTS

When considering the difference quotient $(f(r) - f(x))/(r - x)$, it is often useful to think of r as a second value of x and the difference $(r - x)$ as a *change in x*. The classical notation for such a change is Δx, read "delta x". (Δ is capital δ.) Here Δx is not the product of a number Δ times a number x, but a wholly new numerical variable, called the *increment* in x. Its value is always to be thought of as the *change* in the value of x, from a *first* value to a *second* value, or from an *old* value to a *new* value. The most consistent notation in this situation would be to use subscripts on x, say x_0 and x_1, for the old and new values respectively. Then

$$\Delta x = x_1 - x_0$$

is the change in x in going from the old to the new value. We can also consider the new value as being obtained from the old value by adding the change:

$$x_1 = x_0 + \Delta x.$$

An increment Δx can be of either sign. Thus if $x_0 = 2$ and $\Delta x = -3$, then $x_1 = 2 + (-3) = -1$.

As often as not we discard the zero subscript, so that the old and new values are x and $x + \Delta x(= x_1)$. That is, we use x itself to represent the old value of x. This may not seem quite cricket, but it works out all right.

Now suppose that y is a function of x, $y = f(x)$. When x changes from x to $x + \Delta x$, then y changes from $f(x)$ to $f(x + \Delta x)$. The increment in y thus has the basic formula

$$\Delta y = f(x + \Delta x) - f(x).$$

EXAMPLE 1   If $y = x^2$, then

$$\Delta y = (x + \Delta x)^2 - x^2 = 2x\,\Delta x + (\Delta x)^2 = \Delta x(2x + \Delta x).$$

EXAMPLE 2   When $y = x^3$, we have

$$\begin{aligned}
\Delta y &= f(x + \Delta x) - f(x) = (x + \Delta x)^3 - x^3 \\
&= x^3 + 3x^2\Delta x + 3x(\Delta x)^2 + (\Delta x)^3 - x^3 \\
&= \Delta x(3x^2 + 3x\,\Delta x + (\Delta x)^2).
\end{aligned}$$

These examples show that increment notation may change the way in which a difference factors. Often it makes the simplification more routine in nature. In Example 2 the simplification follows after *expanding the cube* $(x + \Delta x)^3$, whereas before it depended on *factoring* $r^3 - x^3$. Generally speaking, factoring is more elegant, but multiplication is easier and more automatic. And in many other situations the use of increments makes the steps to be taken more or less obvious, whereas the form $f(r) - f(x)$ may require ingenuity in its treatment.

In terms of increments, the difference quotient $(f(r) - f(x))/(r - x)$ becomes

$$\frac{f(x + \Delta x) - f(x)}{\Delta x} = \frac{\Delta y}{\Delta x},$$

which tells us, literally, that the difference quotient is the *change in y divided by the change in x*. The derivative can now be expressed

$$\frac{dy}{dx} = \lim_{\Delta x \to 0} \frac{\Delta y}{\Delta x}.$$

(Although of later origin, the Δ symbolism obviously fits well with Leibniz's symbolism.)

From the above expression for dy/dx we can obtain

$$\lim_{\Delta x \to 0} \Delta y = \lim_{\Delta x \to 0} \left(\frac{\Delta y}{\Delta x}\right) \Delta x$$

$$= \left(\lim_{\Delta x \to 0} \frac{\Delta y}{\Delta x}\right) \cdot \left(\lim_{\Delta x \to 0} \Delta x\right) = \frac{dy}{dx} \cdot 0 = 0.$$

This is just a reformulation in terms of increments of our earlier proof that a differentiable function is continuous.

EXAMPLE 3   We recompute the derivative formula for $y = x^3$ in increment notation. From Example 2,

$$\frac{\Delta y}{\Delta x} = 3x^2 + 3x\Delta x + (\Delta x)^2.$$

Therefore

$$\frac{dy}{dx} = \lim_{\Delta x \to 0} \frac{\Delta y}{\Delta x} = 3x^2 + 3x \cdot 0 + 0^2 = 3x^2.$$

PROBLEMS FOR SECTION 6

Recompute the derivatives of the following functions by the method of increments:

1. $f(x) = x^2$ 2. $f(x) = 1/x$
3. $f(x) = \sqrt{x}$ 4. $f(x) = x^4$
5. $f(x) = x + x^2$ 6. $f(x) = 1/(1 + x)$
7. $f(x) = 1/(1 + x^2)$

8. If $y = x^2$ and x is positive, then y can be interpreted as the area of a square of side x. Then Δy is the *increase in area* when the side is increased by Δx. Draw a figure and interpret the two terms of Δy as areas.

9. If $y = x^3$ and x is positive, then y can be interpreted as the volume of a cube of side x. Then Δy is the *increase in volume* when the side is increased by Δx. Draw a figure and interpret the three terms of Δy as volumes.

10. Write out the proof of Theorem 2 in Section 4 in increment notation. (Note first that if $u = g(x)$, $v = h(x)$, and $y = u + v$, then $\Delta y = \Delta u + \Delta v$.)

7. VELOCITY

Suppose that a particle is moving along a coordinate line. At each instant of time t the particle has a unique position coordinate s, so its position s is a function of t,

$$s = f(t).$$

If the particle is you in your car, and if it takes you half an hour to drive ten miles through a city, then you say that your average velocity is 20 miles per hour through the city, because

$$\frac{\text{Distance travelled}}{\text{Elapsed time}} = \frac{10}{1/2} = 20.$$

In general,

$$\text{Average velocity} = \frac{\text{Distance travelled}}{\text{Elapsed time}} = \frac{\Delta s}{\Delta t},$$

a formula that ought to be in agreement with your intuitive notion of average velocity. Since

$$\frac{\Delta s}{\Delta t} = \frac{f(t + \Delta t) - f(t)}{\Delta t},$$

average velocity is the difference quotient of position as a function of time.

EXAMPLE 1   An object falling from rest falls

$$s = 16t^2$$

feet in t seconds. Here the position function is $f(t) = 16t^2$, measured downward from the point of release. The average velocity of the object during the first second of fall is

$$\frac{\Delta s}{\Delta t} = \frac{f(1) - f(0)}{1} = \frac{16 \cdot 1^2 - 16 \cdot 0^2}{1} = 16 \text{ feet per second.}$$

During the *second* second it is

$$\frac{\Delta s}{\Delta t} = \frac{f(2) - f(1)}{1} = \frac{16 \cdot 2^2 - 16 \cdot 1^2}{1} = \frac{64 - 16}{1} = 48 \text{ ft/sec.}$$

Its average velocity during the *first two seconds* is

$$\frac{\Delta s}{\Delta t} = \frac{16 \cdot 2^2 - 16 \cdot 0^2}{2} = \frac{64}{2} = 32 \text{ ft/sec.}$$

There is also a direct intuitive notion of *instantaneous velocity v*. It is the number that measures how fast something is moving at a given instant. It is the varying reading on a perfect speedometer. In order to obtain a formula for v, let v_0 be its value at time t_0 and let Δt be a very small change in t. Then over the time interval between t_0 and $t_0 + \Delta t$, the velocity v varies only slightly from v_0, so the average velocity $\Delta s/\Delta t$ should be very close to

v_0. That is,

$$\frac{\Delta s}{\Delta t} \approx v_0,$$

where $\approx$ is read "is approximately equal to." For smaller Δt, the difference between initial value and average value should be smaller still. Therefore,

$$v_0 = \lim_{\Delta t \to 0} \frac{\Delta s}{\Delta t} = \frac{ds}{dt}\bigg|_{t_0}.$$

That is, we agree on intuitive grounds that the instantaneous velocity v should be the limit of the average velocity $\Delta s/\Delta t$ as the length of the time interval over which the average is computed tends to zero. So v is given by the derivative

$$v = \frac{ds}{dt}.$$

We thus have a second interpretation of the derivative. For Newton, who was struggling to understand the motions of the planets and other heavenly bodies, this was undoubtedly the major interpretation of the derivative.

EXAMPLE 2 If $s = 3t - 10$, where s is measured in feet and t in seconds, then $v = ds/dt = 3$ feet per second. The particle is moving in *uniform motion* with constant velocity $v = 3$ ft/sec.

EXAMPLE 3 If the motion of the particle is described by $s = t^3 - t$, in the same units as above, then the velocity v is

$$v = \frac{ds}{dt} = (3t^2 - 1) \text{ feet per second.}$$

Thus, at time $t = 2$, its velocity is $v = 3 \cdot 2^2 - 1 = 11$ ft/sec, and at time $t = 0$ it is $v = 3 \cdot 0^2 - 1 = -1$ ft/sec. Note that the velocity is negative at $t = 0$. Thus $\Delta s/\Delta t < 0$ for small Δt and the particle is moving *backward* at time $t = 0$. If the particle was moving and the clock running before the zero setting on the clock, we see that at time $t = -1$ the velocity was $v = 3(-1)^2 - 1 = 2$.

Although the particle in the above example is moving backward and forward on the line, the *graph* of the motion is the graph of the equation $s = t^3 - t$ in the ts-plane.

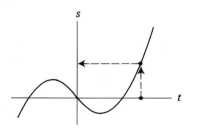

The particle is *not* moving along this curve, but along the *s*-coordinate line, pictured in this figure as the vertical axis. Its position at time *t* is obtained from the graph by the "up-and-over" procedure indicated in the figure. (The graph is called the "world line" of the particle in space–time, space being one-dimensional here.)

We say that a particle is *accelerating* if its velocity is increasing. Its *average acceleration* over an interval of time is given by the formula

$$\text{Average acceleration} = \frac{\text{Change in velocity}}{\text{Elapsed time}} = \frac{\Delta v}{\Delta t}.$$

We also say that a particle is *decelerating* if its velocity is decreasing. In this case, the above formula gives a *negative* average acceleration. Following this clue, we equate *deceleration* to *negative acceleration* and forget about deceleration.

Instantaneous acceleration a is defined to be the limit of the average acceleration as the length of the time interval over which the average is computed tends to zero. The acceleration of the particle is thus the derivative of its velocity:

$$a = \lim_{\Delta t \to 0} \frac{\Delta v}{\Delta t} = \frac{dv}{dt}.$$

In the above example we had $v = 3t^2 - 1$ and so the acceleration is

$$a = \frac{dv}{dt} = 6t.$$

The particle is moving with increasing acceleration.

Since the velocity is so many feet per second, the acceleration, which is the *change in velocity per unit of time*, must be so many (feet per second) per second. We designate this ft/sec^2. Thus in the problem above the acceleration was $6t$ ft/sec^2.

EXAMPLE 4   An object is thrown vertically upward, and its position above the ground (after *t* seconds have elapsed) is given by $s = 50t - 16t^2$ feet. Find its initial velocity, its highest point, and its velocity when it strikes the ground, as well as its acceleration.

Solution. The velocity is

$$v = \frac{ds}{dt} = 50 - 32t \text{ ft/sec.}$$

The initial velocity v_0 is the velocity at time $t_0 = 0$. Thus

$$v_0 = 50 \text{ ft/sec.}$$

The highest point will be reached just when the upward velocity has fallen away to zero. Setting $v = 0$, we have

$$0 = 50 - 32t.$$

$$t = \frac{50}{32} = \frac{25}{16} \text{ sec.}$$

The particle returns to ground at the moment when s becomes 0 again, i.e., when

$$0 = s = 50t - 16t^2.$$

This gives $t = 50/16$ seconds, and then

$$v = 50 - 32t = 50 - 32 \cdot \frac{50}{16}$$

$$= -50 \text{ ft/sec}.$$

The acceleration of the object is

$$a = \frac{dv}{dt} = \frac{d}{dt}(50 - 32t) = -32 \text{ ft/sec}^2.$$

The acceleration is constant and it is *negative*. The velocity is *decreasing* at a constant rate.

PROBLEMS FOR SECTION 7

1. A particle moves along the x-axis and its position as a function of time is given by $x = t^3 - 2t^2 + t + 1$. Find its velocity. At what moments is it standing still? When is it moving forward and when backward?

2. A particle is moving along a horizontal line on which the positive direction is to the right. The equation of motion of the particle is $s(t) = t^3 - 9t^2 + 24t + 1$.

   a) What is the velocity of the particle at $t = 1$?

   b) Is the particle moving to the right or left at $t = 1$?

   c) What is the acceleration of the particle at $t = 1$?

   d) Is the speed of the particle increasing or decreasing at $t = 1$? Explain your reasoning.

   e) At what times, if any, is the velocity of the particle zero?

3. A ball thrown directly upward with a speed of 96 ft/sec moves according to the law

   $$y = 96t - 16t^2,$$

   where y is the height in feet above the ground, and t is the time in seconds after it is thrown. How high does the ball go?

4. The movement of an airplane from the time it releases its brakes until takeoff is governed by the equation $s = 2t^2$, where s is the distance from the starting point in feet and t is the time in seconds since brake release.

   a) If takeoff velocity is 120 mph (176 fps), determine the time elapsed from brake release until takeoff.

   b) What is the distance covered during this time?

5. A coin is thrown straight up from the top of a building which is 300 ft tall. After t seconds, its position is described by $s = -16t^2 + 24t + 300$, where s is its height above the ground in feet. When does the coin begin to descend? What is its velocity when it is 305 feet above the ground?

6. In the first t seconds after an Apollo space shot is launched, the rocket reaches a height of $40t^2$ feet above the earth.

   a) What is the rocket's velocity after 5 seconds?

b) If the speed of sound is 1050 ft/sec, calculate the height at which the space craft attains supersonic speed.

7. Given that the motion of a particle is expressed by the equation

$$s(t) = t^3 - 6t^2 + 2,$$

a) determine the velocity of the particle at $t = 2.3$;

b) determine the acceleration of the particle at 2.3;

c) determine the *average* velocity of the particle in the time interval $t = 1.3$ to $t = 3.3$.

8. The height above the ground of a bullet shot vertically upward with an initial velocity of 320 ft/sec is given by $s(t) = 320t - 16t^2$, where $s(t)$ is the height in ft and t is time in sec.

a) Determine the velocity of the bullet 4 sec after it is fired.

b) Determine the time required for the bullet to reach its maximum height, and the maximum height attained.

9. A ball is thrown down from the top of a 200-ft tower with an equation of motion

$$y = -16t^2 - 40t + 200.$$

What is the *average* velocity of the ball in its trip to the ground? At what time t does the velocity of the ball equal its average velocity?

10. A particle whose velocity is zero during an interval of time is standing still during that time interval. Restate this obvious fact as a theorem about differentiable functions.

11. The distance, rate, and time problems of elementary algebra refer to motions of constant velocity. The formula $d = rt$ would then be restated in our language as

$$\Delta s = r\,\Delta t,$$

where r is the constant velocity. Show that the most general description of such a motion is given by

$$s = rt + c,$$

where c is any constant.

8. THE MEAN-VALUE PRINCIPLE

The following geometric principle is intuitively obvious:

If we draw a secant line through two points on the graph of a differentiable function, then there is at least one point in between where the tangent line is parallel to the secant.

We visualize moving the secant line parallel to itself, and we see that it turns into the tangent line at its last point of contact with the graph.

When we state this analytically as an equality of two slopes, it becomes what is called the mean-value principle.

MEAN-VALUE   *If f is differentiable on the closed interval [a, b], then there is at least one point*
PRINCIPLE   *X strictly between a and b at which*

$$f'(X) = \frac{f(b) - f(a)}{b - a}.$$

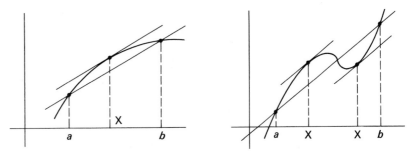

The mean-value principle also has a velocity interpretation. Consider again a particle moving along a coordinate line with position coordinate given by $s = f(t)$. We agreed that its instantaneous velocity is given by $ds/dt = f'(t)$, and that the average velocity over the time interval $[t_0, t_1]$ is

$$\frac{\Delta s}{\Delta t} = \frac{f(t_1) - f(t_0)}{t_1 - t_0}.$$

We assume that the velocity varies continuously with t, so as it varies from above its average value to below (or vice versa), it will cross the average value at some time $t = T$ by the intermediate-value principle. That is, there must be some number T between t_0 and t_1 such that

$$f'(T) = \frac{f(t_1) - f(t_0)}{t_1 - t_0}.$$

(This interpretation explains the phrase *mean-value*, because average values used to be called mean values.)

The mean-value principle thus appears to be true in the contexts of the two principal interpretations of the derivative. On this basis we shall assume it for the time being, i.e., treat it as a temporary axiom. It will be proved in Chapter 12.

Replacing a and b by x and $x + \Delta x$, we have the increment form

$$f'(X) = \frac{f(x + \Delta x) - f(x)}{\Delta x} = \frac{\Delta y}{\Delta x},$$

where X lies strictly between x and $x + \Delta x$. Also useful are the cross-multiplied forms:

$$f(b) - f(a) = f'(X)(b - a)$$
$$\Delta y = f'(X)\Delta x;$$
$$f(x + \Delta x) = f(x) + f'(X)\Delta x.$$

REMARK The cross-multiplied version is trivially true when $b = a$, becoming then $0 = f'(X) \cdot 0$, except for the fine point that X cannot lie *strictly* between a and $b = a$. When we want to include this case, as we do in some applications, we omit the word *strictly*, and interpret *between* as allowing $a \leq X \leq a$.

EXAMPLE 1 When the mean-value principle is applied to the function $f(x) = x^3$, it says that for any two numbers a and b there is a number X lying between a and b, such that
$$b^3 - a^3 = 3X^2(b - a).$$

It is clear from the above figure that there may be more than one such point X. Moreover, the mean-value principle says nothing about where X is, except that it lies somewhere between a and b. One might feel that anything so vague as this couldn't be very helpful, but we shall see that the mean-value principle is extremely useful. In the usual context, $f'(x)$ is known to have some property throughout the interval (a, b), and it therefore has this property at the point X, no matter where it is. So the mere fact that there must *be* such a point X is often enough to get us off the ground. We shall give two important examples below to show how this works, but most of the applications of the mean-value principle will come later.

THEOREM 7   *If f' exists and is everywhere zero on an interval I, then f is a constant function on I. That is, there is a constant c such that $f(x) = c$ for every x in I.*

This is geometrically obvious. If every tangent line to a graph is horizontal, then the graph can never rise or fall, and so must be a horizontal straight line.

Proof. Analytically, the theorem is an immediate corollary of the mean-value principle. Choose any point x_0 in I and set $c = f(x_0)$. For any other x the mean-value principle says that

$$f(x) - f(x_0) = f'(X)(x - x_0),$$

where X is some number between x_0 and x. But since f' is everywhere 0 this shows that $f(x) - f(x_0) = 0$, and so

$$f(x) = f(x_0) = c$$

for all x. ∎

This fact is crucially important when it comes to reversing the differentiation procedure. We shall see a simple example of its use in Section 9.

In the example below we redo one of the examples from Section 5 in order to illustrate how the general form of *any* derivative rule can be obtained by applying the mean-value principle to the special form of that rule.

EXAMPLE 2   We shall prove the general power rule for $k = 3$ in this new way, assuming the special form $dx^3/dx = 3x^2$.

The mean value principle, applied to the function x^3, says that for any two numbers a and b there is some number U lying between a and b such that

$$b^3 - a^3 = 3U^2(b - a).$$

Here we take $b = g(r)$ and $a = g(x)$ and have

$$g^3(r) - g^3(x) = 3U^2[g(r) - g(x)],$$

where U is some number between $g(x)$ and $g(r)$. Thus, if $f(x) = g^3(x)$, then

$$\frac{f(r) - f(x)}{r - x} = \frac{g^3(r) - g^3(x)}{r - x} = 3U^2\frac{g(r) - g(x)}{r - x}.$$

Now let r approach x. Then $g(r) \to g(x)$ since a differentiable function is necessarily continuous, $U \to g(x)$ since U lies between $g(x)$ and $g(r)$, and hence $3U^2 \to 3g^2(x)$. Also $[g(r) - g(x)]/(r - x) \to g'(x)$. The right side therefore has the limit

$$3g^2(x) \cdot g'(x).$$

Thus $f'(x)$ exists and has the above value.

Comparing this example with Example 2 in Section 5, we see that the "mean-value" factor $3U^2$ replaces the more complicated algebraic factor $g^2(r) + g(r)g(x) + g^2(x)$. Each of these factors has the limit $3g^2(x)$ as $r \to x$, and in each case this observation is the critical step in obtaining the general power formula.

The proof in Example 2 works just as well for any other power.

LEMMA 1 *The general power rule*

$$\frac{d}{dx} g^k(x) = kg^{k-1}(x)g'(x)$$

holds for any exponent k for which we have proved the "ordinary" rule $dx^k/dx = kx^{k-1}$.

Proof. Just replace 3 and 2 by k and $k - 1$, respectively, in the proof in Example 2. ∎

When we apply Lemma 1 to the examples and theorems of Sections 2 and 4, we obtain the general power rule for $k = -1$, m, and $1/n$, where m and n are any positive integers. These were just the cases needed in the proof of Theorem 6.

EXAMPLE 3 The mean-value principle for $f(x) = \sqrt{x}$ guarantees that there is a number X between a and b such that

$$\sqrt{b} - \sqrt{a} = \frac{1}{2\sqrt{X}}(b - a).$$

Here we can actually find X by solving this mean-value equation:

$$2\sqrt{X} = \frac{b - a}{\sqrt{b} - \sqrt{a}} = \sqrt{b} + \sqrt{a},$$

$$X = \left(\frac{\sqrt{b} + \sqrt{a}}{2}\right)^2.$$

Note also that if a is replaced by b, then the value of this final expression is increased (since $a < b$) to $(\sqrt{b})^2 = b$. This shows directly that $X < b$. Similarly, it can be directly checked that $a < X$.

PROBLEMS FOR SECTION 8

For each of the following functions and the given values of a and b, find a value for X such that $f'(X) = (f(b) - f(a))/(b - a)$.

1. $f(x) = x^2 - 6x + 5$, $a = 1$, $b = 4$ 2. $f(x) = 3x^2 + 4x - 3$, $a = 1$, $b = 3$

3. $f(x) = x^3$, $a = 0$, $b = 1$ 4. $f(x) = -\dfrac{2}{x}$, $a = 1$, $b = 3$

5. $f(x) = x^{-2}$, $a = 1$, $b = 2$

6. Show that if you average 40 miles per hour on a trip from Chicago to St. Louis, then for at least one instant during your trip your speedometer reads 40 miles per hour.

7. Given that $f(x) = (x + 2)/(x + 1)$ and $a = 1$, $b = 2$, find all values X in the interval $(1, 2)$ such that

$$f'(X) = \frac{f(b) - f(a)}{b - a}.$$

8. Prove that a particle whose velocity is zero during an interval of time is standing still during that time interval.

9. Prove that if $f'(x) = 3x^2$, then $f(x) = x^3 + c$ for some constant c. [*Hint.* What is $(f - g)'$ if $g(x) = x^3$?]

10. Prove that if f and g have the same derivative on an interval I, then $f(x) = g(x) + c$ for some constant c.

11. Prove that if f' is constant, say $f' = m$, then $f(x) = mx + b$.

12. What is the most general function f such that $f''(x) = x$? [*Hint.* Use Problem 10 twice, first finding what f' must be, and then f.]

9. RATE OF CHANGE

So far we have interpreted the derivative as the *slope* of the tangent line to a graph, and as the *velocity* of a moving particle. Now we claim that the increment point of view leads to a generalization of the velocity interpretation: Given a differentiable function $y = f(x)$, we can always interpret $f'(x_0)$ as the *rate of change of y with respect to x as x passes through the value x_0*, or *the rate of change of the function f at the point x_0*. The discussion below is intended to make this interpretation seem plausible.

Consider first a use of the word *rate* that is probably familiar. If f is linear, say

$$y = f(x) = 3x - 5,$$

then

$$\Delta y = f(x_0 + \Delta x) - f(x_0) = [3(x_0 + \Delta x) - 5] - [3x_0 - 5]$$
$$= 3\Delta x,$$

so that the change in y is always exactly 3 times the change in x, no matter what values we use for x_0 and Δx. We say that y changes exactly three times *as fast as x* changes, even though we are not necessarily talking about *motion*. All we mean is that if x is changed, then y will change, and the change in y will always be three units per unit change in x. In this situation it is customary to say that 3 is the constant *rate* at which *y* changes as compared to *x*, or the *constant rate of change of y with respect to x*.

For the general linear function

$$y = ax + b$$

we have, similarly

$$\Delta y = a \Delta x$$

and *a* is the constant rate of change of *y* with respect to *x*. Knowing the rate *a* we can obtain Δy from Δx by the above equation. Knowing Δx and Δy we obtain the rate *a* from

$$\frac{\Delta y}{\Delta x} = a.$$

Note that these two equations in Δx, Δy, and *a* are not quite equivalent, since Δx cannot be zero in the second.

When $f(x)$ is not a linear function, then the increment ratio

$$\frac{\Delta y}{\Delta x} = \frac{f(x_0 + \Delta x) - f(x_0)}{\Delta x}$$

depends on x_0 and Δx. If its value is *r* (for given x_0 and Δx), then the equation

$$\Delta y = r \Delta x$$

says that the total change in *y* is exactly *r* times the total change in *x* over the *x* interval between x_0 and $x_0 + \Delta x$. We therefore say that, over this interval, *y* has been changing *on the average r* times as fast as *x*, and we call *r* the average rate of change of *y* with respect to *x*.

DEFINITION   *The average rate of change of y with respect to x, over the interval between x_0 and $x_1 = x_0 + \Delta x$, is*

$$\frac{\Delta y}{\Delta x} = \frac{f(x_0 + \Delta x) - f(x_0)}{\Delta x}.$$

The derivative

$$\frac{dy}{dx} = \lim_{\Delta x \to 0} \frac{\Delta y}{\Delta x}$$

is now viewed as the limit of the average rate of change as the width of the interval over which the average is computed tends to zero. We therefore interpret

$$\left. \frac{dy}{dx} \right|_{x = x_0} = f'(x_0)$$

as the *true rate of change of y with respect to x at the point x_0, or the true rate of change as x passes through the value $x = x_0$.*

Average velocity and true (or instantaneous) velocity are the special case when *y* is distance and *x* is time. We saw in Section 7 that intuitions about velocity lead pretty directly to the formula $v = ds/dt$ for the true velocity; and this supports the general interpretation of $dy/dx = f'(x)$ as the true rate of change.

The rate of change of any quantity with respect to time is called its *time rate of change*. A time rate of change answers the question of how *fast* something is changing.

EXAMPLE 1   A stone is dropped into a pond and it is observed that the spreading circular ripple has the radius $r = 2t$ ft after t seconds. How fast is the area increasing at the end of 5 seconds?

Solution. We have the formula

$$A = \pi r^2 = \pi(2t)^2 = 4\pi t^2$$

for the area at the end of t seconds. Its rate of change at time t is therefore

$$\frac{dA}{dt} = 8\pi t,$$

and at $t = 5$ the area is increasing at the rate of

$$\left.\frac{dA}{dt}\right|_{t=5} = 40\pi,$$

or about 120 sq. ft/sec.

We now give some further examples of the general rate-of-change interpretation, starting with some words on rate of change as a *coefficient*.

If f is a differentiable function whose rate of change is constant over an interval I, then f must be linear over I,

$$f(x) = mx + b,$$

and its rate of change $f'(x)$ is the coefficient m. (The reason is that if f' has the constant value m, then $f(x) - mx$ has derivative zero and so must be a constant function b, by Theorem 7.)

Because of the above fact, when a rate of change is constant or nearly constant in the applications, it is often called a *coefficient*.

EXAMPLE 2   We consider a metal rod that would be one unit long at zero degrees (centigrade). When it is heated it expands. The rate of change of its length l with respect to the temperature T is called the *coefficient of thermal expansion* for the given metal. This rate of change is the derivative dl/dT, and calling it a *coefficient* implies that it is very nearly a constant c, so that l is given approximately by

$$l = cT + k$$

(where $k = 1$ since $l = 1$ when $T = 0$). Actually, the coefficient of thermal expansion is variable to the extent that its variability is mentioned in tables and handbooks. The notion of a variable coefficient seems paradoxical, but and we keep in mind that what we really are talking about is the derivative dl/dT, the fact that it turns out to be a nonconstant function of T ceases to be troublesome.

EXAMPLE 3 We experiment with an electrical circuit and plot the voltage E required to produce a current I. For simple circuits we find that E is proportional to I, so that

$$E = RI,$$

where the constant of proportionality R is called the *resistance* of the circuit. Of course the rate of change dE/dI then has the constant value R, and R is a true *coefficient* in the above sense.

However, if E and I are the plate voltage and plate current of a vacuum tube (for a given grid voltage), then the graph of E as a function of I looks like this. It is called a *characteristic* of the tube.

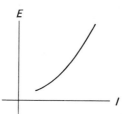

Now we must define the resistance R as the rate of change dE/dI, and we note that it increases as the current I increases.

EXAMPLE 4 In economics the word *marginal* signifies a rate of change. For example, if it costs C dollars to produce x tons of coal, then the *marginal* cost is the increase in total cost C *per extra ton produced*, i.e., the rate of change of C with respect to x, dC/dx. If the relationship between C and x has the simple form

$$C = I + kx,$$

then the marginal cost is the coefficient k. Here I is a fixed initial cost, for machinery etc., and the marginal cost k is the constant running cost per ton, for labor, expended materials like fuel, etc.

Normally the relationship between C and x is more complicated than this. We impose one more condition on our problem, namely, that C is the cost to produce x tons of coal *in a given fixed time interval*, say one week. There will generally be a fixed cost of I dollars per week. But the *extra* running cost per ton, instead of being constant, will probably decrease as production increases, because it is possible to achieve greater internal efficiency with greater volume. The trend will continue only up to a certain point, after which the cost per ton will begin to increase again, because the larger demand will make it necessary to utilize older machinery, pay overtime wages, etc. The marginal cost rises to very high values if the producer strains his capacity in an effort to put out a very large weekly tonnage. Thus a typical cost curve might look like this:

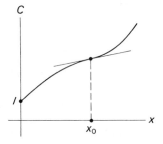

For example, $C = x^3 - 3x^2 + 4x + 1$ has a graph like this. (We shall not try to invent a precise cost situation that would lead to this formula.) The marginal cost is

$$c_M = \frac{dC}{dx} = 3x^2 - 6x + 4.$$

It is variable, and has a minimum value at the point labelled x_0. We shall see later that the minimum value of the marginal cost occurs where *its* derivative dc_M/dx is zero. Since

$$\frac{dc_M}{dx} = 6x - 6 = 6(x - 1),$$

it follows that c_M has its minimum value at $x_0 = 1$, and that its minimum value is

$$\text{Min } c_M = 3(1)^2 - 6 \cdot 1 + 4 = 1.$$

This is the point at which the producer is operating most efficiently, but, because of other factors, it may *not* be the point at which he will choose to operate.

EXAMPLE 5   We continue in the coal vein. Our producer must also consider how many tons of coal he can sell at a given price. Presumably, the lower his price the more coal he can sell (per week), so his *demand curve* is the graph of a decreasing function.

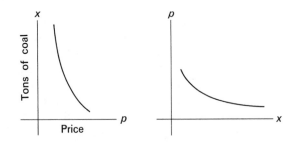

In order to compare this with his cost function, he would plot this relationship with axes interchanged, and so consider p as a function of x.

This function is the producer's *demand function*. Its value $p = d(x)$ is the price the producer must charge in order to sell exactly x tons (per week).

Then $R = px = d(x)x$ is his *total weekly revenue* from selling x tons, and the rate of change of R with respect to x, dR/dx, is called his *marginal revenue*.

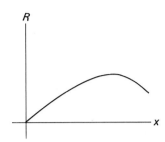

Our scales are all wrong. In order to plot a curve that will react to moderate changes in x, the unit of quantity would probably be a thousand tons, and the unit of money might be $10,000. The revenue R would then be the number of $10,000 units obtained from selling x thousands of tons. (Continued in Chapter 6.)

PROBLEMS FOR SECTION 9

1. a) What is the rate of change of the volume of a sphere with respect to its radius?
   b) What is the rate of change of the radius of a sphere with respect to its volume?

2. Find the rate of change of the volume of a sphere with respect to its radius when the radius is 5 inches. (The answer should be expressed in cubic inches per inch.) Find the rate of change when $r = 10$ inches.

3. A ladder 20 ft long leans against a vertical wall. Its base is the distance from the foot of the ladder to the wall. Find the rate of change of the height of the ladder with respect to its base when the base is 5 feet. (The answer should be expressed as so many feet per foot.)

4. a) What is the rate of change of the area of a circle with respect to its radius?
   b) What is the value of this rate of change when the radius is 5 inches? (The answer should be expressed in square inches per inch.)
   c) What is the rate when $r = 10$ inches?

5. What is the rate of change of the volume of a cube with respect to its edge length, the unit of length being the centimeter? By how much must the volume be increased in order to double this rate of change?

6. A balloon is being filled with air at the rate of 10 cubic feet per minute. How fast is its diameter expanding when its volume is 8 cubic feet? 125 cubic feet?

7. A growing tree increases its diameter at the rate of $\frac{1}{4}$ inch per year, and its height at the rate of 1 foot per year. Assuming that the shape of the tree is approximately conical, at what rate is new wood being added when the tree is 10 years old? 50 years old?

8. A growing cubical crystal increases its edge length at the rate of one millimeter per day. How fast is its surface area increasing at the end of the first week? How fast is its volume increasing then? (Assume it starts at 0 edge length.)

9. If l_0 is the length of a piece of platinum wire at $0°$ centigrade, then its length l_t at $t°C$ is given by

$$l_t = l_0(1 + \alpha t + \beta t^2)$$

where $\alpha = 0.0868 \times 10^{-4}$ and $\beta = 0.013 \times 10^{-7}$. Discuss the sense in which platinum has a coefficient of thermal expansion.

10. A manufacturer finds that it costs him

$$10^{-4}(x^3 - 1500x^2) + 150x + 5000$$

dollars per month to produce x items per month.

   a) What is his marginal cost c_M if he is producing 250 items per month? (The answer should be in dollars per item.) What is his marginal cost if he produces 500 items per month? 1000 items per month?

   b) Now compute his *average* cost per item when his monthly production is 250 items; 500 items; 1000 items.

10. HIGHER DERIVATIVES

The derivative f' may again be a differentiable function and we naturally use the notation f'' for *its* derivative $(f')'$. For example, if $f(x) = x^5$, then

$$f'(x) = 5x^4, \quad \text{and} \quad f''(x) = \frac{d}{dx} f'(x) = \frac{d}{dx} 5x^4 = 20x^3.$$

Then $f'''(x) = 60x^2$, $f''''(x) = 120x$, $f'''''(x) = 120$, and $f''''''(x) = 0$.

When the number of primes begins to be unwieldy, we use a numeral in parentheses instead. Thus,

$$f^{(4)}(x) = f''''(x) = 120x, \quad f^{(5)}(x) = 120, \quad \text{and } f^{(6)}(x) = 0.$$

in the above example.

The Leibniz notation for the second derivative is

$$\frac{d^2 y}{dx^2} = \frac{d}{dx}\left(\frac{dy}{dx}\right).$$

Note that the superscript 2 is on d in the upper part of the term, and on dx in the lower, and that this is consistent with the right side of the equation. In Leibniz notation, the above example is

$$\frac{d}{dx} x^5 = 5x^4, \quad \frac{d^2}{dx^2} x^5 = 20x^3, \quad \frac{d^3}{dx^3} x^5 = 60x^2,$$

$$\frac{d^4}{dx^4} x^5 = 120x, \quad \frac{d^5}{dx^5} x^5 = 120, \quad \frac{d^6}{dx^6} x^5 = 0.$$

The higher-order derivatives of f play an important role in calculus and its applications. The second derivative $d^2 y/dx^2$ is particularly important because it can be directly interpreted. We shall see in Chapter 6 that the sign of $d^2 y/dx^2$ determines which way the graph of $y = f(x)$ is turning. Later on, in Chapter 10, this qualitative meaning of the second derivative will be sharpened into a formula for the *curvature* of a graph.

On the other hand, if $s = f(t)$ is the position coordinate of a particle at time t, then $d^2s/dt^2 = dv/dt$ is the *acceleration* of the particle, and this interpretation is the cornerstone of the mathematical study of motion. Newton's second law of motion says that the acceleration of a body is proportional to the force acting on it (the constant of proportionality being the mass), and this relationship is a *second-order differential equation* that determines the motion from the known forces. This application will be touched on in Chapters 6, 14, and 19.

PROBLEMS FOR SECTION 10

Compute d^2y/dx^2 when:

1. $y = 2x$
2. $y = 2x^2$
3. $y = 1/x$
4. $y = x^{1/3}$
5. $y = x^3 + 3/x$
6. $y = x^n$
7. $y = x^{-3}$
8. $y = x^{-n}$
9. $y = ax + b$
10. $y = x^4 - x^2$
11. $y = \sqrt{x}$
12. $y = 1/(1 - x)$

13. What is d^4y/dx^4 if $y = x^3$?

14. The symbol $n!$ (n factorial) represents the product

$$n(n - 1)(n - 2) \cdots 2 \cdot 1.$$

Thus $4! = 4 \cdot 3 \cdot 2 \cdot 1 \cdot = 24$. Show that

$$\frac{d^n}{dx^n} x^n = n!$$

for $n = 1, 2, 3, 4, 5$

15. Show that if the above formula holds for $n = m$, then it holds for $n = m + 1$. It must therefore hold for every value of n. Why?

16. Find the formula for $d^n y/dx^n$ when $y = 1/x$.

17. If $f(x) = x^{3/2}$, show that

$$f''(x)f(x) = \frac{3}{4}x.$$

18. A body dropped from rest falls $s = 16t^2$ feet in t seconds. What is its acceleration?

19. A particle moves along a line, its position coordinate at time t being

$$s = t^3 - t^2 + 2t.$$

If $t = 0$ is the initial time, what are the initial velocity and initial acceleration of the particle? At what moment does the particle stop decelerating and start accelerating?

20. Suppose that $f(x) = g(x) + h(x)$. Show that if the formula

$$f^{(n)}(x) = g^{(n)}(x) + h^{(n)}(x)$$

holds for some value of n, say $n = m$, then it holds for the next value $n = m + 1$. Then use this result to prove that the formula holds for every value of n.

EXTRA PROBLEMS FOR CHAPTER 3

Calculate the derivatives of the following functions from scratch:

1. $f(x) = x + 1$
2. $f(x) = x^2 + 4$
3. $f(x) = x^3/3$
4. $f(x) = \sqrt{2x}$
5. $g(t) = \sqrt{3t - 2}$
6. $f(y) = 1/(y - 1)$
7. $f(x) = x + (1/x)$
8. $g(x) = (x^2 + 1)/x$
9. $f(s) = 1/(s^2 - 1)$
10. $f(x) = \sqrt{x^2 + 1}$
11. $f(x) = (x - 4)/(x + 1)$
12. $h(x) = 1/\sqrt{x + 1}$
13. $f(x) = (x + 2)^{1/3}$
14. $f(x) = x/(x^2 + 1)$
15. $f(x) = 1/(1 + \sqrt{x})$

16. Find the coordinates of the vertex of the parabola $y = x^2 - 6x + 1$. (Make use of the fact that the slope of the tangent to the curve is zero at the vertex.)

17. Find the points on the curve
$$y = x^3 + x^2$$
where the tangent has slope 1.

18. Find the equation of the line *normal* to the curve $y = x^2 + 3x + 2$ at the point where $x = 3$. (The normal line is perpendicular to the tangent.)

19. Prove that the tangents to the parabola $y = x^2$ at the ends of any chord through $(0, \frac{1}{4})$ are perpendicular to each other.

20. Find the vertex of the parabola
$$y = ax^2 + bx + c.$$
(See Problem 16.)

21. Show that the area of the triangle cut off in the first quadrant by a tangent line to the graph of $f(x) = 1/x$ is always 2, no matter what the point of tangency is. (Find the tangent line at the point $(a, 1/a)$ on the curve, then find the area of the triangle cut off, etc.)

22. Prove that the tangent line to the parabola $y = x^2$ at the point (x_0, y_0) has y-intercept $-y_0(= -x_0^2)$.

23. Show that for any fixed point $(x_0, y_0) = (x_0, x_0^2)$ on the parabola $y = x^2$, the triangle with vertices (x_0, y_0), $(0, -y_0)$, and $(0, 1/4)$ is isosceles. (This is just a distance computation.)

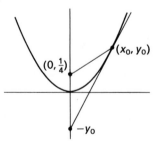

24. Assume that a ray of light reflects off a curve in such a way that the two rays make equal angles with the tangent line at the point of reflection. This is the law of reflection. Also assume the results of Problems 22 and 23. Prove that a point source of light placed at $(0, \frac{1}{4})$ will be reflected off the parabola $y = x^2$ to form a parallel beam of light shining vertically upward. (This is the principle of parabolic reflectors. The point $(0, 1/4)$ is called the focus of the parabola.)

25. Find the equation of the tangent line to the graph of $y = x^3$ at the point $(x_0, y_0) = (x_0, x_0^3)$. Then show that the tangent line intersects the graph again at the point where $x = -2x_0$.

26. Find the lines tangent to the parabola $y = x^2$ through the external point $(2, 3)$. [*Hint:* Equate two expressions for the slope of the tangent line.]

27. Show that a line can be drawn through the point (a, b) tangent to the parabola $y = x^2$ if and only if $b \leq a^2$. Interpret this requirement on (a, b) geometrically.

28. State the limit law (from Section 5 in Chapter 2) needed to make the proof of Theorem 2 logically complete.

29. Prove Theorem 3.

30. The proof of Theorem 4 used the continuity of a certain rather complicated function of r. How do we know this function is continuous?

31. Prove Theorem 4 again, in the manner of Example 5 in Section 2.

32. If $f(x) = x^{1/3}$, show that $f'(x) \to \infty$ as $x \to 0$. Sketch the graph of f.

33. Verify the product law

$$\frac{d}{dx} uv = u\frac{dv}{dx} + v\frac{du}{dx}$$

for the case $u = x^m$, $v = x^n$, where m and n are positive integers.

34. Suppose that the product law stated in the above problem holds for each pair of the four functions f, g, h, and k. Show that it holds for the product

$$(f + g)(h + k).$$

35. Determine the derivative of $f(x) = |x|$, and show that it can be expressed

$$\frac{d}{dx}|x| = \frac{|x|}{x}, \qquad x \neq 0.$$

36. Suppose that $f(x) = xg(x)$ on an interval I about the origin, and that $g(x)$ is continuous at $x = 0$. Prove that $f'(0)$ exists and equals $g(0)$.

37. Suppose that $-x^2 \leq f(x) \leq x^2$ on an interval I about the origin. Show that $f'(0)$ exists and has the value 0.

38. Suppose that $g(x) \leq f(x) \leq h(x)$ on an interval about x_0, and that

$$g(x_0) = h(x_0),$$

$$g'(x_0) = h'(x_0).$$

Show that then $f'(x_0)$ exists and has the same value.

39. A body moving vertically under the influence of gravity has constant acceleration $d^2s/dt^2 = -32$ ft/sec^2 (if the positive s direction is upward).

a) Show that its motion must be of the form

$$s = -16t^2 + bt + c$$

for some constants b and c.

b) Determine the motion explicitly if $s = 0$ and

$$v = \frac{ds}{dt} = 50 \text{ ft/sec}$$

when $t = 0$.

40. If $f(x) = x^{1/3}$, find the unique number X (depending on x) for which

$$f(x) - f(0) = f'(X)(x - 0).$$

41. If $f(x) = x^a (a \neq 1)$, find the unique number X (depending on x) for which

$$f(x) - f(0) = f'(X)(x - 0).$$

42. Show that there is at most one function f defined on the interval $(0, \infty)$ such that

$$f(1) = 0 \qquad \text{and} \qquad f'(x) = 1/x$$

for all x. (Show that if f and g both have these properties, then $f(x) = g(x)$ for all x.)

43. Show that a function f is uniquely determined on an interval I if we know its derivative f' and its value at one point. That is, if $f' = g'$ on I and if $f(x_0) = g(x_0)$, then $f = g$ on I.

44. Prove that the tangent line to the ellipse

$$\frac{x^2}{a^2} + \frac{y^2}{b^2} = 1$$

at the point (x_0, y_0) has the equation

$$\frac{xx_0}{a^2} + \frac{yy_0}{b^2} = 1.$$

45. Assuming the result in Problem 44, show that the normal line to the ellipse at (x_0, y_0) has the x-intercept $c^2 x_0 / a^2$, where $c^2 = a^2 - b^2$.

46. Find the formula for the tangent line to the hyperbola

$$\frac{x^2}{a^2} - \frac{y^2}{b^2} = 1$$

analogous to that given in Problem 44 for the ellipse.

Compute $f'(x)$ when:

47. $f(x) = \sqrt{x + \sqrt{1 - x^2}}$

48. $f(x) = (x^{2/3} + a^{2/3})^{3/2}$

49. $f(x) = \dfrac{1}{x + \sqrt{x^2 - 1}}$

50. $f(x) = 1/(x^2 + 2x + 1)$

51. $f(x) = (x^a + 1)^{1/a}$

52. $f(x) = (1 + g(x))^{1/3}$

53. $f(x) = 1/(x + g(x))$

54. $f(x) = \sqrt{1 - (g(x))^2}$

55. $f(x) = 1/(x^2 + g^2(x))$

56. If $f(x) = (x^{1/3} + a^{1/3})^3$, show that

$$f'(x) = \left(1 + \left(\frac{a}{x}\right)^{1/3}\right)^2.$$

57. Verify by differentiating that $y = (c - x^2)^{1/3}$ satisfies the differential equation

$$3y^2 \frac{dy}{dx} + 2x = 0,$$

for any constant c.

58. Find the rate of change of the volume of a cone of fixed height with respect to the radius.

59. Find the rate of change of the height of a cone of fixed volume with respect to the radius.

60. The marginal cost of production will generally have a graph like the one at the left below. That is, the marginal cost will normally decrease with increasing production until it reaches a minimum value, and then will increase. Show by an intuitive argument that the total cost function $C(x)$ must necessarily look like the graph at the right.

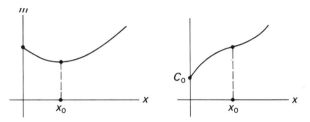

61. A marginal cost graph like the one shown in the preceding problem is approximately a quadratic graph, i.e., approximately the graph of

$$m(x) = ax^2 + bx + c$$

for suitable values of the coefficients a, b, and c.

a) Find this quadratic function if $m(0) = 10$ and if m has the minimum value $m(1000) = 8$ at $x_0 = 1000$. [*Hint:* Cast the general expression for $m(x)$ above into a new form by completing the square.]

b) Find the corresponding cost function $C(x)$, assuming that there is a fixed initial cost of 500 dollars even if no items are produced. (The problem is to find a cubic function $C(x)$ whose derivative is a given quadratic function $m(x)$, and such that $C(0) = 500$.)

62. A manufacturer's average cost per item is $f(x)$ dollars. Show that his marginal cost is

$$f(x) + xf'(x)$$

dollar per item. (This is really a problem for Section 3. You are asked to prove from scratch that if $C(x) = xf(x)$, then $C'(x) = f(x) + xf'(x)$.)

4
The Antiderivative

A major discovery of Newton and Leibniz was that many problems in geometry and physics can be solved by "backwards differentiation," or *anti-differentiation*. We shall see how this comes about in the present chapter. Our antidifferentiation technique will of course be limited to the backwards versions of the rules from Chapter 3, but we shall nevertheless be able to solve many examples of several classical problems.

Important Notice. This chapter is not required for Chapters 5 through 7, and can be postponed beyond any or all of them. It is placed here because of the widespread interest in an early introduction to integration.

1. THE ANTIDERIVATIVE

What do we know about the motion of a particle travelling along a coordinate line with constant velocity 5? In Chapter 3 we decided that if s is the position coordinate of the particle, then its velocity v is given by the derivative ds/dt. So here we are assuming that s is a function of t which has the constant derivative 5. The question is: What does this tell us about the function itself?

You undoubtedly can visualize *one* function of t having the constant derivative 5, namely $5t$. Moreover, adding a constant term doesn't change the derivative, so if s is any one of

$$5t + 3, \qquad 5t - 10, \qquad 5t + \sqrt{2}, \qquad 5t + C,$$

then s is a function of t for which $ds/dt = 5$. Are there others? The answer is *no*:

THEOREM 1 *If h_1 and h_2 are differentiable functions having the same derivative on an interval I, then there is a constant C such that*

$$h_2(x) = h_1(x) + C$$

for all x in I.

Proof. If $h_1' = h_2'$ on I, then $(h_2 - h_1)' = 0$ on I. But any function whose derivative is identically zero must be a constant. (This is Theorem 7 of Chapter 3.) So $h_2 - h_1 = C_1$ and $h_2 = h_1 + C$. ∎

Using this principle we see that if $ds/dt = 5$, then s must differ from $5t$ by a constant, so $s = 5t + C$ for some constant C. (This is all we can conclude without further information.)

The problem above involved finding a function whose derivative is known. A differentiable function F such that $F' = f$ is called an *anti-derivative* of f, and the process of finding F from f is *antidifferentiation*. We saw above that:

A given function f does not have a uniquely determined antiderivative, but if we can find one antiderivative F, then every other antiderivative is of the form F + constant.

For example, $x^3/3$ is one antiderivative of x^2, and every other anti-derivative is then necessarily of the form

$$\frac{x^3}{3} + C.$$

An antiderivative of f is also called an *integral* of f, and antidifferentiation is also called *integration*. The most general integral of x^2 shown above is called its *indefinite integral*, because of the arbitrary constant C. This constant is called the *constant of integration*. The Leibniz notation for the indefinite integral of a function f is

$$\int f(x)\,dx,$$

a notation that will prove to be especially useful in Chapter 8. For example,

$$\int x^2\,dx = \frac{x^3}{3} + C.$$

In this context, x^2 is called the *integrand*. The origin of the Leibniz notation will become apparent in Chapter 9.

The polynomial rules of Chapter 3, read backwards, now become the following integration rules:

$$\int x^r\,dx = \frac{x^{r+1}}{r+1} + C \qquad \text{(if } r \neq -1\text{)},$$

$$\int (af(x) + bg(x))\,dx = a\int f(x)\,dx + b\int g(x)\,dx.$$

We'll check them shortly, but consider first an example of how they are used.

EXAMPLE
$$\int (3x + 2x^3)\,dx = 3\int x\,dx + 2\int x^3\,dx \qquad \text{(By the second rule)}$$

$$= 3\left(\frac{x^2}{2} + C_1\right) + 2\left(\frac{x^4}{4} + C_2\right) \qquad \text{(By the first rule)}$$

$$= \frac{3x^2 + x^4}{2} + (3C_1 + 2C_2).$$

But we can write this as
$$\frac{3x^2 + x^4}{2} + C.$$

because $3C_1 + 2C_2$ is itself just an arbitrary constant.

In practice, when we have to compute the indefinite integral of a sum of functions, we just add up the particular integrals (antiderivatives) that we find, and then add an arbitrary constant at the end. For example, the above calculation would be done like this:

$$\int (3x + 2x^3)\,dx = 3\cdot\frac{x^2}{2} + 2\cdot\frac{x^4}{4} + C$$

$$= \frac{3x^2 + x^4}{2} + C.$$

Proof of the integration rules: Since

$$\frac{d}{dx}\left(\frac{x^{r+1}}{r+1}\right) = \frac{(r+1)x^r}{r+1} = x^r,$$

we see that $x^{r+1}/(r+1)$ is a particular antiderivative of x^r, and the first rule follows.

The second rule is just the integral form of the corresponding derivative identity. For if F and G are antiderivatives of f and g, respectively, then

$$a\int f(x)\,dx + b\int g(x)\,dx = a(F(x) + C_1) + b(G(x) + C_2)$$
$$= aF(x) + bG(x) + C,$$

and this is equal to

$$\int (af(x) + bg(x))\,dx,$$

because

$$(aF + bG)' = aF' + bG' = af + bg.$$

PROBLEMS FOR SECTION 1

Compute the following integrals. Be sure to include the constant of integration in each answer.

1. $\int x^6\,dx$ 2. $\int \sqrt{x}\,dx$ 3. $\int x\,dx$ 4. $\int dx$

5. $\int x^{1/3}\,dx$ 6. $\int (x^2 + x + 1)\,dx$

7. $\int \left(3x^3 - \frac{1}{3x^3}\right)dx$ 8. $\int (t^2 - 2t + 3)\,dt$

9. $\int (2x + 3)(x - 2)\,dx$ 10. $\int x^{2/3}\,dx$

11. $\int \frac{dx}{\sqrt{x}}$ 12. $\int \frac{dx}{\sqrt[3]{x}}$

13. $\int 3ay^2\,dy$ 14. $\int (x^{3/2} - 2x^{2/3} + 5\sqrt{x} - 3)\,dx$

15. $\int \frac{4x^2 - 2\sqrt{x}}{x}\,dx$ 16. $\int \sqrt{x}(3x - 2)\,dx$

17. Find an antiderivative of x^2 that has the value 1 at $x = 2$. [*Hint:* Every antiderivative of x^2 is of the form $y = (x^3/3) + C$. Now see what value C must have to make $y = 1$ when $x = 2$.]

18. Find an antiderivative $F(x)$ of $\sqrt{x}$ such that $F(4) = 4$.

19. Find a function F such that $F'(x) = x$ and $F(1) = 0$. Show that F is uniquely determined by these conditions.

2. THE INITIAL-VALUE PROBLEM

The graphs of the antiderivatives of x^2 are the graphs of the functions

$$y = \int x^2\,dx = \frac{x^3}{3} + C$$

for the various values of the constant C, as shown below. They form a family of curves filling up the plane, exactly one curve going through any given point (x_0, y_0).

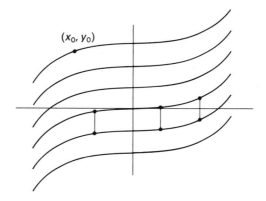

Analytically, this means that if we ask for an antiderivative $F(x)$ of x^2 having a given value at a given point, say $F(-1) = 2$, then there is a unique solution.

We find it as follows. The function F must be of the form

$$F(x) = \int x^2\, dx = \frac{x^3}{3} + C,$$

for some value of the constant C, and C is determined by the extra requirement that $F(-1) = 2$. Thus

$$2 = F(-1) = -\frac{1}{3} + C,$$

so that $C = 2 + 1/3 = 7/3$, giving the unique solution

$$F(x) = \frac{x^3 + 7}{3}.$$

The requirement that $F(-1) = 2$ is called an *initial condition*, and the total problem is called an *initial-value problem*. The above initial-value problem, then, is to determine the unique function F such that

$$F'(x) = x^2, \qquad F(-1) = 2.$$

In pure Leibniz notation it is the problem of determining what function y must be of x if

$$\frac{dy}{dx} = x^2,$$

$$y = 2 \qquad \text{when } x = -1.$$

In this notation the solution would be written out as follows. First,

$$y = \int x^2\, dx = \frac{x^3}{3} + C,$$

where the constant C is to be determined by the initial condition. Substituting the initial values $(x, y) = (-1, 2)$ gives

$$2 = \frac{(-1)^3}{3} + C,$$

so $C = 7/3$, as before.

We now turn to initial-value problems in which we determine the motion of an object from its known velocity or acceleration.

EXAMPLE 1 A particle travels along the number line with constant velocity 5. Find its position s as a function of t if $s = -10$ when $t = 1$.

This is an initial-value problem. From $ds/dt = 5$ we obtain $s = 5t + C$, as before, although we might now write this

$$s = \int 5\,dt = 5t + C.$$

Substituting the initial values $(t, s) = (1, -10)$ gives

$$-10 = 5 \cdot 1 + C,$$

so that $C = -15$ and

$$s = 5t - 15.$$

It is said that in the seventeenth century Galileo experimented by dropping objects from the tower in Pisa, and concluded that, neglecting air resistance, which would slow down light objects more, all falling objects drop with a constant acceleration of 32 ft per sec^2.

If the position of the falling body is measured along a vertical coordinate system with the positive direction upward, then its constant acceleration is -32, because its velocity is becoming increasingly negative. The acceleration due to gravity is usually designated g, so we have $g = -32$ in the present axis system.

EXAMPLE 2 A stone is thrown vertically upward with an initial velocity of 50 ft/sec. What is its velocity two seconds later? How high will it rise?

Solution. Since $dv/dt = g = -32$, we have

$$v = \int g\,dt = \int (-32)\,dt = -32t + C.$$

If we start measuring time at the instant the stone is thrown, then the initial condition is that $v = 50$ when $t = 0$. Thus $50 = -32 \cdot 0 + C$ and $C = 50$, giving the velocity equation

$$v = 50 - 32t.$$

At $t = 2$ we have $v = 50 - 64 = -14$ ft/sec. The stone is already falling.

The highest point in its trajectory will be reached when the stone just stops rising and starts to fall, i.e., when $v = 0$. This occurs when

$$0 = 50 - 32t$$

or $t = 25/16$ seconds. But in order to find how far the stone has risen we have to integrate the equation

$$v = \frac{ds}{dt} = 50 - 32t.$$

We get

$$s = 50t - 16t^2 + C.$$

If we measure s from the point at which the stone is thrown, then the initial condition is $s = 0$ at $t = 0$. This gives $C = 0$, and at $t = 25/16$ we have

$$s = \frac{50 \cdot 25}{16} - 16\left(\frac{25}{16}\right)^2 = \frac{625}{16} \text{ ft,}$$

or approximately 39 ft.

The highest point is the maximum value of s, and can be determined by finding where $ds/dt = 0$ (see Chapter 6). But this is just the condition $v = 0$ that we arrived at intuitively above.

In general, an initial-value problem is the combination of a *differential equation* and one or more *side conditions* that single out a unique solution from among the many solutions to the differential equation. A differential equation is just an equation involving the derivatives of an unknown function. In the first example above, the differential equation is

$$\frac{dy}{dx} = x^2$$

for the unknown function $y = f(x)$. It will be shown in Chapter 5 that the exponential function is the unique solution of the initial-value problem

$$\frac{dy}{dx} = y,$$

$$y = 1 \qquad \text{when } x = 0.$$

PROBLEMS FOR SECTION 2

Solve the following initial-value problems.

1. $f'(x) = x - 3, \qquad f(2) = 9$

2. $g'(x) = x + 3 - 5x^2, \quad g(6) = -20$

3. $f'(y) = y^3 - b^2 y, \quad f(2) = 0$

4. $f'(x) = bx^3 + ax + 4, \quad f(b) = 10$

5. $f'(t) = \sqrt{t} + \dfrac{1}{\sqrt{t}}, \quad f(4) = 0$

6. Given $dy/dx = (2x + 1)$, $y = 7$ when $x = 1$. Find the value of y when $x = 3$.

7. Given $dA/dx = \sqrt{2px}$, $A = p^2/3$ when $x = p/2$. Find the value of A when $x = 2p$.

8. Given the expressions below for acceleration, find the relation between s (displacement) and t if $s = 0$, $v = 20$, when $t = 0$.

   a) $a = 32$ \qquad\qquad\qquad\qquad b) $a = 4 - t$

9. With what velocity will a stone strike the ground if dropped from the top of a building 100 ft high? ($g = 32$ if distance is measured positively in the downward direction.)

10. If the stone in Problem 9 is thrown from the top of the building with an initial velocity of 100 ft/sec downward, what will be its velocity upon hitting the ground? What if it has an *upward* initial velocity of 100 ft/sec?

11. A train leaving a railroad station has an acceleration of

$$\frac{1}{2} + \frac{2}{100}t \text{ ft/sec}^2.$$

How far will the train move in the first 20 sec of motion?

12. What constant acceleration is required to

a) Move a particle 50 ft in 5 sec?

b) Slow a particle from a velocity of 45 ft/sec to a stop in 15 ft?

3. AREAS

One of the unsolved problems of ancient mathematics was the *area* problem. The Greeks knew much about the areas of triangles and circles, and configurations derivable from them, but any other figure represented a new, and generally insoluble, problem. Archimedes was able to apply a method that eventually became known as *exhaustion* to compute the area of a parabolic segment, and to compute a few other particular geometric magnitudes. But for nearly two thousand years this cluster of computations by Archimedes stood as an isolated achievement. During the seventeenth century mathematicians came to understand that the method of exhaustion represented a systematic approach, and eventually it evolved into the calculational procedure that today is associated with the definite integral. But meanwhile Newton and Leibniz discovered that if a quantity could be calculated exactly by exhaustion, then it could be computed *much more easily* from antiderivatives. This remarkable fact, when stated in more precise terms, is called the Fundamental Theorem of Calculus. Our temporary and incomplete version will be that no matter how we consider the area of a figure to be *defined*, we can frequently *calculate* it by antiderivatives.

Suppose, for example, that we want to find the exact area of the region bounded by the half-parabola $y = \sqrt{x}$, the x-axis, and the vertical line $x = 1$, as sketched at the left below.

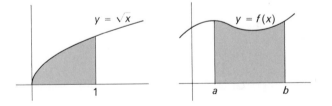

More generally, we will be interested in the area of the region bounded by a graph $y = f(x)$, the x-axis, and two vertical lines

$$x = a \quad \text{and} \quad x = b,$$

as shown at the right. Intuition tells us that any such region *has* an exact area, and our problem is to *find* it.

The key to the Newton–Leibniz approach seems paradoxical. In order to solve this problem, we replace it by an apparently harder problem. Instead of asking what the *fixed* area is, we ask how the area *varies* when we *change* the right-hand boundary line. It is convenient to denote the position of this variable edge by *x*, and the whole figure will be clearer if we use a different independent variable, say *t*, for the graph, as below. The varying area *A* is then a function of the varying right-hand edge coordinate *x*. Let us see how this helps in our parabolic problem.

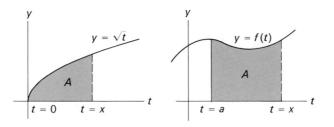

When we give *x* an increment Δx, the area changes by an increment ΔA, as shown below, and ΔA is squeezed between two rectangular areas,

$$\sqrt{x} \cdot \Delta x < \Delta A < \sqrt{x + \Delta x} \cdot \Delta x.$$

But then

$$\sqrt{x} < \frac{\Delta A}{\Delta x} < \sqrt{x + \Delta x}.$$

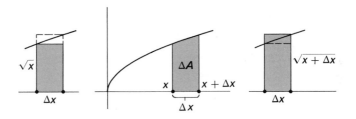

(This assumes Δx positive. If $\Delta x < 0$, then the inequality is reversed.) Since $\sqrt{x + \Delta x} \to \sqrt{x}$ as $\Delta x \to 0$, the difference quotient $\Delta A/\Delta x$ is squeezed between $\sqrt{x}$ and something approaching $\sqrt{x}$ and so must have the same limit $\sqrt{x}$. This shows that *the variable area A is a differentiable function of x and that*

$$\frac{dA}{dx} = \lim_{\Delta x \to 0} \frac{\Delta A}{\Delta x} = \sqrt{x}.$$

Therefore,

$$A = \int x^{1/2}\, dx = \frac{2}{3}x^{3/2} + C,$$

where the constant *C* has a particular value that still has to be determined. To do this we use the fact that $A = 0$ when $x = 0$, which gives

$$0 = \frac{2}{3} \cdot 0 + C$$

and $C = 0$. Thus,

$$A = \frac{2}{3}x^{3/2}.$$

Now we can solve the original problem. Setting $x = 1$, we have $A = 2/3$ for the exact value of the area of the region we began with.

Exactly the same reasoning applies to *any* increasing, positive, continuous function f. When we give x an increment Δx, the area changes by an increment ΔA, as shown below, and ΔA is squeezed between two rectangular areas:

$$f(x) \cdot \Delta x < \Delta A < f(x + \Delta x)\Delta x.$$

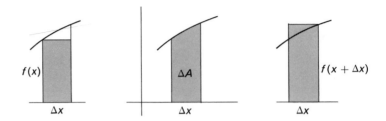

But then

$$f(x) < \frac{\Delta A}{\Delta x} < f(x + \Delta x)$$

(if Δx is positive). Since $f(x + \Delta x) \to f(x)$ as $\Delta x \to 0$, the difference quotient $\Delta A/\Delta x$ is squeezed between $f(x)$ and something approaching $f(x)$, and must have the same limit $f(x)$. This shows that *the variable area A is a differentiable function of x and that*

$$\frac{dA}{dx} = \lim_{\Delta x \to 0} \frac{\Delta A}{\Delta x} = f(x).$$

If f is a *decreasing positive* function, the squeeze inequality is reversed but similar reasoning applies, with the same conclusion.

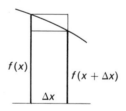

We will not always be dealing with a monotone function. However, if there is a sequence of intervals on which a positive continuous function f is alternately increasing and decreasing, then the conclusion $dA/dx = f(x)$ follows by one or the other of the above arguments, depending on which interval contains x. But it is better to describe this situation in a more general way.

Assuming $\Delta x > 0$, let m and M be the minimum and maximum values of $f(x)$ on the interval $[x, x + \Delta x]$. Then in every case

$$m\,\Delta x \leq \Delta A \leq M\,\Delta x,$$

and

$$m \leq \frac{\Delta A}{\Delta x} \leq M.$$

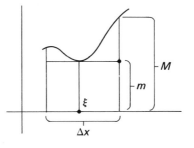

Now m is a function value $f(\xi)$ for some ξ between x and $x + \Delta x$. Since $\xi \to x$ as $\Delta x \to 0$ and since f is continuous, it follows that $m \to f(x)$ as $\Delta x \to 0$. Similarly for M. Thus $\Delta A/\Delta x$ is squeezed to the same limit, and

$$\frac{dA}{dx} = \lim \frac{\Delta A}{\Delta x} = f(x)$$

as before. If Δx is negative, the first inequality is reversed, but it reverses again on dividing by Δx, so

$$m \le \frac{\Delta A}{\Delta x} \le M$$

in this case, too.

It is customary to state this result for an arbitrary continuous function. The trouble is that we don't know enough about what an arbitrary continuous function can be like. If "continuous" brings to mind a graph that smoothly rises and falls, increasing over one interval, then decreasing over a succeeding interval, etc., then our discussion above is adequate. But there are continuous functions that don't behave so nicely. Try to imagine a continuous graph that is everywhere "infinitely crinkly." This may be practically impossible to conceive of, but it can happen. Then the simple pictures that have been supporting our arguments don't apply, and we lose confidence in our intuition. So it is best—and really necessary for logical completeness—to be very explicit about what assumptions are needed to make the above argument work.

First, we assumed that the values of a continuous function on a closed interval $[a, b]$ run from a minimum value m to a maximum value M. This is called the *extreme-value* property of a continuous function; it will be proved in Appendix 4

We also assumed that each region that we describe has a uniquely determined numerical area, and that this area varies with the region in the natural way. For example, we implicitly used the fact that the area of a union of two nonoverlapping regions is exactly the sum of their two areas.

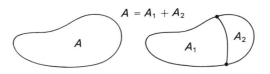

At any rate, subject to these cautionary qualifications, we have established the following theorem.

THEOREM 2   *Let f be a positive continuous function and let A be the area of the region bounded by the graph of $y = f(t)$, the t-axis, and the vertical lines $t = a$ and $t = x$, where $x > a$. Then A is a differentiable function of x, and*

$$\frac{dA}{dx} = f(x).$$

EXAMPLE 1   Find the area of the region below the graph of $f(x) = 2 - x - x^2$ and above the x-axis.

Solution. The graph is shown below. The x-intercepts are the solutions of $2 - x - x^2 = 0$, and are $x = -2$ and $x = 1$, as shown. These are the limits between which we want the area. According to the theorem, the area A from -2 to x is a function of x such that

$$\frac{dA}{dx} = 2 - x - x^2.$$

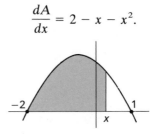

Therefore,

$$A = \int (2 - x - x^2)\,dx = 2x - \frac{x^2}{2} - \frac{x^3}{3} + C,$$

where C has to be determined by the initial condition: $A = 0$ when $x = -2$. This gives

$$0 = 2(-2) - \frac{(-2)^2}{2} - \frac{(-2)^3}{3} + C = -\frac{10}{3} + C,$$

and $C = 10/3$. So

$$A = 2x - \frac{x^2}{2} - \frac{x^3}{3} + \frac{10}{3}.$$

The particular area required was from $x = -2$ to $x = 1$, and is

$$A_1 = 2 - \frac{1}{2} - \frac{1}{3} + \frac{10}{3} = 4\frac{1}{2}.$$

In addition to showing us how to compute an exact area when we can find an antiderivative, Theorem 2 also has an important theoretical consequence.

THEOREM 3   *Every continuous function on a closed interval $[a, b]$ has an antiderivative on $[a, b]$.*

Proof. If f is positive, then the variable area A of Theorem 2 is an antiderivative. If f is not positive, we choose a constant C large enough so that $f(x) + C$ *is* positive and hence has an antiderivative $G(x)$. Then the function $F(x) = G(x) - Cx$ is an antiderivative of f. ∎

This is an *existence* statement. It says that even though we may be unsuccessful in looking for an explicit formula for the antiderivative, nevertheless an antiderivative does exist.

Remember, however, that we have this result only by virtue of assuming certain other properties of a continuous function.

Remember, also, that we have to be able to *find* an antiderivative of f in order to use Theorem 2 to evaluate the area under the graph of f. We know how to do this for simple functions by means of our polynomial rules, but it can be very difficult for more complicated functions f, and we shall spend the whole of Chapter 8 developing the procedures that are available.

EXAMPLE 2 Show that the area A_x^b, with variable *left-hand* edge $t = x$ as shown below, is an antiderivative of $-f(x)$.

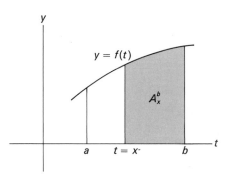

Solution. Choose any fixed number a less than b and consider values of x between a and b. We are using the notation A_a^b for the area under the graph of f from a to b. Thus A_a^x is the area from a to x and A_x^b is the area from x to b. Since

$$A_a^x + A_x^b = A_a^b,$$

where A_a^b is a constant—call it C—we have

$$A_x^b = C - A_a^x,$$

so

$$\frac{d}{dx}(A_x^b) = -\frac{d}{dx}(A_a^x) = -f(x).$$

For the derivative of A_x^b at a point x_0 smaller than our first choice of a, we just repeat the argument with a new value of a to the left of x_0.

These area antiderivatives constitute an important source of new functions. For example, according to the integration formula

$$\int x^k \, dx = \frac{x^{k+1}}{k+1} + C \qquad (k \neq -1),$$

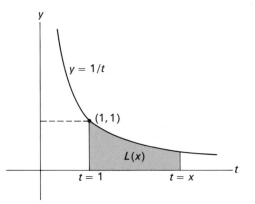

every power of x has an antiderivative that is another power of x (times a constant), *except for the power* $k = -1$. Yet $1/x$ has a perfectly well-defined antiderivative $L(x)$: if $x \geq 1$ we just define $L(x)$ to be the area A_1^x under the graph of $y = 1/t$, from $t = 1$ to $t = x$; if $0 < x < 1$ then $L(x) = -A_x^1$; see Example 2 above. Thus $L(x)$ is defined on the interval $(0, \infty)$, and is the solution there of the initial value problem

$$\frac{dy}{dx} = \frac{1}{x}; \qquad y = 0 \text{ when } x = 1.$$

Clearly, $L(x)$ is a wholly new function, and it turns out to be an important one. Later on, $L(x)$ will be found to be a logarithm function, and its properties will be further investigated. We shall learn a numerical method for computing such area antiderivatives in Chapter 9, and a much better method that can be used for a few of them, including $L(x)$, in Chapter 11.

PROBLEMS FOR SECTION 3

A sketch is an almost necessary part of the solution of a problem involving a geometric magnitude, and you should get in the habit of drawing one as a matter of course, always trying to be reasonably accurate. In these area problems, your sketch should show both the area to be computed and the variable area which lets us use calculus in the computation.

1. Find, by integration, the area of the triangle bounded by the line $y = 2x$, the x-axis, and the line $x = 4$. Verify your answer by using the formula $A = \frac{1}{2}bh$.

2. Find, by integration, the area of the trapezoid bounded by the line $x + y = 15$, the x-axis, and the lines $x = 3$ and $x = 10$. Verify your answer by use of the formula $A = \frac{1}{2}(a + b)h$.

Find the area bounded by the given curve, the x-axis, and the given vertical lines.

3. $y = x^3$; $x = 0, x = 4$
4. $y = x^2 + x + 1$; $x = 2, x = 3$
5. $y = x + 4$; $x = -4, x = -2$
6. $y^2 + 4x = 0$; $x = -1, x = 0$
7. $y = 2x + \dfrac{1}{x^2}$; $x = 1, x = 4$
8. $y = x^2$; $x = 2, x = 5$
9. $y = \dfrac{1}{\sqrt{x + 4}}$; $x = 0, x = 5$
10. $x = 3y^2 - 9$; $x = 0$

11. The variable area A between the graph of $x = f(y)$ and the y-axis, from a to y, is a differentiable function of y satisfying

$$\frac{dA}{dy} = f(y).$$

This can be proved simply by repeating the proof of Theorem 2 in this slightly different context. Instead, show by a geometric argument that the variant above is a corollary of Theorem 2.

Assuming the formula in the preceding problem, find the area of the region bounded by each of the following graphs and the given lines.

12. $x = 9y - y^3$; $y = 0$, $y = 3$ 13. $y^2 = 4x$; $y = 0$, $y = 4$

14. $x = -(y^2 + 4y)$; $x = 0$ 15. $y = 4 - x^2$; $y = 0$, $y = 3$

16. Find the area of the region bounded by the parabola $y = x^2$ and the line $y = x + 2$.

17. The same problem for $y = x^2$ and $y = 2 - x^2$.

4. INTEGRATION BETWEEN LIMITS

The notation $f(x)]_a^b$ stands for $f(b) - f(a)$. For example,

$$x^2]_3^4 = 4^2 - 3^2 = 7.$$

Common variants are $f]_a^b$ and $f|_a^b$. Note that adding a constant to f does not change the result of this "evaluation between limits":

$$[f(x) + C]_a^b = (f(b) + C) - (f(a) + C)$$
$$= f(b) - f(a) = f(x)]_a^b.$$

For this reason,

$$\int f(x)\,dx \bigg]_a^b$$

represents a uniquely determined number (assuming that f is continuous from a to b). Thus, if F is an antiderivative of f on $[a, b]$, then

$$\int f(x)\,dx \bigg]_a^b = F(x) + C \bigg]_a^b = F(x) \bigg]_a^b = F(b) - F(a).$$

EXAMPLE 1

$$\int x^2\,dx \bigg]_a^b = \frac{x^3}{3} \bigg]_a^b = \frac{b^3}{3} - \frac{a^3}{3}.$$

EXAMPLE 2

$$\int (x^2 - x)\,dx \bigg]_1^2 = \frac{x^3}{3} - \frac{x^2}{2} \bigg]_1^2$$

$$= \left(\frac{8}{3} - \frac{4}{2}\right) - \left(\frac{1}{3} - \frac{1}{2}\right)$$

$$= \frac{7}{3} - \frac{3}{2} = \frac{5}{6}.$$

These examples of $\int g(x)\,dx]_a^b$ show how the variable x disappears in the evaluation. It is a *dummy* variable, and any other variable can be used just as well. For example

$$\int t^2\,dt\Bigg]_a^b = \frac{t^3}{3}\Bigg]_a^b = \frac{b^3}{3} - \frac{a^3}{3} = \int x^2\,dx\Bigg]_a^b.$$

When we are using the function symbol f, we can dispense with the dummy variable entirely, and write $\int f]_a^b$. Thus

$$\int f\Bigg]_a^b = \int f(x)\,dx\Bigg]_a^b = \int f(t)\,dt\Bigg]_a^b.$$

The areas that we computed earlier have natural expressions as integrals between limits.

THEOREM 4   *If f is a positive continuous function over the closed interval $[a, b]$, then the area of the region between the graph of f and the x-axis, from $x = a$ to $x = b$, is $\int f]_a^b$.*

Proof. For each number x in $[a, b]$, let $A(x)$ be the area of the region over the interval $[a, x]$. We proved earlier that $A(x)$ is an antiderivative of $f(x)$, so

$$\int f(x)\,dx\Bigg]_a^b = A(b) - A(a).$$

But $A(a) = 0$, since the area $A(x)$ was measured from $x = a$, and $A(b)$ is the area A between $x = a$ and $x = b$ that we want. ∎

This gives a streamlined way to calculate areas. For if we can find *any* antiderivative F of f, then

$$A = \int f\Bigg]_a^b = F(b) - F(a).$$

EXAMPLE 3   The area under (the graph of) $y = x^2$ from 1 to 4 is

$$\int x^2\,dx\Bigg]_1^4 = \frac{x^3}{3}\Bigg]_1^4 = \frac{64}{3} - \frac{1}{3} = 21.$$

EXAMPLE 4   The area under $y = \sqrt{x}$ from 0 to 1 is

$$\int x^{1/2}\,dx\Bigg]_0^1 = \frac{2x^{3/2}}{3}\Bigg]_0^1 = \frac{2}{3}.$$

Any derivative formula becomes an integral formula when reversed. For example, the general power rule in the form

$$\frac{d}{dx}\frac{g^{a+1}(x)}{(a + 1)} = g^a(x) \cdot g'(x)$$

becomes the integration rule

$$\int g^a(x)g'(x)\, dx = \frac{g^{a+1}(x)}{(a+1)} + C \qquad (\text{if } a \neq -1).$$

EXAMPLE 5 What is the area under the graph of $f(x) = 2x\sqrt{1+x^2}$ from $x = 0$ to $x = 1$?

Solution. This is in the above form with $g(x) = 1 + x^2$, so

$$\int \sqrt{1+x^2}\,(2x)\, dx \Big]_0^1 = \frac{(1+x^2)^{3/2}}{3/2}\Big]_0^1 = \frac{2}{3}(2\sqrt{2}-1).$$

EXAMPLE 6 The area under $y = \sqrt{1+x}$ from $x = 3$ to $x = 8$ is

$$\int \sqrt{1+x}\, dx \Big]_3^8 = \frac{2}{3}(1+x)^{3/2}\Big]_3^8 = \frac{2}{3}[27-8] = \frac{38}{3}.$$

EXAMPLE 7 The area under $y = 1/(x^2 - 4x + 4)$ from $x = -1$ to $x = 1$ is

$$\int \frac{dx}{x^2 - 4x + 4}\Big]_{-1}^1 = \int \frac{dx}{(x-2)^2}\Big]_{-1}^1 = -\frac{1}{x-2}\Big]_{-1}^1 = 1 - \frac{1}{3} = \frac{2}{3}.$$

Besides area, there are many other quantities in physics, geometry, and other applications that can be expressed by integrals between limits. The general line of reasoning is always the same.

We want a formula for the amount Q of some quantity (area, volume, mass, charge, etc.) that lies between $x = a$ and $x = b$. To obtain such a formula we first let Q vary, and consider the amount $Q(x)$ of the quantity that lies between a and the variable edge coordinate x. Examination of the increment ΔQ enables us to prove that $Q(x)$ is differentiable, and its derivative

$$\rho(x) = \frac{dQ}{dx}$$

is a function known to us from the conditions of the problem.

The final step is the same in all cases.

THEOREM 5 *In the circumstances above,*

$$Q = \int \rho(x)\, dx \Big]_a^b.$$

Proof. Since $Q'(x) = \rho(x)$, we have

$$\int \rho(x)\, dx \Big]_a^b = Q(b) - Q(a).$$

But $Q(a) = 0$, since $Q(x)$ was measured from $x = a$, and $Q(b)$ is the amount Q of the quantity that lies between $x = a$ and $x = b$, which is what we want. ∎

We can then *compute* Q by finding *any* antiderivative G of ρ and using the fact that

$$\int \rho(x)\, dx \Big]_a^b = G(b) - G(a).$$

The remaining sections in this chapter contain several such applications.

PROBLEMS FOR SECTION 4

Evaluate the following integrals between limits

1. $\displaystyle \int (x - x^2)\, dx \Big]_0^1$

2. $\displaystyle \int (x - x^2)\, dx \Big]_1^2$

3. $\displaystyle \int (x^4 - x^2)\, dx \Big]_{-1}^1$

4. $\displaystyle \int (ax - x^2)\, dx \Big]_0^a$

5. $\displaystyle \int 3t^2\, dt \Big]_x^y$

6. $\displaystyle \int dt/\sqrt{t - 1} \Big]_2^5$

7. $\displaystyle \int (x^3 - x^{1/3})\, dx \Big]_0^1$

8. $\displaystyle \frac{1}{b - a} \int x\, dx \Big]_a^b$

Find the area bounded by the given curve, the x-axis, and the given vertical lines.

9. $y = x^3$; $x = 0, x = 4$

10. $y = x^2 + x + 1$; $x = 2, x = 3$

11. $y = x^3 + 4x^2$; $x = -4, x = -2$

12. $y^2 + 4x = 0$; $x = -1, x = 0$

13. $y = 2x + \dfrac{1}{x^2}$; $x = 1, x = 4$

14. $x = 3y^2 - 9$; $x = 0$

15. $y = \dfrac{1}{\sqrt{x + 4}}$; $x = 0, x = 5$

5. THE VOLUME OF A SOLID OF REVOLUTION

If the first quadrant is rotated about the x-axis, then each point on the graph $y = x^{1/3}$ has a circular path, and the whole graph sweeps out a certain surface, called a *surface of revolution*. The plane region between the graph, the x-axis, and the vertical line $x = 1$ sweeps out a *solid of revolution*. The boundary of the solid lies partly on the surface of revolution and partly on the vertical plane through $x = 1$. We propose to calculate the volume of this solid. More generally, we shall be interested in the volume swept out by rotating about the x-axis the region bounded by a graph $y = f(x)$, the x-axis, and the vertical lines $x = a$ and $x = b$, as shown at the right in the following figure.

Proceeding just as before, we relabel the horizontal axis with t, in order to leave x free to mark the variable right edge. We let $V(x)$ be the volume between $t = 0$ and $t = x$, and we give x an increment Δx. The volume then

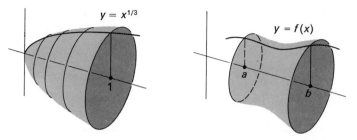

changes by an increment ΔV, as shown below, and ΔV is squeezed between two cylindrical volumes. A circular cylinder with base radius r and altitude h has volume $\pi r^2 h$, so this squeeze on ΔV is the inequality

$$\pi(x^{1/3})^2 \Delta x < \Delta V < \pi((x + \Delta x)^{1/3})^2 \Delta x,$$

or

$$\pi x^{2/3} \Delta x < \Delta V < \pi(x + \Delta x)^{2/3} \Delta x.$$

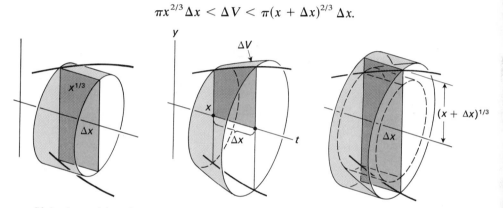

If Δx is positive then

$$\pi x^{2/3} < \frac{\Delta V}{\Delta x} < \pi(x + \Delta x)^{2/3}.$$

If Δx is negative, the inequality is reversed. In either case $\Delta V/\Delta x$ is squeezed between $\pi x^{2/3}$ and something approaching $\pi x^{2/3}$, and so it must approach $\pi x^{2/3}$, too. That is,

$$V'(x) = \lim_{\Delta x \to 0} \frac{\Delta V}{\Delta x} = \pi x^{2/3},$$

by the squeeze limit law. Thus the volume V that we want is given by

$$V = V(1) - V(0) = \int \pi x^{2/3} \, dx \bigg]_0^1 = \frac{3\pi x^{5/3}}{5} \bigg]_0^1 = \frac{3\pi}{5}.$$

If $x^{1/3}$ is replaced by $f(x)$, then the squeeze inequality boxing ΔV between two cylindrical volumes has to be set up in more general terms, but it gives the same conclusion:

$$V'(x) = \lim_{\Delta x \to 0} \frac{\Delta V}{\Delta x} = \pi(f(x))^2.$$

Here $V(x)$ is the volume from $t = a$ to $t = x$. The volume V from $t = a$ to $t = b$ is then given by

$$V = V(b) - V(a) = \int \pi(f(x))^2 \, dx \bigg]_a^b.$$

THEOREM 6   *If f is a positive continuous function over the interval [a, b] and if V is the volume of the solid of revolution generated by the graph of y = f(x) between x = a and x = b, then*

$$V = \int \pi(f(x))^2 \, dx \Big]_a^b.$$

EXAMPLE   Find the volume of the solid of revolution generated by rotating the region under the graph of $y = x^2$, from $x = 1$ to $x = 3$.

Solution.

$$V = \int \pi(x^2)^2 \, dx \Big]_1^3 = \int \pi x^4 \, dx \Big]_1^3 = \frac{\pi x^5}{5} \Big]_1^3$$

$$= \pi \frac{243}{5} - \pi \frac{1}{5} = (48\tfrac{2}{5})\pi.$$

PROBLEMS FOR SECTION 5

1. Prove, by integration, the formula

$$V = \frac{4}{3}\pi r^3$$

for the volume of a sphere. (The sphere is generated by revolving the circle $x^2 + y^2 = r^2$ about a diameter.)

2. Find, by integration, the volume of the truncated cone generated by revolving the area bounded by $y = 6 - x$, $y = 0$, $x = 0$, and $x = 4$ about the x-axis.

Find the volume generated by revolving about the x-axis the regions bounded by the following graphs.

3. $y = x^3$; $y = 0$, $x = 1$ 4. $9x^2 + 16y^2 = 144$

5. $y = x^2 - 6x$; $y = 0$

6. $y^2 = (2 - x)^3$; $y = 0$, $x = 0$, $x = 1$

7. $(x - 1)y = 2$; $y = 0$, $x = 2$, $x = 5$

8. $y = x^{1/3}$; $y = 0$, $x = 0$, $x = 8$

9. $y = \sqrt{1 - x^4}$; $y = 0$ 10. $y = 1/x^2$; $x = 1$, $x = 2$

6. MORE AREAS

We consider now the area *between* two graphs. This slightly generalizes the earlier situation where we considered the area between the x-axis and a function graph.

THEOREM 7   *Let f and g be continuous functions and suppose that g(x) ≤ f(x) over the interval [a, b]. Then the area of the region between the two graphs from x = a to x = b is $\int (f - g)]_a^b$.*

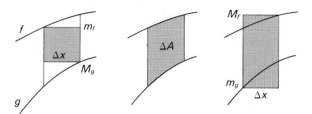

Sketch of proof. The proof depends on squeezing ΔA between the areas of inner and outer rectangles, just as before. The figures here show that

$$(m_f - M_g)\Delta x \leq \Delta A \leq (M_f - m_g)\Delta x,$$

where m and M are the minimum and maximum values on the increment interval. After dividing by Δx, we have an inequality of the form

$$l \leq \frac{\Delta A}{\Delta x} \leq u,$$

where l and u each approach $f(x) - g(x)$ as Δx approaches 0. It follows, as before, that

$$A'(x) = \lim_{\Delta x \to 0} \frac{\Delta A}{\Delta x} = f(x) - g(x),$$

so $\int (f - g)]_a^b$ is the area from a to b, as in Theorem 5.

The missing details in the above proof will be left as an exercise.

EXAMPLE 1 Find the area of the finite region bounded by the graphs of $y = x^2$ and $y = \sqrt{x}$.

Solution. You will generally want to draw a reasonably accurate sketch in a problem of this sort, to ensure that you have the right configuration in mind. Here, for example, we note that $y = \sqrt{x}$ is the *upper* graph between $x = 0$ and the point of intersection of the two curves. You may have noticed that the intersection point is $(1, 1)$—since $(1, 1)$ clearly lies on both graphs—but in any case it can be found algebraically by solving the two equations simultaneously. Here this reduces to solving the equation $x^2 = \sqrt{x}$, or $x^4 = x$, from which we find the solutions $x = 0$ and $x = 1$. These are the left and right edges of the region in question, so our area is

$$A = \int (\sqrt{x} - x^2)\, dx\Big]_0^1 = \left[\frac{2}{3}x^{3/2} - \frac{x^3}{3}\right]_0^1 = \frac{2}{3} - \frac{1}{3} = \frac{1}{3}.$$

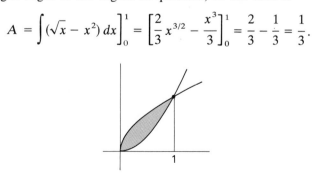

In Theorem 7 there is no requirement that the functions be positive.

EXAMPLE 2   Find the area between the parabola $x = y^2$ and the vertical line $x = 1$.

Solution. Solving for y, we have

$$y = \pm\sqrt{x},$$

and we see that the area lies between the lower function graph $y = -\sqrt{x}$ and the upper function graph $y = \sqrt{x}$. So

$$A = \int [\sqrt{x} - (-\sqrt{x})]\, dx \Big]_0^1 = 2 \cdot \frac{2}{3}x^{3/2} \Big]_0^1 = \frac{4}{3}.$$

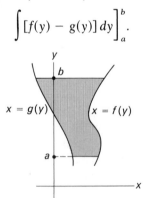

The arguments leading to Theorems 2 and 7 can be carried out just as well with the roles of the axes interchanged. Thus:

THEOREM 7'   *If $g(y) \leq f(y)$ for all y in the interval $[a, b]$, then the area between the graphs $x = g(y)$ and $x = f(y)$, from $y = a$ to $y = b$, is*

$$\int [f(y) - g(y)]\, dy \Big]_a^b.$$

Proof. Instead of running through the same proof in this new configuration, we can reduce the new situation directly to the old as follows. The standard function graphs for the functions of Theorem 7' would have the y-axis horizontal, and then the geometric configurations are exactly the same as for Theorem 7. So the area is

$$\int (f - g) \Big]_a^b = \int [f(y) - g(y)]\, dy \Big]_a^b,$$

by Theorem 7. Now rotate the whole configuration over the 45° line $y = x$, obtaining the configuration of Theorem 7'. The area remains unchanged, so it is still given by $\int (f - g)]_a^b$. ∎

EXAMPLE Solve Example 2 by integration with respect to y.

Solution. Now the region is viewed as extending from $x = y^2$ up to $x = 1$, between the limits -1 and 1. Thus,

$$A = \int (1 - y^2)\, dy\Big]_{-1}^{1} = y - \frac{y^3}{3}\Big]_{-1}^{1} = \left(1 - \frac{1}{3}\right) - \left(-1 + \frac{1}{3}\right) = \frac{4}{3}.$$

PROBLEMS FOR SECTION 6

In each of the following examples, find the area of the finite region bounded by the given graphs.

1. $y = 4x;\quad y = 2x^2$
2. $y^2 = x;\quad y = 4,\ x = 0$
3. $y^2 = 2x;\quad x - y = 4$
4. $y^2 = 6x;\quad x^2 = 6y$
5. $y^2 = 4x;\quad x^2 = 6y$
6. $y^2 = 4x;\quad 2x - y = 4$
7. $y = 4 - x^2;\quad y = 4 - 4x$
8. $y = 6x - x^2;\quad y = x$
9. $y = x^3 - 3x;\quad y = x$
10. $y^2 = 4x;\quad x = 12 + 2y - y^2$
11. $x^2 y = x^2 - 1;\quad y = 1,\ x = 1,\ x = 4$
12. $y = x^2;\quad y = x^3,\ x = 1,\ x = 2$
13. $y = x^2;\quad y = x^3$
14. $w = 2 - x^2;\quad w = x$
15. $s = 1 - t^2;\quad s = t^2$
16. $t = y^4;\quad t = 0,\ y = 1$
17. $y = x^{1/3};\quad y = x/4$
18. $y = \sqrt{x};\quad y = 2x$

7. VOLUMES BY SHELLS

There is a second formula for the volume of a solid of revolution, expressing as an x-integral the volume which is generated by revolving a region about the y-axis. (See the figure below.)

THEOREM 8 *If the region under the graph of f, from $x = a$ to $x = b$, is revolved around the y-axis, then the volume swept out is*

$$\int 2\pi x f(x)\, dx\Big]_{a}^{b}.$$

It is understood that $0 \le a < b$ and that $f \ge 0$ on $[a, b]$, so the back half of the solid can be pictured as in the middle figure below

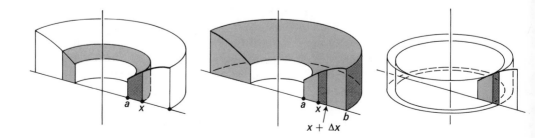

Suppose for example that $f(x) = 2x - x^2$. Then $f(x)$ is above the x-axis between $x = 0$ and $x = 2$, and

$$V = \int 2\pi x(2x - x^2)\,dx \Bigg]_0^2$$

$$= 2\pi \left[\frac{2x^3}{3} - \frac{x^4}{4}\right]_0^2$$

$$= 2\pi \left[\frac{16}{3} - \frac{16}{4}\right] = \frac{8}{3}\pi.$$

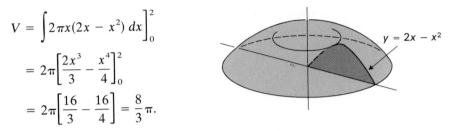

$y = 2x - x^2$

When we use this formula we say that we are computing the volume by *cylindrical shells*, because the incremental volume ΔV associated with the Δx interval $[x, x + \Delta x]$ is approximately the volume of a thin cylindrical shell, obtained by rotating the thin vertical Δx strip about the y-axis, as shown in the figure above.

Proof of theorem. Let $V(x)$ be the volume generated by revolving about the y-axis the region over the interval $[a, x]$. We have to estimate the incremental shell-like volume $\Delta V = V(x + \Delta x) - V(x)$.

The volume of a circular cylinder of altitude h and base radius r is $\pi r^2 h$. So the volume of the cylindrical shell of altitude h and inner and outer radii x and $x + \Delta x$ (supposing Δx positive) is

$$\pi(x + \Delta x)^2 h - \pi x^2 h = \pi(2x + \Delta x)h\,\Delta x.$$

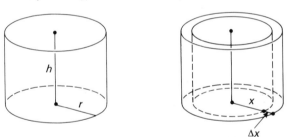

This shell volume will be *less* than ΔV if we take the shell altitude h to be the *minimum* value m that f assumes on the incremental interval $[x, x + \Delta x]$. And the shell volume will be greater than ΔV if h is taken to be the maximum value M of f on $[x, x + \Delta x]$. So

$$\pi(2x + \Delta x)m\,\Delta x \le \Delta V \le \pi(2x + \Delta x)M\,\Delta x,$$

and

$$\pi(2x + \Delta x)m \le \frac{\Delta V}{\Delta x} \le \pi(2x + \Delta x)M.$$

(This final inequality is also correct when Δx and ΔV are negative. In this case the shell volume formula and the first inequality above have to be given minus signs, but the minus sign disappears upon dividing by the positive number $-\Delta x$.) Since m and M both approach $f(x)$ as $\Delta x \to 0$, the difference quotient $\Delta V/\Delta x$ is squeezed to the limit $2\pi x f(x)$. Thus

$$V'(x) = 2\pi x f(x),$$

and the integration between limits formula follows as usual. ∎

PROBLEMS FOR SECTION 7

In the following problems, use the method of shells to compute the volume generated by rotating about the y-axis the region bounded by the given graphs. Draw the plane region in each case.

1. $y = x$; $y = 0$, $x = a$

2. $y = x^2$; $y = 1$, $x = 0$. (Compute this volume also by the earlier formula.)

3. $y^2 = x^3$; $x = 4$ 4. $y = x^2 - 2x$; $y = 2x$

5. $y = x - x^3$; $y = 0$

6. (a) Differentiate $(r^2 - x^2)^{3/2}$. (b) With the preceding formula in mind, find the volume generated by rotating about the y-axis the region to the right of the line $x = a$ and inside the circle $x^2 + y^2 = r^2$ (where $a < r$).

7. (a) Differentiate $(4 - x^2)^{3/2}$. (b) With this result in mind, compute the volume generated by rotating about the y-axis the region outside the hyperbola $x^2 - y^2 = 4$ and inside the line $x = 5$.

8. DENSITY

If Q is a quantity distributed over the x-axis, then the increment quotient $\Delta Q / \Delta x$ can be interpreted as the average amount of Q per unit length lying over the interval $[x, x + \Delta x]$, in which case it is called the *average density* of Q over the increment interval. From this point of view, the limit

$$\rho(x) = \lim \frac{\Delta Q}{\Delta x}$$

is called the *density* of the distribution at x. In these terms, we showed that the area A of the region under the graph of f has the distribution density $f(x)$ at x, and that the volume V generated by rotating this region about the x-axis has the distribution density $\pi[f(x)]^2$ at x. Generally, however, we don't use the density terminology unless the distribution actually lies *on* the axis. Here are some examples.

EXAMPLE 1   Let a piece of wire of varying constitution lie along the x-axis from $x = a$ to $x = b$, and let m be the mass of the segment $[a, x]$. If we give x a positive increment Δx, then the corresponding increment Δm is the mass of the segment $[x, x + \Delta x]$, and the difference quotient $\Delta m / \Delta x$ is thus the average density of the wire throughout the incremental segment. The density at x is defined to be the limit of the average density as Δx tends to zero, and it is thus the derivative dm/dx. It could be measured in ounces per inch, or grams per centimeter, or, for a very fine wire, in milligrams per centimeter.

EXAMPLE 2   A mass distribution on $[0, 1]$ has density $\rho(x) = x - x^2$. By Theorem 5 the amount of mass on $[1/4, 1/2]$, is then

$$\int \rho(x)\, dx \Bigg]_{1/4}^{1/2} = \int (x - x^2)\, dx \Bigg]_{1/4}^{1/2} = \frac{x^2}{2} - \frac{x^3}{3} \Bigg]_{1/4}^{1/2}$$

$$= \left(\frac{1}{8} - \frac{1}{24} \right) - \left(\frac{1}{32} - \frac{1}{192} \right) = \frac{11}{192}.$$

EXAMPLE 3   A distribution of electric charge lies along the interval $[0, 1]$, there being $C(x) = x^2 - x^3$ units of charge in the interval $[0, x]$. What is the charge density at $x = 1/4$? At $x = 1/2$? At $x = 3/4$?

Solution. The charge density $\rho(x)$ is given by

$$\rho(x) = \frac{dC(x)}{dx} = 2x - 3x^2,$$

expressed as units of charge per unit length. Thus,

$$\rho\left(\frac{1}{4}\right) = \frac{5}{16}, \qquad \rho\left(\frac{1}{2}\right) = \frac{1}{4}, \qquad \text{and} \qquad \rho\left(\frac{3}{4}\right) = -\frac{3}{16}.$$

The amount of charge in an incremental interval $[1/4, 1/4 + \Delta x]$ is approximately $\rho(1/4)\,\Delta x = 5\,\Delta x/16$. The interval $[3/4, 3/4 + \Delta x]$ contains negative charge, of approximately $-3\Delta x/16$ units.

EXAMPLE 4   Another important quantity Q of this type is a probability distribution. Suppose we randomly choose points from an interval $[a, b]$ by some mechanism. We assume that for each x there is a number $P = p(x)$ representing the *probability* of a point landing in the subinterval $[a, x]$, where $p(x)$ is an increasing function of x such that $p(a) = 0$ and $p(b) = 1$. Then $p(y) - p(x)$ is the probability of landing in the subinterval $[x, y]$. So if we increase x by Δx, then $\Delta P = p(x + \Delta x) - p(x)$ is the probability of landing in the increment interval $[x, x + \Delta x]$. Thus, probability is a distribution like mass or charge. If $p(x)$ is differentiable, then $\rho(x) = dP/dx$ is the *probability density* function. It is interpreted as the probability per unit length at x, in the sense that

$$\frac{\Delta P}{\Delta x} \approx \rho(x) \qquad \text{and} \qquad \Delta P \approx \rho(x)\,\Delta x.$$

EXAMPLE 5   A probability distribution on the interval $[0, 1]$ is defined by $p(x) = x^2$. Thus, the probability of landing in the interval $[0, 1/2]$ is $p(1/2) = 1/4$. The probability density is $\rho(x) = 2x$, and the probability of landing in the incremental interval $[(3/4), (3/4) + \Delta x]$ is approximately $\rho(x)\,\Delta x = 3\Delta x/2$.

EXAMPLE 6   A probability distribution on $[-1, 1]$ has density $\rho(x) = 3x^2/2$. What is the probability of landing in the interval $[-1/2, 0]$?

Solution. We should first check that we really do have a probability distribution. The basic requirement is that the probability of $[-1, x]$,

$$p(x) = \int \rho(t)\,dt \Big]_{-1}^{x},$$

should increase from 0 to 1 in value as x crosses the interval $[-1, 1]$. Since $p'(x) = \rho(x) = 3x^2/2 \geq 0$, p does increase. Also,

$$p(1) = \int (3x^2/2)\,dx \Big]_{-1}^{1} = \frac{x^3}{2}\Big]_{-1}^{1} = \frac{1}{2} - \left(-\frac{1}{2}\right) = 1,$$

as required. Finally, the probability of $[-1/2, 0]$ is

$$\int \rho(x)\, dx \Big]_{-1/2}^{0} = \frac{x^3}{2} \Big]_{-1/2}^{0} = 0 - \left(-\frac{1}{16}\right) = \frac{1}{16}.$$

We conclude with some remarks about the general notion of density. Consider a quantity Q, such as a mass distribution or a distribution of charge, that is "spread out" in space. We say that the distribution has the constant density ρ if the amount of Q contained in any region R is exactly ρ times the volume V of R. Thus,

$$\rho = \frac{Q(R)}{V(R)},$$

no matter what space region R we consider. If Q is a mass distribution, and if we measure mass in ounces and volume in cubic inches, then the density ρ is measured in ounces per cubic inch. In the metric system density is measured in grams per cubic centimeter.

EXAMPLE 7 We find from a handbook that the density of aluminum is $2.7\ \text{g/cm}^3$. This means that the mass of any aluminum object, measured in grams, is 2.7 times the volume it occupies, measured in cubic centimeters. The density of iron is 7.9 and the density of gold is 19.3.

Unfortunately, a distribution Q will not generally have a constant density, and we have to consider density as something that varies from point to point. In order to define the density of Q at the point p_0, we first choose some small region ΔR around p_0, such as a small ball or a small cube centered at p_0. Let ΔQ be the amount of Q contained in ΔR and let ΔV be the volume of ΔR. The quotient

$$\frac{\Delta Q}{\Delta V}$$

is then called the *average density* of Q throughout ΔR. For a mass distribution in the above units, $\Delta Q/\Delta V$ is the average number of ounces per cubic inch throughout ΔR. *The space density of Q at p_0 is then defined to be the limit*

$$\rho(p_0) = \lim_{\Delta V \to 0} \frac{\Delta Q}{\Delta V},$$

supposing that the limit exists. This is not a derivative in the ordinary sense, and we are in no position to discuss how we might evaluate such a limit. Nevertheless, we can conceive of the limit existing, and if it happened to have the value 5, we would say that the Q density at p_0 is 5 ounces per cubic inch.

Similar remarks apply to a two-dimensional distribution of some quantity Q. We could consider a charge distribution spread out over a two-dimensional sheet, or we could think of a flat sheet of metal with varying constitution as giving us a two-dimensional mass distribution. Now the average plane density in a small region ΔR is the quotient

$$\frac{\Delta Q}{\Delta A},$$

where ΔA is the *area* of ΔR. For a mass distribution with the same units as before, the average density is measured in ounces per *square* inch (or in grams per *square* centimeter). So is the plane point density

$$\rho(p_0) = \lim_{\Delta A \to 0} \frac{\Delta Q}{\Delta A}.$$

Here again we fail to have an ordinary derivative, although we are taking the limit of something like a difference quotient.

The computation of the total amount of a quantity in a plane region R from its density function ρ requires double integration, and is taken up in Chapter 17.

PROBLEMS FOR SECTION 8

1. a) Show that $\rho(x) = x + x^3 - 2x^2$ is a possible density for a mass distribution along the interval $[0, 4]$. (Show that $\rho(x) \geq 0$ in the interval.)

    b) Find the mass of the segment $[1, 2]$.

2. a) A triangular sheet of metal with vertices $(0, 0)$, $(2, 0)$, $(2, 1)$ has constant density 1. What is its mass?

    b) What is the density $\rho(x)$ if the metal triangle is viewed as a mass distribution over the x-axis (or along the x-axis)? Compute its mass by integrating the density $\rho(x)$.

3. A distribution of electric charge along a rod has density $\rho(x) = x - x^2$.

    a) How much charge is on the segment $[0, 1]$?

    b) What is the total charge on the segment $[-1/2, 1]$?

4. a) Show that $\rho(x) = 3(1 - x^2)/2$ is a probability density on the interval $[0, 1]$.

    b) Find the probability that a number will be in the subinterval $[1/2, 1]$.

5. The function $\rho(x) = x^3$ is a probability density on an interval $[0, b]$, What is b?

6. a) The function $\rho(x) = 12x^2$ is a probability density on an interval $[-a, a]$. Find a.

    b) Find the probability of the subinterval $[0, a/2]$.

7. Show that $\rho(x) = 2x/(1 + x^2)^2$ is a probability density on the interval $[0, \infty)$. (First integrate $\rho(x)$ from $x = 0$ to $x = a$, and then take the limit as $a \to \infty$.)

EXTRA PROBLEMS FOR CHAPTER 4

Compute the following integrals. Be sure to include the constant of integration in each answer.

1. $\displaystyle\int (9t^2 + 25t + 14)\, dt$

2. $\displaystyle\int \frac{dx}{x^2}$

3. $\int \dfrac{dx}{\sqrt[3]{x^2}}$

4. $\int (x^3 + 2)^2 3x^2 \, dx$

5. $\int \sqrt{x + 1} \, dx$

6. $\int \dfrac{1}{\sqrt{x + 1}} \, dx$

7. $\int \sqrt{2x + 1} \, dx$

8. $\int (3x + 2)^{-2/3} \, dx$

9. Suppose the rate of change of the volume V of a certain gas with respect to the pressure P of the gas is $-2/P^2$. If $V = 100$ when $P = 1/100$, find the equation of the volume.

10. A swimming pool holds 500 gallons of water. Suppose the water drains so that the rate of change of the volume of the water after t minutes is given by $-(t^2 + 3t + 1)$. How much is left in the pool after 6 minutes?

11. Find the area of the region bounded by the coordinate axes and the line $x + y = 2$.

12. Verify the formula for the area of a triangle in the case of the right triangle with vertices at the origin, the point (b, h) and the point $(b, 0)$.

13. Verify the formula for the area of a trapezoid in the case of the vertices $(0, b_1)$, (h, b_2), and the points on the x-axis under these two points.

14. Find the area of the region under the graph of $y = 1/x^2$ from $x = 1$ to $x = N$. What is its limit as $N \to \infty$? What would you say is the area of the region under the graph of $1/x^2$ extending from $x = 1$ all the way to ∞?

Find the volume generated by revolving about the x-axis the regions bounded by the following graphs.

15. $y = x^3$; $\quad y = 0$, $x = 1$

16. $9x^2 + 16y^2 = 144$

17. $y = x^2 - 6x$; $\quad y = 0$

18. $y^2 = (2 - x)^3$; $\quad y = 0$, $x = 0$, $x = 1$

19. $(x - 1)y = 2$; $\quad y = 0$, $x = 2$, $x = 5$

20. $y = x^{1/3}$; $\quad y = 0$, $x = 0$, $x = 8$

21. $y = \sqrt{1 - x^4}$; $\quad y = 0$

22. $y = (1 + x)^{1/4}$, $y = 0$, $x = 0$, $x = 1$

23. $y = x(1 + x^3)^{1/4}$, $y = 0$, $x = 0$, $x = 1$

24. $y = 1/x^2$; $\quad x = 1$, $x = \infty$. (See Problem 14.)

25. $y = 1/\sqrt{x}$; $\quad x = 1$, $x = \infty$. (See Problem 14.)

26. $y = x/(1 + x^3)^2$; $\quad x = 0$, $x = \infty$

27. Show that

$$\frac{d}{dx} \int f(t) \, dt \bigg]_a^x = f(x).$$

28. Show that

$$\frac{d}{dx} \int f(t) \, dt \bigg]_x^b = -f(x).$$

Compute the derivatives of the following functions.

29. $f(x) = \int t^2 \, dt \bigg]_0^{x^2}$

30. $f(x) = \int \dfrac{dt}{t^2} \bigg]_x^{x^2}$

31. $f(x) = \int \dfrac{dt}{t^3} \bigg]_1^{\sqrt{x}}$

32. $f(x) = \int \sqrt{t} \, dt \bigg]_{x^2}^{x^4}$

33. Show that the volume generated by rotating the ellipse

$$\frac{x^2}{a^2} + \frac{y^2}{b^2} = 1$$

about the x-axis is

$$V = \frac{4}{3}\pi b^2 a.$$

(Note that this generalizes the formula for the volume of a sphere.)

34. Find the limit as $n \to \infty$ of the volume obtained by rotating about the x-axis the region between the x-axis and the graph of $y = 1 + x^n$, from $x = 0$ to $x = 1$. What is the geometric interpretation of this limit?

35. Prove the formula

$$V = \frac{1}{3}\pi r^2 h$$

for the volume of a cone. (Rotate the region bounded by the straight line $y = rx/h$, the x-axis, and the vertical line $x = h$, about the x-axis.)

36. Prove Theorem 6 by imitating the earlier proof of Theorem 2. That is, let m and M be the minimum and maximum of $f(x)$ on the increment interval $[x, x + \Delta x]$, and build the proof around the inequality

$$\pi m^2 \Delta x < \Delta V < \pi M^2 \Delta x.$$

37. Find the formula for the volume of a spherical cap.

38. Find the area of the region between the graph of $y = x^3$ and its tangent line at $x = 1$.

39. Find the area of the region between the circle $x^2 + (y - 1)^2 = 1$ and the semi-parabola $y = \sqrt{x}$.

40. Write out the proof of Theorem 7 in detail.

41. Suppose that we have been studying a quantity Q that is distributed along the x-axis, and have found that the incremental amount ΔQ lying on the increment interval $[x, x + \Delta x]$ satisfies the inequalities

$$l\Delta x \le \Delta Q \le u\Delta x,$$

where $l \le \rho(x) \le u$. Show that then

$$\Delta Q \approx \rho(x)\Delta x,$$

with an error at most $(u - l)\Delta x$ in magnitude.

5

Systematic Differentiation

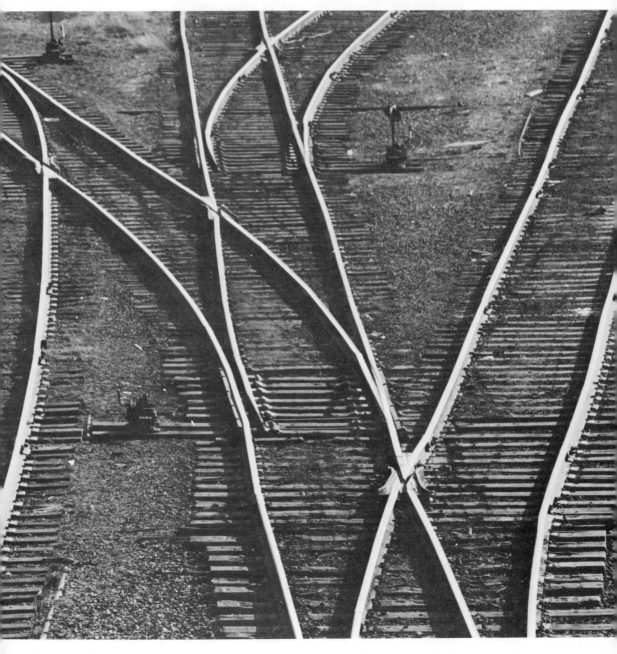

One of the contributions of Newton and Leibniz was their realization that derivatives could be computed systematically. Starting with the derivative formulas for a few basic functions, and with the derivative rules for four or five ways of combining functions, we can compute in a systematic, and almost automatic, way the derivative of practically any function that is likely to come up in everyday mathematical affairs.

What we know so far can be summarized in the formulas

$$\frac{d}{dx} c = 0, \qquad \frac{d}{dx} x = 1,$$

and the rules given in Theorems 2, 3 and 6 of Chapter 3. In pure function notation, they are

$$(f + g)' = f' + g',$$
$$(cf)' = cf',$$
$$(f^k)' = kf^{k-1} \cdot f'.$$

This chapter will complete the mechanism. The remaining rules are the product and quotient rules

$$(fg)' = f \cdot g' + g \cdot f',$$
$$\left(\frac{f}{g}\right)' = \frac{g \cdot f' - f \cdot g'}{g^2},$$

and the chain rule, which we shall state later. One corollary of the chain rule, the inverse-function rule, will be taken up in Chapter 7.

The list of basic functions will be completed with the functions e^x, $\sin x$, and $\cos x$, and their derivative formulas

$$\frac{d}{dx} e^x = e^x, \qquad \frac{d}{dx} \sin x = \cos x, \qquad \frac{d}{dx} \cos x = -\sin x.$$

Although the graph of e^x does not look anything like the graph of $\sin x$ or $\cos x$, these three functions are nevertheless very intimately related. It requires complex numbers and complex variables to show what this relationship really is, but we shall find traces of it, especially in the chapter on power series. Here we shall see certain "structural" similarities: Each function is characterized by a fundamental *addition formula* that it satisfies, and these formulas lead in almost identical ways to the derivative formulas.

The chain rule is put off until after these functions have been studied because it is of little use until then. However, it can be taken up earlier with the other rules if desired.

Finally, there is the problem of differentials. Although modern mathematics develops differentials conceptually, we are better off in a first course with the simpler approach in which differentials are treated in a purely formal, or symbolic, manner. Section 7 takes this symbolic point of view.

1. THE PRODUCT AND QUOTIENT RULES

One might guess offhand that the derivative of the product of two functions ought to be the product of their derivatives. But trying this out on practically any product for which the answer is already known shows that it doesn't work. For instance,

$$x^2 = x \cdot x, \qquad \frac{d(x^2)}{dx} = 2x, \qquad \text{and} \qquad \frac{dx}{dx} = 1,$$

but $2x$ is not $1 \cdot 1$. The quotient rule can't be guessed either.

These rules were stated above in function notation. Here are their Leibniz forms.

THEOREM 1 *If u and v are differentiable functions of x (say, $u = f(x)$ and $v = g(x)$), then so are uv and u/v, and*

$$\frac{d}{dx}(uv) = u\frac{dv}{dx} + v\frac{du}{dx};$$

$$\frac{d}{dx}\left(\frac{u}{v}\right) = \frac{v\dfrac{du}{dx} - u\dfrac{dv}{dx}}{v^2}.$$

It is understood that these formulas hold for those values of x at which both f and g are differentiable and, in the case of f/g, where also $g(x) \neq 0$.

It may be easier to remember "word" versions, such as the following:

The derivative of a product is the first *times the derivative of the* second *plus the* second *times the derivative of the* first.

The derivative of a quotient is the denominator *times the derivative of the* numerator, minus *the* numerator *times the derivative of the* denominator, *all divided by the* denominator squared.

The product and quotient rules enable us to compute derivatives of more complicated functions than we could handle before. We shall practice them now and prove them later.

EXAMPLE 1 Find dy/dx when $y = (x - 3)^9(x + 2)^{15}$. If we insisted on doing this with our "bare hands," so to speak, we could expand the binomials and so express y as a polynomial of degree 24. Then we could write the answer down, using the more elementary rules from Chapter 3. But the product rule lets us do it directly:

$$\begin{aligned}
\frac{dy}{dx} &= (x - 3)^9 \frac{d}{dx}(x + 2)^{15} + (x + 2)^{15}\frac{d}{dx}(x - 3)^9\\
&= (x - 3)^9 \cdot 15(x + 2)^{14} + (x + 2)^{15} \cdot 9(x - 3)^8\\
&= (x - 3)^8(x + 2)^{14}[15(x - 3) + 9(x + 2)]\\
&= (x - 3)^8(x + 2)^{14}[24x - 27]\\
&= 3(x - 3)^8(x + 2)^{14}(8x - 9).
\end{aligned}$$

If we know how to write down the derivatives of the two factors we would not generally bother with the first line above, but would start by writing out the second line.

EXAMPLE 2 Find the derivative of $x^2\sqrt{x-1}$.

Solution. Write $\sqrt{x-1} = (x-1)^{1/2}$ and apply the product rule:

$$\frac{d}{dx}x^2(x-1)^{1/2} = x^2\left(\frac{1}{2}(x-1)^{-1/2}\right) + (x-1)^{1/2}\cdot 2x^\dagger$$

$$= \frac{\dfrac{x^2}{2} + (x-1)\cdot 2x}{(x-1)^{1/2}}$$

$$= \frac{5x^2 - 4x}{2\sqrt{x-1}}.$$

EXAMPLE 3

$$\frac{d}{dx}\frac{(x^2-2x)}{x-1} = \frac{(x-1)\dfrac{d}{dx}(x^2-2x) - (x^2-2x)\dfrac{d}{dx}(x-1)}{(x-1)^2}$$

$$= \frac{(x-1)(2x-2) - (x^2-2x)\cdot 1}{(x-1)^2}$$

$$= \frac{x^2 - 2x + 2}{(x-1)^2}.$$

EXAMPLE 4 Calculate

$$\frac{d}{dx}\left(\frac{x}{\sqrt{x-1}}\right).$$

Solution

$$\frac{d}{dx}\frac{x}{(x-1)^{1/2}} = \frac{(x-1)^{1/2}\cdot 1 - x\cdot\dfrac{1}{2}(x-1)^{-1/2}}{x-1}$$

$$= \frac{(x-1)^{-1/2}\left[(x-1) - \dfrac{x}{2}\right]}{x-1}$$

$$= \frac{x-2}{2(x-1)^{3/2}}.$$

Some students have trouble factoring out a *negative* power because it goes against their feeling that when you "factor something out" there should be *less* of it left behind. For example, when a^2 is factored out of a^3, what is left is a^1:

$$a^3 = a^2\cdot a^1;$$

$\dagger$ This is already a correct answer, but we generally try to express a function in as simple a form as possible, so we go on.

the remaining power is *less* by the amount factored out. But when $a^{-2/3}$ is factored out of $a^{1/3}$ what is left is a^1,

$$a^{1/3} = a^{-2/3} \cdot a^1.$$

This is correct by the law of exponents, but it also fits the above pattern because *when something is decreased by a negative amount it is increased.* Here the remaining power is less than 1/3 by the amount $-2/3$; it is

$$\frac{1}{3} - \left(-\frac{2}{3} \right) = 1.$$

EXAMPLE 5 Work Example 4 by the product rule.

$$\frac{d}{dx} \left(\frac{x}{(x-1)^{1/2}} \right) = \frac{d}{dx} (x(x-1)^{-1/2})$$

$$= x \cdot -\frac{1}{2}(x-1)^{-3/2} + (x-1)^{-1/2} \cdot 1$$

$$= (x-1)^{-3/2} \left[-\frac{x}{2} + (x-1) \right]$$

$$= \frac{x-2}{2(x-1)^{3/2}}$$

EXAMPLE 6 If we don't multiply out the numerator in

$$f(x) = \frac{x^2(x-1)}{x+2}$$

we have a choice as to whether we first take $f(x)$ to be the quotient $[x^2(x-1)]/(x+2)$, or a product, say $x^2[(x-1)/(x+2)]$. The problem can be worked either way. If we start with the quotient rule, we get

$$f'(x) = \frac{(x+2)\dfrac{d}{dx}[x^2(x-1)] - x^2(x-1) \cdot 1}{(x+2)^2},$$

where we still have to compute the product derivative,

$$\frac{d}{dx} x^2(x-1) = x^2 \cdot 1 + (x-1)(2x) = 3x^2 - 2x,$$

and substitute it. We end up with

$$f'(x) = \frac{(x+2)(3x-2)x - x^2(x-1)}{(x+2)^2}$$

$$= \frac{x(2x^2 + 5x - 4)}{(x+2)^2}.$$

It is possible to make such a composite calculation all at once. In applying the quotient rule, the term $v \, du/dx$ involves the above product computation, and this can be written out before going on to complete the quotient rule.

This would give

$$f'(x) = \frac{(x + 2)[x^2 \cdot 1 + (x - 1)2x] - x^2(x - 1) \cdot 1}{(x + 2)^2},$$

which needs only algebraic simplification.

We turn now to the proofs of the product and quotient rules. These proofs can be given in the r,x-notation of Chapter 3, but the special forms of the product and quotient rules are better understood in terms of increments. Recall from Chapter 3 that if $y = f(x)$, then Δy depends on Δx (and x) according to the formula

$$\Delta y = f(x + \Delta x) - f(x).$$

If we add the equation $y = f(x)$, we obtain the equivalent equation

$$y + \Delta y = f(x + \Delta x),$$

which just says that $y + \Delta y$ is the new value of y when $x + \Delta x$ is the new value of x.

Similarly, we can express the new value of any variable u as its old value plus its increment, $u + \Delta u$. This notation is useful because computations frequently become more transparent when all changes are expressed in terms of increments.

EXAMPLE 7   If the variables u and v are given increments Δu and Δv, what is the increment in the product $y = uv$?

Solution. The new values of u and v are $u + \Delta u$ and $v + \Delta v$, so the new value $y + \Delta y$ is given by

$$y + \Delta y = (u + \Delta u)(v + \Delta v).$$

Then

$$\Delta y = (y + \Delta y) - y = (u + \Delta u)(v + \Delta v) - uv,$$

$$= u\Delta v + v\Delta u + \Delta u\Delta v.$$

This expression for $\Delta(uv)$ can be pictured when everything is positive by interpreting uv as the area of a rectangle with sides u and v. Note that the added area is shown in three pieces, corresponding to the three terms in the above sum.

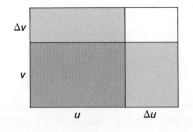

Proof of Theorem 1. We are assuming that the functions $u = g(x)$ and $v = h(x)$ are differentiable. That is if we give x an increment Δx, getting the new values

$$u + \Delta u = g(x + \Delta x) \qquad \text{and} \qquad v + \Delta v = h(x + \Delta x),$$

then our hypothesis is that the limits

$$\lim_{\Delta x \to 0} \frac{\Delta u}{\Delta x} = \frac{du}{dx}, \qquad \lim_{\Delta x \to 0} \frac{\Delta v}{\Delta x} = \frac{dv}{dx}$$

both exist. We computed above that the product $y = uv$ has the increment

$$\Delta y = u\Delta v + v\Delta u + \Delta u\Delta v.$$

Thus,

$$\frac{\Delta y}{\Delta x} = u\frac{\Delta v}{\Delta x} + v\frac{\Delta u}{\Delta x} + \Delta u\frac{\Delta v}{\Delta x}.$$

As $\Delta x \to 0$, the right side of this equation has the limit

$$u\frac{dv}{dx} + v\frac{du}{dx} + 0\frac{dv}{dx}.$$

That is, the limit $dy/dx = \lim_{\Delta x \to 0} \Delta y/\Delta x$ exists, and

$$\frac{dy}{dx} = u\frac{dv}{dx} + v\frac{du}{dx}.$$

Similarly, the quotient $y = u/v$ turns out to have the increment

$$\Delta y = \frac{v\Delta u - u\Delta v}{v(v + \Delta v)}.$$

Therefore,

$$\frac{\Delta y}{\Delta x} = \frac{v\dfrac{\Delta u}{\Delta x} - u\dfrac{\Delta v}{\Delta x}}{v(v + \Delta v)},$$

and taking the limit as $\Delta x \to 0$ we get, similarly,

$$\frac{dy}{dx} = \frac{v\dfrac{du}{dx} - u\dfrac{dv}{dx}}{v^2}. \quad \blacksquare$$

Note in each case how the formula for dy/dx preserves the form of the expression for $\Delta y/\Delta x$.

There is an alternative derivation of the quotient rule that combines the product rule and the power rule with exponent $a = -1$:

$$\frac{d}{dx}\left(\frac{u}{v}\right) = \frac{d}{dx} u \cdot v^{-1} = \cdots$$

(computation to be finished as an exercise). The quotient rule is thus not really a basic rule, but it is so useful that it should be learned anyway.

PROBLEMS FOR SECTION 1

Differentiate the following functions.

1. $y = (x - 1)^3(x + 2)^4$

2. $y = x^m \cdot x^n$ (Do it two ways and compare the answers.)

3. $s = 3t^5(t^2 + 2t)$ (Two ways and compare)

4. $y = x^{21}(1 + x)^{21}$ (Two ways, using $x(1 + x) = x + x^2$)

5. $y = (x^3 + 6x^2 - 2x + 1)(x^2 + 3x - 5)$

6. $h(x) = (x + 1)^2(x^2 + 1)^{-3}$

7. $u = v^{-4}(1 + 2v)^4$ (Two ways)

8. $y = x\sqrt{1 + x^2}$ (Two ways)

9. $u = (x^2 + 2x + 1)^2(3x^2 - 1)^{-1/2}$

10. $s = (t + 1)^2(t + 1)^{-2}$ (Two ways)

11. $y = (x^2 - 1)(x + 1)^{-1}$ (Two ways)

12. $y = x^{1/3}(1 + x)^{1/3}$ (Two ways)

13. $y = (x^2 - 8x + 1/x)^{1/2}(7x^{3/2} - 4x^3 + 1)^{3/4}$

14. $y = (4x + 9x^2 - 32)^{5/2}(x - 8x^4)^{1/2}$

Differentiate the following functions using the quotient rule (although not necessarily as the first step).

15. $y = \dfrac{x}{1 + x}$

16. $s = \dfrac{t}{1 + t^2}$

17. $y = \dfrac{x^2 - 1}{x^2 + 1}$

18. $y = \dfrac{x^3}{1 + 3x^2}$

19. $s = \dfrac{2t}{(1 + t^2)^2}$

20. $y = \dfrac{x + 1}{(x^2 + 2x + 2)^{3/2}}$

21. $u = \dfrac{1 + v^2}{1 + v}$

22. $y = \sqrt{\dfrac{1 + x}{1 - x}}$

23. $y = \dfrac{\sqrt{1 - x^2}}{1 - x}$

24. $w = \dfrac{u}{\sqrt{1 + u^2}}$

25. $y = \dfrac{f(x)}{x}$

26. $y = \left(\dfrac{x^2 - x - 1}{x^2 + 1}\right)^3$

27. $y = \dfrac{x(1/x - 1)^4}{(1 - x)^5(3x - 4)^2}$

2. THE EXPONENTIAL FUNCTION

The exponential function $f(x) = e^x$ is one of the most important functions in mathematics and in the applications of mathematics. For example, we shall see in Chapter 6 that a population grows exponentially, whether it be a population of bacteria growing in a culture, or a bank account growing at continuously compounded interest.

The General Base a

In elementary algebra one learns how to calculate with powers a^m and roots $a^{1/n}$ of a positive "base" number a. Such calculations are governed by the following *laws of exponents*:

$$a^{x+y} = a^x a^y,$$
$$(a^x)^y = a^{xy},$$
$$a^1 = a,$$
$$a^x b^x = (ab)^x.$$

These laws are proved for all rational numbers x and y by purely algebraic manipulations.

We add the fact that if the base a is greater than 1, then a^x is an increasing function of x.

LEMMA 1 *If $a > 1$ and $x < y$, then $a^x < a^y$.*

Proof. If $a > 1$, then $a^{1/n} > 1$ and $a^m > 1$ for any positive integers m and n. It follows that $a^r > 1$ for every positive rational number $r = m/n$ (since $a^r = (a^m)^{1/n}$). So if $x < y$, then $a^y = a^{y-x}a^x > a^x$, because $a^{y-x} > 1$. ∎

To get an idea of how a^x varies with x, we plot 2^x for integer values of x. In the right figure we have filled in the values 2^x for $x = 1/2, 3/2$, etc., using

$$2^{1/2} = \sqrt{2} \approx 1.414, \qquad 2^{3/2} = 2 \cdot 2^{1/2} \approx 2.818, \qquad \text{etc.}$$

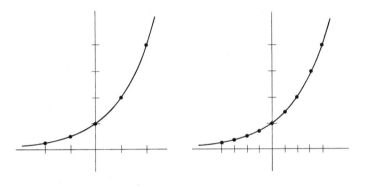

It seems clear that, as we fill in more and more values of 2_x, we are determining the graph of a continuous increasing function. There is nothing special about the base 2 in these considerations, so we conjecture the following theorem.

THEOREM 2 *Given any fixed base number a greater than 1, there is a uniquely determined continuous increasing function f, with domain the whole real line $(-\infty, \infty)$, such that $f(r) = a^r$ for every rational number r.*

Assuming for the moment that this theorem is true, we can use the limit laws and the continuity of a^x to show that the laws of exponents hold for irrational as well as rational values. For example, if we let r and s approach x and y, respectively, through rational values, then $r + s$ approaches $x + y$ through rational values, and

$$a^{x+y} = \lim a^{r+s} = \lim a^r a^s$$
$$= \lim a^r \lim a^s = a^x a^y.$$

The law $(a^x)^y = a^{xy}$ requires knowing that a^x also varies continuously when the base a is varied, which will be shown later. So we have the corollary:

COROLLARY *The function $f(x) = a^x$ of the theorem satisfies the laws of exponents for all real numbers.*

We cannot give a complete proof of Theorem 2 at this point, but at the end of the section we shall show how to determine the value of a^x when x is irrational, and carry the discussion far enough so that the theorem should seem clearly to be true. This is not a critical issue for us, because by taking the exponential function on faith for a while, we will be led ultimately to a much more efficient way of defining and calculating it. (This will be by infinite series, in Chapter 12.)

We shall also see later in this section that the difference quotient

$$\phi(h) = \frac{2^h - 2^0}{h} = \frac{2^h - 1}{h}$$

is an increasing function of h on the interval $[-1, 1]$, omitting the origin, where ϕ is undefined. Furthermore, for positive values of h,

$$\phi(h) - \phi(-h) \to 0 \qquad \text{as } h \to 0.$$

These properties of ϕ imply:

LEMMA 2   *The function $f(x) = 2^x$ is differentiable at the origin.*

Proof.   The above properties of ϕ imply that its graph has the features shown below.

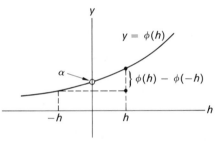

In particular, there is a number α such that

$$\phi(-h) \le \alpha \le \phi(h)$$

for all h between 0 and 1. (From the point of view of the range discussion at the end of Chapter 2, the range of ϕ over the interval $(-1, 0)$ is an open interval of the form $(-\frac{1}{2}, \alpha)$, and over $(0, 1)$ the range is an open interval $(\beta, 1)$. Since ϕ is increasing, the two range intervals cannot overlap and hence $\alpha \le \beta$. In particular,

$$\phi(-h) \le \alpha \le \phi(h)$$

for all h between 0 and 1.) This inequality implies that

$$0 \le \phi(h) - \alpha \le \phi(h) - \phi(-h).$$

Since $\phi(h) - \phi(-h) \to 0$, it follows that $\phi(h) - \alpha \to 0$ as $h \to 0$, by the squeeze limit law. Similarly, $\alpha - \phi(-h) \to 0$. Thus $\phi(h) \to \alpha$ as $h \to 0$, so the function $f(x) = 2^x$ is differentiable at the origin, with $f'(0) = \alpha$. ∎

To get an idea of what kind of number we are talking about, we can use a pocket calculator with a square root key to compute some of the values of $\phi(h)$ and $\phi(-h)$. Starting with 2, and using five successive square roots, we obtain the following table.

h	$\phi(h)$	$\phi(-h)$	$[\phi(h) + \phi(-h)]/2$
1/2	.8284	.5858	.7071
1/4	.7568	.6364	.6966
1/8	.7241	.6640	.6940
1/16	.7084	.6783	.6933
1/32	.7007	.6857	.6932

Since $\phi(-h) < \alpha < \phi(h)$, the first two entries in the last row show that $\alpha \approx 0.69$. The trend of the average values shown in the last column suggests that in fact $\alpha = 0.693$ to the nearest three decimal places. (Look up the value of the natural logarithm of 2 in the table on page 792. This mystery will be explained in Chapter 7.)

We can now use the laws of exponents to prove that $f'(x)$ exists for *all* x and to find a formula for the derivative.

When we are using the increments of only one or two variables, we sometimes forsake the Δ notation if favor of simpler symbols, and it is traditional to set $\Delta x = h$. In this notation we showed above that

$$f'(0) = \lim_{h \to 0} \frac{f(0 + h) - f(0)}{h} = \lim_{h \to 0} \frac{2^h - 1}{h} = \alpha.$$

THEOREM 3   *The function $f(x) = 2^x$ is differentiable for all x, and*

$$\frac{d}{dx} 2^x = \alpha 2^x,$$

where $\alpha = f'(0)$.

Proof

$$f'(x) = \lim_{h \to 0} \frac{f(x + h) - f(x)}{h} = \lim_{h \to 0} \frac{2^{x+h} - 2^x}{h}$$

$$= \lim_{h \to 0} 2^x \cdot \frac{2^h - 1}{h} = 2^x \lim_{h \to 0} \frac{2^h - 1}{h}$$

$$= 2^x f'(0) = \alpha 2^x. \quad \blacksquare$$

The Base e

We saw above that if $f(x) = 2^x$, then $f'(0) = 0.7$ to one decimal place. So if $g(x) = 4^x = f^2(x)$, then the general power rule shows that $g'(0) = 2f'(0) \approx 1.4$. Thus, as we increase the base a from $a = 2$ to $a = 4$, the graphs of the

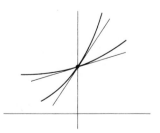

various exponential functions a^x have slopes over $x = 0$ that increase from below 1 to above 1. We therefore expect that there is exactly one base between 2 and 4 for which this slope is 1, that is, exactly one base a such that $da^x/dx\big|_{x=0} = 1$. We now prove this.

THEOREM 4   *There is a unique base, which we shall designate e, such that*

$$\frac{d}{dx}e^x = e^x.$$

Proof. By the intermediate-value principle, *every* positive number k is a value of the function 2^x. In particular, the unknown base e can be expressed in the form $e = 2^b$, where b has to be determined. Then $e^x = (2^b)^x = 2^{bx}$. If we had the general power rule for all exponents, we could now see immediately that

$$\frac{d}{dx}e^x = \frac{d}{dx}2^{bx} = \frac{d}{dx}(2^x)^b = b(2^x)^{b-1} \cdot \alpha 2^x = \alpha b 2^{bx} = \alpha b e^x,$$

so that b must have the value $1/\alpha$ to make the theorem true.

Instead, we start from scratch and rerun the proof of Theorem 3, obtaining

$$\frac{d}{dx}2^{bx} = 2^{bx} \lim_{h \to 0} \frac{2^{bh} - 1}{h}.$$

But

$$\lim_{h \to 0} \frac{2^{bh} - 1}{h} = b \lim_{h \to 0} \frac{2^{bh} - 1}{bh}$$

$$= b \lim_{t \to 0} \frac{2^t - 1}{t} \qquad \text{(where we have set } t = bh\text{)}$$

$$= b\alpha.$$

Therefore, we will have

$$\frac{d}{dx}2^{bx} = 2^{bx}$$

if and only if

$$b\alpha = 1, \qquad \text{or} \qquad b = \frac{1}{\alpha}.$$

That is, $e = 2^{1/\alpha}$ is the unique base such that $de^x/dx = e^x$. ∎

It is because of the simplicity of this law, as compared with $d2^x/dx = \alpha 2^x$, that the base e is always used in calculus.

All we know about the number e at the moment is that it lies between 2 and 4. (Actually, since $e = 2^{1/\alpha} \approx 2^{1.4}$, e is a little less than $2^{1.5} = 2\sqrt{2} \approx 2.8$.) Later on we shall discover how to compute e. Its decimal expansion starts off

$$e = 2.71828 \cdots.$$

The exponential function e^x will be studied further in Chapters 7, 11 and 12.

EXAMPLE 1

$$\frac{d}{dx}e^{2x} = \frac{d}{dx}e^x \cdot e^x = e^x \cdot e^x + e^x \cdot e^x$$

$$= 2e^{2x}.$$

EXAMPLE 2   Show that

$$\frac{d}{dx}\,e^{ax} = ae^{ax}$$

if a is a rational number.

Solution. By the general power rule,

$$\frac{d}{dx}\,e^{ax} = \frac{d}{dx}\,(e^x)^a = a(e^x)^{a-1}\cdot e^x$$

$$= ae^{ax}.$$

EXAMPLE 3   If $y = xe^x$, show that

$$\frac{d^2y}{dx^2} - 2\frac{dy}{dx} + y = 0.$$

Solution. We have

$$\frac{dy}{dx} = xe^x + e^x,$$

$$\frac{d^2y}{dx^2} = xe^x + e^x + e^x$$

$$= xe^x + 2e^x,$$

$$\frac{d^2y}{dx^2} - 2\frac{dy}{dx} + y = (xe^x + 2e^x) - 2(xe^x + e^x) + xe^x = 0.$$

EXAMPLE 4   Show that a differentiable function $y = f(x)$ satisfies the equation

$$\frac{dy}{dx} = y \qquad\qquad (1)$$

over an interval I if and only if $y = ce^x$ on I, for a suitable constant c.

Solution. That the function ce^x satisfies the equation (1) is exactly what the derivative rule $de^x/dx = e^x$ says. Now suppose that f is some other function satisfying (1) on an interval I. Then

$$\frac{d}{dx}\,(f(x)e^{-x}) = f(x)(-e^{-x}) + f'(x)e^{-x}$$

$$= -f(x)e^{-x} + f(x)e^{-x} = 0.$$

Thus,

$$f(x)e^{-x} = c$$

by Theorem 7 of Chapter 3, so $f(x) = ce^x$.

   Thus $f(x) = e^x$ is the unique solution f of (1) satisfying the initial condition

$$f(0) = 1.$$

EXAMPLE 5   Let us find the solution f of (1) that satisfies the initial condition $f(3) = 5$.

Solution. Since $f(x) = ce^x$, the initial condition becomes $5 = f(3) = ce^3$. Then

$$c = 5e^{-3},$$

and

$$f(x) = ce^x = 5e^{-3}e^x = 5e^{x-3}$$

is the unique solution.

The remainder of the section is devoted to some of the details missing from the initial discussion of a^x.

LEMMA 3   *Let r be a positive rational number less than 1. Then*

$$b^r - a^r < r(b - a)$$

if $1 \le a < b$, and

$$b^r - a^r > r(b - a)$$

if $0 < a < b \le 1$.

Proof. We apply the mean-value principle to the function x^r. It guarantees the existence of a number X between a and b such that

$$b^r - a^r = rX^{r-1}(b - a).$$

If $1 \le a < b$ then $X^{r-1} < 1$, because X is larger than 1 and it is raised to a negative power. The first inequality follows. If $a < b \le 1$, we have $X^{r-1} > 1$ and the second inequality. ∎

A first consequence of the lemma is that a^r is a continuous function of the base a (r fixed) when $a > 1$. This fact is needed to help extend the laws of exponents to irrational exponents.

More important here is the following corollary, which gives very tight control of the continuity of 2^r.

LEMMA 4   *If $r < s < N$, then*

$$2^s - 2^r < 2^N(s - r)$$

whenever $s - r < 1$.

Proof. Lemma 3 and the fact that $r \le N$ imply that

$$2^s - 2^r = 2^r(2^{s-r} - 1) < 2^N(s - r)(2 - 1) = 2^N(s - r). ∎$$

A similar result, with a different constant, holds for any base a.

Now consider how we might define 2^u for an irrational number u. To fix our ideas suppose we take $u = \pi$. If r and s are rational numbers such that $r < \pi < s$, then we **want**

$$2^r < 2^\pi < 2^s.$$

So let A be the collection of all values 2^r for $r < \pi$, and let B be the collection of all values 2^s for $\pi < s$. Then we want 2^π to be a "separating number" between the sets A and B.

The existence of such a separating number is a fundamental property of the real numbers that we shall discuss in Appendix 4. Briefly, it is that any two nonoverlapping sets of numbers are separated by a number X. More exactly, if A and B are nonempty sets of numbers, and if A lies below B, in the sense that $a \le b$ for every number a in A and every number b in B, then there is a fixed number X such that

$$a \le X \le b$$

for every a in A and every b in B.

(The number α that we defined earlier is the unique separating number between the set A of all values $\phi(-h)$ for h between 0 and 1, and the set B of all values $\phi(h)$.)

Returning to the present situation, we only have to check that A and B are nonoverlapping, i.e., that $2^r < 2^s$ whenever 2^r is in A and 2^s is in B. But $r < \pi < s$, and hence $2^r < 2^s$ by the monotonicity of 2^x (Lemma 1). So there is a separating number X, by the fundamental property, and we have

$$2^r \le X \le 2^s$$

whenever r and s are rational numbers such that $r < \pi < s$. Now let r and s approach π. Then $s - r \to 0$ so $2^s - 2^r \to 0$ by Lemma 4. Since

$$0 \le X - 2^r \le 2^s - 2^r$$

from the inequality above, it follows that $X - 2^r \to 0$, by the squeeze limit law. Similarly, $2^s - X \to 0$. Thus 2^x approaches the limit X as x approaches π through rational values, and X is the required value for 2^π.

The above reasoning works as well for any irrational number u, and shows that we can define 2^u to be the limit of 2^x as x approaches u through rational values. Lemma 4 then extends to irrational values x and y by applying limit laws, so the extended function 2^x is just as continuous as the unextended function 2^r.

Finally, we have to look at the difference quotient $\phi(h)$. The first inequality of Lemma 3, with $a = 1$, becomes

$$b^r - 1 < r(b - 1).$$

Now let h and k be any two rational numbers such that $0 < h < k$. In the inequality set $b = 2^k$, $r = h/k$, and divide through by h. We then have

$$\frac{2^h - 1}{h} < \frac{2^k - 1}{k}.$$

That is, the difference quotient $\phi(h) = (2^h - 1)/h$ is an increasing function of h for positive rational values of h, and therefore also for all positive *real* values of h, by continuity. We can treat

$$\phi(-h) = \frac{2^{-h} - 1}{-h} = \frac{1 - (1/2)^h}{h}$$

in the same way, using the second inequality of Lemma 3. Finally, note that $\phi(-h) = (1/2)^h \phi(h)$, so

$$\phi(h) - \phi(-h) = h\phi(h)\phi(-h).$$

If $h < 1$, then the right side is less than $h\phi(1)^2 = h$, so $\phi(h) - \phi(-h) \to 0$ as $h \to 0$. This is the last fact we had to establish.

PROBLEMS FOR SECTION 2

Differentiate (assuming the rule from Example 2 in the text).

1. $xe^x - e^x$
2. e^x/x
3. $x^2e^x - 2xe^x + 2e^x$
4. $e^{2x} - e^{-x}$
5. $\sqrt{e^x - 1}$
6. $(e^x - e^{-x})/(e^x + e^{-x})$
7. $e^{x/2}\sqrt{x - 1}$
8. $e^x + e^{2x} + e^{3x}$
9. $(e^x)^a$
10. x/e^x
11. $(e^{3x} + x)^{1/3}$
12. $e^x/(1 - e^x)$
13. $(e^{2x} + 2 + e^{-2x})^{1/2}$
14. $\left(\dfrac{1 - e^x}{1 + e^x}\right)^{1/2}$
15. $\dfrac{xe^{-x}}{1 + x^2}$
16. Show that $(d/dx)e^{x+k} = e^{x+k}$.

In Problems 17 through 20 determine the value or values of k for which $y = e^{kx}$ satisfies the given equation. Use Example 2 in the text.

17. $\dfrac{d^2y}{dx^2} + \dfrac{dy}{dx} - 2y = 0$
18. $\dfrac{d^2y}{dx^2} + 2\dfrac{dy}{dx} + y = 0$
19. $\dfrac{d^2y}{dx^2} + 2\dfrac{dy}{dx} - 3y = 0$
20. $\dfrac{d^2y}{dx^2} - y = 0$

21. If $y = xe^{-2x}$, show that

$$\frac{d^2y}{dx^2} + 4\frac{dy}{dx} + 4y = 0.$$

22. Suppose that the tangent line to the graph of $y = e^x$ at the point x_0 intersects the x-axis at $x = x_1$. Show that $x_0 - x_1 = 1$.

22. Using Example 2, show that if k is rational, then every solution of the equation

$$\frac{dy}{dx} = ky$$

is of the form $y = ce^{kx}$. (Look at Example 4.)

24. Prove from scratch that

$$\frac{d}{dx} e^{x^2} = 2xe^{x^2}.$$

25. Prove that

$$\frac{d}{dx} e^{f(x)} = e^{f(x)} \cdot f'(x).$$

(Use the mean-value principle, as in Example 2 of Section 8 in Chapter 3.)

Find the derivatives of the following functions by using the rule in Problem 25.

26. $e^{\pi x}$ 27. e^{ax} 28. $e^{1/x}$ 29. $e^{\sqrt{x}}$

3. SOME TRIGONOMETRY

Because of their origins in geometry, the basic properties of $\sin t$ and $\cos t$ can be established by geometric reasoning. The crux of the matter is that we

can put a coordinate system on the circumference of the unit circle that
behaves much like the basic coordinate system on a line.

Recall that the coordinate correspondence on a line assigns to each real
number x the uniquely determined point P_x that we get by measuring off the
signed distance x from the origin. That is, P_x is at the distance $|x|$ from the
origin, and is to the right or left of the origin depending on whether x is
positive or negative.

Now we do the same thing on the unit circle $x^2 + y^2 = 1$. The conventional origin here is the point $Q = (1, 0)$ where the circle intersects the
positive x-axis, and the positive direction is the *counterclockwise* direction
along the circle. We assign to each real number t the uniquely determined
point P_t that we get by measuring off the signed distance t along the
circumference from Q. That is, we go in the counterclockwise or clockwise
direction, depending on whether t is positive or negative, and we measure
off the distance $|t|$.

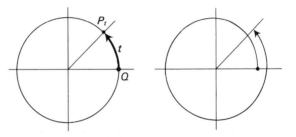

We shall call t *an angular coordinate* of the point P_t. We can't call it *the*
angular coordinate because this coordinate correspondence is not one-to-
one. For example, if $t = 2\pi$, we go counterclockwise around the circle and
get back exactly to the origin point Q, since the length of the full unit
circumference is 2π. If $t = 4\pi$ we go *twice* around and get back exactly to
Q. If $t = -2\pi$, we go clockwise around and come back to Q. Thus
$P_{2\pi} = P_{4\pi} = P_{-2\pi} = P_0 = Q$, so $2\pi, 4\pi, -2\pi$, and 0 are all angular coordinates of Q.

On the other hand, each point P on the unit circle $x^2 + y^2 = 1$ has
exactly one angular coordinate t in the interval $[0, 2\pi)$; t is the length of the
arc from Q around to P in the counterclockwise direction. Then $t + 2n\pi$ is
also an angular coordinate of P, for any integer n, positive or negative.
Moreover, every coordinate of P can be written in this form.

We define cos t (cosine t) and sin t (sine t) as the x and y coordinates of
P_t:

$$\cos t = x\text{-coordinate of } P_t,$$

$$\sin t = y\text{-coordinate of } P_t.$$

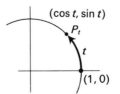

It will be seen shortly that these definitions are substantially the same as others that may be more familiar. Note that cos t and sin t are functions of t, defined for all values of t, and related by the equation of the unit circle:

$$\cos^2 t + \sin^2 t = 1.$$

The graphs of these functions have the general features shown below. Remember that if t is large we may "wrap around" the circle many times before coming to the terminal point $P_t = (x, y)$; and if t is negative then we go in the *clockwise* direction a distance $|t|$. You ought to be able to visualize the x-coordinate of this terminal point P_t oscillating between $+1$ and -1 in the general manner shown, as P_t moves around the circle. In particular, x first becomes 0 when t is the length of one quarter of the full circumference,

$$t = \frac{(2\pi)}{4} = \frac{\pi}{2}.$$

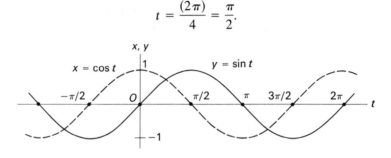

The graphs of sin t and cos t sketched above suggest that

$$\cos(-t) = \cos t,$$
$$\sin(-t) = -\sin t,$$

for all t. These conjectures can be verified from the geometric definitions, as indicated below.

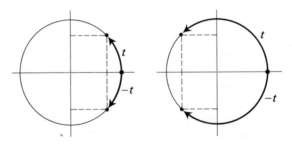

The graphs also show other properties that can be directly checked. Thus, both sine and cosine are *periodic* with *period* 2π, meaning that

$$\sin(t + 2\pi) = \sin t,$$
$$\cos(t + 2\pi) = \cos t,$$

for all t. This follows from our earlier observations that

$$P_{t+2\pi} = P_t.$$

A stronger property of each function suggested by the graphs is that

$$\sin(t + \pi) = -\sin t,$$
$$\cos(t + \pi) = -\cos t.$$

These equations express the fact that $P_{t+\pi}$ is diametrically opposite P_t on the unit circle. Finally, it appears from the graphs that if the cosine graph were slid $\pi/2$ units to the right it would become the sine graph, i.e., that

$$\cos t = \sin(t + \pi/2)$$

for all t. This is true but not quite so easy to prove.

It is very helpful to think geometrically about $\sin t$ and $\cos t$ as we did above, and to regard the general shapes of their graphs as forming a storehouse of various symmetry properties and other relationships that can be retrieved by inspection. But the facts suggested and recalled by such geometric reasoning all have analytic origins in the *addition formulas* for sine and cosine, which we shall come to shortly.

Each point P_t determines a geometric angle which we shall designate θ_t, namely, the angle from the positive x-axis around to the ray OP_t. When θ_t is acute, then the older right-triangle definitions of sine and cosine for θ_t agree with the present ones. For, referring to the right triangle in the next figure, we have

$$\cos t = x = \frac{x}{1} = \frac{\text{Adjacent}}{\text{Hypotenuse}} = \text{cosine of } \theta_t,$$

$$\sin t = y = \frac{y}{1} = \frac{\text{Opposite}}{\text{Hypotenuse}} = \text{sine of } \theta_t.$$

The number t is called the *radian measure* of the angle θ_t. In order to convert from degrees to radians, note that θ_t is a right angle when P_t is one quarter of the way around the circle, i.e., when $t = (1/4)(2\pi) = \pi/2$. This gives the first line in the table below, and the rest of the table follows from it.

θ_t in degrees	t		
90°	$\pi/2$		
1°	$\pi/180$		
$d°$	$(\pi/180)d$		

		$\sin t$	$\cos t$
30°	$\pi/6$	$1/2$	$\sqrt{3}/2$
45°	$\pi/4$	$\sqrt{2}/2$	$\sqrt{2}/2$
60°	$\pi/3$	$\sqrt{3}/2$	$1/2$

The last three lines record the standard facts for a 30°-60°-90° triangle and for an isosceles right triangle (45°-45°-90°).

Now consider an arbitrary point $P = (x, y)$ different from the origin $O = (0, 0)$, and let t be the radian measure of an angle from the positive x-axis around to the ray OP.

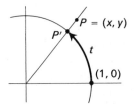

What now is the relationship between x, y, and t? The connecting link is the point P' where the ray OP intersects the unit circle $x^2 + y^2 = 1$. On the one hand,

$$P' = \left(\frac{x}{r}, \frac{y}{r}\right),$$

where $r = \sqrt{x^2 + y^2}$, while on the other hand P' has angular coordinate t. Therefore

$$\cos t = \frac{x}{r}, \qquad \sin t = \frac{y}{r},$$

by the definition of cosine and sine. (For an acute angle, these formulas can also be derived from the right-triangle definitions of $\cos t$ and $\sin t$.) Note that the coordinates (x, y) can be recovered from r and t by the equations

$$x = r \cos t,$$
$$y = r \sin t.$$

The numbers r and t are called *polar coordinates* of P, and the above equations express the change of coordinates for P from polar coordinates to rectangular coordinates. Polar coordinates will be discussed further in Section 4 of Chapter 10.

Finally, if P is on the unit circle $(x - a)^2 + (y - b)^2 = 1$ with center $O' = (a, b)$, then an angle t from the positive x direction around to the ray $O'P$ satisfies

$$\cos t = x - a, \qquad \sin t = y - b,$$

because $x - a$ and $y - b$ are the *new coordinates* of P when we put a new origin at O' (Theorem 7 in Chapter 1). If P is not on the unit circle about O', then, as above,

$$\cos t = \frac{x - a}{r}, \qquad \sin t = \frac{y - b}{r},$$

where $O' = (a, b)$, $P = (x, y)$, and $r = \sqrt{(x - a)^2 + (y - b)^2}$.

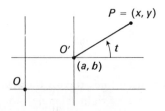

Underlying all the properties of the trigonometric functions (and implicit already in some of our remarks above) is the following basic fact:

If we measure off the signed distance u along the unit circumference, starting from the point P_t, then we end up at the point P_{t+u}.

This should appear geometrically obvious from looking at a few special cases, and we shall simply assume it. It is the analog for angular coordinates of Theorem 1 in Section 1 of Chapter 1: if $t + u$ is replaced by s, so that $u = s - t$, it just says that $s - t$ is a signed distance from P_t to P_s along the unit circle.

According to this basic principle, if we measure off the same signed arc length $(s - t)$ from each of the two initial points P_t and P_0, we obtain the terminal points P_s and P_{s-t}, respectively. And since equal arcs have equal chords, we then have the identity

$$\overline{P_t P_s} = \overline{P_0 P_{s-t}},$$

where we have used the notation $\overline{PQ}$ for the length of the chord from P to Q.

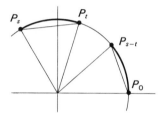

The above identity is equivalent to the trigonometric addition laws! When the chord lengths are expressed by the distance formula, and both sides are squared, the identity becomes

$$(\cos s - \cos t)^2 + (\sin s - \sin t)^2 = (\cos(s - t) - 1)^2 + (\sin(s - t) - 0)^2.$$

Now just simplify algebraically: expand the squares, use the identity $\sin^2 x + \cos^2 x = 1$ three times, and cancel common terms and factors. What is left is the basic law

$$\cos s \cos t + \sin s \sin t = \cos(s - t).$$

In conjunction with the table

	0	$\pi/2$	π
sin	0	1	0
cos	1	0	-1

this law yields *all* the remaining trigonometric identities. A scheme for working these out is suggested in the problems. Meanwhile we shall make use of the addition forms of the above law, namely:

$$\sin(x + y) = \sin x \cos y + \cos x \sin y,$$

$$\cos(x + y) = \cos x \cos y - \sin x \sin y.$$

The functions $\sin x$ and $\cos x$ are the basic trigonometric functions, but there are four others, the tangent, cotangent, secant and cosecant functions, defined as follows:

$$\tan x = \frac{\sin x}{\cos x}, \qquad \cot x = \frac{\cos x}{\sin x} \left(= \frac{1}{\tan x} \right),$$

$$\sec x = \frac{1}{\cos x}, \qquad \csc x = \frac{1}{\sin x}.$$

It follows from these definitions that the graphs of $\tan x$ and $\sec x$ have the general appearance shown below. Some other properties will be derived in the problems.

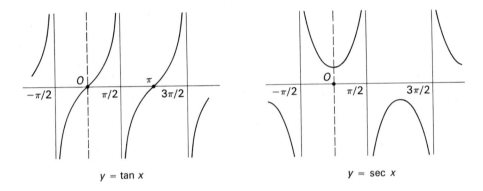

$y = \tan x \qquad\qquad\qquad\qquad y = \sec x$

PROBLEMS FOR SECTION 3

In Problems 1 through 11 below, draw a unit circle, locate the point p_t with the given angular coordinate t, and determine $\cos t$ and $\sin t$ (exactly if t involves π, and approximately otherwise).

1. $t = 3\pi/2$
2. $t = 5\pi/4$
3. $t = -\pi/4$
4. $t = 1$
5. $t = 2$
6. $t = 7\pi$
7. $t = -3\pi$
8. $t = -3\pi/2$
9. $t = 4$
10. $t = 7$
11. $t = -7\pi/4$

In Problems 12 through 20 plot the given point (x, y) on the unit circle, and estimate (or determine exactly if you can) the unique number t in $[0, 2\pi)$ for which $(\cos t, \sin t) = (x, y)$.

12. $(0, -1)$
13. $(-\sqrt{2}/2, -\sqrt{2}/2)$
14. $(1/2, -\sqrt{3}/2)$
15. $(-1/2, \sqrt{3}/2)$
16. $(\sqrt{3}/2, -1/2)$
17. $(4/5, 3/5)$
18. $(3/5, 4/5)$
19. $(-1/\sqrt{5}, 2/\sqrt{5})$
20. $(2/\sqrt{5}, -1/\sqrt{5})$

21. What trigonometric identities are suggested by the following figure?

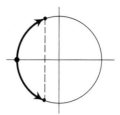

22. What trigonometric identities are suggested by the following figure?

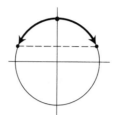

Give the geometric proof (from right triangles) for the identities:

23. $\sin \pi/6 = 1/2$, $\cos \pi/6 = \sqrt{3}/2$.

24. $\sin \pi/4 = \sqrt{2}/2$.

25. Let θ be the angle that the segment from $(1, 2)$ to $(2, 4)$ makes with the direction of the positive x-axis. Compute $\sin \theta$ and $\cos \theta$.

26. Same question for the angle θ that the segment from $(1, -1)$ to $(0, 4)$ makes with the direction of the positive x-axis.

In Problems 27 through 33 estimate the smallest positive angle from the positive x-axis to the ray OP, for the given point P.

27. $P = (1, 1)$ 28. $P = (1, 2)$

29. $P = (-3, 3)$ 30. $P = (2, -4)$

31. $P = (-4, 2)$ 32. $P = (-1/2, -1/2)$

33. $P = (1, -1/2)$

34. Prove the identities:

$$\tan^2 x + 1 = \sec^2 x, \qquad 1 + \cot^2 x = \csc^2 x.$$

It was claimed in the text that the $\cos(s - t)$ law, together with the values of $\cos t$ and $\sin t$ at $t = 0$, $\pi/2$, and π, determine all the trigonometric identities. The order in which things are done is important. Here is one sequence, with some hints:

35. $\cos(-t) = \cos t$ (Take $s = 0$ in the formula for $\cos(s - t)$.)

36. $\cos(s - \pi/2) = \sin s$ (Take $t = \pi/2$.)

37. $\cos(s - \pi) = -\cos s$ (Take $t = \pi$.)

38. $\sin(s - \pi/2) = -\cos s$ (Write $s - \pi = (s - \pi/2) - \pi/2$ in (37) and apply (36).)

39. $\sin(x - y) = \sin x \cos y - \cos x \sin y$ (Use 36, the $\cos(s - t)$ law, 36 and 38, in that order.)

40. $\sin(-y) = -\sin y$ (Take $x = 0$.)

41. $\sin(x + y) = \sin x \cos y + \cos x \sin y$

42. $\cos(x + y) = \cos x \cos y - \sin x \sin y$ (From the $\cos(s - t)$ law, using (35) and (40).)

43. $\tan(x + y) = \dfrac{\tan x + \tan y}{1 - \tan x \tan y}$ (Divide (41) by (42).)

44. $\tan(x - y) = \dfrac{\tan x - \tan y}{1 + \tan x \tan y}$

Double-angle Formulas

45. $\sin 2x = 2 \sin x \cos x$ 46. $\cos 2x = \cos^2 x - \sin^2 x$
$$= 1 - 2 \sin^2 x$$
$$= 2 \cos^2 x - 1$$

47. $\tan 2x = \dfrac{2 \tan x}{1 - \tan^2 x}$

Half-angle Formulas

48. $\cos^2\left(\dfrac{y}{2}\right) = \dfrac{1 + \cos y}{2}$ 49. $\sin^2\left(\dfrac{y}{2}\right) = \dfrac{1 - \cos y}{2}$

50. $\tan\left(\dfrac{y}{2}\right) = \dfrac{\sin y}{1 + \cos y} = \dfrac{1 - \cos y}{\sin y}$ (Verify by setting $y = 2x$ and using double-angle formulas.)

Finally, if we set $x + y = A$ and $x - y = B$, so that

$$A + B = 2x \qquad \text{and} \qquad A - B = 2y,$$

the addition formulas give us:

51. $\sin A - \sin B = 2 \cos\left(\dfrac{A + B}{2}\right) \sin\left(\dfrac{A - B}{2}\right)$

52. $\cos A - \cos B = -2 \sin\left(\dfrac{A + B}{2}\right) \sin\left(\dfrac{A - B}{2}\right)$

53. Show that $a \sin x + b \cos x$ can be written in the form

$$A \sin(x + \theta).$$

Here a and b are arbitrary constants, and A and θ are to be determined from them.

54. What trigonometric identities are suggested by the following figures?

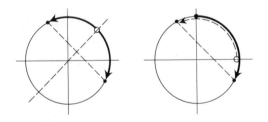

4. THE DERIVATIVES OF sin x AND cos x

The functions $\sin x$ and $\cos x$ are closely related to e^x. The relationship appears here in the similar way in which their derivative formulas can be established; starting from the fact that they are differentiable at the origin, the addition laws can be used to show that they are everywhere differentiable and to calculate their derivative formulas. The starting limits are:

$$\sin'(0) = \lim_{h \to 0} \frac{\sin(0 + h) - \sin(0)}{h} = \lim_{h \to 0} \frac{\sin h}{h},$$

$$\cos'(0) = \lim_{h \to 0} \frac{\cos(0 + h) - \cos(0)}{h} = \lim_{h \to 0} \frac{\cos h - 1}{h}.$$

The geometric definitions of sin t and cos t allow us to establish and evaluate these limits geometrically. The (sin h)/h limit is the basic one.

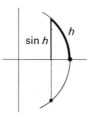

For positive h, the figure above shows that

$$\frac{\sin h}{h} = \frac{2 \sin h}{2h}$$

can be interpreted as the ratio of a chord length to its arc length. It should seem geometrically clear that the chord becomes a very good approximation to the arc when h is small. So intuition suggests the following lemma.

LEMMA 5 $$\frac{\sin h}{h} \to 1 \qquad \text{as } h \to 0.$$

Proof. The geometric proof of this limit is easier in terms of area than it is in terms of arc length. We assume that the area A of a circular sector is proportional to the length of its intercepted arc. Since the area of the full unit circle is π, this proportion is

$$\frac{A}{\pi} = \frac{t}{2\pi},$$

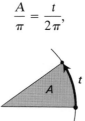

from which we see that $A = t/2$. We note that A lies between the two triangular areas shown below, so

$$\frac{\sin t}{2} < \frac{t}{2} < \frac{\tan t}{2}.$$

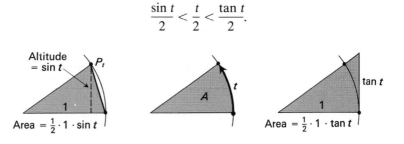

Dividing throughout by $(\sin t)/2$, we obtain the inequality

$$1 < \frac{t}{\sin t} < \frac{1}{\cos t},$$

and this becomes

$$\cos t < \frac{\sin t}{t} < 1$$

upon taking reciprocals. Assuming for the moment that $\cos t$ is continuous at $t = 0$, so that $\lim_{t \to 0} \cos t = \cos 0 = 1$, we conclude that

$$\frac{\sin t}{t} \to 1 \quad \text{as} \quad t \to 0$$

by the squeeze limit law. The proof that $\cos t$ is continuous at $t = 0$ will be left as an exercise.

Since

$$\frac{\sin(-t)}{-t} = \frac{\sin t}{t},$$

we do not need a separate argument for negative h. ∎

The other limit is 0:

$$\cos'(0) = \lim_{h \to 0} \frac{\cos h - 1}{h} = 0.$$

This limit can be derived from the $(\sin h)/h$ limit by using the addition law for $\cos h = \cos(h/2 + h/2)$; it should also seem geometrically obvious because the graph of the cosine function clearly has a horizontal tangent at $t = 0$.

Now the addition law for $\sin(x + h)$ lets us rewrite the difference quotient for $\sin x$ in a form that permits its limit to be evaluated.

THEOREM 5   *The functions* $\sin x$ *and* $\cos x$ *are everywhere differentiable, and*

$$\frac{d}{dx} \sin x = \cos x,$$

$$\frac{d}{dx} \cos x = -\sin x.$$

Proof. We have

$$\frac{\sin(x + h) - \sin x}{h} = \frac{\sin x \cos h + \cos x \sin h - \sin x}{h}$$

$$= \sin x \left(\frac{\cos h - 1}{h} \right) + \cos x \left(\frac{\sin h}{h} \right).$$

We have just seen that the quantities in parentheses have limits 0 and 1 respectively. Thus

$$\frac{d}{dx} \sin x = \lim_{h \to 0} \frac{\sin(x + h) - \sin x}{h}$$

$$= (\sin x) \cdot 0 + (\cos x) \cdot 1$$

$$= \cos x.$$

The proof for cos x goes in exactly the same way, starting from the addition law for $\cos(x + h)$. ∎

The remaining trigonometric functions are defined from sin x and cos x, and their derivatives are computed from these definitions. Here again are the definitions:

$$\tan x = \frac{\sin x}{\cos x}, \qquad \cot x = \frac{\cos x}{\sin x},$$

$$\sec x = \frac{1}{\cos x}, \qquad \csc x = \frac{1}{\sin x}.$$

These are the tangent, secant, cotangent, and cosecant functions.

Dividing the identity $\sin^2 x + \cos^2 x = 1$ first by $\cos^2 x$ and then by $\sin^2 x$ gives the useful identities

$$\tan^2 x + 1 = \sec^2 x,$$

$$1 + \cot^2 x = \csc^2 x.$$

The derivative formulas for these four functions are calculated from the sine and cosine formulas by the quotient rule and/or the general power rule. The remaining formulas are:

$$\frac{d}{dx} \tan x = \sec^2 x,$$

$$\frac{d}{dx} \sec x = \tan x \sec x,$$

$$\frac{d}{dx} \cot x = -\csc^2 x,$$

$$\frac{d}{dx} \csc x = -\cot x \csc x.$$

The tangent and secant derivative formulas come up often enough to be worth memorizing, but they can always be worked out if forgotten.

EXAMPLE 1   Prove that

$$\frac{d}{dx} \tan x = \sec^2 x.$$

Solution. By definition

$$\tan x = \frac{\sin x}{\cos x}.$$

We can therefore compute its derivative by the quotient rule:

$$\frac{d}{dx} \tan x = \frac{\cos x \cdot \cos x - \sin x(-\sin x)}{\cos^2 x}$$

$$= \frac{\cos^2 x + \sin^2 x}{\cos^2 x}$$

$$= \frac{1}{\cos^2 x} = \sec^2 x$$

(since $\sec x = 1/\cos x$).

EXAMPLE 2   If $y = e^x \sin x$, find constants b and c such that

$$y'' + by' + cy = 0$$

(where $y' = dy/dx$ and $y'' = d^2y/dx^2$).

Solution. Since

$$y' = e^x \cos x + e^x \sin x,$$
$$y'' = -e^x \sin x + e^x \cos x + e^x \cos x + e^x \sin x$$
$$= 2e^x \cos x,$$

we see that

$$y'' + by' + cy = (2 + b)e^x \cos x + (b + c)e^x \sin x.$$

This will be 0 if and only if

$$2 + b = 0,$$
$$b + c = 0,$$

i.e., if $b = -2$, $c = 2$. Thus

$$y'' - 2y' + 2y = 0$$

when $y = e^x \sin x$.

We saw at the end of Section 3 that $y = e^x$ is essentially the only function satisfying the equation

$$\frac{dy}{dx} = y,$$

in the sense that every solution of the equation is of the form $y = ce^x$ for some constant c. Since this equation relates an unknown function to one or more of its derivatives, it is called a *differential equation,* and it is of the *first order* because only the first-order derivative occurs.

The functions $\sin x$ and $\cos x$ have a similar property. They are both solutions of the *second-order* differential equation

$$\frac{d^2y}{dx^2} = -y.$$

Obviously neither sine nor cosine can be the unique solution in the above sense, but the two functions together are essentially the only solutions in a similar sense: Every other solution f is of the form

$$f(x) = A \sin x + B \cos x$$

for some constants A and B. A proof of this important fact is outlined below, but the details of the individual steps are left as exercises.

a) If g and h are any two solutions of the differential equation $y'' = -y$, then so is every linear combination of g and h, i.e., every function f of the form

$$f(x) = Ag(x) + Bh(x),$$

where A and B are constants.

b) If f is any solution then

$$f^2 + (f')^2 = \text{constant}.$$

Thus every solution f satisfies an identity of the type

$$\sin^2 x + \cos^2 x = 1.$$

c) It follows from (b) that if ϕ is a solution such that $\phi(0) = \phi'(0) = 0$, then ϕ is the zero function: $\phi(x) = 0$ for all x.

d) If f is any solution, then f must be of the form

$$f(x) = f(0) \cos x + f'(0) \sin x$$

(set $\phi = f - f(0) \cos - f'(0) \sin$ and apply (c)). Thus the general solution of the differential equation $d^2y/dx^2 + y = 0$ is of the form $A \sin x + B \cos x$.

The chain rule will let us show that essentially the same conclusions hold for the differential equation

$$\frac{d^2y}{dx^2} + by = 0,$$

where b is positive. We check that $\sin \sqrt{b}x$ and $\cos \sqrt{b}x$ are solutions, and can then show that

$$A \sin \sqrt{b}x + B \cos \sqrt{b}x$$

is the general solution by repeating the above development. An alternative procedure is to observe that a function f satisfies the new equation if and only if

$$g(x) = f(x/\sqrt{b})$$

satisfies the original equation.

PROBLEMS FOR SECTION 4

Prove the formulas for the derivatives of the remaining trigonometric functions, as follows:

1. $\dfrac{d}{dx} \sec x = \sec x \tan x$ 2. $\dfrac{d}{dx} \cot x = -\csc^2 x$

3. $\dfrac{d}{dx} \csc x = -\csc x \cot x$

Differentiate the following functions:

4. $x \sin x$ 5. $e^x \cos x$

6. $(\sin x)/x$ 7. $\sin^2 x$

8. $\sin x \cos x$ 9. $\cos^2 x + \sin^2 x$

10. $\dfrac{\sin x}{1 + \cos x}$ 11. $\dfrac{1 - \cos x}{\sin x}$

12. $\cos^2 x - \sin^2 x$ 13. $\tan^2 x$

15. $\sec^2 x$ 15. $\sqrt{1 + \cos x}$

16. $\sqrt{1 + \sin^2 x}$ 17. $\sin^3 x \cos^3 x$

18. $\cos^4 x - \sin^4 x$

19. $\sec^2 x - \tan^2 x$

20. $x \sin x - \cos x$

21. $\dfrac{e^x(\sin x + \cos x)}{2}$

22. $x^2 \cos x - 2x \sin x - 2 \cos x$

23. $(1/2)(x + \sin x \cos x)$

24. $y = \tan x \sec x - x$

25. $y = \sin x/(\cos x + x)$

26. Complete the proof of Theorem 5. That is, prove that $d \cos x/dx = -\sin x$ from scratch, using the addition law for $\cos(x + h)$.

27. A particle moves back and forth along the x-axis, its position at time t being given by $x = 3 \sin 2t$. Show that its acceleration a is proportional to its position x but oppositely directed.

28. A particle moves back and forth along the x-axis in a "decaying vibration," its position at time t being given by $x = e^{-t} \sin t$. Express its acceleration a in terms of its velocity v and position x.

5. THE CHAIN RULE

When we start combining the exponential and trigonometric functions with other functions we discover that there are very simple combinations that we can't yet differentiate, such as $\cos \pi x$ and $e^{(x^2)}$. These are examples of functions *inside of* functions, and in order to handle this general method of combination we need one final rule, called the *chain rule*.

If y is a function of u and if u is a function of x, then y is a function of x. For example, if $y = u^{3/2}$ and $u = 3 + x^2$, then $y = (3 + x^2)^{3/2}$. The question is: How does the derivative dy/dx of such a final "composite" function relate to the derivatives dy/du and du/dx of its two "constituent" functions?

It is easy to see from the rate-of-change interpretation of the derivative what the answer ought to be. Consider, first, what happens around $x = 1$ in the above example. If we compute the derivatives at $x = 1$ and at $u = 3 + x^2 = 3 + (1)^2 = 4$, we get

$$\frac{dy}{du} = 3, \qquad \frac{du}{dx} = 2.$$

Thus, as x passes through 1, y is increasing three times as fast as u and u is increasing twice as fast as x. Therefore, altogether y is increasing $3 \cdot 2 = 6$ times as fast as x, so $dy/dx = 6$.

Now consider the general situation. Suppose that at $x = a$ the derivative dy/du has the value r and du/dx has the value s. Thus, as x passes through a, y is increasing r times as fast as u and u is increasing s times as fast as x. Therefore, altogether y is increasing rs times as fast as x. That is, $dy/dx = rs = (dy/dy) (du/dx)$. The following theorem is thus plausible.

THEOREM 6   *If y is a differentiable function of u and if u is a differentiable function of x, then y is a differentiable function of x and*

$$\frac{dy}{dx} = \frac{dy}{du} \cdot \frac{du}{dx}.$$

This is the Leibniz version of the chain rule. Our heuristic derivation needs to be bolstered by a real proof but, as we have done with the other rules, we shall state the chain rule in various ways and practice using it before taking up its proof.

EXAMPLE 1 If
$$y = \sin u \qquad \text{and} \qquad u = 3 + x^2,$$
we have
$$\frac{dy}{du} = \cos u, \qquad \frac{du}{dx} = 2x,$$
and the chain rule gives
$$\frac{dy}{dx} = \frac{dy}{du} \cdot \frac{du}{dx} = (\cos u)2x = 2x \cos(3 + x^2).$$

The intermediate variable u may not be part of the original problem. For instance, the above example would typically arise as the problem of computing dy/dx when $y = \sin(3 + x^2)$. Here we have a function that is too complicated for our earlier rules to handle. But setting $u = 3 + x^2$ breaks the complicated function down into two simpler stages that we do know how to differentiate,
$$y = \sin u \qquad \text{and} \qquad u = 3 + x^2,$$
and we can now compute dy/dx by the chain rule, as above.

Here is the same example without commentary.

EXAMPLE 2 Compute dy/dx when $y = \sin(3 + x^2)$.

Solution. Setting $u = 3 + x^2$, we have $y = \sin u$. Then by the chain rule,
$$\frac{dy}{dx} = \frac{dy}{du} \cdot \frac{du}{dx} = (\cos u) \cdot 2x = 2x \cos(3 + x^2).$$

Here the derivatives dy/du and du/dx are so simple that there is no need to compute them separately before multiplying them in the chain rule. But if one of them has to be worked out, then computing it separately may help to keep track of what is going on.

EXAMPLE 3 If $y = [x/(1 + x)]^{1/3}$, we have a power rule situation, but here we want to apply the chain rule. Setting $u = x/(1 + x)$ breaks it up suitably, into
$$y = u^{1/3} \qquad \text{and} \qquad u = \frac{x}{1 + x}.$$
Then
$$\frac{dy}{du} = \frac{1}{3} u^{-2/3}, \qquad \frac{du}{dx} = \frac{(1 + x) \cdot 1 - x \cdot 1}{(1 + x)^2} = \frac{1}{(1 + x)^2},$$

and so

$$\frac{dy}{dx} = \frac{dy}{du} \cdot \frac{du}{dx} = \frac{1}{3} u^{-2/3} \cdot \frac{1}{(1 + x)^2}$$

$$= \frac{1}{3} \left(\frac{x}{1 + x}\right)^{-2/3} \frac{1}{(1 + x)^2} = \frac{1}{3x^{2/3}(1 + x)^{4/3}}.$$

EXAMPLE 4   We now use the chain rule to prove the power rule for a positive rational exponent $a = m/n$, where m and n are positive integers. The problem is to compute dy/dx when

$$y = x^{m/n} = (x^{1/n})^m.$$

Setting $u = x^{1/n}$, we have

$$y = u^m, \qquad u = x^{1/n}.$$

From Theorems 1 and 4 in Chapter 3,

$$\frac{dy}{du} = mu^{m-1}, \qquad \frac{du}{dx} = \frac{1}{n} x^{(1/n)-1}.$$

Therefore,

$$\frac{dy}{dx} = \frac{dy}{du} \cdot \frac{du}{dx} = mu^{m-1} \cdot \frac{1}{n} x^{(1/n)-1}$$

$$= \frac{m}{n} (x^{1/n})^{m-1} x^{1/n-1}$$

$$= \frac{m}{n} x^{(m/n)-(1/n)+(1/n)-1} = \frac{m}{n} x^{(m/n)-1}$$

$$= ax^{a-1}.$$

With more practice, and in situations where the two derivatives dy/du and du/dx are known, the chain-rule derivative can be written down in one or two steps.

EXAMPLE 5   As we look at

$$y = (\underbrace{3 + x^2}_{u})^{3/2},$$

we can mentally take $3 + x^2$ as u, and see that we then have $y = u^{3/2}$, without writing it down separately. Nor will we need to write down the Leibniz formula. As we put down the righthand side, we first think dy/du, see that it is $(3/2)u^{1/2}$ and write this down, and then think du/dx, see that it is $2x$ and write that down. Thus, we would probably just write

$$\frac{dy}{dx} = \frac{3}{2} u^{1/2} \cdot 2x$$

or

$$\frac{dy}{dx} = \frac{3}{2} (3 + x^2)^{1/2} \cdot 2x.$$

EXAMPLE 6 Of course, a chain-rule calculation may involve other variables, or the same variables used in different ways. Thus, if $s = (3 + t^2)^{3/2}$ and if we set $x = 3 + t^2$, then

$$\frac{ds}{dt} = \frac{ds}{dx} \cdot \frac{dx}{dt} = \frac{3}{2} x^{1/2} \cdot 2t = 3t(3 + t^2)^{1/2}.$$

In the example $y = (3 + x^2)^{3/2} = u^{3/2}$, we can replace y by what it is equal to and write this case of the chain rule as

$$\frac{d}{dx} \underbrace{(3 + x^2)}_{u}^{3/2} = \frac{d}{du} u^{3/2} \cdot \frac{du}{dx} = \frac{3}{2} u^{1/2} \frac{du}{dx}.$$

In general, if $y = g(u)$ and u is a function of x then the chain rule can be rewritten

$$\frac{d}{dx} g(u) = \frac{d}{du} g(u) \cdot \frac{du}{dx} = g'(u) \frac{du}{dx}.$$

This form is useful enough to be given special status.

THEOREM 6' *If $y = g(u)$ and u is a function of x, and if both functions are differentiable, then y is a differentiable function of x and*

$$\frac{dy}{dx} = g'(u) \cdot \frac{du}{dx},$$

or

$$\frac{d}{dx} g(u) = g'(u) \frac{du}{dx}.$$

The "Built-in" Chain Rule

When we take g in the chain rule

$$\frac{d}{dx} g(u) = g'(u) \frac{du}{dx}$$

to be one of the functions u^a, e^u, $\sin u$, and $\cos u$, we obtain the *general* differentiation formulas

$$\frac{d}{dx} u^a = au^{a-1} \frac{du}{dx},$$

$$\frac{d}{dx} e^u = e^u \frac{du}{dx},$$

$$\frac{d}{dx} \sin u = \cos u \frac{du}{dx},$$

$$\frac{d}{dx} \cos u = -\sin u \frac{du}{dx}.$$

The first of these formulas is the general power rule from Chapter 3.

Some people like to learn all of the basic derivative formulas in this general form, incorporating the chain rule. Then most routine applications of the chain rule occur automatically. For example

$$\frac{d}{dx}\sin(x^{1/3}) = [\cos(x^{1/3})]\frac{d}{dx}x^{1/3}$$

is just an application of the general sine formula, and needs no special thought of the chain rule. This procedure is very efficient, but it can also be costly if it leads one to lose sight of the fundamental chain rule itself and so possibly to flounder when a less routine application of it is needed.

The other possible approach, and the one we have been following in this section, is to make every application of the chain rule a conscious act. Then we start with the formulas

$$\frac{d}{dx}x^a = ax^{a-1}, \qquad \frac{d}{dx}e^x = e^x, \qquad \frac{d}{dx}\sin x = \cos x, \qquad \frac{d}{dx}\cos x = -\sin x,$$

etc., and combine any one of them with the chain rule whenever it is necessary.

Composition of Functions

In Theorem 6', u is still some unspecified function of x. If $u = h(x)$, so that $du/dx = h'(x)$, then we can replace all mention of u and obtain the pure function form

$$\frac{d}{dx}g(h(x)) = g'(h(x))h'(x).$$

This allows the most precise statement of the chain rule:

THEOREM 6″   *If the function h is differentiable at the point $x = x_0$ and if g is differentiable at the point $u_0 = h(x_0)$, then the function $f(x) = g(h(x))$ is differentiable at x_0 and*

$$f'(x_0) = g'(u_0) \cdot h'(x_0)$$
$$= g'(h(x_0))h'(x_0).$$

The function

$$f(x) = g(h(x))$$

is called the *composition product* of g and h. As a function of x it is formed by applying the outside function g to the inside function of x, $h(x)$.

EXAMPLE 7   The function $f(x) = (3 + x^2)^{3/2}$ is the composition product of the outside function $g(u) = u^{3/2}$ and the inside function $h(x) = 3 + x^2$.

The composition product of g with h is sometimes designated by the special notation $g \circ h$. This means that $g \circ h$ is the function defined by

$$(g \circ h)(x) = g(h(x)).$$

This notation allows us to write the function form of the chain rule without any variables at all:

$$(g \circ h)' = (g' \circ h)h'.$$

Proof of Theorem 6'. We are considering

$$y = g(u),$$

where u is a differentiable function of x, and we shall assume that g has a continuous derivative. An increment Δx in x produces an increment Δu in u, and this in turn produces the increment

$$\Delta y = g(u + \Delta u) - g(u)$$

in y. We rewrite the right side of this equation by the mean-value principle, and have

$$\Delta y = g'(U) \, \Delta u,$$

where U is some point between u and $u + \Delta u$. Now divide by Δx,

$$\frac{\Delta y}{\Delta x} = g'(U) \frac{\Delta u}{\Delta x},$$

and let $\Delta x \to 0$. Then $\Delta u/\Delta x \to du/dx$ by hypothesis. Also, $\Delta u \to 0$ because a differentiable function is continuous, $U \to u$ because U lies between u and $u + \Delta u$, and $g'(U) \to g'(u)$ because g'. is continuous. The right side above thus approaches $g'(u)(du/dx)$ as Δx approaches 0. Thus the limit

$$\frac{dy}{dx} = \lim_{\Delta x \to 0} \frac{\Delta y}{\Delta x}$$

exists, and

$$\frac{dy}{dx} = g'(u) \frac{du}{dx}. \quad \blacksquare$$

We have thus proved Theorem 6', but under the extra hypothesis that g' is continuous in an interval about u_0. This will cover practically any situation that comes up. For the theorem as stated we can still proceed in essentially the above way because we can concoct a function $G(\Delta u)$ that plays the same role that $g'(U)$ played above. We define G artificially as follows:

$$G(\Delta u) = \frac{g(u_0 + \Delta u) - g(u_0)}{\Delta u} \qquad \text{when } \Delta u \neq 0,$$

$$G(0) = g'(u_0).$$

It follows from this definition that

$$\Delta y = g(u_0 + \Delta u) - g(u_0) = G(\Delta u) \cdot \Delta u$$

for all values of Δu, and we can then proceed exactly as before, using the fact that $G(\Delta u) \to g'(u_0)$ as $\Delta u \to 0$.

PROBLEMS FOR SECTION 5

Prove the following differentiation rules:

1. $\dfrac{d}{dx} \sin(ax + b) = a \cos(ax + b)$ 2. $\dfrac{d}{dx} \cos(ax + b) = -a \sin(ax + b)$

3. $\dfrac{d}{dx} e^{ax+b} = ae^{ax+b}$

Compute the derivatives of the following functions.

4. e^{x^2}

5. $\sin \sqrt{x}$

6. $\sqrt{\sin x}$

7. $\cos (e^x)$

8. $e^{(\cos x)}$

9. $e^{(e^x)}$

10. $e^{ax} \sin bx$

11. $\sin (x^3 - x)$

12. $\sin (e^{ax})$

13. Suppose that there is a differentiable function f such that $e^{f(x)} = x$ for all positive x. Show that $f'(x) = 1/x$.

14. Suppose that there is a differentiable function f such that $\tan f(x) = x$ for all x. Show that $f'(x) = 1/(1 + x^2)$.

15. Suppose that f and g are differentiable functions such that $f(g(x)) = x$ for all x. Show that $g'(x) = 1/f'(u)$, where $u = g(x)$.

16. Show that

$$\frac{d}{dx} \sin (f(x)) = \cos (f(x))$$

only if $f(x)$ is of the form

$$f(x) = x + b.$$

17. From the sketches of graphs of even and odd functions decide how such functions behave under differentiation, and then prove your conjecture from the chain rule.

18. Show that $y = e^{kx}$ satisfies the differential equation

$$y'' + ay' + by = 0$$

if and only if

$$k^2 + ak + b = 0.$$

(Here $y' = dy/dx$ and $y'' = d^2y/dx^2$.)

19. Show that $y = xe^{kx}$ satisfies the above equation if and only if $a^2 = 4b$ and $k = -a/2$.

20. Show that $y = e^{mx} \sin rx$ satisfies the second-order differential equation

$$\frac{d^2y}{dx^2} - 2m\frac{dy}{dx} + (m^2 + r^2)y = 0.$$

Differentiate:

21. $y = e^{\sqrt{1 - x^2}}$

22. $s = \sqrt{1 - e^{2t}}$

23. $y = \sin (1/x)$

24. $y = \cos (\tan \theta)$

25. $y = x^2 \sin (1/x)$

26. $w = \dfrac{e^{(u^2)}}{u}$

27. $y = e^{-x^2}(x + 1)$

28. $y = \sin (\sin x)$

29. $y = \tan (\cos (x))$

30. $y = \sin (1/(x^2 + 1))$

31. $y = \sin (\cos (\sin x))$

32. $y = e^{-1/x^2}$

6. FUNCTIONS DEFINED IMPLICITLY

If n is an odd integer, or if n is even and y is positive, then the equation

$$y^n = x$$

defines y as a function of x. In this case we can solve for y,

$$y = f(x) = x^{1/n},$$

and we computed the derivative of this function in Theorem 4 of Chapter 3. However, the chain rule lets us find a formula for the derivative *without* solving for y. We just differentiate the original equation *with respect to x, treating y as a differentiable function of x.* Here this involves computing dy^n/dx, and hence involves the chain rule (general power rule). Thus, differentiating

$$y^n = x$$

with respect to x, we get

$$ny^{n-1} \frac{dy}{dx} = 1;$$

so

$$\frac{dy}{dx} = \frac{1}{ny^{n-1}} = \frac{1}{n} y^{1-n}.$$

The answer involves the *dependent* variable y. Here we know what y is in terms of x, so we can recapture our earlier formula:

$$\frac{dy}{dx} = \frac{1}{n} y^{1-n} = \frac{1}{n} (x^{1/n})^{1-n} = \frac{1}{n} x^{(1/n)-1} = ax^{a-1}.$$

If an equation determines y as a function of x but cannot be solved for y, then the above procedure is the only one available to give us a formula for the derivative dy/dx.

EXAMPLE 1   Suppose that the equation

$$e^y = x$$

determines y as a differentiable function of x. Find a formula for dy/dx.

Solution. We differentiate the equation with respect to x, treating y as a function of x. This involves the chain rule (or the exponential formula with "built-in" chain rule). We get

$$e^y \frac{dy}{dx} = 1,$$

$$\frac{dy}{dx} = \frac{1}{e^y} = \frac{1}{x}.$$

Here again (and this will always happen), the answer first appears involving the dependent variable y. When we can't solve the original equation for y in terms of x, we won't generally be able to express the derivative dy/dx entirely in terms of x. Here we were lucky.

It is possible for a function $y = f(x)$ to satisfy an equation in x and y without being uniquely determined by the equation. For example, the function

$$y = f(x) = \sqrt{1 - x^2}$$

satisfies the equation

$$x^2 + y^2 = 1.$$

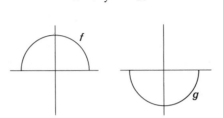

However, the function graph is only the upper half of the equation graph, and there is a second function satisfying the same equation whose graph is the lower half of the equation graph. This is, of course, the negative of the above function,

$$y = g(x) = -\sqrt{1 - x^2}.$$

Since both functions satisfy the same equation, we can't say that either one of them is uniquely determined by the equation. We can call them *function solutions* of the equation.

If we allow discontinuity, then we can find still more function solutions of $x^2 + y^2 = 1$. For example, the function h whose graph is drawn below is one. This function h is made up artificially by choosing part of the function f above and part of g. Such extraneous solution functions h can be eliminated by requiring that a solution function be continuous.

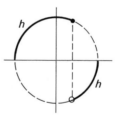

The functions defined *implicitly* by an equation are simply its solution functions. We shall consider the domain of an implicitly defined function to be an interval, although this slightly oversimplifies a rather fussy question.

The point of our discussion is that we can obtain derivative formulas for implicitly defined functions by the same procedure that we used above for an equation defining y as a function of x.

EXAMPLE   Suppose that y is a differentiable function of x defined implicitly by the equation

$$x^2 + y^2 = 1.$$

Differentiating with respect to x, we get

$$2x + 2y\frac{dy}{dx} = 0,$$

and so

$$\frac{dy}{dx} = -\frac{x}{y}.$$

Note that this is *always* correct. If $y = f(x) = \sqrt{1 - x^2}$, it gives

$$\frac{dy}{dx} = \frac{-x}{\sqrt{1 - x^2}};$$

while if $y = g(x) = -\sqrt{1 - x^2}$, it gives

$$\frac{dy}{dx} = \frac{x}{\sqrt{1 - x^2}}.$$

EXAMPLE 3 Find a formula for du/dx, supposing that u is a differentiable function of x defined implicitly by the equation

$$u^3 + ux^2 + x^3 = 1.$$

Solution. Differentiating the equation with respect to x, we have

$$3u^2\frac{du}{dx} + \left[u \cdot 2x + x^2\frac{du}{dx} \right] + 3x^2 = 0,$$

$$\frac{du}{dx} = -\frac{2xu + 3x^2}{3u^2 + x^2}.$$

If $u = h(x)$ is the function we are talking about, then this equation becomes

$$h'(x) = -\frac{2xh(x) + 3x^2}{3(h(x))^2 + x^2}.$$

Higher derivatives of implicitly defined functions can also be computed. The following example illustrates the process.

EXAMPLE 4 Compute d^2y/dx^2 when y is a differentiable function of x implicitly defined by the equation

$$\frac{x^2}{a^2} - \frac{y^2}{b^2} = 1.$$

Solution. We first compute dy/dx implicitly as in our earlier examples. We have

$$\frac{2x}{a^2} - \frac{2y\dfrac{dy}{dx}}{b^2} = 0,$$

$$\frac{dy}{dx} = \frac{b^2x}{a^2y}.$$

Differentiating this equation with respect to x gives

$$\frac{d^2y}{dx^2} = \frac{b^2}{a^2} \left[\frac{y - x\dfrac{dy}{dx}}{y^2} \right].$$

Now all we have to do is substitute the formula already obtained for dy/dx in the right side above. Thus

$$\frac{d^2y}{dx^2} = \frac{b^2}{a^2} \left[\frac{y - x\left(\dfrac{b^2 x}{a^2 y}\right)}{y^2} \right]$$

$$= \frac{b^2}{a^2} \left[\frac{a^2 y^2 - b^2 x^2}{a^2 y^3} \right]$$

$$= \frac{b^2}{a^2} \left[-\frac{a^2 b^2}{a^2 y^3} \right] = \frac{-b^4}{a^2 y^3}.$$

In order to reach the last line above we used the equation

$$a^2 y^2 - b^2 x^2 = -a^2 b^2,$$

which is another form of the equation of the hyperbola

$$\frac{x^2}{a^2} - \frac{y^2}{b^2} = 1.$$

We saw above how to calculate the derivative of a function h if we assume that h is a differentiable function defined implicitly by an equation in x and y. Such an equation can always be written in the form

$$f(x, y) = 0,$$

where f is a function of *two* variables. Then $y = h(x)$ is a function satisfying the identity

$$f(x, h(x)) \equiv 0.$$

In Chapter 16 it is shown that if $f(x, y)$ has continuous *partial* derivatives $\partial f/\partial x$ and $\partial f/\partial y$, and if $\partial f/\partial y \neq 0$ near a point (x_0, y_0) such that $f(x_0, y_0) = 0$, then *near that point* the equation $f(x, y) = 0$ has such a smooth solution function $y = h(x)$. The theorem in question is naturally called the *implicit function theorem*.

PROBLEMS FOR SECTION 6

For Problems 1 through 9 find (dy/dx) by:

    a) solving explicitly for y and then differentiating;

    b) using implicit differentiation.

Show that your two answers are equivalent.

1. $2xy + 5 = 0$ 2. $y^3 = 4x^2 + 2x + 1$

3. $x^2 + y^2 = 16$ 4. $y^2 = 2x + 1$

5. $3x^2 + 2x + y^2 = 10$ 6. $y^5 = x^3$

7. $y^2 = \dfrac{x^2 - 1}{x^2 + 1}$ 8. $x^2 + 2xy = 3y^2$

9. $x^{3/2} + y^{3/2} = 2$

In Problems 10 through 17 find (dy/dx) by implicit differentiation.

10. $2x^3 + 3y^3 + 6 = 0$ 11. $x^2y - xy^2 + x^2 + y^2 = 0$

12. $x^4y^3 - 5xy = 100$ 13. $\cos y = x$

14. $xy + x - 2y - 1 = 0$ 15. $\sin y = xy$

16. $e^y = x^2$ 17. $y^3 + y = \sin x$

18. $\tan y = x$ 19. $\sec^2 y = xy$

20. Write the equation of the line tangent to the curve $xy + y^2 - 2x = 0$ at the point $(1, 1)$.

21. Find y'' in terms of x and y when $x^2 + 4y^2 = 25$.

22. Find the equation of the tangent line to the curve $x^3 - 3xy^3 + xy^2 = xy + 6$ at the point $(1 - 1)$.

23. Find the equation of the normal line to $2x^2 - y^2 = 1$ at the point $(1, 1)$.

24. Find y' and y'' when $2xy - x^2 + 3y^2 + 6 = 0$.

7. THE DIFFERENTIAL FORMALISM

Although we have insisted that the Leibniz symbol dy/dx is not a quotient, nevertheless the chain rule

$$\frac{dy}{dt} = \frac{dy}{dx} \cdot \frac{dx}{dt}$$

looks as though it were obtained by cancelling the two occurrences of dx on the right, and this formal cancellation is a good way of remembering the rule.

A similar situation will be discovered in Chapter 8, when an integration is performed by changing variables. If $x = g(t)$ and if an integral has the form

$$\int \cdots g'(t)\, dt,$$

then we are tempted to make the formal cancellation

$$g'(t)dt = \frac{dx}{dt}\, dt = dx,$$

and so to write the integral as

$$\int \cdots dx.$$

This formal cancellation is found to be correct, again because of the chain rule.

Thus the chain rule permits us to treat dy/dx as though it were a quotient in several important situations. We could let it go at that, but you are likely to run into differentials such as dx and dy in your reading, and we should say something about them. They will be treated here in a purely formal way, as a useful symbolism. In fact, they are just going to be new variables that are related to each other by certain equations.

Suppose that we had been using only function notation up to this point. In particular, we would then have the chain rule only in the following form:

If $y = f(x)$ and $x = g(s)$, and if we set $h(s) = f(g(s))$, then

$$h'(s) = f'(g(s)) \cdot g'(s)$$
$$= f'(x) \cdot g'(s).$$

We now reintroduce the Leibniz symbols, but in a different way. Suppose there is some underlying independent variable, say t. We associate with t a new independent variable dt, called the differential of t. Like Δt, this is not a product of something d times something t, but a wholly new variable, independent of t, but to be used along with t. We can give dt any value we wish, and in practice we often set $dt = \Delta t$.

Now suppose that x is a differentiable function of t, say

$$x = g(t).$$

Then the differential dx is also a dependent variable. It is the function of the two variables t and dt defined by

$$dx = g'(t)\, dt.$$

The differential of any other dependent variable is defined in the same way. Thus if

$$y = h(t),$$

then

$$dy = h'(t)\, dt.$$

Now, let us suppose that in the above situation y also happens to be a differentiable function of x, say

$$y = f(x).$$

Since $x = g(t)$, we then have $y = f(g(t))$, so

$$h(t) = f(g(t)),$$

and

$$h'(t) = f'(x)g'(t)$$

by the chain rule. Therefore

$$dy = h'(t)dt = f'(x)g'(t)\, dt = f'(x)\, dx.$$

We thus see that the equation

$$dy = f'(x)\, dx$$

holds if $y = f(x)$, regardless of whether x is independent or both x and y depend on some other variable t. So long as $y = f(x)$, then also $dy = f'(x)\, dx$, no matter what variable ultimately turns out to be the underlying independent variable.

Moreover, if $dx \neq 0$, then we can divide by it and have

$$\frac{dy}{dx} = f'(x),$$

where now the Leibniz symbol dy/dx *is* a quotient, a quotient of two related differentials.

The Differential Rules

The derivative rules now acquire very useful differential formulations. For example, if we multiply the product rule

$$\frac{d}{dx}(uv) = u\frac{dv}{dx} + v\frac{du}{dx}$$

by dx, it becomes the differential product rule

$$d(uv) = u\,dv + v\,du.$$

This form has the advantage of being "independent of the independent variable." Moreover, it turns out to be just as valid for functions of several variables as it is for functions of one variable.

Here is our basic list in differential form:

$$d(u^a) = au^{a-1}du \quad \text{(for any rational exponent } a),$$
$$d(e^u) = e^u\,du,$$
$$d(\sin u) = \cos u\,du,$$
$$d(\cos u) = -\sin u\,du,$$
$$d(u + v) = du + dv,$$
$$d(cu) = c\,du,$$
$$d(uv) = u\,dv + v\,du,$$
$$d\left(\frac{u}{v}\right) = \frac{v\,du - u\,dv}{v^2}.$$

We recover the derivative forms of these rules by dividing through by the differential of the independent variable.

PROBLEMS FOR SECTION 7

Find the following differentials.

1. $d(x^3 + 2x + 3)$

2. $d(3x + 2)$

3. $d(x^2 + 2)(2x + 1)^3$

4. $d\left(\frac{x-3}{x+2}\right)$

5. de^{x^2}

6. $d\sqrt{1 + \sqrt{1 + x}}$

7. $d\left(\frac{x^3 + 2x + 1}{x^2 + 3}\right)$

8. $d(\cos^2 2x + \sin 3x)$

9. $d(x^{-1} + 2x + 5)$

10. $d\left(\frac{\sin x}{\cos x}\right)$

Using the method of differentials, find the derivatives of the following functions.

11. $y = 2 + 2x$

12. $y = \frac{1}{(2 + 3x)^3}$

13. $y = (x + 2)^{2/3}(x - 1)^{1/3}$ 14. $y = \dfrac{x^{2/3}}{(x + 1)^{2/3}}$

Use differentials to find the derivative dy/dx in each of the following situations, assuming in each case that the equation defines y as a differentiable function of x. Also find the derivative dx/dy, assuming that the equation defines x as a differentiable function of y.

15. $e^y = x^2$ 16. $y^3 + x = \sin xy$

17. $y^3 + xy - x^5 = 5$

18. $2x^2 + xy - y^2 + 2x - 3y + 5 = 0$

EXTRA PROBLEMS FOR CHAPTER 5

Differentiate:

1. $w = (z + 1)^{1/2}(z - 1)^{-1/2}$ 2. $y = (ax^2 + bx + c)\sqrt{2ax + b}$

3. $y = (1 - x)^{2/3}(1 + x)^{1/3}$ 4. $y = x^3(x - 1)^4(x + 2)^2$

5. Given that $f(x) = u(x) \cdot v(x) \cdot w(x)$, show that $f' = uvw' + uv'w + u'vw$.

Use the formula derived in Problem 5 to find the derivatives of the following functions:

6. $s = (2t^2 + t^{-2})(t^2 - 3)(4t + 1)$

7. $f(x) = (x - 3)(x + 1)(x + 2)$

Differentiate:

8. a) $y = \dfrac{1}{1 - x^2}$ b) $y = \dfrac{x^2}{1 - x^2}$ (Compare the answers and explain.)

9. $y = \dfrac{6x^2 - 2}{(x^2 + 1)^3}$ 10. $y = \left(\dfrac{1 - x}{1 + x}\right)^2$

11. $y = \left(\dfrac{1 - x}{1 + x}\right)^n$ 12. $y = \dfrac{1 - f(x)}{1 + f(x)}$

13. $y = \left(\dfrac{f(x)}{1 + f(x)}\right)^a$ 14. $y = \dfrac{f(x) - g(x)}{f(x) + g(x)}$

15. $y = \dfrac{1 + x}{1 - x}(3x + 2)$ 16. $y = \dfrac{x\sqrt{1 + x}}{1 + x^2}$

17. $y = \dfrac{x}{(1 + 3x)\sqrt{1 + x^2}}$

18. Prove from the product rule that if it is true that

$$(f^m)' = mf^{m-1}f',$$

then it is true that

$$(f^{m+1})' = (m + 1)f^m f'.$$

19. Verify that the above formula is true for $m = 1$. Then show how its truth for an arbitrary positive integer p can be established by repeatedly applying the result in the above problem, starting from $m = 1$. This gives a new way of approaching the general power rule.

20. Assuming that f and $\sqrt{f}$ are both differentiable, prove the formula

$$(\sqrt{f})' = \frac{1}{2\sqrt{f}} \cdot f'$$

from the product rule.

21. Assuming that f and $1/f$ are both differentiable, prove the formula

$$\left(\frac{1}{f}\right)' = -\frac{1}{f^2} \cdot f'$$

from the product rule.

22. If $f(x) = g(x)h(x)$, show that

$$f'' = g''h + 2g'h' + gh''; \qquad f''' = g'''h + 3g''h' + 3g'h'' + gh'''.$$

Find d^2y/dx^2 when:

23. $y = \sqrt{1 - x^2}$,   24. $y = 1/(1 + x^2)$.

25. Call $(m - n)$ the degree of the rational function

$$\frac{a_m x^m + a_{m-1}x^{m-1} + \cdots + a_0}{b_n x^n + b_{n-1}x^{n-1} + \cdots + b_0},$$

where a_m and b_n are both nonzero. Show that if $r(x)$ is a rational function with *nonzero* degree p, then $r'(x)$ is of degree $(p - 1)$.

26. Show by a simple example that if $r(x)$ is a rational function of degree zero, then the degree of $r'(x)$ can be any negative integer except -1.

27. Prove the quotient rule from the product rule and the general power rule $(n = -1)$.

28. Prove the formula for Δy when $y = u/v$.

29. If $y = uv$, what is the formula for $\Delta y/y$ (when u and v are given increments Δu and Δv, respectively)?

30. If $y = uvw$, find the formula for $\Delta y/y$, when u, v, and w are given the increments Δu, Δv, and Δw, respectively.

31. Formalize the proof of the product law given in the text by quoting the limit law from Section 5 of Chapter 2 that justifies each step in the limit computation. (You will also have to quote Theorem 5 of Chapter 3.)

32. Formalize the proof of the quotient law in the same way.

33. Compute $f'(a)$ if $f(x) = (x - a)g(x)$. Can this result be true if g is not differentiable at $x = a$? Discuss.

34. Show by using the product law (and not the mean-value principle) that the derivative of $(x - a)^m(x - b)^n$ is zero at a point between a and b.

35. Suppose that f is a differentiable function such that

$$tf'(t) = nf(t).$$

Show that $f(t) = ct^n$. (Compute the derivative of $f(t)/t^n$.)

36. Write out the proof of the product rule using function notation.

Find dx/dy when:

37. $y = [x/(e^x + x)]^{1/2}$ 38. $y = e^x(e^x + x)^{1/2}$

39. $y = 1/\sqrt{e^x}$ 40. $y = (e^x + x/2e^{2x})/x$.

41. If $f(x) = a^x$, where a is any positive base, prove that $f'(x) = f'(0)f(x)$. (Repeat the proof of Theorem 3.)

42. Use the above result to show that if a is any positive base different from 1, then $f'(x) = d(a^x)/dx$ is an increasing function of x.

43. We can now show, without using a calculator, that $1/2 < \alpha < 1$, where α is the value at the origin of $d(2^x)/dx$. Prove this, by first applying the mean-value principle to the chords of the graph $y = 2^x$ over the pairs of points $x = -1, 0$, and $x = 0, 1$, and then using Problem 42.

44. Prove that if $g(x) = 4^x$, then $1 < g'(0) < 2$, using the steps outlined in the above problem.

45. Assuming the result in Example 4, Section 2, in the text, find the solution of the differential equation $dy/dx = y$ that satisfies the initial condition $f(1) = 1$.

46. a) Compute

$$e^{-kx}\left(\frac{d}{dx}\, e^{kx}f(x)\right).$$

b) Assume that if $g'(x) = e^{rx}$, then $g(x) = (e^{rx}/r) + c$. Keeping this in mind, together with the result of (a), solve the equation

$$f'(x) + 3f(x) = e^{5x}.$$

47. A particle moves along the x-axis from the origin toward $x = 1$, its position at time t being given by $x = 1 - e^{-t}$. Show that its acceleration is the negative of its distance from 1.

48. A particle moves along the x-axis, its position at time t being given by $x = te^{-t}$.

    a) How far to the right does it get? (This should be its position when its velocity has fallen to zero.)

    b) Show that its position x, velocity v, and acceleration a are related by

$$a + 2v + x = 0.$$

Use the $\sin(x + y)$ and $\cos(x + y)$ laws stated in the text, together with the special values for $\sin x$ and $\cos x$ at $x = \pm\pi/2, \pi$ (these can be read off from a diagram), to prove the following laws:

49. $\sin(x + \pi/2) = \cos x$ 50. $\sin(x - \pi/2) = -\cos x$

51. $\cos(x + \pi/2) = -\sin x$ 52. $\cos(x - \pi/2) = \sin x$

53. $\sin(x + \pi) = -\sin x$ 54. $\cos(x + \pi) = -\cos x$

55. Prove the identities:

$$\sin(x + \pi/4) = \frac{\sqrt{2}}{2}(\sin x + \cos x),$$

$$\cos(x + \pi/4) = \frac{\sqrt{2}}{2}(\cos x - \sin x).$$

56. Prove the identities:

$$\sin(x + \pi/3) = \frac{1}{2}(\sin x + \sqrt{3}\cos x),$$

$$\cos(x + \pi/3) = \frac{1}{2}(\cos x - \sqrt{3}\sin x).$$

57. Show that $\tan x$ is periodic with period π. That is, prove that

$$\tan(x + \pi) = \tan x$$

for all x.

58. Show that if t is (the measure of) a positive acute angle, then $\tan t$, $\cot t$, $\sec t$, and $\csc t$ have the right-triangle characterizations

$$\tan t = \frac{\text{opposite}}{\text{adjacent}}, \qquad \cot t = \frac{\text{adjacent}}{\text{opposite}},$$

$$\sec t = \frac{\text{hypotenuse}}{\text{adjacent}}, \qquad \csc t = \frac{\text{hypotenuse}}{\text{opposite}}.$$

59. Show that if $g(x)$ and $h(x)$ are any two solutions of the differential equation

$$\frac{d^2y}{dx^2} = -y, \tag{I}$$

then so is any linear combination

$$f(x) = Ag(x) + Bh(x),$$

where A and B are constants.

60. Show that $f^2 + (f')^2$ is constant for any solution f of (I). Conclude that if ϕ is a solution such that $\phi(0) = \phi'(0) = 0$, then ϕ is identically zero.

61. Show that every solution f of (I) is of the form

$$f(x) = A \cos x + B \sin x.$$

[*Hint:* Set $\phi(x) = f(x) - f(0) \cos x - f'(0) \sin x$ and apply the above problems.]

62. Show that $e^x \sin x$ satisfies the second-order differential equation

$$\frac{d^2y}{dx^2} - 2\frac{dy}{dx} + 2y = 0.$$

63. If $y = f(x)$ is any solution of the above differential equation, show that $u = e^{-x} f(x) = e^{-x} y$ satisfies the differential equation

$$\frac{d^2u}{dx^2} + u = 0,$$

and conversely.

Using the most general solution of this equation, as determined in Problem 61, write down the most general solution of the equation in Problem 62.

64. Show that $y = x \sin x$ satisfies the fourth-order differential equation

$$\frac{d^4y}{dx^4} + 2\frac{d^2y}{dx^2} + y = 0.$$

65. a) If $f(x) = Ae^x \sin x + Be^x \cos x$, show that

$$f'(x) = (A - B)e^x \sin x + (A + B)e^x \cos x.$$

b) Without differentiating any further, write down what $f''(x)$, $f'''(x)$, $f^{(4)}(x)$, and $f^{(5)}(x)$ are.

Some more derivatives:

66. $\dfrac{\sin x + \cos x}{\sin x - \cos x}$

67. $\sqrt{\dfrac{1 - \cos x}{1 + \cos x}}$

68. $xe^x \sin x$

69. $\sqrt{1 - \cos^2 x}$

70. $\dfrac{\tan x}{1 + \tan^2 x}$

71. $\dfrac{1 - \sin 2x}{1 + \sin 2x}$

72. $(1/5)\sin^5 x - (2/3)\sin^3 x + \sin x$ (Simplify the answer.)

73. $(1/3) \tan^3 x - \tan x + x$ (Simplify the answer.)

74. $(1/5) \tan^5 x + (2/3) \tan^3 x + \tan x$ (Simplify.)

75. We saw in the text that $0 < \sin t < t$ when t is positive. Use this inequality to prove that $\cos t \to 1$ as $t \to 0$. (There are a couple of ways to start this. For one thing, $\sin t = \sqrt{1 - \cos^2 t}$.)

76. Show that the unit circle about the origin lies entirely below the graph of $\cos x$ except for the point of tangency of the two graphs at $x = 0$.

77. Use the identity for $\sin A - \sin B$ (Problem 51 in Section 3) to give another proof that the derivative of $\sin x$ is $\cos x$.

78. The six trigonometric functions can be grouped into three pairs: $(\sin x, \cos x)$, $(\tan x, \cot x)$, $(\sec x, \csc x)$. In a given pair, each function is called the cofunction of the other. Thus, $\cot x$ is the cofunction of $\tan x$, but also $\tan x$ is the cofunction of $\cot x$: $\tan x = \mathrm{co}(\cot x)$. This relationship is carried out to combinations of functions. For example, $\sin x + \cot x$ is the cofunction of $\cos x + \tan x$, and $\csc x \cot x$ is the cofunction of $\sec x \tan x$:

$$\csc x \cot x = \mathrm{co}(\sec x \tan x).$$

   With the above remarks in mind, show that the derivative formulas of the six trigonometric functions satisfy the following rule:

$$\frac{d}{dx} \mathrm{co} \, (f(x)) = -\mathrm{co}\!\left(\frac{d}{dx} f(x)\right).$$

The hyperbolic sine of x ($\sinh x$) and the hyperbolic cosine of x ($\cosh x$) are defined as follows:

$$\sinh x = \frac{e^x - e^{-x}}{2}, \qquad \cosh x = \frac{e^x + e^{-x}}{2}.$$

The remaining hyperbolic functions are defined from $\sinh x$ and $\cosh x$ in the standard way. Show that the hyperbolic functions have the following properties.

79. $\dfrac{d}{dx} \sinh x = \cosh x$

80. $\dfrac{d}{dx} \cosh x = \sinh x$

81. $\dfrac{d}{dx} \tanh x = \mathrm{sech}^2 x$

82. $\dfrac{d}{dx} \mathrm{sech}\, x = -\mathrm{sech}\, x \tanh x$

83. $\cosh(-x) = \cosh x$

84. $\sinh(-x) = -\sinh x$

85. $\sinh(x + y) = \sinh x \cosh y + \cosh x \sinh y$.

86. $\cosh(x + y) = \cosh x \cosh y + \sinh x \sinh y$.

Differentiate:

87. $y = xe^{1/x}$

88. $x = a \sin b\theta + b \sin a\theta$

89. $y = \sqrt{a + \dfrac{b}{x}}$

90. $y = \sqrt{\sin x} - \sqrt{\cos x}$

91. $s = \sqrt{1 + \sin^2(\sqrt{t})}$

92. $y = e^{a\sqrt{x}}$

93. $y = \dfrac{1 + \sin(x^2)}{1 + \cos(x^2)}$

94. $y = e^{|x|}$

95. $y = f(x^2)$

96. $y = f(e^x)$

97. $y = e^{2f(x)}$

98. $y = f(x + g(x))$

99. $y = f(e^{g(x)})$

100. $y = f([g(x)]^n)$

101. If $f(x) = x^2 \sin(1/x)$ when $x \neq 0$ and $f(0) = 0$, then f is everywhere differentiable. (See Problem 37 in the Extra Problems for Chapter 3.) Prove that $f'(x)$ is not continuous at $x = 0$.

102. Prove that a function $y = f(x)$ satisfies the differential equation

$$\frac{d^2y}{dx^2} + b^2 y = 0,$$

where b is a nonzero constant, if and only if $f(x)$ is of the form

$$y = A \sin bx + B \cos bx$$

for some constants A and B. (Show that $g(x) = f(x/b)$ satisfies $g'' + g = 0$, and apply Problem 61.)

103. Prove by implicit differentiation that the tangent line to the ellipse

$$\frac{x^2}{a^2} + \frac{y^2}{b^2} = 1$$

at the point (x_0, y_0) has the equation

$$\frac{xx_0}{a^2} + \frac{yy_0}{b^2} = 1.$$

104. Find the equation for the tangent line to the hyperbola

$$\frac{x^2}{a^2} - \frac{y^2}{b^2} = 1$$

at the point (x_0, y_0) in a form analogous to the one given above for the ellipse.

In the following problems compute dy/dx and the second derivative d^2y/dx^2 by implicit differentiation.

105. $y = e^{(x+y)}$

106. $y^3 + y = x$

107. $\cos x = \sin y$

6

Applications of
the Derivative

1. ELEMENTARY GRAPH SKETCHING

If we draw some random graphs of differentiable functions and a few tangent lines to them, we are led to certain conclusions about the relationship between the behavior of f' and the behavior of f.

First, it seems clear that a graph with positive slope is rising and that a graph with negative slope is falling.

Second, it seems clear that a graph can change from sloping up to sloping down only by going over a summit point where the slope is zero. Similarly, it can change from sloping down to sloping up only by going through a bottom point where the slope is zero.

Combining these two observations we conclude that if the slope is never zero then the graph cannot change directions, and is either everywhere rising or everywhere falling. A function whose graph behaves this way is said to be *monotone*. Thus f is monotone if f is either everywhere increasing or everywhere decreasing. Our combined conclusion is therefore:

If f is a differentiable function on an interval I and if f' is never zero on I, then f is monotone on I.

This "shape principle" tells us that the values of x for which $f'(x) = 0$ are the critical values in determining the overall behavior of f, since $f(x)$ moves in a constant direction on each of the intervals marked off by such values. This idea enables us to find the general overall shape of a graph with a minimum of calculation.

EXAMPLE 1   Consider $f(x) = (x^3 - 3x + 2)/3$. The critical points of f are the values of x for which $f'(x) = 0$. Since

$$f'(x) = \frac{3x^2 - 3}{3} = x^2 - 1,$$

the critical points of f are the roots of $x^2 - 1 = 0$, i.e., the two points $x = -1, +1$.

We make a little table showing the values of f at the critical points and we add the behavior of f at $\pm\infty$. We have

$$f(-1) = \frac{-1 + 3 + 2}{3} = \frac{4}{3} \quad \text{and} \quad f(1) = \frac{1 - 3 + 2}{3} = 0.$$

Also, since $f(x)$ behaves like its highest power $x^3/3$ for large x, we see that $f(x) \to -\infty$ as $x \to -\infty$, and $f(x) \to +\infty$ as $x \to +\infty$. Our table is thus

x	$-\infty$	-1	1	$+\infty$
$f(x)$	$-\infty$	$4/3$	0	$+\infty$

We now apply the shape principle. Since $f'(x)$ is never zero in the interval $(-1, 1)$, the graph of f must be either always rising over $(-1, 1)$ or always falling over $(-1, 1)$. We see from our table that it must be falling since $f(x)$ goes from 4/3 to 0 as x runs from -1 to $+1$. In the same way, we see from the shape principle and the table that the graph is rising over the whole interval $(-\infty, -1)$, and also over $(1, \infty)$. We can add the y-intercept by calculating $f(0) = 2/3$. Schematically, the graph therefore has the features shown below. (The scales are slightly different on the two axes.)

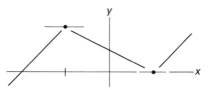

Now consider how we would actually sketch the graph from this information. There aren't going to be any sharp corners; the graph is going to be as smooth and simple as possible subject to the above conditions. It changes from rising to falling as it passes over $x = -1$, and so is smoothly turning downward there. (In Section 2 we shall consider what this really means.) Similarly, it is smoothly turning upward as it passes over $x = +1$. Its overall shape must therefore be something like this.

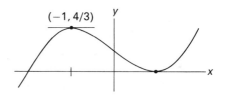

Note that the lefthand x-intercept is shown at $x = -2$. This is easily checked to be correct, since $f(-2) = 0$, but finding it in the first place requires knowing how to factor a cubic (or making a lucky guess). Normally, we will not know the x-intercepts, because locating the roots of $f(x) = 0$ can be very difficult. As a matter of fact, we sometimes graph a function to help us guess approximately where the roots are, as a preliminary step to a calculation procedure for the roots.

EXAMPLE 2   Sketch the graph of $f(x) = x^4 - 2x^2$ in the above way.

Solution. We have $f'(x) = 4x^3 - 4x = 4x(x^2 - 1)$, giving the critical points $x = -1, 0, +1$. When $|x|$ is very large, $f(x)$ behaves like x^4 and so approaches $+\infty$ as x approaches $-\infty$ or $+\infty$. We thus have the following determining table.

x	$-\infty$	-1	0	$+1$	∞
$f(x)$	∞	-1	0	-1	∞

         decr.   incr.   decr.   incr.

In this case also we can find the x-intercepts, i.e., the solution of $x^4 - 2x^2 = 0$, which are $-\sqrt{2}$, 0, $\sqrt{2}$. Plotting the critical points and intercepts, we get the graph below.

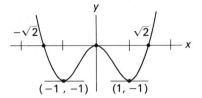

Another method for discovering which way $f(x)$ is going on one of its intervals of monotonicity is to compute the value of $f'(x)$ *at one point* in the interval, because that will tell us the constant sign of f' over the whole interval and hence whether f is increasing or decreasing there.

Consider the above example from this point of view.

We need the values of f at its critical points in order to draw the graph, and they already determine which way f is moving on $(-1, 0)$ and on $(0, +1)$. But instead of finding how f behaves at $-\infty$ we can calculate

$$f'(-2) = 4x(x^2 - 1)|_{x=-2} = -8 \cdot 3 = -24.$$

Since this is negative, $f'(x)$ is negative on the whole interval $(-\infty, -1)$ and the graph of f is falling there. Similarly, the evaluation $f'(2) = 24$ shows that f must be increasing over $(1, \infty)$.

The shape principle probably seems just as obvious geometrically as the mean-value principle. Logically, it is a consequence of the mean-value principle and properties of continuous functions. Here is the first step.

THEOREM 1   *If f' exists and is everywhere positive on an interval I, then f is increasing on I.*

Proof. By the mean-value principle and the hypothesis that f' is everywhere positive,

$$\frac{f(x_2) - f(x_1)}{x_2 - x_1} = f'(X) > 0$$

for any two distinct points x_1 and x_2 in I. Thus $f(x_2) - f(x_1)$ is positive whenever $x_2 - x_1$ is positive, so f is increasing on I. ∎

Similarly, f is decreasing on I if f' is everywhere negative on I. We thus have the first of the two observations that were combined in the shape principle.

In practice it will always be the case that if f' exists on an interval I then f' is continuous there. Then f' cannot change sign without going through the value 0, by the intermediate-value principle, and the shape principle follows as before. But there do exist differentiable functions having discontinuous derivatives, and for them we have to work harder to show that f' cannot change sign without going through the value 0. See Chapter 12, Section 1.

Some authors define a function f to be *increasing at the point* x_0 if $f'(x_0) > 0$. However, this does *not* imply that f is increasing on *any* interval about x_0, because of the same possibility: f' can be discontinuous at x_0 and assume negative values arbitrarily close to x_0. (See Problem 4, Section 1, Chapter 12.) The following result is all that can be proved from the single hypothesis that $f'(x_0) > 0$.

THEOREM 2   If $f'(x_0) > 0$, then $f(x) - f(x_0)$ changes sign from $-$ to $+$ as x crosses the point x_0. That is, there is a small interval about x_0 on which we have

$$f(x) < f(x_0) \qquad \text{if } x < x_0,$$
$$f(x) > f(x_0) \qquad \text{if } x > x_0.$$

(If $f'(x_0) < 0$, the f inequalities are interchanged.)

Proof. Since $f'(x_0)$ is the limit of the difference quotient $(f(x) - f(x_0))/(x - x_0)$ as $x \to x_0$, it follows that if $f'(x_0)$ is positive then so is the difference quotient for all x sufficiently close to x_0. (See the limit law **L6** in Chapter 2, Section 5.) And since a quotient is positive exactly when its numerator and denominator have the same sign, we conclude that

$$f(x) - f(x_0) \qquad \text{and} \qquad x - x_0$$

have the same sign when x is close to x_0. ∎

There remains one fine point that should be mentioned. If we wish, we can always throw in the endpoints of an interval of monotonicity, because if f is continuous on $[a, b]$ and increasing on (a, b) then it is automatically increasing on $[a, b]$ (see Theorem 5 in Chapter 2, Section 7). For example, in Example 1 we concluded from the shape principle that f was decreasing on the open interval $(-1, 1)$. So f was in fact decreasing on $[-1, 1]$, even though $f' = 0$ at the two endpoints.

PROBLEMS FOR SECTION 1

Use calculus to sketch the graphs of the following equations:

1. $y = x^2 - x$

2. $y = 2 - x - x^2$

3. $y = x^2 - 4x + 4$

4. $y = x^2 - 2x + 2$

5. $s = t^3 - t$

6. $y = x^4 - 3x^2 + 2$

7. $u = 3v^4 + 4v^3$

8. $y = x^3 - 4x^2 - 3x$

9. $y = x^3 - 3x^2 + 3x$

10. $y = x^3 - 4x^2 + 3x$

11. $y = x + \dfrac{1}{x}$.   (Be careful around the origin.)

12. $y = x - \sqrt{x}$.   (Same warning.)

13. $3y = 3x^5 - 10x^3 + 15x + 3$

14. $y = 3x^4 - 8x^3 + 6x^2 + 1$

15. Discuss the graph of

$$y = ax^2 + bx + c$$

in light of the shape principle.

16. Sketch the graph of a function f defined for positive x and having the properties

$$f(1) = 0,$$
$$f'(x) = 1/x, \qquad \text{all } x > 0.$$

2. CONCAVITY

One of the most distinctive features of a graph is the direction in which it is turning or curving. The first two patches of graphs shown below are *concave up* (turning upward) and the second two are *concave down* (turning downward). Note that the graph of a decreasing function can be turning in *either* direction (first and third patches), as can the graph of an increasing function (second and fourth patches).

Knowing the concavity of a graph contributes more to getting a correct overall picture than anything else except locating the critical points. Knowing how the graph is curving is really what tells us its shape. For example, if we know that the three points shown below lie on a graph that is concave down then we know how the graph must look.

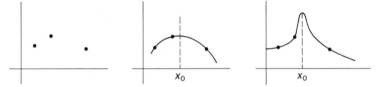

But if we don't know that the graph is concave down then it may look entirely different, as at the right. Roughly speaking, *constant* concavity rules out oscillatory, or wiggly, behavior.

Most graphs will be concave up over some intervals and concave down over other intervals. Points across which the direction of concavity changes are called *inflection points*. In the following figure the graph on page 201 is relabelled for concavity.

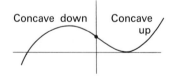

The only point of inflection is at $x = 0$, the curve being concave down over $(-\infty, 0]$ and concave up over $[0, \infty)$.

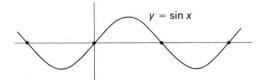

The graph of sin x has a point of inflection at every multiple of π. It is concave down whenever it lies above the x-axis and concave up whenever it lies below.

Geometrically, it appears that a graph is *concave up* if its tangent line rotates *counterclockwise* and its slope *increases* as the point of tangency runs from left to right along the curve. You should be able to "see" the slope increasing. If you visualize a succession of tangent lines as in the figure below, then you see that each slope is greater than the one before. Or you can visualize holding a straightedge against the bottom of the curve and "rolling" it counterclockwise along the curve, in which case the tangent segments above are successive positions of the rolling line.

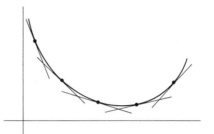

We make this property our definition.

DEFINITION *A function f is concave up over the interval I if f' exists and is an increasing function on I. Similarly, f is concave down on I if f' is decreasing on I.*

In view of the above definition, the shape principles of Section 1 relate the *concavity* of f to its *second* derivative f'', as follows.

1. If f'' exists and is never zero on an interval I, then f is of constant concavity on I (since then f' is monotone on I). The intervals of constant concavity of f (the intervals of monotonicity of f') are thus marked off by the values of x where $f''(x) = 0$, i.e., by the critical points of f'.

Of course, if we happen to know the sign of f'', then we know which concavity we have:

2. If $f''(x) > 0$ on an interval I, then f is concave up on I (since then f' is increasing on I), and if $f''(x) < 0$, then f is concave down on I.

EXAMPLE 1 Since $d^2 \sin x/dx^2 = -\sin x$, it follows that the graph of sin x is concave down whenever it is above the x-axis and concave up whenever it is below. We have been drawing the graph this way, but until now we couldn't be absolutely certain that there wasn't some extra wiggle we were overlooking.

EXAMPLE 2 Let us sketch the graph of

$$f(x) = x^4 - 2x^3 = x^3(x - 2),$$

using the concavity shape principle. We have

$$f'(x) = 4x^3 - 6x^2 = 2x^2(2x - 3),$$
$$f''(x) = 12x^2 - 12x = 12x(x - 1).$$

The intervals of constant concavity for f are marked off by the values of x for which $f''(x) = 0$ (the critical points of f') as we noted above, and we see that these values are $x = 0$ and $x = 1$. The intervals of constant concavity are thus $(-\infty, 0]$, $[0, 1]$, and $[1, \infty)$. In order to determine which way the graph of f is concave over these intervals (i.e., whether f' is increasing or decreasing), we compute the values of f' at the ends of the intervals and the behavior of f' at infinity, and get the table

x	$-\infty$	0	1	∞
$f'(x)$	$-\infty$	0	-2	∞

Thus f is concave up over $(-\infty, 0]$. The reason, again, is that we already know that f' is monotone on this interval because it has no critical points between $-\infty$ and 0, and now we see that it is increasing, since its values run from $-\infty$ to 0. Similarly, f is concave down on $[0, 1]$, because f' runs from 0 to -2 and hence is decreasing there. Finally f is concave up over $[1, \infty)$.

We also need the critical points of f, and find that $f'(x) = 0$ at $x = 0$ and $x = 3/2$. In Section 1 we then went on to determine whether f is increasing or decreasing on each of the intervals these two points determine. However, this is unnecessary when we know the concavity of f. For example, a curve that is concave up must be increasing just after a horizontal tangent.

We compute the values of f at all these special points and end up with the following combined table.

Concavity	Up		Down	Up	
x	$-\infty$	0	1	3/2	$+\infty$
$f(x)$	$+\infty$	0	-1	$-27/16$	$+\infty$
$f'(x)$	$-\infty$	0	-2	0	$+\infty$
$f''(x)$		0	0		

These data determine the shape and position of the graph.

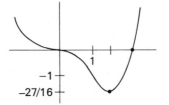

EXAMPLE 3 We shall sketch the graph of

$$f(x) = \frac{1}{1 + x^2}.$$

The first two derivatives are

$$f'(x) = \frac{-2x}{(1 + x^2)^2},$$

$$f''(x) = \frac{(1 + x^2)^2(-2) - (-2x)2(1 + x^2)(2x)}{(1 + x^2)^4}$$

$$= \frac{-2(1 + x^2) + 8x^2}{(1 + x^2)^3}$$

$$= \frac{2(3x^2 - 1)}{(1 + x^2)^3}.$$

This time we go directly to the combined table. Remember, we want the critical points of both f and f' (the zeros of both f' and f'') and their values at infinity. You may be able to do the arithmetic mentally, or you may want to check it with pencil and paper.

Concavity	Up		Down		Up
x	$-\infty$	$-1/\sqrt{3}$	0	$1/\sqrt{3}$	$+\infty$
$f(x)$	0	$3/4$	1	$3/4$	0
$f'(x)$	0	$9/8\sqrt{3}$	0	$-9/8\sqrt{3}$	0
$f''(x)$		0		0	

The direction of concavity depends on whether f' is increasing or decreasing, and this is determined by comparing the values of f' at the endpoints of its critical intervals.

In plotting the above values we use the approximation $1/\sqrt{3} \approx 3/5$. As x tends to infinity, the graph approaches the x-axis. Two curves related this way appear to be "tangent at infinity." We say they are *asymptotic*.

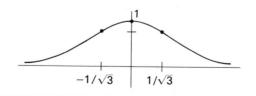

$-1/\sqrt{3}$ $1/\sqrt{3}$

The Notation of Concavity

Probably you would agree that a concave upward graph has all four of the properties suggested by the following figures.

These properties are:

1. *The slope $f'(x)$ of the tangent line is an increasing function of x $(f'(x_1) < f'(x_2)$ if $x_1 < x_2)$.*

2. *The graph of f lies strictly above each tangent (except for the point of tangency).*

3. *Each arc of the graph of f lies strictly below its chord (except for the two endpoints).*

4. *The second of two nonoverlapping chords always has the larger slope.*

But then a graph can't be considered to be concave up unless it has all four of these properties. That is, the intuitive notion of upward concavity is equivalent to the sum total of these properties. Fortunately, at least for those who like simple definitions, there is a theorem to the effect that these four properties are all logically equivalent to each other. So if the graph of a differentiable function has any one of these properties, then of necessity it has them all. It is by virtue of this theorem that the definition of concavity given earlier is adequate.

PROBLEMS FOR SECTION 2

1. Show in the manner of Example 1 that we have been drawing the graph of $y = e^x$ with the right general shape.

Determine the intervals of constant concavity and the critical points for the graph of each of the following equations. Then sketch the graph.

2. $x^2 - 4 = y$ 3. $3y = 8x^3 - 6x + 1$

4. $y = 8x^3 - 6x^2 + 1$ 5. $y = x^3 + x$

6. $y = \sin 3x$ 7. $y = x^3 - 3x + 2$

8. $y = x^2 - 2x + 3$ 9. $y = x^3 - (21/4)x^2 + 9x - 4$

10. $y = x^4 - 2x^2$ 11. $9y = (x + 2)(x - 2)^3$

12. $y = (x - 1)(x + 1)^3$ 13. $y = (1/3)x^3 - 4x^2 + 12x - 8$

14. $y = xe^{1-x}$, over the interval $[0, \infty)$ (Use the approximate value $e = 2.7$.)

15. $y = x - x^3$, over the interval $[0, 1]$ (Be sure to have the right slopes at the endpoints.)

16. $y = x/2 - \sin x$, over the interval $[0, 2\pi]$

17. $y = 2 \sin x + \sin 2x$, over $[0, 2\pi]$

     a) Do this first by sketching the graphs of $\sin 2x$ and $2 \sin x$ and visually adding them.

     b) Then apply the methods of this section.

3. SINGULAR POINTS

According to Theorem 5 of Chapter 3, a differentiable function is necessarily continuous. This theorem does not have a converse, however. A continuous function can fail to be differentiable, and this can happen at isolated points even for functions that are otherwise very smooth. For example, if $f(x) = x^{1/3}$, then

$$f'(x) = \frac{1}{3}x^{-2/3} = \frac{1}{3x^{2/3}}$$

except at $x = 0$. Since $f'(x) \to \infty$ as $x \to 0$, we visualize the tangent line rotating into a vertical position as the point of tangency approaches the origin. We therefore expect that at the origin the graph has a vertical tangent. In fact, at $x = 0$, we have the secant slope

$$\frac{f(r) - f(0)}{r - 0} = \frac{r^{1/3} - 0}{r - 0} = r^{-2/3},$$

and therefore

$$\lim_{r \to 0} \frac{f(r) - f(0)}{r - 0} = \lim_{r \to 0} r^{-2/3} = +\infty,$$

as predicted. However, $f(x) = x^{1/3}$ is continuous *everywhere*. Here are the graphs:

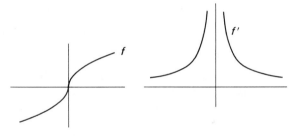

   The functions $f(x) = x^{2/3}$ is somewhat similar, except that here the derivative

$$f'(x) = \frac{2}{3}x^{-1/3} = \frac{2}{3x^{1/3}}$$

approaches $-\infty$ as x approaches 0 from the left ($x \to 0^-$), while $f'(x) \to +\infty$ as $x \to 0^+$. The graph of $f(x)$ has a *cusp* at the origin.

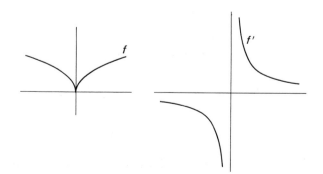

   Another way in which a continuous function can fail to be differentiable is shown by $f(x) = |x|$. Again we have an everywhere continuous function, with graph as shown below on the left. When $x < 0$, we have $f(x) = -x$ and

$f'(x) = -1$. If $x > 0$, then $f(x) = x$ and $f'(x) = 1$. The graph of f' is thus as shown on the right.

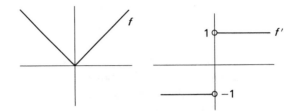

At $x = 0$ we can obviously expect trouble. The secant slope (difference quotient) there is

$$\frac{f(r) - f(0)}{r - 0} = \frac{|r|}{r} \begin{cases} = \dfrac{-r}{r} = -1, & \text{if } r < 0. \\[2mm] = \dfrac{r}{r} = +1, & \text{if } r > 0. \end{cases}$$

The difference quotient thus has the *one-sided* limits -1 on the left and $+1$ on the right, and we could say that f has *one-sided derivatives*

$$f'(0^-) = -1, \qquad f'(0^+) = +1.$$

But $f'(0)$, the two-sided limit, does not exist.

A more interesting function with the same behavior at $x = 0$ is $f(x) = |x| + x^2$. Its graph looks like this:

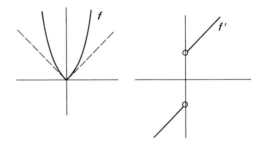

Each of the above functions has an *isolated singularity* at the origin. The function

$$f(x) = x^{2/3} + (x - 1)^{2/3} - 1$$

has isolated singularities at $x = 0$ and $x = 1$.

In the example above f was always defined and continuous at a singular point. This is not essential to the idea. For example, the function $f(x) = 1/x$ is considered to have an isolated singularity at the origin, a point where f is not defined. Thus a function f may or may not be defined at a singularity; the sole criterion is that f' fails to exist there.

DEFINITION   *A function f has an isolated singularity at x_0 if $f'(x_0)$ fails to exist but f and f' are defined at all other points of an interval about x_0.*

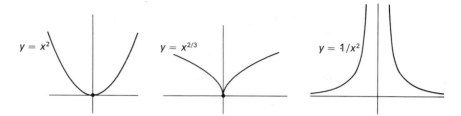

When we compare functions like $f(x) = x^2$, $g(x) = x^{2/3}$, and $h(x) = 1/x^2$, graphed above, we see that a singularity can be just as critical for the behavior of a graph as a point where $f'(x) = 0$. We therefore must list singularities along with critical points as points across which the behavior of f may change.

EXAMPLE   The function

$$f(x) = \frac{1}{x^2 - 1}$$

has the added complication of infinite singularities at ± 1. Now we have to determine whether $f(x)$ approaches $+\infty$ or $-\infty$ as x approaches one of these points from the left and from the right. But again concavity considerations save us from calculating these facts, as we shall see in a minute. We have

$$f'(x) = \frac{-2x}{(x^2 - 1)^2},$$

$$f''(x) = \frac{(x^2 - 1)^2(-2) - (-2x)2(x^2 - 1)2x}{(x^2 - 1)^4}$$

$$= \frac{-2(x^2 - 1) + 8x^2}{(x^2 - 1)^3}$$

$$= \frac{2(3x^2 + 1)}{(x^2 - 1)^3}.$$

Notice that f' has *no* critical points, but has singularities at $x = \pm 1$. The intervals of constant concavity are thus $(-\infty, -1)$, $(-1, 1)$, $(1, \infty)$. This time we shall determine the direction of concavity from the sign of f'' at one point in each of these intervals. Then we have the following table, including points where the function is not defined (nd):

Concavity		Up			Down		Up	
x	$-\infty$	-2	-1	0	$+1$	$+2$	$+\infty$
$f(x)$	0		nd	-1	nd		0
$f'(x)$			nd	0	nd		
$f''(x)$		$+$	nd	$-$	nd	$+$	

We now know that the graph is concave down over $(-1, 1)$ and that it blows up at each endpoint. It therefore must approach $-\infty$ at both endpoints. Similarly it is concave up over $(-\infty, -1)$ and blows up at -1, and so must approach $+\infty$ there. We thus have the graph shown below. The graph is asymptotic to the x-axis and to the two vertical lines at $x = 1$ and $x = -1$. Note how these three asymptotes combine with the known concavity to govern the overall shape of the graph.

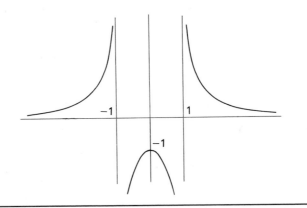

PROBLEMS FOR SECTION 3

Sketch the graphs of the following functions, using the procedures of Section 1. In each case the new feature will be one or more points where the graph has a vertical tangent, or two different one-sided tangents.

1. $f(x) = x^{1/3}$

2. $f(x) = x^{2/3}$

3. $f(x) = x^{2/3}\left(\dfrac{5}{2} - x\right)$

4. $f(x) = x^{1/3} + x^{2/3}$

5. $f(x) = 1/(x^2 - x)$

6. $f(x) = 1/(x^3 - x^2)$

7. $f(x) = \sin|x|$

8. $f(x) = (x - 1)^{1/3} + (x + 1)^{1/3}$

9. $f(x) = (x - 1)^{2/3} - (x + 1)^{2/3}$

10. $y(x - 2) = 1$

11. $x^2 y - 2x^2 - 16y = 0$

12. $xy - y - x - 2 = 0$

13. $x^2 y - x^2 + 4y = 0$

14. $x^2 y - x - 4y = 0$

15. $x^2 y + y - 4x = 0$

16. Define a function f such that $f'(x) = |x|$, and draw its graph. This is a continuous function with a continuous derivative, for which f'' is singular at the origin.

17. Let f be any function at all satisfying the inequality

$$x \le f(x) \le x + x^2.$$

Prove that $f'(0)$ exists and has the value 1.

4. FINDING A MAXIMUM OR MINIMUM VALUE

We turn now to a simple but important application of the derivative to the problem of finding the maximum (or minimum) value that a varying quantity can have, and "where" it occurs. Historically, this was one of the very first successes of calculus.

Suppose, for example, that a rancher has one mile of fence that he can use to fence off a rectangular grazing area along the bank of a straight river. He doesn't need to fence the bank of the river itself, so his fence will run along only three sides of the rectangle. His problem is to choose the dimensions of the rectangle so as to maximize its area. Note that if he chose a *square*, then each of the 3 fenced sides would be 1/3 of a mile long and the area would be 1/9 square mile; but he can do better than that.

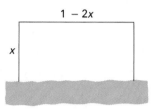

If x is the fence length *perpendicular* to the river, then the length available to run parallel to the river is $1 - 2x$ and the area enclosed is

$$A = x(1 - 2x) = x - 2x^2.$$

The physical limitation on x is that $0 \leq x \leq 1/2$. The graph of A as a function of x over this interval must look something like this figure,

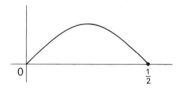

because $A = 0$ when $x = 0$ or when $x = 1/2$, and $A > 0$ in between. It seems clear that A will have its maximum value at a point where the tangent line is horizontal, i.e., where $dA/dx = 0$. Assuming that this is correct, we have

$$\frac{dA}{dx} = 1 - 4x,$$

and $dA/dx = 0$ when $x = 1/4$. The maximum area is thus

$$A = x - 2x^2 \Big|_{x=1/4} = \frac{1}{4} - \frac{1}{8} = \frac{1}{8} \, \text{sq. mi.}$$

Before considering more examples, we shall show that the simple procedure used above is correct.

Suppose that a quantity to be maximized is called y and that we have managed to express y as a continuous function of some other variable x, $y = f(x)$, where the conditions of the problem restrict x to an interval I.

The following diagrams illustrate the possible ways in which $f(x)$ can achieve a maximum value over an interval I:

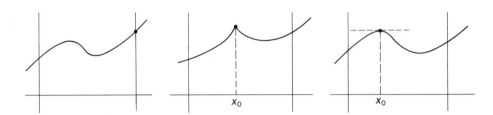

1. At an endpoint of the interval; or
2. At an interior point where the derivative is not defined; or
3. At an interior point x_0 where the derivative exists.

In the third case it probably seems geometrically clear that $f'(x_0)$ must be 0, and this is easily verified analytically.

THEOREM 3   *If a function f assumes a maximum (or minimum) value at an interior point x_0 in its domain, and if $f'(x_0)$ exists, then $f'(x_0) = 0$.*

Proof.   Theorem 2 of Section 1 shows that if $f'(x_0)$ is *not* zero, then $f(x_0)$ is *not* a maximum value. So if $f(x_0)$ is the maximum value, then $f'(x_0)$ must be zero. ∎

The possibilities noted above suggest extending the notion of a critical point of a continuous function to include singularities.

DEFINITION   *A **critical point** of a continuous function f is an interior point x in its domain at which either $f'(x) = 0$ or $f'(x)$ is not defined.*

A singularity at which f itself is undefined is not called a critical point; we want a critical point to correspond to a point on the graph of f. Thus $f(x) = 1/x$ has an isolated singularity at the origin, but this is not called a critical point of f.

In view of this definition and the remark at the end of Section 1, the shape principle can be reformulated as follows:

SHAPE PRINCIPLE   *If f is continuous on an interval I and without critical points interior to I then f is monotone on I.*

We saw above that if f has a maximum value on an interval I, then it can occur only at a critical point or at an endpoint of I. To find the maximum value for f, we should therefore calculate the values of f at all such points and choose the largest of them. This might be impossible if there were infinitely many critical points, and there also might be trouble if f were not defined at an endpoint. However, supposing that the number of critical points is finite, and that the domain of f is a closed interval, then the shape principle implies that the above procedure does give the maximum value of f. The following figure shows how this goes.

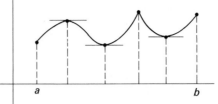

Here the maximum value of f on the interval $[a, b]$ is the largest of the six key values shown. Any other value $f(x)$ lies *between* two adjacent values in this list, because f is monotone between any two adjacent key points.

THEOREM 4   *Let f be a continuous function on a closed interval $[a, b]$ and suppose that f has only a finite number of critical points, say n. Let M be the maximum of the $(n + 2)$ values of f at the critical points and at the two endpoints. Then M is the maximum value of f on the whole interval $[a, b]$. Similarly, the minimum of these $(n + 2)$ values is the minimum value of f on $[a, b]$.*

Proof. The n critical points divide $[a, b]$ up into $(n + 1)$ nonoverlapping closed subintervals. If I is one of these subintervals, then f has no critical point interior to I, and hence f is monotone on I, by the shape principle. This means that the values of f at the two endpoints of I are the maximum and minimum values of f on I. Therefore, the largest and smallest of the values of f at all $(n + 2)$ of these endpoints are the maximum and minimum values of f on $[a, b]$. ∎

If there is only *one* critical point x_0 in I, then it may not be necessary to check the endpoint values at all. Suppose we know that f' changes sign from plus to minus as x crosses the single critical point x_0. Then we know that f is increasing just before x_0 and therefore on the whole of I up to x_0, and that f is decreasing just after x_0 and therefore on the whole of I after x_0. Therefore we know that $f(x_0)$ is the maximum value of f on I. Similarly, $f(x_0)$ must be the minimum value of f if f' changes sign from $-$ to $+$ as x crosses x_0. This reasoning doesn't even require that the endpoints of I belong to the domain of f, and I might perfectly well extend to infinity. The situation is particularly simple if $f''(x_0)$ exists and is not zero.

THEOREM 5   *Suppose that f is a differentiable function with exactly one critical point x_0 on an interval I. Then $f(x_0)$ is the maximum value of f on I if $f''(x_0)$ exists and is negative, and $f(x_0)$ is the minimum value of f if $f''(x_0)$ is positive.*

Proof. Suppose first that $f''(x_0) > 0$. Thus the function f' is zero at x_0 and its derivative is positive there. Then Theorem 2 in Section 1 says exactly that $f'(x)$ changes sign from $-$ to $+$ as x crosses x_0. Therefore $f(x_0)$ must be the minimum value of f on I, as we saw above. Similarly, if $f''(x_0) < 0$, then f' changes from $+$ to $-$ as x crosses x_0 and $f(x_0)$ is the maximum value. ∎

EXAMPLE 1   Find the maximum and minimum values of $f(x) = x^4 - 4x$ on the interval $[0, 2]$.

Solution. The derivative $f'(x) = 4x^3 - 4 = 4(x^3 - 1)$ is zero when $x^3 - 1 = 0$, i.e., when

$$x^3 = 1, \quad \text{or} \quad x = 1.$$

Thus $x = 1$ is the only critical point of f. Since the second derivative $f''(x) = 12x^2$ is positive at $x = 1$, Theorem 5 says that $f(1) = -3$ is the absolute minimum value of f over its whole domain. In particular, this is the minimum value of f on the interval $[0, 2]$.

Theorem 4 says that the maximum value of f on $[0, 2]$ is either this critical point value or one of the two endpoint values, $f(0) = 0$ and $f(2) = 16 - 8 = 8$. So the maximum value is the right-hand endpoint value 8.

EXAMPLE 2   Find the maximum and minimum values of $f(x) = x^3 - x^2$ on the interval $[1, 4]$.

Solution. Since

$$f'(x) = 3x^2 - 2x = x(3x - 2),$$

the critical points of f are $x = 0, \frac{2}{3}$. Neither of these critical points belongs to the interval $[1, 4]$, so the maximum and minimum values of f on this interval are the two endpoint values,

$$f(1) = 0 \quad \text{and} \quad f(4) = 48,$$

by Theorem 4.

EXAMPLE 3   Apply Theorem 5 to $f(x) = x + (1/x)$ on the interval $(0, \infty)$.

Solution. Since $f'(x) = 1 - (1/x^2)$, the only critical point of f in the interval $(0, \infty)$ is at $x = 1$. At this point $f''(x) = 2/x^3$ is positive. Therefore $f(1) = 2$ is the minimum value of f, by Theorem 5.

PROBLEMS FOR SECTION 4

Use Theorem 4 to find the maximum and minimum values of each of the following functions over the specified interval. (There may be none.)

1. $f(x) = 3 + x - x^2$; $[0, 2]$
2. $f(x) = x^3 - x^2 - x + 2$; $[0, 2]$
3. $F(x) = x^3 5x^2 - 8x + 20$; $[-1, 5]$
4. $g(x) = x\sqrt{x + 3}$; $[-3, 3]$
5. $h(x) = x^{2/3}(x - 5)$; $[-1, 1]$
6. $G(x) = \sqrt[3]{x^2 - 2x}$; $(0, 2)$
7. $y = x\sqrt{2x - x^2}$; $[0, 2]$
8. $y = x^{3/2}(x - 8)^{-1/2}$; $[10, 16]$
9. $f(x) = 2x^4 + x$; $[-1, 1]$
10. $f(x) = \sin x + \cos x$; $[0, \pi]$
11. $f(x) = x^2 e^{-x}$; $[0, 4]$
12. $f(x) = xe^x$; $[-2, 0]$
13. $f(x) = \sin x + \cos^2 x$; $[0, 2\pi]$

Use Theorem 5 to find the maximum or minimum value of $f(x)$ in each of the following situations.

14. $f(x) = 2x^2 + x$
15. $f(x) = 3 + x - x^2$
16. $f(x) = x^3 - 3x$, over $[-1, 2]$.
17. $f(x) = x^3 - x^2 - x + 2$, over $[-1, 0]$.

18. $f(x) = x\sqrt{x + 3}$, over $[-3, 3]$.   19. $f(x) = 2x^4 + x$

20. $f(x) = \sin x + \cos x$, over $[0, \pi]$.   21. $f(x) = x^2 e^{-x}$, over $[0, 4]$.

22. $f(x) = 2 \sec x + \tan x$, over $(-\pi/2, \pi/2)$.

5. MAXIMUM–MINIMUM PROBLEMS

EXAMPLE 1   We continue with our coal producer from Examples 4 and 5 in Section 9 of Chapter 3. Suppose that his revenue function is

$$R = 9x - 2x^2,$$

and that his cost function is

$$C = x^3 - 3x^2 + 4x + 1.$$

Then

$$P = R - C = -x^3 + x^2 + 5x - 1$$

is his profit, and, being selfish, he proposes to operate so as to maximize his profit. Since

$$\frac{dP}{dx} = -3x^2 + 2x + 5 = -(3x - 5)(x + 1),$$

and since his production x is always positive, we see that the only critical point for his profit occurs at $x = 5/3$. The corresponding value of his profit works out to be $P = 148/27 \approx 5\frac{1}{2}$. It seems clear that this must be his maximum profit, and this is easily checked. For one thing, $d^2P/dx^2 = -6x + 2$ is negative at $x = 5/3$, so Theorem 5 guarantees that this is the maximum point. In the units we chose before, his maximum profit will be \$5500 per week (he is probably a corporation) and this will be achieved by producing 1667 tons of coal per week.

EXAMPLE 2   In general, any entrepreneur will experience losses when his production is too low, because of fixed costs, and also when his production is too high, because of very high marginal costs. Unless he can operate profitably at some in-between production he won't be in business at all, so we can suppose his profit curve looks like this:

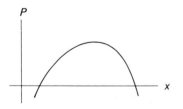

Then his maximum profit will always occur at a critical point, i.e., supposing P to be a differentiable function of x, it will occur at a point where $dP/dx = 0$.

Now $P = R - C$, so

$$0 = \frac{dP}{dx} = \frac{dR}{dx} - \frac{dC}{dx}$$

if and only if $dR/dx = dC/dx$. That is,

Profit will be maximum when the marginal revenue equals the marginal cost.

EXAMPLE 3   It will generally happen, in setting up a maximum problem, that the variable to be maximized (or minimized) is most naturally expressed in terms of more than one independent variable; but then the conditions of the problem will allow all but one of these variables to be eliminated, as the present example will show.

Consider the problem of finding the most efficient shape for a one-cubic-foot cylindrical container, the most efficient cylinder being the one using the least material, i.e., having the *smallest total surface area*. Now the volume of a cylinder is $V = \pi r^2 h$, and its total surface area is $A = 2\pi r^2 + 2\pi rh$. The problem, therefore, is so to determine r and h that A has the smallest possible value subject to the requirement that $V = 1$. The restriction

$$\pi r^2 h = 1$$

allows us to solve for either h or r as a function of the other, say

$$h = \frac{1}{\pi r^2};$$

and when this is substituted in the area formula, the area becomes a function of r alone,

$$A = 2\pi r^2 + \frac{2}{r}.$$

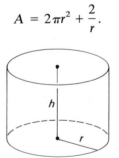

The physical limitation on r is that it be positive. To find the critical points we compute dA/dr and set it equal to 0,

$$\frac{dA}{dr} = 4\pi r - \frac{2}{r^2} = 0,$$

which gives $4\pi r^3 = 2$. Thus $r = (1/2\pi)^{1/3}$ is the only critical point. Since

$$\frac{d^2 A}{dr^2} = 4\pi + \frac{4}{r^3} > 0,$$

we know from Theorem 5 that A has its minimum value at the critical point.

Since we asked for the *shape* of the most efficient container, we want h also, and we compute

$$h = \frac{1}{\pi r^2} = \frac{(2\pi)^{2/3}}{\pi} = \frac{2}{(2\pi)^{1/3}} = 2r.$$

The most efficient shape is therefore that for which the altitude h equals the diameter $2r$ of the base. The can has a square cross section along its vertical axis.

Here are some general suggestions for tackling problems of maxima and minima. Unfortunately, they do not guarantee success, but it is difficult to imagine getting anywhere without them. (That is, they are *necessary* but not *sufficient* conditions for success.)

1. In a problem that involves relationships among geometric magnitudes, draw a reasonably accurate figure. Be sure to show the *general* configuration. For instance, if the problem involves a general triangle, don't draw one that looks isosceles, because that might lead you astray in seeing what is going on.
2. Label the figure with appropriate constants and variables, being sure that you know what is being held fixed and what is varying.
3. Derive from known geometric relationships and formulas the equations connecting the variables involved in the problem.
4. If Q is the quantity to be maximized or minimized, then these equations should determine Q as a function of *one* of the variables. Then apply Theorem 4 or Theorem 5.

Sometimes an angle can be taken as the independent variable, as in the following example.

EXAMPLE 4 A corridor of width a meets a corridor of width b at right angles. Workmen wish to push a heavy beam of length c on dollies around the corner, but they want to be sure it will be able to make the turn before starting. How long a beam will go around the corner (neglecting the width of the beam)?

Solution. This is essentially the problem of minimizing the length l of the segment cut off by the corridor on a varying line through the corner point P. The beam will go around the corner if its length c is less than the minimum of l.

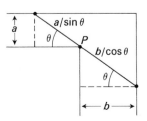

In terms of the angle θ shown in the figure, the length l is given by

$$l = \frac{a}{\sin \theta} + \frac{b}{\cos \theta}.$$

The domain of θ is $(0, \pi/2)$, and l approaches $+\infty$ as θ approaches 0 or $\pi/2$. The minimum of l will therefore occur at a point where $dl/d\theta = 0$. We see that

$$\frac{dl}{d\theta} = -\frac{a \cos \theta}{\sin^2 \theta} + \frac{b \sin \theta}{\cos^2 \theta}$$

$$= \frac{b \sin^3 \theta - a \cos^3 \theta}{\sin^2 \theta \cos^2 \theta},$$

so $dl/d\theta = 0$ if and only if

$$b \sin^3 \theta - a \cos^3 \theta = 0 \qquad \text{or} \qquad \tan \theta = \left(\frac{a}{b}\right)^{1/3}.$$

The corresponding value of l then works out to be

$$l = (a^{2/3} + b^{2/3})^{3/2}.$$

The beam will go around the corner if its length is not greater than this value of l.

REMARK   Strictly speaking, this example requires that Theorem 4 be modified so as to apply to a function f defined on an interval that is missing one or both endpoints. We replace any such missing endpoint value of f by the limit of $f(x)$ as x approaches that end of the interval, and take the maximum of this modified collection of numbers. If this maximum is an actual value of f, then we have found the maximum value of f. But if the maximum of the modified collection occurs only as the limit of $f(x)$ at a missing endpoint, then f has no maximum value on the interval. Three such situations are pictured below.

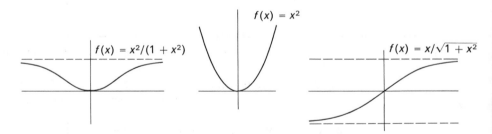

In the first example, $\lim_{x\to\infty} f(x) = 1 = \lim_{x\to-\infty} f(x)$, and this is larger than the value 0 at the only critical point, $x = 0$; f has no maximum value. In the second case, $\lim_{x\to\infty} f(x) = \infty = \lim_{x\to-\infty} f(x)$, which is larger than the value of f at the only critical point 0. Again, no maximum. In the third case, there are *no* critical points *or* endpoint values, and hence no maximum or minimum value.

Going back to Example 3, we had

$$A = 2\pi r^2 + \frac{2}{r}$$

on the domain $(0, \infty)$, and we see that $A \to \infty$ as $r \to 0$ and as $r \to \infty$. Thus, the minimum of A must be at a critical point, by the modified Theorem 4; the same is true for Example 4.

The Method of Auxiliary Variables

It was noted earlier that the variable to be maximized or minimized may be most naturally expressed as a function of two or more other variables, but that the conditions of the problem would always allow the elimination of all these variables but one. We now treat this situation in another way.

EXAMPLE 5   In Example 3 we wanted to minimize

$$A = 2\pi r^2 + 2\pi rh = 2\pi[r^2 + rh]$$

subject to the condition that
$$V = \pi r^2 h = 1.$$

This restriction determines either h or r as a function of the other, and what we did before was to solve for h, substitute in the first equation, and go on.

The new procedure will be to treat h as a function of r that is determined by the second equation, but leave both h and r in both equations. We differentiate the two equations with respect to r, keeping in mind that h is a function of r:

$$\frac{dA}{dr} = 2\pi\left[2r + r\frac{dh}{dr} + h\right]$$

$$\frac{dV}{dr} = \pi\left[r^2\frac{dh}{dr} + 2rh\right] = 0.$$

Now solve the second equation for dh/dr and substitute in the first. We get $dh/dr = -2h/r$ and

$$\frac{dA}{dr} = 2\pi\left[2r + r\left(\frac{-2h}{r}\right) + h\right]$$

$$= 2\pi[2r - h].$$

The critical-point equation $dA/dr = 0$ now appears as $2r - h = 0$, or

$$h = 2r.$$

This example typifies what one can expect from the new approach. There are two things to notice. First, the answer appears as a relationship among the supporting variables, and for some problems this is really all we want. The above problem, for instance, asked for the most efficient shape and this is exactly what the answer $h = 2r$ gives us. If we want to know what the minimum area is, and for what value of r it occurs, then we have to substitute $h = 2r$ in the equation $\pi r^2 h = 1$, and solve for r. The second point about this method is that it really only gives us the critical point configuration, and does not (at least not without further work) let us conclude with assurance that this configuration is where the maximum (or minimum) value occurs.

EXAMPLE 6   Find the rectangle of largest area (with sides parallel to the axes) that can be inscribed in the ellipse

$$\frac{x^2}{a^2} + \frac{y^2}{b^2} = 1.$$

Solution. If (x, y) is the rectangle vertex lying in the first quadrant, then the area of the rectangle is

$$A = 4xy,$$

where

$$\frac{x^2}{a^2} + \frac{y^2}{b^2} = 1.$$

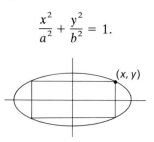

The second equation determines y as a function of x, and we can therefore differentiate both equations with respect to x:

$$\frac{dA}{dx} = 4x\frac{dy}{dx} + 4y,$$

$$\frac{2x}{a^2} + \frac{2y\,dy/dx}{b^2} = 0.$$

We can then solve for dy/dx from the second equation and substitute in the first, obtaining

$$\frac{dy}{dx} = -\frac{b^2 x}{a^2 y},$$

$$\frac{dA}{dx} = 4\left(-\frac{b^2 x^2}{a^2 y} + y\right).$$

The critical-point equation $dA/dx = 0$ is thus

$$0 = \frac{-b^2 x^2 + a^2 y^2}{a^2 y},$$

from which we conclude that

$$\frac{y}{x} = \frac{b}{a}$$

at the critical point. That is, the critical rectangle has sides proportional to the axes of the ellipse. In order to find the rectangular area, we have to substitute $y = bx/a$ in the equation of the ellipse and solve for x and y. We find that

$$x = \frac{a}{\sqrt{2}}, \qquad y = \frac{b}{\sqrt{2}},$$

so the critical (maximum) area is

$$A = 4xy = 2ab.$$

PROBLEMS FOR SECTION 5

1. In a certain physical situation, it is found that a variable quantity Q can be expressed in the form $Q = x + y$, where x and y are related by

$$x^2 + y^2 = 1.$$

   Find the maximum value of Q.

2. In a certain situation it is found that a variable quantity Q that is to be maximized is expressed in terms of variable quantities x and y by $Q = xy^2$, where x and y are related by

$$x^2 + y^2 = 1.$$

   Find the maximum value of Q.

3. Find the maximum value of $Q = xy$, if x and y are related by the condition

$$2x^2 + y^2 = 1.$$

4. A quantity Q is given by $Q = x^3 + 2y^3$, where x and y are positive variables related by the equation

$$x + y = 1.$$

   Show that the minimum value of Q is $6 - 4\sqrt{2}$, and that this is greater than $1/3$.

5. Find the point on the graph $y = x^3 - 6x^2 - 3x$ at which the tangent line has minimum slope.

6. Find the maximum value of $x - 2y$ on the unit circle $x^2 + y^2 = 1$.

7. Find the minimum value of $3x + y^3$ on the circle $x^2 + y^2 = 2$.

8. Find the rectangle of maximum area that can be inscribed in the circle

$$x^2 + y^2 = 1.$$

9. Find the circular cylinder of maximum volume that can be inscribed in the cone of altitude H and base radius R.

10. Find the circular cylinder of maximum volume that can be inscribed in a sphere of radius r.

11. A rectangular garden is to be laid out with one side adjoining a neighbor's lot, and is to contain 48 square yards. If the neighbor pays for half the dividing fence, what dimensions of the garden will minimize your cost for the fence?

12. The post office places a limit of 120 in. on the combined length and girth of a package. What are the dimensions of a rectangular box with square cross section, that will contain the largest mailable volume?

13. Find the proportions for a rectangle of given area A that will minimize the distance from one corner to the midpoint of a nonadjacent side.

14. You are the owner of an 80-unit motel. When the daily charge for a unit is $20, all units are occupied. If the daily charge is increased by d dollars then $3d$ of the units become vacant. Each occupied unit requires $3 daily for service and repairs. What should be your daily charge to realize the most profit?

15. A wire 24 in. long is cut in two, and then one part is bent into the shape of a circle and the other into the shape of a square. How should it be cut if the sum of the areas of the circle and the square is to be a *minimum*?

16. Find the positive number for which the sum of its reciprocal and four times its square is the smallest possible.

17. Find the shortest distance from the point $(0, 2)$ to the hyperbola $x^2 - y^2 = 1$.

18. Find the rectangle of maximum perimeter than can be inscribed in the ellipse $4x^2 + 9y^2 = 36$. The sides of the rectangle are parallel to the axes of the ellipse.

19. The strength of a wooden beam of rectangular cross section is proportional to the width of the beam and the square of its depth. Determine the proportions of the strongest beam that can be cut from a circular log.

20. A small loan company is limited by law to an 18% interest charge on any loan. The amount of money available for loans is proportional to the interest rate the company will pay investors. If the company can loan out all the money that is invested with it, what interest rate should it pay its investors in order to maximize profits?

6. THE CLASSIFICATION OF ISOLATED CRITICAL POINTS

If x_0 is an isolated critical point of a continuous function f, and if the interval $I = [u, v]$ has been chosen small enough about x_0 so that it contains no other critical point of f, then f is strictly monotone on each half interval $[u, x_0]$ and $[x_0, v]$. We therefore have only the four possibilities shown schematically below.

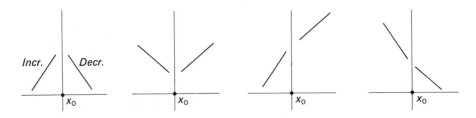

If f is increasing on $[u, x_0]$ and decreasing on $[x_0, v]$ then $f(x_0)$ is the maximum value of f on the interval $[u, v]$. This may not be the absolute maximum value of f because f may very well have larger values outside of $[u, v]$. We therefore say that f has a *local maximum*, or a *relative maximum* at x_0.

Similarly, if f is decreasing on $[u, x_0]$ and increasing on $[x_0, v]$, then x_0 is a *relative minimum* point for f.

If f has neither a local maximum nor a local minimum at the isolated critical point x_0, then f is monotone in the interval $[u, v]$ about x_0. That is, f is then monotone "in the neighborhood of x_0." This can happen whether $f'(x_0)$ is zero or is undefined, as in the following figures.

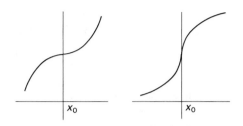

The derivative f' presumably has only isolated critical points (or singular points) also, and therefore we can further restrict the size of the interval $[u, v]$ about x_0 so that f' is monotone in each half interval $[u, x_0]$ and $[x_0, v]$. Now these are also intervals of *constant concavity* for f.

If $f'(x_0) = 0$, then the four possibilities for concavity before and after x_0 imply the same fourfold classification for the behavior of f in the neighborhood of x_0 that we found earlier, except that the ordering of the cases is different. For example, if the concavity changes from up to down at x_0, then x_0 is a horizontal point of inflection, and this is the earlier case where f is decreasing in the neighborhood of the isolated critical point x_0. Or, if $f'(x_0) = 0$ and f is concave down in an interval about x_0, then f has a relative maximum at x_0.

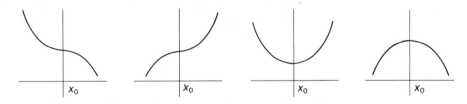

If x_0 is a singular point ($f'(x_0)$ is not defined), then there are many more possibilities. For example, each of the four concavity possibilities can now occur around a relative maximum point.

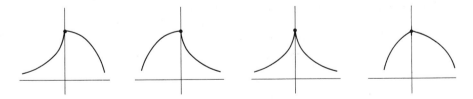

PROBLEMS FOR SECTION 6

1. By applying Theorem 5 to $f'(x)$, prove the following *second derivative test for a relative minimum:*

   If $f'(x_0) = 0$ and $f''(x_0) > 0$, then f has a relative minimum at x_0.

2. Similarly, prove that if $f'(x_0) = 0$ and $f''(x_0) < 0$, then f has a relative maximum at x_0.

3. Show that if $f'(x_0) = f''(x_0) = 0$ but $f'''(x_0) > 0$, then f has a horizontal point of inflection at x_0 and is increasing on an interval about x_0. (This time apply Theorem 5 to $f''(x)$, or apply Problem 1 to $f'(x)$.)

4. Show that if $f'(x_0) = f''(x_0) = f'''(x_0) = 0$ but $f''''(x_0) > 0$, then f has a relative minimum at x_0. (Apply Theorem 5 to $f'''(x)$ and work backward, or apply Problem 3 to $f'(x)$.)

5. Reread Problems 1, 3, and 4. Now state and prove the next in this chain of results.

6. Draw figures to show that if x_0 is a singular point, then the four concavity possibilities on the two sides of x_0 are all consistent with f having a relative minimum at x_0.

Find and classify all critical points of the following functions, using the higher derivative tests from Problems 1 through 4.

7. $y = x^3 - 3x^2$

8. $y = x^3 - x^2 - x + 2$

9. $y = (x - 1)^2(x + 1)$

10. $y = x^4 - 2x^3 + 3$

11. $y = 2x^3 - 24x$

12. $y = 3x^4 - 4x^3 - 6x^2 + 4$

13. $y = x^3(x + 3)^2$

14. $y = x^4 - 4x^3 + 16x$

15. $y = x^6 - 2x^3 + 1$

16. $y = (x^3 - 1)^2$

7. RELATED RATES

Suppose that y is a function of x, say $y = f(x)$, and that x varies with time t. Because y depends on x, it will also vary with time. That is, if y is a function of x and x is a function of t, then y is a function of t. The chain rule says that

$$\frac{dy}{dt} = f'(x)\frac{dx}{dt}.$$

Thus the rate of change of y is related to the rate of change of x because of the relationship between y and x.

EXAMPLE 1 Suppose that a circular ripple is spreading out over a pond, and that the radius r is increasing at the rate of 2 feet per second at the moment when r goes through the value $r = 5$ feet. How fast is the disturbed area increasing at that moment?

Solution. The disturbed area A and the radius r are varying with time, but at all times they are related by the equation

$$A = \pi r^2.$$

We take the time t to be the underlying independent variable, and differentiate this identity with respect to t:

$$\frac{dA}{dt} = (2\pi r)\frac{dr}{dt}.$$

At the moment in question we are given $dr/dt = 2$ and $r = 5$. Therefore, at that moment

$$\frac{dA}{dt} = 2\pi5 \cdot 2 = 20\pi,$$

or approximately 63 square feet per second.

EXAMPLE 2   A tank in the shape of an inverted cone with equilateral cross section is being filled with water at the constant rate of 10 cubic feet per minute. How fast is the water rising when its depth is 5 feet?

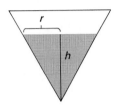

Solution. The amount of water in the cone is related to its depth h and its surface radius r by the formula for the volume of a cone:

$$V = \frac{1}{3}\pi r^2 h.$$

Here V, r, and h are all increasing with time as water pours in, but are always related by the above equation. We are given $dV/dt = 10$, and we want to know the value of dh/dt when $h = 5$. Our general principle is to *relate, by an equation that holds throughout the variation, the quantity whose rate is wanted to the quantity whose rate is given, and then differentiate with respect to t.* The above volume equation is such an equation, but it isn't quite right because it contains the extra variable r. However, we can eliminate r by using the fact that the tank cross section is equilateral: We have

$$h^2 + r^2 = 4r^2$$

by the Pythagorean theorem, so

$$h^2 = 3r^2.$$

Therefore

$$V = \frac{1}{3}\pi\frac{h^2}{3} \cdot h = \frac{\pi}{9}h^3,$$

$$\frac{dV}{dt} = \frac{\pi}{9} \cdot 3h^2 \cdot \frac{dh}{dt}.$$

We are given $dV/dt = 10$, and we want dh/dt when $h = 5$. At that moment

$$10 = \frac{\pi 25}{3}\frac{dh}{dt},$$

and

$$\frac{dh}{dt} = \frac{6}{5\pi},$$

or approximately 2/5 ft/min.

PROBLEMS FOR SECTION 7

1. Two variable quantities Q and R are found to be related by the equation
$$Q^3 + R^3 = 9.$$
What is the rate of change dQ/dt at the moment when $Q = 2$, if $dR/dt = 3$ at that moment?

2. Two variable quantities x and y are found to be related by the equation
$$x^2 + y^2 = 25.$$
If x increases at the constant rate of 3 units per unit of time, what is the rate at which y is increasing at the moment when $x = 0$? When $x = 3$? When $x = 5$?

3. A ladder 26 ft long leans against a vertical wall. If the lower end is being moved away from the wall at the rate of 5 ft/sec, how fast is the top descending when the lower end is 10 ft from the wall?

4. A conical funnel is 8 in. across the top and 12 in. deep. A liquid is flowing in at a rate of 60 cu. in./sec and flowing out at a rate of 40 cu. in/sec. Find how fast the surface is rising when the liquid is 6 in. deep.

5. A rope 32 ft long is attached to a weight and passed over a pulley 16 feet above the ground. The other end of the rope is pulled away along the ground at the rate of 3 ft/sec. At what rate does the weight rise at the instant when the end being pulled is 12 ft from its initial point?

6. A man 6 ft tall walks away from a street light 15 ft high at the rate of 3 mi/hr.

   a) How fast is the far end of his shadow moving when he is 30 ft from the light pole?

   b) How fast is his shadow lengthening?

7. Two cars start from the same point at the same time. One travels north at 25 mi/hr, and the other travels east at 60 mi/hr. How fast is the distance between them increasing as they start out? At the end of an hour?

8. A revolving beacon located 3 miles from a straight shore line makes 2 revolutions/min. Find the speed of the spot of light along the shore when it is two miles away from the point on the shore nearest the light.

9. A drawbridge with two 10 ft spans is being raised at a rate of 2 radians per minute. How fast is the distance increasing between the ends of the spans when they both are at an elevation of $\pi/4$ radians?

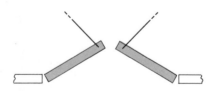

10. A bridge is 30 ft above a canal. A motor boat traveling at 10 ft/sec passes under the center of the bridge at the same instant that a man walking 5 ft/sec reaches that point. How rapidly are they separating 3 sec later?

11. If, in Problem 10, the man reaches the center of the bridge 5 sec before the boat passes under it, find the rate at which the distance between them is changing 4 sec later.

12. A conical reservoir 8 ft across and 4 ft deep is filled at a rate of 2 cu. ft/sec. A leak allows fluid to run out at a rate depending on the depth of the water, the loss being at the rate of $h^4/8$ cubic feet per minute when its depth is h feet. At what depth will the surface level stabilize?

13. A hemispherical bowl of diameter 18 inches is being filled with water. If the depth of the water is increasing at the rate of 1/8 in/sec when it is 8 in. deep, how fast is the water flowing in? (The volume of a segment of a sphere of radius r is

$$\pi h^2 \left(r - \frac{h}{3} \right),$$

where h is the height of the segment.)

14. A light is at the top of a tower 80 ft high. A ball is dropped from the same height from a point 20 ft from the light. Assuming that the ball falls according to the law $s = 16t^2$, how fast is the shadow of the ball moving along the ground two seconds after release?

8. GROWTH AND DECAY

We saw in Chapter 5, Section 3, Example 4, that "up to a multiplicative constant," $y = e^x$ is the unique function satisfying the differential equation

$$\frac{dy}{dx} = y.$$

That is, if f is a function such that

$$f' = f,$$

then $f(x) = ce^x$ for some constant c.

If we carry through the same argument, using e^{kx} instead of e^x, we find that:

THEOREM 6   *A function $y = f(x)$ satisfies the equation*

$$\frac{dy}{dx} = ky$$

if and only if $y = ce^{kx}$ for some constant c.

The equation

$$\frac{dy}{dt} = ky$$

is the basic law of population growth. Neglecting special inhibiting and stimulating factors, a population normally reproduces itself at a rate proportional to its size, and this is exactly what the above equation says. Bacteria colonies grow this way as long as they have normal environment, and so does a bank balance with a fixed rate of continuously compounded interest. Theorem 6 shows that a population normally grows exponentially, and if you will look back at the shape of an exponential graph (page 157) you will see what "population explosion" is all about.

The constant k is sometimes called the *rate of exponential growth*. This is not the rate of change of the population size, which is

$$\frac{dy}{dt} = ky,$$

but the constant that y must be multiplied by to get its rate of change. It is thus a different use of the word *rate*. It is like the *interest rate* paid by a bank. If the interest rate is 0.05, we do not mean that your bank balance y is growing at the rate of 0.05 dollars per year but at the rate of $0.05y$ dollar per year. We therefore express the rate as 5% per year, rather than 0.05 dollars per year. We could say that the rate is 0.05 dollars *per dollar* per year. (Where the interest is compounded continuously, the interest rate is a true exponential growth rate.)

When a quantity Q is growing exponentially,

$$Q = ce^{kt},$$

the time T it takes to double depends only on the exponential growth rate k. Conversely, the growth rate k can be found from the doubling time T.

In working out a problem concerning exponential growth we often end up with an equation of the form $e^x = c$ that has to be solved for x. These solutions are given in the table on page 792. The number x such that $e^x = a$ is called the natural logarithm of a and is designated

$$x = \log a.$$

The calculus properties of natural logarithms are taken up in the next chapter.

EXAMPLE Suppose that a culture of bacteria doubles its size from 0.15 grams to 0.3 grams during the seven-hour period from $t = 15$ hrs to $t = 22$ hrs. What is its exponential growth rate k?

Solution. Assuming normal exponential growth

$$y = ce^{kt},$$

our data are

$$0.15 = ce^{k \cdot 15},$$
$$0.30 = ce^{k \cdot 22}.$$

Dividing the second equation by the first gives us

$$2 = e^{k(22-15)} = e^{k \cdot 7}.$$

Thus,

$$k \cdot 7 = \log 2,$$
$$k = \frac{\log 2}{7}.$$

The table on page 792 gives $\log 2 = 0.6931$ to four decimal places. So the growth rate k is approximately 1.

In general the exponential growth rate k and the doubling time T are related by

$$kT = \log 2 \approx 0.6931,$$

so that either determines the other. The proof is like the example above.

If the exponential growth rate is *negative*, say $-k$ where k is positive, then the equation $dy/dt = -ky$ shows y to be *decreasing* as a function of time, and the solution $y = ce^{-kt}$ shows it to be decreasing exponentially. This is called exponential *decay*. A radioactive element behaves this way; it disintegrates at a rate proportional to the amount present. In this context, we have the *half-life* of the element instead of the doubling time. This is the time it takes for $y = ce^{-kt}$ to decrease to half of an original value. That is, if

$$y_0 = ce^{-kt_0},$$

$$\frac{y_0}{2} = ce^{-kt_1},$$

then $T = t_1 - t_0$ is the half-life. Again we find that

$$kT = \log 2,$$

so that the half-life T depends only on the decay rate k. In particular, it is independent of the initial population size y_0 and the initial time t_0.

If we know the exponential growth (or decay) rate k, then the remaining constant c is determined by a single initial condition (t_0, y_0), as before.

PROBLEMS FOR SECTION 8

1. A colony of bacteria has an exponential growth rate of 1% per hour. What is its doubling time?

2. A culture of bacteria is found to increase by 41% in 12 hours. What is its exponential growth rate per hour? (Note that $1.41 \approx \sqrt{2}$.)

3. An exponential growth rate of 100% per day corresponds to what growth rate per hour?

4. A bank advertises that it compounds interest continuously and that it will double your money in 10 years. What is its annual interest rate?

5. A certain radioactive material has a half-life of 1 year. When will it be 99% gone? ($\log 10 \approx 2.3$)

6. A quantity Q_1 grows exponentially with a doubling time of one week. Q_2 grows exponentially with a doubling time of three weeks. If the initial amounts of Q_1 and Q_2 are the same, when will Q_1 be twice the size of Q_2?

7. There is nothing sacred about doubling time. Any two measurements of an exponentially growing population $y = ce^{kt}$ will determine both parameters k and c. Show that if y has the values y_0 and y_1 at times t_0 and t_1, then

$$k = \frac{\log(y_1/y_0)}{t_1 - t_0}.$$

8. In a chemical reaction a substance A decomposes at a rate proportional to the amount of A present. It is found that 8 pounds of A will reduce to 4 pounds in 3 hours. At what time will there be only 1 pound left?

9. In a chemical reaction a substance B decomposes at a rate proportional to the amount of B present It is found that 8 pounds of B will reduce to 7 pounds in one hour. Make a rough calculation to show that more than four hours are needed for 8 pounds of B to reduce to 4 pounds.

10. The temperature T of a cooling body drops at a rate that is proportional to the difference $T - C$, where C is the constant temperature of the surrounding medium. Use Theorem 6 to show that

$$T = ae^{-kt} + C.$$

11. The temperature of a cup of freshly poured coffee is 200° and the room temperature is 70°. The coffee cools 10° in 5 minutes. How much longer will it take to cool 20° more? (Use the above formula.)

12. If a cold object is placed in a warm medium whose temperature is held constant at C°, the object warms up according to the same general principle. State a general law of temperature change that includes both the warming and cooling situations.

9. SIMPLE HARMONIC MOTION

When a weight suspended at the end of a coiled spring is disturbed, it bobs up and down with a definite frequency. The motion dies out as time passes, due to air resistance and internal friction in the spring. Presumably, if the motion were to occur in a perfect vacuum, and if the spring were perfectly elastic, then the oscillatory motion would continue undiminished forever.

The motion described above is a simple example of *elastic vibration*, and a lot can be learned about the nature of this universal phenomenon by studying such a simple mass–spring prototype. At the moment we do not have the technique to handle friction, but the ideal frictionless model is still very instructive and useful.

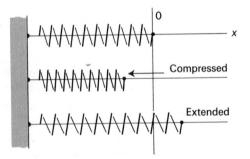

Consider, then, a coiled spring lying along a frictionless x-axis, with one end held fixed and the free end at the origin. When we compress the spring it resists and pushes back. When we extend the spring it resists and pulls back. That is, the spring always exerts a "restoring force" when it is deformed. Moreover, we find after careful measurement that the restoring force F is exactly proportional to the amount of deformation x (within the so-called elastic limits). Since F is directed against the deformation, it is given by

$$F = -sx,$$

where s is the constant of proportionality. This is Hooke's Law. If we use a stronger spring, s will be larger, and we call s the "*stiffness* of the spring."

If we now attach a mass m to the end of the spring and disturb the system somehow, then our ideal oscillatory motion will occur. In order to study this motion mathematically, we need to know how a mass m moves when it is acted on by a force F. We find that m is *accelerated* by F, and that its acceleration d^2x/dt^2 is proportional to F:

$$F = k\frac{d^2x}{dt^2}.$$

Moreover, the constant of proportionality k is a measure of the mass m: if we double the mass, then it requires double the force to produce the same acceleration. This means that we can write the force law as

$$F = ma = m\frac{d^2x}{dt^2}$$

(when suitable units are used for all these quantities). This is Newton's second law of motion.

The laws of Newton and Hooke combine to show that the motion of the model mass–spring system satisfies the differential equation

$$m\frac{d^2x}{dt^2} = F = -sx, \qquad \text{or} \qquad \frac{d^2x}{dt^2} + \frac{s}{m}x = 0.$$

We already know the general solution of this equation from Section 4 in Chapter 5. It is

$$x = A \sin \omega t + B \cos \omega t,$$

where $\omega = \sqrt{s/m}$, and A and B are constants of integration. Without knowing A and B, we don't know the motion exactly, but we do know its frequency: since sine and cosine are periodic with period 2π, it follows that x is a periodic function of t with period

$$T = \frac{2\pi}{\omega} = 2\pi \sqrt{\frac{m}{s}}.$$

The frequency of the motion is the number of cycles (periods) per unit time:

$$f = \frac{1}{T} = \frac{1}{2\pi} \sqrt{\frac{s}{m}}.$$

These equations show exactly how the period and frequency depend on the physical constants of the elastic system. The frequency varies as the square root of the stiffness of the spring, and is *inversely proportional* to the square root of the mass.

A particular motion is generally determined by known "initial conditions." We shall neglect mentioning units in the examples below.

EXAMPLE 1 A particle of mass $m = 4$ is attached to a spring of stiffness $s = 2$ in the configuration discussed above. Starting at the equilibrium position, the particle is struck and given an initial velocity $v_0 = -3$. What is the motion?

Solution. Here $\omega = \sqrt{s/m} = 1/\sqrt{2}$, and the initial conditions are $x = 0$ and $dx/dt = -3$ when $t = 0$. Substituting these values in the equations

$$x = A \sin \omega t + b \cos \omega t,$$

$$\frac{dx}{dt} = \omega A \cos \omega t - \omega B \sin \omega t,$$

we see that

$$0 = A \cdot 0 + B \cdot 1,$$

$$-3 = \frac{1}{\sqrt{2}} A \cdot 1 - \frac{1}{\sqrt{2}} B \cdot 0,$$

so that $B = 0$ and $A = -3\sqrt{2}$. Thus

$$x = -3\sqrt{2} \sin(t/\sqrt{2})$$

is the equation of motion. The particle oscillates with frequency

$$f = \frac{\omega}{2\pi} = \frac{1}{2\sqrt{2}\,\pi} \approx \frac{1}{9} \qquad \text{cycles per unit time,}$$

back and forth across the interval $[-3\sqrt{2}, 3\sqrt{2}]$. Its maximum displacement $3\sqrt{2}$ is called the *amplitude* of its motion.

EXAMPLE 2 Show that any particular solution

$$x = A \sin \omega t + B \cos \omega t$$

can be written uniquely in the form

$$x = a \sin(\omega t + \alpha,)$$

where a is positive and $0 \le \alpha < 2\pi$.

Solution. Expanding the desired form by the sine addition law,

$$x = a[\cos \alpha \sin \omega t + \sin \alpha \cos \omega t],$$

and comparing with the given equation

$$x = A \sin \omega t + B \cos \omega t,$$

we see that the requirements on α and a reduce to

$$a \cos \alpha = A,$$

$$a \sin \alpha = B.$$

Then $A^2 + B^2 = a^2$, so a and α are determined by

$$a = \sqrt{A^2 + B^2},$$

$$(\cos \alpha, \sin \alpha) = (A/a, B/a).$$

If we consider $x = \sin \omega t$ to be the normal mode of oscillation, then the motion $x = \sin(\omega t + \alpha)$ *leads* the normal oscillation by the fixed angle α. We call α the *leading phase angle*. Recapitulating:

The motion of the ideal frictionless elastic system is always sinusoidal. If it is given by

$$x = A \sin \omega t + B \cos \omega t,$$

then its *frequency* is

$$f = \frac{\omega}{2\pi},$$

its *amplitude* is

$$a = \sqrt{A^2 + B^2},$$

and its *leading phase angle* α is determined by

$$(\cos \alpha, \sin \alpha) = \left(\frac{A}{a}, \frac{B}{b}\right).$$

The motion

$$x = a \sin(\omega t + \alpha)$$

is called *simple harmonic motion*. The development above shows it to be the basic type of periodic motion occurring when elastic objects vibrate.

PROBLEMS FOR SECTION 9

In each of the following motions calculate the period, amplitude, and phase angle. (Rewrite each in the form $x = a \sin(\omega t + \alpha)$, computing a exactly, but possibly only estimating α.)

1. $x = \sin(3t) - \sqrt{3} \cos(3t)$ 2. $x = \cos t + \sin t$

3. $x = 5 \cos t - 5 \sin t$ 4. $x = \sin 2t + 2 \cos 2t$

5. A particle is moving in simple harmonic motion with amplitude a and frequency f (in cycles per second). Show that its velocity at $x = 0$ is $\pm 2\pi af$.

6. A particle of mass 10 is attached to a spring of stiffness 4 and oscillates freely in simple harmonic motion. Suppose that at time $t = 0$, the position and velocity of the particle are

$$x_0 = 2, \qquad v_0 = -3.$$

Find the motion.

7. The same system is given a new motion with initial conditions $x_0 = 2$, $v_0 = 3$. Find the new motion.

8. Show that a mass m hanging at the bottom of a vertical spring of stiffness k executes simple harmonic motion about its rest position. (Gravity acts with the

constant downward force $mg = 32.2m$. Calculate the total force, and then put a new origin at the point where the total force is zero.)

A pendulum consists of a pendulum bob (a weight) suspended at the end of a light rigid rod, and allowed to swing back and forth under the action of gravity. We consider it *ideally* to be a point mass m at the end of a weightless rod of length l. The acceleration caused by gravity in a freely falling body is $g \approx 32.2$ ft/sec^2, so by Newton's law $F = ma$, the force exerted by gravity on a body of mass m is $mg \approx 32.2m$. In the case of the pendulum, most of the force of gravity is counteracted by the support of the rod, but a small amount of sidewise force is left over to make the pendulum oscillate. Assuming that forces combine according to the parallelogram law, this residual force is $mg \sin \theta$. Newton's second law $F = ma$ thus becomes

$$-mg \sin \theta = m \frac{d^2 s}{dt^2} = m \frac{d^2 (l\theta)}{dt^2} \quad \text{or} \quad \frac{d^2 \theta}{dt^2} = -\left(\frac{g}{l}\right) \sin \theta.$$

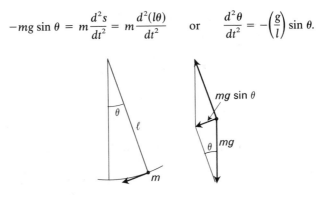

Because of the appearance of $\sin \theta$ rather than θ on the right, we cannot solve this equation. However, when θ is small,

$$\sin \theta \approx \theta$$

(since $\sin \theta / \theta \approx 1$), and small-amplitude motions are therefore governed by the "linearized" equation

$$\frac{d^2 \theta}{dt^2} = -\left(\frac{g}{l}\right) \theta.$$

9. The above discussion is important to clockmakers. A one-second pendulum takes one second for each swing, so its period is 2 sec. Show that the length l of such a *second-pendulum* is 39.2 inches.

10. Find the length of a one-half-second pendulum.

11. Find the length of a two-second pendulum.

12. Show that the period of a pendulum is proportional to the square root of its length, and calculate the constant of proportionality.

13. On Planet X the force of gravity is one-third what it is on earth. Find the length of a one-second pendulum on X.

10. SOME ELEMENTARY ESTIMATION

We shall now apply the mean-value principle directly to some elementary calculations. In practical computations we almost always replace each number by a finite decimal that approximates it. For example, we know that

$$3.14 < \pi < 3.15,$$

and on the basis of this inequality we know that each of 3.14 and 3.15 approximates π with an error less than 0.01, i.e., with an error less than 1 in the second decimal place.

Let us be clear about what we are saying here. When we regard a number x as an *approximation of a number a*, we call the difference $(a - x)$ the *error in the approximation*; it is *how much we miss the mark by*. The choice of sign is arbitrary, and some people would say that $(x - a)$ is the error; but since we are usually concerned with the *magnitude* of the error, $|a - x|$, this ambiguity in sign is seldom important.

Now if we subtract 3.14 from all terms of the inequality

$$3.14 < \pi < 3.15,$$

we obtain the inequality

$$0 < \pi - 3.14 < 0.01,$$

and this says *exactly* that the error in the approximation of π by 3.14 is positive and less than .01. If we subtract 3.15 instead, we get

$$-.01 < \pi - 3.15 < 0,$$

So this time the error is *negative* but less than .01 in magnitude. These conclusions depend on the basic inequality law:

If $x < y$, then $x + c < y + c$, for any number c.

Other inequality laws also will be needed in these examples. They are mostly in accord with common sense, and will be applied without comment, but can be looked up in Appendix 1 in case of doubt.

Actually, 3.14 is the two-place decimal *closest* to π, i.e., the *best* approximation of π by a two-place decimal. This means that

$$|\pi - 3.14| < 0.005,$$

and it is a consequence of the fact that

$$3.141 < \pi < 3.142.$$

(This is what we mean when we write $\pi = 3.141 \ldots$.) This inequality shows, in particular, that

$$3.14 < \pi < 3.142,$$

so 3.14 approximates π with an error less than 0.002.

We don't want to discuss "rounding off" here, so we shall adopt the following temporary definition:

The number x approximates the number a accurately to two decimal places if $|x - a| \leq 0.005$. More generally, the approximation is accurate to n decimal places if

$$|x - a| \leq \frac{10^{-n}}{2} \qquad (= 5 \cdot 10^{-(n+1)}).$$

Most of the following examples can also be worked out by algebra, but this will not be possible for our later applications of the mean-value principle to computations.

EXAMPLE 1   Show that $(3.14)^3$ approximates π^3 with an error less than 1 in the first decimal place, i.e., with an error less than 0.1. Use the inequality $\pi - 3.14 < 0.002$ that we noted above.

Solution. We use the mean-value principle with $f(x) = x^3$ and the x-values 3.14 and π. It guarantees a number X lying between these two x values such that

$$\pi^3 - (3.14)^3 = 3X^2(\pi - 3.14).$$

Since $X < 4$ and $\pi - 3.14 < 0.002$, we can conclude that $\pi^3 - (3.14)^3 < 48(0.002) = 0.096 < .1$, as we claimed.

EXAMPLE 2   A table gives $a = 1.414$ as the value of $\sqrt{2}$ to the nearest three decimal places. Justify this, using the value $(1.414)^2 = 1.999396$.

Solution. We use $g(x) = \sqrt{x}$ and the x-values 2 and 1.999396. Then there is a number X lying between these two x values such that

$$\sqrt{2} - 1.414 = g'(X)[2 - 1.999396]$$

$$= \frac{1}{2\sqrt{X}}(0.000604) < \frac{0.000604}{2.8} < 0.0003,$$

where we have used the fact that $\sqrt{X} > 1.4$. This meets our criterion for three-decimal-place accuracy.

EXAMPLE 3   Show that $1/3.14$ approximates $1/\pi$ accurately to three decimal places. Use the fact that $\pi - 3.14 < 0.002$.

Solution. We use the mean-value principle for $f(x) = 1/x$ and $f'(x) = -1/x^2$. Thus

$$\frac{1}{3.14} - \frac{1}{\pi} = \frac{1}{X^2}(\pi - 3.14) < \frac{0.002}{X^2} < \frac{0.002}{9} < 0.0003.$$

Since this is less than 0.0005, the approximation is accurate to three decimal places.

EXAMPLE 4   How closely should we approximate π by a number r in order to ensure that r^3 approximates π^3 accurately to two decimal places?

Solution. We want to know how small to take $|\pi - r|$ in order to ensure that

$$|\pi^3 - r^3| < 0.005.$$

But if $f(x) = x^3$, then

$$\pi^3 - r^3 = 3X^2(\pi - r)$$

for some X between π and r, by the mean-value principle. If we restrict r to the interval $[3, 4]$, then $3X^2 < 3(4)^2 = 48$, so

$$|\pi^3 - r^3| < 48|\pi - r|.$$

Therefore, in order to ensure that $|\pi^3 - r^3| < 0.005$, it is sufficient to require that

$$48|\pi - r| < 0.005 \quad \text{or} \quad |\pi - r| < \frac{0.005}{48}.$$

Since

$$\frac{0.005}{48} > \frac{0.005}{50} = 0.0001,$$

it is sufficient to take r so that

$$|\pi - r| < 0.0001,$$

i.e., to take r as an approximation to π that is accurate to within 1 in the fourth decimal place. You may know that $\pi = 3.1415 \cdots$, which means that

$$3.1415 < \pi < 3.1416.$$

We can thus take r as either 3.1415 or 3.1416, and be certain that r^3 approximates π^3 accurately to two decimal places.

Example 4 is harder than the preceding examples, where we were asked only to verify that $f(x) - f(r)$ is as small as we claimed for two given numbers x and r. Here r is *not* given, and we have to *find* it: we have to find how close r should be taken to x in order to ensure that $f(x) - f(r)$ is as small as required. This kind of calculation answers the question of *how continuous* f is at x. We noted in Theorem 5 of Chapter 3 that any differentiable function is automatically continuous, but here we use the mean-value principle to compute *how* small $(x - r)$ has to be in order to ensure that the error $f(x) - f(r)$ lies within preassigned bounds.

PROBLEMS FOR SECTION 10

1. We know that 3.1416 approximates π with an error less than 10^{-5} in magnitude. Show, then, that $(3.1416)^5$ approximates π^5 to the nearest two decimal places. (Use the fact that $3.15 < \sqrt{10}$.)

2. Assume again that 3.1416 approximates π with an error less than 10^{-5} in magnitude. We find by long division that $1/3.1416 = 0.31831$ with an error less than 10^{-6}. Show, therefore, that 0.31831 approximates $1/\pi$ to the nearest five decimal places.

3. Show that 10 approximates $(100,001)^{1/5}$ to the nearest four decimal places.

4. Show that, if r is the closest eight-place decimal to π, then $r^{1/3}$ approximates $\pi^{1/3}$ with an error less than 1 in the ninth decimal place.

5. Show that, if r is the closest n-place decimal to π, then $r^{1/3}$ approximates $\pi^{1/3}$ with an error less than 1 in the $(n + 1)$st decimal place.

6. a) How closely should we approximate a positive number a by a finite decimal r $(r < a)$ in order to ensure that r^3 approximates a^3 with an error less than 0.01? The answer will be in terms of a.

   b) Same question if the tolerable error in the approximation of a^3 by r^3 is e.

7. How closely should the finite decimal r approximate π in order to ensure that $1/r$ approximates $1/\pi$ with an error at most e?

8. Prove from the mean-value principle that

$$x^n - 1 > n(x - 1)$$

if $x > 1$ and $n > 1$.

9. Prove that if $b > 1$ then $b^n \to \infty$ as $n \to \infty$. [*Hint:* Use the preceding problem, with x fixed and n variable, to show that b^n is larger than any preassigned number M when n is chosen large enough.]

10. Prove that if $0 < a < 1$, then $a^n \to 0$ as $n \to \infty$. (Use the preceding problem.)

11. Prove that if $x > 0$, then

$$\sqrt{1 + x} < 1 + \frac{x}{2}.$$

12. Prove that if $x > 1$, then

$$x^{1/5} < \frac{4}{5} + \frac{x}{5}.$$

11. THE TANGENT LINE APPROXIMATION

The line tangent to the graph of

$$y = f(x)$$

at the point $(a, f(a))$ has the equation

$$y_t = f(a) + f'(a)(x - a).$$

This is the point–slope equation of the tangent line, slightly modified by throwing the y_0 term to the right side. We have used the different dependent variable y_t so that we can compare the two graphs for all values of x.

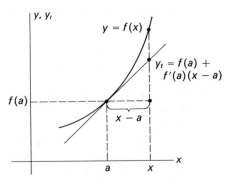

The tangent line has the direction of the graph at the point of tangency, and therefore the tangent line is a good approximation to the graph near the point of tangency. In analytic terms, the linear function $f(a) + f'(a)(x - a)$ is a good approximation to $f(x)$ near $x = a$:

$$f(x) \approx f(a) + f'(a)(x - a).$$

This tangent-line approximation is related to the calculations in the last section, but here the approximation occurs at the next stage. There we approximated $f(x)$ by $f(a)$. The error was $f(x) - f(a)$, and we used the mean-value principle to estimate it. Here we are using the better approximation $f(a) + f'(a)(x - a)$ for $f(x)$. This is equivalent to approximating $f(x) - f(a)$ by $f'(a)(x - a)$, so *here we are, in effect, approximating the former error term.* The error in *this* approximation is

$$[f(x) - f(a)] - f'(a)(x - a),$$

and when we use the mean-value principle to estimate this "second generation" error, it will be the *second* derivative f'' that will be involved.

For the moment, let us put aside the fundamental problem of estimating the error and simply use the above approximate equality as a rough estimate of $f(x)$ when x is near a. Since the new estimating formula contains $(x - a)$, it will be most easily used in situations where we know $x - a$ exactly, and the following examples are of this sort. This temporarily steers us away from the problems in the last section, where we only had a bound on $x - a$. More on this later.

EXAMPLE 1   When $f(x) = \sqrt{x}$ and $a = 25$, the approximation

$$f(x) \approx f(a) + f'(a)(x - a)$$

becomes

$$\sqrt{x} \approx \sqrt{25} + \frac{1}{2\sqrt{25}}(x - 25).$$

If $x = 27$, this gives us the rough estimate

$$\sqrt{27} \approx 5 + \frac{2}{10} = 5.2.$$

This example shows the circumstances under which the above approximation is most easily used. *If we know or can easily find the values of f and f' at a point $x = a$, then the formula estimates the value of f at a nearby point x.*

EXAMPLE 2   Estimate $63^{1/3}$.

Solution. Since $64^{1/3} = 4$, we set $f(x) = x^{1/3}$ and evaluate the formula at $a = 64$ and $x = 63$. Here $f'(x) = \frac{1}{3}x^{-2/3}$, and the approximation becomes

$$63^{1/3} \approx 64^{1/3} + \frac{1}{3 \cdot 64^{2/3}} \cdot (-1) = 4 - \frac{1}{48} \approx 4 - \frac{1}{50} = 3.98.$$

(When we replace 1/48 by 1/50 we increase the error by

$$\frac{1}{48} - \frac{1}{50} = \frac{2}{48 \cdot 50} = \frac{1}{1200} < 0.001.)$$

The estimate above is correct to two decimal places, although we have no way of showing this at the moment.

If we set $\Delta x = x - a$, then $f(x) - f(a) = \Delta y$, and the approximate equality

$$f(x) \approx f(a) + f'(a)(x - a)$$

turns into

$$\Delta y \approx f'(a)\Delta x$$

upon subtracting $f(a)$ from both sides. This form also can be used to make rough estimates.

EXAMPLE 3   By approximately how much does the volume of a sphere increase when its radius is increased from 10 to 11 feet?

Solution. Here $V = (4/3)\pi r^3$ and $dV/dr = 4\pi r^2$, so the formula is

$$\Delta V \approx 4\pi a^2 \Delta r.$$

When $a = 10$ and $\Delta r = 11 - 10 = 1$, we have the estimate

$$\Delta V \approx 4\pi \cdot 100 \cdot 1 = 400\pi \approx 1200 \text{ cu. ft.}$$

This is just an order-of-magnitude estimate, but such rough approximations can be very useful in giving a feel for what is happening. The actual value of ΔV here is

$$\Delta V = \frac{4}{3}\pi(11^3 - 10^3)$$

$$= \frac{4}{3}\pi[331] = 441\frac{1}{3}\pi \text{ cu. ft.}$$

Now let us consider the error E in these approximations. It is the difference $E = y - y_t$, for which we have the two expressions

$$E = f(x) - [f(a) + f'(a)(x - a)]$$
$$= \Delta y - f'(a)\Delta x.$$

Note that the second form of E is obtained from the first by rebracketing. The error E has a very simple geometric interpretation. Of course, $E \to 0$ as $\Delta x \to 0$, but it appears from the figure that $E = \Delta y - f'(a)\Delta x$ becomes small *even in comparison to* Δx, and thus approaches 0 *faster than* Δx.

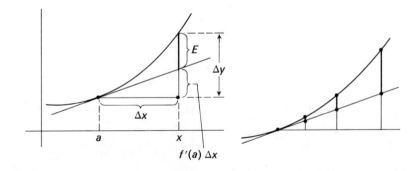

Let us calculate E explicitly for $y = f(x) = x^3$. Then

$$\Delta y = (a + \Delta x)^3 - a^3 = 3a^2\Delta x + 3a(\Delta x)^2 + (\Delta x)^3,$$

$$f'(a)\Delta x = 3a^2\Delta x,$$

$$E = \Delta y - f'(a)\Delta x = (\Delta x)^2[3a + \Delta x].$$

Such calculations make it appear that $E = \Delta y - f'(a)\Delta x$ approaches 0 faster than Δx by virtue of having $(\Delta x)^2$ as a factor. But something else is involved too.

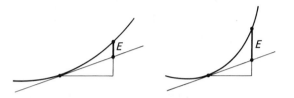

It is apparent that E will be comparatively small for a graph that bends away from its tangent line slowly, and our experience with concavity suggests that this will occur if the *second* derivative $f''(x)$ is small near $x = a$.

The following theorem verifies these reasonable conjectures. Its proof will be given in Chapter 10.

THEOREM 7 *If f'' exists on an interval I, then for any two points a and $x = a + \Delta x$ in I, the error E in the tangent line approximation can be written in the form*

$$E = \frac{f''(X)}{2}(\Delta x)^2,$$

where X is some point lying strictly between a and x.

EXAMPLE 4 We reconsider Example 1. Taking $f(x) = \sqrt{x}$, $a = 25$, and $\Delta x = 2$, we had

$$\sqrt{27} \approx 5 + \frac{1}{2}\frac{1}{\sqrt{25}} \cdot 2 = 5.2,$$

but now we can go on. Since $f''(x) = -1/(4x^{3/2})$, the theorem says that the error E in this approximation can be written

$$E = -\frac{1}{8X^{3/2}}(2)^2 = -\frac{1}{2X^{3/2}},$$

where X lies between 25 and 27. In particular,

$$|E| \leqq \frac{1}{2(25)^{3/2}} = \frac{1}{250} = 0.004,$$

so

$$\sqrt{27} \approx 5.20,$$

correct to the nearest two decimal places.

Consider now one of the problems from the last section, say the problem of estimating π^3 starting from the estimate 3.14 for π. We now have

$$\pi^3 \approx (3.14)^3 + 3(3.14)^2(\pi - 3.14),$$

with an error

$$E = \frac{6X}{2}(\pi - 3.14)^2 < 10(.002)^2 = 4 \cdot 10^{-5}.$$

The trouble is that in order to use the above estimate we have to approximate the factor $\pi - 3.14$ on the right side. If we use the approximation $\pi \approx 3.1416$, then $\pi - 3.14 \approx 0.0016$ with an error $< 10^{-5}$. However, by the time the outside factor is multiplied in, this error bound increases to about $30 \cdot 10^{-5} = 3 \cdot 10^{-4}$ and becomes the major error. On the other hand, by the simpler method of the last section, $\pi^3 \approx (3.1416)^3$, with an error

$$3X^2(\pi - 3.1416) < 30 \cdot 10^{-5} = 3 \cdot 10^{-4}$$

of the same order of magnitude, and with about the same amount of arithmetic needed to multiply everything out. So the tangent line approximation offers no great advantage in this problem.

PROBLEMS FOR SECTION 11

Use the tangent-line approximation to estimate each of the following values.

1. $\sqrt{65}$ 2. $\sqrt[3]{124}$

3. $\sqrt[4]{80}$ 4. $\sqrt[5]{31}$

5. $(.98)^{-1}$ 6. $(17)^{1/4}$

7. $26^{1/3}$ 8. $\sqrt{24}$

9. $1/\sqrt{15}$ 10. $1/80$

11. Find the approximate change in the volume of a sphere of radius r caused by increasing the radius by 2%.

12. A metal cylinder is found by measurement to be 2 ft in diameter and 5 ft long. What, approximately, is the error in the computed volume of this cylinder if the error in measuring the diameter was 1/2 in?

13. Let V and A be the volume and surface area of a sphere of radius r. If the radius is changed by a small amount Δr, show that

$$\Delta V \approx A \cdot \Delta r.$$

14. Let V and A be the volume and surface area of a cube of edge length x. If x is given a small increment Δx, show that

$$\Delta V \approx A \frac{\Delta x}{2}.$$

15. Solve Problem 1 again, along the lines of Example 4 in the text. That is, obtain an expression for the error E by applying Theorem 7, and then state the tangent-line estimate in decimal form, with some statement about E.

16. Same for Problem 2.   17. Same for Problem 3.

18. Same for Problem 4.   19. Same for Problem 5.

20. Same for Problem 6.

12. NEWTON'S METHOD

Assuming that we know $f(a)$ and $f'(a)$, the approximate equality

$$f(x) \approx f(a) + f'(a)(x - a)$$

was used in the last section to estimate the value of $f(x)$ in terms of x. But it can also be used in the opposite direction to estimate x from $f(x)$. That is, we can estimate a root x of the equation $f(x) = k$, supposing that we know a reasonably close initial estimate a. (The *existence* of the root x will always follow from the intermediate-value principle.)

The solution takes a slightly simpler form if we want a root of the equation $f(x) = 0$, and problems can always be set up this way. In this case the approximate equation becomes

$$0 \approx f(a) + f'(a)(x - a),$$

and solving for x gives

$$x \approx a - \frac{f(a)}{f'(a)}.$$

Thus,

$$x_1 = a - \frac{f(a)}{f'(a)}$$

is our new estimate of the root x of the equation $f(x) = 0$ in terms of an initial rough estimate a.

EXAMPLE 1 Estimate the positive solution of the equation $x^2 = 27$.

Solution. We want a root of $x^2 - 27 = 0$, so here $f(x) = x^2 - 27$. Since 27 lies between $5^2 = 25$ and $6^2 = 36$, it is reasonable to take $a = 5$ as the initial rough estimate. So

$$x_1 = a - \frac{f(a)}{f'(a)} = 5 - \frac{-2}{10} = 5.2$$

is our new estimate of the positive root of the equation $x^2 = 27$.

Note that we are again approximating $\sqrt{27}$, obtaining the same estimate as in the last section, but by a different method. This time we are approximating the root r of the equation

$$f(x) = 0$$

by the solution $x = x_1$ of the equation

$$f(a) + f'(a)(x - a) = 0.$$

Again, there is a very simple geometric interpretation. The root r and its approximation x_1 are simply the x-values where the graph of f and its tangent line at $(a, f(a))$ intersect the x-axis, as pictured below.

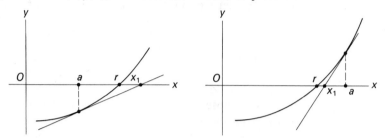

In practice, a will be a preliminary, very rough estimate of the root r, and x_1 will be an improved estimate. The figures above show that *if the graph of f is increasing and concave up, then the estimate x_1 will always be larger than the root r, regardless of whether the initial estimate a is larger or smaller than r.*

We can now *iterate* the approximation, replacing the initial estimate a by x_1, and thus obtaining a "second generation" estimate

$$x_2 = x_1 - \frac{f(x_1)}{f'(x_1)}.$$

It appears from the figure that x_2 is an exceedingly close estimate of r.

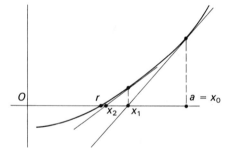

If we go on in this way, we obtain a sequence $x_0, x_1, x_2, x_3, \ldots$, of rapidly improving approximations, i.e., an infinite sequence that converges very fast on the true root $x = r$. This iterative procedure is called *Newton's method*. We shall see that it can be extraordinarily accurate at even the second step.

EXAMPLE 2   Compute a "second-generation" approximation to the solution of $x^2 = 27$ by starting from the estimate x_1 obtained in Example 1.

Solution.

$$x_2 = x_1 - \frac{f(x_1)}{f'(x_1)}$$

$$= 5.2 - \frac{(5.2)^2 - 27}{2(5.2)}$$

$$= 5.2 - \frac{0.04}{10.4}$$

$$= 5.2 - 0.0038461 \cdots.$$

Thus

$$x_2 = 5.196154$$

is our improved estimate. We carried the long division .04/10.4 out to six decimal places because it can be shown that *this second estimate is correct to within two units in the sixth decimal place.* That is, the error $x_2 - r$ is less than $2(10)^{-6}$.

Estimating the Error

We shall now find a bound for the error $E = r - x_1$, where r and x_1 are defined by

$$f(r) = 0$$

$$f(a) + f'(a)(x_1 - a) = 0.$$

Then

$$f(x_1) - f(r) = f(x_1) - f(a) - f'(a)(x_1 - a).$$

We now apply the mean-value principle on the left and Theorem 7 on the right, and conclude that

$$f'(X)(x_1 - r) = \frac{f''(Y)}{2}(x_1 - a)^2,$$

where Y is some number between a and x_1, and X is some number between x_1 and r. Now suppose that we know that $|f'(x)| \geq L > 0$ and $|f''(x)| \leq B$ for all the x-values in question. Then we have

$$|x_1 - r| \leq \frac{B}{2L}(x_1 - a)^2$$

as a bound for the error $E = r - x_1$ in the estimate of r by x_1. In applying this formula, remember:

a) a is a preliminary, rough estimate of the root r of the equation $f(x) = 0$;

b) x_1 is the new estimate obtained from a by one application of Newton's method;

c) B is an upper bound for $|f''(x)|$ over the values of x in question;

d) L is a *lower* bound for $|f'(x)|$ over these values of x.

EXAMPLE 3   Use this formula to estimate the error in Example 2.

Solution. Here $f(x) = x^2 - 27$, and we are operating with values of x larger than 5. Therefore $f'(x) = 2x \geq 10$, while $f''(x) \equiv 2$. Using these constants in the error formula above, we get

$$|E| \leq \frac{2}{2 \cdot 10}(x_2 - x_1)^2 = \frac{(x_2 - x_1)^2}{10}.$$

(We have replaced a and x_1 by x_1 and x_2.) Since $x_1 - x_2 = 0.0038 \cdots < 0.004$, we see that

$$|E| < \frac{(0.004)^2}{10} = 0.0000016 = 1.6(10)^{-6}.$$

Therefore

$$\sqrt{27} \approx 5.196154,$$

with an error less than 2 in the sixth decimal place.

PROBLEMS FOR SECTION 12

In each of the following problems, locate the nearest integer a to the solution of $f(x) = 0$ by sketching the graph of f, and then estimate the root by one application of Newton's formula.

1. $f(x) = x^2 - 10, x > 0$ 2. $f(x) = x^2 - 5, x > 0$
3. $f(x) = x^3 + x - 1$ 4. $f(x) = x^3 + 2x - 5$

5. a) In Problem 4 above the answers are $a = 1$, $x_0 = 1\frac{2}{5} = 1.4$. Now use this estimate for x_0 as a new value of a, and obtain the second-generation estimate

$$x_1 = 1.331.$$

   b) Use the error formula, with $f''(x) \le 6(1.4) = 8.4$ and $f'(x) \ge 5$, to conclude that $x = 1.33$ is the solution of the equation $x^3 + 2x = 5$ to the nearest two decimal places.

6. The answers in Problem 3 above are $a = 1$, $x_0 = 3/4$. Now show that

$$x = 0.686$$

   is the solution of the equation $x^3 + x = 1$ with an error at most 5 in the third decimal place. (Use x_0 as a new starting estimate a, to obtain a second generation estimate x_1, and then apply the error formula.)

7. Show that the positive solution of the equation

$$\sin x = \frac{2}{3} x$$

   is given by $x = 3/2$, with an error at most 0.01 in magnitude. (Take $a = \pi/2$ and use the approximation once.)

8. Show by two figures that if the graph of f is increasing and concave down, then Newton's estimate of the solution $f(x) = 0$ is always too small, regardless of whether the initial estimate is too small or too large.

9. Starting from the nearest integer solution of the equation $x^2 - 10 = 0$, and applying the reverse tangent line approximation twice (Newton's method) show that

$$\sqrt{10} = 3.16228,$$

   correct to five decimal places. (The error is less than 5 in the sixth decimal place.)

10. Since $(2.2)^2 = 4.84$ and $(2.3)^2 = 5.29$, we know that $2.2 < \sqrt{5} < 2.3$. Using 2.2 as the initial estimate, apply Newton's method once. Show that your answer is accurate to three decimal places.

11. Compute $\sqrt{101}$ to three decimal places by a single application of Newton's method. (Start with $a = 10$.)

In the following problems apply Newton's method twice, starting from the given rough approximation. Estimate the error in the final approximation.

12. $\sqrt{3} \approx 2$.

13. $\sqrt{2} \approx \dfrac{3}{2}$.

14. $\sqrt{5} \approx \dfrac{5}{2}$.

15. $\sqrt{3} \approx \dfrac{7}{4}$.

EXTRA PROBLEMS FOR CHAPTER 6

Sketch the graphs of the following equations (concavity, critical points, asymptotes).

1. $y = x^3 + x^2 - x - 1$

2. $y = x^5 - 5x^4$

3. $y = x + \sin x \ [0, 3\pi]$

4. $y = x(x - 4)^2(x + 4)^2$

5. $y = e^{-x^2/2}$

6. $y = x^2/(1 + x^2)$

7. $y = x^3/(1 + x^2)$

8. $y = xe^x$

9. $y = x^2 e^x$

10. $y = \tan x$, over $(-\pi/2, \pi/2)$. Your results should justify the general features shown in Section 3 of the last chapter.

11. $y = \sec x$, over $(-\pi/2, 3\pi/2)$.

12. $y = \dfrac{1}{x} + x$. Note that this graph is asymptotic to the line $y = x$, since the difference between the two y-coordinates $(1/x)$ approaches 0 as $x \to \infty$.

13. $y = \dfrac{1}{x} - x$

14. $y = \dfrac{1}{x^2} + 2x$

15. $y = \dfrac{1}{x} + \sqrt{x}$. (This graph is asymptotic to the half parabola $y = \sqrt{x}$.)

16. $f(x) = (\cos x)^{2/3}$

17. $f(x) = (4 - 4x^2)^{1/2}$

18. A function graph turns in the same direction whether it is traced forward or backward, so $f(x)$ and $f(-x)$ should show the same concavity behavior. State this fact more carefully and verify it analytically.

19. Show that the cubic graph

$$y = x^3 + ax^2 + bx + c$$

has three possible shapes, depending on whether $a^2 < 3b$, $a^2 = 3b$, or $a^2 > 3b$. Sketch the three possible shapes.

In the following four problems we consider a differentiable function f over an interval I and show that the various possible definitions for f being *concave up* are all equivalent.

20. Show that if $f'(x)$ is increasing, then the graph of f lies strictly above each tangent line (except for the point of tangency). [*Hint*: Try to use the fact that if $g'(x) > 0$, then g is increasing, for a suitably chosen function g.]

21. Show that if the graph of f lies strictly above each of its tangent lines, then the second of any pair of nonoverlapping chords always has the larger slope.

22. Show that if the graph of f has the property that the second of any pair of nonoverlapping chords always has the larger slope, then each arc of the graph of f lies strictly below its chord (except for the endpoints).

23. Show, finally, that if each arc of the graph of f lies strictly below its chord, then $f'(x)$ is increasing.

24. Show that a rational function of degree at most 1 has an asymptote as x approaches infinity.

A polygonal function is a continuous function whose graph is made up of a number of straight line segments.

25. Show that $f(x) = |x - 2| + (x - 2)$ is a polygonal function.

26. Show that if f is a polygonal function whose graph has just one vertex, at $x = 2$, then $f(x)$ can be expressed in the form

$$f(x) = a|x - 2| + bx + c.$$

27. Show that a polygonal function with just one vertex, at $x = x_0$, is of the form

$$f(x) = a|x - x_0| + bx + c.$$

28. Show that

$$f(x) = |x + 1| + |x - 1| - 1$$

is a polygonal function. (Show, for example, that if x lies in the interval $(-\infty, -1)$ then $f(x) = -(x + 1) - (x - 1) - 1 = -2x - 1$). A polygonal function can be graphed exactly from its vertices and its two slopes before the first vertex and after the last vertex. Graph the above function in this manner.

29. Show that any function of the form

$$f(x) = a_1|x - x_1| + a_2|x - x_2| + \cdots + a_n|x - x_n| + bx + c$$

is a polygonal function.

Graph the following polygonal functions

30. $f(x) = \dfrac{1}{2}(x + |x|)$.   31. $f(x) = |x + 1| - |x - 1|$.

32. $f(x) = |x + 1| + |x - 1| - 2x$.

Find a formula for each of the polygonal functions graphed below.

33.

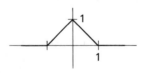

34.

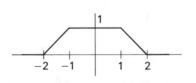

35.

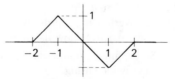

36. A rectangular box with a square base and a cover is to be built to contain 640 cu. ft. If the cost per square foot for the bottom is 15¢ and for the top and sides 10¢, what are the dimensions for a minimum cost?

37. Find the minimum distance from the point $(1, 4)$ to the parabola $x^2 = 3y$.

38. In connecting a water line to a building at A (see figure below), the contractor finds that he must connect to a certain point C on the water main which lies under the paved parking lot of a shopping center. It will cost him $20 a foot to dig, lay pipe, fill, and resurface the parking lot, but only $12 a foot to lay pipe along the edge. Find the distance from the store water inlet to the point B, at which he should turn the water line and go directly to point C in order to minimize his cost.

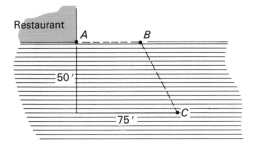

39. A frame for a cylindrically shaped lamp shade is made from a piece of wire 16 ft. long. The frame consists of two equal circles, two diametral wires in the upper circle, and four equal wires from the upper to the lower circle. For what radius will the volume of the cylinder be a maximum?

40. A rectangular box is to be made from a sheet of tin 20 in. by 20 in. by cutting a square from each corner and turning up the sides. Find the edge of this square which makes the volume a maximum.

41. Two towers 40 ft apart are 30 and 20 ft high respectively. A wire fastened to the top of each tower is guyed to the ground at a point between the towers, and is tightened so that there is no sag. How far from the taller tower will the wire touch the ground if the length of the wire is a minimum?

42. A steel mill is capable of producing x tons per day of a low-grade steel and y tons per day of a high-grade steel, where

$$y = \frac{40 - 5x}{10 - x}.$$

If the fixed-market price of low-grade steel is half that of the high-grade steel, find the amount of low-grade steel to be produced each day that will yield maximum receipts.

43. A window in the shape of a rectangle surmounted by a semicircle has a perimeter of 24 ft. If the amount of light entering through the window is to be as large as possible, what should the dimensions of the window be?

44. Two roads intersect at right angles, and a spring is located in an adjoining field 10 yds from one road and 5 yds from the other. How should a straight path just passing the spring be laid out from one road to the other so as to cut off the least amount of land? How much land is cut off?

45. Find the shortest path in the above situation.

46. Which is the more efficient container, a cube, or a circular cylinder? (Find the most efficient cylinder and then compare with the cube. See page 218.)

47. Solve the Norman window problem (Problem 43) if colored glass is used for the semicircle that admits only half as much light per unit area as does the clear glass in the rectangle.

48. A silo consists of a circular cylinder with a hemispherical top. Find its most efficient shape. That is, find the relative dimensions that maximize the volume for a given total area (base area plus lateral area of the cylinder, plus the area of the hemisphere).

49. Find the most economical shape for a silo like that in the above problem if the construction of the hemisphere costs twice as much per unit area as the cylinder.

50. A smooth graph not passing through the origin always has a point (x_0, y_0) closest to the origin. Show that the segment from the origin to (x_0, y_0) is perpendicular to the graph.

51. A tank has hemispherical ends and a cylindrical center. Find its most efficient shape. That is, find the proportions of the cylinder that will maximize the volume for a given surface area.

52. If the hemispherical ends to the tank in the above problem cost twice as much (per unit area) as the cylindrical center, find its most economical shape.

53. It costs a manufacturer $x^3 - 3x^2 + 4x + 1$ hundreds of dollars to produce x thousands of an item per week, and he can sell them for $10x$ hundreds of dollars. How many should he sell per week in order to maximize his profit, and what is this maximum profit?

54. A manufacturer's cost function is $4\sqrt{x} + 1$, and his revenue function is $9x - x^2$, both in thousands of dollars per thousand items. Find his maximum profit.

55. The illumination from a light source is inversely proportional to the square of the distance from the light and directly proportional to the sine of the angle of incidence. How high should a light be placed on a pole in order to maximize the illumination on the ground along the circumference of a circle of radius 25 feet?

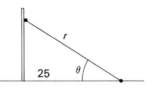

56. A balloon is rising vertically over a point A on the ground at the rate of 15 ft/sec. A point B on the ground is level with and 30 ft from A. The angle of elevation is being measured at B. At what rate is the elevation changing when the balloon is $30\sqrt{3}$ ft above A?

57. Water is being pumped into a trough 8 ft long with a triangular cross section 2 ft × 2 ft × 2 ft at the rate of 16 cu. ft/min. How fast is the surface level rising when the water is 1 ft deep?

58. Gas is escaping from a spherical balloon at the rate of 1/2 cu. in./sec. At what rate is the surface area decreasing when the radius is 6 in.?

59. If $uv = 1$ and if v is increasing at the constant rate of 2 units per unit time, what is the rate of change of u when $v = 4$?

60. A particle moves along the parabola $y = 3x^2$ in such a way that its x-coordinate is increasing at the constant rate of 5 inches per second (the inch being the unit of distance in the coordinate plane). What is the rate at which its y-coordinate is changing when $x = -1$? When $x = 0$? When $x = 1/3$?

61. A particle moves along the parabola $y = x^2$, and dy/dt is found to have the value 3 when $x = 2$. How fast is x changing at that moment?

62. A circular wheel of radius 5 feet lies in the xy-plane, with its center at the origin, and revolves steadily in the counterclockwise direction at the rate of 25 revolutions per minute. How fast is the y-coordinate of a particle on the wheel increasing as the particle goes through the point $(3, 4)$? The same question for its x-coordinate.

63. Referring to the above problem, suppose that the wheel is rotating at the constant rate of one radian per minute. Show that then $dy/dt = x$, for all positions of the particle.

64. Suppose a particle traces a circle about the origin in such a way that $dy/dt = x$. Show that the particle is moving along the circle at the constant rate of 1 radian per unit of time.

65. A rolling snowball picks up new snow at a rate proportional to its surface area. Assuming that the snowball always remains spherical, show that its radius is increasing at a constant rate.

66. A ladder 20 feet long leans against a wall 10 feet high. If the lower end of the ladder is pulled away from the wall at the rate of 3 feet per second, how fast is the angle between the ladder and the top of the wall changing at the moment when the angle is $45° = \pi/4$ radians?

Find and classify all critical points of the following functions, using the higher derivative tests.

67. $y = (x^2 - 3x)^2$ 　　　　　　　　　　　　　68. $y = x^4(x + 1)^2$

69. Find the approximate change in the surface area of a sphere if the radius is increased from 10 in. to 10.1 in.; if the volume is decreased from 27π in^3 to 26π in^3.

70. Using the fact that $(9/2)^2 = 20\frac{1}{4}$, show that

$$\sqrt{20} = 4.4722\ldots$$

with an error less than 1 in the fourth decimal place.

71. Using the fact that $(3/2)^4 = 5\frac{1}{16}$, and assuming that $5^{3/4} > 3$, show that

$$5^{1/4} \approx \frac{3}{2} - \frac{1}{216}$$

with an error less than $1/10 \cdot 2^{12}$, and hence less than $3(10)^{-5}$.

72. Show that

$$\sin t \approx t$$

with an error less than $t^3/2$. (Use the fact that $|\sin t| \le |t|$.)

73. Show that

$$1 - \cos x \le \frac{x^2}{2}$$

for all x.

74. Prove for any positive number x that

$$\sqrt{1 + x} \approx 1 + \frac{x}{2}$$

with an error less than $x^2/8$.

75. Show that if $h > 0$ and $a < 2$, then

$$(1 + h)^a \approx 1 + ah$$

with an error at most $|a(a - 1)h^2/2|$ in magnitude.

76. Draw the figure for Theorem 7 showing E as the length of a vertical line segment for the case when Δx is negative (and f'' is positive).

77. Theorem 7 shows that the error E has the same sign as f'' (supposing that f'' has constant sign). The figures in the text and Problem 76 show this geometrically for the concave up case. Draw the corresponding figures for the case that E and f'' are both negative.

78. By comparing $\sqrt{10}$ to $\sqrt{9} = 3$, and referring to the above problem, show that $\sqrt{10} \approx 19/6$, with an error that is negative and less than $1/200$ in magnitude. Conclude that

$$3.16 < \sqrt{10} < 3.17.$$

79. By comparing $\sqrt{3}$ to $\sqrt{25/9} = 5/3$, show that $\sqrt{3} \approx 26/15$, with an error that is negative and less than $1/750$ in magnitude. Conclude that

$$1.732 < \sqrt{3} < 1.734.$$

80. In Chapter 5 we defined the differential dy in terms of a new independent variable dx (supposing that $y = f(x)$). Look up that definition, and show:

If we set $dx = \Delta x$, then $\Delta y - dy$ is the error E in the tangent-line approximation.

Draw the figure for the error E and relabel it so as to exhibit the identity $E = \Delta y - dy$.

7
Inverse Functions

1. THE NATURAL LOGARITHM FUNCTION

The logarithm function $\log x$ is derived from the exponential function e^x in the same way that the cube root function $x^{1/3}$ is derived from the cubing function x^3, and a brief review of how this simpler function arises may be helpful.

Every number y has a uniquely determined cube root $y^{1/3}$, defined to be the unique number x whose cube is y:

$$y^{1/3} = \text{the unique number } x \text{ such that } y = x^3.$$

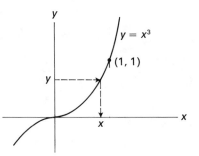

A look at the graph of $y = x^3$ shows this to be a sensible definition: the graph shows that each number y is a value of the cubing function $f(x) = x^3$ for one and only one number x. Actually, it is the intermediate-value property of $f(x) = x^3$ that we are visualizing. For example, as x runs across the interval $[1, 2]$, we visualize the value $y = x^3$ running across the interval $[1^3, 2^3] = [1, 8]$ on the y-axis *without skipping over any numbers*. In particular, as $y = x^3$ moves from below 2 to above 2 it must go *through* the value *2*, so $x^3 = 2$ for a certain number x in $[1, 2]$, and $2^{1/3}$ exists. This was discussed at the end of Chapter 2. There is *only* one solution to the equation $2 = x^3$ because the function $f(x) = x^3$ is strictly increasing. So the cube root function $x = g(y) = y^{1/3}$ is well-defined because the cubing function $f(x) = x^3$ is increasing and runs through all intermediate values.

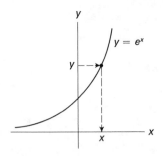

In the same way, the graph of $y = e^x$ shows that each positive number y is a value of the exponential function $f(x) = e^x$ for one and only one number x. This unique x is called the *natural logarithm* of y, and is designated $\log y$. Thus, for each positive number y,

$$\log y = \text{the unique number } x \text{ such that } y = e^x.$$

Again it is the intermediate-value principle that we are visualizing, this time for $f(x) = e^x$. As x runs across the interval $(-\infty, \infty)$, the value of

$f(x) = e^x$ increases from arbitrarily small positive values, *through all intermediate values*, to arbitrarily large values, so the range of $f(x) = e^x$ is the interval $(0, +\infty)$. (The general theorem to this effect was given in the last section of Chapter 2, along with other properties of increasing functions that we shall use here.)

The graph also shows that log y increases with y. Formally, this reduces to the *reverse increasing* property of the increasing function e^x, namely, that

$$\text{if } e^{x_1} < e^{x_2} \quad \text{then } x_1 < x_2.$$

Setting $y_1 = e^{x_1}$ and $x_1 = \log y_1$, etc., this turns into the fact that

$$\text{if } y_1 < y_2 \quad \text{then } \log y_1 < \log y_2.$$

Thus log y is an increasing function, with domain $(0, +\infty)$ and range $(-\infty, +\infty)$. It was shown at the end of Chapter 2 that any such function is automatically continuous. It will be shown below that in fact log y is differentiable.

Logarithms to the base 10 are often taught in secondary school, starting from the definition of $\log_{10} y$ as that number x such that $y = 10^x$. Comparing this definition with the definition of log y given above, we see that log y is the logarithm of y to the base e, $\log_e y$. From the point of view of calculus, e will turn out to be the natural base, which is why log $y = \log_e y$ is called the natural logarithm of y. This explains a third notation in common usage, namely, ln y. Thus

$$\log y = \log_e y = \ln y.$$

Of course it doesn't matter what variables we use to define the logarithm function. We could define log t as the unique number s such that $t = e^s$. In particular, and since we are going to concentrate on this new function for a while, we shall interchange the earlier roles of x and y and write:

$$\log x = \text{the unique number } y \text{ such that } x = e^y.$$

Then $y = \log x = \log(e^y)$ and $x = e^y = e^{\log x}$. Therefore

$$\log(e^x) = x \text{ for every number } x,$$
$$e^{\log x} = x \text{ for every positive number } x.$$

These equations show that the functions e^x and log x cancel each other. Each undoes what the other has done. Two functions related this way are said to be *inverse* to each other. For the cubing and cube root functions we have the similar cancellation identities

$$(x^3)^{1/3} = x,$$
$$(x^{1/3})^3 = x,$$

showing these two functions to be mutually inverse. Section 3 contains a general discussion of such inverse pairs of functions.

The graph of the equation $y = \log x$ is the graph of the equivalent equation $x = e^y$, and it is therefore the mirror image of the graph of $y = e^x$ in the line $y = x$. (This was discussed in Chapter 1, Section 6, and Chapter 2, Section 2.) That is, the graph of log x is obtained geometrically from the graph of e^x by reflecting in, or rotating about, the 45° line $y = x$.

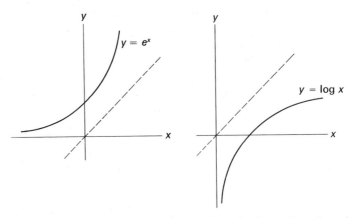

The cancellation identities in effect define the logarithm function from the exponential function, so we should be able to use them to deduce the logarithm law from the law of exponents. This is short but tricky. We have

$$\log uv = \log(e^{\log u}e^{\log v}) = \log(e^{(\log u + \log v)})$$
$$= \log u + \log v,$$

where we have used both cancelling identities and the law of exponents. The law

$$\log uv = \log u + \log v$$

is the characteristic property of a logarithm, and the above proof shows why the inverse of any exponential function is a logarithm function.

Also useful are the properties:

$$\log 1 = 0,$$
$$\log \frac{1}{u} = -\log u.$$

(Apply the logarithm law to the identities $1 \cdot 1 = 1$, $u(1/u) = 1$. The details are left to the reader.)

Repeated applications of the logarithm law show that

$$\log(u_1 u_2 \cdots u_n) = \log u_1 + \log u_2 + \cdots + \log u_n$$

for any finite collection of positive numbers $u_1, \ldots, u_n$. In particular,

$$\log u^n = n \log u$$

for every positive integer n.

We shall now use the mean-value principle to prove that $y = \log x$ is differentiable. In the usual increment notation ($y + \Delta y = \log(x + \Delta x)$, etc.), we have

$$\Delta x = (x + \Delta x) - x = e^{y + \Delta y} - e^y = e^Y \Delta y,$$

the final equation being the mean-value principle for e^y. Here Y is some number between y and $y + \Delta y$, so the number $X = e^Y$ lies between $x = e^y$ and $x + \Delta x = e^{y + \Delta y}$. Thus

$$\frac{\Delta y}{\Delta x} = \frac{1}{X},$$

where X is some number between x and $x + \Delta x$. But then $X \to x$ as $\Delta x \to 0$, by the squeeze limit law, so

$$\lim_{\Delta x \to 0} \frac{\Delta y}{\Delta x} = \lim_{\Delta x \to 0} \frac{1}{X} = \frac{1}{x}.$$

That is, $y = \log x$ is differentiable and its derivative is $1/x$.

Section 3 contains the general version of this proof, roughly to the effect that differentiable functions always have differentiable inverses. If we assume this general result, then the formula for the derivative of a particular inverse function can always be found from the chain rule. Thus, differentiating the cancellation identity

$$x = e^{\log x},$$

we get

$$1 = \frac{d}{dx}(e^{\log x}) = e^{\log x} \frac{d}{dx} \log x = x \frac{d}{dx} \log x,$$

so the derivative formula for $\log x$ is

$$\frac{d}{dx} \log x = \frac{1}{x}.$$

With the built-in chain rule, this formula becomes

$$\frac{d}{dx} \log u = \frac{1}{u} \frac{du}{dx},$$

$$d(\log u) = \frac{du}{u}.$$

EXAMPLES

1. $\dfrac{d}{dx} \log(1 + x^2) = \dfrac{1}{1 + x^2} \dfrac{d}{dx} (1 + x^2) = \dfrac{2x}{1 + x^2}.$

2. $\dfrac{d}{dx} \log kx = \dfrac{k}{kx} = \dfrac{1}{x}.$

3. $\dfrac{d}{dx} \log(\cos x) = \dfrac{1}{\cos x} \cdot (-\sin x) = -\tan x.$

4. $\dfrac{d}{dx} \log(\log x) = \dfrac{1}{\log x} \left(\dfrac{1}{x}\right) = \dfrac{1}{x \log x}.$

5. $\dfrac{d}{dx} (x \log x - x) = x \cdot \dfrac{1}{x} + \log x \cdot 1 - 1 = \log x.$

6. $\dfrac{d}{dx} \log\left(\dfrac{1 - x}{1 + x}\right) = \dfrac{1}{\left(\dfrac{1 - x}{1 + x}\right)} \cdot \dfrac{(1 + x)(-1) - (1 - x) \cdot 1}{(1 + x)^2}$

$$= \dfrac{-2}{(1 - x)(1 + x)} = -\dfrac{2}{1 - x^2}.$$

7. Since $\log[(1 - x)/(1 + x)] = \log(1 - x) - \log(1 + x)$,

$$\dfrac{d}{dx} \log\left(\dfrac{1 - x}{1 + x}\right) = \dfrac{1}{(1 - x)} (-1) - \dfrac{1}{(1 + x)} = \dfrac{-2}{(1 - x)(1 + x)}.$$

REMARK The calculation of the derivative of $\log x$ is really an implicit differentiation. To see this, remember that the function $\log x$ exists only because the equation

$$x = e^y$$

determines y as a function of x. If we leave y in this equation, instead of setting $y = \log x$, then the calculation given earlier amounts to differentiating this equation implicitly:

$$1 = \frac{d}{dx} e^y = e^y \frac{dy}{dx} = x \frac{dy}{dx},$$

and

$$\frac{dy}{dx} = \frac{1}{x}.$$

REMARK A function f is uniquely determined if we know its derivative and its value at one point. In particular, $\log x$ is the unique function $f(x)$ satisfying the equations

$$f'(x) = \frac{1}{x} \quad \text{and} \quad f(1) = 0.$$

Therefore, the addition law

$$\log ax = \log a + \log x$$

must somehow be implied by these equations. See Problem 18 in the extra problems at the end of the chapter.

We conclude with the following remarkable identity for the exponential function: for every number x,

$$e^x = \lim_{n \to \infty} \left(1 + \frac{x}{n}\right)^n.$$

Proof. We know that

$$\frac{\log(1 + h)}{h} \to 1$$

as $h \to 0$, because the left side is just the difference quotient for $\log x$ at $x = 1$. If we multiply by x and set $r = x/h$, we see that

$$r \log\left(1 + \frac{x}{r}\right) \to x$$

as $r \to \infty$. This holds in particular as $r \to \infty$ through integer values n, and since $n \log u = \log u^n$, we have

$$\log\left(1 + \frac{x}{n}\right)^n \to x$$

as $n \to \infty$. Now just apply the exponential function to both sides, and use the identity $e^{\log t} = t$ on the left. ∎

PROBLEMS FOR SECTION 1

Reformulate the following expressions by using the logarithm law.

1. $\log(x - 1) + \log(x + 1)$ 2. $2 \log x + 3 \log(x - 2)$

3. $\log(x^2 - 2x + 1)$

4. $\log(x - 1) - \log(x + 1) + \log(x^2 + x + 1) - \log(x^2 - x + 1)$

5. Prove the more general formula

$$\frac{d}{dx} \log |x| = \frac{1}{x} \quad \text{(if } x \neq 0\text{)}.$$

Differentiate the following functions.

6. $\log(x + 1)$ 7. $\log(x^2 + 1)$ 8. $\log \sqrt{x^2 + 4}$ 9. $x \log x$

10. $e^x \log x$ 11. $(\log x)/x$ 12. $\log(e^x + 1)$ 13. $\log(e^x)^2$

14. $x^2 \log x - x^2/2$ 15. $\dfrac{x^{n+1}}{n + 1}\left(\log x - \dfrac{1}{n + 1}\right)$

16. $(\log x)^2$

Prove the following formulas.

17. $\dfrac{d}{dx} \log(\sin x) = \cot x$ 18. $\dfrac{d}{dx} \log(\sec x) = \tan x$

19. $\dfrac{d}{dx} \log(\sec x + \tan x) = \sec x$ 20. $\dfrac{d}{dx} \log\left(\dfrac{1 + \sin x}{1 - \sin x}\right) = 2 \sec x$

21. $\dfrac{d}{dx} \log(x + \sqrt{1 + x^2}) = \dfrac{1}{\sqrt{1 + x^2}}$ 22. $\dfrac{d}{dx} \log(x + \sqrt{x^2 - 1}) = \dfrac{1}{\sqrt{x^2 - 1}}$

23. $\dfrac{d}{dx} \log\left(\dfrac{1 + \sqrt{1 + x^2}}{x}\right) = -\dfrac{1}{x\sqrt{1 + x^2}}$

24. Prove that $a \log x = \log x^a$ for any rational number a. [*Hint:* Show that both sides have the same derivative.]

2. THE GENERAL EXPONENTIAL FUNCTION

In Chapter 5 we briefly discussed the exponential function a^x with a general positive base a, but since then we have concentrated on the base e. We can now reduce other bases to the base e by using the cancelling law $a = e^{\log a}$, which gives us the identity

$$a^x = e^{x \log a}.$$

We always use this identity to reduce problems involving other bases to the known rules for the base e.

EXAMPLE 1   We start by proving the "mixed" law

$$\log a^x = x \log a.$$

Proof. $\log(a^x) = \log(e^{x \log a}) = x \log a$, by the cancellation law. ∎

EXAMPLE 2 $\dfrac{d}{dx}(2^x) = \dfrac{d}{dx} e^{(x \log 2)}$

$$= e^{x \log 2} \cdot \frac{d}{dx}(x \log 2)$$

$$= 2^x \cdot \log 2.$$

Thus the constant α that we met in Chapter 5 when we first considered the derivative of 2^x is $\alpha = \log 2$.

EXAMPLE 3   At last we can prove the power rule for an *arbitrary* exponent a. We have

$$\frac{d}{dx}\, x^a = \frac{d}{dx}\, e^{a\log x}$$

$$= e^{a\log x} \frac{d}{dx}\,(a\log x)$$

$$= x^a \cdot \frac{a}{x}$$

$$= a x^{a-1}.$$

EXAMPLE 4

$$\frac{d}{dx}\, x^x = \frac{d}{dx}\, e^{x\log x}$$

$$= e^{x\log x} \frac{d}{dx}\,(x\log x)$$

$$= x^x[1 + \log x].$$

The function

$$[f(x)]^{g(x)} = e^{g(x)\log f(x)}$$

is defined only when $f(x)$ is positive; it is then differentiable whenever f and g are differentiable. One way of computing its derivative is illustrated in the examples above. There is another procedure, called *logarithmic differentiation*, where we compute the derivative of the logarithm of the function in question.

EXAMPLE 5   Find the derivative of $y = x^x$.

Solution. Since

$$\log y = \log x^x = x\log x,$$

differentiating with respect to x gives

$$\frac{1}{y}\frac{dy}{dx} = x \cdot \frac{1}{x} + \log x \cdot 1 = 1 + \log x,$$

so

$$\frac{dy}{dx} = y[1 + \log x] = x^x[1 + \log x],$$

as before.

EXAMPLE 6   Products and quotients can be attacked by logarithmic differentiation. We shall illustrate this by finding the rule for the differential of a product with three factors. If

$$y = uvw,$$

then

$$\log|y| = \log|u| + \log|v| + \log|w|.$$

Applying the differential rules and Problem 5 in Section I, we see that

$$\frac{dy}{y} = \frac{du}{u} + \frac{dv}{v} + \frac{dw}{w},$$

or

$$dy = uvw\left(\frac{du}{u} + \frac{dv}{v} + \frac{dw}{w}\right).$$

PROBLEMS FOR SECTION 2

Differentiate the following.

1. $x^{\sqrt{x}}$ 2. 3^{2x} 3. 9^x 4. $x^{\log x}$

5. $(x^a)^x$ 6. $x^{(a^x)}$ 7. $(\sin x)^x$ 8. $(\log x)^{\log x}$

9. $10^{(ax^2+bx+c)}$ 10. $(e^e)^x$ 11. $x^{\tan x}$ 12. $(x^2)^{x^2}$

13. We define $\log_a x$ as the function inverse to a^x. Show that

$$\log_a x = \frac{\log x}{\log a}.$$

14. Strictly speaking, the equation

$$a^x = e^{x \log a}$$

should be taken as the *definition* of a^x. Prove, then, that:

$$a^{x+y} = a^x a^y, \quad (a^x)^y = a^{xy}, \quad \text{and} \quad a^1 = a.$$

In the following problems, compute dy/dx by logarithmic differentiation.

15. $y = \sqrt{1 + x^2}$ 16. $y^3 = x(x - 1)$ 17. $y^n = x^m$

18. $y = x^{(e^x)}$ 19. $(xy)^x = 4$ 20. $y = (\sin x)^x$

21. $y = x^{\tan x}$ 22. $y^3 + y = x^3$

23. $y = \sqrt{\left(\frac{1-x}{1+x}\right)}$ 24. $y = \dfrac{(1+x)^{1/3}(1-x)^{2/3}}{(1+x^2)^{1/6}}$

3. THE INVERSE FUNCTION RULE

The reasoning used in Section 1 to define the logarithm function and prove its differentiability can be copied almost verbatim for the inverse of an arbitrary increasing function.

We start with a function f that is continuous and increasing on an interval I, and proceed in a number of steps.

I.   *The range of f over I is an interval J.* This was shown at the end of Chapter 2 for two cases. If the domain is the closed interval $[a, b]$, then the intermediate-value principle says exactly that the range is $[f(a), f(b)]$. If the domain is the open interval (a, b), and if A and B are the limits of $f(x)$ as x approaches a and b, respectively, then the first case is used to show that the range is the open interval (A, B). Here a and/or A can be $-\infty$, etc.

II.   Since f is increasing, each number y in the range interval J is assumed as a value at only one point x in I. We can therefore define the *inverse function g* on J as follows: For each number y in J,

$$g(y) = \text{the unique number } x \text{ in } I \text{ such that } y = f(x).$$

Reversing the roles of x and y, we can also write:

$$g(x) = \text{the unique number } y \text{ such that } x = f(y).$$

III.   *The function g is increasing and continuous, with domain J and range I.* In order to see this, we start with the *reverse increasing* property of f:

$$\text{if } f(y_1) < f(y_2) \qquad \text{then } y_1 < y_2.$$

Setting $y_1 = g(x_1)$, $x_1 = f(y_1)$, etc., turns this into

$$\text{if } x_1 < x_2 \qquad \text{then } g(x_1) < g(x_2).$$

Thus g is increasing. By definition, g has domain J and range I, and it follows (Theorem 6 in Chapter 2, Section 7) that g is automatically continuous.

IV.   By definition, $g(x) = y$ if and only if $x = f(y)$. Therefore $y = g(x) = g(f(y))$ and $x = f(y) = f(g(x))$, and we have the cancellation identities

$$g(f(x)) = x \text{ for every } x \text{ in the domain of } f;$$

$$f(g(x)) = x \text{ for every } x \text{ in the domain of } g.$$

These identities characterize mutually inverse pairs of functions. Each function undoes what the other has done.

At this point we have the following theorem.

THEOREM 1   *Let f be continuous and increasing on an interval I. Then the range of f over I is an interval J, and f has a continuous, increasing inverse function g with domain J and range I.*

V.   Since the equations $y = g(x)$ and $x = f(y)$ are equivalent, they have the same graph in the standard x,y-coordinate plane. On the other hand, the equation $y = f(x)$ is obtained from $x = f(y)$ by interchanging the variables, so *its* graph is the symmetric image of the above common graph in the line $y = x$. Therefore:

The graphs of the inverse functions g(x) and f(x) in the standard x,y-plane are mirror images of each other in the line y = x.

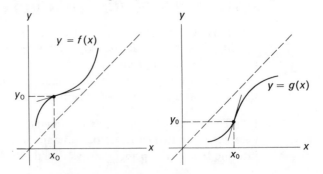

VI. Finally, we come to the derivative rule for inverse functions.

THEOREM 2 **The inverse function rule.** *If f has a continuous, everywhere positive derivative on I then g is differentiable on J, and*

$$g'(x) = \frac{1}{f'(g(x))}.$$

Proof. In the usual increment notation (setting $y = g(x)$, $y + \Delta y = g(x + \Delta x)$, etc.) we have

$$\Delta x = (x + \Delta x) - x = f(y + \Delta y) - f(y) = f'(Y)\Delta y,$$

the final equation being the mean-value principle for f. So

$$\frac{\Delta y}{\Delta x} = \frac{1}{f'(Y)},$$

where Y is some number between y and $y + \Delta y$. Now let $\Delta x \to 0$. Then $\Delta y \to 0$ because $y = g(x)$ is continuous, $Y \to y$ by the squeeze limit law, and $f'(Y) \to f'(y)$ because f' is continuous. Thus

$$\frac{\Delta y}{\Delta x} \to \frac{1}{f'(y)} \qquad \text{as } \Delta x \to 0.$$

That is, $y = g(x)$ is differentiable and its derivative is $1/f'(y) = 1/f'(g(x))$. ∎

REMARK 1 The results above all hold equally well for a *decreasing* function having an everywhere *negative* derivative. The inverse of a decreasing function is decreasing. The whole argument can be repeated, but it is simpler to derive the decreasing results as a corollary of the increasing results. Just note that if $f(x)$ is a *decreasing* function with an everywhere *negative* derivative, then $h(x) = f(-x)$ is an *increasing* function with an everywhere *positive* derivative. Moreover, if $k(x)$ is the inverse of $h(x)$ given by the above development, then $g(x) = -k(x)$ can be checked to be the inverse of $f(x)$.

REMARK 2 We now know that if f is a monotone function on an interval I, with a continuous derivative that is never zero, then f has a differentiable inverse. There is a very simple partial converse which explains why we have been looking only at monotone functions in our search for inverses.

If f is continuously differentiable on the interval I and if f has a differentiable "left inverse" g,

$$g(f(x)) = x \text{ for every } x \text{ in } I,$$

then f is necessarily monotone.

Proof. By the chain rule

$$g'(f(x)) \cdot f'(x) = 1.$$

Therefore f' is never zero, and must have constant sign. Therefore f is monotone. ∎

REMARK 3 The proof of Theorem 2 used the prior result that the inverse function g is continuous. This can be avoided, and the proof made entirely self-contained, as follows. Suppose first that the domain and range intervals are both closed, say $I = [a, b]$ and $J = [f(a), f(b)] = [A, B]$. Then the positive continuous function f' has a minimum value m on I (the extreme-value principle), and $m > 0$. Adding this fact in the middle of the proof, we have

$$0 < \frac{\Delta y}{\Delta x} = \frac{1}{f'(Y)} \le \frac{1}{m}.$$

So $|\Delta y| \le m |\Delta x|$. This shows that g is continuous, and we are back where we were. If I and J are open, the result for closed intervals shows that g is differentiable on every closed subinterval $[A, B]$ in J and hence everywhere.

PROBLEMS FOR SECTION 3

The following corollary of Theorems 1 and 2 depends on the fact that a function is increasing on a closed interval $[a, b]$ if it is known to be continuous on $[a, b]$ and increasing on (a, b). (See Chapter 2, Section 7, Theorem 5.)

COROLLARY *If f is continuous on the interval I, and if $f'(x)$ is everywhere positive (with the possible exception of a finite number of points), then f is invertible over I, and its inverse function g is of the same type. That is, g is continuous, and has an everywhere positive derivative (except possibly at a finite number of points).*

Some of the functions in Problems 1 through 15 are already known to be invertible. But give a (new) proof in each case by applying the corollary above.

1. $f(x) = x^3$
2. $f(x) = x^2$, on the interval $[0, \infty)$
3. $f(x) = x^5$
4. $f(x) = x^3 + x$
5. $f(x) = 2x - x^2$, on $[-1, 1]$
6. $f(x) = x^3 + x^2$, on $[0, \infty)$
7. $f(x) = x^3 - 3x^2 + 3x$
8. $f(x) = \frac{x^5}{5} - \frac{x^4}{2} + \frac{x^3}{3}$
9. $f(x) = x^3 - x^2 + x$
10. $f(x) = x\sqrt{1 + x^2}$
11. $f(x) = \frac{x}{\sqrt{1 + x^2}}$
12. $f(x) = \sin x$, on $[-\pi/2, \pi/2]$
13. $f(x) = \tan x$, on $(-\pi/2, \pi/2)$
14. $f(x) = x \cos x$, on $[-\pi/4, \pi/4]$
15. $f(x) = \sin^3 x$, on $[-\pi/2, \pi/2]$
16. Prove that $f(x) = x^3/(1 + x^2)$ is invertible, and show that its inverse function is everywhere defined, i.e., its domain is $(-\infty, \infty)$.

In general, a function is invertible if and only if it is *one to one*, meaning that

$$\text{if } x_1 \ne x_2 \quad \text{then } f(x_1) \ne f(x_2).$$

This is just a way of saying that each y in the range of f comes from only one x in the domain, so the equation $y = f(x)$ determines x as a function of y. The function $x = g(y)$ defined this way,

$$g(y) = \text{the unique } x \text{ such that } y = f(x),$$

is called the inverse of f. Since the two equations $y = f(x)$ and $x = g(y)$ are then equivalent, we can substitute from each into the other, converting them into the cancellation identities

$$g(f(x)) = x \text{ for every } x \text{ in the domain of } f,$$
$$f(g(y)) = y \text{ for every } y \text{ in the domain of } g.$$

We take these identities to be the formal definition of a pair of inverse functions.

17. Show that $f(x) = (x + 4)/3$ and $g(x) = 3x - 4$ are mutually inverse functions by computing $f(g(x))$ and $g(f(x))$.

18. Show that $f(x) = (4 - x)/(1 + x)$ is its own inverse by computing $f(f(x))$.

19. Compute $f(g(x))$ and $g(f(x))$, where

$$f(x) = \frac{x}{\sqrt{1 + x^2}} \quad \text{and} \quad g(x) = \frac{x}{\sqrt{1 - x^2}}.$$

What is your conclusion?

20. Find the values of a and b such that the linear function $f(x) = ax + b$ is self-inverse. There is more than one solution. Graph each type.

The following problems explore the inverse function relationship when the equation $y = f(x)$ is simple enough to be explicitly solvable for x in terms of y. Show that each function f has an inverse function g by solving the equation for x in terms of y to get an equivalent equation of the form $x = g(y)$.

21. $y = f(x) = 3x - 5$

22. $y = f(x) = 1/x$

23. $y = f(x) = x^{3/5}$

24. $y = f(x) = 1 + x^{1/7}$

25. $y = f(x) = \dfrac{1 - x}{1 + x}$

26. $y = f(x) = x^3 + 3x^2 + 3x + 1$

27. $y = f(x) = \dfrac{x}{\sqrt{1 + x^2}}$

4. THE ARCSINE FUNCTION

We think of the functions $f(x) = \sqrt{x}$ and $g(x) = x^2$ as being inverse to each other, but this is only partly correct. The function $\sqrt{x}$ is strictly increasing and has a differentiable inverse, but its inverse is only the "right-hand half" of $g(x) = x^2$. This is clear geometrically from the reflection principle. It must also show up in the cancelling identities. We find that one is all right,

$$g(f(x)) = (\sqrt{x})^2 = x,$$

but the other is not, for

$$f(g(x)) = \sqrt{x^2} = |x|.$$

Thus we have $f(g(x)) = x$ only when $x \geq 0$.

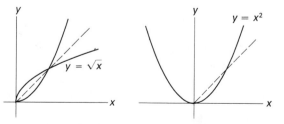

It was shown in the last section that if g is a differentiable function whose domain is an interval I, then g has a differentiable inverse if and only if g is monotone. From this viewpoint, the function $g(x) = x^2$ in its entirety doesn't have an inverse because it isn't monotone, and what we have done above, in order to get an inverse for x^2, is to restrict its domain to an interval on which it *is* monotone, choosing as large an interval as possible.

Now consider $y = g(x) = \sin x$. In order to get an inverse function we have to restrict $\sin x$ to an interval on which it is monotone. There are many possible choices, but the standard one is $[-\pi/2, \pi/2]$. The inverse of this restricted sine function is called the *arcsine* function, and its value at y is designated arcsin y. Thus, for each number y in the closed interval $[-1, 1]$, arcsin y is defined to be *that number x in $[-\pi/2, \pi/2]$ such that $y = \sin x$*. When the roles of x and y are interchanged, the definition is

arcsin x = the unique y in $[-\pi/2, \pi/2]$ such that $x = \sin y$.

The domain of arcsin x is the closed interval $[-1, 1]$.

The graph of arcsin x is the reflection in the line $y = x$ of the graph of $\sin x$ (restricted to $[-\pi/2, \pi/2]$).

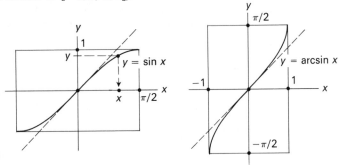

The cancellation identities are

$$\sin(\arcsin x) = x \qquad \text{(for } x \text{ in } [-1, 1]),$$
$$\arcsin(\sin x) = x \qquad \text{(for } x \text{ in } [-\pi/2, \pi/2]).$$

According to Theorem 2, $y = \arcsin x$ is everywhere differentiable in $[-1, 1]$ except at the two endpoints. To compute its derivative, we can differentiate the first cancellation identity. The algebra is a little simpler if instead we differentiate the equation

$$x = \sin y$$

implicitly, knowing that it determines y as a differentiable function of x. This gives

$$1 = \frac{d}{dx} \sin y = \cos y \cdot \frac{dy}{dx}.$$

Therefore

$$\frac{dy}{dx} = \frac{1}{\cos y} = \frac{1}{\pm\sqrt{1 - \sin^2 y}} = \frac{1}{\pm\sqrt{1 - x^2}}.$$

Moreover, $\cos y > 0$ when $-\pi/2 < y < \pi/2$ so the plus sign before the radical is correct. Thus

$$\frac{d}{dx} \arcsin x = \frac{1}{\sqrt{1 - x^2}},$$

$$\frac{d}{dx} \arcsin u = \frac{1}{\sqrt{1 - u^2}} \cdot \frac{du}{dx},$$

$$d(\arcsin u) = \frac{du}{\sqrt{1 - u^2}}.$$

EXAMPLES   1. If a is positive, then

$$\frac{d}{dx} \arcsin \frac{x}{a} = \frac{1}{\sqrt{1 - \left(\frac{x}{a}\right)^2}} \cdot \frac{1}{a} = \frac{1}{\sqrt{a^2 - x^2}}.$$

2. $$\frac{d}{dx}(\sqrt{1 - x^2} \arcsin x - x) = \frac{\sqrt{1 - x^2}}{\sqrt{1 - x^2}} + \arcsin x \cdot \frac{-x}{\sqrt{1 - x^2}} - 1$$

$$= -\frac{x \arcsin x}{\sqrt{1 - x^2}}.$$

3. $$\frac{d}{dx} \arcsin \frac{1}{x} = \frac{1}{\sqrt{1 - (1/x)^2}} \cdot \frac{-1}{x^2} = \frac{-\sqrt{x^2}}{x^2 \sqrt{x^2 - 1}} = \frac{-1}{|x|\sqrt{x^2 - 1}}.$$

The domain of this function consists of two separate intervals: $(-\infty, -1]$ and $[1, \infty)$.

The inverse of $\cos x$, called $\arccos x$, is similar to $\arcsin x$. We have to choose an interval on which $\cos x$ is monotone, and the usual choice is $[0, \pi]$, so that this time we are dealing with *decreasing* functions. The graph follows.

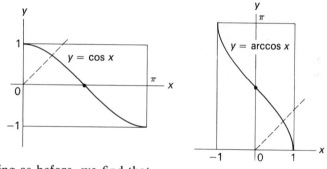

Proceeding as before, we find that

$$\frac{d}{dx} \arccos x = -\frac{1}{\sqrt{1 - x^2}}.$$

REMARK   Therefore the derivative of $\arcsin x + \arccos x$ is zero, and the sum is a constant! Once our attention has been directed to this fact, we find that we can establish it directly from the definition of the two functions. It depends on the identity

$$\cos(\pi/2 - y) = \sin y$$

for all y, and the fact that $\pi/2 - y$ runs from π to 0 as y runs from $-\pi/2$ to $\pi/2$. Thus, if x is the common value in the above identity, then

$$\arcsin x = y,$$

$$\arccos x = \frac{\pi}{2} - y,$$

and

$$\arcsin x + \arccos x = \frac{\pi}{2}.$$

The derivative formula for $\arccos x$ can now be obtained by differentiating $\pi/2 - \arcsin x$.

PROBLEMS FOR SECTION 4

Find the derivatives of the following functions.

1. $\arcsin \sqrt{x}$ 2. $\arcsin kx$ 3. $\arcsin x^2$

4. $\arcsin(\sin^2 x)$ 5. $\arcsin(\cos x)$ 6. $\arcsin(1 - 2x)$

7. $x \arcsin x + \sqrt{1 - x^2}$ 8. $\arcsin x + \sqrt{1 - x^2}$

9. $\arcsin x + x\sqrt{1 - x^2}$ 10. $\arcsin\left(\dfrac{1 - x}{1 + x}\right)$

11. $\arcsin\left(\dfrac{\cos x}{1 + \sin x}\right)$ 12. $\cos^2(\arcsin x)$

13. $\cos(\arcsin x) - \sqrt{1 - x^2}$ 14. $\arcsin\dfrac{x}{\sqrt{1 + x^2}}$

15. If we don't restrict the domain of $g(x) = x^2$, then g fails to be the inverse of $f(x) = \sqrt{x}$, as was noted in the text. But g is a *left inverse* of f, in the sense that $g(f(x)) = x$ for all x in the domain of f. What breaks down is that g fails to be a right inverse of f. Show, similarly, that $\sin x$ is a left inverse of $\arcsin x$ but not a right inverse.

16. Show that $\pi - \arcsin x$ is another right inverse of $\sin x$.

17. Describe the most general differentiable right inverse of $\sin x$ with domain $[-1, 1]$,

18. Draw the graph of $\arcsin(\sin x)$ for $-2\pi \le x \le 2\pi$.

19. Do Problems 1 and 6, and compare their answers. Try to figure out what the explanation is.

20. Show that

$$\arccos x = \arcsin \sqrt{1 - x^2}$$

for all x in the interval $[0, 1]$.

21. Derive the formula for the derivative of $\arccos x$ by differentiating the right side of the above identity.

5. THE ARCTANGENT FUNCTION

To invert the tangent function we use the same restricted domain as for the sine function, except that now the endpoints are missing. Thus, if x is restricted to the open interval $(-\pi/2, \pi/2)$, then $f(x) = \tan x$ is an increasing function with an everywhere positive derivative, and so has a differentiable inverse, by Theorems 1 and 2. This new inverse function is called the *arctangent* function, and its value at x is designated $\arctan x$. Thus,

$$\arctan x = \text{the unique } y \text{ in } (-\pi/2, \pi/2) \text{ such that } x = \tan y.$$

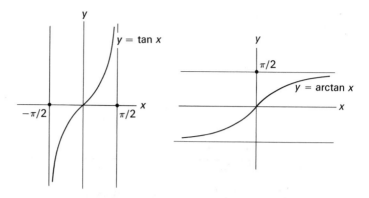

The domain of arctan x *is the whole real-number system* $(-\infty, \infty)$*.* This fact makes arctan x a more generally useful function than arcsin x.

The cancellation identities are:

$$\tan(\arctan x) = x \quad \text{for every } x,$$
$$\arctan(\tan x) = x \quad \text{for } x \text{ in } (-\pi/2, \pi/2).$$

We can find the derivative of arctan x by differentiating the equation

$$x = \tan y$$

implicitly, knowing that it determines y as a differentiable function of x. We get

$$1 = \frac{d}{dx}(\tan y) = sec^2 y \frac{dy}{dx},$$

so

$$\frac{dy}{dx} = \frac{1}{sec^2 y} = \frac{1}{1 + \tan^2 y} = \frac{1}{1 + x^2}.$$

Thus,

$$\frac{d}{dx}\arctan x = \frac{1}{1 + x^2},$$

$$\frac{d}{dx}\arctan u = \frac{1}{1 + u^2}\frac{du}{dx}$$

$$d \arctan u = \frac{du}{1 + u^2}.$$

EXAMPLES 1.
$$\frac{d}{dx}\arctan\frac{x}{a} = \frac{1}{1 + \left(\dfrac{x}{a}\right)^2}\cdot\frac{1}{a} = \frac{a}{a^2 + x^2}.$$

2.
$$\frac{d}{dx}(1 + x^2)\arctan x = \frac{1 + x^2}{1 + x^2} + (\arctan x)2x$$

$$= 2x \arctan x + 1.$$

3. If $x \neq -1$, then

$$\frac{d}{dx} \arctan \frac{x-1}{x+1} = \frac{1}{1 + \dfrac{(x-1)^2}{(x+1)^2}} \cdot \frac{(x+1) - (x-1)}{(x+1)^2}$$

$$= \frac{2}{(x+1)^2 + (x-1)^2} = \frac{1}{x^2 + 1}.$$

See Extra Problem 26 for what is behind this interesting result.

PROBLEMS FOR SECTION 5

Differentiate the following functions.

1. $\arctan x^2$ 2. $\arctan 3x$

3. $\arctan ax$ 4. $\arctan \sqrt{x}$

5. $\arctan \sqrt{x^2 - 1}$ 6. $\arctan\left(\dfrac{\sin x}{1 + \cos x}\right)$

7. $\arctan x + \log \sqrt{1 + x^2}$

Prove the following identities.

8. $\dfrac{d}{dx} [x \arctan x - \log\sqrt{1 + x^2}] = \arctan x$

9. $\dfrac{d}{dx} [\sqrt{x} - \arctan \sqrt{x}] = \dfrac{\sqrt{x}}{2(1 + x)}$ 10. $\dfrac{d}{dx} \arctan\left(\dfrac{e^x - e^{-x}}{2}\right) = \dfrac{2}{e^x + e^{-x}}$

11. If we invert $\cot x$ on the interval $(0, \pi)$, show that

$$\operatorname{arccot} x = \frac{\pi}{2} - \arctan x,$$

and hence that

$$\frac{d}{dx} \operatorname{arccot} x = -\frac{1}{x^2 + 1}.$$

12. Show that for positive values of x,

$$\operatorname{arccot} x = \arctan \frac{1}{x}.$$

Then compute the derivative of $\operatorname{arccot} x$ from this identity, and compare with the above problem.

13. If we invert $x = \sec y$ on the interval $[0, \pi]$, excluding $\pi/2$, show that

$$\frac{d}{dx} \operatorname{arcsec} x = \frac{1}{|x|\sqrt{x^2 - 1}}.$$

14. Prove similarly that

$$\frac{d}{dx} \operatorname{arccsc} x = -\frac{1}{|x|\sqrt{x^2 - 1}}.$$

Differentiate (using the formulas in Problems 11 through 14 when necessary).

15. $\operatorname{arccot}(\tan x)$ 16. $\operatorname{arcsec} \sqrt{x^2 + 1}$

17. $\arcsin \dfrac{x}{\sqrt{1 + x^2}}$ 18. $\operatorname{arcsec} \dfrac{1}{x}$

19. $\operatorname{arccot}(\log x)$

6. CONCAVITY AND INVERSE FUNCTIONS

The graphs below suggest what happens to concavity when a graph is reflected in, or rotated over, the 45° line $y = x$. It appears that *an increasing graph reverses its concavity when reflected*, and that *a decreasing graph preserves concavity*.

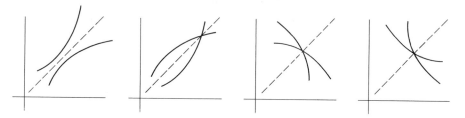

These conjectures are easy to prove. We saw earlier that two functions related this way are mutually inverse and, in particular, satisfy

$$f(g(x)) = x.$$

Then $f'(g(x))g'(x) = 1$ and

$$g'(x) = \frac{1}{f'(g(x))},$$

by the chain rule. We have been through all of this before. Differentiating once more, and then substituting from the above equation, we have

$$g''(x) = -\frac{1}{[f'(g(x))]^2} \cdot f''(g(x)) \cdot g'(x)$$

$$= -\frac{f''(g(x))}{[f'(g(x))]^3} = -\frac{f''(y)}{[f'(y)]^3},$$

where $y = g(x)$. We are supposing that f' is never 0. Thus if f is increasing, then f' is everywhere positive and the formula above shows that g'' and f'' have opposite signs, so that concavity reverses. Similarly, if f is decreasing, then f' and $(f')^3$ are everywhere negative, and the formula shows that concavity is preserved.

EXTRA PROBLEMS FOR CHAPTER 7

Differentiate the following functions:

1. $x \log x - x$
2. $2x^2 \log x - x^2$
3. $\log(\cos x)$
4. $\log(e^x + e^{-x})$
5. $\log(\log x)$
6. $x^{(x^n)}$
7. $x^{1/x}$
8. $(x^x)^x$
9. $x^{(x^x)}$
10. $\arcsin(e^x)$
11. $\arcsin(\tan x)$
12. $\arctan(\sin x)$
13. $\arctan(e^x)$
14. $\arctan\left[\dfrac{e^x - e^{-x}}{2}\right]$
15. $\operatorname{arcsec}\left[\dfrac{e^x + e^{-x}}{2}\right]$

16. Prove the identities

$$\log 1 = 0,$$
$$\log(1/u) = -\log u$$

from the addition law $\log(ab) = \log a + \log b$.

17. Use the cancellation identities to show that the law of exponents follows from the logarithm law.

18. Prove that if f is a function defined for positive x such that $f'(x) = 1/x$ and $f(1) = 0$, then
$$f(ab) = f(a) + f(b)$$
for all positive a and b. [*Hint:* What is $f'(ax)$?]

19. Using the notation $u' = du/dx$, etc., show that
$$\frac{d}{dx} \log(uv) = \frac{u'}{u} + \frac{v'}{v},$$
$$\frac{d}{dx} \log(uvw) = \frac{u'}{u} + \frac{v'}{v} + \frac{w'}{w}.$$

20. Prove that $a \log x < x^a - 1$ when $x > 1$ for any positive number a. (Use the mean-value principle.)

21. Using the above inequality, show that for any positive number b,
$$\log x < x^b$$
when x is large enough. (Take $a = b/2$ and manipulate a little.)

22. Show then that
$$x^n < e^x$$
when x is large enough.

23. Show from the preceding problem that
$$x^m e^{-x} \to 0$$
as $x \to +\infty$, for any positive number m.

24. Find the minimum value of $f(x) = x \log x$.

25. Find the minimum value of $f(x) = x^2 \log x$.

26. It appeared from Example 3 in Section 5 that
$$\arctan \frac{x - 1}{x + 1} = \arctan x + C$$
over the interval $(-1, \infty)$. Show directly that this is so and that $C = -\pi/4$.

27. Draw the graph of $\arctan(\tan x)$ over $[-2\pi, 2\pi]$.

28. Show that
$$\arctan x = \arcsin \frac{x}{\sqrt{1 + x^2}}.$$

29. Show that
$$\arcsin x = \arctan \frac{x}{\sqrt{1 - x^2}}.$$

30. Find the maximum value of
$$f(x) = \arctan(x + 1) - \arctan(x - 1).$$

31. Find the value of x where
$$f(x) = \arctan 2x - \arctan x$$
has its maximum value on the interval $[0, \infty)$.

32. Sketch the graph of
$$y = 2 \arctan \frac{x}{2} - \arctan x$$
(critical points, concavity, behavior at infinity).

33. Sketch the graph of
$$y = \arctan 2x - \arctan x.$$

34. A statue a feet tall stands on a pedestal whose top is b feet above an observer's eye level. If the observer stands x feet away from the pedestal, show that the angle θ subtended by the statue at his eyes is given by
$$\theta = \arctan\frac{a+b}{x} - \arctan\frac{b}{x}.$$

35. Find where the observer must stand in order to maximize the angle of vision θ given by the above formula.

36. Show that if f is increasing and equal to its own inverse, then $f(x) = x$.

37. Suppose that an equation E in x and y determines each of x and y as a function of the other. Show that these mutually inverse functions are the same function if and only if interchanging x and y in the equation E yields an equivalent equation.

38. Keeping in mind the principles stated in Problems 36 and 37, draw the graph of a self-inverse function f that is decreasing and concave up. Find an equation for such a function.

39. Draw a graph and find an equation for a decreasing self-inverse function f that is concave down.

40. What can you say about the function solution of the equation
$$x^{1/3} + y^{1/3} = 2?$$
Sketch its graph.

41. a) Suppose f is a differentiable function defined on the interval $(-1, 1)$ such that $f'(x) = 1/\sqrt{1-x^2}$. Show that f is invertible, and that its inverse function g necessarily satisfies the equation
$$g^2 + (g')^2 = 1.$$

    b) Prove from this equation that $g'' = -g$. (Remember that $g' \neq 0$.) What can you then say about g?

42. Compute $f(g(x))$ and $g(f(x))$, where
$$f(x) = x^2 - 2x + 1, \qquad g(x) = 1 + \sqrt{x}.$$
Show that f and g are mutually inverse if the domain of f is suitably restricted.

43. Show that
$$f(x) = (x^2 + 1)^{1/4} \qquad \text{and} \qquad g(x) = \sqrt{(x^2+1)(x^2-1)}$$
are mutually inverse if their domains are suitably restricted.

44. Show that $f(x) = \sqrt{1 - x^2}$ is its own inverse, provided that its domain is suitably restricted.

45. a) Show that if $g(f(x)) = x$ for every x in the domain of f, then the domain of f is included in the range of g and the range of f is included in the domain of g.

    b) Show therefore that if f and g are mutually inverse, then
$$\text{domain } f = \text{range } g,$$
$$\text{range } f = \text{domain } g.$$

46. Show that if f and g are both invertible, then so is their composition product $f \circ g$, and
$$\text{inv}\,(f \circ g) = (\text{inv } g) \circ (\text{inv } f).$$

47. Assuming that $ad \neq bc$, show that

$$f(x) = \frac{ax + b}{cx + d}$$

is self-inverse if and only if:

a) $d = -a$,   or

b) $d = a, c = b = 0$.

48. Show that if f and g are mutually inverse, then so are $f(ax)$ and $g(x)/a$, for any nonzero constant a.

Instead of defining $\log x$ as the inverse function of e^x as in Section 1, it is possible, and from some points of view preferable, to define e^x as the inverse of $\log x$. The following Problems 49–54 outline this alternative development.

49. Define $L(x)$ on the interval $(0, \infty)$ as follows: If $x \geq 1$ then

$L(x)$ = the area under the graph of $y = 1/t$ from $t = 1$ to $t = x$.

If $0 < x < 1$, then

$L(x)$ = the negative of the area under $y = 1/t$ from $t = x$ to $t = 1$.

We know from Chapter 4 that then $L(x)$ is an antiderivative of $1/x$ on $(0, \infty)$,

$$L'(x) = \frac{1}{x},$$

and that $L(1) = 0$.

Use these properties of L to prove that L is a logarithm, i.e., that

$$L(ab) = L(a) + L(b)$$

for any two positive numbers a and b. [*Hint:* show by the chain rule that $L'(ax) = L'(x)$.]

50. Since L has a positive derivative it is an increasing function. Show that its range is $(-\infty, \infty)$. [*Hint:* Show that $L(2) > L(1) = 0$ and $L(2^n) = nL(2)$ for every positive integer n.]

51. Assuming the above conclusion, use the results in Section 3 to show that there is a differentiable function $E(x)$, with domain $(-\infty, \infty)$ and range $(0, \infty)$, such that

$$E(L(x)) = x \text{ for every positive number } x,$$
$$L(E(x)) = x \text{ for every number } x.$$

Also, that

$$E'(x) = E(x),$$
$$E(0) = 1.$$

52. Assuming the above cancellation identities and the logarithm law $L(ab) = L(a) + L(b)$, prove that E is an exponential function, i.e., that

$$E(a + b) = E(a)E(b)$$

for all numbers a and b.

53. Give an alternate proof that E is an exponential function by using the derivative law $E' = E$ and the initial condition $E(0) = 1$. [*Hint:* Compute the derivative of $E(a - x)E(x)$. It should come out to be 0.]

54. Assume only the exponential law. Setting $e = E(1)$, show that

$$E(r) = e^r$$

for every rational number r. [*Hint:* Show first that $E(na) = [E(a)]^n$ for every a and every positive integer n.]

8

Finding Antiderivatives

If we start with the five functions x, e^x, log x, sin x, and arcsin x, and build new functions by using the algebraic operations and composition, possibly repeatedly and in combination with each other, we generate the class of *elementary functions*. Such functions are said to have *closed form* because they can be explicitly written down.

The functions cos x and arctan x were omitted from the basic list because they arise as simple compositions:

$$\cos x = \sin(x + \pi/2),$$

$$\arctan x = \arcsin \frac{x}{\sqrt{1 + x^2}}.$$

The derivative of an elementary function is always an elementary function, and it can be found in a systematic way by applying the differentiation rules. When we first turn to antidifferentiation, we probably expect the same pattern; that is, we expect every elementary function to have an antiderivative in "closed form" that we can find in an explicit and systematic manner. We saw in Chapter 4 that polynomials behave this way; in fact, polynomials can be integrated as easily as they can be differentiated.

This is not so, however, in general. There is, in general, no systematic integration procedure that can be followed step by step to a guaranteed answer. There may not even *be* an answer, at least until we invent a new function for the purpose. For example, such a simple looking function as $f(x) = e^{-x^2}$ has no antiderivative at all within the class of elementary functions. It does have an antiderivative: If we define $F(x)$ to be the area under the graph of $y = e^{-t^2}$ from $t = 0$ to $t = x$, then we know from Chapter 4 that $F(x)$ is an antiderivative of e^{-x^2}, and we shall learn how to calculate $F(x)$ in Chapter 11. But it can be proved that there is no way of representing F in terms of functions we already know about. That is, $F(x)$ is not an elementary function.

Despite such uncertainties, it is very useful to find antiderivatives explicitly in terms of known functions when it is possible and feasible. That this process must be viewed as an art rather than a systematic routine should not be discouraging. Many students find, after they get the hang of the three or four basic tricks of the trade, that integrating is more interesting than differentiating because it is more like solving puzzles.

We recall that systematic differentiation involves:

a) The derivative formulas for the functions x^a, e^x, sin x, cos x, log x, arcsin x, and arctan x;
b) The basic differentiation rules:
   1. linearity, $(af + bg)' = af' + bg'$;
   2. the product rule, $(fg)' = fg' + gf'$;
   3. the chain rule, $(f \circ g)' = (f' \circ g) \cdot g'$.

The quotient rule is omitted because it really is a secondary rule derived from the above: f/g is the product $f(1/g)$, and $1/g$ is the composition g^{-1}, a special case of the general power function g^a.

Integration involves these same formulas and rules, but run backwards. We start with the following formulas:

$$\int x^a \, dx = \frac{x^{a+1}}{a + 1} + C, \qquad \text{provided } a \neq -1;$$

$$\int \frac{dx}{x} = \log x + C \qquad \text{(the case } a = -1\text{)};$$

$$\int e^x \, dx = e^x + C;$$

$$\int \sin x \, dx = -\cos x + C;$$

$$\int \cos x \, dx = \sin x + C;$$

$$\int \frac{dx}{1 + x^2} = \arctan x + C;$$

$$\int \frac{dx}{\sqrt{1 - x^2}} = \arcsin x + C.$$

Then more complicated functions are attacked by the backwards rules. We have already considered linearity in Chapter 4, and this leaves only the chain rule and the product rule. The backwards chain rule is called *integration by substitution* or *change of variable*. The backwards product rule is called *integration by parts*.

In addition, there is one integration procedure that does not correspond to a differentiation rule: Occasionally we can subdue a balky integrand by algebraically manipulating it into a more docile form. The systematic integration of rational functions is accomplished this way (in conjunction with the rules), and there are a couple of other situations in which we can benefit by such tactics.

However, no matter how clever we are, there will always be more functions that we cannot integrate than ones we can, and some technique of numerical integration is a necessary companion to the art of integrating. The next chapter takes up this *estimation* approach to integration.

The logarithm integral has a complication that we must discuss. Although the integrand $1/x$ is defined for all x except $x = 0$, the formula $\int dx/x = \log x + c$ is valid only for $x > 0$, since negative numbers don't have logarithms. However, if $x < 0$, then $\log(-x)$ exists, and

$$\frac{d}{dx} \log(-x) = \frac{1}{-x} \cdot (-1) = \frac{1}{x}.$$

Therefore,

$$\int \frac{dx}{x} = \log(-x) + C \qquad \text{over the interval } (-\infty, 0).$$

It is customary to combine these two formulas and write

$$\int \frac{dx}{x} = \log |x| + C, \qquad x \neq 0.$$

But this is a dangerous formula, and it must be used very carefully. It seems to say that $\log |x| + C$ is the most general antiderivative of $1/x$ over the full domain of $1/x$. This is false. The function

$$f(x) = \begin{cases} \log |x|, & \text{when } x < 0, \\ \log |x| + 2, & \text{when } x > 0, \end{cases}$$

is an antiderivative which is not in the above form. The theorem guaranteeing that $F(x) + C$ is the most general antiderivative was only established over an *interval* lying in the domains of both F and $f = F'$. It fails for $\log|x|$ and $d \log|x|/dx = 1/x$, because their domain is not an interval. Therefore the standard formula

$$\int \frac{dx}{x} = \log|x| + C$$

must be understood *subject to this correction*, and must be applied very carefully. The same considerations apply to

$$\int dx/x^2 = -1/x + C,$$

and to other negative powers of x.

1. CHANGE OF VARIABLE: DIRECT SUBSTITUTION

So far the dx in Leibniz's notation $\int f(x)dx$ has played no role, but now we shall find it useful to take dx at its face value as the differential of x.

Consider the integral

$$\int (\sin x)^2 \cos x \, dx.$$

If we set $u = \sin x$, then $du = \cos x \, dx$, and the integral takes on the new form

$$\int u^2 du.$$

Since

$$\int u^2 du = \frac{u^3}{3} + C,$$

and since $u = \sin x$, the answer we are looking for ought to be

$$\int (\sin x)^2 \cos x \, dx = \frac{(\sin x)^3}{3} + C.$$

We see that it is, by differentiating. This is a chain-rule differentiation, and our change of variables from x to u can thus be considered as a backward application of the chain rule.

The general pattern of the above calculation is this. We have a complicated function $f(x)$ that can be seen to be in the form $f(x) = g(h(x))h'(x)$, and we want to compute the integral

$$\int f(x)dx = \int g(h(x))h'(x)dx.$$

If we set $u = h(x)$, then $du = h'(x)dx$, and the integral takes on the new form

$$\int g(u)du.$$

We suppose that we can find an antiderivative G of g, so

$$\int g(u)du = G(u) + C.$$

Then, since $u = h(x)$, the answer we are looking for ought to be

$$\int f(x)dx = G(h(x)) + C.$$

We see that it is, by the chain rule:

$$\frac{d}{dx} G(h(x)) = G'(h(x))h'(x) = g(h(x))h'(x) = f(x).$$

Thus, if we take Leibniz's notation at its face value, as involving a *differential*, then it automatically gives us the correct integration procedure that corresponds to the chain rule.

One might think that this device requires an integrand of such special form that it would hardly ever be applicable, but it is surprising how often it can be used. Here are some further examples:

EXAMPLE 1 $\quad\displaystyle\int \frac{2x\,dx}{1 + x^2} = \int \frac{du}{u}\qquad\qquad \left(\begin{array}{l}\text{set } u = 1 + x^2; \\ \text{then } du = 2x\,dx\end{array}\right)$

$$= \log u + C = \log(1 + x^2) + C.$$

EXAMPLE 2 $\quad\displaystyle\int \frac{2x\,dx}{(1 + x^2)^2} = \int \frac{du}{u^2} = -\frac{1}{u} + C = -\frac{1}{1 + x^2} + C.$

EXAMPLE 3 $\quad\displaystyle\int \frac{\log x}{x}\,dx = \int u\,du\qquad\qquad \left(\begin{array}{l}u = \log x, \\ du = \dfrac{dx}{x}\end{array}\right)$

$$= \frac{u^2}{2} + C = \frac{(\log x)^2}{2} + C.$$

EXAMPLE 4 Frequently we will have to multiply by a constant to get $h'(x)dx$. Thus, in

$$\int xe^{x^2}\,dx,$$

we try setting $u = x^2$. Then $du = 2x\,dx$, which we don't quite have. However, we do have

$$x\,dx = \frac{du}{2},$$

and this is good enough because a constant can be factored out:

$$\int xe^{x^2}\,dx = \int e^{x^2}(x\,dx) = \int \frac{e^u\,du}{2} = \frac{1}{2}\int e^u\,du$$

$$= \frac{1}{2}e^u + C = \frac{1}{2}e^{x^2} + C.$$

The simplest substitution of all occurs when we have the derivative of a known function except for an added constant inside.

EXAMPLE 5
$$\int \cos(x+3)\,dx = \int \cos u\,du \qquad \left(\begin{array}{l} u = x+3, \\ du = dx \end{array}\right)$$
$$= \sin u + C = \sin(x+3) + C.$$

EXAMPLE 6
$$\int \frac{dx}{x+a} = \int \frac{du}{u} \qquad \left(\begin{array}{l} u = x+a, \\ du = dx \end{array}\right)$$
$$= \log|u| + C = \log|x+a| + C.$$

EXAMPLE 7   A multiplicative constant inside is about as simple,

$$\int \cos 2x\,dx = \frac{1}{2}\int \cos u\,du \qquad \left(\begin{array}{l} u = 2x, \\ du = 2dx, \\ dx = \dfrac{du}{2} \end{array}\right)$$
$$= \frac{1}{2}\sin 2x + C.$$

Note that we could have viewed this substitution as setting $x = u/2$, $dx = du/2$. In the next example this is the *natural* procedure. (See Section 4.)

EXAMPLE 8
$$\int \frac{dx}{\sqrt{4-x^2}} = \int \frac{du}{\sqrt{1-u^2}} \qquad \left(\begin{array}{l} x = 2u, \\ dx = 2du \end{array}\right)$$
$$= \arcsin u + C = \arcsin\frac{x}{2} + C.$$

EXAMPLE 9
$$\int \frac{dx}{(x+1)^2 + 2} = \frac{1}{\sqrt{2}}\int \frac{du}{u^2+1} \qquad \left(\begin{array}{l} x+1 = \sqrt{2}\,u, \\ dx = \sqrt{2}\,du \end{array}\right)$$
$$= \frac{1}{\sqrt{2}}\arctan u = \frac{1}{\sqrt{2}}\arctan\frac{x+1}{\sqrt{2}} + C.$$

After a little practice it is perfectly feasible to make simple substitutions mentally and just write the answer down, as in
$$\int \cos(x+3)\,dx = \sin(x+3) + C.$$

PROBLEMS FOR SECTION 1

Integrate the following.

1. $\int \sin 2x\,dx$
2. $\int e^{x/2}\,dx$
3. $\int \sin^2 x \cos x\,dx$
4. $\int x \cos x^2\,dx$
5. $\int t^2 e^{-t^3}\,dt$
6. $\int \sqrt{ax+b}\,dx$

7. $\int \cos(2 - x)dx$

8. $\int e^x \cos(e^x)dx$

9. $\displaystyle\int \frac{\sin x}{1 + \cos x}\, dx$

10. $\int \tan x \, dx$

11. $\displaystyle\int \frac{\log x \, dx}{x}$

12. $\displaystyle\int \frac{dx}{x \log x}$

13. $\displaystyle\int \frac{2x + 1}{x^2 + x - 1}\, dx$

14. $\displaystyle\int \frac{\sin \theta}{\cos^2 \theta}\, d\theta$

15. $\int y \sqrt[3]{a + y}\, dy$

16. $\displaystyle\int \frac{x^{1/2}\, dx}{1 + x^{3/4}}$

17. $\displaystyle\int \frac{(x + 2)dx}{x\sqrt{x - 3}}$

18. $\int (x^3 + 2)^{1/3} x^2 \, dx$

19. $\displaystyle\int \frac{8x^2 \, dx}{(x^3 + 2)^3}$

20. $\displaystyle\int \frac{dx}{4 - 2x}$

21. $\displaystyle\int \frac{e^{\sqrt{x}}}{\sqrt{x}}\, dx$

22. $\displaystyle\int \frac{dx}{(1 + x)^{3/2} + (1 + x)^{1/2}}$

23. $\int (e^x + 1)^3 e^x \, dx$

24. $\displaystyle\int \frac{x \, dx}{x^2 - 1}$

25. $\displaystyle\int \frac{(\log x)^2 \, dx}{x}$

26. $\displaystyle\int \frac{x \, dx}{1 + x^4}$

2. RATIONAL FUNCTIONS

In this section and the next, we shall look at two general situations in which an integrand can be algebraically manipulated into a form that can be integrated.

First, we look at some simple rational functions. The basic device employed here—called a *partial fraction decomposition*—can be used to integrate any rational function whatsoever. We shall discuss this systematic process further in Section 6.

EXAMPLE 1

$$\int \frac{x \, dx}{(x - 1)(x + 3)}$$

The crucial fact is that there are constants a and b such that

$$\frac{x}{(x - 1)(x + 3)} = \frac{a}{x - 1} + \frac{b}{x + 3}.$$

In order to determine these constants, first cross-multiply on the right:

$$\frac{x}{(x - 1)(x + 3)} = \frac{a(x + 3) + b(x - 1)}{(x - 1)(x + 3)}.$$

Then the two numerators have to be equal:

$$x = (a + b)x + (3a - b).$$

But two polynomials are equal only if their corresponding coefficients are equal, so

$$1 = a + b, \qquad 0 = 3a - b.$$

Solving these equations simultaneously, we find that $a = 1/4$ and $b = 3/4$.

Therefore,

$$\int \frac{x\,dx}{(x-1)(x+3)} = \frac{1}{4} \int \frac{dx}{x-1} + \frac{3}{4} \int \frac{dx}{x+3}$$

$$= \frac{1}{4} \log|x-1| + \frac{3}{4} \log|x+3| + C$$

$$= \log|x-1|^{1/4} |x+3|^{3/4} + C.$$

EXAMPLE 2   In order to integrate

$$\int \frac{x\,dx}{(x-1)^2}$$

we have to modify the above scheme slightly. This time we look for constants a and b such that

$$\frac{x}{(x-1)^2} = \frac{a}{(x-1)} + \frac{b}{(x-1)^2}.$$

Proceeding as in the first example we find that $a = b = 1$, so

$$\int \frac{x\,dx}{(x-1)^2} = \int \frac{dx}{(x-1)} + \int \frac{dx}{(x-1)^2}$$

$$= \log|x-1| - \frac{1}{(x-1)} + C.$$

An alternative procedure here is to write $x = (x-1) + 1$ in the numerator:

$$\frac{x}{(x-1)^2} = \frac{(x-1)+1}{(x-1)^2} = \frac{1}{(x-1)} + \frac{1}{(x-1)^2}.$$

EXAMPLE 3   In order for the above device to work, the integrand must be a *proper* rational function; that is, the degree of its numerator must be less than the degree of its denominator. If it isn't proper, we use long division to reduce to the proper case. For example, we find by long division that

$$\frac{x^3 - x + 2}{x^2 - x} = x + 1 + \frac{2}{x^2 - x}.$$

Therefore,

$$\int \frac{x^3 - x + 2}{x^2 - x}\,dx = \frac{x^2}{2} + x + 2 \int \frac{dx}{x(x-1)},$$

and the latter integrand is then computed as above. Thus

$$\frac{1}{x(x-1)} = -\frac{1}{x} + \frac{1}{(x-1)},$$

and the final answer is

$$\frac{x^2}{2} + x + 2 \log \left| \frac{x-1}{x} \right| + C = \frac{x^2}{2} + x + \log \left(1 - \frac{1}{x} \right)^2 + C.$$

EXAMPLE 4

$$\int \frac{x \, dx}{x^2 + 2x - 3} = \int \frac{x \, dx}{(x-1)(x+3)},$$

which is the same as the first example.

EXAMPLE 5

$$\int \frac{dx}{x^2 + 2x + 3}$$

Here we can't factor the denominator, and in fact this is an arctangent integral rather than a logarithm. In order to see what is going on, we complete the square in the denominator, and have

$$\int \frac{dx}{x^2 + 2x + 3} = \int \frac{dx}{(x+1)^2 + 2} = \frac{1}{\sqrt{2}} \int \frac{du}{u^2 + 1}$$

(where we have set $x + 1 = \sqrt{2} \, u$, $dx = \sqrt{2} \, du$). This is the arctan integral

$$\frac{1}{\sqrt{2}} \arctan u + C = \frac{1}{\sqrt{2}} \arctan \left(\frac{x+1}{\sqrt{2}} \right) + C.$$

Note that if the numerator in the above example were $(2x + 2)$, then the answer would be $\log|x^2 + 2x + 3|$, by simple substitution. If the numerator were $2x$, we would rewrite it as $(2x + 2) - 2$, and so have a combination of a logarithm and arctangent.

The device of completing the square could be used even for a quadratic that can be factored, such as in Example 4:

$$x^2 + 2x - 3 = (x+1)^2 - 4$$
$$= [(x+1) - 2][(x+1) + 2]$$
$$= (x-1)(x+3).$$

EXAMPLE 6 We have just seen that there are three distinct possibilities for a proper rational function with a quadratic denominator, depending on how the denominator factors. If the degree of the denominator is three, then there will be three unknown coefficients to determine in the partial-fraction decomposition, and we consider three different possible setups for them.

1. *Three different linear factors*:

$$\frac{1}{(x-1)(x+2)(x+3)} = \frac{a}{x-1} + \frac{b}{x+2} + \frac{c}{x+3}.$$

2. *Two different linear factors,* but *one repeated*:

$$\frac{1}{(x-1)^2(x+3)} = \frac{a}{(x-1)} + \frac{b}{(x-1)^2} + \frac{c}{(x+3)}.$$

3. *One linear factor* and *one irreducible quadratic factor*:

$$\frac{1}{(x-1)(x^2+1)} = \frac{a}{(x-1)} + \frac{bx+c}{(x^2+1)}.$$

The coefficients are computed in the same way as before. Thus (3) goes on:

$$\frac{1}{(x-1)(x^2+1)} = \frac{a(x^2+1) + (bx+c)(x-1)}{(x-1)(x^2+1)};$$

$$1 = (a+b)x^2 + (c-b)x + (a-c);$$

$$\left.\begin{array}{l} a + b = 0 \\[2mm] c - b = 0 \\[2mm] a - c = 1 \end{array}\right\} \quad \begin{array}{l} a = \dfrac{1}{2}, \\[2mm] b = -\dfrac{1}{2}, \\[2mm] c = -\dfrac{1}{2}. \end{array}$$

$$\int \frac{dx}{(x-1)(x^2+1)} = \frac{1}{2}\int \frac{dx}{(x-1)} - \frac{1}{2}\int \frac{x\,dx}{x^2+1} - \frac{1}{2}\int \frac{dx}{x^2+1}$$

$$= \frac{1}{2}\log|x-1| - \frac{1}{4}\log(x^2+1) - \frac{1}{2}\arctan x + C$$

$$= \frac{1}{2}\left[\log\frac{|x-1|}{\sqrt{x^2+1}} - \arctan x\right] + C.$$

In each of the above three cases, the numerator 1 can be replaced by any polynomial of degree less than 3 (the quotient must be proper) without changing the procedure. For example, if we set

$$\frac{x^2+2}{(x-1)(x^2+1)} = \frac{a}{x-1} + \frac{bx+c}{x^2+1},$$

we just change the final set of three equations in the three unknowns *a*, *b*, and *c*. When we equate numerators this time, we have

$$x^2 + 2 = (a+b)x^2 + (c-b)x + (a-c),$$

so

$$\left.\begin{array}{l} a + b = 1 \\[2mm] c - b = 0 \\[2mm] a - c = 2 \end{array}\right\} \quad \begin{array}{l} a = \dfrac{3}{2}. \\[2mm] b = -\dfrac{1}{2}, \\[2mm] c = -\dfrac{1}{2}. \end{array}$$

The general rational function is treated in essentially the same way, but the complications can be greater. Further discussion will be postponed to Section 6.

PROBLEMS FOR SECTION 2

Integrate

1. $\int \dfrac{dx}{x^2 - 9}$

2. $\int \dfrac{dx}{x^2 + 9}$

3. $\int \dfrac{dx}{x^2 - 9x}$

4. $\int \dfrac{dx}{x^2 + x - 2}$

5. $\int \dfrac{dx}{x^2 + 2x - 2}$

6. $\int \dfrac{dx}{x^2 + 2x + 2}$

7. $\int \dfrac{dx}{x^2 + 3x + 2}$

8. $\int \dfrac{dx}{x^2 + 2x + 1}$

9. $\int \dfrac{dx}{4x^2 + 9}$

10. $\int \dfrac{dx}{2x^2 + x - 3}$

11. $\int \dfrac{x\,dx}{x^2 + 4x - 5}$

12. $\int \dfrac{4\,dx}{x^3 + 4x}$

13. $\int \dfrac{x^2\,dx}{x^2 + 2x + 1}$

14. $\int \dfrac{y^2 + 1}{y^2 + y + 1}\,dy$

15. $\int \dfrac{(x + 1)\,dx}{(x^2 + 4x - 5)}$

16. $\int \dfrac{2x^3 + x + 3}{(x^2 + 1)}\,dx$

17. $\int \dfrac{2x^2 + x + 2}{x(x - 1)^2}\,dx$

18. $\int \dfrac{dx}{x^3 + 8}$

19. $\int \dfrac{\cos\theta\,d\theta}{1 - \sin^2\theta}$

20. $\int \dfrac{\sin\theta\,d\theta}{\cos^2\theta + \cos\theta - 2}$

21. $\int \dfrac{e^t\,dt}{e^{2t} + 3e^t + 2}$

22. $\int \dfrac{dx}{x^3 + x^2 + 5x}$

23. $\int \dfrac{dx}{x^3 - 4x}$

24. $\int \dfrac{dx}{x^3 - 4x^2}$

3. TRIGONOMETRIC POLYNOMIALS

Trigonometric polynomials can also be integrated in a systematic way. The integrating device will again be algebraic manipulation of the integrand, this time by using the trigonometric identities:

$$\cos^2 x + \sin^2 x = 1$$
$$\cos^2 x - \sin^2 x = \cos 2x \qquad \text{(from the addition law for } \cos(x + x)\text{).}$$

Adding and subtracting these two laws gives the "power-reducing" formulas:

$$\cos^2 x = \frac{1 + \cos 2x}{2} \qquad \text{and} \qquad \sin^2 x = \frac{1 - \cos 2x}{2}.$$

A trigonometric polynomial is a sum of products of trigonometric functions and, by using the definitions of these functions in terms of sine and cosine, any such product can be reduced to the form

$$\sin^m x \cos^n x,$$

where m and n are integers (positive, zero, or negative). The problem is therefore to integrate any such reduced product. We consider three cases.

I. If m and n are *even positive* integers, we use the power-reducing formulas.

EXAMPLE 1
$$\int \cos^2 x\,dx = \int \frac{(1 + \cos 2x)}{2}\,dx = \frac{x}{2} + \frac{\sin 2x}{4} + C.$$

Since $\sin 2x = 2 \sin x \cos x$ (by the addition law for $\sin(x + x)$), this answer can be rewritten

$$\frac{x + \sin x \cos x}{2} + C.$$

EXAMPLE 2 Two applications of the power-reducing identity yield

$$\cos^4 x = \left(\frac{1 + \cos 2x}{2}\right)^2 = \frac{1}{4} + \frac{1}{2} \cos 2x + \frac{\cos^2 2x}{4}$$

$$= \frac{1}{4} + \frac{\cos 2x}{2} + \frac{1}{4}\left(\frac{1 + \cos 4x}{2}\right) = \frac{3}{8} + \frac{\cos 2x}{2} + \frac{\cos 4x}{8}.$$

Therefore

$$\int \cos^4 x \, dx = \frac{3x}{8} + \frac{\sin 2x}{4} + \frac{\sin 4x}{32} + C.$$

EXAMPLE 3 The above two examples combine to give

$$\int \sin^2 x \cos^2 x \, dx = \int (1 - \cos^2 x)\cos^2 x \, dx$$

$$= \int \cos^2 x \, dx - \int \cos^4 x \, dx$$

$$= \frac{x}{2} + \frac{\sin 2x}{4} - \frac{3x}{8} - \frac{\sin 2x}{4} - \frac{\sin 4x}{32} + C$$

$$= \frac{x}{8} - \frac{\sin 4x}{32} + C.$$

In this way we can attack any product $\sin^m x \cos^n x$, where m and n are *even positive integers*.

II. If $\cos x$ occurs to an *odd* power, then the integration can be performed by simple substitution, by the following device:

$$\cos^{2p+1} x \, dx = \cos^{2p} x \cos x \, dx = (1 - \sin^2 x)^p \cos x \, dx$$

$$= (1 - u^2)^p du \qquad\qquad (u = \sin x).$$

EXAMPLE 4 $$\int \cos^3 x \, dx = \int (1 - \sin^2 x)\cos x \, dx = \int \cos x \, dx - \int \sin^2 x \cos x \, dx$$

$$= \sin x - \frac{\sin^3 x}{3} + C.$$

The same idea lets us integrate any product $\sin^m x \cos^n x$ where m and n are integers and at least one is odd. But when the odd integer is *negative*, the final integrand is a rational function, as in the following example.

EXAMPLE 5

$$\int \frac{dx}{\cos x} = \int \frac{\cos x \, dx}{\cos^2 x} \qquad \left(\begin{matrix} \text{set } u = \sin x, \\ du = \cos x \, dx \end{matrix}\right)$$

$$= \int \frac{du}{1 - u^2} = \frac{1}{2} \int \left[\frac{1}{1 - u} + \frac{1}{1 + u} \right] du$$

$$= \frac{1}{2} \log \left[\frac{1 + u}{1 - u} \right] + C = \log \left[\frac{1 + \sin x}{1 - \sin x} \right]^{1/2} + C.$$

This answer can be simplified by multiplying top and bottom inside the brackets by $(1 + \sin x)$. It becomes

$$\log \left[\frac{1 + \sin x}{|\cos x|} \right] + C = \log |\sec x + \tan x| + C.$$

EXAMPLE 6

$$\int \frac{dx}{\cos^5 x} = \int \frac{\cos x \, dx}{\cos^6 x} = \int \frac{du}{(1 - u^2)^3}.$$

Here the partial-fraction decomposition is

$$\frac{1}{(1 - u)^3 (1 + u)^3} = \frac{a}{1 - u} + \frac{b}{(1 - u)^2} + \frac{c}{(1 - u)^3}$$

$$+ \frac{d}{1 + u} + \frac{e}{(1 + u)^2} + \frac{f}{(1 + u)^3},$$

and a lot of algebra is required.

EXAMPLE 7 If there is a *positive* odd power, then the other exponent does not have to be an integer.

$$\int \sqrt{\sin x} \, \cos^3 x \, dx = \int \sqrt{\sin x} \, (1 - \sin^2 x) \cos x \, dx$$

$$= \int (u^{1/2} - u^{5/2}) du = \frac{2}{3} u^{3/2} - \frac{2}{7} u^{7/2} + C$$

$$= \frac{2}{3} (\sin x)^{3/2} - \frac{2}{7} (\sin x)^{7/2} + C.$$

EXAMPLE 8

$$\int \frac{\sin x \, dx}{\cos^2 x} = - \int \frac{du}{u^2} = \frac{1}{u} + C = \frac{1}{\cos u} + C.$$

The above example shows how we easily recapture the formula

$$\int \tan x \sec x \, dx = \sec x + C$$

in case we have forgotten that

$$\frac{d}{dx} \sec x = \sec x \tan x.$$

But the formula $d \tan x/dx = \sec^2 x$ doesn't reappear in this way. As an integral in $\sin x$ and $\cos x$, it is

$$\int \frac{dx}{\cos^2 x} = \frac{\sin x}{\cos x} + C,$$

and this represents the one case of $\sin^m x \cos^n x$ that we haven't covered so far.

III. *Both m and n are even and at least one of them negative.* In order to handle such integrals we add the formulas

$$\int \sec^2 x \, dx = \tan x + C, \qquad \int \csc^2 x \, dx = -\cot x + C$$

to our basic list, and we convert the identity $\sin^2 x + \cos^2 x = 1$ to the equivalent forms

$$\tan^2 x + 1 = \sec^2 x,$$
$$1 + \cot^2 x = \csc^2 x,$$

in order to apply these additional integrals.

EXAMPLE 9

$$\int \frac{dx}{\cos^4 x} = \int \sec^4 x \, dx = \int (1 + \tan^2 x) \sec^2 x \, dx$$

$$= \int (1 + u^2) du \qquad\qquad (u = \tan x)$$

$$= u + \frac{u^3}{3} + C = \tan x + \frac{\tan^3 x}{3} + C.$$

EXAMPLE 10

$$\int \tan^2 x \, dx = \int (\sec^2 x - 1) dx = \tan x - x + C.$$

EXAMPLE 11

$$\int \frac{dx}{\sin^2 x \cos^2 x} = \int \left(\frac{1}{\sin^2 x} + \frac{1}{\cos^2 x} \right) dx$$

$$= \int (\sec^2 x + \csc^2 x) dx = \tan x - \cot x + C.$$

PROBLEMS FOR SECTION 3

Integrate the following.

1. $\int \sin^2 x \, dx$

2. $\int \sin^3 x \, dx$

3. $\int \sin^4 x \, dx$

4. $\cos^2 3x \, dx$

5. $\int \sin^3 6x \cos 6x \, dx$

6. $\int \cos^3 2\theta \sin 2\theta \, d\theta$

7. $\int \cos^3 x \sin^{-4} x \, dx$

8. $\int \frac{\sin^5 y \, dy}{\sqrt{\cos y}}$

9. $\int \cos^6 \theta \, d\theta$

10. $\int \cos^3 \left(\frac{x}{2} \right) \sin^2 \left(\frac{x}{2} \right) dx$

11. $\int \frac{\sin^3 2x}{\sqrt[3]{\cos 2x}} dx$

12. $\int \sin^3 x \cos^3 x \, dx$

13. $\displaystyle\int \frac{dx}{\sin x \cos x}$ 14. $\int \sin 3x \cot 3x \, dx$ 15. $\int \csc^4 x \cot^2 x \, dx$

16. $\int \tan^4 x \, dx$ 17. $\int \csc^4 \dfrac{x}{4} \, dx$ 18. $\int \sec^3 x \, dx$

19. $\int \tan^3 x \sec^{5/2} x \, dx$ 20. $\int \tan x \sqrt{\sec x} \, dx$

21. $\int \tan^3 x \, dx$. Do this in two ways:
   a) By substitution, since $\sin x$ (and $\cos x$ also) occurs to an odd power;
   b) By using $\tan^2 x = \sec^2 x - 1$.

22. $\int \sec x \, dx$. Do this in two ways:
   a) By substitution, since $\cos x$ occurs to an odd power;
   b) By tricky substitution, multiplying top and bottom by $(\sec x + \tan x)$.

23. $\int \tan^5 x \, dx$ (Do it the easier way.)

24. $\displaystyle\int \frac{\cos^5 x}{\sin^5 x} \, dx$ (Ditto.)

25. In Example 5 the absolute-value sign was not introduced inside the logarithm until the very end. Why was this permissible?

4. INVERSE SUBSTITUTIONS

In making the direct substitutions of Section 1, we set $u = h(x)$ where $h(x)$ was a part of the integrand. But then we also had to find $du = h'(x)dx$ as part of the integrand, and this meant that altogether the integrand had to be in a rather special form.

A much easier way to change variables in the integral $\int f(x)dx$ is simply to set $x = k(u)$ and $dx = k'(u)du$, where $k(u)$ is some function that is suggested by the form of the integrand.

EXAMPLE 1   In order to integrate

$$\int \sqrt{4 - x^2} \, dx,$$

we try setting $x = 2 \sin u$, which will at least eliminate the radical. Then

$$dx = 2 \cos u \, du,$$
$$\sqrt{4 - x^2} = 2\sqrt{1 - \sin^2 u} = 2 \cos u,$$
$$\int \sqrt{4 - x^2} \, dx = 4 \int \cos^2 u \, du = 4 \int \frac{1 + \cos 2u}{2} \, du \qquad \text{(as in Section 3)}$$
$$= 2u + \sin 2u + C$$
$$= 2 \sin u \cos u + 2u + C.$$

Now we have to write the answer in terms of the original variable x, which means that we have to know the inverse of the substitution function. In this case $u = \arcsin x/2$, so

$$\int \sqrt{4 - x^2} \, dx = \frac{1}{2} x\sqrt{4 - x^2} + 2 \arcsin x/2 + C$$

is the final answer. Note that $x = 2 \sin u$ must be restricted to the domain $[-\pi/2, \pi/2]$ to make it invertible.

This process is called *inverse substitution.* Instead of setting $h(x) = u$ as we did in Section 1, we now set $x = k(u)$. This makes for a very easy transformation of the integral, but we have to know an inverse function $u = h(x)$ in order to express the u answer in terms of the original variable x.

There are two other useful substitutions similar to the one used in Example 1:

1. *to rationalize* $\sqrt{a^2 + x^2}$, *set* $x = a \tan u$, *with* $a > 0$;

2. *to rationalize* $\sqrt{x^2 - a^2}$, *set* $x = a \sec u$.

EXAMPLE 2

$$\int \frac{dx}{\sqrt{a^2 + x^2}}$$

We set $x = a \tan u$, $dx = a \sec^2 u \, du$. Then

$$\int \frac{dx}{\sqrt{a^2 + x^2}} = \int \sec u \, du = \int \frac{\cos u \, du}{\cos^2 u}$$

$$= \int \frac{dv}{1 - v^2} = \frac{1}{2} \int \left[\frac{1}{1 - v} + \frac{1}{1 + v} \right] dv$$

$$= \log \left[\frac{1 + v}{1 - v} \right]^{1/2} + C = \log \left[\frac{1 + \sin u}{1 - \sin u} \right]^{1/2} + C$$

$$= \log |\sec u + \tan u| + C \qquad \text{(as in Section 3).}$$

There remains the question: What is sec u when tan $u = x/a$? We could write

$$\sec u = \sec(\arctan x/a)$$

and have an answer. But this is more complicated than necessary, since any such composition of a trigonometric function with an inverse trigonometric function can always be simplified. If we label a right triangle in the simplest way possible consistent with tan $u = x/a$, we see that when tan $u = x/a$, then

$$\sec u = \sqrt{a^2 + x^2}/a.$$

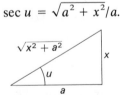

This could also be worked out directly,

$$\sec u = \sqrt{\sec^2 u} = \sqrt{1 + \tan^2 u} = \sqrt{1 + \left(\frac{x}{a} \right)^2},$$

but the right-triangle device is easier. In any case, the final answer is

$$\int \frac{dx}{\sqrt{a^2 + x^2}} = \log\left[\frac{\sqrt{a^2 + x^2} + x}{a}\right] + C = \log\left[\sqrt{a^2 + x^2} + x\right] + C.$$

EXAMPLE 3
$$\int \frac{dx}{\sqrt{x^2 - a^2}}$$

Set $x = a \sec u$, $dx = a \sec u \tan u$. Then

$$\int \frac{dx}{\sqrt{x^2 - a^2}} = \int \frac{a \sec u \tan u \, du}{a \tan u} = \int \sec u \, du$$

$$= \log|\sec u + \tan u| + C$$

$$= \log\left|\frac{x + \sqrt{x^2 - a^2}}{a}\right| + C$$

$$= \log|x + \sqrt{x^2 - a^2}| + C.$$

EXAMPLE 4 The same substitution and diagram is used here:

$$\int \frac{\sqrt{x^2 - a^2} \, dx}{x} = \int \frac{a^2 \sec u \tan^2 u}{a \sec u} du = a \int \tan^2 u \, du$$

$$= a \int (\sec^2 u - 1) du = a \tan u - au + C$$

$$= \sqrt{x^2 - a^2} - a \arctan \frac{\sqrt{x^2 - a^2}}{a} + C.$$

EXAMPLE 5
$$\int \sqrt{2x - x^2} \, dx.$$

We complete the square under the radical, and then proceed as in Example 1.

$$\int \sqrt{2x - x^2} \, dx = \int \sqrt{1 - (1 - x)^2} dx \qquad \left(\begin{array}{l}\text{set } 1 - x = \sin u \\ dx = -\cos u \, du\end{array}\right)$$

$$= -\int \cos^2 u \, du = \frac{u + \sin u \cos u}{2} + C$$

$$= -\frac{\arcsin(1 - x) + (1 - x)\sqrt{2x - x^2}}{2} + C.$$

The computations above have an important feature that we have overlooked. It was careless to write

$$\cos u = \sqrt{1 - \sin^2 u}$$

without checking signs, because $\cos u$ is negative as often as it is positive. However, $u = \arcsin x$ here, so u is restricted to the interval $[-\pi/2, \pi/2]$, and $\cos u$ is positive, as was assumed.

Similarly, $\sec u$ is positive over the interval $(-\pi/2, \pi/2)$. Since this is the range of the inverse function $u = \arctan x$, it is correct to write

$$\sqrt{1 + x^2} = \sqrt{1 + \tan^2 u} = \sqrt{\sec^2 u} = \sec u.$$

However, if we take the domain of $\operatorname{arcsec} x$ to be $[0, \pi]$ minus the point $\pi/2$, so that

$$\operatorname{arcsec} x = \arccos (1/x),$$

then

$$\sqrt{x^2 - 1} = \sqrt{\sec^2 u - 1} = \sqrt{\tan^2 u} = \tan u$$

is correct only when u is in $[0, \pi/2)$, because $\tan u < 0$ on $(\pi/2, \pi]$. This restriction was therefore implicit in Examples 3 and 4. However, see Problems 25 and 26.

Here are a few more rationalizing substitutions.

EXAMPLE 6

$$\int \frac{x \, dx}{(1 + x)^{1/3}}$$

We would like to set $u = (1 + x)^{1/3}$ in order to eliminate the radical. Fortunately, we can easily solve the substitution equation for x in terms of u,

$$x = u^3 - 1,$$
$$dx = 3u^2 \, du,$$

so

$$\int \frac{x \, dx}{(1 + x)^{1/3}} = \int \frac{(u^3 - 1)3u^2 \, du}{u} = 3 \int (u^4 - u) \, du$$

$$= 3 \left(\frac{u^5}{5} - \frac{u^2}{2} \right) + C = \frac{3u^2}{10}(2u^3 - 5) + C$$

$$= \frac{3}{10}(1 + x)^{2/3}(2x - 3) + C.$$

EXAMPLE 7

$$\int \frac{dx}{x^{1/2}(1 + x^{1/3})} = \int \frac{5u^5 \, du}{u^3(1 + u^2)} \qquad \left(\begin{array}{l} x = u^6, \\ dx = 5u^5 \, du \end{array} \right)$$

$$= 5 \int \frac{u^2 \, du}{1 + u^2} = 5 \int \left[1 - \frac{1}{1 + u^2} \right] du$$

$$= 5[u - \arctan u] + C$$

$$= 5[x^{1/6} - \arctan x^{1/6}] + C.$$

EXAMPLE 8   We can do Example 4 in this way, too.

$$\int \frac{\sqrt{x^2 - a^2}\,dx}{x} = \int \frac{\sqrt{x^2 - a^2}\,x\,dx}{x^2} \qquad \left(\text{set } u^2 = x^2 - a^2,\atop u\,du = x\,dx\right)$$

$$= \int \frac{u \cdot u\,du}{u^2 + a^2}$$

$$= \int \left[1 - \frac{a^2}{u^2 + a^2}\right] du$$

$$= u - a \arctan \frac{u}{a} + C$$

$$= \sqrt{x^2 - a^2} - a \arctan \frac{\sqrt{x^2 - a^2}}{a} + C.$$

This device can be used whenever one of the radicals

$$\sqrt{x^2 - a^2}, \qquad \sqrt{a^2 - x^2}, \qquad \sqrt{a^2 + x^2}$$

is combined with an *odd* power of x.

We have not yet justified the inverse substitution process. Here is a general description of what we have been doing. Setting $x = k(u)$ and $dx = k'(u)du$, we write

$$\int f(x)\,dx = \int f(k(u))k'(u)\,du = \int g(u)\,du,$$

where $g(u) = f(k(u))k'(u)$. We now find an antiderivative $G(u)$ of $g(u)$, and claim that then

$$\int f(x)\,dx = G(h(x)) + C.$$

where h is the inverse of the function k.

In order to see that this is correct we simply differentiate:

$$\frac{d}{dx} G(h(x)) = G'(h(x))h'(x) = g(h(x))h'(x)$$

$$= f(k(h(x)))k'(h(x))h'(x),$$

where the final complicated expression comes from the definition of g. But k and h are inverse functions, so

$$k(h(x)) = x \qquad \text{and} \qquad k'(h(x))h'(x) = 1.$$

The complicated answer above therefore collapses to just $f(x)$, so

$$\frac{d}{dx} G(h(x)) = f(x),$$

as claimed.

PROBLEMS FOR SECTION 4

In the following integrations a labeled right triangle will often be helpful when substituting back, at the end.

1. $\displaystyle\int \frac{x^3\,dx}{\sqrt{4 - x^2}}$ 2. $\displaystyle\int \frac{dx}{x\sqrt{x^2 - a^2}}$ 3. $\displaystyle\int \frac{dx}{(x^2 + a^2)^{3/2}}$

4. $\displaystyle\int \frac{dx}{(x^2 - 9)^{3/2}}$ 5. $\displaystyle\int x\sqrt{16 - x^2}\,dx$ 6. $\displaystyle\int \frac{x\,dx}{\sqrt{(x^2 + a^2)^3}}$

7. $\displaystyle\int \frac{dx}{x\sqrt{x^2 + 1}}$ 8. $\displaystyle\int \frac{\sqrt{9 - 4x^2}}{x}\,dx$ 9. $\displaystyle\int \frac{dx}{x\sqrt{25 - x^2}}$

10. $\displaystyle\int \frac{dy}{y^2\sqrt{y^2 - 7}}$

The following problems would normally be tackled in a different way, but this time do them using the trigonometric rationalizing substitutions.

11. $\displaystyle\int \frac{x\,dx}{\sqrt{1 - x^2}}$ 12. $\displaystyle\int \frac{dx}{x^2 + 4}$ 13. $\displaystyle\int \frac{x\,dx}{\sqrt{1 + x^2}}$ 14. $\displaystyle\int \frac{dx}{x^2 - a^2}$

15. $\displaystyle\int \frac{x\,dx}{\sqrt{x^2 - a^2}}$ 16. $\displaystyle\int \frac{dx}{(x^2 + a^2)^2}$ 17. $\displaystyle\int \frac{dy}{\sqrt{a^2 - y^2}}$

Integrate by finding a suitable substitution

18. $\displaystyle\int \frac{dx}{1 + x^{1/3}}$ 19. $\displaystyle\int \frac{dx}{1 + \sqrt{x}}$

20. $\displaystyle\int \frac{1 + x}{\sqrt{1 - x}}\,dx$ 21. $\displaystyle\int \frac{1 + x^{1/2}}{1 + x^{1/3}}\,dx$

22. Show that a right triangle as labeled in the text for reading off the values of the trigonometric functions when $x = a \tan u$ will give the right signs for these functions when x is negative $(-\pi/2 < u < 0)$.

23. Show that a right triangle labeled to read off the values of the trigonometric functions when $x = a \sin u$ will give the right signs of these functions when x is negative $(-\pi/2 \le u < 0)$.

24. The right triangle labeled to show the substitution $x = a \sec u$ does not give the correct signs for the other trigonometric functions when x is negative $(\pi/2 < u < \pi)$. Show that the signs come out right for x negative if one leg is labeled $-\sqrt{x^2 - a^2}$.

25. a) Example 4 in the text was worked under the assumption that x is positive. Show that the answer is also correct for x negative.
   b) Work Example 4 out again for the case of negative x. [*Hint:* The proper procedure is related to the conclusion in Problem 24.]

26. a) Example 3 in the text was worked out under the assumption that x is positive. Show by differentiation that the conclusion is correct for negative x.
   b) Work Example 3 out again for the case that x is negative.

5. INTEGRATION BY PARTS

We make the product rule

$$\frac{d}{dx}\,uv = u\frac{dv}{dx} + v\frac{du}{dx}$$

into an integration rule by solving for the term $u\, dv/dx$ and integrating:

$$\int u\frac{dv}{dx}\, dx = uv - \int v\frac{du}{dx}\, dx.$$

Although the indefinite integral of $d(uv)/dx$ is $uv + C$, we have suppressed the constant C since a constant of integration is contained implicitly in the remaining integrals.

The above formula states a very special integration procedure called *integration by parts*. We can try this device whenever an integrand is a product of two functions, *provided that we already know how to integrate one of them.* For example

$$\int x \cos x\, dx$$

can be considered to be in the form

$$\int u\frac{dv}{dx}\, dx$$

with $u = x$, $dv/dx = \cos x$. We can then use the formula because we know how to integrate $dv/dx = \cos x$ to find v. Any antiderivative is suitable, and we generally (but not always) choose the simplest. So here we set $v = \sin x$. Since $du/dx = 1$, the formula gives

$$\int x \cos x\, dx = x \sin x - \int \sin x \cdot 1 \cdot dx$$
$$= x \sin x + \cos x + C.$$

In terms of differentials, the integration-by-parts formula takes the neater form

$$\int u\, dv = uv - \int v\, du.$$

For instance, in the above example we have $x = u$ and $\cos x\, dx = dv$, from which we can conclude that $du = dx$ and $v = \sin x$.

In addition to the requirement that we be able to integrate one factor, we also need to have the resulting new integral simpler than the original, or at least more amenable. This doesn't necessarily happen. For example, if we had reversed the roles of x and $\sin x$ in the above example, we would get

$$\int (\sin x)x \cdot dx = (\sin x)\frac{x^2}{2} - \int \frac{x^2}{2}\cos x\, dx,$$

which has made things worse rather than better. Thus, when we know how to integrate *both* factors of the integrand, and accordingly have a choice as to how to apply the integration-by-parts formula, we can expect that only one of the possibilities will be useful (and maybe neither will).

As we have said, integration by parts is a very special device, but sometimes it works when nothing else at all will.

EXAMPLE 1

$$\int \log x \, dx = x \log x - \int x \frac{dx}{x}$$

$$= x \log x - \int dx$$

$$= x \log x - x + C.$$

EXAMPLE 2   In Section 3 we had the most trouble with even powers of sin x and cos x. These can successfully be tackled by parts.

$$\int \cos^2 x \, dx = \int \cos x \cdot \cos x \, dx$$

$$= \cos x \sin x - \int \sin x (-\sin x \, dx)$$

$$= \cos x \sin x + \int \sin^2 x \, dx$$

$$= \cos x \sin x + \int (1 - \cos^2 x) dx$$

$$= \cos x \sin x + x + C - \int \cos^2 x \, dx.$$

Solving for $\int \cos^2 x \, dx$, we have

$$\int \cos^2 x dx = \frac{\cos x \sin x + x}{2} + C.$$

EXAMPLE 3

$$\int \tan^2 x \, dx = \int \frac{\sin^2 x}{\cos^2 x} \, dx = \int \sin x \cdot \frac{\sin x \, dx}{\cos^2 x}$$

$$= \sin x \left(\frac{1}{\cos x}\right) - \int \cos x \left(\frac{1}{\cos x}\right) dx$$

$$= \tan x - x + C.$$

EXAMPLE 4   In order to integrate $\int x^2 \cos x \, dx$ we have to apply the integration-by-parts process twice.

$$\int x^2 \cos x \, dx = x^2 \sin x - \int \sin x \, 2x \, dx$$

$$\int x \sin x \, dx = x(-\cos x) - \int (-\cos x) dx$$

$$= -x \cos x + \sin x + C;$$

$$\int x^2 \cos x \, dx = x^2 \sin x - 2[-x \cos x + \sin x + C]$$

$$= x^2 \sin x + 2x \cos x - 2 \sin x + C.$$

Note that we have replaced $2C$ by C. We should probably use something like C' or C_1, but we say to ourselves that $2C$ is just an arbitrary constant, and so relabel it and call it C.

EXAMPLE 5 Higher even powers of $\sin x$ and $\cos x$ can be approached as in Example 2, but more than one step will be needed.

$$\int \cos^4 x \, dx = \int \cos^3 x \cos x \, dx$$

$$= \cos^3 x \sin x - \int \sin x \cdot 3 \cos^2 x (-\sin x) \, dx$$

$$= \cos^3 x \sin x + \int 3 \cos^2 x \sin^2 x \, dx$$

$$= \cos^3 x \sin x + \int 3 \cos^2 x (1 - \cos^2 x) \, dx$$

$$= \cos^3 x \sin x - 3 \int \cos^4 x \, dx + 3 \int \cos^2 x \, dx.$$

Solving in this equation for $\int \cos^4 x \, dx$, we get

$$\int \cos^4 x \, dx = \frac{\cos^3 x \sin x}{4} + \frac{3}{4} \int \cos^2 x \, dx,$$

which reduces to the integral in Example 2. Similarly, $\int \cos^6 x \, dx$ would involve two such reducing steps.

EXAMPLE 6 In this trigonometric context, when the odd power occurred in the denominator we got involved with rational functions, possibly quite complicated ones. Here, also, integration by parts may lead to a reducing step. For example,

$$\int \frac{dx}{\cos^5 x} = \int \sec^5 x \, dx = \int \sec^3 x \cdot \sec^2 x \, dx$$

$$= \sec^3 x \tan x - \int \tan x \cdot 3 \sec^2 x \cdot \sec x \tan x \, dx$$

$$= \sec^3 x \tan x - 3 \int \sec^5 x \, dx + 3 \int \sec^3 x \, dx.$$

Therefore

$$\int \sec^5 x \, dx = \frac{1}{4} \sec^3 x \tan x + \frac{3}{4} \int \sec^3 x \, dx.$$

PROBLEMS FOR SECTION 5

Integrate.

1. $\int x \sin x \, dx$
2. $\int e^x \cos x \, dx$
3. $\int x\sqrt{x+1} \, dx$
4. $\int x \log x \, dx$
5. $\int y^2 \sin y \, dy$
6. $\int xe^{ax} \, dx$
7. $\int x^2 e^{-x} \, dx$
8. $\int x \cos ax \, dx$
9. $\int \sin^2 x \, dx$
10. $\int \arctan x \, dx$
11. $\int \arcsin x \, dx$
12. $\int x^n \log x \, dx$
13. $\int \dfrac{\log x \, dx}{x}$
14. $\int \sec^3 x \, dx$
15. $\int \sin \sqrt{x} \, dx$
16. $\int \sin(\log x) dx$
17. $\int \log(1 + x^2) dx$
18. $\int \dfrac{xe^x dx}{(1+x)^2}$

In the following examples find degree-reducing formulas analogous to the special cases worked out in Examples 5 and 6.

19. $\int \sec^n x \, dx$
20. $\int \cos^n x \, dx$
21. $\int (\log x)^n dx$

6. RATIONAL FUNCTIONS (CONCLUDED)

Partial-fraction decompositions depend on a lemma from polynomial algebra that we shall assume.

LEMMA *If $r(x) = p(x)/q(x)$ is a proper rational function, and if its denominator $q(x)$ can be written in the form*

$$q(x) = q_1(x)q_2(x),$$

where $q_1(x)$ and $q_2(x)$ are polynomials that have no factor in common (i.e., are relatively prime), then $r(x)$ can be written as the sum of two proper rational functions having denominators $q_1(x)$ and $q_2(x)$, respectively:

$$\frac{p(x)}{q_1(x)q_2(x)} = \frac{p_1(x)}{q_1(x)} + \frac{p_2(x)}{q_2(x)}.$$

The lemma doesn't tell us how to find the new numerators $p_1(x)$ and $p_2(x)$, but this is easy once we know that the decomposition is possible.

Consider, for example, the rational function $r(x) = x/(x-1)(x+3)$, where $q_1(x)$ and $q_2(x)$ are each of the form $x + c$. A rational function with such a denominator will be proper only if its numerator is a constant, i.e., a polynomial of degree zero. The lemma therefore guarantees that there are constants a and b such that

$$\frac{x}{(x-1)(x+3)} = \frac{a}{x-1} + \frac{b}{x+3},$$

and we proceed as in Section 2.

Consider next

$$r(x) = \frac{x^2 - 3}{(x-1)(x^2+1)}.$$

A rational function with $x^2 + 1$ in the denominator is proper only if its numerator has the form $bx + c$, this being the most general polynomial of degree less than 2. The lemma therefore guarantees constants a, b, and c, such that

$$\frac{x^2 - 3}{(x - 1)(x^2 + 1)} = \frac{a}{x - 1} + \frac{bx + c}{x^2 + 1},$$

again a situation that we worked out in Section 2.

If $q_2(x)$ in turn could be factored into relatively prime factors, then the lemma could be applied again to the rational function $p_2(x)/q_2(x)$. Proceeding in this way, the lemma extends to denominators having many relatively prime factors $q_1(x)$, $q_2(x)$, ..., $q_n(x)$, and shows that we can always find polynomials $p_1(x)$, ..., $p_n(x)$ such that

$$\frac{p(x)}{q(x)} = \frac{p_1(x)}{q_1(x)} + \frac{p_2(x)}{q_2(x)} + \cdots + \frac{p_n(x)}{q_n(x)},$$

each quotient being a proper rational function.

For example, we know that there are constants a, b, and c, such that

$$\frac{x - 2}{x(x + 1)(x + 2)} = \frac{a}{x} + \frac{b}{x + 1} + \frac{c}{x + 2},$$

another type considered in Section 2.

Determining the constants in the above examples falls under the general heading of methods involving *undetermined coefficients*. In the first example, the answer we are looking for is a *linear combination* of $1/(x - 1)$ and $1/(x + 3)$, i.e., a sum of constants times these two functions. We start off with a linear combination having undetermined coefficients a and b, and then determine the coefficients by algebra.

The second example also involves a linear combination having undetermined coefficients, this time a linear combination

$$a\left(\frac{1}{x - 1}\right) + b\left(\frac{x}{x^2 + 1}\right) + c\left(\frac{1}{x^2 + 1}\right)$$

of the three functions

$$\frac{1}{x - 1}, \qquad \frac{x}{x^2 + 1}, \qquad \frac{1}{x^2 + 1}.$$

If a denominator factor is repeated there is more to do. For example, the lemma guarantees constants a, b, and c, such that

$$\frac{3x^2 + 2}{x(x - 1)^2} = \frac{a}{x} + \frac{bx + c}{(x - 1)^2},$$

because x and $(x - 1)^2$ are relatively prime. We thus end up with a linear combination of the three functions

$$\frac{1}{x}, \qquad \frac{x}{(x - 1)^2}, \qquad \frac{1}{(x - 1)^2}.$$

We can write down the integrals of the first and third of these functions, but not the second. However,

$$\frac{x}{(x - 1)^2} = \frac{(x - 1) + 1}{(x - 1)^2} = \frac{1}{x - 1} + \frac{1}{(x - 1)^2},$$

and now we can integrate.

Instead of resorting to this second change of form, it is customary to incorporate it into the original linear combination decomposition. That is, we look for a, b, and c, such that

$$\frac{3x^2 + 2}{x(x - 1)^2} = \frac{a}{x} + \frac{b}{x - 1} + \frac{c}{(x - 1)^2},$$

again as in Section 2. Similarly, if $(x^2 + 1)^2$ is a denominator factor, we look for two terms

$$\frac{ax + b}{x^2 + 1} + \frac{cx + d}{(x^2 + 1)^2},$$

instead of one term

$$\frac{ax^3 + bx^2 + cx + d}{(x^2 + 1)^2}.$$

We then have a linear combination of the following four functions:

$$\frac{x}{x^2 + 1}, \qquad \frac{1}{x^2 + 1}, \qquad \frac{x}{(x^2 + 1)^2}, \qquad \frac{1}{(x^2 + 1)^2}.$$

The first three can be directly integrated, but the last falls into the one category of elementary partial fractions that we don't yet know how to integrate:

$$\int \frac{dx}{(x^2 + 1)^n},$$

where n is an integer ≥ 2.

What is involved is a step-by-step reduction process, each step requiring an integration by parts. The procedure runs parallel to the treatment of the even powers of cos x that we discussed at the end of the last section. As a matter of fact, the substitution $x = \tan \theta$ reduces the present problem to the former one. The direct attack is illustrated below.

EXAMPLE   We consider the case $n = 3$. Since

$$\int \frac{x \, dx}{(x^2 + 1)^3} = -\frac{1}{4} \frac{1}{(x^2 + 1)^2}$$

by simple substitution, we try to get this integrand as dv in the integration-by-parts formula. This can be done by writing

$$\frac{1}{(x^2 + 1)^3} = \frac{-x^2 + x^2 + 1}{(x^2 + 1)^3} = \frac{-x^2}{(x^2 + 1)^3} + \frac{1}{(x^2 + 1)^2}.$$

Then

$$\int \frac{dx}{(x^2 + 1)^3} = -\int x \cdot \frac{x \, dx}{(x^2 + 1)^3} + \int \frac{dx}{(x^2 + 1)^2}$$

$$= \frac{1}{4}\left[\frac{x}{(x^2 + 1)^2} - \int \frac{dx}{(x^2 + 1)^2} \right] + \int \frac{dx}{(x^2 + 1)^2}$$

$$= \frac{1}{4} \frac{x}{(x^2 + 1)^2} + \frac{3}{4} \int \frac{dx}{(x^2 + 1)^2}.$$

The power of $(x^2 + 1)$ has been reduced by 1. The next reduction gives arctan x.

There is a theorem, called the Fundamental Theorem of Algebra, which asserts that every polynomial with real coefficients can be expressed as a product of linear polynomials such as $x - 1$ and quadratic polynomials such as $x^2 + 1$ and $x^2 + 2x + 2$ which cannot be factored into linear polynomials with real coefficients. Therefore, in theory, every rational function has a denominator like those in the examples we have been working, except that the number of factors might be large. This is why, in theory, every rational function can be integrated.

In practice, finding the linear and irreducible quadratic factors of a polynomial may be very difficult. For example,

$$x^4 + 4 = (x^2 + 2x + 2)(x^2 - 2x + 2),$$

but this factorization is by no means obvious, and a more complicated fourth-degree polynomial would be even more difficult to factor.

PROBLEMS FOR SECTION 6

Integrate.

1. $\int \dfrac{dx}{x^4 - x^3}$
2. $\int \dfrac{(x^2 + 1)dx}{x^4 - x^3}$
3. $\int \dfrac{dx}{x^4 - x^2}$

4. $\int \dfrac{x^2 + x + 1}{(x^4 - x^2)}\, dx$
5. $\int \dfrac{dx}{x^4 + x^2}$
6. $\int \dfrac{x^3 + 1}{x^4 + x^2}\, dx$

7. $\int \dfrac{dx}{(1 - x^2)^2}$
8. $\int \dfrac{x^3 + x}{(1 - x^2)^2}\, dx$
9. $\int \dfrac{dx}{(x^2 - 1)(x^2 - 4)}$

10. $\int \dfrac{x^2 + 2}{(x^2 - 1)(x^2 - 4)}\, dx$
11. $\int \dfrac{dx}{(x^2 + 1)(x^2 + 4)}$
12. $\int \dfrac{x^2 + 2}{(x^2 + 1)(x^2 + 4)}\, dx$

7. SYSTEMATIC INTEGRATION

We shall now describe several other classes of functions which can be integrated systematically in closed form. The main point here is to be able to recognize these functions. The integration procedures themselves can get involved enough to lead us to look for shortcuts (or to use integral tables). Then we are back to ingenuity, but with the important difference that now we *know* the function has a closed-form integral that can be systematically calculated, and we are merely challenged to find a more direct route to it.

A *monomial* in two variables u and v is a product of a constant times a power of u times a power of v, such as $4u^2v^3$ and $-u^{19}v^5$. A *polynomial* in u and v, $p(u, v)$, is a sum of such monomials. A *rational function* $R(u, v)$ in u and v is a quotient $p(u, v)/q(u, v)$ of two polynomials in u and v. Finally, a rational function in sin x and cos x, $R(\sin x, \cos x)$, is obtained by setting $u = \sin x$ and $v = \cos x$ in a rational function $R(u, v)$. For example, if

$$R(u, v) = \frac{uv^2}{u^3 - 1},$$

then

$$R(\sin x, \cos x) = \frac{\sin x \cos^2 x}{\cos^3 x - 1}.$$

EXAMPLE The function $\tan^3 x - \sec x$ is of the form $R(\sin x, \cos x)$:

$$\tan^3 x - \sec x = \frac{\sin^3 x}{\cos^3 x} - \frac{1}{\cos x}$$

$$= \frac{\sin^3 x - \cos^2 x}{\cos^3 x} = R(\sin x, \cos x),$$

where $R(u, v) = (u^3 - v^2)/v^3$.

We shall use $R(u)$ and $R(u, v)$ for arbitrary rational functions in one and two variables. In this terminology, here are some classes of functions that can be systematically integrated in closed form:

a) $R(e^x)$ b) $R(\sin x, \cos x)$

c') $R(x, \sqrt{1 - x^2})$ c'') $R(x, \sqrt{1 + x^2})$ c''') $R(x, \sqrt{x^2 - 1})$

d) $R(x, \sqrt{ax^2 + bc + c})$ e) $R(x, (ax + b)^{1/n})$

EXAMPLES

1. $\dfrac{x^3}{\sqrt{1 - x^2}}$ is of type (c'), with $R(u, v) = \dfrac{u^3}{v}$.

2. $\dfrac{e^x + e^{-x}}{e^x - e^{-x}}$ is of type (a), with $R(u) = \dfrac{u + 1/u}{u - 1/u} = \dfrac{u^2 + 1}{u^2 - 1}$.

3. $\dfrac{x^2 - 2x}{x\sqrt{x^2 +}} \dfrac{+ 1}{x + 1}$ is of type (d), with $R(u, v) = \dfrac{u^2 - 2u + 1}{uv}$.

Finally, a word about integral tables. Because finding an antiderivative is such an uncertain business, often requiring an unrealistic combination of ingenuity and stamina to push a calculation through, it has been found worthwhile to gather together the answers that are reasonable and/or important into tables, where they are classified by the form of the integrand and can simply be looked up. You ought to spend a little time examining such a table, to see how the classification goes and what kinds of functions are treated. A short table of integrals is given on pages 796–801.

EXTRA PROBLEMS FOR CHAPTER 8

1. $\int (e^x)^2 dx$

2. $\int \dfrac{x^3 + 1}{x - 1} dx$

3. $\int \dfrac{\sin x}{\cos^3 x} dx$

4. $\int \dfrac{e^{\sqrt{x}}}{\sqrt{x}} dx$

5. $\int e^{(e^x)} e^x dx$

6. $\int \dfrac{\log \sqrt{x}}{x} dx$

7. $\displaystyle\int \frac{\log x}{\sqrt{x}}\,dx$

8. $\displaystyle\int \sqrt{x}\log x\,dx$

9. $\displaystyle\int \frac{x^2\,dx}{\sqrt{1-x^2}}$

10. $\displaystyle\int \frac{dx}{\sqrt{x}\sqrt{x+1}}$

11. $\displaystyle\int x^{n-1}\sqrt{a+bx^n}\,dx$

12. $\displaystyle\int \frac{x^{n-1}\,dx}{\sqrt{a+bx^n}}$

13. $\displaystyle\int \frac{dx}{x^2\sqrt{x^2+4}}$

14. $\displaystyle\int \frac{\arcsin x\,dx}{\sqrt{1-x^2}}$

15. $\displaystyle\int \frac{dx}{x(\log x)^3}$

16. $\displaystyle\int e^{2x}\sin(e^x)\,dx$

17. $\displaystyle\int \frac{x^3}{\sqrt{x^2+1}}\,dx$

18. $\displaystyle\int xe^{\sqrt{x}}\,dx$

19. $\displaystyle\int \frac{dx}{\sqrt{x}\cos\sqrt{x}}$

20. $\displaystyle\int \frac{x\,dx}{(a^2-x^2)^{3/2}}$

21. $\displaystyle\int \sin x\log(\cos x)\,dx$

22. $\displaystyle\int \frac{dx}{x\cos(\log x)}$

23. $\displaystyle\int \frac{x^2+1}{x^3-4x^2}\,dx$

24. $\displaystyle\int \frac{dx}{x^{1/2}(x-1)}$

25. $\displaystyle\int \frac{dx}{x^{1/2}(x^{1/3}+1)}$

26. $\displaystyle\int \tan^4 x\sec^4 x\,dx$

27. $\displaystyle\int \tan^6 x\,dx$

28. $\displaystyle\int \sec^6 x\,dx$

29. $\displaystyle\int \frac{\arctan t}{1+t^2}\,dt$

30. $\displaystyle\int \frac{dx}{x^{1/2}(1+x)}$

31. $\displaystyle\int \frac{\sin\sqrt{x}}{\sqrt{x}}\,dx$

32. $\displaystyle\int \frac{\cos(\log x)\,dx}{x}$

33. $\displaystyle\int \frac{\cos t}{\sqrt{1-\sin t}}\,dt$

34. $\displaystyle\int \frac{x^2}{\sqrt{1-x}}\,dx$

35. $\displaystyle\int \sqrt{x^4+x^2}\,dx$

36. $\displaystyle\int \frac{e^{2t}\,dt}{1+e^{2t}}$

37. $\displaystyle\int \frac{e^t}{1+e^{2t}}\,dt$

38. $\displaystyle\int \tan\theta\sec^2\theta\,d\theta$

39. $\displaystyle\int \tan^n\theta\sec^2\theta\,d\theta$

40. $\displaystyle\int \sec^n\theta\tan\theta\,d\theta$

41. $\displaystyle\int \sin^n\theta\cos\theta\,d\theta$

42. $\displaystyle\int \frac{\sin t\,dt}{(a+b\cos t)^n}$

43. $\displaystyle\int \frac{dx}{x\sqrt{9+4x^2}}$

44. $\displaystyle\int \frac{\sqrt{16-t^2}}{t^2}\,dt$

45. $\displaystyle\int \frac{x^2\,dx}{(a^2-x^2)^{3/2}}$

46. $\displaystyle\int \frac{x^2\,dx}{\sqrt{2x-x^2}}$

47. $\displaystyle\int \frac{dx}{(x^2-2x-3)^{3/2}}$

48. $\displaystyle\int \frac{\sqrt{1-x}}{1+x}\,dx$

49. $\displaystyle\int \frac{\cos t}{2-\sin^2 t}\,dt$

50. $\displaystyle\int \frac{\sin t\,dt}{1+\cos^2 t}$

51. $\displaystyle\int \sqrt{\frac{1-x}{1+x}}\,dx$

52. $\displaystyle\int \frac{x^3\,dx}{a^2-x^2}$

53. $\displaystyle\int \frac{x^3\,dx}{\sqrt{a^2-x^2}}$

54. $\displaystyle\int \frac{\sqrt{x}}{1+x}\,dx$

55. $\displaystyle\int \frac{x^{1/3}}{1+x}\,dx$

56. $\displaystyle\int x\arctan x\,dx$

57. $\displaystyle\int x\arcsin x\,dx$

58. $\displaystyle\int \frac{\log x}{(x+1)^2}\,dx$

59. $\displaystyle\int x^3\sin x\,dx$

60. $\displaystyle\int \arctan\sqrt{x}\,dx$

61. $\displaystyle\int x^2\log(x+1)\,dx$

62. $\displaystyle\int e^{ax}\cos bx\,dx$

63. $\displaystyle\int (\log x)^2\,dx$

64. a) Show that

$$\frac{x^3}{(x^2+1)^2} = \frac{ax+b}{(x^2+1)} + \frac{cx+d}{(x^2+1)^2}$$

for suitable constants a, b, c, and d.

b) Using the result in (a), compute

$$\int \frac{x^3}{(x^2 + 1)^2}\,dx.$$

65. Show that

$$\frac{Ax^3 + Bx^2 + Cx + D}{(x^2 + 1)^2}$$

can always be written in the form

$$\frac{ax + b}{x^2 + 1} + \frac{cx + d}{(x^2 + 1)^2}.$$

66. Find the degree-reducing formula for

$$\int \frac{dx}{(x^2 + a^2)^n},$$

using integration-by-parts as in the example in the text.

67. Find the degree-reducing formula for

$$\int \frac{dx}{(x^2 + a^2)^n},$$

by using the substitution $x = a \tan \theta$.

68. Prove that the area of the ellipse

$$\frac{x^2}{a^2} + \frac{y^2}{b^2} = 1$$

is πab.

69. One arch of the curve $y = \sin x$ is revolved about the x-axis. Compute the volume of the region thus generated.

70. Compute the volume of the solid generated by rotating about the x-axis the region under the graph of $y = \log x$, from $x = 1$ to $x = 2$.

71. Calculate the area under the graph of $f(x) = \arctan x$ from $x = 0$ to $x = 1$.

72. A torus is a doughnut-like solid formed by rotating a circular disk about an axis not touching it. If the radius of the circle is r and the distance from the center of the circle to the axis is R $(R > r)$, show that the volume of the torus is

$$V = 2\pi^2 r^2 R.$$

73. A cylindrical core of radius a is removed from a spherical apple of radius r. Find the volume of the cored apple.

In each of the problems below, find the volume generated by revolving about the y-axis the region bounded by the given graphs. In each case, draw the plane region to be rotated.

74. $y = \sin x,$ $y = 0,$ $x = 0,$ $x = \pi.$
75. $y = \log x,$ $y = 0,$ $x = 1,$ $x = 2.$
76. $y = e^{-x^2},$ $y = 0,$ $x = 0,$ $x = \infty.$
77. $y = e^x,$ $y = -e^x,$ $x = 0,$ $x = a.$

78. Calculate the area of the region between the graph of $y = \log x$, its tangent line at $x = 1$, and the line $x = 2$.

79. Find the volume swept out when the above area is rotated about the y-axis.

80. Same problem when the region is rotated about the x-axis.

81. Find the formula for the volume of a sphere by the shell method.

The Definite Integral

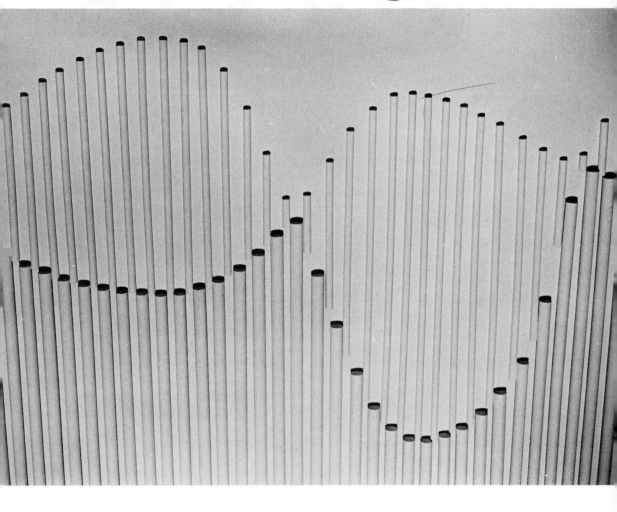

In Chapter 4 we took it as geometrically obvious that plane regions have exact numerical areas and space regions have exact numerical volumes, and our goal was to discover how to compute such quantities. We found an unexpectedly successful method using antiderivatives. The present chapter develops the older method of exhaustion, based here on Riemann sum approximations and the definite integral. Of course, it must be shown that these two wholly different computational procedures always give the same answer (the Fundamental Theorem of Calculus).

A definite integral is usually specified as the limit of a *sequence* of Riemann sums, so the chapter starts with a brief discussion of sequences and sequential convergence. Section 2 then introduces the definite integral process and the Fundamental Theorem, on the basis of properties of area. The remainder of the chapter is a mixture of applications and further development of the integral.

1. CONVERGENT SEQUENCES

EXAMPLE 1 If P_n is the regular polygon of n sides inscribed in the unit circle, then its perimeter p_n is an approximation to the circumference of the circle that improves as n is taken larger, and approaches the circumference as its limit as n tends to infinity. We thus have an *infinite sequence*

$$p_3, p_4, \ldots, p_n, \ldots$$

of numbers, each the perimeter of a polygon, such that

$$p_n \to 2\pi \qquad \text{as } n \to \infty.$$

In this chapter we are going to be concerned with similar approximating sequences that approach areas, volumes, and other quantities as their limits, and we therefore start with a brief discussion of sequential convergence.

The notion of an infinite sequence arises very naturally from examples like the one above. Some procedure or formula determines a number a_n for every integer n from a certain point on. It might seem desirable always to start with a "first" term a_1, but the sequence above starts naturally at $n = 3$ because a polygon presumably has at least three sides. On the other hand, in the sequence

$$1, 2, 4, 8, \ldots, 2^n, \ldots$$

of powers of 2, we would probably like to think of $8 = 2^3$ as the third term a_3, and 2^n as the nth term a_n, in which case the beginning number $1 = 2^0$ is the zeroth term a_0. The three dots $\cdots$ can be loosely read "and so on."

We shall be especially interested in sequences that approach limits, in the way that $p_n \to 2\pi$ in the first example above. For a much simpler example, consider the sequence

$$1, \frac{1}{2}, \frac{1}{3}, \frac{1}{4}, \ldots,$$

where the nth term is $1/n$ for every positive integer n. Then

$$\frac{1}{n} \to 0 \qquad \text{as } n \to \infty,$$

in exactly the same way that

$$\frac{1}{x} \to 0 \qquad \text{as } x \to \infty.$$

It is clear that sequence limits are much like function limits, the analogy being with the limit of $f(x)$ as x tends to infinity.

DEFINITION   *If $\{a_n\}$ is an infinite sequence of numbers, then we say that a_n approaches the limit l as n approaches infinity if we can make the difference $l - a_n$ as small as we wish merely by taking n large enough.*

We then write
$$a_n \to l \qquad \text{as } n \to \infty,$$
or
$$\lim_{n \to \infty} a_n = l.$$

EXAMPLE 2   If $a_n = 1/n$, then
$$a_n \to 0 \qquad \text{as } n \to \infty,$$
as noted above. Technically, if we want a_n closer to 0 than a given small positive number ϵ, i.e., if we want

$$\frac{1}{n} < \epsilon,$$

then we only have to take n larger than $1/\epsilon$.

EXAMPLE 3   Later in the chapter we will meet a type of sequence $\{a_n\}$ such that

$$|a_n - l| < \frac{K}{n}$$

for all n where K is a constant. So if we want a_n to be closer to its limit l than $1/100$, i.e., if we want

$$|a_n - l| < \frac{1}{100},$$

it is sufficient to take n large enough so that

$$\frac{K}{n} \leq \frac{1}{100},$$
or
$$n \geq 100\,K.$$

A sequence having the limit l is said to *converge* to l. A sequence that doesn't converge to any l is said to *diverge*. Divergence can occur in two "pure" forms: *oscillatory divergence*, as in

$$1, \quad -1, \quad 1, \quad -1, \quad 1, \quad \ldots,$$

where $a_n = (-1)^n$, and *divergence to* ∞, as in

$$1, \quad 4, \quad 9, \quad 16, \quad 25, \quad \ldots,$$

where $a_n = n^2$. Generally a divergent sequence would exhibit a mixture of these two types of divergent behavior.

The computation of sequential limits is governed by the same algebraic limit laws that control function limits:

A. *If $a_n \to a$ and $b_n \to b$ as $n \to \infty$, then*

$$a_n + b_n \to a + b,$$

$$a_n b_n \to ab,$$

$$\frac{a_n}{b_n} \to \frac{a}{b} \qquad \left(\begin{array}{c} \textit{if the denominators} \\ \textit{are nonzero} \end{array} \right).$$

B. *If f is continuous at $x = a$ and if $x_n \to a$ as $n \to \infty$, then*

$$f(x_n) \to f(a) \qquad \textit{as } n \to \infty.$$

C. (Squeeze limit law). *If $0 \le a_n \le b_n$ for all n, and if $b_n \to 0$ as $n \to \infty$, then $a_n \to 0$ as $n \to \infty$.*

Finally, there is the "sign-preserving" property of a convergent sequence.

D. *If $a_n \to a$ as $n \to \infty$ and if $a_n \ge 0$ for all n, then $a \ge 0$.*

All of these limit laws should seem to be correct on purely intuitive grounds. For example, if a_n is very close to a and b_n is very close to b, then surely $a_n + b_n$ is very close to $a + b$.

EXAMPLE 4 Let $a_n = p_n/2$, where p_n is the perimeter from Example 1. We know that $a_n \to \pi$ as $n \to \infty$. Then (B) tells us that

$$a_n^2 \to \pi^2 \qquad \text{and} \qquad e^{a_n} \to e^{\pi}.$$

Also, $\cos a_n \to -1$.

EXAMPLE 5 Starting with the fact that $1/n \to 0$ as $n \to \infty$, and applying rules from (A), we see that

$$\lim_{n \to \infty} \frac{4 - n}{1 + 3n} = \lim_{n \to \infty} \frac{\dfrac{4}{n} - 1}{\dfrac{1}{n} + 3} = \frac{0 - 1}{0 + 3} = -\frac{1}{3}.$$

Also,

$$\lim_{n \to \infty} \frac{n^3 - 5n^2}{1 + n^3} = \lim_{n \to \infty} \frac{1 - \dfrac{5}{n}}{\dfrac{1}{n^3} + 1} = \frac{1 + 0}{0 + 1} = 1.$$

EXAMPLE 6 $\sin\dfrac{1}{n} \to 0$ as $n \to \infty$ by the squeeze limit law (C), because

$$0 < \sin\frac{1}{n} < \frac{1}{n}$$

and $1/n \to 0$.

We use the words *increasing* and *decreasing* in a slightly weaker sense when we are talking about sequences. A sequence $\{a_n\}$ is said to be *increasing* if

$$a_{n+1} \geq a_n \qquad \text{for all } n.$$

If $a_{n+1} > a_n$ for all n, then we say that $\{a_n\}$ is *strictly increasing*. The meaning of decreasing is similarly modified.

EXAMPLE 7 $a_n = 1/n$ is strictly decreasing.

EXAMPLE 8 If a_n is the n-place decimal truncation of the expansion of π, so that $a_1 = 3.1$ and $a_5 = 3.14159$, then $\{a_n\}$ is an increasing sequence. However, we can't claim that it is strictly increasing, because presumably there will be a zero somewhere along the infinite-decimal expansion of π, and if the zero occurs in the jth place we will have

$$a_{j-1} = a_j.$$

(Decimal expansions are discussed at the beginning of Chapter 11.)

PROBLEMS FOR SECTION 1

Evaluate the limit as $n \to \infty$ of the following sequences.

1. $2 - \dfrac{3}{n} + \dfrac{4}{n^2}$

2. $\dfrac{n+1}{n-1}$

3. $\dfrac{n}{n^2+1}$

4. $\dfrac{100n - n^2}{2n^2+1}$

5. $\dfrac{5n+2}{3-n}$

6. $\dfrac{(2-n)(3-n)}{(1+2n)(1+3n)}$

7. $\dfrac{2n}{\sqrt{4+n^2}}$

8. $\sqrt{n+1} - \sqrt{n}$

9. $\sqrt{n^2+n} - n$

10. $\dfrac{e^n - e^{-n}}{e^n + e^{-n}}$

Here is a principle related to (B) (it can be proved from (B)).

$$\text{If } \lim_{x \to 0} f(x) = l, \qquad \text{then} \qquad \lim_{n \to \infty} f\left(\frac{1}{n}\right) = l.$$

11. Use the new principle stated above to prove that $\lim n \sin(1/n = 1$.

Use the limit laws, including the new principle above, to find the limits of the following sequences.

12. $n^2\left(1 - \cos\dfrac{1}{n}\right)$ 13. $n\log\left(1 + \dfrac{1}{n}\right)$

14. $\dfrac{\log n}{n}$ 15. $n(1 - e^{1/n})$

16. Use an algebraic limit law from (A) and the squeeze limit law (C) to prove the more general squeeze limit law:

   If $a_n \le b_n \le c_n$ for all n, and if $a_n \to l$ and $c_n \to l$ as $n \to \infty$, then $b_n \to l$ as $n \to \infty$.

2. THE IDEA OF THE DEFINITE INTEGRAL; THE FUNDAMENTAL THEOREM

The definite integral has roughly the same array of interpretations and applications as the antiderivative procedure developed in Chapter 4, and again the area interpretation serves as a particularly intuitive starting point.

   The following figure shows how the new procedure estimates the area under a function graph. It shows that the area A under the graph of f from $x = a$ to $x = b$ is approximately equal to a sum of certain rectangular areas. We shall describe these areas in a minute, but the figure clearly shows what they are. The error in this approximation is the amount of shaded area lying between the graph and the tops of the rectangles. It is obvious that this shaded area is greatly reduced by doubling the number of rectangles. In fact, if we examine what happens when a single rectangle is replaced by two, we see that the error is roughly cut in half. A roughly triangular error area is replaced by two out of four roughly equal triangular quarters.

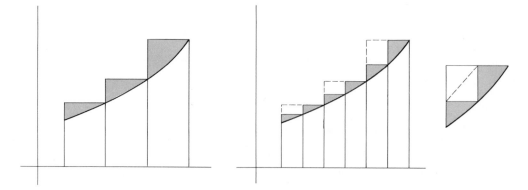

   Here is the algebraic description of this configuration. First we subdivide the interval $[a, b]$ into N equal subintervals by inserting $N - 1$ equally spaced subdividing points $x_1, x_2, \ldots, x_{N-1}$.

For convenience we set $x_0 = a$ and $x_N = b$. Note that the common width Δx of the N subintervals is given by

$$\Delta x = \frac{b - a}{N},$$

and that the subdividing points can be written in terms of a and Δx as follows:

$$x_1 = x_0 + \Delta x = a + \Delta x,$$
$$x_2 = x_1 + \Delta x = a + 2\Delta x,$$
$$x_3 = x_2 + \Delta x = a + 3\Delta x, \text{etc.}$$

Taking the first subinterval as base, we erect the rectangle with altitude $f(x_1)$ and area $bh = \Delta x f(x_1)$. The second rectangle has altitude $f(x_2)$ and area $\Delta x f(x_2)$. And so on. The sum S_N of these N rectangular areas can thus be written

$$S_N = \Delta x[f(x_1) + f(x_2) + \cdots + f(x_N)].$$

It is called the nth *Riemann sum* for f over the interval $[a, b]$.

EXAMPLE 1   Compute S_8 for $f(x) = x^2$ over the interval $[0, 1]$, and determine the error in the approximation $S_8 \approx A$.

Solution. Here $\Delta x = (b - a)/N = (1 - 0)/8 = 1/8$. Also, $x_1 = \Delta x = 1/8$, $x_2 = 2\Delta x = 2/8$,   etc.,   so $f(x_1) = x_1^2 = (\Delta x)^2 = 1/64$, $f(x_2) = x_2^2 = (2\Delta x)^2 = 4/64$, etc. Thus

$$S_8 = \Delta x[f(x_1) + \cdots + f(x_8)]$$
$$= \frac{1}{8}\left[\frac{1}{64} + \frac{4}{64} + \frac{9}{64} + \frac{16}{64} + \frac{25}{64} + \frac{36}{64} + \frac{49}{64} + \frac{64}{64}\right]$$
$$= \frac{1}{8} \cdot \frac{204}{64} = \frac{51}{128}.$$

Since $A = \int x^2\, dx]_0^1 = 1/3$, the error has magnitude

$$S_8 - \frac{1}{3} = \frac{51}{128} - \frac{1}{3} = \frac{153 - 128}{384} = \frac{25}{384} < \frac{25}{375} = \frac{1}{15}.$$

EXAMPLE 2   We compute S_{16} for the same area. Now $\Delta x = 1/16$, and

$$S_{16} = \Delta x[f(x_1) + f(x_2) + \cdots + f(x_{16})]$$
$$= \frac{1}{16}\left[\left(\frac{1}{16}\right)^2 + \left(\frac{2}{16}\right)^2 + \cdots + \left(\frac{16}{16}\right)^2\right]$$
$$= \frac{1}{16}\left[\frac{1}{256} + \frac{4}{256} + \cdots + \frac{256}{256}\right]$$
$$= \frac{1}{16} \cdot \frac{1496}{256} \quad \text{(by pocket calculator)}$$
$$= \frac{187}{512}.$$

The error has magnitude

$$S_{16} - \frac{1}{3} = \frac{187}{512} - \frac{1}{3} = \frac{49}{1536} < \frac{50}{1500} = \frac{1}{30}.$$

Note that the error has been approximately halved by doubling the number of intervals.

Since the error $A - S_N$ is roughly halved each time N doubles, it appears that $A - S_N$ approaches the limit 0 as N increases, i.e., that

$$A = \lim_{N \to \infty} S_N.$$

Assuming for the moment that this is so, it constitutes a second formula for the area: A can be computed *either* as $F(b) - F(a)$, where $F' = f$, as discussed in Chapter 4, *or* as the limit of the finite sum approximation $S_N = \Delta x[f(x_1) + \cdots + f(x_N)]$ as $N \to \infty$ (and $\Delta x \to 0$). But now we can drop A out. These two formulas, each equal to A, must necessarily be equal to *each other*, and the resulting equation

$$F(b) - F(a) = \lim \Delta x[f(x_1) + \cdots + f(x_N)]$$

contains no reference to area, or to any other interpretation of its terms. It is an identity asserting simply that two different calculation procedures give the same answer. As such it is a form of the Fundamental Theorem of Calculus.

Now observe that the Riemann sum formula

$$S_N = \Delta x[f(x_1) + f(x_2) + \cdots + f(x_N)]$$

makes sense for *any* function f defined on $[a, b]$, positive or not, increasing or not. In view of our tentative conclusions above, two questions then arise:

1. For how general a function f can it be proved that the limit $\lim_{N \to \infty} S_N$ exists?
2. If the limit exists, does it equal $F(b) - F(a)$, i.e., does the Fundamental Theorem hold for f?

If the Riemann sum limit does exist then f is said to be *integrable* on $[a, b]$, and $\lim S_N$ is called the *definite integral of f from a to b*. It is designated $\int_a^b f$, or $\int_a^b f(x)\,dx$. Thus

$$\int_a^b f = \lim_N S_N,$$

by definition. (A more inclusive, formal definition will be stated in Section 6. Until then the partial definition just given is adequate.)

The basic theorems of integration theory, answering the above questions, are the following:

THEOREM I   *If f is a continuous on $[a, b]$, then f is integrable there.*

THEOREM II   *If f is continuous on $[a, b]$, then f has an antiderivative there.*

THEOREM III   *(Fundamental Theorem of Calculus) If f is integrable on [a, b] and if F is any antiderivative of f on [a, b], then*

$$\int_a^b f = F(b) - F(a).$$

REMARK   In the notation of Chapter 4, the Fundamental Theorem asserts that

$$\int_a^b f = \left[\int f \right]_a^b.$$

We shall only partially prove these theorems. The situation at the moment is as follows: Chapter 4 contained area proofs of Theorem II and part of Theorem III, to the effect that the area A_a^b is given by $F(b) - F(a)$. In this section it will be shown, again by an area argument, that if f is continuous, positive, and increasing, then f is integrable and $\int_a^b f = A_a^b$. This will complete the area proofs of Theorems I–III for this class of functions. In Section 6 it will be shown that these results extend to any function f that can be written as a *difference* $f = g - h$ of two increasing functions. Moreover, it will be shown that every "normal" continuous function is of this sort. At that point we will have an adequate working theory of integration, based on geometric arguments.

Proofs that are independent of geometry are harder to come by but are necessary for theoretical reasons. Some of this theory will be found in Section 6 and in Appendix 4, but the complete proof of Theorem I is left to more advanced courses.

It will be convenient to accept and use Theorems I–III as stated, even though weaker versions would suffice in practice. This will save us from having constantly to check that our functions are "normal," although in practice they always will be.

The Fundamental Theorem is central for most applications of the definite integral. It is used in two opposite ways.

I. In applied problems certain quantities Q are seen to be given approximately by finite sum estimates of the form

$$Q \approx \Delta x [f(x_1) + f(x_2) + \cdots + f(x_N)],$$

where the function f is suggested by the context. If it also seems clear that the error in this approximation approaches 0 as $N \to \infty$, then we conclude that Q is given exactly by the definite integral $\int_a^b f$. Now the Fundamental Theorem tells us that Q can be computed much more simply as $\int f]_a^b = F(b) - F(a)$ (provided, of course, that we can find an antiderivative F). This will be illustrated in Section 4 and later sections.

II. In the reverse direction, the Fundamental Theorem allows us to estimate function values by Riemann sums. For example, since

$$\frac{d}{dt} \log t = \frac{1}{t},$$

it follows from the Fundamental Theorem that

$$\log \frac{3}{2} = \log \frac{3}{2} - \log 1 = \int_1^{3/2} \frac{dt}{t},$$

and therefore $\log(3/2)$ can be estimated with as small an error as desired by Riemann sums for the function $f(x) = 1/x$ over the interval $[1, 3/2]$. More on this later.

So far we have proved nothing. The example below shows how the Fundamental Theorem can be established for some functions by directly evaluating the limit of the Riemann sum. After that, we shall verify our original conjecture by giving an area proof that if f is continuous, positive, and increasing over $[a, b]$, then f is integrable there, and $\int_a^b f$ is the area A.

EXAMPLE 3   It can be proved (by induction) that

$$1^2 + 2^2 + 3^2 + \cdots + n^2 = \frac{n(n + 1)(2n + 1)}{6}$$

for any positive integer n. Use this identity to prove the Fundamental Theorem for $f(x) = x^2$ over the interval $[0, b]$.

Solution. We are asked to prove that if $f(x) = x^2$, then

$$F(b) - F(0) = \lim_{N \to \infty} \Delta x[f(x_1) + \cdots + f(x_N)],$$

where F is any antiderivative of f (say $F(x) = x^3/3$), $\Delta x = b/N$, and the subdividing points are $x_1 = \Delta x$, $x_2 = 2\Delta x$, $x_3 = 3\Delta x$, etc. Thus, $f(x_1) = x_1^2 = (\Delta x)^2$, $f(x_2) = x_2^2 = (2\Delta x)^2$, etc., so

$$\lim_{N \to \infty} \Delta x[f(x_1) + f(x_2) + \cdots + f(x_N)]$$

$$= \lim_{N \to \infty} \Delta x[(\Delta x)^2 + (2\Delta x)^2 + \cdots + (N\Delta x)^2]$$

$$= \lim_{N \to \infty} (\Delta x)^3[1^2 + 2^2 + \cdots + N^2]$$

$$= \lim_{N \to \infty} \left(\frac{b}{N}\right)^3 \left[\frac{N(N + 1)(2N + 1)}{6}\right]$$

$$= \lim_{N \to \infty} \frac{b^3}{6}\left(1 + \frac{1}{N}\right)\left(2 + \frac{1}{N}\right) = \frac{b^3}{6} \cdot 2 = \frac{b^3}{3}.$$

Since this is $F(b) - F(0)$, we are done.

We return now to the case of a continuous, positive, increasing function f, as illustrated earlier in the figure on page (312). Note again that the region under the graph of f between a and b is wholly covered by the rectangles whose areas make up the Riemann sum S_N, so $A < S_N$.

Because S_n involves the right-hand endpoints, it will now be renamed the nth *right* Riemann sum. The nth *left* Riemann sum, designated s_n, will use the left-hand endpoints instead, as shown in the following figure for the same function and sixfold subdivision. The general expression for s_n is

$$s_n = \Delta x[f(x_0) + f(x_1) + \cdots + f(x_{n-1})].$$

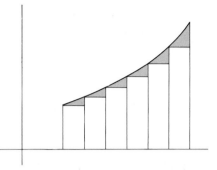

Note that when f is increasing (as pictured) the terms of this sum are the areas of rectangles lying wholly *below* the graph of f, so $s_n < A$. Thus s_n approximates A from below with an error $A - s_n$ that is again represented by the shaded area.

When we combine the diagrams for S_n and s_n a remarkable fact appears. The difference $S_n - s_n$, represented as the sum of the areas of the small shaded rectangles in the figure below, is exactly $\Delta x[f(b) - f(a)]$.

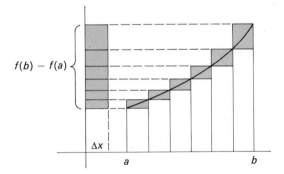

This is because these difference rectangles can be slid left and stacked to form exactly a single rectangle of base width Δx and altitude $f(b) - f(a)$. Note that since $\Delta x = (b - a)/n$, we can also write $S_n - s_n = K/n$, where K is the constant $(b - a)[f(b) - f(a)]$.

Altogether, we have established the following estimate:

Let f be continuous, positive, and increasing over the interval $[a, b]$. Let s_n and S_n be the nth left and right Riemann sums for f over $[a, b]$, and let A be the area of the region between the graph of f and the x-axis, from $x = a$ to $x = b$. Then

$$s_n < A < S_n,$$

and

$$S_n - s_n = \frac{K}{n},$$

where $K = (b - a)[f(b) - f(a)]$. Thus each of s_n and S_n approximates A with an error less than K/n in magnitude.

It follows that we can approximate A by the Riemann sum S_n (or s_n) as closely as we wish by taking n sufficiently large. That is, A is the *limit* of the right Riemann sum S_n as n tends to infinity. Similarly for s_n.

To make the formal calculation, we combine the two statements in the estimate and have

$$0 \le S_N - A \le S_N - s_N = \frac{K}{n}.$$

Since $K/n \to 0$ as $n \to \infty$, it follows from the squeeze limit law that $S_N - A \to 0$ as $N \to \infty$, i.e., that

$$A = \lim_{N \to \infty} S_N.$$

This proves that f is integrable and that $\int_a^b f = A$. In Chapter 4 it was shown that if the region under the graph of f is given a variable right-hand edge $t = x$, then its (variable) area A_a^x is an antiderivative of f. So Theorem II holds for f. It was also shown that then $A = A_a^b = F(b) - F(a)$, for any antiderivative F or f, completing the proof of III for such functions f.

We have thus proved (geometrically):

THEOREM 1   *If f is continuous, positive, and increasing on the interval $[a, b]$, then:*

f is integrable on $[a, b]$;
f has an antiderivative over $[a, b]$;
if F is any antiderivative of f on $[a, b]$, then

$$\int_a^b f = F(b) - F(a).$$

Since $\int_a^b f$ is probably not exactly halfway between s_n and S_n, we can expect it to be closer to one of these numbers than half the distance $S_n - s_n$ between them. In general we can't tell which Riemann sum gives this closer estimate, but we can if we know that f has constant concavity over $[a, b]$. Suppose, for example, that f is increasing and concave down. The error rectangle over a single subdivision interval is divided by the graph of f into the upper error area for the right Riemann sum S_n (shaded in the figure), and the lower error area for s_n (unshaded).

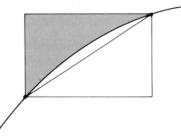

Since the graph lies above its chord when f is concave down, the upper error area is less than half the total rectangle area. This holds for all the individual errors, and therefore for their sums. So we have the sharper estimate:

If f is positive, increasing, and concave down, then the right Riemann sum S_n approximates $\int_a^b f$ from above with an error of magnitude less than

$$\frac{1}{2}(S_n - s_n) = \frac{1}{2}\Delta x[f(b) - f(a)].$$

REMARK 1 If f is increasing and concave *up*, we have the same conclusion except that now $\int_a^b f$ is closer to the *left* Riemann sum s_n.

REMARK 2 The whole discussion so far has concerned a *strictly* increasing function. If f is only known to be *weakly* increasing ($f(x) \leq f(y)$ whenever $x < y$), then there is always the possibility that f might in fact be a constant function, in which case the inequalities all become equalities. So in this case our conclusions have to be stated as weak inequalities, such as $s_n \leq \int_a^b f \leq S_n$.

REMARK 3 If f is decreasing, then the roles of the left and right Riemann sums are interchanged throughout. For example, S_n is now the sum of areas of *inscribed* rectangles. Also:

> *If f is decreasing and concave up, then S_n approximates $\int_a^b f$ from below with an error less than $\frac{1}{2}|f(b) - f(a)| \Delta x$.*

REMARK 4 The identity

$$S_n - s_n = \Delta x[f(b) - f(a)] = K/n$$

is a purely algebraic fact that is true for *any* function defined over $[a, b]$: we just note that if the Riemann sums S_n and s_n are written out, then the terms in $S_n - s_n$ cancel in pairs except for the last term of S_n and the first term of s_n, leaving

$$\Delta x\, f(x_n) - \Delta x\, f(x_0) = \Delta x[f(b) - f(a)].$$

So $\int_a^b f$ can always be evaluated as the limit of the left Riemann sums s_n (when it exists).

We are now ready to illustrate how the Fundamental Theorem can be used to estimate the values of a function.

EXAMPLE 4 Use a fourfold subdivision of $[1, 3/2]$ to estimate $\log (3/2)$.

Solution. Since

$$\frac{d}{dx} \log x = \frac{1}{x},$$

the Fundamental Theorem says that

$$\log \frac{3}{2} = \log \frac{3}{2} - \log 1 = \int_1^{3/2} \frac{dx}{x},$$

and therefore that $\log(3/2)$ can be approximated by the Riemann sums of $f(x) = 1/x$. Since $1/x$ is decreasing and concave up, we use the right

Riemann sum S_4 to get the better estimate. We have

$$S_4 = \frac{1}{8}\left[\frac{8}{9} + \frac{8}{10} + \frac{8}{11} + \frac{8}{12}\right]$$

$$= \left[\frac{1}{9} + \frac{1}{10} + \frac{1}{11} + \frac{1}{12}\right]$$

$$= [0.1111\cdots + 0.1 + 0.0909\cdots + 0.0833\cdots]$$

$$= 0.385\cdots.$$

(The first section of Chapter 11 contains a review of decimal expansions.) We know that S_4 approximates $\log(3/2)$ from below with an error less than

$$\frac{1}{2}|f(b) - f(a)|\Delta x = \frac{1}{2}\cdot\frac{1}{3}\cdot\frac{1}{8} = \frac{1}{48} = 0.0208\cdots.$$

Therefore,

$$0.385 < \log\frac{3}{2} < 0.386 + 0.021 = 0.407,$$

so

$$\log\frac{3}{2} \approx 0.40,$$

with an error less than 2 in the second decimal place. The table on p. 792 gives $\log\frac{3}{2} = 0.4055$ to four places.

From now on we shall use the summation notation, involving Σ, the Greek letter capital sigma. Consider again the sixfold subdivision of the earlier figures. We can indicate the six equations defining $x_1, x_2, \ldots, x_6$ by writing that

$$x_i = a + i\Delta x$$

for $i = 1, \ldots, 6$. That is, the subscript letter i is used as a variable that takes on integer values, as in sequences, and as i runs from 1 to 6 in value, the equation above represents the six equations

$$x_1 = a + \Delta x, \qquad x_2 = a + 2\Delta x, \ldots, x_6 = a + 6\Delta x.$$

Then we indicate the sum

$$f(x_1) + \cdots + f(x_6)$$

by writing

$$\sum_{i=1}^{6} f(x_i),$$

which is read "the sum of the numbers $f(x_i)$ from $i = 1$ to $i = 6$." In the same way, the sixth *left* Riemann sum is

$$s_6 = \Delta x\sum_{i=1}^{6} f(x_{i-1}).$$

In general, $\sum_{i=p}^{q} a_i$ is the sum of the terms a_1 from $i = p$ to $i = q$ (it being understood that $p \leq q$):

$$\sum_{i=p}^{q} a_i = a_p + a_{p+1} + \cdots + a_q.$$

EXAMPLE 5   The sixth left Riemann sum can also be written

$$s_6 = \Delta x \sum_{i=0}^{5} f(x_i).$$

EXAMPLE 6   We rewrite three of the lines in Example 3, using the subscript k instead of i:

$$\lim_{N \to \infty} \Delta x \sum_{k=1}^{N} f(x_k)$$

$$= \lim_{N \to \infty} \Delta x \sum_{k=1}^{N} (k \, \Delta x)^2$$

$$= \lim_{N \to \infty} (\Delta x)^3 \sum_{k=1}^{N} k^2.$$

PROBLEMS FOR SECTION 2

1. Calculate S_4 for $\int_{1/2}^{1} x \, dx$ and find the error in the approximation of the integral by S_4.

2. Show that the error in approximating $\int_{1/2}^{1} x \, dx$ by S_n is cut exactly in half by going from S_4 to S_8.

3. Use a pocket calculator to calculate S_6 for $\int_{1}^{2} \sqrt{x} \, dx$ to three decimal places. Calculate the integral to three decimal places and the error in the approximation to two places.

4. We know that $\log 2 = \int_{1}^{2} dt/t$. Estimate $\log 2$ by calculating S_6 to three decimal places. Look up $\log 2$ in the tables at the end of the book, and so obtain the approximation error to two places.

Each of the following identities can be proved by mathematical induction.

$$1 + 2 + 3 + \cdots + n = \frac{n(n + 1)}{2}$$

$$1^2 + 2^2 + 3^2 + \cdots + n^2 = \frac{n(n + 1)(2n + 1)}{6}$$

$$1^3 + 2^3 + 3^3 + \cdots + n^3 = \frac{n^2(n + 1)^2}{4}$$

Use these identities in the manner of Example 3 to verify the Fundamental Theorem in each of the following cases.

5. $\int_{0}^{b} x \, dx$ 6. $\int_{0}^{b} x^3 \, dx$ 7. $\int_{a}^{b} x \, dx$

8. $\int_{a}^{b} x^2 \, dx$ (expand the squares in the sum S_n)   9. $\int_{a}^{b} x^3 \, dx$

10. It can be proved that $1^4 + 2^4 + 3^4 + \cdots + n^4$ is a fifth degree polynomial in n with leading term $n^5/5$. Using just this fact, verify the Fundamental Theorem for $\int_0^b x^4 \, dx$.

11. Show that if $[1, \frac{3}{2}]$ is divided into eight equal subintervals, then the left Riemann sum for $1/x$ is

$$\sum_{k=0}^{7} \frac{1}{16 + k}.$$

12. In the manner of the above problem, write down a formula for the right Riemann sum for $1/x$ obtained by dividing $[1, 2]$ into 1000 equal subintervals.

13. a) Write down the fourth Riemann sum S_4 for $f(x) = x$, over the interval $[1, \frac{3}{2}]$, leaving its terms as fractions.
    b) Compute $\int_1^{3/2} x \, dx$ and show that the error in using S_4 as an estimate of the integral is exactly $1/32$.
    c) Show therefore that the sharper error estimate for a concave function given in the text cannot be improved.

14. a) Write down the fourth left Riemann sum s_4 for $f(x) = x^2$ over the interval $[1, \frac{3}{2}]$, leaving its terms as fractions.
    b) Compute $\int_1^{3/2} x^2 \, dx$ and show that the actual error is less than the sharper error bound for a concave function.

15. We wish to use a Riemann sum S_n for $f(x) = 1/(1 + x^2)$ over the interval $[0, 1]$ to compute

$$\frac{\pi}{4} = \arctan 1 = \int_0^1 \frac{dx}{1 + x^2}.$$

Show that S_n will approximate $\pi/4$ accurately to the nearest two decimal places if $n = 100$ by using an appropriate estimate from the text.

16. Compute the fourth Riemann sums for $f(x) = 1/(1 + x^2)$ over the interval $[0, 1]$. Show that

$$2.88 < \pi < 3.39.$$

17. Suppose we wish to compute $\pi = 4\int_0^1 dx/(1 + x^2)$ to the nearest four decimal places, i.e., with an error less than $5(10)^{-5}$, by using a Riemann sum. How fine a subdivision would have to be used?

18. Show from the Fundamental Theorem that

$$\frac{d}{dx} \int_a^x f(t) \, dt = f(x).$$

19. Show that

$$\frac{d}{dx} \int_x^b f(t) \, dt = -f(x).$$

Compute the derivatives of the following functions:

20. $f(x) = \displaystyle\int_0^{x^2} t^2 \, dt$

21. $f(x) = \displaystyle\int_{-x}^{x} \cos t \, dt$

22. $f(x) = \displaystyle\int_1^{\sqrt{x}} \frac{dt}{t}$

23. $f(x) = \displaystyle\int_0^{\log x} e^t \, dt$

3. PROPERTIES OF THE DEFINITE INTEGRAL

The final solution to a problem involving integration is frequently expressed as a definite integral that must be evaluated, or at least estimated. In this connection, and also in using the definite integral as a theoretical tool, it is important to know its principal properties. Some of these properties are merely restatements in definite integral form of the rules we have been using to find antiderivatives, but one or two properties belong uniquely to the definite integral.

We assume here that all continuous functions are integrable. The properties of the definite integral $\int_a^b f$ are then most easily verified in terms of antiderivatives, by way of the Fundamental Theorem. The basic fact that every continuous function *has* an antiderivative (Theorem II) was proved in Chapter 4 by an area argument and will be proved again in Appendix 4 by purely analytic means. In a situation where the integrability of some function is being proved, then the Riemann sum definition has to be used. See, for example, Theorem 4 in Section 6.

We start with linearity:

i)
$$\int_a^b [cf(x) + dg(x)]\,dx = c\int_a^b f(x)\,dx + d\int_a^b g(x)\,dx.$$

This merely reflects the linearity of antidifferentiation. We know that if F and G are antiderivatives of f and g, respectively, then $cF + dG$ is an antiderivative of $cf + dg$. The above equation is thus simply the evaluation identity

$$cF(x) + dG(x)\Big]_a^b = c\Big[F(x)\Big]_a^b + d\Big[G(x)\Big]_a^b.$$

Next, there is the additivity of the integral with respect to changing limits of integration:

ii)
$$\int_a^b f + \int_b^c f = \int_a^c f.$$

The hypothesis here is that the limits a, b, and c all belong to an interval I on which f is defined and continuous. Then f has an antiderivative F on I, and the equation above reduces to the identity

$$[F(b) - F(a)] + [F(c) - F(b)] = F(c) - F(a).$$

If $a > b$, it is convenient to adopt the definition

$$\int_a^b f = -\int_b^a f.$$

The Fundamental Theorem then remains true, and the antiderivative identity above shows that (ii) is then true for *any* ordering of the points a, b, c.

The mean-value principle can be given the integral form:

iii)
$$\int_a^b f = f(X)(b - a),$$

for some X between a and b. This just restates the mean-value principle for an antiderivative F: $F(b) - F(a) = F'(X)(b - a)$. In particular,

iv)
$$\int_a^b f > 0 \qquad \text{if } f > 0 \text{ between } a \text{ and } b, \text{ and}$$

$$\int_a^b g \leqq \int_a^b f \qquad \text{if } g \leq f \text{ on the interval } [a, b].$$

The second inequality follows from applying the mean-value principle (or the first inequality) to the function $f - g$. An important special case:

$$m(b - a) \leq \int_a^b f \leq M(b - a)$$

if $m \leq f(x) \leq M$ for all x between a and b.

Now consider substitution. When we are looking for an antiderivative $\int f(x)\,dx$, we can substitute $x = k(u)$, $dx = k'(u)\,du$, but it is necessary that k be invertible (over the x interval in question), because we have to substitute back the inverse function $u = h(x)$ to get the final answer. This was the gist of Section 4 in the last chapter. But, in the computation of a definite integral $\int_a^b f$, the final step is unnecessary, and we can substitute with almost no worries at all.

THEOREM 2   *If the integral $\int f(x)\,dx$ transforms into the integral $\int g(u)\,du$ under the substitution $x = k(u)$, $dx = k'(u)\,du$, all functions being continuous over the intervals in question, then*

v)
$$\int_a^b f(x)\,dx = \int_A^B g(u)\,du$$

whenever A and B are such that $a = k(A)$ and $b = k(B)$.

Of course, we have to be able to find the new limits A and B, but this is frequently easier than finding the whole inverse function. In fact, the transformation function k isn't required to be invertible here.

EXAMPLE 1   In Section 4 we found that

$$\int \frac{dx}{\sqrt{a^2 + x^2}} = \int \sec u\,dx = \int \frac{dv}{1 - v^2} = \log\left[\frac{1 + v}{1 - v}\right]^{1/2},$$

where the substitutions were

$$x = a\tan u, \qquad v = \sin u.$$

We then had to substitute back for the answer. However,

$$\int_0^a \frac{dx}{\sqrt{a^2 + x^2}} = \int_0^{\pi/4} \sec u\,du = \int_0^{\sqrt{2}/2} \frac{dv}{1 - v^2} = \log\left[\frac{1 + \sqrt{2}/2}{1 - \sqrt{2}/2}\right]^{1/2},$$

and we are done, except possibly for algebraic simplification. We have used the fact that $a\tan(\pi/4) = a$ and that $\sin(\pi/4) = \sqrt{2}/2$. The limits of integration here are obviously rather special ones in relation to the integrand, but such special limits occur frequently in applications.

Incidentally, the answer given above can be simplified by "rationalizing the denominator" to give, finally

$$\int_0^a \frac{dx}{\sqrt{a^2 + x^2}} = \log(\sqrt{2} + 1).$$

Proof of theorem. By hypothesis,

$$g(u) = f(k(u))k'(u).$$

If F is an antiderivative of f, then $F(k(u))$ is an antiderivative of $g(u)$, by the chain rule. (This is just *direct* substitution.) Therefore,

$$\int_A^B g = F(k(B)) - F(k(A)).$$

But if $a = k(A)$ and $b = k(B)$, then

$$F(k(B)) - F(k(A)) = F(b) - F(a)$$

$$= \int_a^b f.$$

Therefore, $\int_a^b f = \int_A^B g$, as claimed. ∎

EXAMPLE 2 If we were careless about the continuity of the new integrand, we might set $x = 1/y$, $dx = -dy/y^2$, and conclude that

$$2 = \int_{-1}^1 dx = -\int_{-1}^1 \frac{dy}{y^2}.$$

This is nonsense, since $f(y) = 1/y^2$ has no antiderivative at all on $[-1, 1]$. In fact, the integral on the right must be $-\infty$ under any reasonable definition of its value, because the area under the graph of $1/y^2$ from $y = -1$ to $y = 1$ can be shown to be infinite.

The integration-by-parts formula, in its definite-integral garb, offers no surprises. It is:

vi)
$$\int_a^b f(x)g'(x)\,dx = f(x)g(x)\Big]_a^b - \int_a^b f'(x)g(x)\,dx,$$

or
$$\int_a^b fg' = f(b)g(b) - f(a)g(a) - \int_a^b f'g.$$

The power of this formula may be surprising, though. In many advanced areas of mathematics it is a crucial tool.

The following examples illustrate, and briefly discuss, the notion of an improper integral.

EXAMPLE 3   Find the total area under the graph of $y = 1/x^2$ from $x = 1$ to $x = \infty$.

Solution. As worded, the problem doesn't make sense, since the only area formula we know is for a finite region. However, the area from $x = 1$ to $x = b$ is

$$\int_1^b \frac{dx}{x^2} = -\frac{1}{x}\bigg]_1^b = 1 - \frac{1}{b}.$$

Then

$$\lim_{b\to\infty}[\text{area from 1 to } b] = \lim_{b\to\infty}\left(1 - \frac{1}{b}\right) = 1.$$

This limit is what we *mean* by the area from 1 to ∞. So here we have an example of an infinitely long region with a finite area.

EXAMPLE 4   Show that the region under the graph of $y = 1/\sqrt{x}$ from $x = 1$ to ∞ has infinite area.

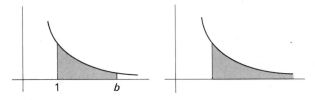

Solution. By definition the area A from 1 to ∞ is the limit as $b \to \infty$ of the area from 1 to b. So

$$A = \lim_{b\to\infty}\int_1^b \frac{dx}{\sqrt{x}} = \lim_{b\to\infty} 2\sqrt{x}\bigg]_1^b$$
$$= \lim_{b\to\infty}(2\sqrt{b} - 2) = \infty.$$

An integral of the form $\int_a^\infty f(x)\,dx$ is an example of an *improper* integral. It is defined as the limit

$$\int_a^\infty f(x)\,dx = \lim_{b\to\infty}\int_a^b f(x)\,dx.$$

If the limit exists (as a finite number l) we say that the improper integral *converges*, and that its value is l. If the limit does not exist, we say that the improper integral *diverges*.

EXAMPLE 5   Find the area under the graph of $y = 1/\sqrt{x}$ from $x = 0$ to $x = 1$.

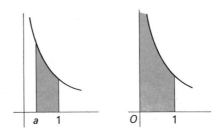

Solution. This region is infinitely long along the y-axis, rather than along the x-axis. In other words, the integral that should give its area, $\int_0^1 dx/\sqrt{x}$, is improper because the integrand tends to infinity as x approaches 0, rather than because it is integrated over an infinitely long interval. However, we can follow essentially the same scheme. We first compute the area over the interval $[a, 1]$, where a is a small positive number. Over this interval the integrand $1/\sqrt{x}$ is continuous and there is no problem. Then we can let a tend to zero and see what happens "in the limit." We have:

$$A = \int_0^1 \frac{dx}{\sqrt{x}} = \lim_{a \to 0} \int_a^1 \frac{dx}{\sqrt{x}} = \lim_{a \to 0} 2\sqrt{x} \Big]_a^1$$

$$= \lim_{a \to 0} (2 - 2\sqrt{a}) = 2.$$

So the improper integral is convergent, and its value is 2. The area we were looking for is thus 2.

EXAMPLE 6   The improper integral $\int_0^1 dx/x^2$ is divergent:

$$\int_0^1 \frac{dx}{x^2} = \lim_{a \to 0} \int_a^1 \frac{dx}{x^2} = \lim_{a \to 0} -\frac{1}{x} \Big]_a^1$$

$$= \lim_{a \to 0} \left(\frac{1}{a} - 1 \right) = \infty.$$

PROBLEMS FOR SECTION 3

Integrate.

1. $\displaystyle\int_0^{\pi/2} \cos^3 \theta \, d\theta$ 2. $\displaystyle\int_0^{\pi} \cos^2 x \, dx$ 3. $\displaystyle\int_0^{\pi/3} \tan^3 x \sec^2 x \, dx$

4. $\displaystyle\int_0^{\pi/4} \tan^2 x \sec^2 x \, dx$ 5. $\displaystyle\int_0^{\pi/6} \tan^3 \theta \sec \theta \, d\theta$ 6. $\displaystyle\int_0^{\pi} \frac{dx}{1 + \sin x}$

7. $\displaystyle\int_{-1}^{1} \frac{dx}{4 - x^2}$ 8. $\displaystyle\int_{-1}^{1} \frac{dx}{x^2 + 2x + 5}$ 9. $\displaystyle\int_0^{\log 2} \frac{dx}{1 + e^x}$

10. $\displaystyle\int_0^1 \arctan x \, dx$ 11. $\displaystyle\int_0^{1/2} \arctan \sqrt{x} \, dx$ 12. $\displaystyle\int_0^{\pi/3} x \sec^2 x \, dx$

13. $\displaystyle\int_a^{2a} x^3 \sqrt{x^2 - a^2} \, dx$ 14. $\displaystyle\int_0^1 \frac{dx}{(1 + x^2)^{3/2}}$ 15. $\displaystyle\int_a^{\sqrt{3}a} \frac{dx}{x^2 \sqrt{a^2 + x^2}}$

Calculate the following integrals.

16. $\int_0^1 \left[\int_0^{\sqrt{x}} t^3 \, dt \right] dx$ 17. $\int_{-1}^2 \left[\int_{x^2}^{2+x} t \, dt \right] dx$

Determine whether each of the following improper integrals is convergent or divergent, and calculate its value if convergent.

18. $\int_1^\infty \dfrac{dx}{x^3}$ 19. $\int_1^\infty \dfrac{dx}{x^{3/2}}$ 20. $\int_1^\infty \dfrac{dx}{x^{2/3}}$

21. $\int_0^\infty \dfrac{dx}{1+x^2}$ 22. $\int_0^\infty \dfrac{dx}{1+x}$ 23. $\int_0^1 \dfrac{dx}{x^{2/3}}$

24. $\int_0^1 \dfrac{dx}{x^3}$ 25. $\int_0^1 \dfrac{dx}{\sqrt{1-x^2}}$ 26. $\int_0^{\pi/2} \sec x \tan x \, dx$

27. $\int_0^{\pi/2} \sec^2 x \, dx$ 28. $\int_{-1}^1 \dfrac{dx}{x^2}$ 29. $\int_{-1}^1 \dfrac{dx}{x^{2/3}}$

30. We know (Problem 20) that the region under the graph of $y = x^{-2/3}$ from $x = 1$ to ∞ has infinite area. Yet the volume generated by revolving this region about the x-axis is finite. Prove this, by calculating the volume.

4. APPROXIMATING OTHER QUANTITIES BY RIEMANN SUMS; APPLYING THE FUNDAMENTAL THEOREM

Suppose that $T = f(t)$ is the temperature at time t recorded at a weather station on a certain day. The station uses a 24-hour clock, so the domain of f is the time interval $[0, 24]$. And we assume f to be continuous.

In order to find the average temperature for the given day, we might take six temperature readings at 4-hour intervals, starting at midnight: $T_0 = f(0)$, $T_1 = f(4), \ldots, T_5 = f(20)$. The average reading would then be the sum of these six readings divided by 6:

$$
\begin{aligned}
T_{av} &= \frac{T_0 + T_1 + T_2 + T_3 + T_4 + T_5}{6} \\
&= \frac{f(0) + f(4) + f(8) + f(12) + f(16) + f(20)}{6} \\
&= \frac{1}{6} \sum_{k=0}^5 f(4k).
\end{aligned}
$$

This computation of the average temperature probably does not give the right answer. For example, suppose it is a hot summer day, and that at 2:00 in the afternoon (hour 14 on the 24-hour clock) there is a short thunderstorm that cools the air for an hour. This temporary dip in temperature wouldn't even show in the average computed above. And there might be other fluctuations that wouldn't be taken into account adequately by reading temperatures at 4-hour intervals.

Clearly we would do better by taking 24 readings on the hour, starting at midnight and using their average

$$
T_{av} = \frac{f(0) + f(1) + \cdots + f(23)}{24} = \frac{1}{24} \sum_{k=0}^{23} f(k).
$$

But this probably wouldn't be exactly right either, for the same reason that small temporary fluctuations may not be adequately represented in the average. Still better would be the average of 48 readings taken on the half-hour,

$$T_{av} = \frac{f(0) + \cdots + f(23\frac{1}{2})}{48} = \frac{1}{48} \sum_{k=0}^{47} f\left(\frac{k}{2}\right),$$

and so on.

We thus have various ways of computing an average daily temperature, none of them exactly right. Our intuition tells us that there is a *true average temperature* $\overline{T}$ that these approximating averages come close to, and that by taking a sufficiently large number of equally spaced temperature readings, we could, in principle, make the approximation as good as we wished.

On the other hand, the above finite sums look somewhat like Riemann sums for f. In fact, each of them is the corresponding Riemann sum divided by 24. For example, the first sum can be written

$$\frac{1}{6} \sum_{k=0}^{5} f(4k) = \frac{1}{24}\left[4 \sum_{k=0}^{5} f(4k)\right] = \frac{s_6}{24},$$

where s_6 is the sixth left Riemann sum for f over the interval $[0, 24]$. Now we know that $\int_0^{24} f(t)\, dt$ is the unique number approximated arbitrarily closely by the Riemann sums s_n themselves. So each of the numbers $\overline{T}$ and $(\int_0^{24} f)/24$ is the unique number approximated arbitrarily closely by $s_n/24$ as n increases, and hence

$$\overline{T} = \frac{1}{24} \int_0^{24} f(t)\, dt.$$

We can reason this way about the average value of any continuous function $y = f(x)$ over a closed interval $[a, b]$, and conclude:

If f is continuous on the closed interval $[a, b]$, then its average value over the interval is

$$\frac{1}{b-a} \int_a^b f(x)\, dx.$$

We can now compute the average value by the Fundamental Theorem (provided we can find an antiderivative of f).

EXAMPLE   The average value of $f(x) = x^2$ over the interval $[0, 2]$ is

$$\frac{1}{2} \int_0^2 x^2\, dx = \frac{1}{2} \frac{x^3}{3}\bigg]_0^2 = \frac{4}{3}.$$

Note that although the value of $f(x)$ increases from 0 to 4, we wouldn't expect the average value to be 2 because we see from the following graph that $f(x)$ is less than 2 more than half the time.

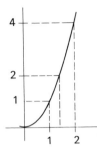

We now have a new procedure, complementing and reinforcing the one followed in Chapter 4, for showing that quantities Q in geometry, physics, and other disciplines are expressible by integrals. Here is how it goes. We observe that Q can be estimated by a finite sum of a certain type, and we notice that this finite sum is also a Riemann sum for a certain function g over the interval $[a, b]$. Then Q and $\int_a^b g(x)\,dx$ are *both* approximated as closely as we wish by the *same* number, and this means that

$$Q = \int_a^b g(x)\,dx.$$

Finally, Q can be computed as $\int g]_a^b$, by the Fundamental Theorem. For example, the figure below shows how the volume V of a solid of revolution can be approximated by a finite sum of volumes of thin circular cylinders:

$$V \approx \sum \pi r_i^2 \Delta x = \sum \pi f(x_i)^2 \Delta x.$$

But this sum is also a Riemann sum $\sum g(x_i)\Delta x$ for the function $g(x) = \pi f(x)^2$. Therefore, V is given exactly by $V = \int_a^b \pi f(x)^2\,dx = \int \pi f(x)^2\,dx]_a^b$.

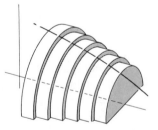

We have glossed over the important step of proving that the Riemann sum actually does estimate the volume, with an error that approaches zero as n tends to infinity. The approximation seemed intuitively clear on the basis of a picture, and we let it go at that. Such minimal evidence will be accepted throughout much of the chapter. It will be intuitively clear that a quantity Q can be estimated arbitrarily closely by a finite sum that turns out to be a Riemann sum for a function g, and we accept the evaluation of Q by $\int_a^b g$ on this basis.

PROBLEMS FOR SECTION 4

Find the average value of the following functions over the specified interval.

1. $y = 2x^3$; $[-1, 1]$

2. $y = 4 - x^2$; $[-2, 2]$

3. $y = x^2 - x + 1$; $[0, 2]$

4. $y = (x/2) + 1$; $[2, 6]$

5. $y = 2x + 1$; $[-1, 3]$

6. $y = x^n$; $[0, 1]$. (Why is the answer reasonable?)

7. A man travels 20 miles an hour for $\frac{1}{2}$ hour, 30 miles per hour for 2 hours, and 40 miles per hour for $\frac{1}{2}$ hour. What is the average velocity with respect to time?

8. A typist's speed over a four-hour interval increases as she warms up and decreases as she tires. Her speed in words per minute can be approximated by $w(t) = 6[4^2 - (t - 1)^2]$. Find her speed at the beginning of the interval, the end of the interval, her maximum speed, and her average speed over the 4-hour period.

9. If a particle is moving along a coordinate line, and if its position at time t is given by $s = f(t)$, then we concluded in Chapter 3 that its average velocity over the interval $[t_0, t_1]$ is given by

$$\frac{s_1 - s_0}{t_1 - t_0} = \frac{f(t_1) - f(t_0)}{t_1 - t_0},$$

and that its instantaneous velocity at time t is

$$V = \frac{ds}{dt} = f'(t).$$

Show now that the average velocity really is the average value of the instantaneous velocity, according to the notion of average value developed in this section.

10. If $y = f(x)$ and if $f'(x_0)$ is interpreted as the rate of change of y with respect to x at x_0, show that

$$\frac{\Delta y}{\Delta x} = \frac{f(x + \Delta x) - f(x)}{\Delta x}$$

is the average rate of change of y with respect to x over the interval $[x, x + \Delta x]$.

11. A particle moves along a straight line with varying velocity $v = g(t)$. Give a direct justification, in terms of finite-sum approximations, for the fact that the distance travelled during the time interval $[a, b]$ is given by

$$s = \int_a^b v \, dt = \int_a^b g(t) \, dt.$$

Start with the basic idea that over a very short time interval from t_0 to $t_0 + \Delta t$ the velocity is nearly constant, with value $v_0 = g(t_0)$, so the distance travelled is approximately $v_0 \Delta t = g(t_0) \Delta t$.

12. Give a direct justification, in terms of finite-sum approximations, for the fact that $\int_a^b f$ is the area under the graph of f from a to b. Start with the basic idea that, over a very short interval from x_0 to $x_0 + \Delta x$, the height $f(x)$ is nearly constant, so the area over this incremental interval is approximately a rectangular area.

5. VOLUMES BY SLICING

Suppose we put a coordinate line along beside a solid body S (or, possibly, skewering S). We shall call the coordinate line the x-axis. We want to consider how the solid intersects various planes perpendicular to the x-axis.

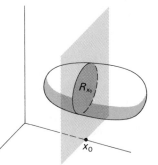

If the plane perpendicular to the x-axis at $x = x_0$ slices the solid, then the cross section of the solid in this plane will be a plane region R_{x_0}. Nearby cross sections can be projected onto the same plane (by moving them parallel to the x-axis), so we can compare how the cross section R_x changes as we move the slicing plane. In general, the cross section R_x will vary with x. If it does not, then we have a *cylinder*.

DEFINITION   *If the solid S runs from the plane at $x = a$ to the plane at $x = b$, and if all its cross sections R_x at points x between a and b are the same plane region R, then we say that S is a cylinder, with base R and altitude $h = b - a$.*

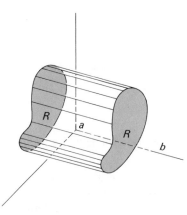

The volume of a cylinder is the product of its base area and its altitude:

$$V = Ah,$$

where A is the area of the base and h is the altitude. We shall assume this formula, and use it to derive a very general formula for the volume of a solid body.

Suppose the first and the last planes to touch the solid S are at $x = a$ and $x = b$, and consider a subdivision $a = x_0 < x_1 < \cdots < x_n = b$ of the interval $[a, b]$ into n equal subintervals of common length $\Delta x = (b - a)/n$. Then the planes $x = x_i$ slice the solid into incremental slabs ΔS_i. (Imagine slicing a loaf of French bread. The following figure shows a quarter of a slab.) Each slab ΔS_i is approximately a cylinder of base area $A(x_i)$ and altitude Δx, where $A(x)$ is the cross-sectional area. So

$$\Delta V_i \approx A(x_i)\Delta x$$

for each i by the cylinder volume formula, and

$$V = \sum_{i=1}^{n} \Delta V_i \approx \Delta x \sum_{i=1}^{n} A(x_i).$$

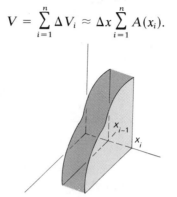

Our geometric intuition tells us that this approximation to V improves with finer subdivision (finer slicing). It is also a Riemann sum for $A(x)$. Thus:

THEOREM 3   *Let S be a solid whose cross section R_x varies continuously with x as x runs from a to b. Let $A(x)$ be the area of the cross section R_x. Then the volume of the slab between $x = a$ and $x = b$ is given by*

$$V = \int_a^b A(x)\,dx.$$

EXAMPLE 1   The solid of revolution generated by rotating the graph of f about the x-axis has the cross-sectional area $A(x) = \pi[f(x)]^2$. Therefore,

$$V = \int_a^b A(x)\,dx = \int_a^b \pi[f(x)]^2\,dx,$$

and we recover the formula of Section 5, Chapter 4 as a special case of the above "volumes-by-slicing" formula and the Fundamental Theorem.

EXAMPLE 2   The figure below shows a solid whose cross sections perpendicular to the x-axis are all isosceles right triangles with one vertex on the half-parabola

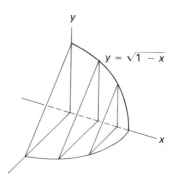

$y = \sqrt{1 - x}$. The area of the triangle at x is

$$\frac{1}{2} bh = \frac{1}{2}\sqrt{1 - x}\sqrt{1 - x} = \frac{1 - x}{2}.$$

This is the cross-sectional area $A(x)$, and the volume of the solid between $x = 0$ and $x = 1$ is therefore

$$V = \int_0^1 A(x)\, dx$$

$$= \int_0^1 \frac{1 - x}{2}\, dx = \frac{x}{2} - \frac{x^2}{4}\Big]_0^1 = \frac{1}{4}.$$

The formula $V = \int_a^b A(x)\, dx$ is important and deserves more of a proof than was provided above. We go back to the examination of a single increment. Let ΔS be the incremental slab of width Δx that is sliced from the solid by the planes $x = x_0$ and $x = x_0 + \Delta x$, and let ΔV be the volume of ΔS. We assume that the cross section R_x varies continuously with x, in the following sense. The incremental slab ΔS completely includes a cylinder C' whose base area A' is only slightly less than $A(x_0)$, and is completely included in a cylinder C'' whose base area A'' is only slightly greater than $A(x_0)$. That is, the incremental slab ΔS is "squeezed between" two cylinders C' and C'', whose base areas A' and A'' can be made as close to $A(x_0)$ as desired by making Δx sufficiently small. (The inner and outer cylinders C' and C'' will not generally be circular cylinders. The figure below is drawn with circular cross sections for simplicity and clarity.)

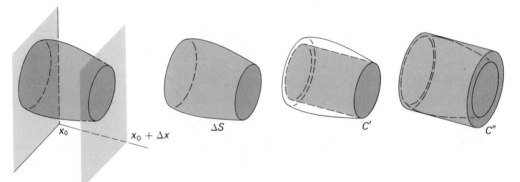

Consequently,

$$\text{Volume of } C' \leq \text{Volume of } \Delta S \leq \text{Volume of } C'',$$

or

$$A'\Delta x \leq \Delta V \leq A''\Delta x,$$

or

$$A' \leq \frac{\Delta V}{\Delta x} \leq A''.$$

Then, since A' and A'' both approach $A(x_0)$ as Δx approaches 0, $\Delta V/\Delta x$ is squeezed to the same limit. Thus $dV/dx = A(x)$, so

$$V = \int A(x)\,dx \Big]_a^b = \int_a^b A(x)\,dx$$

(by Theorem 5 in Chapter 4).

PROBLEMS FOR SECTION 5

Find the volume of each of the solids described below. You may assume that the cross section R_x varies continuously with x and that Theorem 3 can therefore be applied. Cross-section continuity is one of those annoying conditions that are obviously true but fussy to prove.

In the first four of these problems, the base of the solid is the unit circular disk $x^2 + y^2 \leq 1$, and its cross section perpendicular to the x-axis is the given figure:

1. A semicircle.   2. A square.

3. An equilateral triangle.

4. A rectangle of height $|y^3|$ (where $x^2 + y^2 = 1$).

5. The base of the solid is the square centered at the origin with one vertex at $(1, 1)$, and its cross sections perpendicular to the x-axis are triangles of altitude $(1 - x^2)$.

6. The same base, with triangular cross sections of altitude $1 + x$.

7. The base of the solid is the square centered at the origin with one vertex at $(1, 0)$, and its cross sections perpendicular to the x-axis are triangles with altitude $1 - x^4$.

8. The same base, with triangular cross sections of altitude 1.

9. The pyramid with square base of side a, and altitude h.

10. The pyramid whose base is an equilateral triangle of side a and whose altitude is h.

11. A cone with an irregular base region R is formed by drawing line segments from boundary points of R to a fixed vertex P. Prove that its volume is

$$V = \frac{1}{3}Ah,$$

where A is the area of R and h is the altitude of the cone. (Assume that similar plane figures have areas proportional to the squares of their diameters. The diameter of a plane figure is the length of the longest segment that can be drawn between two of its points.)

6. MORE GENERAL RIEMANN SUMS

We continue in the context of Section 2. It is not necessary to use regular (evenly spaced) subdivisions to get Riemann sum approximations to the area A. The following figure shows that unequal spacing leads to the same kind of area estimate for an increasing function. We allow *any* set of subdividing points $\{x_i\}_{i=0}^n$ such that

$$a = x_0 < x_1 < \cdots < x_n = b,$$

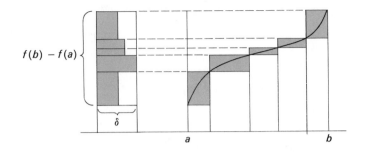

thus partitioning $[a, b]$ into n subintervals of (possibly) varying lengths

$$\Delta x_1 = x_1 - x_0, \quad \Delta x_2 = x_2 - x_1, \ldots, \quad \Delta x_n = x_n - x_{n-1}.$$

The maximum of these n subinterval lengths $\Delta x_1, \ldots, \Delta x_n$ is called the *mesh* of the subdivision.

No longer do we have a single subdivision for each positive integer n, so a different notation is necessary. We shall use Δ to stand for the arbitrary subdivision described above. The left and right Riemann sums for Δ are now given by

$$s_\Delta = f(x_0)\Delta x_1 + \cdots + f(x_{n-1})\Delta x_n = \sum_{i=1}^{n} f(x_{i-1})\Delta x_i$$

and

$$S_\Delta = f(x_1)\Delta x_1 + \cdots + f(x_n)\Delta x_n = \sum_{i=1}^{n} f(x_i)\Delta x_i.$$

These are again sums of rectangular areas, and the figure shows, just as before, that

$$s_\Delta \le A \le S_\Delta.$$

Moreover, just as before, we can stack the n "error" rectangles whose areas make up the difference $S_\Delta - s_\Delta$ inside a single rectangle, this time having as base width the number δ that is the mesh of Δ (the maximum of the subinterval lengths $\Delta x_1, \ldots, \Delta x_n$). Thus,

$$S_\Delta - s_\Delta \le \delta[f(b) - f(a)].$$

The two inequalities above combine to show that

$$0 \le A - s_\Delta \le S_\Delta - s_\Delta \le K\delta,$$

where K is the constant $f(b) - f(a)$. Thus the left Riemann sum s_Δ approximates A from below with an error at most $K\delta$. Similarly for S_Δ, except that it approximates A from above.

Finally, note that we get area approximations that seem even better by evaluating f at *intermediate* points of the intervals of a subdivision. Thus, we choose an *arbitrary* evaluation point ξ_k in the kth subinterval, for $k = 1, 2, \ldots, n$, and form the sum

$$R = f(\xi_1)\Delta x_1 + f(\xi_2)\Delta x_2 + \cdots + f(\xi_n)\Delta x_n$$

$$= \sum_{k=1}^{n} f(\xi_k)\Delta x_k.$$

This new sum is called a *general Riemann sum* for the function f over the interval $[a, b]$.

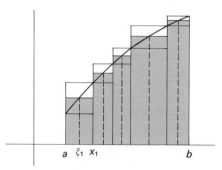

It is clear from the figure that if f is increasing then a general Riemann sum R lies between the left and right Riemann sums for the associated subdivision Δ:

$$s_\Delta \le R \le S_\Delta.$$

Since R and the area A *both* lie between s_Δ and S_Δ, it follows that R and A differ from each other by at most $S_\Delta - s_\Delta$, so

$$|R - A| \le S_\Delta - s_\Delta \le \delta[f(b) - f(a)].$$

Altogether, we have proved (geometrically) the following estimate.

Let f be continuous, positive, and increasing on the interval $[a, b]$, and let A be the area of the region under the graph of f, over the interval $[a, b]$. Let R be any Riemann sum for f over $[a, b]$, and let δ be the mesh of the associated subdivision. Then

$$|A - R| \le K\delta,$$

where $K = f(b) - f(a)$. That is, the Riemann sum R approximates the area A with an error at most $K\delta$ in magnitude.

Because of the error estimate, we can say that A is the limit of the general Riemann sum R as δ approaches 0. Here R is *not* a function of δ, and we are using the limit notion in a new way. But it is perfectly meaningful and appropriate.

In review, for any set of interval widths $\Delta x_1, \ldots, \Delta x_n$ adding to $b - a$, and any associated set of evaluation points $\xi_1, \ldots, \xi_n$, we form the two numbers

$$R = \sum_{k=1}^{n} f(\xi_k)\Delta x_k,$$

$$\delta = \max\{\Delta x_1, \ldots, \Delta x_n\}.$$

The error estimate shows that we can cause the difference $A - R$ to be as small as we please in magnitude merely by requiring that δ be suitably small. This is what we *mean* by:

$$R \to A \qquad \text{as } \delta \to 0.$$

We continue following the pattern of Section 2, noting now, just as before, that the general Riemann sum formula makes sense for an *arbitrary* function f. The same question then arises:

For how general a function f does the Riemann sum limit $\lim_{\delta \to 0} R$ exist?

Any function for which this general Riemann sum limit exists is said to be integrable.

DEFINITION   *Let f be an arbitrary function defined over the interval $[a, b]$. Let R be a general Riemann sum for f associated with a subdivision Δ of $[a, b]$, and let δ be the mesh of Δ. Then f is said to be integrable on $[a, b]$ if the limit $\lim_{\delta \to 0} R$ exists in the above sense, and $\int_a^b f$ is defined to be this limit:*

$$\int_a^b f = \lim_{\delta \to 0} R.$$

The estimate established above can now be restated as follows:

Let f be a continuous, positive, increasing function defined on the closed interval $[a, b]$. Then f is integrable over $[a, b]$ (and $\int_a^b f = A$). Moreover,

$$\left| \int_a^b f - R \right| \le \delta[f(b) - f(a)],$$

where R is any Riemann sum for f over $[a, b]$, and δ is the mesh of the associated subdivision.

This is essentially the same as Theorem 1 in Section 2, except that the meaning of $\int_a^b f$ is now taken more broadly.

Any continuous function that comes up in practice will almost certainly be *piecewise monotone*, that is, it will alternately increase for a while and then decrease for a while, and will change directions only a finite number of times on any given closed interval in its domain.

DEFINITION   *The function f is piecewise monotone on the closed interval $[a, b]$ in its domain if there is a subdivision*

$$a = x_0 < x_1 < \cdots < x_n = b,$$

such that f is monotone on each subinterval $[x_{i-1}, x_i]$.

For example, any function having only a finite number of critical points is piecewise monotone, as we saw in Chapter 6, Section 4.

The next figures show how a piecewise monotone function f can be expressed as a sum $f = g + h$ of a positive (weakly) increasing function g and a negative (weakly) decreasing function h. First we draw (define) g. Over each subinterval where f is *increasing* we simply copy the graph of f, but raise it or lower it by a suitable constant amount ($g = f + \text{constant}$). Over each interval where f is *decreasing* we draw a horizontal line segment ($g = \text{constant}$). The constants are successively chosen so that each new piece of the graph of g starts at the point where the preceding piece ended. The first constant is chosen to make $g(a) \ge 0$ (and also $g(a) \ge f(a)$). We then set $h = f - g$, and note that h is constant wherever f is increasing and is

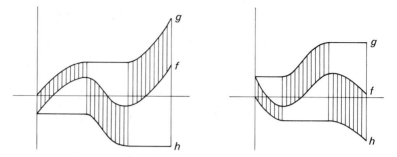

decreasing where f is decreasing, so h is a weakly decreasing function. Finally, the identity

$$f = g - (-h)$$

expresses f as a difference of two increasing functions.

A function that is piecewise monotone on $[a, b]$ need not be everywhere differentiable there. For example, $f(x) = x^{1/3}$ is monotone on $[-1, 1]$ and $g(x) = x^{2/3}$ is piecewise monotone on $[-1, 1]$, but each function fails to be differentiable at the origin.

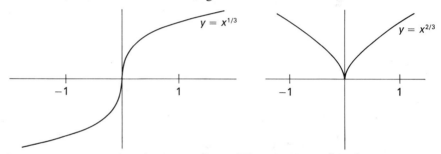

Suppose, however, that f *is* everywhere differentiable on $[a, b]$, and that f' is bounded there, i.e., that there is a constant B such that $|f'(x)| < B$ for every x in $[a, b]$. In this case also, it is easy to show that f is a difference of two increasing functions: the function

$$h(x) = Bx - f(x)$$

has a positive derivative and hence is increasing, and

$$f(x) = Bx - (Bx - f(x))$$
$$= Bx - h(x),$$

a difference of two increasing functions. (To ensure that both functions are positive we can add a suitably large constant C to each, and write $f(x) = (Bx + C) - (h(x) + C)$. Similarly, the functions g and h in the first decomposition can be made *strictly* increasing by adding x to them both.)

In view of the above remarks, the following theorem shows that every "normal" continuous function is integrable.

THEOREM 4 *Let f be a continuous function on $[a, b]$ which either is piecewise monotone or has a bounded derivative. Then f is integrable on $[a, b]$.*

Proof. Let $f = g - h$ where g and h are continuous, positive, and increasing. Let R_f be any Riemann sum for f over $[a, b]$,

$$R_f = f(\xi_1)\Delta x_1 + \cdots + f(\xi_n)\Delta x_n.$$

Then R_f is the difference of the corresponding Riemann sums (same subdivision Δ and evaluation points ξ_i) for g and h. This is because $f(\xi_i) = g(\xi_i) - h(\xi_i)$ for $i = 1, 2, \ldots, n$, so R_f splits into two parts, giving

$$R_f = R_g - R_h.$$

This lets us go back to the estimates for the increasing functions g and h. Thus, setting $I = \int_a^b g - \int_a^b h$, we have

$$
\begin{aligned}
|R_f - I| &= \left| \left(R_g - \int_a^b g \right) - \left(R_h - \int_a^b h \right) \right| \\
&\leq \left| R_g - \int_a^b g \right| + \left| R_h - \int_a^b h \right| \\
&\leq \delta[g(b) - g(a)] + \delta[h(b) - h(a)] \\
&= K\delta,
\end{aligned}
$$

where the number δ is the mesh of the subdivision Δ, and $K = [g(b) - g(a)] + [h(b) - h(a)]$ is the total increase of the two functions g and h over the interval $[a, b]$. It follows that R_f can be made as close to I as desired by taking δ suitably small, which proves that I is the integral $\int_a^b f$. ∎

REMARK 1   The proof showed that

$$\left| R - \int_a^b f \right| \leq K\delta,$$

where R is any Riemann sum for f over $[a, b]$, and δ is the mesh of the associated subdivision. As noted earlier, it is proved in more advanced courses that every continuous function is integrable. In particular, the error $E = \int_a^b f - R$ in the approximation of $\int_a^b f$ by the Riemann sum R can be made as small as desired merely by taking the subdivision mesh δ sufficiently small. However, the relationship between E and δ is not generally as simple as $|E| \leq K\delta$.

REMARK 2   The proof of Theorem 4 also showed that $\int_a^b f = \int_a^b g - \int_a^b h$. With some change in detail it becomes the direct proof in terms of Riemann sum approximations that a linear combination $f(x) = cg(x) + dh(x)$ of two integrable functions g and h is integrable, and that

$$\int_a^b f = c \int_a^b g + d \int_a^b h.$$

This identity was proved in Section 3 in terms of antiderivatives, under the assumption that all the functions involved are integrable and that the Fundamental Theorem holds. Here we have to go back to the Riemann sum definition of $\int_a^b f$ because we are establishing the integrability of f.

Theorems II and III for the "normal" continuous function $f = g - h$ follow directly from Theorem 1 for the functions g and h, plus (for III) the fact that

$$\int_a^b f = \int_a^b g - \int_a^b h.$$

This completes our geometric discussion of the integrability of "normal" continuous functions.

We conclude with an analytic proof of the Fundamental Theorem, in order to show a type of argument that is frequently useful. The idea is to establish some relationship (equality or inequality) for each subinterval of a subdivision, and then add them up to obtain a result about Riemann sums and/or the definite integral.

LEMMA 1 *If F is an antiderivative of f over* $[a, b]$, *and if* Δ *is any subdivision of* $[a, b]$, *then f has a particular Riemann sum* R_F *associated with* Δ *such that* $R_F = F(b) - F(a)$.

Proof. Suppose Δ is given by the subdividing points

$$a = x_0 < x_1 < \cdots < x_n = b.$$

By the mean-value principle there is a point ξ_1 lying strictly between x_0 and x_1 such that $F(x_1) - F(x_0) = F'(\xi_1)\Delta x_1 = f(\xi_1)\Delta x_1$. There is a corresponding equation for each subinterval of the subdivision Δ, so we have the following n equations:

$$F(x_1) - F(x_0) = f(\xi_1)\Delta x_1$$
$$F(x_2) - F(x_1) = f(\xi_2)\Delta x_2$$
$$F(x_3) - F(x_2) = f(\xi_3)\Delta x_3$$
$$\cdot$$
$$\cdot$$
$$\cdot$$
$$F(x_n) - F(x_{n-1}) = f(\xi_n)\Delta x_n.$$

When we add this stack of n equations, all the terms on the left cancel in pairs except for two, and we have

$$F(x_n) - F(x_0) = \sum_{j=1}^n f(\xi_j)\Delta x_j.$$

The sum on the right is a Riemann sum associated with Δ, so we are done. **I**

THEOREM III *If f is integrable on* $[a, b]$ *and if F is an antiderivative of f on* $[a, b]$ *then*

$$\int_a^b f = F(b) - F(a).$$

Proof. By Lemma 1, for every subdivision Δ of $[a, b]$ there is an associated Riemann sum R_F for f such that $R_F = F(b) - F(a)$. On the other hand, $R_F \to \int_a^b f$ as the mesh of Δ approaches 0, by the definition $\int_a^b f$. But this can be true only if $F(b) - F(a) = \int_a^b f$.

PROBLEMS FOR SECTION 6

Let

$$\bar{S}_n = \Delta x \sum_{k=1}^{n} f(\bar{x}_k)$$

be the Riemann sum obtained by evaluating f at the *midpoints* of n equal subintervals of length $\Delta x = (b - a)/n$ each. This midpoint-evaluation Reimann sum is generally a much better approximation of $\int_a^b f$ than the simple Riemann sum S_n, as the following estimate shows.

If the second derivative of f is bounded by B on the interval $[a, b]$, then the nth midpoint-evaluation Riemann sum $\bar{S}_n$ approximates $\int_a^b f$ with an error at most

$$\frac{B}{24}(b - a)(\Delta x)^2.$$

The following exercises center around this estimate. First we use it; then we prove it.

1. Show that

$$\log 2 \approx 2\left[\frac{1}{17} + \frac{1}{19} + \frac{1}{21} + \frac{1}{23} + \frac{1}{25} + \frac{1}{27} + \frac{1}{29} + \frac{1}{31}\right]$$

   with an error less than $(1/768)$ in magnitude.

2. If we wish to use the midpoint-evaluation Riemann sum $\bar{S}_n$ to approximate $\log 2 = \int_1^2 dx/x$ accurately to the nearest two decimal places, show that it is sufficient to take $n = 5$.

3. Show geometrically that $\bar{S}_n < \int_a^b f$ if f is decreasing and concave up. (Draw a typical subinterval and compare the errors in the area estimate on the two halves of the interval.)

4. Show that

$$\log\frac{3}{2} \approx \left[\frac{2}{17} + \frac{2}{19} + \frac{2}{21} + \frac{2}{23}\right]$$

   with an error less than $(1/1536)$. The estimate is too small, by Problem 3. Divide out the fractions to four decimal places and hence show that

$$\log\frac{3}{2} \approx 0.405$$

   correct to three decimal places.

5. Show that

$$\frac{\pi}{4} \approx \frac{1}{5}\left[\frac{1}{1.01} + \frac{1}{1.09} + \frac{1}{1.25} + \frac{1}{1.49} + \frac{1}{1.81}\right]$$

   with an error less than $(1/300)$. (Assume the fact that $|f''(x)| \le 2$ on $[0, 1]$, where $f(x) = 1/(1 + x^2)$.)

6. Show that the rectangular area $f(\bar{x})\Delta x$ is the same as the trapezoidal area under the tangent line at $\bar{x}$, as suggested by the following figure, where $\bar{x}$ is the midpoint of the increment integral. Show therefore that

$$\left(\int_x^{x+\Delta x} f\right) - f(\bar{x})\Delta x$$

   is equal to the integral

$$\int_x^{x+\Delta x} [f(t) - f(\bar{x}) - f'(\bar{x})(t - \bar{x})]dt.$$

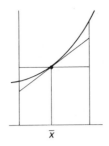

$\bar{x}$

7. Use the tangent-line error formula (Theorem 7, Chapter 6) to show that if f'' is bounded by B between x and $x + \Delta x$, then the above integral is at most

$$\frac{B}{24}(\Delta x)^3$$

in magnitude. This bounds the error over each subdivision interval. Now add these errors up and hence prove the midpoint Riemann sum estimate.

8. Prove that if f' exists and $|f'(x)| \le B$ on $[a, b]$, then

$$s_n \approx \int_a^b f,$$

with an error at most

$$\frac{1}{2}B(b - a)\Delta x$$

in magnitude. Here s_n is the nth regular left Riemann sum, and $\Delta x = (b - a)/n$. (Follow the scheme of Lemma 1, but use the tangent line error formula from the last section in Chapter 6 instead of the mean-value principle. Thus

$$F(x_1) - F(x_0) = f(x_0)\Delta x + \tfrac{1}{2}f'(\xi_1)(\Delta x)^2,$$

etc. Show that the sum of these n equations can be cast into the above form.)

9. Prove that the same result holds for the nth regular *right* Riemann sum S_n.

10. Prove that if $f(x) = Bx$ then

$$\int_a^b f = s_n + \frac{1}{2}B(b - a)\Delta x.$$

This shows that the error bound given in the above problems cannot be improved.

11. By slightly modifying the scheme suggested in Problem 8, show that if $|f'(x)| \le B$ on $[a, b]$, then for any Riemann sum R

$$\left| R - \int_a^b f \right| \le \frac{1}{2}B(b - a)\delta,$$

where δ is the mesh of the subdivision associated with R.

7. ARC LENGTH

In high school geometry the length s of a circular arc is characterized roughly as follows:

we inscribe a polygonal arc composed of n segments, compute its length s_n, and then obtain s as the limit of s_n as n tends to infinity. We shall now show that this procedure works for the graph of any function that is *smooth* in the sense of having a *continuous* derivative.

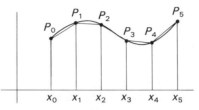

Consider, then, the graph of a smooth function f from $x = a$ to $x = b$, and inscribe a polygonal arc, with vertices $P_0, P_1, \ldots, P_n$. The x-coordinates of the vertices, $x_0, x_1, \ldots, x_n$ subdivide the interval $[a, b]$ into n subintervals of lengths

$$\Delta x_k = x_k - x_{k-1},$$

where $k = 1, 2, \ldots, n$.

Consider, to begin with, the kth polygon side. We isolate and magnify it in the figure below.

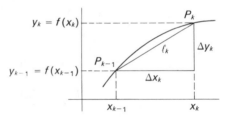

Its length l_k is the distance between P_{k-1} and P_k,

$$l_k = \sqrt{(\Delta x_k)^2 + (\Delta y_k)^2}.$$

By the mean-value principle, there is a point ξ_k between x_{k-1} and x_k such that

$$\Delta y_k = f'(\xi_k)\Delta x_k,$$

and, when this is substituted in the radical above, we end up with

$$l_k = \sqrt{1 + [f'(\xi_k)]^2}\, \Delta x_k$$

as the length of the segment $P_{k-1}P_k$.

The length of the inscribed polygon is thus

$$l = \sum_{k=1}^{n} l_k = \sum_{k=1}^{n} \sqrt{1 + [f'(\xi_k)]^2}\, \Delta x_k.$$

But this is simply a Riemann sum for the function

$$g(x) = \sqrt{1 + [f'(x)]^2},$$

and therefore approaches the limit

$$\int_a^b \sqrt{1 + [f'(x)]^2}\, dx$$

as the mesh of the subdivision of $[a, b]$ approaches 0. We have thus not only shown that the arc length characterization works, but also have derived the formula

$$\int_a^b \sqrt{1 + [f'(x)]^2}\, dx$$

for the length of the graph of f over the interval $[a, b]$. If $y = f(x)$, this can be written

$$\int_a^b \sqrt{1 + \left(\frac{dy}{dx}\right)^2}\, dx.$$

This formula is consistent with, in fact equivalent to, a very intuitive property of arc length. First note that the length s from $t = a$ to $t = x$,

$$s = \int_a^x \sqrt{1 + [f'(t)]^2}\, dt,$$

is an antiderivative of the integrand (by the Fundamental Theorem):

$$\frac{ds}{dx} = \sqrt{1 + \left(\frac{dy}{dx}\right)^2}.$$

Then both $\Delta s / \Delta x$ and

$$\frac{l}{\Delta x} = \frac{\sqrt{(\Delta x)^2 + (\Delta y)^2}}{\Delta x} = \sqrt{1 + \left(\frac{\Delta y}{\Delta x}\right)^2}$$

have the same limit $\sqrt{1 + (dy/dx)^2}$ as $\Delta x \to 0$, so their quotient has the limit 1. Thus

$$\lim \frac{\Delta s}{l} = \lim \frac{\Delta s / \Delta x}{l / \Delta x} = 1.$$

That is,

$$\frac{\text{arc length}}{\text{chord length}} \to 1,$$

as chord length $\to 0$, which is the intuitive property referred to above.

The arc length formula leads to complicated integrands and there aren't very many functions for which we can work out the integral. Even simple changes of scale can affect the complexity of the calculation. For example, we can easily find the length of the graph of $f(x) = (e^x + e^{-x})/2$, but not $g(x) = e^x + e^{-x}$!

EXAMPLE 1   Find the length of the graph of

$$f(x) = \frac{e^x + e^{-x}}{2}$$

from $x = 0$ to $x = a$.

Solution. Here $f'(x) = (e^x - e^{-x})/2$, and

$$1 + [f'(x)]^2 = 1 + \frac{e^{2x} - 2 + e^{-2x}}{4} = \frac{e^{2x} + 2 + e^{-2x}}{4}$$

$$= \left[\frac{e^x + e^{-x}}{2}\right]^2.$$

The arc length is therefore

$$\int_0^a \sqrt{1 + [f'(x)]^2}\, dx = \int_0^a \frac{e^x + e^{-x}}{2}\, dx = \frac{e^x - e^{-x}}{2}\bigg]_0^a$$

$$= \frac{e^a - e^{-a}}{2}.$$

EXAMPLE 2   Find the length of the graph of $\log x$ from $x = 1$ to $x = 2$.

Solution. We have to integrate

$$\int \sqrt{1 + [f'(x)]^2}\, dx = \int \sqrt{1 + \left(\frac{1}{x}\right)^2}\, dx = \int \frac{\sqrt{x^2 + 1}}{x}\, dx.$$

Setting $x = \tan\theta$, $dx = \sec^2\theta\, d\theta$, this becomes

$$\int \frac{\sec^3\theta\, d\theta}{\tan\theta} = \int \frac{\sin\theta\, d\theta}{\sin^2\theta \cos^2\theta} = -\int \frac{du}{(1 - u^2)u^2}$$

$$= -\int \left[\frac{1}{2(1 + u)} + \frac{1}{2(1 - u)} + \frac{1}{u^2}\right] du$$

$$= \frac{1}{2} \log \frac{1 - u}{1 + u} + \frac{1}{u}.$$

Retracing the substitutions $u = \cos\theta$, $x = \tan\theta$, we obtain $u = 1/\sqrt{x^2 + 1}$. The arc length is then

$$\left[\frac{1}{2}\log\left(\frac{\sqrt{x^2 + 1} - 1}{\sqrt{x^2 + 1} + 1}\right) + \sqrt{x^2 + 1}\right]_1^2 = \left[\log\left(\frac{\sqrt{x^2 + 1} - 1}{x}\right) + \sqrt{x^2 + 1}\right]_1^2$$

$$= \log \frac{\sqrt{5} - 1}{2(\sqrt{2} - 1)} + \sqrt{5} - \sqrt{2}.$$

Problem 13 asks for this computation to be redone using the definite integral substitution rule from Section 3. Keep this rule in mind for all the problems.

PROBLEMS FOR SECTION 7

Find the lengths of the graphs of the following equations, over the given x intervals.

1. $y = \log \cos x$, $[0, \pi/4]$

2. $y = x^{1/2} - x^{3/2}/3$, $[1, 2]$

3. $y = ax^{3/2}$, $[0, 1]$

4. $3y = 2(x^2 + 1)^{3/2}$, $[0, a]$

5. $y^3 = x^2$, $[0, 8]$

6. $y = (1 - x^{2/3})^{3/2}$, $[0, 1]$

7. $y = e^x$, $[0, 1]$ 8. $y = ax^2$, $[0, 1]$

9. Find the length of the graph of $y = mx + k$ from $x = a$ to $x = b$.

10. Show that the length of a quarter circle of radius r is $\int_0^r r \, dx/\sqrt{r^2 - x^2}$.

11. Find the length of the graph of $3y = 2x^{3/2}$ from $x = 0$ to $x = 2$.

12. Find the length of the graph of $8y = x^4 + 2/x^2$ from $x = 1$ to $x = 2$.

13. Redo Example 2, using the substitution rule for definite integrals from Section 3.

8. CENTER OF MASS

Let us consider the x-axis as a rigid, weightless, horizontal rod pivoting at the point p, and suppose that we place masses m_i at the positions x_i. According to the principle of the lever, or the "seesaw" principle, the system will exactly balance if

$$\sum m_i(x_i - p) = 0.$$

In any case, the sum $\sum m_i(x_i - p)$ measures the tendency of the system to turn about p. It is called the *moment* of the system about p, and if the moment is zero, then the system is in equilibrium. If we are given the masses m_i sitting at positions x_i, and if we are free to move the pivot point p, then we can find a unique position $p = \bar{x}$ such that the moment about $\bar{x}$ is zero. This holds if and only if

$$\sum m_i(x_i - \bar{x}) = 0,$$

or

$$\sum m_i x_i - \bar{x} \sum m_i = 0,$$

so

$$\bar{x} = \frac{\sum m_i x_i}{\sum m_i}$$

is the zero-moment point. This balancing point is called the *center of mass* of the system. The formula above says that the center of mass is obtained as the quotient

$$\bar{x} = \frac{\text{moment of system about the origin}}{\text{total mass of the system}}.$$

Now consider a rod with variable density $\rho(x)$ extending along the x-axis from $x = a$ to $x = b$. Our intuition tells us that there will be a unique pivot position $\bar{x}$ where the rod will exactly balance, and which we can therefore call the center of mass of the rod. The question is how to compute it.

What we do is approximate the rod by a finite collection of point masses. We subdivide the rod $[a, b]$ by the points

$$a = x_0 < x_1 < \cdots < x_n = b,$$

and we choose an arbitrary point ξ_i in the ith subrod $[x_{i-1}, x_i]$ for each i. Then the mass of this subrod is approximately

$$m_i = \rho(\xi_i)\Delta x_i,$$

and we consider it to be concentrated at the point ξ_i (We discussed the relationship between density and mass in Chapter 4.) We thus end up with the finite collection of masses m_i at the points ξ_i as an approximation of the rod. The moment of the rod about p should then be approximately the moment of this finite system of point masses,

$$\sum m_i(\xi_i - p) = \sum \rho(\xi_i)(\xi_i - p)\Delta x_i.$$

This approximation should improve when we break the rod up into a larger number of smaller pieces and "in the limit" should give the moment exactly. On the other hand, the sum above is a Riemann sum for the function $\rho(x)(x - p)$ over the interval $[a, b]$ and hence approaches $\int_a^b \rho(x)(x - p)\, dx$ as its limit. We therefore conclude:

> *If a rod with density $\rho(x)$ lies along the interval $[a, b]$, then its moment about the point p is $\int_a^b (x - p)\rho(x)\, dx$.*

Going on, we define the center of mass of the rod as the point $\bar{x}$ about which the moment is zero, and we can solve for $\bar{x}$ as before:

$$0 = \int_a^b (x - \bar{x})\rho(x)\, dx = \int_a^b x\rho(x)\, dx - \bar{x}\int_a^b \rho(x)\, dx,$$

$$\bar{x} = \frac{\displaystyle\int_a^b x\rho(x)\, dx}{\displaystyle\int_a^b \rho(x)\, dx} = \frac{\text{moment about } 0}{\text{total mass}}.$$

EXAMPLE 1   A rod extends along the x-axis from $x = 0$ to $x = 1$ and has density $\rho(x) = x^2$. Find its center of mass $\bar{x}$.

Solution

$$\bar{x} = \frac{\displaystyle\int_0^1 x\rho(x)\, dx}{\displaystyle\int_0^1 \rho(x)\, dx} = \frac{\displaystyle\int_0^1 x^3\, dx}{\displaystyle\int_0^1 x^2\, dx} = \frac{1/4}{1/3} = \frac{3}{4}.$$

We gave above an argument that is typical of the way scientists pass from laws and formulas governing finite discrete systems to their continuous analogues involving definite integrals. Such arguments are not mathematical proofs, and their conclusions are physical laws subject to verification.

We could make the mathematics completely sound by starting from a set of postulates for moments. This would be an improvement, but it does not guarantee a physically correct conclusion. It just shifts the onus to the postulates. That is, if the result of a completely logical argument is found to be in disagreement with experience, then some premise must already be out of line.

Here is such a postulate system. We fix the pivot point p and assume that every finite mass distribution along the line has a uniquely determined moment about p satisfying the following two axioms:

1. The moment is additive. That is, if two mass distributions are superimposed to form a single "sum" distribution, then the moment of the sum is the sum of the moments.

2. If the mass distribution lies entirely between $x = x_1$ and $x = x_2$ (where $x_1 < x_2$), then

$$m(x_1 - p) \le \text{moment} \le m(x_2 - p),$$

where m is the total mass.

It follows from (2) that a mass m at a point x has the moment $m(x - p)$. Then (2) says that if we concentrate all the mass at a point to the left of the actual distribution, then we decrease the moment, and if we concentrate all the mass at a point to the right then we increase the moment.

Starting from these axioms it is possible to *prove* the formula that we obtained heuristically. (See Problem 17.)

PROBLEMS FOR SECTION 8

In each of the following problems, a rod with variable density $\rho(x)$ lies along the given interval. Find its center of mass.

1. $\rho(x) = 1 + x$; $[0, 2]$ 2. $\rho(x) = x^2$; $[1, 2]$
3. $\rho(x) = 2x - x^2$; $[0, 2]$ 4. $\rho(x) = x^3$; $[0, 1]$
5. $\rho(x) = x^4$; $[1, 2]$ 6. $\rho(x) = x^4$; $[0, 2]$
7. $\rho(x) = \sin x$; $[0, \pi]$ 8. $\rho(x) = e^x$; $[0, 1]$
9. $\rho(x) = \log x$; $[1, 2]$

10. A rod with density $\rho(x) = x$ lies over the interval $[0, 2]$. If the rod is broken in two at $x = 1$, find the mass and center of mass of the whole rod and of each of its two pieces. If these are respectively m, $\bar{x}$, m_1, $\bar{x}_1$, m_2, $\bar{x}_2$, show that

$$\bar{x} = \frac{m_1 \bar{x}_1 + m_2 \bar{x}_2}{m_1 + m_2}.$$

That is, the center of mass of the whole rod can be computed by replacing each of its two pieces by a point mass located at the center of mass of the piece, and then computing the moment of this system of two point masses.

11. A rod of varying density $\rho(x)$ lies over the interval $[a, c]$. If it is broken into two pieces at the point $x = b$, and if these pieces have masses m_1 and m_2, and centers of mass $\bar{x}_1$ and $\bar{x}_2$, show that

$$\bar{x} = \frac{m_1 \bar{x}_1 + m_2 \bar{x}_2}{m_1 + m_2}$$

is the center of mass of the whole rod.

12. A finite number of point masses m_i sit at positions x_i on the x-axis. Let M be the moment of this system about the pivot point $x = p$, as given by the formula

in the text. Let $\bar{x}$ be the center of mass of the system, and m its total mass. Show that in the moment computation the system behaves as though it consisted of a single point mass m at $x = \bar{x}$. That is, show that

$$M = m(\bar{x} - p).$$

13. A rod with variable density $\rho(x)$ has total mass m, center of mass $\bar{x}$, and moment M about the pivot point $x = p$, all given by the integral formulas in the text. Show that, as far as the computation of M is concerned, the rod behaves as though it were a point mass m at $x = \bar{x}$. That is, show that $M = m(\bar{x} - p)$.

14. A rod with variable density $\rho(x)$ lies over the interval $[a, c]$, and is broken into two pieces at $x = b$. If M, M_1, and M_2 are, respectively, the moments about $x = p$ of the whole rod and its two pieces, as given by the integral formula in the text, show that

$$M = M_1 + M_2.$$

15. Use the results of Problems 13 and 14 to give a new proof of Problem 11.

16. If the moment axiom (1) is applied repeatedly it acquires the following more general formulation:

1'. *If a mass distribution is broken up into a finite number of pieces, then the moment of the whole is equal to the sum of the moments of its pieces.*

Show that the formula $M = \sum m_i(x_i - p)$ for the moment about p of a system of point masses m_i at positions x_i is a consequence of the moment axioms (1') and (2).

17. Assuming the moment axioms (1) and (2), we can use the methods of Chapter 4 to prove the formula

$$M = \int_a^b \rho(x)(x - p)\, dx$$

for the moment about p of a continuous mass distribution along the interval $[a, b]$. What we have to show is that if ΔM is the moment of the piece lying over the increment interval $[x, x + \Delta x]$, then

$$\lim_{\Delta x \to 0} \frac{\Delta M}{\Delta x} = \rho(x)(x - p).$$

Do this, as follows. First show, using axiom (2), that if Δm is the mass on the increment interval, then

$$(x - p)\Delta m \le \Delta M \le (x + \Delta x - p)\Delta m.$$

Then finish off, as usual, by the squeeze limit law.

9. THE CENTROID OF A PLANE LAMINA

We can view a region G in the plane as representing a thin sheet of material of constant density k. That is, the mass of any part of G is k times the area of that part. When G is interpreted this way, it is often called a *homogeneous lamina*. We may as well suppose that $k = 1$, so that the mass is just the area.

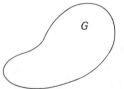

Suppose now that we view the coordinate plane as lying horizontally and that we interpret the region between the graph of a function f and the x-axis as a horizontal homogeneous lamina. We fix a line $x = p$ parallel to the y-axis and we consider whether or not the lamina G will "balance" on this new axis. That is, we consider the moment of the lamina G with respect to the axis $x = p$.

The seesaw principle for mass distributions in the plane has exactly the same formula as before: If masses m_i are placed at the points (x_i, y_i) on a rigid, weightless, horizontal sheet, then the system will exactly balance about the axis at $x = p$ if

$$\sum m_i(x_i - p) = 0.$$

That is, the moment of a mass m_i depends only on its distance from the axis at $x = p$, and hence is independent of its y-coordinate. We therefore treat the homogeneous lamina by dividing it into thin strips parallel to the balancing axis, each strip then having an approximately constant distance from the balancing axis. As usual, this slicing up of the lamina is discussed in terms of a subdivision

$$a = x_0 < x_1 < \cdots < x_n = b$$

of the interval $[a, b]$.

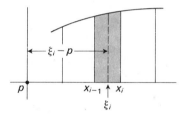

If Δm_i is the mass of the ith strip, lying between $x = x_{i-1}$ and $x = x_i$, and if an evaluation point ξ_i is chosen from this interval, then the moment of this strip about the axis $x = p$ is approximately

$$\Delta m_i(\xi_i - p).$$

We are assuming that the lamina has constant density equal to 1, so Δm_i is the area ΔA_i of the strip. Also, $\Delta A_i \approx f(\xi_i)\Delta x_i$ by the usual rectangular approximation to the area. We thus end up with

$$(\xi_i - p)f(\xi_i)\Delta x_i$$

as an approximation to the moment of the ith strip, and the total moment is thus approximately

$$\sum (\xi_i - p)f(\xi_i)\Delta x_i.$$

As the subdivision is taken finer we assume that this approximation will improve and have the exact moment as its limit. On the other hand, the above sum is a Riemann sum for the function $(x - p)f(x)$, so its limit is known to be the definite integral of this function. Thus,

$$M = \int_a^b (x - p)f(x)\, dx$$

is the exact moment of the lamina about the axis $x = p$. Here, again, the moment axioms can be used to make the mathematics precise, but the physical conclusion is a law in the sense of physics, subject to constant verification.

This is the same formula that we obtained for the rod with variable density. The point is that since only distances in the x-direction count in determining this moment, the contribution of the whole strip over $[x_{i-1}, x_i]$ is the same as though it were a mass distribution on the x-axis of density approximately $f(\xi_i)$.

To get the axis $p = \bar{x}$ where the lamina will balance, we set the moment equal to zero and find

$$\bar{x} = \frac{\displaystyle\int_a^b xf(x)\,dx}{\displaystyle\int_a^b f(x)\,dx} = \frac{\text{moment about } y\text{-axis}}{\text{area of region}},$$

just as before. For a region bounded by two function graphs, say above by f and below by g, we just replace $f(x) = f(x) - 0$ by $f(x) - g(x)$ in the above computations. Thus,

$$\bar{x} = \frac{\displaystyle\int_a^b x(f(x) - g(x))\,dx}{\displaystyle\int_a^b (f(x) - g(x))\,dx}$$

is the more general form of the formula for $\bar{x}$.

Actually, this more general form is a direct consequence of the earlier special form and the addition principle for moments (Axiom (1)), because of the identity

$$\int_a^b x(f(x) - g(x))\,dx = \int_a^b xf(x)\,dx - \int_a^b xg(x)\,dx.$$

This expresses the integral on the left as the difference of the moments about the y-axis of the regions R and R_2 below, and this is the moment of R_1, by Axiom (1).

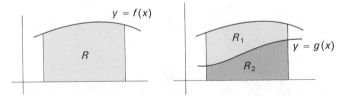

A plane lamina G will also have a balancing axis $y = \bar{y}$ parallel to the x-axis, and then $(\bar{x}, \bar{y})$ is the balancing *point* for G. This may not be obvious, but it can be proved. The balancing point $(\bar{x}, \bar{y})$ of a plane lamina G is called its *center of mass*. In the case of a homogeneous lamina, such as we are considering here, the point $(\bar{x}, \bar{y})$ is also called the *centroid* of the region G.

EXAMPLE 1 Find the centroid of a semicircular disk.

Solution. Let G be the right half-disk bounded by the circle $x^2 + y^2 = r^2$ and the y-axis.

Then

$$\bar{x} = \frac{\int_0^r x \cdot 2\sqrt{r^2 - x^2}\, dx}{\frac{1}{2}\pi r^2}$$

$$= \frac{-\frac{2}{3}(r^2 - x^2)^{3/2}\big]_0^r}{\frac{1}{2}\pi r^2}$$

$$= \frac{\frac{2}{3}r^3}{\frac{1}{2}\pi r^2} = \frac{4r}{3\pi}.$$

By symmetry $\bar{y}$ must be 0, so $(4r/3\pi, 0)$ is the centroid.

If a region G is not symmetric about the x-axis, then $\bar{y}$ must be computed. When G can be described as lying between the y-axis and the graph of $x = h(y)$, from $y = c$ to $y = d$, then exactly the same reasoning as before will now lead to the corresponding formula for $\bar{y}$:

$$\bar{y} = \frac{\int_c^d y h(y)\, dy}{\text{area of } G} = \frac{\text{moment about } x\text{-axis}}{\text{area of } G}.$$

EXAMPLE 2 If G is the region between the y-axis, the parabola $y = x^2$, and the line $y = 4$, then we can compute $\bar{x}$ as before:

$$\bar{x} = \frac{\int_0^2 x(4 - x^2)\, dx}{\int_0^2 (4 - x^2)\, dx} = \frac{2x^2 - \dfrac{x^4}{4}\Big]_0^2}{4x - \dfrac{x^3}{3}\Big]_0^2} = \frac{4}{16/3} = \frac{3}{4}.$$

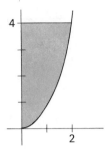

But G can also be described as lying between the y-axis and the graph of $x = \sqrt{y}$, from $y = 0$ to $y = 4$. So we can compute $\bar{y}$ by the corresponding y integral:

$$\bar{y} = \frac{\displaystyle\int_c^d yh(y)\,dy}{\text{area}} = \frac{3}{16}\int_0^4 y^{3/2}\,dy = \frac{3}{16} \cdot \frac{2}{5} y^{5/2}\Big]_0^4 = \frac{12}{5}.$$

Unfortunately, a region that can easily be described when it is viewed as lying over the x-axis may be hard to describe when viewed as lying over the y-axis. One wonders therefore if it is possible to compute $\bar{y}$ by some kind of an x integral. It is, and in Problems 25 and 26 a proof is sketched for the following result:

If G is bounded above and below by the graphs of $y = f(x)$ and $y = g(x)$, and runs from $x = a$ to $x = b$, then its moment about the x-axis is given by

$$\frac{1}{2}\int_a^b (f^2(x) - g^2(x))\,dx.$$

Then $\bar{y}$ is found by dividing this moment by the area of G.

EXAMPLE 3   Find the y coordinate of the centroid of the region bounded by the graphs of $y = e^x$ and $y = 2e^x$, the y-axis and the line $x = 1$.

Solution. The area of G is

$$\int_0^1 (2e^x - e^x)\,dx = e^x\Big]_0^1 = e - 1.$$

According to the new formula above, the moment of G about the x-axis is

$$\frac{1}{2}\int_0^1 (4e^{2x} - e^{2x})\,dx = \frac{3}{4}e^{2x}\Big]_0^1 = \frac{3}{4}(e^2 - 1).$$

Then

$$\bar{y} = \frac{\text{moment about } x\text{-axis}}{\text{area}}$$

$$= \frac{\dfrac{3}{4}(e^2 - 1)}{e - 1} = \frac{3}{4}(e + 1).$$

PROBLEMS FOR SECTION 9

Find the center of mass of the homogeneous plane lamina G (centroid of the plane region G) where:

1. G is the rectangle with two sides along the axes and a vertex at (a, b).
2. G is an arbitrary rectangle with sides parallel to the axes.
3. G is bounded by the graph $y = |x|$ and the line $y = 2$.
4. G is bounded by the lines $y = x$, $y = 2$, and the y-axis.
5. G is bounded by the y-axis and the parabola $y^2 = 4 - x$.
6. G is the triangle formed by the coordinate axes and the line $x + 2y = 2$.
7. G is bounded by the parabola $y = x^2$ and the line $y = x$.
8. G is bounded by the two parabolas $y = x^2$, $x = y^2$.
9. G is the region in the first quadrant bounded by the cubic $y = x^3$ and the line $y = 2x$.
10. G is bounded by the curve $y^2 = x^3$ and the line $x = 2$.
11. G is bounded by the curves $y = e^x$, $y = -e^x$, $x = 0$, and $x = a$.
12. G is the part of the circular disk of radius r about the origin lying in the first quadrant.
13. G is the right half of the interior of the ellipse

$$\frac{x^2}{a^2} + \frac{y^2}{b^2} = 1.$$

14. G is the region $0 \le y \le \log x$, $x \le 2$.
15. G is the region $x \ge 0$, $-e^{-x} \le y \le e^{-x}$. (Here G extends to infinity and the integrals in the formula for $\bar{x}$ have to be evaluated as $\lim_{a \to \infty} \int_0^a \cdots$.)
16. G is the triangle with vertices at $(a, 0)$, $(b, 0)$, and $(0, c)$, where $a < b$ and $c > 0$. [*Hint:* The numerator in the formula for $\bar{x}$ (the moment about the y-axis) can conveniently be evaluated as the sum of two moments.]
17. G is bounded by the parabola $y = x^2$ and the line $y = x + 2$.

10. THE TRAPEZOIDAL APPROXIMATION

By now the reader has experienced some of the difficulties one can meet in evaluating a definite integral $\int_a^b f$, and can better appreciate the need for numerical estimates. If f is a function, like $f(x) = e^{-x^2}$, for which no elementary antiderivative exists, then a direct numerical evaluation is our only recourse. But even if we can integrate f, the antiderivative F that we find may be very complicated, and the evaluation

$$\int_a^b f = F(b) - F(a)$$

may be more difficult than using some direct numerical estimate.

The only numerical method that we have considered so far is the approximation of $\int_a^b f$ by a Riemann sum. However, this approximation is inefficient, at least for hand computation, since it requires considerable labor to guarantee even a moderately good estimate. The trouble is that we are, in effect, approximating f locally by a constant function, and this takes no account at all of the shape of the graph of f.

When f is very smooth, in the sense of having several derivatives, then unexpected fluctuations are impossible and the values of f at successive subdivision points ought to determine the whole graph to within very narrow margins. Therefore, it should be possible to compute the integral $\int_a^b f$ from the subdivision values $f(x_i)$ very accurately.

The most obvious way to make some use of the shape of the graph is to replace each rectangle associated with a Riemann sum by the corresponding trapezoid, as shown below. The area of a trapezoid is $h(l_1 + l_2)/2$, where l_1 and l_2 are the lengths of the parallel sides, i.e., the two base lengths.

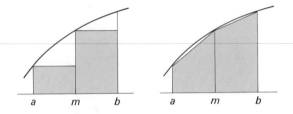

Thus, the figure on the right above suggests the estimates

$$\int_a^m f \approx \Delta x \frac{f(a) + f(m)}{2},$$

$$\int_m^b f \approx \Delta x \frac{f(m) + f(b)}{2},$$

and so

$$\int_a^b f \approx \Delta x \left[\frac{f(a)}{2} + f(m) + \frac{f(b)}{2}\right].$$

We expect from the look of the geometric figures that this estimate will be much better than our original Riemann-sum estimate

$$\int_a^b f \approx \Delta x[f(a) + f(m)].$$

In general, we subdivide the interval $[a, b]$ into n equal subintervals of common length $\Delta x = (b - a)/n$ the subdividing points being $a = x_0$, $x_1, \ldots, x_n = b$.

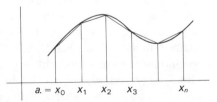

Then we add up all the individual trapezoidal estimates

$$\int_{x_{i-1}}^{x_i} f \approx \Delta x \frac{f(x_{i-1}) + f(x_i)}{2},$$

from $i = 1$ to $i = n$, and end up with the approximation

$$\int_a^b f \approx \Delta x \left[\frac{f(x_0)}{2} + f(x_1) + f(x_2) + \cdots + f(x_{n-1}) + \frac{f(x_n)}{2} \right]$$

$$= \Delta x \left[\frac{f(a) + f(b)}{2} + \sum_{k=1}^{n-1} f(x_k) \right].$$

Again, this is the analytic estimate corresponding to the approximation of the area under the graph of f by the sum of all the trapezoidal areas associated with the subdivision. This formula is called the *trapezoidal rule*. It is a much better estimate of $\int_a^b f$, as the following error formula shows.

Error formula for trapezoidal rule. *If the second derivative of f exists and is bounded by K over the interval $[a, b]$, then the error in using the trapezoidal rule to estimate $\int_a^b f$ is at most*

$$\frac{K(b - a)}{12} (\Delta x)^2$$

in magnitude.

We shall skip the proof. However, Chapter 12 contains the proof of a similar result.

EXAMPLE 1   We use the same fourfold subdivisions of $[1, 3/2]$ as in Example 4 of Section 2 to estimate $\log(3/2)$ by the trapezoidal rule. We have

$$\log \frac{3}{2} = \int_1^{3/2} \frac{dt}{t} \approx \frac{1}{8} \left[\frac{1}{2} + \frac{8}{9} + \frac{4}{5} + \frac{8}{11} + \frac{1}{2} \cdot \frac{2}{3} \right]$$

$$= \frac{1}{16} + \frac{1}{9} + \frac{1}{10} + \frac{1}{11} + \frac{1}{24}.$$

Here $f''(x) = 2/x^3$, which is bounded by $K = 2$ over the interval $[1, 3/2]$. The error is thus at most

$$\frac{K}{12} (b - a)(\Delta x)^2 = \frac{2}{12} \cdot \frac{1}{2} \cdot \left(\frac{1}{8} \right)^2 = \frac{1}{768}.$$

In Section 2 we found that the simple Riemann sum estimates $\log(3/2)$ with an error at most $1/48$. So here we are guaranteed a better estimate by a factor of 16.

Since $1/768$ is less than $1/500 = 0.002$, so we can say that the error in our trapezoidal estimate of $\log(3/2)$ is less than 2 in the third decimal place.

In order to see how large the error actually is we can compare our estimate with the value from a log table. Dividing out the fractions in the trapezoidal estimate shows that its decimal expansion begins 0.4061. (We haven't discussed decimal expansions yet but probably this is familiar.) From a four-place table of natural logarithms we find that

$$\log \frac{3}{2} = 0.4055$$

to the nearest four decimal places. The actual error in our trapezoidal estimate of log 3/2 is thus less than $0.4062 - 0.4054 = 0.0008$ and so is less than 1 in the third decimal place.

We shall see in Chapter 12 that the same fourfold subdivision of [1, 3/2] leads to still closer estimates of log 3/2 when we take into account higher derivatives. Thus, with essentially the same amount of arithmetic as required by a Riemann sum, but more sophistication, we can get very good approximations to a definite integral.

PROBLEMS FOR SECTION 10

1. Write down the trapezoidal approximation to $\log 2$ arising from a tenfold subdivision of $[1, 2]$. Leave the answer as a sum of rational numbers, as in Example 1 in the text. Show that this sum approximates $\log 2$ with an error less than 0.002.

2. How fine a subdivision do we need to take in order to ensure that the trapezoidal sum will approximate log 2 to the nearest four decimal places (i.e., with an error less than $0.00005 = 5(10)^{-5}$)?

3. Write down the trapezoidal approximation to

$$\frac{\pi}{4} = \int_0^1 dx/(1 + x^2)$$

arising from a tenfold subdivision of $[0, 1]$. Leave the answer as a sum of rational numbers, as in Example 1 in the text. Show that the error is less than 0.002.

4. Show that the trapezoidal approximation is exact (i.e., has zero error) for a linear integrand $f(x) = Ax + B$.

5. Compute the trapezoidal approximation to $\int_0^b x^2\, dx$ for the trivial subdivision of $[0, b]$ (no subdividing points). Compare with the value of the integral, and hence show that the error formula cannot be improved.

6. Compute the trapezoidal approximation to $\int_0^1 x^3\, dx$ for the fivefold subdivision of $[0, 1]$. Show that the actual error is exactly one half the error bound provided by the error formula.

EXTRA PROBLEMS FOR CHAPTER 9

1. $\displaystyle\int_0^1 \frac{dx}{(2 - x^2)^{3/2}}$

2. $\displaystyle\int_0^a \frac{x\,dx}{\sqrt{a^2 - x^2}}$

3. $\displaystyle\int_0^r \sqrt{r^2 - x^2}\, dx$

4. $\displaystyle\int_0^1 \sqrt{1 + x^2}\, dx$

5. $\displaystyle\int_1^2 \sqrt{x^2 - 1}\, dx$

6. $\displaystyle\int_0^\infty \frac{dx}{1 + x^2}$

7. $\displaystyle\int_0^\infty \frac{dx}{4 + x^2}$

8. $\displaystyle\int_0^\infty \frac{dx}{(1 + x^2)^{3/2}}$

9. a) Show that

$$\int_0^a f(-x)\,dx = \int_{-a}^0 f(x)\,dx$$

for any function f.

b) Show that therefore

$$\int_{-a}^a f(x)\,dx = 0$$

if f is an odd function on the interval $[-a, a]$, and that

$$\int_{-a}^a f(x)\,dx = 2\int_0^a f(x)\,dx$$

if f is even on $[-a, a]$.

10. Show that

$$\int_A^B f(cx + d)\,dx = \frac{1}{c}\int_{cA+d}^{cB+d} f(x)\,dx.$$

11. The "wood chopper's wedge" pictured below, is cut from a cylinder of radius r by a plane through a diameter of the base circle, making an angle of $30° = \pi/6$ radians with the base plane. Compute the volume using plane sections perpendicular to the wedge edge.

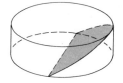

12. Solve the above problem by sections parallel to the wedge edge.

13. A tent is made by stretching canvas from a circular base of radius r to a semicircular rib, erected at right angles to the base at the ends of a diameter. Compute the volume of the tent.

14. Show that the volume of the ellipsoid

$$\frac{x^2}{a^2} + \frac{y^2}{b^2} + \frac{z^2}{c^2} = 1$$

is $(4/3)\pi abc$. [Hint: The cross section of this solid perpendicular to the z-axis at the point $z = z_0$ is the ellipse

$$\frac{x^2}{a^2} + \frac{y^2}{b^2} = \left(1 - \frac{z_0^2}{c^2}\right).$$

The area of the standard ellipse

$$\frac{x^2}{a^2} + \frac{y^2}{b^2} = 1$$

is πab, by Extra Problem 68, Chapter 8.]

15. Two circular cylinders intersect to form a solid S. The cylinders have the same radius a and have perpendicular intersecting axes. Find the volume of S.

16. Let f be a smooth decreasing function whose graph runs from $(0, 1)$ to $(1, 0)$. Prove that the length of the graph of f necessarily lies between $\sqrt{2}$ and 2. [*Hint:* Show that the length of any inscribed polygonal arc necessarily lies between these numbers.]

17. In the same way, show, for any two positive numbers a and b, that if the graph of f decreases from $(0, b)$ to $(a, 0)$, then the length of the graph necessarily lies between $\sqrt{a^2 + b^2}$ and $a + b$. What would be the corresponding statement for an increasing graph?

18. Over the interval $[0, 1]$, the graph of $f(x) = x^n$ increases smoothly from 0 to 1. Let a_n be the point where f has the value $\sqrt{1/n}$: $f(a_n) = \sqrt{1/n}$.

    a) Prove that $a_n > 1 - \sqrt{1/n}$. [*Hint:* Show that if $a_n \le 1 - \sqrt{1/n}$, then the area under the graph must be larger than $1/n$. But compute the area.]

    b) Show that the length of the graph of f over $[0, 1]$ is greater than $2 - 2\sqrt{1/n}$ (because there is a simple inscribed polygon with length greater than this).

19. Show, partly on the basis of the above result, that the bounds 2 and $\sqrt{2}$ in Problem 16 cannot be improved upon.

20. Let f and g be mutually inverse functions with domain I and J, respectively. Show that the length of the graph of g over J equals the length of the graph of f over I.

21. Prove the following Theorem of Pappus.

    Theorem. *If a plane region G is revolved about a line in the plane that does not cut into G, then the volume generated is equal to the area of G times the length of the circle traced by the centroid of G.*

    Suggestion. Taking the axis of revolution as the y-axis, the center of mass $(\bar{x}, \bar{y})$ traces a circle of length $2\pi\bar{x}$ and the formula to be proved is

    $$V = 2\pi\bar{x}A,$$

    where A is the area of G and V is the volume of the solid of revolution generated by rotating G. In order to prove this formula, write down the formula for $\bar{x}$ and note that its numerator is very close to being the formula for computing V by shells.

22. We know the formulas for the area of a circle and the volume of a sphere. Use the Theorem of Pappus to compute the centroid of a semicircular disk.

23. Compute the formula for the volume of a torus by the Theorem of Pappus. Take the torus to be the solid generated by rotating the circle

    $$(x - R)^2 + y^2 = r^2$$

    about the y-axis, where $0 < r < R$.

24. The area of an ellipse is πab, where a and b are its semiaxes. Compute the volume of the "elliptical torus" generated by revolving the ellipse

    $$\frac{(x - R)^2}{a^2} + \frac{y^2}{b^2} = 1$$

    about the y-axis, where $a < R$.

25. Use the Theorem of Pappus to compute the volume of a cylindrical shell with rectangular cross section.

26. The sides of a square S lie along lines having slopes ± 1. Its left vertex is the point $(a, 0)$ and its area is A. Show that the volume generated by rotating S

about the y-axis is

$$V = 2\pi A\left(a + \sqrt{\frac{A}{2}}\right).$$

27. Prove from the integral formula for moments that the moment of a region G about the axis $x = p$ is $A(\bar{x} - p)$, where A is the area of G and $\bar{x}$ is the x-coordinate of its centroid.

28. Consider the region G lying between the graphs $y = f(x)$ and $y = g(x)$ and running from $x = a$ to $x = b$, where $a < b$, and $g(x) \leq f(x)$ over the interval $[a, b]$.

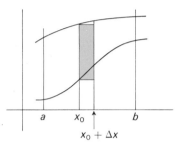

The incremental slice of G between x_0 and $x_0 + \Delta x$ is approximately a rectangle of altitude $f(x_0) - g(x_0)$. Show that the moment of this rectangle about the x-axis is

$$\frac{1}{2}[f^2(x_0) - g^2(x_0)]\Delta x.$$

(Assume the moment formula analogous to the one in Problem 27.)

29. Continuing the above problem, write down an approximation for the moment of G about the x-axis as a sum of moments of rectangles, and then conclude that the exact moment is

$$\frac{1}{2}\int_a^b (f^2(x) - g^2(x))\,dx.$$

Find the centroid of each of the following regions G.

30. G is bounded by the x-axis and the first arch of the graph of $\cos x$.

31. G is bounded by the y-axis and the lines $y = 2x$, $y = x + 2$.

32. G is bounded by the parabolas $y = x^2$, $y = 2 - x^2$.

33. G is bounded by the parabolas $y = 2x^2$, $y = x^2 + 1$.

34. For a region G running from the x-axis up to the graph of $y = f(x)$ the moment M given by the formula in Problem 29 has the simpler expression

$$M = \frac{1}{2}\int_a^b f^2(c)\,dx.$$

Show that if $0 \leq g(x) \leq f(x)$, then the more general formula of Problem 29 follows from the above special case and the addition Axiom (1) for moments.

35. Assuming the Theorem of Pappus (Problem 21), show that the above moment formula (in Problem 34) follows from the formula for the volume of a solid of revolution.

36. Show that

$$\int_a^b (x - a)(b - x)\, dx = \frac{(b - a)^3}{6}.$$

37. Show that

$$\int_a^b (x - a)^2(b - x)^2\, dx = \frac{(b - a)^5}{30}.$$

38. Show that if $f(a) = f(b) = 0$, and if f'' is continuous over $[a, b]$, then

$$\int_a^b f(x)\, dx = \frac{1}{2} \int_a^b (x - a)(b - x)f''(x)\, dx.$$

[*Hint:* Start with the right side and integrate by parts twice. The boundary term will be zero at each step.]

39. Prove that if f has a continuous fourth derivative on $[a, b]$, and if $f(a)$, $f(b)$, $f'(a)$, and $f'(b)$ are all zero, then

$$\int_a^b f(x)\, dx = \frac{1}{24} \int_a^b (x - a)^2(x - b)^2 f^{(4)}(x)\, dx.$$

[*Hint:* Start with the right side and repeatedly integrate by parts.]

40. Prove the following result.

If $f(a)$, $f'(a)$, $f(b)$, and $f'(b)$ are all 0, and if the fourth derivative $f^{(4)}(x)$ is continuous on $[a, b]$, then

$$\int_a^b f = \frac{f^{(4)}(X)}{720}(b - a)^5$$

for some number X in $[a, b]$.

(Let m and M be the minimum and maximum of $f^{(4)}(x)$ on $[a, b]$, and set

$$I = \frac{1}{24} \int_a^b (x - a)^2(x - b)^2\, dx.$$

Use Problem 39 and Property (iv) for definite integrals from Section 3 to show that

$$mI \leq \int_a^b f \leq MI.$$

Conclude that $\int_a^b f = f(X)I$ for some X in $[a, b]$, and then use Problem 37.)

41. The reasoning used in Problem 40 comes up frequently. Prove the following general formulation:

The weighted mean-value theorem. *If f and g are continuous on $[a, b]$, and if $g > 0$ on (a, b), then there is a number X in $[a, b]$ such that*

$$\int_a^b fg = f(X) \int_a^b g.$$

10 Paths and Polar Coordinates

1. PATHS

EXAMPLE 1   It is often useful to view a graph as a path that is traced in some manner. Consider the two equations

$$x = t^3 - 3t,$$
$$y = t^2.$$

For any given value of t, the corresponding values of x and y, taken together, give a point in the xy-coordinate plane. For instance, when $t = -2$, we have $(x, y) = (-2, 4)$. As t varies, the point

$$(x, y) = (t^3 - 3t, t^2)$$

varies and traces out the curve in the xy-plane shown below. That is, the collection of all points (x, y) of the form $(x, y) = (t^3 - 3t, t^2)$ makes up this curve. We shall see later how we know the graph looks like this.

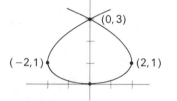

Note that we don't plot the variable t, but plot coordinates x and y that are determined by t. Thus, t is the independent variable, but it doesn't appear in the geometric picture. We call the equations $x = t^3 - 3t$, $y = t^2$ a *parametric representation* of the curve, and t is the *parameter*.

Any pair of functions f and g having a common domain interval can be used together to define a curve parametrically in the above way. The parametric equations are

$$x = g(t),$$
$$y = h(t),$$

and the curve is made up of the points $(x, y) = (g(t), h(t))$. We imagine the parameter t changing, and visualize the varying point $(x, y) = (g(t), h(t))$ tracing the curve. One of the most important interpretations comes from thinking of t as time and $(x, y) = (g(t), h(t))$ as the position at time t of a moving point or particle. Then the path of this moving particle lies along the curve. Because of this interpretation, a parametric representation of a curve is often called a *path*. Sometimes we call the curve itself a path.

The example above shows that a path does not necessarily lie along the graph of a function. In fact, one reason for introducing parametric representations is to obtain a new way of discussing curves that are more general than function graphs.

Sometimes we can eliminate the parameter t from the parametric equations

$$x = g(t),$$
$$y = h(t),$$

and obtain a single equation in x and y for the graph of the path. In the above example,

$$x = t^3 - 3t,$$
$$y = t^2,$$

if we square x, then the right side involves only powers of $t^2 = y$, and we obtain the equation

$$x^2 = y^3 - 6y^2 + 9y.$$

Thus, the point $(x, y) = (t^3 - 3t, t^2)$ moves along the graph of this equation.

Often we can recognize a path by eliminating the parameter in the above way. In each of the following examples, the parameter is easily eliminated to get an equation in x and y, and in each case we can easily sketch the graph of the equation. Then the path of the point along this known curve can be indicated by adding arrows.

EXAMPLE 2 $x = t^3$; $y = t^2$.

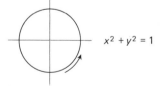

$y = x^{2/3}$

This example is closely related to the first example.

EXAMPLE 3 $x = \cos t$; $y = \sin t$.

$x^2 + y^2 = 1$

EXAMPLE 4 $x = \cos t^3$; $y = \sin t^3$.

$x^2 + y^2 = 1$

The point of this last example is the use of the word *path*. Some people would say that Examples 3 and 4 involve the same path, although it is traced somewhat differently in the two situations. Others feel that the idea of a path includes the way it is traced, and that these examples are, therefore, *different* paths. This ambiguity in the meaning of "path" causes no difficulty. We just have to be careful to be clear in cases where it matters.

In Example 2, even though the coordinate functions

$$x = g(t) = t^3 \qquad \text{and} \qquad y = h(t) = t^2$$

are everywhere continuously differentiable, the graph itself has a kink (called a *cusp*) at the origin, and the path direction along the graph takes a sudden 180° turn there. It will be seen later that such a discontinuity in path direction can be ruled out by requiring path smoothness in the following sense.

DEFINITION   *A path* $(x, y) = (g(t), h(t))$ *is smooth if g and h have continuous derivatives and if* g' *and* h' *are never simultaneously zero.*

We shall need a formula for the slope of the path in terms of g' and h'. In order to find such a formula, we consider a general path point $(x, y) = (g(t), h(t))$, and give the parameter t a (nonzero) increment Δt, obtaining a new path point

$$(x + \Delta x, y + \Delta y) = (g(t + \Delta t), h(t + \Delta t)).$$

The secant line between these two path points has the slope $\Delta y/\Delta x$, and the tangent slope m is the limit of the secant slope as the second point approaches the first point along the path, i.e., as Δt approaches 0. Thus

$$m = \lim_{\Delta t \to 0} \frac{\Delta y}{\Delta x} = \lim_{\Delta t \to 0} \frac{\dfrac{\Delta y}{\Delta t}}{\dfrac{\Delta x}{\Delta t}} = \frac{h'(t)}{g'(t)}.$$

This calculation is valid only if $g'(t) \neq 0$. Then $\Delta x \neq 0$ when Δt is sufficiently small (Theorem 2, Chapter 6), and there is no division by zero. However, the formula is still correct in the case $g'(t) = 0$, $h'(t) \neq 0$, provided it is interpreted as saying that the slope is infinite and the tangent line is vertical. For in this case the above calculation can be repeated to show that the *reciprocal* slope is zero,

$$\lim_{\Delta t = 0} \frac{\Delta x}{\Delta y} = \frac{g'(t)}{h'(t)} = 0.$$

And the fact that $\Delta x/\Delta y$ approaches zero means geometrically that the secant line approaches a vertical line as $\Delta t \to 0$. We have proved the following theorem

THEOREM 1   *A smooth path* $(x, y) = (g(t), h(t))$, *has a tangent line at every point, with slope m given by*

$$m = \frac{h'(t)}{g'(t)}.$$

If $g'(t) = 0$, *the formula is interpreted as saying that the slope is infinite and the tangent line is vertical.*

Since the slope h'/g' and the reciprocal slope g'/h' are both continuous functions of t wherever defined, it follows that the tangent line varies continuously.

We can sketch smooth paths much as we sketched graphs in Chapter 6. When we first considered function graphs, we decided that a graph could change from rising to falling (or vice versa) only by going through a point where the tangent is *horizontal*. A path like the first one drawn above has a complementary property: It can change from moving forward to moving backward (or vice versa) only by going through a point where the tangent is *vertical*.

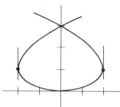

So the points of vertical tangency (the points where $g'(t) = 0$) divide the path into pieces, each of which is traced steadily in one direction over the x-axis, either to the right, like a function graph, or to the left, like a function graph traced backwards. The proof of this observation (Theorem 2) depends on showing that each of these pieces actually does lie along the graph of a function F, but F plays only a theoretical role and rarely has to be found explicitly.

Each function-graph piece of the path in turn is divided into *monotone* pieces by its *horizontal* tangents, as in Chapter 6. A smooth path thus has two types of critical points.

DEFINITION *A critical point of a smooth path is a point where the tangent line is either vertical or horizontal, i.e., a point where either $g'(t) = 0$ or $h'(t) = 0$.*

It was concluded above that the critical points of a smooth path divide it up into monotone-function-graph pieces, traced forward or backward. These pieces are thus qualitatively characterized by their four possible "quadrant directions": to the right and up, to the right and down, to the left and up, to the left and down.

EXAMPLE 5 We know that the path

$$x = \cos t,$$

$$y = \sin t,$$

traces the unit circle $x^2 + y^2 = 1$ counterclockwise. Let us see, however, what the critical point analysis by itself would tell us about the path.

The path critical points are the points where either derivative is zero. We find

$$\frac{dx}{dt} = -\sin t = 0 \qquad \text{at } t = 0, \pi, 2\pi, \text{etc.},$$

$$\frac{dy}{dt} = \cos t = 0 \qquad \text{at } t = \pi/2, 3\pi/2, \text{etc.}$$

We compute the corresponding path points, and get the following table.

t	0	$\pi/2$	π	$3\pi/2$	2π	$\cdots$
x	1	0	-1	0	1	$\cdots$
y	0	1	0	-1	0	$\cdots$

These points are in consecutive order on the path, which runs along a monotone graph between each pair. The general configuration must therefore be like this:

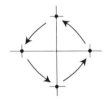

The slope formula $m = g'(t)/h'(t)$ shows whether the critical point tangent is vertical or horizontal. However, it can also be decided geometrically. For example, the tangent at the critical point $(1, 0)$ must be vertical because the path doubles back from that point. The simplest curve that we can draw that will meet these requirements is some kind of oval, and we thus get the right general shape from these path considerations.

EXAMPLE 6   We now show that we have been using the correct figure for our first path

$$x = g(t) = t^3 - 3t,$$
$$y = h(t) = t^2.$$

First,

$$\frac{dx}{dt} = 3(t^2 - 1) = 0 \qquad \text{at } t = \pm 1,$$

$$\frac{dy}{dt} = 2t = 0 \qquad \text{at } t = 0,$$

so that altogether these critical parameter values break the t-axis up into four intervals.

We then make up the corresponding critical point table.

t	$-\infty$	-1	0	1	∞
x	$-\infty$	2	0	-2	∞
y	$+\infty$	1	0	1	∞

Note also that $x = t^3 - 3t = 0$ when $t = 0, \pm\sqrt{3}$, so we can add the y-intercept at $y = (\sqrt{3})^2 = 3$. Plotting these points gives the framework shown at the left below, and we draw as smooth a path as we can, following these directions.

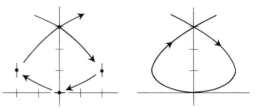

PROBLEMS FOR SECTION 1

In each of the following problems eliminate the parameter to obtain an equation in x and y for the graph along which the path runs. Sketch and identify (if possible) the graph, and indicate by one or more arrows how it is being traced. Compute the path slope dy/dx as a function of t and determine the path critical points (points where $dy/dx = 0$ or ∞, or is undefined).

1. $x = t, y = t^2$
2. $x = t - 1, y = 2t + 3$
3. $x = t - 1, y = t^3$
4. $x = 3 \sec t, y = 2 \tan t$
5. $x = 2t^{1/3}, y = 3t^{1/3}$
6. $x = t^2 - 1, y = t + 1$
7. $x = 2 \sin t, y = 3 \cos t$
8. $x = t + 1, y = t(t + 4)$
9. $x = 2 \cos t, y = 4 \sin t$
10. $x = a \cos^3 t, y = a \sin^3 t$
11. $x = 2 \cos t, y = 2 \sin t$
12. $x = a \cos t, y = b \sin t$

Parametric equations can be used to great advantage in a practical way when analyzing the motion of projectiles. In this case, the motion may be represented by:

$$x = x_0 + v_0 t \cos \theta,$$

$$y = y_0 + v_0 t \sin \theta - \frac{1}{2} gt^2,$$

where v_0 is the initial velocity, (x_0, y_0) is the initial point, θ is the initial angle of inclination from the horizontal, and g is the acceleration of gravity (32 ft/sec^2).

13. A shell is fired with an initial velocity of 100 ft/sec and $\cos \theta = 3/5$ and $\sin \theta = 4/5$. Find the time t when the shell hits the ground, and the distance between the gun and the spot where the shell lands; also, find the maximum height reached by the shell.

14. A ball is thrown off the top of a building 100 ft high with a velocity of 60 ft/sec and at an angle of 45° above the horizontal. Find

a) The maximum height attained by the ball.

b) The distance from the building to the point where the ball hits the ground.

Analyze each of the following paths following the scheme used in Example 6. Include the x and y intercepts in your table if you can determine them, and indicate any singular points that you find (points where the path is not smooth).

15. $x = t^2, y = t^2$

16. $x = t^2, y = t^2 - 2t$

17. $x = t^3 - 3t, y = t^3 - 3t^2$

18. $x = t^2, y = 2t^3 - 9t^2 + 12t$

2. THE MISSING DETAILS

If a path $(x, y) = (g(t), h(t))$ runs for a while along the graph of a differentiable function F, then the coordinate-parameter functions

$$x = g(t) \quad \text{and} \quad y = h(t)$$

are related by the equation

$$y = F(x),$$

or

$$h(t) = F(g(t)),$$

along this part of the path. Therefore

$$\frac{dy}{dt} = F'(x)\frac{dx}{dt}$$

along this piece, provided g is differentiable. This derivative relationship has two consequences.

First, if the path is smooth, then dx/dt can never be zero along this piece. For if $dx/dt = 0$, then the formula shows that $dy/dt = 0$ also, and this contradicts the definition of a smooth path. Second, if $dx/dt \neq 0$, then

$$F'(x) = \frac{dy/dt}{dx/dt} = \frac{h'(t)}{g'(t)},$$

along this piece. Since $F'(x)$ is the slope m of the graph of F, we thus recover this case of the formula

$$m = \frac{h'(t)}{g'(t)}$$

for the slope of a path in terms of the parameter functions.

Theorem 2 is a partial converse of the above remarks.

THEOREM 2   *If f and g are differentiable functions on an interval I and if $g'(t)$ is never zero in the interior of I, then as t runs across I the path $(x, y) = (g(t), h(t))$ runs along the graph of a continuous function F, forwards if $x = g(t)$ is increasing, and backwards if g is decreasing. Moreover, F is differentiable for t in the interior of I.*

Proof. Since the function $x = g(t)$ has no critical points interior to I, it is monotone on I, and has a monotone, continuous inverse function $t = k(x)$, by the inverse function theorem (Chapter 7, Theorem 1). Then

$$y = h(t) = h(k(x))$$

while t is in I. So if we set $F(x) = h(k(x))$, then the equation

$$y = F(x)$$

holds along the path while t is crossing I. If $x = g(t)$ is increasing, then the path runs along the graph of F in the normal left-to-right direction. If $x = g(t)$ is decreasing, then the path traces the graph of F backwards.

Since g' is never zero interior to I, its inverse k is differentiable, and $k'(x) = 1/(g'(t))$ at the corresponding values of x, again by the inverse-function rule. Then the composite function $F(x) = h(k(x))$ is differentiable, by the chain rule. This completes the proof of the theorem. ∎

Along an interval I containing no path critical points both derivatives $g'(t)$ and $h'(t)$ are different from zero. Since $g' \neq 0$, the path runs along the graph of a function F. Since $h' \neq 0$, the formula $F'(x) = h'(t)/g'(t)$ shows that $F' \neq 0$, and therefore that F is monotone. Thus, between successive critical points, a smooth path runs along a monotone function graph, as claimed earlier.

3. PATH LENGTH

We saw in the last chapter that we could define the length of a smooth function graph as the limit of lengths of inscribed polygons, and that this limit had a natural formulation as a definite integral.

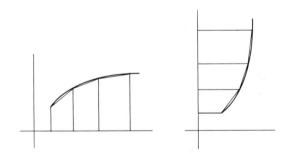

If we rotate the plane over the $y = x$ diagonal, such a function graph rotates to a nonstandard function graph along which x is a function of y. But the inscribed polygonal lengths remain the same and they necessarily have the same limit, so the *length* of a piece of a graph remains unchanged by such a rotation. However, the description of the length calculation for the new graph proceeds with the roles of x and y interchanged, so the integral formula for the length is now

$$\int_a^b \sqrt{1 + \left(\frac{dx}{dy}\right)^2}\, dy.$$

We have seen that a smooth path can be broken up into a succession of arcs, each of which is one of the above two types. The length of a smooth path is therefore a sum of lengths, each of which can be computed by one of these two formulas. But this is a piecemeal way to deal with path length, and we clearly would like a single integral formula with the parameter as variable.

The most straightforward way to approach such a path length formula would be to start over again with inscribed polygons and express their lengths in terms of the parameter t. This works all right, but it involves an extra complication, namely, that the resulting finite t sums are not quite Riemann sums and a new integral estimate has to be proved.

Instead, we shall adopt an antiderivative approach. We know that for small arcs and their chords,

$$\frac{\text{arc length}}{\text{chord length}} \to 1$$

as chord length $\to 0$. (See Section 7 in Chapter 9.)

So let s be the length of the smooth path $(x, y) = (g(t), h(t))$ measured from the fixed point (x_0, y_0) at $t = t_0$. Then:

THEOREM 3   *The path length s is a differentiable function of t, and*

$$\left(\frac{ds}{dt}\right)^2 = \left(\frac{dx}{dt}\right)^2 + \left(\frac{dy}{dt}\right)^2.$$

Proof. We assume that $t > t_0$, so that s is an increasing function of t and $\Delta s/\Delta t$ is positive. If we give t a positive increment Δt, then the chord length l for the corresponding arc increment is

$$l = \sqrt{(\Delta x)^2 + (\Delta y)^2}.$$

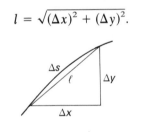

Therefore,

$$\frac{\Delta s}{\Delta t} = \frac{\Delta s}{l} \cdot \frac{l}{\Delta t} = \frac{\Delta s}{l} \sqrt{\left(\frac{\Delta x}{\Delta t}\right)^2 + \left(\frac{\Delta y}{\Delta t}\right)^2}.$$

Now let $\Delta t \to 0$. Then Δx and Δy both approach 0 (differentiable functions being continuous), so $l \to 0$ and $\Delta s/l \to 1$. Thus $ds/dt = \lim_{\Delta t \to 0}(\Delta s/\Delta t)$ exists and

$$\frac{ds}{dt} = \sqrt{\left(\frac{dx}{dt}\right)^2 + \left(\frac{dy}{dt}\right)^2}.$$

If Δt is negative, then we have to keep track of minus signs in the above calculation, but the result is the same. However, if $t < t_0$, so that s is a decreasing function of t, then the length of the incremental arc is $-\Delta s$ for positive Δt, and this extra minus sign leads to

$$\frac{ds}{dt} = \lim_{\Delta t \to 0} \frac{\Delta s}{\Delta t} = -\sqrt{\left(\frac{dx}{dt}\right)^2 + \left(\frac{dy}{dt}\right)^2}.$$

In every case, however,

$$\left(\frac{ds}{dt}\right)^2 = \left(\frac{dx}{dt}\right)^2 + \left(\frac{dv}{dt}\right)^2. \quad \blacksquare$$

REMARK   In terms of differentials, this equation becomes

$$(ds)^2 = (dx)^2 + (dy)^2,$$

which is easy to remember by thinking of the Pythagorean theorem for an "infinitesimal" right triangle.

REMARK   A function graph traced normally is a particular path with parameter $t = x$:

$$x = x,$$
$$y = f(x).$$

For this special path, our new path formula reduces to the original function formula

$$\frac{ds}{dx} = \sqrt{1 + \left(\frac{dy}{dx}\right)^2}.$$

The new integral formula is of course

$$s = \int_{t_1}^{t_2} \sqrt{\left(\frac{dx}{dt}\right)^2 + \left(\frac{dy}{dt}\right)^2}\, dt.$$

EXAMPLE 1   The path $(x, y) = (\cos \theta, \sin \theta)$ traces the unit circumference once as θ runs from 0 to 2π. The length formula should therefore give the answer 2π. We check that it does:

$$\int_0^{2\pi} \sqrt{\left(\frac{dx}{d\theta}\right)^2 + \left(\frac{dy}{d\theta}\right)^2}\, d\theta = \int_0^{2\pi} \sqrt{\sin^2\theta + \cos^2\theta}\, d\theta$$

$$= \int_0^{2\pi} 1 \cdot d\theta$$

$$= \theta \Big]_0^{2\pi}$$

$$= 2\pi.$$

EXAMPLE 2   The path $(x, y) = (\cos(\theta^2), \sin(\theta^2))$ traces the unit circumference once as θ runs from 0 to $\sqrt{2\pi}$. The answer should again be 2π.

$$\int_0^{\sqrt{2\pi}} \sqrt{\left(\frac{dx}{d\theta}\right)^2 + \left(\frac{dy}{d\theta}\right)^2}\, d\theta = \int_0^{\sqrt{2\pi}} \sqrt{4\theta^2 \sin^2\theta^2 + 4\theta^2 \cos^2\theta^2}\, d\theta$$

$$= \int_0^{\sqrt{2\pi}} 2\theta\, d\theta$$

$$= \theta^2 \Big]_0^{\sqrt{2\pi}}$$

$$= 2\pi.$$

EXAMPLE 3   The path

$$x = t^3 - 3t,$$

$$y = 3t^2,$$

is the same as our original example of a path except that the y-coordinates are multiplied by 3. This makes a big difference in the length computation. There is no obvious way of doing it for the original path, but it is very easy for this modified path. Here it is:

$$s = \int \sqrt{\left(\frac{dx}{dt}\right)^2 + \left(\frac{dy}{dt}\right)^2}\, dt$$

$$= \int \sqrt{9(t^2 - 1)^2 + 36t^2}\, dt$$

$$= \int \sqrt{9(t^2 + 1)^2}\, dt$$

$$= \int 3(t^2 + 1)\, dt$$

$$= t^3 + 3t + C.$$

The closed loop in the path runs from $t = -\sqrt{3}$ to $t = \sqrt{3}$, and its length is

$$\int_{-\sqrt{3}}^{\sqrt{3}} = (t^3 + 3t)\Big]_{-\sqrt{3}}^{\sqrt{3}} = 12\sqrt{3}.$$

PROBLEMS FOR SECTION 3

Find the derivative of the path length (ds/dt) for each of the following paths.

1. $x = t, y = t^2$ 2. $x = \cos t, y = \sin t$
3. $x = 3 \sin t, y = \cos t$ 4. $x = t^3, y = t^2$

Compute the lengths of the following paths over the given parameter intervals.

5. $3x = 4t^{3/2}, 2y = t^2 - 2t; [0, a]$
6. $x = a \cos^3 t, y = a \sin^3 t; [0, \pi/2]$
7. $x = t^3 + 3t^2, y = t^3 - 3t^2;$ a) $[0, 1]$ b) $[0, 2]$
8. $x = t^3, y = 2t^2; [0, 1]$
9. $x = e^t \cos t, y = e^t \sin t; [0, 1]$
10. $x = t^2 \cos t, y = t^2 \sin t; [0, 1]$
11. $x = \cos t + t \sin t, y = \sin t - t \cos t; [0, 1]$

4. POLAR COORDINATES

Sometimes a graph seems to have a special affinity for the origin. It loops about and seems to be constantly pulled inward toward the origin.

Such a curve will not be a function graph in a Cartesian coordinate system, and its equation may be complicated and unrevealing of what is going on. When a point traces such a curve, the most natural description may be in terms of its distance from and orientation to the origin, i.e., in terms of its *polar coordinates.*

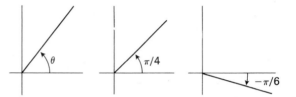

Recall that any number θ, considered as an angular coordinate, determines a unique ray, or half-line, drawn from the origin. Then a nonnegative number r determines the unique point P on the ray at the distance r from the origin. The numbers r and θ together determine the point P uniquely, and are called *polar coordinates* of P.

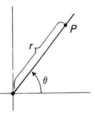

The rectangular coordinates of P are obtained from the polar coordinates r, θ by the change-of-coordinates equations

$$x = r \cos \theta, \qquad y = r \sin \theta.$$

We considered all of this in Chapter 5, but did not pursue the matter there.

The polar-coordinate correspondence

$$(r, \theta) \to P$$

is not one-to-one. That is, it is not true, conversely, that P determines a unique pair of polar coordinates. An angle θ can always be modified by adding a multiple of 2π without changing the ray it determines, so P determines its polar coordinates only "up to" the addition of a multiple of 2π to its angular coordinate θ.

So far it has been implicitly assumed that r is positive or zero. But the equations $x = r \cos \theta$, $y = r \sin \theta$, define a point P for *any* numbers r and θ. If r is negative, then P is obtained geometrically by first measuring off the angle θ to obtain a ray and then proceeding along this ray *backward across the origin* a distance $|r|$ on the other side. Such negative values of r are not normally used, but we shall see that some polar graphs seem incomplete unless negative r is allowed.

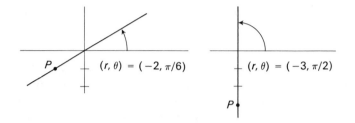

$$(r, \theta) = (-2, \pi/6) \qquad (r, \theta) = (-3, \pi/2)$$

Any equation in r and θ has a *polar graph*, consisting of all points in the plane whose polar coordinates satisfy the equation. The polar graph of a function f is the polar graph of the equation

$$r = f(\theta).$$

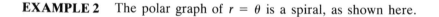

EXAMPLE 1   The polar graph of $r = 1$ is the unit circle about the origin.

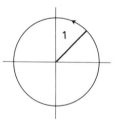

EXAMPLE 2   The polar graph of $r = \theta$ is a spiral, as shown here.

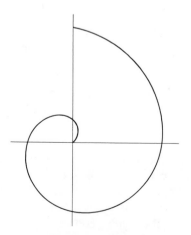

EXAMPLE 3 The graph of $r = \cos \theta$ is another circle, but this is not so obvious. We can try to get an idea of the shape of an unknown polar graph by computing and plotting a few selected points. We do this below for half the graph, and see that the graph is probably some kind of *oval*.

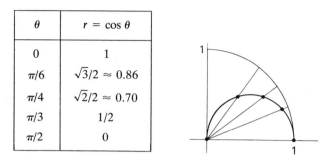

θ	$r = \cos \theta$
0	1
$\pi/6$	$\sqrt{3}/2 \approx 0.86$
$\pi/4$	$\sqrt{2}/2 \approx 0.70$
$\pi/3$	$1/2$
$\pi/2$	0

A better procedure, when it works, is to find and recognize the Cartesian equation of the graph. Here we use the change-of-coordinate equation $x = r \cos \theta$ together with the given equation $r = \cos \theta$, to get, in order,

$$r = \cos \theta = x/r,$$

$$r^2 = x,$$

$$x^2 + y^2 = x,$$

$$x^2 - x + y^2 = 0,$$

$$\left(x - \frac{1}{2}\right)^2 + y^2 = \frac{1}{4},$$

and finally we recognize that the graph is the circle of radius 1/2 about the center $(1/2, 0)$.

EXAMPLE 4 If e is a positive number less than 1, show in a similar way that the polar graph of

$$r = \frac{1}{1 - e \cos \theta}$$

is an ellipse.

Solution. Setting $\cos \theta = x/r$ and cross-multiplying, the equation becomes, in turn,

$$r\left(1 - e\frac{x}{r}\right) = 1,$$

$$r = 1 + ex,$$

$$x^2 + y^2 = r^2 = (1 + ex)^2 = 1 + 2ex + e^2 x^2,$$

$$x^2(1 - e^2) - 2ex + y^2 = 1,$$

$$(1 - e^2)\left(x - \frac{e}{1 - e^2}\right)^2 + y^2 = 1 + \frac{e^2}{1 - e^2} = \frac{1}{1 - e^2},$$

$$\frac{(x - \alpha)^2}{a^2} + \frac{y^2}{b^2} = 1,$$

where $a = 1/(1 - e^2)$, $b = 1/\sqrt{1 - e^2}$, $\alpha = e/(1 - e^2)$.

If $e = 4/5$, then $a = 25/9$, $b = 5/3$, $\alpha = 20/9$, and the ellipse is shown below.

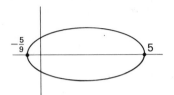

EXAMPLE 5   If negative r is not allowed, then the graph of $r = \cos 2\theta$ is as shown at the left below. At the right is the "full" graph, using negative r.

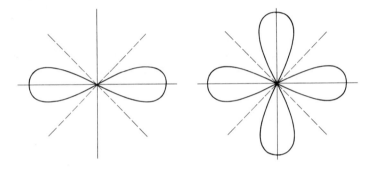

PROBLEMS FOR SECTION 4

Sketch the polar graphs of each of the following equations.

1. $r = \cos 3\theta$ 2. $r = \sin 2\theta$
3. $r = 2 \cos 5\theta$ 4. $r = 2 \sin 4\theta$
5. $r^2 = \cos 2\theta$ 6. $r^2 = \cos \theta$
7. $r = 2 \sin \theta$ (Find the x, y equation first.)
8. $r = 1 + \cos \theta$
9. $r = 1 - \cos \theta$ 10. $r = 1 - 2 \cos \theta$
11. Show that $r \sin \theta = 2$ is the equation of the horizontal straight line two units above the x-axis.
12. In view of the above problem, consider the equation

$$r \sin (\theta - \theta_0) = 2.$$

It has the above form if we set $\phi = \theta - \theta_0$. On this basis, show by a geometric argument that it must be the equation of the line which is at a distance 2 from the origin and which makes the angle θ_0 with the direction of the x-axis.

13. Show that the polar graph of the equation

$$r = \frac{1}{1 - \cos \theta}$$

is a parabola. (Follow Example 4.)

5. AREA IN POLAR COORDINATES

Polar area is measured between given values of the *angular* coordinate, and the geometric picture is thus different. The incremental approximating area is now in the form of a circular sector with vertex angle $\Delta\theta$, instead of a rectangle with base Δx, and this change is reflected in the new formula below.

THEOREM 4 *Let f be a positive continuous function and let A be the area of the region bounded by the polar graph of the function f and the two rays $\theta = a$ and $\theta = b$. Then*

$$A = \frac{1}{2} \int_a^b f^2(\theta)\, d\theta.$$

EXAMPLE 1 Find the area swept over in the first revolution of the spiral $r = \theta$.

Solution. Here $r = f(\theta) = \theta$, and θ runs from 0 to 2π. Therefore,

$$A = \frac{1}{2} \int_0^{2\pi} f(\theta)^2 d\theta = \frac{1}{2} \int_0^{2\pi} \theta^2 d\theta = \frac{1}{2}\left[\frac{\theta^3}{3}\right]_0^{2\pi} = \frac{4}{3}\pi^3,$$

by the theorem.

The form of the integrand in the theorem stems from the formula for the area of a circular sector, $A = (1/2)r^2\alpha$. In a given circle the area of a sector is proportional to its central angle, so if the sector central angle is α, then the sector area A is the fraction $\alpha/2\pi$ of the area of the whole circle,

$$A = \left(\frac{\alpha}{2\pi}\right)\pi r^2 = \frac{1}{2} r^2 \alpha.$$

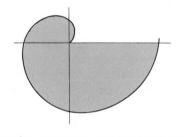

We could prove Theorem 4 in the manner of Chapter 4 or we could proceed as in the earlier sections of Chapter 9. Actually, there is a sort of combination of these two methods that often can be used to strengthen the Riemann-sum approach and we shall illustrate it here.

Proof of theorem. We start as usual with a subdivision of the angular interval $[a, b]$, $a = \theta_0 < \theta_1 < \cdots < \theta_n = b$. But now, for each k, let m_k and M_k be the minimum and maximum values of $f(\theta)$ on the kth subinterval $[\theta_{k-1}, \theta_k]$. Also, let ΔA_k be the area lying "over" this incremental subinterval, i.e., the area of the incremental region bounded by the polar graph of $r = f(\theta)$ and the rays $\theta = \theta_{k-1}$ and $\theta = \theta_k$.

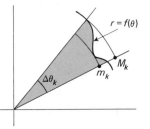

We have pictured one of these incremental regions here. It includes an "inscribed" circular sector of radius $r = m_k$ and is included in a "circumscribed" circular sector of radius $r = M_k$. Therefore,

$$\text{area of inside sector} \le \Delta A_k \le \text{area of outside sector},$$

or

$$\frac{1}{2} m_k^2 \Delta\theta_k \le \Delta A_k \le \frac{1}{2} M_k^2 \Delta\theta_k.$$

Summing these inequalities from $k = 1$ to $k = n$, we have the inequality

$$\sum_{k=1}^n \frac{1}{2} m_k^2 \Delta\theta_k \le A \le \sum_{k=1}^n \frac{1}{2} M_k^2 \Delta\theta_k.$$

Each of these two sums is a Riemann sum for the function $(1/2)f^2(\theta)$, so each of them approaches the limit $(1/2)\int_a^b f^2(\theta)\, d\theta$ as we vary the subdivision in the manner discussed in Section 6 of Chapter 9. Finally, since A is squeezed between these varying sums, it must equal their limit. ∎

EXAMPLE 2   Find the area enclosed by the cardioid $r = 1 + \cos\theta$.

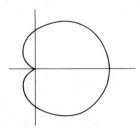

Solution

$$A = \frac{1}{2} \int_0^{2\pi} r^2 d\theta = \frac{1}{2} \int_0^{2\pi} (1 + \cos \theta)^2 d\theta$$

$$= \frac{1}{2} \int_0^{2\pi} [1 + 2 \cos \theta + \cos^2 \theta] \, d\theta.$$

$$= \frac{1}{2} \int_0^{2\pi} \left[1 + 2 \cos \theta + \frac{1}{2} + \frac{\cos 2\theta}{2}\right] d\theta$$

$$= \left[\frac{3}{4} \theta + \sin \theta + \frac{\sin 2\theta}{8}\right]_0^{2\pi}$$

$$= \frac{3}{4} 2\pi$$

$$= \frac{3\pi}{2}.$$

PROBLEMS FOR SECTION 5

1. Find the area swept out in the first revolution of the spiral $r = \theta^2$.
2. The area formula of Theorem 4 should be consistent with the formula for the area of a circle. Show that it is. (Use Theorem 4 to retrieve the value πr^2 as the area of a circle of radius r.)
3. Find the area of the region bounded by the polar graph $r = \cos \theta$ and the rays $\theta = 0$, $\theta = \pi/2$.
4. Find the area bounded by the polar graph $r = \sec \theta$ and the rays $\theta = 0$, $\theta = \pi/4$.
5. Show that the area cut off by the spiral $r = e^\theta$ in successive quadrants increases by the constant factor e^π.
6. Find the area of the region bounded on the outside by the cardioid $r = 1 + \cos \theta$ and on the inside by $r = 2 \cos \theta$.

In each of the following problems, calculate the area of the region bounded by the given polar graph or graphs.

7. $r = 1 - \sin \theta$
8. $r = 2 + \cos \theta$
9. $r = 3 - 2 \sin \theta$
10. $r^2 = 2 \cos 2\theta$
11. $r^2 = a \cos n\theta$ (one loop)

6. ANGLE AS PARAMETER

When a curve loops around the origin, it is frequently convenient to express it as a path with an angle θ as the parameter. However, the angle θ may or may not be the angular coordinate of the path point.

EXAMPLE 1   The central ellipse

$$\frac{x^2}{a^2} + \frac{y^2}{b^2} = 1$$

has the important parametric representation

$$x = a \cos \theta$$
$$y = b \sin \theta.$$

But θ is not the angular coordinate of the point $(x, y) = (a \cos \theta, b \sin \theta)$ on this ellipse. The geometric meaning of θ here is shown in the accompanying figure.

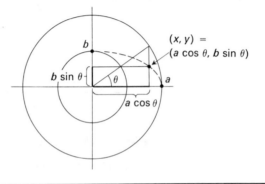

The angular coordinate θ is the natural parameter when the polar graph of a function f is interpreted as a path, for along the polar graph the x and y coordinates,

$$x = r \cos \theta,$$
$$y = r \sin \theta,$$

satisfy the equation

$$r = f(\theta),$$

so that

$$x = f(\theta) \cos \theta,$$
$$y = f(\theta) \sin \theta$$

along the graph. But these are the equations of a path with the angular coordinate of the path point as parameter.

EXAMPLE 2   The spiral $r = \theta$ is traced by the path

$$x = \theta \cos \theta,$$
$$y = \theta \sin \theta.$$

EXAMPLE 3   The cardioid $r = 1 + \cos \theta$ is traced by the path

$$x = \cos \theta + \cos^2 \theta,$$
$$y = \sin \theta + \sin \theta \cos \theta.$$

The expression of the polar graph

$$r = f(\theta)$$

as the path

$$x = f(\theta) \cos \theta \qquad \text{and} \qquad y = f(\theta) \sin \theta$$

gives us an easy derivation of the arc-length formula in polar coordinates. We have

$$\frac{dx}{d\theta} = f'(\theta) \cos \theta - f(\theta) \sin \theta,$$

$$\frac{dy}{d\theta} = f'(\theta) \sin \theta + f(\theta) \cos \theta,$$

and when we square each equation and add, we find that

$$\left(\frac{dx}{d\theta}\right)^2 + \left(\frac{dy}{d\theta}\right)^2 = [f'(\theta)]^2 + [f(\theta)]^2$$

$$= r^2 + \left(\frac{dr}{d\theta}\right)^2.$$

The parametric arc-length formula

$$s = \int_a^b \sqrt{\left(\frac{dx}{d\theta}\right)^2 + \left(\frac{dy}{d\theta}\right)^2}\, d\theta$$

thus becomes the formula

$$s = \int_a^b \sqrt{[f(\theta)]^2 + [f'(\theta)]^2}\, d\theta$$

$$= \int_a^b \sqrt{r^2 + \left(\frac{dr}{d\theta}\right)^2}\, d\theta$$

for the length of the polar graph of f from $\theta = a$ to $\theta = b$.

EXAMPLE 4   Find the length of the cardioid $r = 1 + \cos \theta = 2 \cos^2(\theta/2)$.

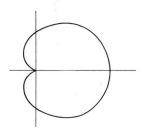

Solution

$$s = \int_0^{2\pi} \sqrt{r^2 + \left(\frac{dr}{d\theta}\right)^2}\, d\theta$$

$$= \int_0^{2\pi} \sqrt{4 \cos^4(\theta/2) + 4 \cos^2(\theta/2) \sin^2(\theta/2)}\, d\theta$$

$$= 2 \int_0^{2\pi} \sqrt{\cos^2 \theta/2} \; d\theta = 2 \int_0^{2\pi} |\cos(\theta/2)| \; d\theta$$

$$= 4 \int_0^{\pi} \cos(\theta/2) \; d\theta = 8 \sin(\theta/2) \Big]_0^{\pi}$$

$$= 8.$$

EXAMPLE 5   Find the length of the first revolution of the spiral $r = \theta$.

Solution. By the formula above, this length is

$$s = \int_0^{2\pi} \sqrt{r^2 + \left(\frac{dr}{d\theta}\right)^2} \; d\theta = \int_0^{2\pi} \sqrt{\theta^2 + 1} \; d\theta.$$

This is a difficult integral. Substituting first $\theta = \tan u$ and then $\sin u = v$, we get:

$$\int \sqrt{\theta^2 + 1} \; d\theta = \int \sec^3 u \; du = \int \frac{\cos u \; du}{(1 - \sin^2 u)^2}$$

$$= \int \frac{dv}{(1 - v^2)^2}.$$

We then express $1/(1 - v^2)^2$ as the sum of its partial fractions. Skipping the details, the answer is:

$$\frac{1}{(1 - v^2)^2} = \frac{1}{4}\left[\frac{1}{(1 - v)^2} + \frac{1}{(1 + v)^2} + \frac{1}{1 - v} + \frac{1}{1 + v}\right].$$

Therefore,

$$\int \frac{dv}{(1 - v^2)^2} = \frac{1}{4}\left[\frac{1}{1 - v} - \frac{1}{1 + v} - \log(1 - v) + \log(1 + v)\right]$$

$$= \frac{1}{4}\left[\frac{2v}{1 - v^2} + \log\left(\frac{1 + v}{1 - v}\right)\right].$$

The v limits of integration are found to be 0 and $2\pi/\sqrt{1 + 4\pi^2}$, and when these are substituted the answer simplifies to

$$s = \int_0^{2\pi} \sqrt{\theta^2 + 1} \; d\theta = \pi\sqrt{4\pi^2 + 1} + \frac{1}{2}\log[\sqrt{4\pi^2 + 1} + 2\pi].$$

PROBLEMS FOR SECTION 6

Find the lengths of the following polar graphs.

1. The first revolution of the spiral $r = \theta^2$ 2. $r = \cos \theta$, from $\theta = 0$ to $\theta = \pi/2$
3. $r = \sec \theta$, from $\theta = 0$ to $\theta = \pi/4$ 4. $r = e^\theta$, from $\theta = 0$ to $\theta = a$
5. $r = 1/\theta$, from $\theta = \pi/2$ to $\theta = \pi$

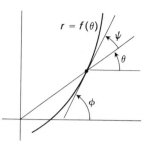

It seems reasonable to specify the direction of a polar graph $r = f(\theta)$ by the angle ψ (Greek psi) it makes with the polar ray, as in the above figure. That this is the natural angle to use is shown by the simplicity of the formula:

$$\cot \psi = \frac{f'(\theta)}{f(\theta)} = \frac{1}{r} \frac{dr}{d\theta}.$$

The following problems center around this formula.

6. Derive the identity

$$\cot(\alpha - \beta) = \frac{1 + \tan \alpha \tan \beta}{\tan \alpha - \tan \beta}$$

from the trigonometric addition formulas (Chapter 5, Section 3).

7. Referring to the above figure again, the slope of the polar graph is $\tan \phi$, where $\phi = \psi + \theta$. Thus,

$$\tan \phi = \frac{dy}{dx} = \frac{dy/d\theta}{dx/d\theta} = \frac{f'(\theta) \sin \theta + f(\theta) \cos \theta}{f'(\theta) \cos \theta - f(\theta) \sin \theta},$$

where the values on the right have been copied from our earlier calculations in the text. Now expand

$$\cot \psi = \cot(\phi - \theta)$$

by the formula in Problem 6, substitute the above value for $\tan \phi$ and simplify. Your answer should be the simple formula for $\cot \psi$ stated earlier (after Problem 5).

8. Prove the identities

$$\tan x = \frac{\sin 2x}{1 + \cos 2x} \qquad \text{and} \qquad \cot x = \tan \left(\frac{\pi}{2} - x \right)$$

from the trigonometric addition formulas.

Compute $\cot \psi$ for each of the following polar graphs. In each case, try to interpret the answer geometrically. (Problem 8 can help.)

9. $r = e^{\theta}$ 10. $r = \sec \theta$

11. $r = \cos \theta$ 12. $r = 1 + \cos \theta$

13. Show, as claimed in the text, that the path $x = a \cos \theta$, $y = b \sin \theta$ does lie along the ellipse $x^2/a^2 + y^2/b^2 = 1$.

7. CURVATURE

DEFINITION   *The curvature of a curve is the amount it bends per unit length, or its change in direction per unit length, where the change in direction between two points is* *the angle between the two tangent lines. The curvature is thus the rate of change of the angle ϕ with respect to the arc length s,*

$$\text{curvature} = \frac{d\phi}{ds},$$

where ϕ is the angle from the x-axis to the tangent line.

We shall first investigate the curvature of a function graph. We saw earlier that sign of the second derivative of a function f tells us which way the graph of f is turning. The magnitude of the second derivative ought to give us an indication of how fast the graph is changing direction, i.e.; of its curvature. But d^2y/dx^2 is not exactly what we want, because $dy/dx = \tan \phi$ and

$$\frac{d^2y}{dx^2} = \frac{d}{dx}\left(\frac{dy}{dx}\right) = \frac{d \tan \phi}{dx},$$

whereas we want $d\phi/ds$. The circle is a good test of suitability. The circle is certainly turning at a constant rate, i.e., has constant curvature. However, from the equation $y = \sqrt{1 - x^2}$ for the upper half of the unit circle we get

$$\frac{dy}{dx} = -\frac{x}{\sqrt{1 - x^2}} \quad \text{and} \quad \frac{d^2y}{dx^2} = -\frac{1}{(1 - x^2)^{3/2}},$$

which is not constant.

The formula for the curvature $d\phi/ds$ depends on the formula for the arc-length derivative ds/dx and the equations

$$\phi = \arctan \frac{dy}{dx}, \qquad \left(\text{from } \tan \phi = \frac{dy}{dx}\right)$$

$$\frac{d\phi}{dx} = \frac{d\phi}{ds} \cdot \frac{ds}{dx}.$$

Then

$$\frac{d\phi}{ds} = \frac{\dfrac{d\phi}{dx}}{\dfrac{ds}{dx}} = \frac{\dfrac{d}{dx}\left(\arctan \dfrac{dy}{dx}\right)}{\sqrt{1 + \left(\dfrac{dy}{dx}\right)^2}} = \frac{\dfrac{1}{1 + \left(\dfrac{dy}{dx}\right)^2} \cdot \dfrac{d^2y}{dx^2}}{\sqrt{1 + \left(\dfrac{dy}{dx}\right)^2}}.$$

Thus:

THEOREM 5   *If $y = f(x)$ is a twice differentiable function, then the curvature of its graph is given by*

$$\text{curvature} = \frac{\dfrac{d^2y}{dx^2}}{\left[1 + \left(\dfrac{dy}{dx}\right)^2\right]^{3/2}} = \frac{f''(x)}{[1 + (f'(x))^2]^{3/2}}.$$

There is still a sign problem. If you visualize a point running along a curve, you will see that if the tangent line rotates counterclockwise as the curve is traced in one direction, then it rotates clockwise when the curve is traced in the opposite direction. In other words, the curvature $d\phi/ds$ can have either sign, depending on which way we measure arc length along the curve. In the calculation above, this ambiguity lurks in the ambiguous sign of ds/dx. We have derived the formula for the case when s increases with x, but if s happened to be measured in the opposite direction we would have acquired a minus sign. What this means is that curvature is a property of a *directed*, or *oriented*, curve. In order to get a definite sign for the curvature, we must be told which way to trace the curve. For undirected curves, we use the *absolute* curvature

$$\kappa = \left| \frac{d\phi}{ds} \right| = \frac{\left| \dfrac{d^2 y}{dx^2} \right|}{\left[1 + \left(\dfrac{dy}{dx} \right)^2 \right]^{3/2}}.$$

EXAMPLE The function $f(x) = x^2$ has a constant second derivative. But its graph is a parabola, and we are familiar enough with its general shape to realize that it flattens out, or becomes less curved, as $|x|$ increases. We thus expect the curvature to approach 0 as $x \to \infty$. In fact

$$\kappa = \frac{\left| \dfrac{d^2 y}{dx^2} \right|}{\left(1 + \left(\dfrac{dy}{dx} \right)^2 \right)^{3/2}} = \frac{2}{(1 + 4x^2)^{3/2}}.$$

The curvature of a path is computed in the same way. This time we start with

$$\tan \phi = \text{slope of path} = \frac{\dfrac{dy}{dt}}{\dfrac{dx}{dt}}.$$

The proof then follows exactly the same pattern as before, although the computations are a little more complicated because of the quotient in the expression for $\tan \phi$. The result is

$$\text{curvature} = \frac{\dfrac{d^2 y}{dt^2} \cdot \dfrac{dx}{dt} - \dfrac{d^2 x}{dt^2} \cdot \dfrac{dy}{dt}}{\left[\left(\dfrac{dx}{dt} \right)^2 + \left(\dfrac{dy}{dt} \right)^2 \right]^{3/2}}.$$

Newton's Dot Notation

Newton used the following very efficient notation for derivatives with respect to time:

$$\dot{x} = \frac{dx}{dt}, \qquad \ddot{x} = \frac{d^2 x}{dt^2}, \qquad \text{etc.}$$

In particular, velocities and accelerations were always expressed this way.

The dot notation is also useful for paths where the parameter isn't necessarily interpreted as time. For example, in this notation, the path curvature formula is

$$\text{curvature} = \frac{\ddot{y}\dot{x} - \ddot{x}\dot{y}}{[\dot{x}^2 + \dot{y}^2]^{3/2}},$$

which is much simpler and neater than the Leibniz form given above.

REMARK The computation of path curvature that we outlined above is incomplete, in that it fails for a path point at which the tangent line is vertical ($dx/dt = 0$, $\tan \phi = \infty$). Near such a point, we have to use the alternate formula

$$\cot \phi = \frac{dx/dt}{dy/dt}.$$

The calculation is then substantially the same, and gives the same curvature formula.

The trouble here is that in each case we are working with a slope formula, and there is always one direction in which a slope becomes infinite and the slope formula becomes useless. Later we shall learn to work with a variable pair $(\Delta x, \Delta y)$ rather than a variable quotient $\Delta y/\Delta x$. This new "vector" technique avoids these "blowing up" difficulties and provides a more satisfactory way to approach such geometric notions as direction and curvature. Chapter 14 contains a very brief introduction to such "vector calculus of paths."

The path curvature formula can be used to find points of inflection. These are the points across which the curvature changes sign—the curve changes from turning to the right to turning to the left, or vice versa—so they are the points where the *numerator* in the curvature formula changes sign. In particular, they are among the points where the numerator is zero.

EXAMPLE Find the points of inflection on the path $(x, y) = (t^2 + t, t^3)$.

Solution. We have

$$\ddot{y}\dot{x} - \ddot{x}\dot{y} = 6t(2t + 1) - 2(3t^2) = 6(t^2 + t),$$

which is zero at $t = 0$ and $t = -1$. Moreover, $t^2 + t$ changes sign at each of these points. Therefore the corresponding path points $(0, 0)$ and $(-1, 0)$ are points of inflection.

PROBLEMS FOR SECTION 7

Compute the curvature of the graph of each of the following functions.

1. $y = x^2$ 2. $y = x^{1/2}$

3. $y = \log x$ 4. $y = \log(\cos x)$ $(-\pi/2 < x < \pi/2)$

5. $y = mx + b$ 6. $y = \sqrt{1 - x^2}$

7. Show that the curvature of $y = \cos x$ oscillates between $+1$ and -1, with period 2π.

8. Find the maximum value of the absolute curvature κ of the graph $y = e^x$.

9. Find the minimum value of the absolute curvature κ of the path

$$x = a \cos^3 t, \qquad y = a \sin^3 t,$$

over the t interval $[0, \pi/2]$.

10. Show that the circle

$$(x, y) = (r \cos \theta, r \sin \theta)$$

has constant curvature $\kappa = 1/r$.

11. Prove that a function graph of zero curvature is a straight line.

12. Consider the function $f(x) = x^a$ on the positive x-axis $(0, \infty)$, where the exponent a can be any real number. Show that the curvature of the graph of f has a finite limit as $x \to 0$ if and only if

$$a \leq \frac{1}{2} \qquad \text{or} \qquad a = 1 \qquad \text{or} \qquad a \geq 2.$$

13. Compute the curvature of the ellipse

$$x = a \cos t, \qquad y = b \sin t.$$

Find its minimum and maximum values.

14. Carry out the proof of the parametric curvature formula.

Find the points of inflection (if any) on the following paths.

15. $(x, y) = (t^3, t^3 - 1)$

16. $(x, y) = (t^3 - 3t, t^2)$

17. $(x, y) = (t^3 + 3t, t^4)$

18. $(x, y) = (t^3 - 3t, t^4)$

19. $(x, y) = (e^{-t}, \log t)$

8. THE PARAMETRIC MEAN-VALUE PRINCIPLE

There is a generalization of the mean-value principle that seems just as obvious geometrically. Suppose we have a smooth path $(x, y) = (g(t), h(t))$ running from (x_0, y_0) to (x_1, y_1). Then the secant line through these two points must be parallel to the tangent line at some in-between point. If the lines are not vertical, this is equivalent to their slopes being equal:

$$\frac{y_1 - y_0}{x_1 - x_0} = \text{slope of tangent line.}$$

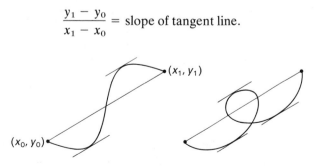

And we saw in Section 1 that the tangent-line slope is $h'(t)/g'(t)$. The following theorem is thus intuitively obvious on geometric grounds.

THEOREM 6   *Let the functions g and h be differentiable on the closed interval $[a, b]$, with $g(a) \neq g(b)$, and $g'(t)$ and $h'(t)$ never simultaneously zero. Then there is at least one point T strictly between a and b at which*

$$\frac{h(b) - h(a)}{g(b) - g(a)} = \frac{h'(T)}{g'(T)}.$$

The proof will be given in Chapter 12. Here we shall illustrate its usefulness by proving a theorem called l'Hôpital's rule, and by proving the tangent-line error formula used in Chapter 6. Then, in Chapter 12, it will be used to prove a much more general error formula. First, we state a special case of the theorem that better fits these applications.

THEOREM 7   *Let g and h be differentiable functions on an interval I containing $x = a$, and suppose that $g(a) = h(a) = 0$ and that $g'(x) \neq 0$ when $x \neq a$. Then for each x different from a, there is a point ξ lying strictly between a and x such that*

$$\frac{h(x)}{g(x)} = \frac{h'(\xi)}{g'(\xi)}.$$

As a first application of Theorem 7, we let $h(x)$ be the error $E(x)$ in the tangent-line approximation, as discussed in Chapter 6:

$$E(x) = f(x) - f(a) - f'(a)(x - a).$$

Note that $E(a)$ and $E'(a)$ are both zero, and that $E''(x) = f''(x)$. Let $g(x) = (x - a)^2$. Then $g(a)$ and $g'(a)$ are also both zero. We can then apply Theorem 7 twice in a row, and conclude that

$$\frac{E(x)}{(x - a)^2} = \frac{E'(\xi)}{2(\xi - a)} = \frac{E''(X)}{2} = \frac{f''(X)}{2},$$

where ξ lies between a and x, and X lies between a and ξ. Thus,

COROLLARY   *Let f be twice differentiable on an interval I containing the point a. Then for every other point x in I, the error*

$$E = f(x) - [f(a) + f'(a)(x - a)]$$

in the tangent-line approximation to f at a can be written in the form

$$E = \frac{f''(X)}{2}(x - a)^2,$$

where X is some number lying strictly between a and x.

It follows from the above formula that if f'' is bounded by K on an interval I about a, then

$$|E(x)| \leq \frac{K}{2}(x - a)^2$$

for each x in I.

Theorem 7 can also be used to answer a natural question about paths, and, in the process, yields a result that is useful in contexts having nothing to do with paths. We saw earlier that a smooth path always has tangent line, with slope $m = h'(t)/g'(t)$. The tangent line could be vertical, in which case

$g'(t) = 0$, $h'(t) \neq 0$, and m is infinite. Now suppose the path is smooth except for one (singular) point $t = a$, where $h'(a) = g'(a) = 0$. The slope formula then takes the meaningless form $0/0$, and we say that $h'(t)/g'(t)$ is *indeterminate* at $t = a$. But the path may still have a tangent line there.

EXAMPLE 1 If

$$x = g(t) = t^3,$$

$$y = h(t) = t - \sin t,$$

then $g'(0) = h'(0)$, and $t = 0$ is a singular point where the slope formula $h'(t)/g'(t)$ is indeterminate. Yet the path has slope 1/6 there, as we shall soon see.

Let $(x, y) = (g(t), h(t))$ be any path through the origin, and suppose that $(0, 0) = (g(a), h(a))$. The path slope there is

$$m = \lim_{\Delta t \to 0} \frac{\Delta y}{\Delta x} = \lim_{\Delta t \to 0} \frac{h(a + \Delta t) - h(a)}{g(a + \Delta t) - g(a)} = \lim_{t \to a} \frac{h(t)}{g(t)}.$$

We can't evaluate this limit by the quotient limit law (the limit of a quotient is the quotient of the limits), because each of the functions g and h has the limit 0 at $t = a$, so the quotient of the limits is the indeterminate $0/0$. The *quotient limit law is valid only when the limit of the denominator is not zero.* We avoided this problem for smooth paths by Theorem 1, which proved that the limit is then $h'(a)/g'(a)$. But what if the path is singular at $t = a$, so that this quotient is also the indeterminate $0/0$?

The solution to this dilemma lies in Theorem 7 which tells us that for any $t \neq a$, there is a number T lying strictly between a and t such that

$$\frac{h(t)}{g(t)} = \frac{h'(T)}{g'(T)}.$$

Suppose that $h'(T)/g'(T)$ approaches a limit l as T approaches a, and consider what happens in the above equation as t approaches a. Then T must also approach a, since T lies between a and t. The righthand side thus approaches the limit l, as $t \to a$, and hence so must the lefthand side. We have proved the following corollary of Theorem 7.

THEOREM 8 *Suppose that the functions g and h are differentiable on an interval I containing $x = a$, and suppose that $g(a) = h(a) = 0$ and $g'(x)$ is not equal to zero on I except possibly at $x = a$. Then*

$$\lim_{x \to a} \frac{h(x)}{g(x)} = \lim_{x \to a} \frac{h'(x)}{g'(x)},$$

in the sense that if the derivative limit exists then the function limit exists and has the same value.

This formula is called l'Hôpital's rule. (Rather, it is one of a *collection* of similar results that all go under this name.)

EXAMPLE 2 We return to our original problem, where

$$g(x) = x^3, \quad \text{and} \quad h(x) = x - \sin x.$$

By the theorem,

$$\lim_{x \to 0} \frac{x - \sin x}{x^3} = \lim_{x \to 0} \frac{1 - \cos x}{3x^2},$$

provided the righthand limit exists. We are still in trouble, however, because the righthand quotient is indeterminate at $x = 0$, so we cannot apply the quotient-limit law. However, the numerator and denominator functions in the righthand quotient satisfy the requirements for h and g in the theorem, so we can apply the theorem once more, and conclude that

$$\lim_{x \to 0} \frac{1 - \cos x}{3x^2} = \lim_{x \to 0} \frac{\sin x}{6x},$$

provided the righthand limit exists. This new quotient is again indeterminate at $x = 0$, but we can apply the theorem a third time, to conclude that

$$\lim_{x \to 0} \frac{\sin x}{6x} = \lim_{x \to 0} \frac{\cos x}{6},$$

provided the righthand limit exists. And finally it *does* exist because we now have a quotient of two continuous functions with the denominator nonzero. Thus

$$\lim_{x \to 0} \frac{\cos x}{6} = \frac{\cos 0}{6} = \frac{1}{6},$$

and, backtracking through the above steps in reverse order, we see that each lefthand limit now does exist and has the value 1/6, by the theorem.

The above example was complicated because of its origin as a path problem to which Theorem 1 could not be applied. It is perhaps typical of the limits evaluated by the theorem that the theorem must be applied more than once, but this doesn't always occur.

EXAMPLE 3 Does the limit of $\sin x/x$ exist as $x \to 0$?

Solution. We know, of course, that it does, because we had to establish the limit of this quotient, which is simply the difference quotient for $\sin'(0)$, before we could even prove that $\sin x$ is differentiable. However, it is interesting to note that we can apply the theorem and conclude that

$$\lim_{x \to 0} \frac{\sin x}{x} = \lim_{x \to 0} \frac{\cos x}{1} = \frac{\cos 0}{1} = 1.$$

PROBLEMS FOR SECTION 8

Evaluate the following limits, mostly by l'Hôpital's rule.

1. $\lim_{t \to 0} \dfrac{\cos t - 1}{t}$ 2. $\lim_{x \to 0} \dfrac{e^x - 1}{x}$ 3. $\lim_{\theta \to 0} \theta \csc \theta$ 4. $\lim_{x \to 0} x \cot x$

5. $\lim\limits_{x\to 0}\dfrac{x-\log(1+x)}{x^2}$

6. $\lim\limits_{t\to 0}\dfrac{1-\cos t}{t^2}$

7. $\lim\limits_{\theta\to 0}\dfrac{1-\cos\theta}{\sin^2\theta}$

8. $\lim\limits_{t\to 0}\dfrac{1-\cos^2 t}{t^2}$

9. $\lim\limits_{x\to 0}\dfrac{x-\sin x}{x-\tan x}$

10. $\lim\limits_{x\to 0}\dfrac{e^x-x-1}{x^2}$

11. $\lim\limits_{t\to 0}\dfrac{2\sqrt{1+t}-t-2}{t^2}$

12. $\lim\limits_{x\to 0}\dfrac{2\cos x+x^2-2}{x^4}$

13. $\lim\limits_{x\to 0}\dfrac{\log(1+x)}{x}$

14. $\lim\limits_{x\to 0}\log(1+x)^{1/x}$

15. $\lim\limits_{x\to 0} x\log x$

16. $\lim\limits_{x\to 0} x\log(\sin x)$

17. $\lim\limits_{t\to 0}\dfrac{\cos t}{1+t^2}$

18. $\lim\limits_{x\to 0}\dfrac{\cos x}{x^2}$

19. $\lim\limits_{x\to 0} x\sin\dfrac{1}{x}$

EXTRA PROBLEMS FOR CHAPTER 10

1. The function f has the graph shown below.

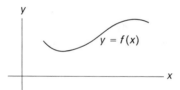

Sketch the following paths.

a) $x = t,\ y = f(t)$

b) $x = -t,\ y = f(-t)$

c) $x = f(t),\ y = t$

2. Sketch the path $x = -t,\ y = f(t)$, where f is the function of Problem 1.

3. Sketch the path $x = t^2,\ y = t^4$, including some arrows to show how it is traced.

4. Do the same for the path $x = t^3 - 3t,\ y = (t^3 - 3t)^2$.

Analyze the following paths by the scheme used in Example 6, Section 1.

5. $x = t^3 - 3t^2 + 3t,\ y = t^3 - 3t^2$ 6. $x = t^2 - t,\ y = t^4 - 2t^2$

7. $x = 4t^3 - 3t^2,\ y = t^4 - 2t^2$

8. Elsewhere we have called a function f smooth if it is continuously differentiable. Show that a function is smooth if and only if its graph is a smooth path when it is traced in the standard way (with the independent variable as parameter).

9. It is reasonable to call a *curve* smooth if there is a smooth path running along it. It is always possible to trace a smooth curve by a different path that fails to meet the smoothness requirement at one or more points. (Later we shall see that this amounts to tracing the curve by a particle that comes to a stop one or

more times.) Show that Examples 3 and 4 in Section 1 in the text illustrate this possibility.

Compute the lengths of the following paths over the given parameter intervals.

10. $x = t^2, y = \dfrac{1}{3}(t^3 - 3t); [1, 3]$

11. $x = \dfrac{4}{n+2} t^{n/2+1}, y = \dfrac{1}{n+1}(t^{n+1} - (n+1)t); [1, n+1], n \geq 0$

12. A smooth path runs along the graph of a function $y = f(x)$. Show that if x is increasing with t, then the formulas

$$\frac{ds}{dx} = \sqrt{1 + \left(\frac{dy}{dx}\right)^2}, \qquad \frac{ds}{dt} = \sqrt{\left(\frac{dx}{dt}\right)^2 + \left(\frac{dy}{dt}\right)^2}$$

are equivalent, by the chain rule.

13. Continue the line of reasoning begun in the above problem, and thus give an alternative proof of the parametric path-length formula

$$s = \int_a^b \sqrt{\left(\frac{dx}{dt}\right)^2 + \left(\frac{dy}{dt}\right)^2}\, dt.$$

14. Show that the formula

$$s = \int_c^d \sqrt{1 + \left(\frac{dx}{dy}\right)^2}\, dy$$

for the length of a graph $x = g(y)$ is a special case of the parametric formula for path length.

15. Show that the polar graph of the equation

$$r = \frac{1}{1 - \epsilon \cos \theta}$$

is a hyperbola if $\epsilon > 1$.

16. Show more generally that for any positive constants e and d the polar graph of

$$r = \frac{d}{1 - e \cos \theta}$$

is an ellipse if $e < 1$, a parabola if $e = 1$, and a hyperbola if $e > 1$.

17. Consider the locus traced by a point whose distance from the origin is a constant e times its distance from the line $x = -k$. Assuming the above problem, show that this locus is an ellipse if $e < 1$, a parabola if $e = 1$, and a hyperbola if $e > 1$.

Find the area bounded by each of the following polar graphs.

18. $r = a \cos 3\theta$ (one loop)

19. $r = a \cos n\theta$ (one loop)

20. $r = 1 + 2 \cos \theta$ (outside loop)

21. $r = 1 + 2 \cos \theta$ (inside loop)

22. $r = 1/\sqrt{1 + \theta}, \theta = 0, \theta = \pi$

23. $r = 1/(1 + \theta), \theta = 0, \theta = \pi$

24. $r = 3 \sec \theta, r = 4 \cos \theta$

25. $r = \sec \theta, r = 2 \cos \theta$

26. $r = 1 + \cos \theta, r = 3/2$

27. $r = 1/\sin \theta, \theta = \pi/4, \theta = \pi/2$

28. $r = \sin \theta - \cos \theta, \theta = 0, \theta = 2\pi$

29. Use the parametric representation

$$x = a \cos \theta,$$
$$y = b \sin \theta$$

for the ellipse

$$x^2/a^2 + y^2/b^2 = 1,$$

to prove the following theorem.

Theorem. *Let F and F′ be the points $(c, 0)$ and $(-c, 0)$ on the x-axis, where $c = \sqrt{a^2 - b^2}$. Let $P = (x, y)$ be any point on the ellipse. Then the segments PF and PF′ have the lengths*

$$\overline{PF} = a - ex \qquad \text{and} \qquad \overline{PF'} = a + ex,$$

where $e = c/a$.

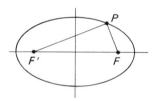

This shows that $\overline{PF} + \overline{PF'} = 2a$; the ellipse is the locus of a point P moving in such a way that the sum of its distance from the two fixed points F and F' is constant and equal to $2a$.

30. Show that the hyperbola

$$\frac{x^2}{a^2} - \frac{y^2}{b^2} = 1$$

has the parametric representation

$$x = a \sec \theta,$$
$$y = b \tan \theta.$$

31. Let F and F' be the points $(c, 0)$ and $(-c, 0)$, where $c = \sqrt{a^2 + b^2}$. Let P be any point on the hyperbola of Problem 30. Using the above parametric representation, show that the segments PF and PF' have the lengths

$$PF = |ex - a|, \qquad PF' = |ex + a|,$$

where $e = c/a$. [*Hint.* Set $\tan^2 \theta = \sec^2 \theta - 1$ at an appropriate moment.]

32. Assuming the above equations show that

$$|PF - PF'| = 2a.$$

(Consider separately the cases $x > 0$, $x < 0$. Note that $e > 1$ and that $|x| \geq 1$ on the graph.) Restate this equation in words as a locus characterization of the hyperbola, similar to that for the ellipse in Problem 29.

33. Carry out the computation showing that

$$\left(\frac{dx}{d\theta}\right)^2 + \left(\frac{dy}{d\theta}\right)^2 = r^2 + \left(\frac{dr}{d\theta}\right)^2,$$

where $x = r \cos \theta$, $y = r \sin \theta$, and r is a function of θ.

Use the parametric curvature formula to compute the curvatures of the following.

34. $x = t^3 + 3t^2, y = t^3 - 3t^2$

35. $x = \cos t + t \sin t, y = \sin t - t \cos t$

36. The formula for the curvature in polar coordinates is

$$\text{curvature} = \frac{r^2 + 2\left(\dfrac{dr}{d\theta}\right)^2 - r\dfrac{d^2r}{d\theta^2}}{\left[r^2 + \left(\dfrac{dr}{d\theta}\right)^2\right]^{3/2}}.$$

Prove this formula, starting from the parametric form of the formula given in the text, and using the expressions for $dx/d\theta$ and $dy/d\theta$ from the last section.

Using the above formula, compute the curvature of the following polar graphs.

37. $r = 2a \cos \theta$ 38. $r = a(1 + \cos \theta)$

39. $r = e^{a\theta}$ 40. $r = a\theta$

Evaluate the following limits

41. $\displaystyle\lim_{x \to 0} \frac{x - \tan x}{x^3}$ 42. $\displaystyle\lim_{x \to 0} \frac{\arctan x}{x}$

43. $\displaystyle\lim_{x \to 0} \frac{x - \arctan x}{x^3}$ 44. $\displaystyle\lim_{x \to +\infty} x \sin \frac{1}{x}$

45. Prove the weighted integral mean-value theorem (Extra Problem 41 in Chapter 9) from the parametric mean-value theorem. (Let F be an antiderivative of fg and let G be an antiderivative of g. Consider the quotient $[F(b) - F(a)]/[G(b) - G(a)]$.)

46. Show that

$$\lim_{x \to 0} \frac{f(x) - f(-x)}{x} = 2f'(0).$$

47. Show that

$$\lim_{x \to 0} \frac{f(x) + f(-x) - 2f(0)}{x^2} = f''(0).$$

48. Prove that

$$\lim_{x \to 0} (1 + x)^{1/x} = e.$$

(Look first at Problem 14 in Section 8.)

49. Prove l'Hôpital's second rule:

If $\displaystyle\lim_{x \to \infty} g(x) = \lim_{x \to \infty} h(x) = 0$, and if $g'(x) \neq 0$ for all sufficiently large x, then

$$\lim_{x \to \infty} \frac{h(x)}{g(x)} = \lim_{x \to \infty} \frac{h'(x)}{g'(x)},$$

in the same sense as before.

[*Hint:* Set $H(y) = h(1/y)$, $G(y) = g(1/y)$, and try to apply Theorem 8, Section 8 to H/G.]

11

Infinite Series

It will be shown later in the chapter that the derivative sum rule remains true for certain "infinitely long polynomials":

$$If \quad f(x) = \sum_{n=0}^{\infty} a_n x^n, \quad then \quad f'(x) = \sum_{n=1}^{\infty} n a_n x^{n-1}.$$

Before considering the calculus of such infinite polynomials, however, it is necessary to learn something about the nature of the infinite sum operation and its arithmetic. The early sections lay this groundwork.

We start with the infinite decimal representation of a positive real number, a familiar and important description of a real number that exhibits some of the phenomena we have to analyze.

1. DECIMAL REPRESENTATIONS

Ordinary arithmetic is concerned with *rational* numbers, i.e., with quotients of integers, such as 5/3, −2/7, 1/100, 4 = 4/1, and 3.1 = 31/10. A real number that is *not* rational is called *irrational*. For example it can be shown that $\sqrt{2}$, $\sqrt{3}$, π, and e are all irrational. Proofs of irrationality for these numbers are of varying degrees of difficulty. If p is a positive prime integer, such as 2, 3, 5, or 7, then it is relatively easy to show that $\sqrt{p}$ is irrational. But the proofs that π and e are irrational are much harder.

A rational number that can be written as a fraction with a power of 10 as denominator is called a *decimal fraction*, and is normally expressed as a *finite decimal*. For example,

$$\frac{31}{10} = 3.1,$$

$$\frac{141}{100} = 1.41,$$

$$\frac{2}{1000} = 0.002.$$

Finite decimals are simpler to handle than general rational numbers because the rules for adding and multiplying fractions reduce, in this case, to the simpler rules for integers (plus keeping track of the decimal point).

The crucial fact we wish to discuss is that *every real number can be represented by an infinite decimal expansion*. A decimal fraction has two such decimal expansions, but for any other number the infinite decimal representation is unique. In order to see how this goes, suppose that x is some particular number in the interval [1, 2] and suppose that x is not itself a finite decimal. The following diagram shows how the one-place decimals divide the number line into intervals of length 1/10.

Note that we have written the integers 1 and 2 as the one-place decimals 1.0 and 2.0. The number x must lie in the interior of one of these ten subintervals of [1, 2], i.e., between a uniquely determined pair of adjacent one-place decimals. In the figure, x is shown as lying between 1.6 and 1.7. *The decimal expansion of x conveys this location by starting off* $1.6\cdots$.

The two-place decimals divide each of the above intervals into ten subintervals, each of length 1/100. For example, [1.60, 1.61] is the first of the ten subintervals of [1.6, 1.7], and [1.64, 1.65] is the fifth. Again, exactly one of these ten subintervals contains x. It appears from the figure that this might be [1.64, 1.65], in which case the next digit in the decimal expansion of x is 4 and the expansion starts off $1.64\cdots$. Then the three-place decimals divide [1.64, 1.65] into ten subintervals of length $1/1000 = 10^{-3}$, and the one containing x is recorded by the digit in the third decimal place in the expansion of x. And so on. Thus, *the first n places in the decimal expansion of x are given by the n-place decimal a_n uniquely determined by the fact that*

$$a_n < x < a_n + 10^{-n}.$$

This is for a number x that is not a finite decimal. In general we have to allow for equality: the number x represented by a decimal expansion is related to the first n places in the expansion by the *weak* inequalities

$$a_n \leq x \leq a_n + 10^{-n}.$$

When these bracketing inequalities are rewritten

$$0 \leq x - a_n \leq 10^{-n},$$

they provide the formal justification (by the squeeze limit law) for the obvious fact that $x - a_n \to 0$ and $a_n \to x$ as $n \to \infty$.

The decimal expansion of x can be viewed as the base ten "address" of x on the number line. By thinking in a general way about how a particular number x must sit amongst the finite decimals, we have seen that x must *have* such a base ten address, but *finding* the expansion for x is another matter.

We can find the decimal expansion of a rational number by continued long division. For example, to find the expansion of 27/88, we start dividing 88 into 27, and get

$$\frac{27}{88} = 0.306818181\cdots,$$

where the three dots indicate that the sequence of digits goes on forever. This is a very simple infinite decimal expansion because the two-digit block 81 repeats forever.

The decimal expansion of $\sqrt{2}$ can be obtained by a process (algorithm) for extracting square roots that is often taught in school. It is more complicated than long division, but similar in being nonterminating (unless the beginning number happens to be a finite decimal that is a perfect square, such as $\sqrt{1.44} = 1.2$). If we apply this process to compute $\sqrt{2}$, we get

$$\sqrt{2} = 1.414\cdots,$$

where again the three dots indicate that the sequence of digits goes on forever. This time, though, there isn't any block of digits that forever repeats.

The decimal expansion representing π begins

$$\pi = 3.14159\cdots.$$

The calculation of this expansion is harder still, but there are various ways of going about it.

Whether a number is rational or irrational shows up in its infinite decimal expansion. An expansion of a rational number is *always* a repeating decimal, as in the example above. After a certain point it consists entirely of repetitions of a certain block of digits. Here is a brief indication of why this is so.

We obtain the decimal expansion of m/n by continued long division. Suppose we have gone far enough so that only zeros are being brought down. Suppose also that no subsequent remainder is ever zero; this rules out m/n being equal to a decimal fraction. The only possible remainders are then the $n-1$ numbers $1, \ldots, n-1$, so during the next n steps there must be a repetition of a remainder r. Then this remainder r and the block of steps between the two occurrences of r will repeat forever, as will the block of quotient digits arising from these steps.

It is convenient to indicate the repeated block by an overhead bar, and in this notation the first example is

$$\frac{27}{88} = 0.30\overline{681}.$$

Conversely, we shall see in the next section how any repeating decimal represents a rational number. Thus, *the irrational numbers are exactly those numbers having nonrepeating infinite decimal expansions.*

Since the finite decimal 3.14 can be obtained by cutting off the decimal expansion of π after two places, we call 3.14 the two-place *truncation* of the expansion of π. What it says about the location of π was discussed above.

The following examples illustrate the "positioning" role of such truncations. Their solutions depend on the laws for inequalities, which are reviewed in Appendix 1.

EXAMPLE 1   A classical approximation to π is 22/7. Assuming that the expansion of π begins $3.141\cdots$, show that $\pi < 22/7$ and that $22/7 - \pi < 0.002$.

Solution. Dividing out 22/7, we see that its decimal expansion begins $3.142\cdots$. Therefore,

$$3.141 < \pi < 3.142 < \frac{22}{7} < 3.143.$$

This shows that π is less than 22/7 and that

$$\frac{22}{7} - \pi < 3.143 - 3.141 = 0.002.$$

EXAMPLE 2 The decimal expansion of e begins $2.71 \cdots$. Show from this that $\sqrt{e} < 5/3$.

Solution. Since $(5/3)^2 = 25/9 = 2.77 \cdots$, we have

$$\left(\frac{5}{3}\right)^2 > 2.77 > 2.72 > e,$$

and hence

$$\frac{5}{3} > \sqrt{e}.$$

EXAMPLE 3 Knowing that the decimal expansions of π and $\sqrt{2}$ start 3.14 and 1.41 respectively, what can we say about $\sqrt{2}\pi$?

Solution. We are assuming that

$$3.14 < \pi < 3.15,$$

$$1.41 < \sqrt{2} < 1.42.$$

Multiplying these two inequalities together, we find that

$$4.4274 < \sqrt{2}\pi < 4.4730.$$

In particular,

$$4.4 < \sqrt{2}\pi < 4.5,$$

so the expansion of $\sqrt{2}\pi$ is 4.4, to one decimal place. (But we don't know from this calculation which of 4.4 and 4.5 is the *closest* one-place decimal to $\sqrt{2}\pi$.)

Passing from the decimal expansion of x to the *closest n-place decimal* is called *rounding off* to n places. Thus 3.14 is both the two-place truncation of the expansion of π,

$$3.14 < \pi < 3.15,$$

and also the two-place round-off

$$|3.14 - \pi| < 0.005.$$

On the other hand, the expansion of π through five places is $3.14159 \cdots$, so the four-place truncation is 3.1415 while the four-place round-off is 3.1416.

As this example suggests, we can generally obtain the n-place round-off if we know the expansion through $n + 1$ places.

The number 0.325 represents a rarely occurring special case, lying exactly halfway between 0.32 and 0.33, and therefore not determining a unique closest two-place decimal. Here we adopt the artificial convention of rounding to the even digit. Thus 0.325 rounds to 0.32, while 0.335 rounds to 0.34.

Tables contain rounded-off values. The table on page 792 gives 0.6931 as the value for log 2, which means that

$$|\log 2 - 0.6931| < 5(10)^{-5},$$

or

$$\log 2 \approx 0.6931$$

with an error less than $5(10)^{-5}$ in magnitude.

The following examples explore the relationship between the two types of error inequalities occurring in truncation and rounding off.

EXAMPLE 4   Show that the inequalities

$$4.43 < x < 4.49 \qquad \text{and} \qquad |x - 4.46| < 0.03$$

are equivalent.

Solution. We can rewrite the first inequality as

$$4.46 - 0.03 < x < 4.46 + 0.03,$$

or

$$-0.03 < x - 4.46 < 0.03,$$

which is equivalent to

$$|x - 4.46| < 0.03.$$

EXAMPLE 5   Show that an inequality of the form

$$a < x < b$$

can always be rewritten in the form

$$|x - c| < e.$$

Solution. Let c be the midpoint of the interval $[a, b]$, and let e be half the interval width $b - a$.

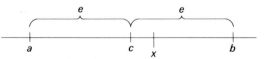

Then it should seem geometrically clear that the inequalities

$$a < x < b \qquad \text{and} \qquad |x - c| < e$$

are equivalent. Algebraically, the proof is as follows. We have defined c and e as

$$c = \frac{b + a}{2},$$

$$e = \frac{b - a}{2}.$$

It follows that $c + e = b$ and $c - e = a$. We thus have, in order,

$$a < x < b,$$
$$c - e < x < c + e,$$
$$-e < x - c < e,$$
$$|x - c| < e,$$

where at each step an inequality is replaced by an equivalent one.

When we are viewing a number c as an approximation to a number x, the difference $x - c$ is called the *error* in the approximation. The inequality

$$|x - c| < e$$

can thus be read: *c approximates x with an error less than e in magnitude.*

Finally, we should discuss a little snag that comes up in finding decimal approximations. At the end of Chapter 6, in Example 3 of Section 10, it was shown that

$$\frac{1}{3.14} - \frac{1}{\pi} < 0.0003.$$

So it seems reasonable to expect that by using long division to obtain the decimal expansion of $1/3.14$ we can read off the three-place decimal closest to $1/\pi$. This works out all right. We find (by hand or by pocket calculator) that

$$\frac{1}{3.14} = 0.31847\cdots,$$

so

$$0.31847\cdots - 0.0003 < \frac{1}{\pi} < 0.31847\cdots,$$

and

$$0.3181 < \frac{1}{\pi} < 0.3185.$$

This shows that 0.318 is *both* the three-place truncation of the expansion of $1/\pi$ *and* the three-place round-off. But this was a lucky accident, as the next example shows.

EXAMPLE 6   The table on page 794 gives the value 0.438 for sin 26°. What can be said about $(\sin 26°)^2$?

Solution. The hypothesis is that

$$|\sin 26° - 0.438| < 0.0005,$$

or

$$0.4375 \le \sin 26° \le 0.4385.$$

By simply squaring this inequality, or by using the mean-value principle for the function x^2, as in Section 9 of Chapter 6, we find that

$$0.1913 < (\sin 26°)^2 < 0.1923.$$

That is,

$$(\sin 26°)^2 \approx 0.1918$$

with an error less than 0.0005, which looks like three-place accuracy. But this approximation tells us *neither* the three-place truncation *nor* the three-place round-off for $(\sin 26°)^2$. The simplest conclusion is that

$$(\sin 26°)^2 \approx 0.192$$

with an error less than 0.001 in magnitude.

EXAMPLE 7   The decimal expansion of a number x begins $0.547\ldots$. What can we say about x^2?

Solution.   By assumption

$$0.547 \le x \le 0.548.$$

Squaring, we find that

$$0.299209 \le x^2 \le 0.300304.$$

Therefore $x^2 \approx 0.300$, with an error less than 0.001. But we don't know whether the decimal expansion of x^2 begins $0.299\ldots$ or $0.300\ldots$, or which is the three-place round-off.

We can do just this well, but no better, in general. (However, see Problem 17.)

LEMMA 1   *Suppose we have found a rational number r approximating a number x "to the nearest n decimal places," in the sense that the error is at most $10^{-n}/2$ in magnitude. Then we can find an n-place decimal d that approximates x to within an error of 1 in the nth decimal place:*

$$|x - d| \le 10^{-n}.$$

Proof.   We simply compute the decimal expansion of r by long division, and so determine the n-place decimal d that is *closest* to r:

$$|r - d| \le \frac{10^{-n}}{2}.$$

(This will be either the first n places in the expansion of r, or the number one greater in the last place, whichever is closer to r.) Then

$$|x - d| = |(x - r) + (r - d)|$$

$$\le |x - r| + |r - d| \le \frac{10^{-n}}{2} + \frac{10^{-n}}{2} = 10^{-n}. \quad \blacksquare$$

We saw earlier how each positive number x determines an infinite decimal expansion that records the exact location of x on the number line. Conversely, each infinite decimal defines a sequence of intervals, each being one-tenth of the preceding one, and altogether defining an exact location on the number line. The explicit assumption that such a squeezing down process always produces a point on the line is called the *axiom of continuity* in geometry. When we convert the geometric line to the number line, the point in question acquires a coordinate, namely, the number whose infinite decimal expansion we began with. Thus, when we state the axiom of continuity as a property of the real number system, we are asserting that *every infinite decimal is the expansion of a real number.* This property is called the *completeness* of the real number system. It says that every "position" amongst the finite decimals is occupied by a number.

For flexibility, we need a more general formulation of this property. For example, if we compute in base two arithmetic, then we end up specifying a real number by a different type of interval sequence, where each interval is one of the two *halves* of the preceding interval. So we need the fact that *any* squeezing-down sequence of closed intervals determines a number.

DEFINITION   *A sequence of closed intervals $I_n = [a_n, b_n]$ is nested if each interval I_n includes the next interval I_{n+1} (and hence includes all later intervals I_{n+j}).*

This just means that $a_n \leq a_{n+1}$ and $b_n \geq b_{n+1}$ for each n (and, of course, $a_n \leq b_n$), so it is a simple way of saying simultaneously that $\{a_n\}$ is an increasing sequence, that $\{b_n\}$ is a decreasing sequence, and that $a_n \leq b_n$ for all n.

The Nested-interval Principle. *Suppose that $\{I_n\}$ is a nested sequence of closed intervals whose lengths approach 0 as n tends to infinity. Then there is a unique real number x lying in them all.*

If $I_n = [a_n, b_n]$ in the nested-interval principle, then of course $a_n \to x$ as $n \to \infty$. The formal proof first converts the inequality $a_n \leq x \leq b_n$ to the form

$$0 \leq x - a_n \leq b_n - a_n.$$

Then, since $b_n - a_n \to 0$, it follows that $x - a_n \to 0$ by the squeeze limit law. Therefore, $a_n \to x$. It is also clear that $b_n \to x$ as $n \to \infty$, and this also has a simple formal proof.

The completeness of the real numbers is the basic fact underlying all the theory needed for calculus. Appendix 4 contains some other forms of this property and derives some theoretical consequences.

PROBLEMS FOR SECTION 1

1. Archimedes proved that

$$3\frac{10}{71} < \pi < 3\frac{1}{7}.$$

Show from this that 22/7 approximates π with an error less than 1/497.

2. Find what the above double inequality tells us about the decimal expansion of π.

3. From a classical partial-fraction expansion of e, we find that

$$2\frac{334}{465} < e < 2\frac{385}{536}.$$

Show therefore that either of these rational numbers approximates e with an error less than $1/249{,}000$.

4. Compute e from the above data to the nearest five decimal places.

Here are some decimal expansions through two decimal places:

$$\pi = 3.14\cdots, \qquad \sqrt{2} = 1.41\cdots, \qquad e = 2.71\cdots,$$

$$\log 3 = 1.09\cdots, \qquad \sin\sqrt{2} = 0.98\cdots.$$

Determine what can be said about the one-place decimal approximations to each of the following products. In each case the answer should be *either* the decimal expansion through one decimal place, *or* the closest one-place decimal approximation (or both).

5. $e\sqrt{2}$ 6. $\pi\log 3$ 7. $\sqrt{2}\sin\sqrt{2}$ 8. e^2

9. $\pi\sin\sqrt{2}$ 10. $e\log 3$ 11. $\log 3\sin\sqrt{2}$

12. If $|x - 1.4849| < 0.0001$, how close must x be to 1.485?

13. If $|x - 1.099| < 0.005$, how close must x be to 1.1?

14. If $|x - 4.473| < 0.001$, how close must x be to 4.47?

15. If $|x - 2.152| < 0.003$, how close must x be to 2.15? 2.16?

16. If $|x - 2.289| < 0.001$, how close must x be to 2.28? 2.29?

17. In the situation of Lemma 1, if we start with a rational r even closer to a positive number x, then we can get a better answer. Show that if we first find a rational r such that

$$|x - r| < \frac{10^{-n}}{4},$$

and then find the closest n-place decimal to r,

$$|r - d| \le \frac{10^{-n}}{2},$$

then we can determine *either* the n-place decimal closest to x (it will be d), *or* the first n places in the decimal expansion of x (it will be d or $(d - 10^{-n})$). [*Hint:* Consider separately the cases $|r - d| \le 10^{-n}/4$; $d - r > 10^{-n}/4$; $r - d > 10^{-n}/4$.]

2. THE SUM OF AN INFINITE SERIES

It is surprising and very important that we can sometimes calculate what the sum of an infinite collection of numbers ought to be even though we can't actually carry out the infinite number of additions seemingly required. For example, we shall see below that there is a sense in which the sum of all the powers of 1/2, starting from $1 = (1/2)^0$, is exactly equal to 2,

$$1 + \frac{1}{2} + \frac{1}{4} + \frac{1}{8} + \cdots = 2,$$

even though the actual calculation of the infinitely many additions indicated on the left is impossible.

This is an example of a *geometric series*, a fundamental type of infinite sum that is sometimes taken up in high school. In general, we are given a number t between -1 and 1, and we want to calculate the infinite sum of all the powers of t,

$$1 + t + t^2 + \cdots + t^n + \cdots.$$

We clearly can't perform all the additions that are indicated above, but we *can* start adding and see what happens as we keep on going. The sum of the terms from 1 through t^n is called the nth *partial sum*, and is designated s_n. Thus,

$$s_1 = 1 + t,$$
$$s_2 = 1 + t + t^2,$$
$$\cdot \qquad \cdot$$
$$\cdot \qquad \cdot$$
$$\cdot \qquad \cdot$$
$$s_n = 1 + t + t^2 + \cdots + t^n.$$

If there is an exact infinite sum s, it should seem reasonable that, as we keep on adding, this accumulating partial sum s_n comes closer and closer to s. That is, we ought to be able to compute the exact infinite sum s as the limit:

$$s = \lim_{n \to \infty} s_n.$$

What makes this computation possible for the geometric series is that we can find a simple explicit formula for the partial sum s_n. It comes from the factorization formula,

$$1 - t^{n+1} = (1 - t)(1 + t + t^2 + \cdots + t^n).$$

We can write this as

$$1 - t^{n+1} = (1 - t)s_n,$$

so

$$s_n = \frac{1 - t^{n+1}}{1 - t}.$$

But we know that $t^{n+1} \to 0$ as $n \to \infty$, because $|t| < 1$. Therefore,

$$s_n = \frac{1 - t^{n+1}}{1 - t} \to \frac{1}{1 - t}$$

as $n \to \infty$. Thus, supposing that there is an exact infinite sum s of the geometric series $1 + t + \cdots + t^n + \cdots$, we can calculate it as the limit of the accumulating partial sum s_n:

$$s = \lim_{n \to \infty} s_n = \frac{1}{1 - t}.$$

EXAMPLE 1

$$1 + \frac{3}{7} + \frac{9}{49} + \cdots + \left(\frac{3}{7}\right)^n + \cdots = \frac{1}{1 - (3/7)} = \frac{7}{4}.$$

Now comes a slight shift in point of view. Further investigation will not turn up any other way to calculate the exact infinite sum than as the limit $s = \lim s_n$, and we therefore *define* the infinite sum to be the number calculated in this way. We also say that the series converges to s. Therefore:

THEOREM 1   If $|t| < 1$, then the geometric series $1 + t + t^2 + \cdots + t^n + \cdots$ converges, and its sum is $1/(1 - t)$.

The sum of any other infinite series

$$a_1 + a_2 + \cdots + a_n + \cdots$$

is investigated in the same way. We form the sum s_n of the terms through a_n, and then see what happens as we keep on adding, i.e., as n increases. *If the "lengthening" partial sum s_n approaches a limit s, then we define s to be the exact sum of the infinite series, and we say that the series converges to s.*

EXAMPLE 2   The finite decimals $3.1, 3.14, 3.141, \cdots$, obtained by truncating the infinite decimal expansion of π, are just the partial sums of the infinite series

$$3 + \frac{1}{10} + \frac{4}{(10)^2} + \frac{1}{(10)^3} + \frac{5}{(10)^4} + \frac{9}{(10)^5} + \cdots.$$

Therefore, the fact that π is given exactly by its infinite decimal expansion can now be reinterpreted as the fact that this series converges to π.

Any other infinite decimal expansion represents a convergent infinite series in the same way. However, these examples are entirely different from the geometric series. There we were able to compute the sum $s = \lim s_n$ by an explicit formula. Here we know the sum exists because of the general property of completeness for the real numbers, but the sum may not have any description other than that provided by the series, i.e., by the infinite decimal expansion. This difference will be clearer after we have looked at the comparison test for convergence (Theorem 4).

When an infinite series fails to have a sum we say that it *diverges*. This can happen in two quite different ways. Suppose that we consider the infinite series

$$1 + 1 + \cdots + 1 + \cdots,$$

i.e., the series

$$a_1 + a_2 + \cdots + a_n + \cdots$$

for which all the terms a_n have the value 1. For this series, the partial sum s_n is just the integer n, which becomes arbitrarily large and doesn't approach any number s. We say that s_n tends to $+\infty$ (plus infinity), and that the series *diverges* to $+\infty$.

A more interesting example of a series that diverges to $+\infty$ is the so-called *harmonic* series,

$$1 + \frac{1}{2} + \frac{1}{3} + \frac{1}{4} + \cdots + \frac{1}{n} + \cdots.$$

Here, the fact that the series diverges isn't so obvious, principally because we don't have a simple formula for the partial sum s_n that we can examine. However, by gathering the terms into groups we see that

$$1 + \frac{1}{2} + \overbrace{\frac{1}{3} + \frac{1}{4}} + \overbrace{\frac{1}{5} + \frac{1}{6} + \frac{1}{7} + \frac{1}{8}} + \cdots$$

$$> 1 + \frac{1}{2} + \underbrace{\frac{1}{4} + \frac{1}{4}} + \underbrace{\frac{1}{8} + \frac{1}{8} + \frac{1}{8} + \frac{1}{8}} + \cdots$$

$$= 1 + \frac{1}{2} \quad + \frac{1}{2} \quad\quad + \frac{1}{2} \quad\quad + \cdots.$$

Thus, by going out along the harmonic series far enough, we can make the partial sum s_n larger than $1 + m(1/2)$, no matter how large we have taken the integer m. Therefore, the harmonic series diverges to $+\infty$.

The other kind of divergent behavior is illustrated by taking $a_n = (-1)^n$ so that the series (starting with a_0) is

$$1 + (-1) + 1 + (-1) + \cdots + (-1)^n + \cdots.$$

After we have added n terms, we have either 0 or 1, depending on whether n is even or odd. The partial sum doesn't grow arbitrarily large, but neither does it settle down near one number. Instead it oscillates forever.

These two "pure" forms of divergence can be combined, as in

$$1 + (-2) + 3 + (-4) + \cdots + n(-1)^{n+1} + \cdots.$$

The following theorem states a simple condition that is occasionally useful.

THEOREM 2 *If the series $a_0 + a_1 + \cdots$ converges, then $a_n \to 0$ as $n \to \infty$. Therefore, if a_n does not have the limit zero, then the series $a_0 + a_1 + \cdots$ diverges.*

Proof. Suppose that the series converges, with sum s. That is, the partial sum s_n approaches the limit s as $n \to \infty$. Arguing informally, we can see that since s_n and s_{n-1} are both very close to s, their difference, which is a_n, must be close to zero.

The formal argument would be that since we have both $s_n \to s$ and $s_{n-1} \to s$ as $n \to \infty$, then $a_n = s_n - s_{n-1} \to s - s = 0$. ∎

The harmonic series shows that this theorem cannot be reversed: it is a divergent series and yet the general term $1/n$ has the limit zero. The condition that $a_n \to 0$ is thus *necessary*, but not *sufficient*, for the convergence of the series $a_1 + a_2 + \cdots + a_n + \cdots$.

In working with the sums of convergent series we need to use a few principles that will probably seem intuitively obvious. For example, we shall see later on that

$$e = 1 + 1 + \frac{1}{2} + \frac{1}{3!} + \frac{1}{4!} + \cdots + \frac{1}{n!} + \cdots,$$

and that

$$1/e = 1 - 1 + \frac{1}{2} - \frac{1}{3!} + \frac{1}{4!} + \cdots + (-1)^n \frac{1}{n!} + \cdots.$$

Here 4! (read "4 factorial") is $4 \cdot 3 \cdot 2 \cdot 1$, and

$$n! = n(n - 1)(n - 2) \cdots 3 \cdot 2 \cdot 1.$$

Assuming these two equations, we would certainly expect to be able to add the two series term by term, and to conclude that

$$e + 1/e = 2 + 2\left(\frac{1}{2}\right) + 2\left(\frac{1}{4!}\right) + \cdots + 2\left(\frac{1}{(2m)!}\right) + \cdots.$$

Then we would like to be able to multiply throughout by 1/2, ending up with

$$\frac{e + 1/e}{2} = 1 + \frac{1}{2!} + \frac{1}{4!} + \cdots + \frac{1}{(2m)!} + \cdots.$$

We would also like to be able to prove that $e < 3$ by the following argument:

$$e = 1 + 1 + \frac{1}{2} + \frac{1}{3!} + \frac{1}{4!} + \cdots + \frac{1}{n!} + \cdots$$

$$= 1 + 1 + \frac{1}{2} + \frac{1}{2 \cdot 3} + \frac{1}{2 \cdot 3 \cdot 4} + \cdots$$

$$< 1 + 1 + \frac{1}{2} + \frac{1}{2 \cdot 2} + \frac{1}{2 \cdot 2 \cdot 2} + \cdots$$

$$= 1 + \left[1 + \frac{1}{2} + \left(\frac{1}{2}\right)^2 + \left(\frac{1}{2}\right)^3 + \cdots + \left(\frac{1}{2}\right)^n + \cdots\right]$$

$$= 1 + \frac{1}{1 - 1/2} = 1 + 2 = 3.$$

The rules that let us do things like this have to be derived from the rules about sequential limits, because a series sum *is* a sequence limit. Before looking at these laws we shall introduce the sigma (Σ) notation for infinite series. Remember that $\sum_{j=1}^{n} a_j$ is the sum of the terms a_j from $j = 1$ to $j = n$:

$$\sum_{j=1}^{n} a_j = a_1 + a_2 + \cdots + a_n.$$

For example,

$$\sum_{j=1}^{5} \frac{1}{j^2} = 1 + \frac{1}{2^2} + \frac{1}{3^2} + \frac{1}{4^2} + \frac{1}{5^2} = 1 + \frac{1}{4} + \frac{1}{9} + \frac{1}{16} + \frac{1}{25}.$$

It is then natural to denote the sum of a convergent infinite series by

$$\sum_{j=1}^{\infty} a_j \qquad \text{or} \qquad \sum_{j=0}^{\infty} a_j.$$

If we start at the mth term, the sum is $\sum_{j=m}^{\infty} a_j$. We sometimes designate the series itself by

$$\sum a_j \qquad \text{or} \qquad \sum_{j=1} a_j,$$

the rationale being that since we are not indicating the sum to any particular point we are looking at the "unsummed" series.

The general laws for combining series can be discussed conveniently in the $\sum$ notation, although this is by no means essential. In the statements of these laws, any equation of the form $\sum_1^{\infty} x_j = X$ is to be read:

The infinite series $\sum_1^{\infty} x_j$ is convergent and its sum is X.

THEOREM 3 *If $\sum_1^{\infty} a_j = A$ and $\sum_1^{\infty} b_j = B$, then $\sum_1^{\infty} (a_j + b_j) = A + B$. Also, for any constant c, $\sum_1^{\infty} (ca_j) = cA$.*

Proof. We are assuming that the partial sums $\sum_1^n a_j$ and $\sum_1^n b_j$ have the limits A and B, respectively, as n tends to ∞. Since we can rearrange a finite sum in any way, it then follows that

$$\sum_1^n (a_j + b_j) = \sum_1^n a_j + \sum_1^n b_j \to A + B$$

as $n \to \infty$, by the law for the limit of the sum of two sequences. Since an infinite sum is by definition the limit of its partial sums, this shows that

$$\sum_{j=1}^{\infty} (a_j + b_j) = A + B.$$

Similarly,

$$\sum_1^n ca_j = c\sum_1^n a_j \to cA$$

as $n \to \infty$, so that

$$\sum_{j=1}^{\infty} ca_j = cA.$$

This completes the proof of the theorem. ∎

Whether a series converges or not is independent of any given initial block of terms, because we always have the identity

$$\sum_1^{\infty} a_j = \sum_1^N a_j + \sum_{N+1}^{\infty} a_j.$$

This means that if either of the two infinite series converges, then so does the other, with the equation holding between the sums. It can be considered a special case of the theorem if each of the two incomplete sums on the right is considered to be a full infinite series with zeros in all the missing positions.

The first term on the right above is the partial sum s_N for the given series. The second term, the series sum from the $(N + 1)$st term on, is called the Nth *remainder series*, and is frequently designated r_N. Thus, if the series converges, then the equation

$$s = s_N + r_N$$

expresses the infinite sum s as the sum of the first N terms plus the remainder. Note that then

$$r_N \to 0$$

as $N \to \infty$, since $r_N = s - s_N$ and $s_N \to s$.

EXAMPLE 3   If we assume the formula

$$1 + t + \cdots + t^n + \cdots = \frac{1}{1 - t},$$

then we can recover the partial-sum formula algebraically, as follows:

$$\frac{1}{1 - t} = s_n + r_n = s_n + [t^{n+1} + t^{n+2} + \cdots]$$

$$= s_n + t^{n+1}[1 + t + \cdots]$$

$$= s_n + t^{n+1} \cdot \frac{1}{1 - t},$$

and

$$s_n = \frac{1}{1 - t} - \frac{t^{n+1}}{1 - t} = \frac{1 - t^{n+1}}{1 - t}.$$

EXAMPLE 4   We can now see why a repeating decimal always represents a rational number: it is essentially a geometric series from a certain point on. Consider our early example, $0.30\overline{681}$. We group together the terms in the repeating block, and write the infinite decimal as the series sum,

$$\frac{306}{1000} + \frac{1}{1000}\left[\frac{81}{100} + \frac{81}{(100)^2} + \cdots\right]$$

$$= \frac{306}{1000} + \frac{1}{1000} \cdot \frac{81}{100} \cdot \sum_{k=0}^{\infty} \frac{1}{100^k}$$

$$= \frac{306}{1000} + \frac{1}{1000} \cdot \frac{81}{100} \cdot \frac{100}{99} = \frac{306}{1000} + \frac{1}{1000} \cdot \frac{9}{11}$$

$$= \frac{306 \cdot 11 + 9}{11,000} = \frac{3,375}{11,000} = \frac{27}{88}.$$

EXAMPLE 5 This geometric series interpretation also explains why each finite decimal has two decimal expansions. Thus,

$$0.1999\cdots = 0.1\overline{9} = \frac{1}{10} + \frac{9}{(10)^2} + \frac{9}{(10)^3} + \cdots$$

$$= \frac{1}{10} + \frac{9}{100} \sum_0^\infty \frac{1}{10^j}$$

$$= \frac{1}{10} + \frac{9}{100} \cdot \frac{10}{9} = \frac{2}{10} = 0.2\overline{0}.$$

PROBLEMS FOR SECTION 2

Find the sum of each of the following series.

1. $1 + \dfrac{1}{4} + \dfrac{1}{16} + \dfrac{1}{64} + \cdots + \left(\dfrac{1}{4}\right)^n + \cdots$

2. $1 - \dfrac{1}{6} + \dfrac{1}{36} - \dfrac{1}{216} + \cdots + \left(-\dfrac{1}{6}\right)^n + \cdots$

3. $1 - \dfrac{2}{3} + \dfrac{4}{9} - \dfrac{16}{27} + \cdots + \left(-\dfrac{2}{3}\right)^n \cdots$

4. $1 + \dfrac{3}{4} + \dfrac{9}{16} + \dfrac{27}{64} + \cdots + \left(\dfrac{3}{4}\right)^n + \cdots$

5. $2 + \dfrac{2}{3} + \dfrac{2}{9} + \dfrac{2}{27} + \cdots + 2\left(\dfrac{1}{3}\right)^n + \cdots$

6. $3 + \dfrac{6}{5} + \dfrac{12}{25} + \dfrac{24}{125} + \cdots + 3\left(\dfrac{2}{5}\right)^n + \cdots$

7. $\dfrac{1}{4} - \dfrac{1}{8} + \dfrac{1}{16} + \cdots + \dfrac{1}{4}\left(-\dfrac{1}{2}\right)^n + \cdots$

8. $\dfrac{7}{6} + \dfrac{35}{36} + \dfrac{175}{216} + \dfrac{875}{1296} + \cdots + \dfrac{7}{6}\left(\dfrac{5}{6}\right)^n + \cdots$

9. $\dfrac{1}{4} + \dfrac{1}{6} + \dfrac{1}{9} + \dfrac{2}{27} + \cdots + \dfrac{1}{4}\left(\dfrac{2}{3}\right)^n + \cdots$

10. $\sum_{n=0}^\infty \dfrac{2^n}{3^n}$
11. $\sum_{n=3}^\infty \dfrac{3^n}{4^n}$

12. $\sum_0^\infty \dfrac{2^n}{5^{n/2}}$

Suppose that each of the following partial sums continues as a geometric series, possibly multiplied by a constant k, so that the nth term is kr^n. Find its sum.

13. $3 + \dfrac{9}{4} + \cdots$
14. $\dfrac{1}{9} + \dfrac{2}{27} + \cdots$

15. $\dfrac{1}{5} - \dfrac{1}{25} + \cdots$
16. $\dfrac{1}{2} + \dfrac{1}{3} + \cdots$

17. Show that a geometric series $\sum r^n$ diverges if $|r| \geq 1$. (Apply Theorem 2.)

18. Suppose that $a_n \geq 1/n$ for every n. Assuming that the harmonic series diverges to $+\infty$, show that the series $\sum a_n$ diverges to $+\infty$.

Assuming the above fact (and possibly using Theorem 3), show that each of the following series is divergent to $+\infty$.

19. $\dfrac{1}{2} + \dfrac{1}{4} + \cdots + \dfrac{1}{2n} + \cdots$ (Suppose it converges and multiply by 2.)

20. $1 + \dfrac{1}{3} + \dfrac{1}{5} + \cdots + \dfrac{1}{2n - 1} + \cdots$ 21. $\dfrac{1}{4} + \dfrac{1}{7} + \dfrac{1}{10} + \cdots + \dfrac{1}{3n - 1} + \cdots$

22. Prove the following divergence criteria from Theorem 3.
   a) If $\sum a_n$ diverges and $c \neq 0$, then $\sum c a_n$ diverges.
   b) If $\sum (a_n + b_n)$ diverges, then either $\sum a_n$ diverges or $\sum b_n$ diverges.

Express each of the following repeating decimals as an infinite series and compute its sum.

23. $0.232323 \cdots$ 24. $0.\overline{63}$ 25. $0.\overline{315}$

26. $0.012012012 \cdots$ 27. $2.0\overline{1}$ 28. $4.162162162 \cdots$

29. Show that $s_n = 1 - 1/(n + 1)$ if s_n is the nth partial sum of the following series:

$$\dfrac{1}{2} + \dfrac{1}{6} + \dfrac{1}{12} + \cdots + \dfrac{1}{n(n + 1)} + \cdots.$$

[*Hint*: $(1/6) = (1/2) - (1/3)$.]

30. Consider the series

$$1 + \dfrac{1}{4} + \dfrac{1}{9} + \dfrac{1}{16} + \cdots + \dfrac{1}{n^2} + \cdots.$$

   a) Show that its sequence of partial sums s_n is strictly increasing.
   b) Show that $t_n = s_n + 1/n$ is a decreasing sequence.
   c) Conclude that the present series is convergent and that its sum is less than 2.

Show that each of the following series converges, and find its sum.

31. $\dfrac{1}{1 \cdot 3} + \dfrac{1}{3 \cdot 5} + \dfrac{1}{5 \cdot 7} + \cdots + \dfrac{1}{(2n - 1)(2n + 1)} + \cdots$

   [*Hint*: $2/(3 \cdot 5) = (1/3) - (1/5)$.]

32. $\dfrac{1}{2 \cdot 4} + \dfrac{1}{4 \cdot 6} + \dfrac{1}{6 \cdot 8} + \cdots$

33. $\dfrac{1}{1 \cdot 3} + \dfrac{1}{2 \cdot 4} + \dfrac{1}{3 \cdot 5} + \cdots + \dfrac{1}{n(n + 2)}$

34. Find the sum of the series

$$\dfrac{1}{1 \cdot 4} + \dfrac{1}{2 \cdot 5} + \dfrac{1}{3 \cdot 6} + \cdots + \dfrac{1}{n(n + 3)}.$$

35. A ball bearing is dropped from 10 feet onto a heavy metal plate. Being very elastic, but not *perfectly* elastic, the ball bounces each time to a height that is 19/20 of its preceding height. How far does the bounding ball travel altogether?

36. A man walks a mile, from A to B, at a steady pace of 3 miles per hour. According to an ancient paradox attributed to the Greek Zeno, he will never get to B, because he first must pass the halfway point, then the 3/4 point, then the 7/8 point, and so on, so that he must pass an infinite number of distance markers on the way, which is clearly impossible. Our modern answer to Zeno is that the man does get to B, in 20 minutes, because two different geometric series each have a finite sum. Explain.

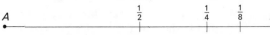

3. THE COMPARISON TEST

The next theorem is the crux of the whole subject. It allows us to establish the convergence of a series by comparing its terms with those of a known convergent series, and is called the *comparison test for convergence*.

THEOREM 4 *If $0 \le a_n \le b_n$ for all n, and if $\sum b_j$ is convergent, then the series $\sum a_j$ is convergent, and*

$$\sum_{j=1}^{\infty} a_j \le \sum_{j=1}^{\infty} b_n.$$

If $0 \le a_n \le b_n$ for all n, and if $\sum a_n$ diverges, then $\sum b_n$ diverges.

EXAMPLE 1 We saw earlier that the series

$$1 + 1 + \frac{1}{2!} + \frac{1}{3!} + \cdots + \frac{1}{n!} \cdots$$

is term-by-term dominated, in the sense of the theorem, by a geometric series whose sum is 3. Therefore, the displayed series is convergent and its sum is at most 3.

The series

$$1 + \frac{1}{\sqrt{2}} + \cdots + \frac{1}{\sqrt{n}} + \cdots$$

diverges, because $1/\sqrt{n} \ge 1/n$ for all n and $\sum 1/n$ is known to diverge.

Proof of Theorem. Let s_n be the nth partial sum of the series $\sum a_j$, let $r_n = \sum_{n+1}^{\infty} b_j$ be the nth *remainder* of the series $\sum b_j$, and let $t_n = s_n + r_n$. Then:

1) the sequence $\{s_n\}$ is increasing, because

$$s_n - s_{n-1} = a_n \ge 0;$$

2) the sequence $\{t_n\}$ is decreasing, because

$$t_n - t_{n-1} = (s_n - s_{n-1}) + (r_n - r_{n-1}) = a_n - b_n \le 0;$$

3) $t_m - s_n \to 0$, because $t_n - s_n = r_n$ and the sequence of remainders of a convergent series approaches 0.

The intervals $[s_n, t_n]$ thus form a nested sequence with lengths approaching 0, so there is a unique number s common to them all, by the nested-interval principle. Then $s_n \to s$ as $n \to \infty$, so the series $\sum a_j$ converges to the sum s. Moreover, $s \le t_1 \le \sum_1^{\infty} b_j$. The second statement in the theorem follows from the first. ∎

If a number s is the sum of an infinite series, we may be able to use the partial sums s_n to estimate s. The error $s - s_n$ in the approximation is the sum of the nth remainder series

$$r_n = a_{n+1} + a_{n+2} + \cdots,$$

and we may be able to show that $s - s_n$ is small by comparing this remainder series with a convergent series whose sum is known.

EXAMPLE 2   This is the ultimate explanation of our original interpretation of an infinite decimal expansion. We now regard the equation

$$\pi = 3.14159 \cdots$$

as meaning that π is the sum of the infinite series

$$\pi = 3 + \frac{1}{10} + \frac{4}{10^2} + \frac{1}{10^3} + \frac{5}{10^4} + \frac{9}{10^5} \cdots.$$

The second remainder series,

$$r_2 = \frac{1}{10^3} + \frac{5}{10^4} + \frac{9}{10^5} + \cdots,$$

is term-by-term dominated by the series

$$\frac{9}{10^3} + \frac{9}{10^4} + \frac{9}{10^5} + \cdots = \frac{9}{10^3}\left[\sum_{k=0}^{\infty}\frac{1}{10^k}\right] = \frac{9}{10^3} \cdot \frac{10}{9} = \frac{1}{10^2}.$$

Therefore $r_2 < 0.01$, and

$$3.14 = s_2 < \pi = s_2 + r_2 < 3.14 + 0.01 = 3.15.$$

That is, the bracketing inequality

$$3.14 < \pi < 3.15$$

is now seen to be a consequence of the comparison test for series convergence, applied to the second remainder series r_2.

EXAMPLE 3   Assuming that e is the sum of the series

$$e = 1 + 1 + \frac{1}{2!} + \cdots + \frac{1}{n!} + \cdots,$$

we can compute e by comparing its remainder series with the geometric series. Suppose, for example, that we compare the remainder series r_6 term-by-term with the geometric series that its first terms suggest. We have

$$r_6 = \frac{1}{7!} + \frac{1}{8!} + \frac{1}{9!} + \cdots$$

$$= \frac{1}{7!}\left[1 + \frac{1}{8} + \frac{1}{8 \cdot 9} + \frac{1}{8 \cdot 9 \cdot 10} + \cdots\right]$$

$$< \frac{1}{7!}\left[1 + \frac{1}{8} + \frac{1}{8^2} + \frac{1}{8^3} + \cdots\right] = \frac{1}{7!}\left(\frac{1}{1 - (1/8)}\right) = \frac{1}{7!}\frac{8}{7}$$

$$< \frac{1}{7!}\frac{7}{6} = \frac{1}{6} \cdot \frac{1}{6!}.$$

Thus,

$$s_6 < e = s_6 + r_6 < s_6 + \frac{1}{6} \cdot \frac{1}{6!},$$

and we have e pinned down to an interval of width $1/6 \cdot 6!$. Therefore, each endpoint of the interval, s_6 and $s_6 + 1/6 \cdot 6!$, approximates e with an error less than the interval width $1/6 \cdot 6!$. A quick check shows this to be of the order of magnitude of $1/4{,}000 = 0.00025$, so s_6 approximates e with three-place accuracy.

The computation of s_6 goes very simply as follows. (We start with 7 decimal places in order to be ready for a later improvement.)

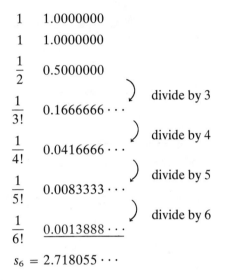

$$
\begin{array}{ll}
1 & 1.0000000 \\
1 & 1.0000000 \\
\dfrac{1}{2} & 0.5000000 \\
\dfrac{1}{3!} & 0.1666666\cdots \\
\dfrac{1}{4!} & 0.0416666\cdots \\
\dfrac{1}{5!} & 0.0083333\cdots \\
\dfrac{1}{6!} & \underline{0.0013888\cdots} \\
s_6 = & 2.718055\cdots
\end{array}
$$

divide by 3
divide by 4
divide by 5
divide by 6

The interval width $1/6 \cdot 6!$ is $1/6$ of the last line before adding, so

$$\frac{1}{6 \cdot 6!} = 0.000231 \cdots < 0.000232.$$

Since

$$s_6 < e < s_6 + \frac{1}{6 \cdot 6!},$$

we have

$$2.718055 < e < 2.718056 + 0.000232 = 2.718288.$$

Thus, $e \approx 2.718$, correct to three decimal places.

Actually, the upper bound $s_6 + 1/6 \cdot 6! = 2.718287\cdots$ is much closer to e than is the partial sum s_6. This is because

$$s_6 + \left(\frac{1}{7!} + \frac{1}{8!} + \frac{1}{9!} \right) = s_9 < e < s_6 + \frac{1}{6 \cdot 6!},$$

so the error in using $s_6 + 1/6 \cdot 6!$ as the estimate is less than

$$\frac{1}{6 \cdot 6!} - \left(\frac{1}{7!} + \frac{1}{8!} + \frac{1}{9!}\right) = \frac{1}{6!}\left[\frac{1}{6} - \frac{1}{7} - \frac{1}{7 \cdot 8} - \frac{1}{7 \cdot 8 \cdot 9}\right]$$

$$= \frac{1}{6!}\left[\frac{1}{6 \cdot 7} - \frac{1}{7 \cdot 8} - \frac{1}{7 \cdot 8 \cdot 9}\right]$$

$$= \frac{1}{6!}\left[\frac{2}{6 \cdot 7 \cdot 8} - \frac{1}{7 \cdot 8 \cdot 9}\right]$$

$$= \frac{1}{6!}\left[\frac{12}{6 \cdot 7 \cdot 8 \cdot 9}\right] = \frac{2}{9!} < 6 \cdot 10^{-6}.$$

Thus 2.718288 is larger than e, but by less than 8 in the sixth decimal place, so

$$e = 2.71828 \cdots.$$

It may be necessary to multiply a comparison series by a constant before making the comparison.

EXAMPLE 4   Another series that can be shown to be convergent by directly evaluating its partial sum s_n is $\sum 1/n(n + 1)$. (See Problem 29, Section 2.) Since

$$\frac{1}{n^2} \le \frac{2}{n(n + 1)}$$

for all positive n, it follows from Theorem 4 that

$$\sum \frac{1}{n^2}$$

is convergent. This is a convenient series to use for further comparisons.

EXAMPLE 5   Show that the series

$$\frac{1}{3} + \frac{1}{8} + \frac{1}{15} + \cdots + \frac{1}{n(n - 2)} + \cdots$$

is convergent by comparing with (a constant time) the above convergent series

$$\frac{1}{9} + \frac{1}{16} + \frac{1}{25} + \cdots + \frac{1}{n^2} + \cdots,$$

where each series starts with $n = 3$.

Solution. Again we can't compare directly because the inequalities go the wrong way. However, we can find a positive constant k such that

$$\frac{1}{n(n - 2)} \le k\frac{1}{n^2}$$

for all n from some point on. You may be able to see by inspection that

$k = 2$ will do, but in any case we can work it out. Cross-multiplying, we want

$$n^2 \le k \, n(n - 2) = kn^2 - 2kn,$$

or

$$2kn \le (k - 1)n^2,$$

or

$$\frac{2k}{k - 1} \le n.$$

If $k = 2$, this holds for all $n \ge 4$.

Retracing the steps, we see that the given series, starting from its second term, is term-by-term dominated by the known convergent series $2 \sum 1/n^2$, and hence itself converges, by Theorem 4.

In some situations, the comparison between $\sum a_n$ and $\sum b_n$ can be made most easily by showing that the sequence a_n/b_n has a limit.

THEOREM 5 *Suppose that $\sum a_n$ and $\sum b_n$ are series with positive terms such that $\lim_{n \to \infty} (a_n/b_n)$ exists. Then $\sum a_n$ converges if $\sum b_n$ converges. Also, $\sum b_n$ diverges if $\sum a_n$ diverges.*

Proof. The difference between a_n/b_n and its limit l can be made as small as desired by taking n large enough. In particular, the difference is less than 1 in magnitude for all n from some particular value N on, and

$$\frac{a_n}{b_n} < l + 1 \qquad \text{for all} \qquad n \ge N.$$

Thus

$$a_n < (l + 1)b_n$$

for $n \ge N$, and $\sum a_n$ converges by comparison with $\sum (l + 1)b_n = (l + 1)\sum b_n$.

EXAMPLE 6 The comparison in Example 5 is probably more easily made this way. We have

$$\frac{a_n}{b_n} = \frac{\dfrac{1}{n(n - 2)}}{1/n^2} = \frac{n^2}{n^2 - 2n} = \frac{1}{1 - (2/n)} \to 1,$$

so $\sum 1/n(n - 2)$ converges by comparison with $\sum 1/n^2$, by virtue of Theorem 5.

EXAMPLE 7 We could apply the theorem in the same way for the series $\sum 1/n(n + 2)$, but here it is easier to make the direct comparison

$$\frac{1}{n(n + 2)} < \frac{1}{n^2}.$$

PROBLEMS FOR SECTION 3

Use the comparison test, including Theorem 5, to determine whether the following series converge or diverge. For convergence, compare with a geometric series or the series $\sum 1/n^2$ (Problem 30, Section 2). For divergence, compare with the harmonic series $\sum 1/n$. In some instances, you may have to multiply by a constant before comparing.

1. $\sum \dfrac{1}{n+1}$ 2. $\sum \dfrac{1}{n^3}$ 3. $\sum \dfrac{1}{2\sqrt{n}}$ 4. $\sum \dfrac{1}{n^2+1}$

5. $\sum \dfrac{1}{n^2-1}$ 6. $\sum \dfrac{n-1}{n^2+1}$ 7. $\sum \dfrac{1}{3n+5}$ 8. $\sum \dfrac{3n+5}{n^3}$

9. $\sum \dfrac{\log n}{n^3}$ 10. $\sum \dfrac{1}{2^n-1}$ 11. $\sum \sin \dfrac{1}{n^2}$ 12. $\sum \dfrac{1}{n!}$

13. $\sum \dfrac{n!}{n^n}$ 14. $\sum \dfrac{1}{\sqrt{n(n-1)}}$ 15. $\sum \dfrac{3n+5}{n^3}$ 16. $\sum \dfrac{1}{n(n-2)}$

17. $\sum \dfrac{1}{3n+5}$ 18. $\sum \dfrac{\log n}{n^3}$ 19. $\sum \sin \dfrac{1}{n^2}$ 20. $\sum \dfrac{1}{\sqrt{n^2+n}}$

21. Carry out the proof of the error formula

$$E < \frac{1}{(n+3)!}\left(1+\frac{6}{n}\right)$$

for the approximation of e by $s_n + 1/(n \cdot n!)$, following the calculation given in the text for the special case $n = 6$.

22. Carry out the computation of e given in Example 3 in Section 3 through three more factorials (i.e., through $1/9!$). Compute $S_9 + 1/(9 \cdot 9!)$ through ten decimal places, and use the error formula given in Problem 21 to show that this estimation of e is accurate through eight decimal places.

23. Using the infinite-series interpretation of the decimal expansion $\pi = 3.1415\cdots$, prove that

$$3.141 < \pi < 3.142.$$

4. THE RATIO TEST

We shall need other corollaries of the comparison test. The first is the "ratio test," which describes a class of series that can be shown to be convergent by comparison with the geometric series. First an example.

EXAMPLE 1   In the series

$$1\left(\frac{1}{2}\right) + 2\left(\frac{1}{2}\right)^2 + 3\left(\frac{1}{2}\right)^3 + \cdots + n\left(\frac{1}{2}\right)^n + \cdots,$$

a pair of successive terms a_n and a_{n+1} have the ratio

$$\frac{a_{n+1}}{a_n} = \frac{(n+1)\left(\dfrac{1}{2}\right)^{n+1}}{n\left(\dfrac{1}{2}\right)^n} = \left(\frac{n+1}{n}\right)\left(\frac{1}{2}\right) = \left(1+\frac{1}{n}\right)\frac{1}{2}.$$

This "test ratio" a_{n+1}/a_n varies with n, but it has the limit $1/2$ as $n \to \infty$. Therefore, if we choose a number between $1/2$ and 1, say $3/4$, then we will have

$$\frac{a_{n+1}}{a_n} < \frac{3}{4}$$

for all integers n beyond some value N. We don't have to know what N is, but in this example we can easily find its value by solving the inequality

$$\frac{a_{n+1}}{a_n} = \left(1 + \frac{1}{n}\right)\frac{1}{2} < \frac{3}{4},$$

obtaining

$$1 + \frac{1}{n} < \frac{3}{2}, \quad \frac{1}{n} < \frac{1}{2}, \quad n > 2.$$

Thus,

$$\frac{a_{n+1}}{a_n} < \frac{3}{4} \quad \text{or} \quad a_{n+1} < \frac{3}{4}\,a_n,$$

for all $n \geq 3$. So

$$a_4 < \frac{3}{4}\,a_3,$$

$$a_5 < \frac{3}{4}\,a_4 < \left(\frac{3}{4}\right)^2 a_3,$$

$$a_6 < \frac{3}{4}\,a_5 < \left(\frac{3}{4}\right)^3 a_3,$$

$$\cdot$$
$$\cdot$$
$$\cdot$$

$$a_n < \frac{3}{4}\,a_{n-1} < \left(\frac{3}{4}\right)^{n-3} a_3.$$

Therefore, if we sum starting with the third term, the comparison test shows that $\sum a_j$ is convergent and that

$$\sum_{n=3}^{\infty} a_n < a_3 \sum_{k=0}^{\infty} \left(\frac{3}{4}\right)^k = a_3 \left(\frac{1}{1 - 3/4}\right) = 4a_3.$$

Here is the theorem.

THEOREM 6 *Suppose that $\sum a_n$ is a series of positive terms such that*

$$\frac{a_{n+1}}{a_n} \to l$$

as $n \to \infty$. Then $\sum a_n$ converges if $l < 1$ and diverges if $l > 1$.

Proof. Suppose that $l < 1$ and choose a number t between l and 1: $l < t < 1$. Since $a_{n+1}/a_n \to l$, we will have $a_{n+1}/a_n < t$, i.e.,

$$a_{n+1} < ta_n,$$

for all indices n starting at some value N. Thus

$$a_{N+1} < ta_N,$$

$$a_{N+2} < ta_{N+1} < t^2 a_N,$$

$$a_{N+3} < ta_{N+2} < t^3 a_N,$$

$$\cdot$$
$$\cdot$$
$$\cdot$$

$$a_{N+m} < ta_{n+m-1} < t^m a_N.$$

Therefore, if we start at the Nth term, the series $\sum a_n$ converges by comparison with the geometric series $a_N \sum t^m$.

If $l > 1$, then it follows as above that $a_{n+1} > a_n$ for all n beyond some value N. Thus the series terms a_j form an increasing sequence from N on. In particular the sequence $\{a_j\}$ does not converge to 0, so $\sum a_j$ diverges, by Theorem 2. ∎

EXAMPLE 2   Consider the series

$$\frac{1}{2} + \frac{2^2}{2^2} + \frac{3^2}{2^3} + \cdots + \frac{n^2}{2^n} + \cdots .$$

The test ratio is

$$\frac{a_{n+1}}{a_n} = \frac{(n+1)^2/2^{n+1}}{n^2/2^n} = \left(\frac{n+1}{n}\right)^2 \frac{1}{2}$$

$$= \left(1 + \frac{1}{n}\right)^2 \Big/ 2.$$

Since $1/n \to 0$ as $n \to \infty$, the test ratio approaches the limit $l = (1 + 0)^2/2 = 1/2$. The series therefore converges, by the theorem.

The theorem says nothing about the case $l = 1$. This is because nothing *can* be said: the harmonic series is a *divergent* series for which the test ratio has the limit 1, whereas the series $\sum 1/n^2$ is a *convergent* series for which $l = 1$. (Section 5 will treat the general case $\sum 1/n^p$.)

PROBLEMS FOR SECTION 4

Use the ratio test to conclude what you can about the convergence or divergence of each of the following series.

1. $\sum \dfrac{n+4}{n2^n}$ 2. $\sum nx^n$ 3. $\sum \dfrac{n^3}{2^n}$

4. $\sum n(n-1)x^{n-2}$ 5. $\sum \dfrac{1}{n^3}$ 6. $\sum \dfrac{n^4}{n!}$

7. $\sum \dfrac{1}{\sqrt{n}}$ 8. $\sum \dfrac{n!}{n^n}$ $\left(\text{Use: } \left(1 + \dfrac{1}{n}\right)^n \to e.\right)$ 9. $\sum \dfrac{2^n + n}{3^n}$

Test for convergence or divergence the series whose general term (from a certain point on) is:

10. $\dfrac{1}{\sqrt{n^4 - 1}}$

11. $\dfrac{1}{n + \sqrt{n}}$

12. $\dfrac{1}{\sqrt{n!}}$

13. $\dfrac{n^{10}}{10^n}$

14. $\dfrac{10^n}{n!}$

15. $\sin \dfrac{1}{n}$

16. $\dfrac{1}{100n + 10^6}$

17. $\dfrac{1}{\log n}$

18. $\dfrac{(n!)^2}{(2n)!}$

19. $\dfrac{2^n (n!)^2}{(2n)!}$

20. Prove that $n!/n^n \to 0$ as $n \to \infty$.   21. Prove that $2^n/n! \to 0$ as $n \to \infty$.

5. THE INTEGRAL TEST

At the moment we cannot say anything about the convergence of the series

$$1 + \frac{1}{2\sqrt{2}} + \frac{1}{3\sqrt{3}} + \cdots + \frac{1}{n\sqrt{n}} + \cdots .$$

However, look at the following figure. We see that the sum of the above series from 2 to n is the sum of the areas of rectangles lying under the graph of $y = x^{-3/2}$ between $x = 1$ and $x = n$.

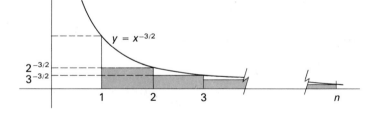

So if A_n is the area under this graph from $n - 1$ to n, then

$$\frac{1}{n\sqrt{n}} < A_n$$

for $n \geq 2$. But

$$\sum_2^n A_k = \int_1^n x^{-3/2} = -2x^{-1/2}\Big]_1^n = 2 - \frac{2}{\sqrt{n}},$$

which has the limit 2 as $n \to \infty$. The series $\sum A_k$ thus converges, with sum 2, and hence $\sum_2^\infty 1/n\sqrt{n}$ converges by comparison, with sum ≤ 2. When we add on the first term that was missing in the above comparison we end up with

$$\sum_{k=1}^\infty k^{-3/2} \leq 3.$$

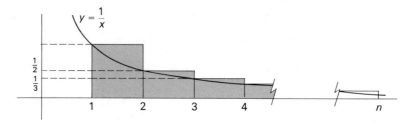

In the same way, but using upper rectangles rather than lower rectangles, we obtain a new proof that the harmonic series $\sum 1/n$ diverges. The figure above shows that

$$\log n = \int_1^n \frac{dx}{x} < \sum_{k=1}^{n-1} \frac{1}{k},$$

where both sides are interpreted as areas. Since $\log n \to \infty$ as $n \to \infty$, it follows that the partial sums of the harmonic series form a sequence diverging to ∞, and hence that $\sum 1/n$ is divergent.

In order to state the theorem covering these two situations, we recall that, by definition,

$$\int_a^\infty f(x)\, dx = \lim_{b \to \infty} \int_a^b f(x)\, dx.$$

If the limit exists (as a finite number), we say that $\int_a^\infty f$ is a *convergent* improper integral. If the limit does not exist the improper integral is divergent.

THEOREM 7   *Let f be a positive decreasing continuous function defined on the interval $[1, \infty)$. Then:*

a) *if the improper integral $\int_1^\infty f$ converges, then the series $\sum f(n)$ converges, and*

$$\sum_2^\infty f(n) \le \int_1^\infty f(x)\, dx \le \sum_1^\infty f(n);$$

b) *if the improper integral $\int_1^\infty f(x)\, dx$ diverges then the series $\sum f(n)$ diverges.*

Proof. Left to the reader. Just write down in general terms the arguments from the two examples. The second inequality of part (a) follows from the same argument used to prove part (b). ∎

REMARK   Since a series converges if and only if a remainder series $\sum_N a_k$ converges, we can restate the theorem in terms of the improper integral $\int_N^\infty f$.

PROBLEMS FOR SECTION 5

Determine the convergence or divergence of the series whose general term is:

1. $\dfrac{1}{n^p}\ (p > 1)$ 2. $\dfrac{1}{n^p}\ (0 < p \le 1)$

3. ne^{-n^2} 4. ne^{-n}

5. $\dfrac{1}{n \log n}$ 6. $\dfrac{1}{n(\log n)^p}$ $(p > 1)$

7. Show that $\sum_1^\infty 1/(n^2) < 2$. (Apply Theorem 7)

8. Show that $\sum_1^\infty 1/(n^2 + 1) < \dfrac{\pi}{2}$.

9. Show that $\sum_1^\infty 1/(n^3) < \dfrac{3}{2}$.

We have been comparing series with series, and series with integrals. Prove the following inequalities by comparing integrals with integrals.

10. $\displaystyle\int_0^\infty \dfrac{dx}{x^2 + x + 1} < \dfrac{\pi}{2}$.   11. $\displaystyle\int_2^\infty \dfrac{x \, dx}{x^3 + 1} < \dfrac{1}{2}$.

12. $\displaystyle\int_2^\infty \dfrac{\sqrt{x} \, dx}{x^2 + 1} < \sqrt{2}$.   13. $\displaystyle\int_1^\infty e^{-x^2} \, dx < \dfrac{1}{2e}$.

14. The finite sum $\log(n!) = \log 2 + \log 3 + \cdots + \log n$ can be compared with $\int \log x \, dx$ between appropriate limits, as in the proof of the integral test. Show in this way that

$$\log(n - 1)! < n \log n - n + 1 < \log(n!)$$

for $n \geq 2$. Conclude that

$$(n - 1)! < en^n e^{-n} < n!$$

15. Using the trapezoidal approximation in a similar way, show that

$$\sqrt{n}(n - 1)! < en^n e^{-n},$$

or

$$n! < en^n e^{-n} n^{1/2}.$$

16. Finally, interpret $\log k$ as the area of a trapezoid below the tangent line to the graph of $\log x$ at $x = k$, the trapezoid extending from $x = k - 1/2$ to $x = k + 1/2$. Thus $\log(n!)$ is a sum of trapezoidal areas that *include* the area under the graph of $\log x$, except for sticking out too far by $1/2$ on the right, and not reaching quite far enough (by $1/2$) on the left. Show therefore that

$$\log n! + \dfrac{1}{4} - \dfrac{1}{2} \log n > n \log n - n + 1,$$

and conclude that

$$n! > e^{7/8} n^n e^{-n} n^{1/2}.$$

6. ABSOLUTE CONVERGENCE AND CONDITIONAL CONVERGENCE

We shall have to know something about the convergence of a series $\sum a_n$ whose terms have both signs. If the sum $\sum |a_n|$ of their magnitudes (absolute values) converges, it should seem probable that the series $\sum a_n$ itself must converge, and to a smaller sum, because of the cancelling that goes on as we add together the positive and negative terms.

EXAMPLE 1   Consider the geometric series with ratio $t = -(1/2)$. Theorem 1 shows that

$$1 - \frac{1}{2} + \frac{1}{4} - \frac{1}{8} + \cdots + \left(-\frac{1}{2}\right)^n + \cdots$$

has the sum

$$\frac{1}{1 - \left(-\frac{1}{2}\right)} = \frac{1}{1 + 1/2} = \frac{2}{3}.$$

This is much smaller than the sum of the powers of $1/2$,

$$1 + \frac{1}{2} + \frac{1}{4} + \cdots = 2,$$

and the reason is the cancelling that goes on as we add. We can see exactly the cancelling that occurs by dividing up the series into its positive and negative parts:

$$P = 1 + \frac{1}{4} + \frac{1}{16} + \cdots = \frac{1}{1 - (1/4)} = \frac{4}{3},$$

$$N = -\left(\frac{1}{2} + \frac{1}{8} \cdots\right)$$

$$= \left(-\frac{1}{2}\right)\left[1 + \frac{1}{4} + \cdots\right]$$

$$= \left(-\frac{1}{2}\right)P = -\frac{2}{3},$$

$$\Sigma = P + N = \frac{4}{3} + \left(-\frac{2}{3}\right) = \frac{2}{3}.$$

THEOREM 8   *If $|a_n| \le b_n$ for all n and if $\sum b_n$ converges, with sum B, then $\sum a_n$ converges, and*

$$\left|\sum_1^\infty a_n\right| \le B.$$

Proof. We simply divide up $\sum a_n$ into two series, one containing all the positive (and zero) terms a_n, and the other containing all the negative (and zero) terms a_n. Specifically, we define a_n^+ as a_n if $a_n > 0$, and as zero otherwise. Then $0 \le a_n^+ \le b_n$ for all n, and so $\sum a_n^+$ converges, with sum $\le B$, by Theorem 4. Next we define a_n^- as $-a_n$ if $a_n < 0$ and as 0 otherwise, and get, similarly, $0 \le \sum_1^\infty a_n^- \le B$. But $a_n = a_n^+ - a_n^-$ for all n. Therefore, $\sum a_n$ converges, with $\sum_1^\infty a_n = \sum_1^\infty a_n^+ - \sum_1^\infty a_n^-$, by Theorem 3. And since each sum on the right lies between 0 and B, their difference lies between $-B$ and B. That is, $|\sum_1^\infty a_n| \le B$. ∎

DEFINITION   *When we have a series $\sum a_n$ such that $\sum |a_n|$ converges, we say that $\sum a_n$ converges absolutely.*

Theorem 8 is then equivalent to the following:

If a series $\sum a_n$ converges absolutely, then it converges, and

$$\left| \sum_1^\infty a_n \right| \leq \sum_1^\infty |a_n|.$$

Theorems 6 and 8 have the following useful corollary:

COROLLARY   *Suppose $|a_{n+1}/a_n| \to l$ as $n \to \infty$. Then $\sum a_n$ converges absolutely if $l < 1$, and diverges if $l > 1$.*

When the terms of a series *alternate* in sign, the following special test is available.

THEOREM 9   *If $\{a_n\}$ is a strictly decreasing sequence of positive numbers such that $a_n \to 0$ as $n \to \infty$, then the series*

$$a_0 - a_1 + a_2 + \cdots + (-1)^n a_n + \cdots$$

is convergent. Moreover, its sum s is approximated by its nth partial sum s_n with an error less than the next term a_{n+1}:

$$|s - s_n| < a_{n+1}.$$

EXAMPLE 2   The series

$$1 - \frac{1}{2} + \frac{1}{3} - \frac{1}{4} + \cdots + (-1)^{n+1} \frac{1}{n} + \cdots$$

is convergent, by the theorem. However it is not absolutely convergent because the harmonic series

$$1 + \frac{1}{2} + \frac{1}{3} + \cdots + \frac{1}{n} + \cdots$$

diverges.

A series that is convergent but *not* absolutely convergent is said to converge *conditionally*. Conditionally convergent series are of little interest to us here. They converge so slowly that their partial sums are of little use in estimating their sums. Moreover, we are not allowed to handle them algebraically with freedom. For example, we cannot express the above sum as the difference of the sums of the positive and negative parts,

$$s = \left(1 + \frac{1}{3} + \frac{1}{5} + \cdots\right) - \left(\frac{1}{2} + \frac{1}{4} + \frac{1}{6} + \cdots\right),$$

because each of these series diverges to infinity and we have the nonsensical

$$s = \infty - \infty.$$

So for us the importance of alternating series is the error estimate $|s - s_n| < a_{n+1}$ in situations where we do have rapid, absolute convergence.

Proof of theorem. The odd partial sums s_{2m+1} form an *increasing* sequence, as shown by the following bracketing:

$$s_{2m+1} = (a_0 - a_1) + (a_2 - a_3) + \cdots + (a_{2m} - a_{2m+1}).$$

Each term in parentheses is positive, and the odd partial sums s_{2m+1} are the partial sums of this bracketed series. Similarly, the even partial sums s_{2m} form a *decreasing* sequence:

$$s_{2m} = a_0 - (a_1 - a_2) - \cdots - (a_{2m-1} - a_{2m}).$$

Moreover,

$$s_{2m} > s_{2m-1}$$

for all m, since $s_{2m} - s_{2m-1} = a_{2m} > 0$. Therefore, the intervals

$$I_m = [s_{2m-1}, s_{2m}]$$

form a nested sequence of closed intervals with lengths a_{2m} tending to 0. It follows from the nested-interval principle that the two endpoint sequences converge as $m \to \infty$, and to the same limit s. That is, $s_n \to s$ as n runs to ∞ through the even integers, and also as n runs to ∞ through the odd integers. Therefore, $s_n \to s$ as $n \to \infty$.

Finally,

$$s_{2m-1} < s < s_{2m},$$

since the endpoint sequences are strictly monotone, so

$$s - s_{2m-1} < s_{2m} - s_{2m-1} = a_{2m}.$$

Also $s_{2m+1} < s < s_{2m}$, so

$$s_{2m} - s < s_{2m} - s_{2m+1} = a_{2m+1}.$$

Thus in all cases

$$|s - s_n| < a_{n+1}. \quad \blacksquare$$

EXAMPLE 3   *All* the hypotheses of the theorem have to be met to ensure convergence. For example, the series

$$3 - \frac{1}{2} + \frac{3}{3} - \frac{1}{4} + \frac{3}{5} - \cdots + \frac{1 + (-1)^{n+1}2}{n} + \cdots$$

diverges, even though $a_n \to 0$ and the signs alternate. What allows divergence here is that a_n is not a *decreasing* sequence. The proof of divergence is left as a problem.

PROBLEMS FOR SECTION 6

The series in Problems 1 through 12 alternate in sign. Determine in each case whether the series converges absolutely, conditionally, or not at all.

1. $\sum (-1)^n/n^2$ 2. $\sum (-1)^n/\sqrt{n}$ 3. $\sum (-1)^n 2^n/n^2$

4. $\sum (-1)^n/\log n$ 5. $\sum (-1)^n(n-1)/n$ 6. $\sum (-1)^n(\sqrt{n}-1)/n$

7. $\sum (-1)^n (n-1)/n^2$ 8. $\sum (-1)^n (\sqrt{n} - 1)/n^2$ 9. $\sum (-1)^n \log n/n$

10. $\sum (-1)^n \log n/\log 2n$ 11. $\sum (-1)^n \log n/\log n^2$ 12. $\sum (-1)^n \log n/\log(n^n)$

13. $\sum (-1)^n \sin (n\pi/6)/n$

14. Show that the series in Example 3 is divergent.

15. Construct a divergent alternating series with terms that decrease in absolute value.

16. Prove Theorem 8 as follows. Set $c_n = a_n + b_n$ and show that $0 \le c_n \le 2b_n$. Then use the comparison test, etc.

7. POWER SERIES

We can now show that the elementary transcendental functions e^x, $\sin x$, etc., can be expressed as "infinite polynomials," in the sense that each is an infinite sum of terms $a_n x^n$. For example, the series that we have been using for e will be established by showing more generally that

$$e^x = 1 + x + \frac{x^2}{2!} + \frac{x^3}{3!} + \cdots + \frac{x^n}{n!} + \cdots$$

for all x. Such a series is called a *power series* because its nth term is a constant times x^n. This remarkable equation for e^x, and the others like it that we shall establish, all depend on a few simple facts about power series.

This whole subject is important for computation. For example, most of the numerical tables for transcendental functions, such as the tables for the natural logarithm, sine and cosine, are computed from the power-series expansions of these functions.

A power series is an infinite series whose nth term is a constant a_n times x^n:

$$a_0 + a_1 x + \cdots + a_n x^n + \cdots.$$

Such a series cannot be said to converge or diverge as it stands, since it is not an infinite series of numbers. However, it becomes a numerical series if we give x a numerical value, and we can expect that it may converge for some values of x and diverge for other values. Here is the basic fact.

THEOREM 10 *If a power series $\sum a_n x^n$ converges for a particular value of x, say $x = c$, then the series converges absolutely for every x that is smaller in absolute value, i.e., for every x in the interval $(-|c|, |c|)$.*

Proof. We are assuming that $\sum a_n c^n$ converges. Then $a_n c^n \to 0$ as $n \to \infty$, by Theorem 2, so $|a_n c^n| \le 1$ from some point on, say from the integer N on. We can then see that $\sum a_n x^n$ converges absolutely for all x in the interval $(-|c|, |c|)$, by comparison with the geometric series. For suppose that $|x| < |c|$ and set $t = |x/c|$. Then $0 \le t < 1$, and

$$|a_n x^n| = |a_n c^n t^n| = |a_n c^n| \cdot t^n \le t^n,$$

when $n \ge N$. Therefore the remainder series

$$\sum_{N}^{\infty} |a_n x^n|$$

converges by comparison with the geometric series $\sum_N t^n$. This proves that $\sum a_n x^n$ converges absolutely for every x in the interval $(-|c|, |c|)$. ∎

It can be proved, as a corollary of Theorem 10 and the completeness property of the real numbers, that a power series $\sum a_n x^n$ converges exactly on an interval centered at the origin. This is a theoretical result, and does not provide any way of determining the interval of convergence. However, the ratio test often provides a direct computation of the convergence interval, and thus bypasses all such theoretical considerations.

There will still be the question of convergence at the interval endpoints, and other tests have to be used for this.

EXAMPLE 1   In order to determine whether the series

$$x + 2x^2 + 3x^3 + \cdots + nx^n + \cdots$$

converges for a given number x we try the ratio test. The test ratio for $\sum |nx^n|$ is

$$\left| \frac{(n+1)x^{n+1}}{nx^n} \right| = \left(\frac{n+1}{n} \right) |x| = \left(1 + \frac{1}{n} \right) |x|,$$

which approaches $|x|$ as n approaches infinity. Therefore, we know (from the corollary of Theorem 8) that $\sum nx^n$ will converge absolutely if $|x| < 1$ and diverge if $|x| > 1$. The domain of convergence is thus the interval $(-1, 1)$, with the possible addition of one or both endpoints. Since the ratio test fails when the limit $|x|$ has the value 1, the series behavior at the endpoints $x = \pm 1$ has to be determined by other tests. Here the nth terms in the endpoint series are $\pm n$, so these two series both diverge. The open interval $(-1, 1)$, is thus the exact domain of convergence.

EXAMPLE 2   The series

$$(x - 3) + 2(x - 3)^2 + \cdots + n(x - 3)^n + \cdots$$

is not of the type we are interested in here, but it has a feature worth noticing. We apply the ratio test exactly as in the above example:

$$\left| \frac{(n+1)(x-3)^{n+1}}{n(x-3)^n} \right| = \left(1 + \frac{1}{n} \right) |x - 3| \to |x - 3|.$$

We conclude as above that the series converges absolutely if $|x - 3| < 1$ and diverges if $|x - 3| \geq 1$, so the interval of convergence is defined by the inequality

$$-1 < x - 3 < 1, \quad \text{or} \quad 2 < x < 4.$$

It is thus the interval $(2, 4)$, centered about the point $x = 3$.

EXAMPLE 3   Consider the series $\sum x^n/n$. The ratio test shows that it converges if $|x| < 1$ and diverges if $|x| > 1$, but it leaves up in the air the behavior of the series at $x = \pm 1$.

At $x = 1$ the series becomes the harmonic series, which we know to be divergent.

At $x = -1$ it is the *alternating harmonic* series

$$-1 + \frac{1}{2} - \frac{1}{3} + \frac{1}{4} + \cdots + \frac{(-1)^n}{n} + \cdots,$$

which converges, by Theorem 9.

The domain of convergence is thus the set of numbers x such that $-1 \leq x < 1$, that is, the interval $[-1, 1)$, closed on the left and open on the right.

EXAMPLE 4   For the series

$$1 + x + \frac{x^2}{2!} + \frac{x^3}{3!} + \cdots + \frac{x^n}{n!} + \cdots,$$

we have the test ratio

$$\left| \frac{x^{n+1}/(n+1)!}{x^n/n!} \right| = \frac{|x|}{n+1},$$

which approaches 0 as $n \to \infty$, no matter what value x has. The series therefore converges absolutely for all x, and the interval of convergence is $(-\infty, \infty)$.

If the power series $\sum a_n x^n$ converges on the interval $(-c, c)$, then its sum depends on x and hence is a function f of x,

$$f(x) = a_0 + a_1 x + a_2 x^2 + \cdots + a_n x^n + \cdots.$$

The major question we have to answer is this: Does the law that the derivative of a sum is equal to the sum of the derivatives hold for *infinite* sums? In other words, must the function f above necessarily have a derivative f' that is given by

$$f'(x) = a_1 + 2a_2 x + \cdots + n a_n x^{n-1} + \cdots$$

on the interval $(-c, c)$?

It is really remarkable that the answer to this question is *yes*, and that we can therefore apply calculus to the study of such infinite-sum functions. We first show that the sum of the derivatives exists.

THEOREM 11   *If the power series $\sum a_n x^n$ converges for each x in the open interval $(-c, c)$, then it and its term-by-term differentiated series $\sum n a_n x^{n-1}$ both converge absolutely for every x in the interval $(-c, c)$.*

Proof. Consider any x in $(-c, c)$. Then choose a positive number u such that $|x| < u < c$. This number u will now play the role that c played in Theorem 9. Thus, since $\sum a_n u^n$ converges by hypothesis, it follows from Theorem 9 that $\sum a_n x^n$ is absolutely convergent.

For the differentiated series we have to go back to the line of reasoning we used in Theorem 9. Since $\sum a_n u^n$ converges, it follows that $|a_n u^n| \leq 1$ for all n from some integer N on. Assuming $x \neq 0$, we set $t = |x/u|$. Then

$$|na_n x^n| = n|a_n u^n| \cdot t^n \leq nt^n$$

for $n \leq N$. Since $0 < t < 1$, the series $\sum nt^n$ converges by the ratio test. Then $\sum na_n x^n$ converges absolutely by comparison. Finally, we can multiply this series by $1/x$ and conclude that $\sum na_n x^{n-1}$ is absolutely convergent (Theorem 3). Of course, every power series converges trivially at $x = 0$, so the proof is complete. ∎

This brings us to the major theorem.

THEOREM 12   *If $\sum a_n x^n$ converges on the interval $(-c, c)$, then its sum function*

$$f(x) = a_0 + a_1 x + a_2 x^2 + \cdots + a_n x^n + \cdots$$

is differentiable and

$$f'(x) = a_1 + 2a_2 x + \cdots + na_n x^{n-1} + \cdots.$$

We shall give the proof of this theorem later on. It is not really hard, but it takes longer than anything we have met so far.

EXAMPLE 5   As a special case of the geometric series we have

$$\frac{1}{1 - x^2} = 1 + x^2 + x^4 + \cdots = \sum_0^\infty x^{2n}.$$

Therefore

$$\frac{2x}{(1 - x^2)^2} = 2x + 4x^3 + \cdots = \sum_1^\infty 2nx^{2n-1}$$

by Theorem 12, so

$$\frac{1}{(1 - x^2)^2} = 1 + 2x^2 + 3x^4 + \cdots = \sum_1^\infty nx^{2n-2}$$

$$= \sum_0^\infty (n + 1)x^{2n}.$$

The final series can also be obtained by squaring the first series, but we have not taken up how to do this.

Theorem 12 suggests that perhaps we also can integrate a power series term by term. As a matter of fact, if

$$f(x) = a_0 + a_1 x + \cdots + a_n x^n + \cdots$$

on $(-r, r)$, then the integrated series

$$a_0 x + \frac{a_1 x^2}{2} + \cdots + \frac{a_n x^{n+1}}{n + 1} + \cdots$$

also converges on $(-r, r)$, because, after we factor x out, its terms are smaller and we can apply the comparison test. Thus

$$F(x) = a_0 x + \frac{a_1 x^2}{2} \cdots + \frac{a_n x^{n+1}}{n+1} + \cdots$$

is defined on $(-r, r)$, and $F'(x) = f(x)$, by Theorem 12. Since obviously $F(0) = 0$, we have proved:

THEOREM 13 *If the power series*

$$f(x) = a_0 + a_1 x + \cdots + a_n x^n + \cdots$$

converges on $(-r, r)$, then so does its term-by-term integrated series, and the sum $F(x)$ of the integrated series is the antiderivative of $f(x)$ which has the value 0 at 0.

COROLLARY 1 *For every x in the interval $(-1, 1)$,*

$$\log(1 + x) = x - \frac{x^2}{2} + \frac{x^3}{3} + \cdots + (-1)^{n+1} \frac{x^n}{n} + \cdots.$$

Proof. Since

$$1 - x + x^2 + \cdots + (-1)^n x^n + \cdots = \frac{1}{1+x} \left(= \frac{1}{1 - (-x)} \right)$$

on $(-1, 1)$, the sum of the integrated series

$$x - \frac{x^2}{2} + \frac{x^3}{3} + \cdots + (-1)^n \frac{x^{n+1}}{n+1} + \cdots$$

is the antiderivative of $1/(1 + x)$ which has the value 0 at $x = 0$, and hence is just $\log(1 + x)$, without any constant of integration. ∎

COROLLARY 2 *For every x in the interval $(-1, 1)$,*

$$\arctan x = x - \frac{x^3}{3} + \frac{x^5}{5} - \frac{x^7}{7} + \cdots + (-1)^m \frac{x^{2m+1}}{2m+1} + \cdots.$$

Proof. Left to the reader. [*Hint.* Set $y = -x^2$ in $1/(1 + x^2)$ and use the geometric series expansion.] ∎

PROBLEMS FOR SECTION 7

In each problem below determine the *open* interval of convergence by the ratio test. The test will fail at every endpoint. Then determine endpoint behavior by other tests.

1. $1 + x + \dfrac{x^2}{2} + \dfrac{x^3}{3} + \cdots + \dfrac{x^n}{n} + \cdots$

2. $1 + \dfrac{x}{2} + \dfrac{x^2}{4} + \cdots + \dfrac{x^n}{2^n} + \cdots$

3. $1 + x + \dfrac{x^2}{2} + \dfrac{x^3}{2 \cdot 3} + \dfrac{x^4}{2 \cdot 3 \cdot 4} + \cdots + \dfrac{x^n}{n!} + \cdots$

4. $1 + 3x + 9x^2 + \cdots + 3^n x^n + \cdots$

5. $1 + 3x + 5x^2 + \cdots + (2n + 1)x^n + \cdots$

6. $1 + \dfrac{x^2}{2} + \dfrac{x^4}{4} + \dfrac{x^6}{8} + \cdots + \dfrac{x^{2n}}{2^n} + \cdots$

7. $x - \dfrac{x^3}{3} + \dfrac{x^5}{5} - \dfrac{x^7}{7} + \cdots + (-1)^n \dfrac{x^{2n+1}}{2n + 1} \cdots$

8. $1 - \dfrac{x}{1 \cdot 2} + \dfrac{x^2}{2 \cdot 4} - \dfrac{x^3}{3 \cdot 8} + \cdots + (-1)^n \dfrac{x^n}{n2^n} + \cdots$

9. $\dfrac{(-1)^n x^n}{n}$

10. $\dfrac{x^n}{n2^n}$

11. $\dfrac{x^n}{\sqrt{n}}$

12. $\dfrac{(x - 3)^n}{n^2}$

13. $\dfrac{(-1)^n (x - 2)^n}{n2^n}$

14. $\dfrac{n(x - 1)^n}{2^n}$

15. $\dfrac{(x - 1)^n}{n3^n}$

16. $\dfrac{(x - 1)^n}{n^2 3^n}$

17. $\sum \dfrac{(x - 2)^n}{2^n}$

18. $\sum \dfrac{(x - 1)^n}{2^n}$

19. $\sum 2^n (x - 1)^n$

20. $\sum \dfrac{(x - 3)^n}{n2^n}$

21. $\sum \dfrac{(x - 3)^n}{n^2 3^n}$

The following problems center around the geometric series

$$\frac{1}{1 - x} = 1 + x + x^2 + \cdots + x^n + \cdots.$$

22. a) Using Theorem 12, find a power series whose sum is $1/(1 - x)^2$.
    b) Compute the exact sum of the series $\sum_1^\infty n/2^n$.

23. Compute the exact sum of the series $\sum_1^\infty n^2/2^n$ by some device like that used in the problem above.

Use Theorems 12 and 13 to identify the following functions.

24. $f(x) = 1 + \dfrac{x}{2} + \dfrac{x^2}{3} + \cdots + \dfrac{x^n}{n - 1} + \cdots$

25. $f(x) = 1 + 2x + 3x^2 + \cdots + (n + 1)x^n + \cdots$

26. $f(x) = 2 + 3x + 4x^2 + 5x^3 + \cdots + (n + 2)x^n + \cdots$

27. $f(x) = 1 + 4x + 9x^2 + \cdots + (n + 1)^2 x^n + \cdots$

28. $f(x) = 1 + x + \dfrac{x^2}{1 \cdot 2} + \dfrac{x^3}{2 \cdot 3} + \cdots + \dfrac{x^n}{(n - 1)n} + \cdots$

8. THE MACLAURIN SERIES OF A FUNCTION

It is difficult to overemphasize the importance of Theorem 12. We saw above how it determines power series expansions for $\log(1 + x)$ and arctan x. We show next how it determines the series coefficients a_n in terms of the function derivatives $f^{(n)}(0)$.

THEOREM 14   *If*

$$f(x) = a_0 + a_1x + a_2x^2 + \cdots + a_nx^n + \cdots$$

over an interval $(-r, r)$, *then f has derivatives of all orders in this interval and*

$$a_n = \frac{f^{(n)}(0)}{n!}$$

for every n, where $f^{(n)}$ *is the nth derivative of f. The series can thus be rewritten*

$$f(x) = f(0) + f'(0)x + \frac{f''(0)x^2}{2!} + \cdots + \frac{f^{(n)}(0)x^n}{n!} + \cdots.$$

REMARK   In order to interpret $f(0)$ as $f^{(0)}(0)/0!$ we adopt the convention that

$$0! = 1.$$

Proof of theorem. By Theorems 11 and 12, we can differentiate the above series as often as we wish. Thus

$$
\begin{aligned}
f(x) &= a_0 + a_1x + a_2x^2 + a_3x^3 + \cdots + a_nx^n + \cdots\\
f'(x) &= a_1 + 2a_2x + 3a_3x^2 + \cdots + na_nx^{n-1} + \cdots\\
f''(x) &= 2a_2 + 3 \cdot 2a_3x + \cdots + n(n-1)a_nx^{n-2} + \cdots\\
f'''(x) &= 3!a_3 + \cdots + n(n-1)(n-2)a_nx^{n-3} + \cdots\\
&\quad\vdots\\
f^{(n)}(x) &= n!a_n + \cdots
\end{aligned}
$$

Setting $x = 0$ in these equations reduces them to

$$
\begin{aligned}
f(0) &= a_0\\
f'(0) &= a_1\\
f''(0) &= 2a_2\\
f'''(0) &= 3!a_3\\
&\vdots\\
f^{(n)}(0) &= n!a_n.
\end{aligned}
$$

Thus $a_j = f^{(j)}(0)/j!$ for all j. ∎

DEFINITION   *The series*

$$\sum \frac{f^{(n)}(0)}{n!} x^n$$

is called the Maclaurin *series for f.*

Let us be clear about what we have proved here. We have proved that *if f is the sum of a power series on an interval* $(-r, r)$, *then f is infinitely differentiable and the series is necessarily the Maclaurin series for f.* We have *not* proved that if f is infinitely differentiable, then it is the sum of its Maclaurin series. Two things can go wrong for such a function.

1. Its Maclaurin series $\sum f^{(n)}(0)x^n/n!$ may not converge for any x except $x = 0$.
2. The Maclaurin series may converge to a function different from f.

Let us explore the second possibility a little further. Suppose that f is infinitely differentiable on $(-r, r)$ and that its Maclaurin series $\sum f^n(0)x^n/n!$ converges to a function g on $(-r, r)$, with g different from f. Then the series also has to be the Maclaurin series for g, by Theorem 14. That is, its coefficients must also have the form $g^{(n)}(0)/n!$, so

$$f^{(n)}(0) = g^{(n)}(0)$$

for all n. Then $(f - g)^{(n)}(0) = 0$ for all n. Thus $f - g$ is an infinitely differentiable function on $(-r, r)$ that is not the zero function but has the property that all of its derivatives are zero at $x = 0$. This may seem paradoxical, and yet we can concoct just such a function. If we define h by

$$h(x) = e^{-1/|x|} \qquad \text{if } x \neq 0,$$
$$h(0) = 0,$$

then it can be proved that h is infinitely differentiable and that all the derivatives of h are 0 at $x = 0$. Yet the graph of h is like this:

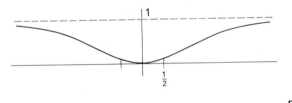

It is decreasing on $(-\infty, 0]$, increasing on $[0, \infty)$, concave up on $\left[-\dfrac{1}{2}, \dfrac{1}{2}\right]$ and concave down elsewhere.

The Maclaurin series for h has only 0 coefficients and its sum is the 0 function, not h.

We now establish the power-series representations of the elementary transcendental functions. In each case, the role of Theorem 14 is only to identify the series we should be looking at. The proof that the series represents the function then depends on a calculus characterization of the individual function.

THEOREM 15   *For every real number x,*

$$e^x = 1 + x + \frac{x^2}{2!} + \frac{x^3}{3!} + \cdots + \frac{x^n}{n!} + \cdots = \sum_{j=0}^{\infty} \frac{x^j}{j!}.$$

Proof. Since $f(x) = e^x$ is its own derivative, we see that $f(0), f'(0), f''(0), \ldots$ all have the same value $e^0 = 1$. The above series is therefore the Maclaurin series for e^x, and hence is the only series that can possibly represent e^x. But does it?

We noted earlier that this series converges for all x, by the ratio test, and therefore defines *some* function g,

$$g(x) = 1 + x + \frac{x^2}{2!} + \frac{x^3}{3!} + \cdots.$$

By Theorem 12, $g(x)$ is everywhere differentiable and $g'(x)$ is the sum of the term-by-term differentiated series. But here the differentiated series turns out to be the original series:

$$0 + 1 + \frac{2x}{2!} + \frac{3x^2}{3!} + \frac{4x^3}{4!} + \cdots = 1 + x + \frac{x^2}{2!} + \frac{x^3}{3!} + \cdots,$$

and so $g'(x) = g(x)$. Also $g(0) = 1$. But we know from Chapter 5 (page 161), that e^x is the unique function f satisfying the differential equation $f' = f$ and the "initial condition" $f(0) = 1$. Therefore $g(x) = e^x$. ∎

THEOREM 16 *For every real number x,*

$$\cos x = 1 - \frac{x^2}{2!} + \frac{x^4}{4!} - \frac{x^6}{6!} + \cdots + (-1)^m \frac{x^{2m}}{(2m)!} + \cdots.$$

Proof. This proof goes exactly like the one above, and we shall only list the steps, leaving their verification to the reader.

1. The series is the Maclaurin series for $\cos x$ and so is the only power series that can possibly represent $\cos x$.
2. The series converges for all x (by the ratio test) and so defines a function g.
3. The function g satisfies the differential equation $g'' = -g$ and the initial conditions $g(0) = 1$, $g'(0) = 0$.
4. But $\cos x$ is the unique function satisfying this differential equation and these initial conditions (Chapter 5, Section 4). ∎

THEOREM 17 *For every real number x,*

$$\sin x = x - \frac{x^3}{3!} + \frac{x^5}{5!} - \frac{x^7}{7!} + \cdots + (-1)^m \frac{x^{2m+1}}{(2m+1)!} + \cdots$$

Proof. Just the same as for Theorem 16 (or by Theorem 12 applied to Theorem 16). ∎

THEOREM 18 *For any fixed real number α,*

$$(1 + x)^\alpha = 1 + \alpha x + \frac{\alpha(\alpha - 1)}{2!} x^2 + \cdots$$

$$+ \frac{\alpha(\alpha - 1) \cdots (\alpha - n + 1)}{n!} x^n + \cdots$$

on the interval $(-1, 1)$.

This proof is left as an exercise. The steps are:

1. Use the ratio test to find the interval of convergence.
2. Show that if g is the sum of the series, then

$$g'(x) = \frac{\alpha g(x)}{(1 + x)}, \qquad g(0) = 1.$$

3. Show that the unique solution of this initial-value problem is the function $(1 + x)^\alpha$.

It was mentioned earlier that most mathematical tables are computed from power series expansions. The simplest situation concerns alternating series, where the error is always less than the next term.

EXAMPLE 1 The Maclaurin series for sin x is an alternating series in this sense if $|x| \leq 2$. Therefore $x - x^3/3!$ approximates sin x with an error less than $x^5/5!$; for example,

$$\sin 0.2 \approx 0.2 - 0.008/6$$
$$= 0.2 - 0.001333 \cdots$$
$$= 0.198666 \cdots$$

with an error less than

$$\frac{(0.2)^5}{120} = \frac{0.00032}{120} < 0.000003.$$

The partial sums of an alternating series are alternately too large and too small, so here

$$0.198666 < \sin 0.2 < 0.198670,$$

and

$$\sin 0.2 \approx 0.19867$$

is the closest five-place decimal approximation to sin 0.2.

EXAMPLE 2 Determine a range of x over which the first three terms in the Maclaurin series for sin x are accurate as an approximation to sin x to within an error of 1 in the fifth decimal place.

Solution. The error is less than $x^7/7!$ so it will be sufficient to restrict x so that

$$\frac{x^7}{7!} < 10^{-5},$$

or

$$x^7 < (7!)10^{-5} = 5040 \cdot 10^{-5} = 0.0504$$

or

$$x < (0.0504)^{1/7}.$$

Solving this requires logarithm tables (or a pocket calculator) but you can check that $(0.6)^7 < 0.03$ so the desired approximation holds at least over the interval $(0, 0.6)$.

PROBLEMS FOR SECTION 8

Compute the Maclaurin series for each of the following functions f by evaluating its derivatives $f^{(n)}(0)$.

1. $1/(1 - x)$ 2. $\log(1 - x)$

3. x^3 4. $1/(1 + x)$

5. $x^3 - 2x + 5$ 6. $\sqrt{1 + x}$

Assuming Theorems 13 through 18, express each of the following functions as a sum of a power series. Also give the interval of convergence in each case.

7. e^{-x^2} 8. $\cos \sqrt{x}$

9. $\arctan x^3$ 10. $\int_0^x \cos \sqrt{t}\, dt$

11. $\int_0^x e^{-t^2/2}$

12. Complete the proof of Theorem 16 by giving the details for steps 1, 2, and 3.

13. Prove Theorem 17.

14. Show that if $g'(x) = \alpha g(x)/(1 + x)$, then $g(x) = c(1 + x)^\alpha$ for some constant c. (How can you show that $g(x)/(1 + x)^\alpha$ is a constant?)

15. a) Show that the infinite series defined in Theorem 18 converges on $(-1, 1)$.
    b) Show that its sum $g(x)$ has the property that $g'(x)(1 + x) = \alpha g(x)$.
    c) Assuming the result of Problem 14, show that $g(x) = (1 = x)^\alpha$.

16. We know that $x^n e^{-x} \to 0$ as $x \to \infty$ for any positive integer n (see Chapter 7, Extra Problems 20 through 23). Show therefore that

$$\frac{e^{-1/x^2}}{x^m} \to 0$$

as $x \to 0$, for any positive integer m. It follows then that

$$p\left(\frac{1}{x}\right) e^{-1/x^2} \to 0$$

as $x \to 0$ for any polynomial in $1/x$:

$$p\left(\frac{1}{x}\right) = \frac{a_n}{x^n} + \frac{a_{n-1}}{x^{n-1}} + \cdots + \frac{a_1}{x} + a_0.$$

17. Show that the derivative of a function of the form

$$f(x) = p\left(\frac{1}{x}\right) e^{-1/x^2}$$

is again of this form, with a different polynomial. It is sufficient to show this for one of the terms of $f(x)$, that is, for a "monomial" term $ae^{-1/x^2}/x^m$.

18. Now define $F(x)$ by

$$F(x) = e^{-1/x^2} \qquad \text{if } x \neq 0,$$
$$F(0) = 0.$$

a) Show that F is everywhere differentiable, with $F'(0) = 0$ and with $F'(x)$ of the form $p(1/x)e^{-1/x^2}$, when $x \neq 0$, where p is a polynomial. (Use Problems 16 and 17.)

b) Then show, continuing to use Problems 16 and 17, that all the derivatives of F exist and that $F^{(n)}(0) = 0$ for every n.

The function F is thus an infinitely differentiable function such that $F(x) \neq 0$ when $x \neq 0$, and yet its Maclaurin-series coefficients are all 0.

19. Show that for any value of x the Maclaurin series for $\sin x$ is alternating from some term on.

20. Compute $\sin 0.2$ to eight decimal places. (First show that the error after the first three terms is less than 3×10^{-9}.)

21. Compute $\sin 0.5$ to three decimal places.

22. Prove that

$$\int_0^x e^{-t^2/2}\, dt = x - \frac{x^3}{3 \cdot 2} + \frac{x^5}{5 \cdot 2^2 \cdot 2!} - \frac{x^7}{7 \cdot 2^3 \cdot 3!} + \cdots$$

$$+ \frac{(-1)^n x^{2n+1}}{(2n+1)2^n n!} + \cdots .$$

Use the above series to compute $\int_0^1 e^{-t^2/2}$ to three decimal places.

9. PROOF OF THEOREM 12

We come finally to the postponed proof of the basic theorem on differentiating power series. It does not involve any new principles, but is a little more complicated than anything we have done so far in this chapter.

THEOREM 12   *If*

$$f(x) = a_0 + a_1 x + a_2 x^2 + \cdots + a_n x^n + \cdots$$

on an open interval $I = (-c, c)$, *then f is differentiable and*

$$f'(x) = a_1 + 2a_2 x + \cdots + n a_n x^{n-1} + \cdots$$

on I.

Proof. We know from Theorem 11 that the term-by-term differentiated series $\sum n a_n x^{n-1}$ converges on the same interval $(-c, c)$. Call its sum $g(x)$. We want to prove that $g(x) = f'(x)$, that is, that

$$\frac{f(x + \Delta x) - f(x)}{\Delta x} - g(x)$$

approaches 0 as Δx approaches 0. According to Theorem 3, this algebraic combination is the sum of the series

$$\sum_{n=1}^{\infty} a_n \left[\frac{(x + \Delta x)^n - x^n}{\Delta x} - n x^{n-1} \right].$$

We shall show below that this series is dominated, term by term, by Δx times the twice-differentiated series $\sum n(n-1) a_n r^{n-2}$ (for a suitable value of r). We know that this twice-differentiated series is absolutely convergent, by Theorem 11 again, and if we set

$$s = \sum_{n=2}^{\infty} n(n-1) |a_n| r^{n-2},$$

then it will follow directly from Theorem 8 that the two sums satisfy the inequality

$$\left| \frac{f(x + \Delta x) - f(x)}{\Delta x} - g(x) \right| \le s |\Delta x|.$$

The left side, therefore, has the limit zero as $\Delta x \to 0$. That is,

$$g(x) = \lim_{\Delta x \to 0} \frac{f(x + \Delta x) - f(x)}{\Delta x},$$

or $g(x) = f'(x)$, as we claimed.

Our proof will thus be complete when we have shown that

$$\left| \frac{(x + \Delta x)^n - x^n}{\Delta x} - nx^{n-1} \right| \le n(n - 1)r^{n-2} \cdot |\Delta x|,$$

provided that r is larger than both $|x|$ and $|x + \Delta x|$. But this is a direct consequence of the tangent-line error formula in Chapter 6, Section 10, which tells us that

$$(x + \Delta x)^n - x^n - nx^{n-1}\Delta x = \frac{n(n - 1)X^{n-2}(\Delta x)^2}{2}$$

for some number X between x and $x + \Delta x$. Then $|X| < r$, and we obtain the above inequality upon dividing by $|\Delta x|$. ∎

EXTRA PROBLEMS FOR CHAPTER 11

Test for convergence the series having the following general terms, distinguishing between absolute convergence and conditional convergence.

1. $\dfrac{1}{n^2 + n}$

2. $\dfrac{1}{n^3 - 50}$

3. $\dfrac{1}{n^2 - n}$

4. $\dfrac{(-1)^n + \sqrt{n}}{n}$

5. $\dfrac{n^2}{2^n}$

6. $\dfrac{1 + (-1)^n\sqrt{n}}{n^2}$

7. $\dfrac{n^2}{(n - 2)(n - 1)(n + 1)(n + 2)}$

8. $\dfrac{n}{1 + (-1)^n n^2}$

9. $\dfrac{(-1)^n}{n \log n}$

10. $\dfrac{1}{n\sqrt{n - 1}}$

11. $\dfrac{2^n}{n^2}$

12. $\dfrac{(-1)^n}{1 + \sqrt{n}}$

13. $\dfrac{1}{\sqrt{n^3 - 5}}$

14. $(-1)^n \sin \dfrac{1}{n}$

15. $\dfrac{\log n}{n^2}$

16. $r^{(-1)^n}$

17. $\dfrac{e^n}{n!}$

18. $\dfrac{1}{(n - 2)^2}$

19. $\sin \dfrac{(-1)^n}{n}$

20. $\cos \dfrac{(-1)^n}{n}$

21. $\dfrac{1}{1 + (-1)^n n^{3/2}}$

22. $\dfrac{n}{1 + (-1)^n n^{3/2}}$

23. $\sin \left(\dfrac{1}{n^{3/2}} \right)$

24. $\dfrac{1 + (-1)^n\sqrt{n}}{n}$

25. $\dfrac{1}{(n \log n)^2}$

26. $\dfrac{(-1)^n n}{5n + 10}$

27. $\dfrac{\sin(x + n\pi)}{n}$

28. $\dfrac{(-1)^n n!}{n^n}$

29. $\dfrac{(-1)^{n(n+1)/2}}{n}$

30. $\dfrac{\log n}{n\sqrt{n}}$

31. $r^{(n^2)}$

32. $e^{(-1)^n n}$

33. $\dfrac{5^n (n!)^2}{(2n)!}$

34. $\dfrac{x^{(n^2)}}{2n}$

35. $\dfrac{n!}{\sqrt{(2n)!}}$

36. $\dfrac{3^n (n!)^2}{(2n)!}$

37. $n^{(-1)^n}$

38. e^{-n^2}

39. $r^{(n^{3/2})}$

40. $\dfrac{(\log n)^a}{n^b}$

41. $e^{-\sqrt{n}}$

42. $\dfrac{(n!)^2 x^n}{(2n)!}$

43. $1 - \dfrac{1}{2} + \dfrac{1}{4} - \dfrac{1}{5} + \dfrac{1}{7} - \dfrac{1}{8} + \cdots$

   (harmonic series, omitting every third term, and alternating)

44. $1 - \dfrac{1}{2} - \dfrac{1}{3} + \dfrac{1}{4} - \dfrac{1}{5} - \dfrac{1}{6} + \dfrac{1}{7} - \dfrac{1}{8} - \dfrac{1}{9} + \cdots$

   (harmonic series, alternating unevenly in the pattern one positive term followed by two negative terms)

45. $1 + \dfrac{1}{2} - \dfrac{1}{3} - \dfrac{1}{4} + \dfrac{1}{5} + \dfrac{1}{6} - \dfrac{1}{7} - \dfrac{1}{8} + \cdots$

   (harmonic series, with signs changing in blocks of two)

46. The series for $\log(1 + x)$ in Theorem 13, Corollary 1 is not useful for computation because it converges too slowly. Show that it can be modified to give the following improved version:

$$\log \frac{1 + x}{1 - x} = 2\left[x + \frac{x^3}{3} + \frac{x^5}{5} + \cdots + \frac{x^{2n-1}}{2n - 1} + \cdots \right].$$

   (First write down the $\log(1 - x)$ series and then combine.)

47. In the above series, show that the error after n terms is at most

$$\frac{2}{2n + 1} x^{2n+1}\left(\frac{1}{1 - x^2} \right).$$

   (Compare the remainder with a geometric series.)

48. Compute $\log 3/2$ to four decimal places by using the above series. (First show that the error after the first three series terms, i.e., through the x^5 term, is less than 10^{-5}.) Note how much more efficient this computation is than the Riemann-sum method of Chapter 9.

49. Compute $\log 3/2$ to eight decimal places.

50. Compute $\log 2$ to five decimal places.

51. The arctan x series

$$\arctan x = x - \frac{x^3}{3} + \frac{x^5}{5} + \cdots + (-1)^n \frac{x^{2n+1}}{2n + 1} + \cdots$$

provides the standard method of computing π, but here again a little artistic modification pays enormous dividends. Prove first that, if $\tan \theta = 1/5$, then

$$\tan\left(4\theta - \frac{\pi}{4}\right) = \frac{1}{239}.$$

(Use the formulas for $\tan(x + y)$ and $\tan(x - y)$ from Chapter 5. Compute $\tan 2\theta$, $\tan 4\theta$, and $\tan(4\theta - \pi/4)$, in that order.) Conclude that

$$\pi = 16 \arctan \frac{1}{5} - 4 \arctan \frac{1}{239}.$$

This remarkable identity has been known for nearly 300 years.

52. We start by seeing what we can do with only the first term of the series for arctan $1/239$. Using just the crude estimate $239 > 200$ in the error term, show that $4/239$ approximates $4 \arctan 1/239$ with an error less than $2(10)^{-7}$.

53. The decimal expansion of $1/239$ (and hence that of $4/239$) repeats in blocks of 7. Carry out the long division far enough to determine this expansion. Then use the result from the above problem, and the inequality arctan $x < x$, to show that

$$0.01673620 < 4 \arctan \frac{1}{239} < 0.01673641.$$

54. Now compute $16 \arctan 0.2$ through the $x^9/9$ term, and show that

$$3.15832895 < 16 \arctan 0.2 < 3.15832899.$$

55. Combine the above two results using Problem 14, and so prove that

$$3.1415925 < \pi < 3.1415928.$$

Thus $\pi = 3.141593$ to the nearest six decimal places, and the expansion of π begins $3.141592 \cdots$

56. Show that if arctan $1/239$ is replaced by its series expansion through the second term (the $x^3/3$ term) in the formula

$$\pi = 16 \arctan \frac{1}{5} - 4 \arctan \frac{1}{239},$$

and if arctan $1/5$ is replaced by its expansion through the eighth term (the $x^{15}/15$ term), then the right side approximates π with an error less than 5×10^{-12}. (Use the crude estimates $239 > 200$ and $2^{13} = 8192 < 10^4$.)

12
Polynomial Approximation

1. THE MEAN-VALUE THEOREM

In this section we shall prove the mean-value principle and related results from the extreme-value and intermediate-value principles. These two results, in turn, are proved in Appendix 4. The reader may wonder why it is now necessary to prove facts that so far have been accepted as self-evident on the basis of geometric intuition. There seems to be little point in proving the obvious. The reason lies in the uncertain and tenuous nature of the obvious. Something that appeared obvious yesterday may seem less so today, and the present proofs become relevant once one has seen how geometric intuition can be misleading. For example, consider again our (correct) geometric observation that a positive slope must imply an increasing graph. Similar considerations might in the beginning have led us to conjecture conversely that if f is strictly increasing over an interval I, then f' is positive everywhere in I. But the example $f(x) = x^3$ shows that this isn't so. Here f is a strictly increasing function, and yet $f'(x) = 3x^2$ is zero at the origin.

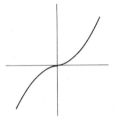

This conjecture, and the "counterexample" showing the conjecture to be false, together illustrate why geometric intuition cannot be trusted absolutely. Something may seem to be true principally because we have not yet noticed, and cannot imagine, a situation in which it is false. If we can't imagine how something might fail to happen, we are tempted to conclude that it must always happen. Of course, according to this principle of reasoning, the poorer our imagination the more facts we could establish!

A more serious example concerns the nature of a derivative. Every derivative we have met so far is continuous on its domain. That is, every function f that we have met so far has the property that if f is differentiable on an interval I then f' is continuous on I. This is true whether f is given by a formula or by a graph that we draw: we find it impossible to draw an everywhere differentiable graph (over I) for which the tangent line varies discontinuously. So it would be reasonable to conjecture, on the basis of intuition bolstered by experience, that a derivative is *necessarily* continuous on its domain.

Yet in Problem 4 we shall concoct a function f with the property that f' exists everywhere on $(-1, 1)$ and is discontinuous at the origin! So our conjecture is false. It can be proved that the graph of any such function must contain an infinite number of wiggles, and this is why such a graph cannot be drawn.

This conjecture and counterexample affect geometric intuition in the following way. Since any differentiable graph we draw is of necessity smooth, it follows that a geometric property, like the mean-value principle, that we have inferred from looking at such a graph may actually depend on the *continuity* of f', and not just the *existence* of f'.

So we are led to investigate what (if any) versions of the geometric principle we really can prove.

We shall prove a slight refinement of the mean-value principle that is standard and occasionally useful. Note that the mean-value property can be spoiled by a single interior singularity. For example, consider $f(x) = x^{2/3}$ over the interval $[a, b] = [-1, 1]$. The secant slope $[f(1) - f(-1)]/(1 - (-1))$ is then 0, but there is no point where the derivative $f'(x) = 2/3x^{-1/3}$ is 0. This failure of the mean-value property occurs because f has a singularity at the interior point where we would expect f' to be zero, namely, at the origin. On the other hand, the figure on the right shows that a singularity at an endpoint of $[a, b]$ does no damage, provided f is continuous there.

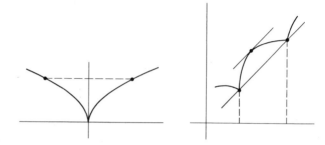

THEOREM 1 **The Mean-Value Theorem.** *If f is continuous on the closed interval $[a, b]$ and differentiable everywhere except possibly at one or both endpoints, then there is an interior point X at which*

$$f'(X) = \frac{f(b) - f(a)}{b - a}.$$

To prove the mean-value theorem we have to find a point x interior to the interval $[a, b]$ at which $f'(x) = m$, where m is the secant slope. But $f'(x) = m$ just when the derivative of $f(x) - mx$ is zero. So what we have to find is an interior critical point of the function $\phi(x) = f(x) - mx$. A sketch makes it seem obvious that ϕ has such an interior critical point; the secant line for ϕ is clearly horizontal so the ϕ configuration is just the horizontal case of the figure for the mean-value principle. This special case of the mean-value theorem is called Rolle's Theorem, and we prove it first.

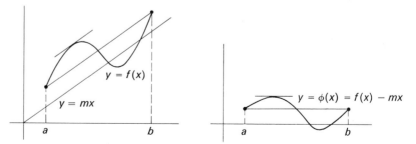

THEOREM 2 **Rolle's Theorem.** *Suppose that f is continuous on $[a, b]$, and differentiable on (a, b), and that $f(a) = f(b)$. Then there is an interior point X at which $f'(X) = 0$.*

Proof. By the extreme-value principle, f assumes maximum and minimum values on $[a, b]$. If the maximum value of f on $[a, b]$ is not the common endpoint value $f(a) = f(b)$, then a point X where the maximum value is assumed must necessarily be an interior point, and so $f'(X) = 0$, by Theorem 4 in Chapter 6. In the same way, if the minimum value of f is not the common endpoint value, then we get an interior point W where $f'(W) = 0$. The only remaining possibility is that the common endpoint value $f(a) = f(b)$ is both the maximum and minimum value of f on $[a, b]$. In this case, f is the constant function $f(x) = f(a)$, and $f'(x) = 0$ at *all* interior points.

Proof of the mean-value theorem. Determine the constant m so that the function

$$\phi(x) = f(x) - mx$$

has equal values at a and b (so that Rolle's Theorem can be applied):

$$f(a) - ma = f(b) - mb.$$

Solving for m gives

$$m = \frac{f(b) - f(a)}{b - a},$$

which is the secant slope, as of course it should be. Since then $\phi(a) = \phi(b)$, it follows from Rolle's Theorem that $\phi'(X) = 0$ for some point X interior to $[a, b]$. That is,

$$f'(X) = m = \frac{f(b) - f(a)}{b - a}. \quad \blacksquare$$

The refined form of the increasing-function theorem is an immediate corollary:

THEOREM 3   *If f is continuous on the interval I and if f' exists and is positive throughout the interior of I, then f is increasing on I.*

Proof. If x_1 and x_2 are any two distinct points of I then, by the mean-value theorem, there exists a point X strictly between x_1 and x_2 (and hence in the interior of I) such that

$$\frac{f(x_2) - f(x_1)}{x_2 - x_1} = f'(X) > 0.$$

Thus $f(x_2) - f(x_1)$ always has the same sign as $x_2 - x_1$, so f is increasing. $\quad \blacksquare$

THEOREM 4   ***The parametric mean-value theorem.*** *Let g and h be continuous on $[a, b]$ and differentiable at every interior point. Suppose that $g(a) \neq g(b)$, and that $g'(t)$ and $h'(t)$ are never simultaneously zero. Then there is at least one interior point X at which*

$$\frac{h'(X)}{g'(X)} = \frac{h(b) - h(a)}{g(b) - g(a)}.$$

Proof. The proof is essentially the same as for the mean-value theorem. We try to determine the constant m so that the function

$$\phi(x) = h(x) - mg(x)$$

has equal values at a and b:

$$h(a) - mg(a) = h(b) - mg(b).$$

When we solve this equation for m, we get

$$m = \frac{h(b) - h(a)}{g(b) - g(a)}$$

as the required value. This is possible because $g(a) \neq g(b)$ by hypothesis. Since now $\phi(a) = \phi(b)$, Rolle's Theorem guarantees the existence of a number X lying strictly between a and b such that $\phi'(X) = 0$. That is,

$$h'(X) = mg'(X).$$

If $g'(X) = 0$, then also $h'(X) = 0$, contradicting the hypothesis of the theorem. Therefore $g'(X) \neq 0$ and we can divide by it to get the equation of the theorem. ∎

Consider, finally, the zero-crossing principle for f'. It seems geometrically obvious that a graph can change from sloping upward to sloping downward only by passing through a point where its slope is zero. Now, try to examine your visualization of the situation to see *why* this conclusion is obvious. There seem to be two ways to think about it. First, one can visualize a varying tangent line and watch its inclination changing as the point of tangency moves along the curve. If it starts with a positive slope and rotates, as it slides along, to a position with negative slope, then it had to go through a position of zero slope. When we think this way, we are visualizing the tangent line changing position *continuously*, and our conclusion is that if $f'(x)$ changes continuously from a positive value to a negative value, then it has to go through a zero value in the process. So for a smooth function the underlying principle is the intermediate-value principle.

There is another way to think about what is happening. When we start with a graph sloping upward and end with the graph sloping downward it seems clear that somewhere in between the graph had to go over a highest point, and that at this peak point the slope must be zero.

Such reasoning can be developed into a formal proof of the zero-crossing property for f' from the extreme-value property for f. (See Problem 5.) This is a more general result than that given by our first line of argument (using the intermediate-value property of f'), since it doesn't require the continuity of f'. In practice, this is an empty generality, because in practice a differentiable function will always have a continuous derivative. But in theory, the stronger proof is needed; there *do exist* functions having discontinuous derivatives. (See Problem 4.)

We are now down to the intermediate-value and extreme-value principles for a continuous function. We might feel that these properties are inherently a part of our notion of continuity, and are less in need of analytic justification than the other properties we have discussed. Indeed, until about a hundred years ago they were so taken for granted that they weren't even separately formulated. To Newton and other early workers in calculus, a continuous graph was the path traced by a moving particle, and that such a graph had these properties went without saying.

Nevertheless, this was another case of equating intuitive plausibility with proof. Something was accepted as being true because it was unimaginable that it be false. And if we have agreed that a weak imagination is not a suitable basis for mathematical truth, then the intermediate- and extreme-value principles need analytic proof, just as everything else does.

These proofs were discovered only relatively recently (during the last hundred years or so). They turn out to involve more sophisticated techniques than are needed elsewhere in beginning calculus, and they really belong to the foundations of a larger discipline, called analysis, in which calculus is imbedded. The proofs are given in Appendix 4.

The critical nature of these two properties becomes clearer if we note that even one discontinuity may cause them both to fail. The function graphed below is continuous everywhere on $[0, 1]$ except at $x = 1/2$, but it has no maximum value, nor does it assume the intermediate value of $1/2$. Thus we have to know that f is continuous *at every single point* of the interval $[a, b]$ in order to apply these properties.

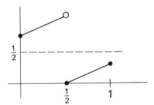

Moreover, we can't drop an endpoint from $[a, b]$ without losing, possibly, an extreme value. The function $f(x) = x$ loses its maximum value on $[0, 1]$ if we drop off $x = 1$, and it loses its minimum if we drop $x = 0$. It has neither a maximum value nor a minimum value on the *open* interval $(0, 1)$.

PROBLEMS FOR SECTION 1

Along with Theorem 3, there is of course its "change-of-sign" counterpart:

Theorem 3′. *If f is continuous on I and f′ is negative at every interior point of I, then f is decreasing on I.*

1. Prove Theorem 3′ from the mean-value theorem.

2. Prove Theorem 3′ directly from Theorem 3.

3. Prove the mean-value theorem from the zero-crossing property for $f′$ and the two monotone properties Theorems 3 and 3′. [*Hint:* Suppose there is no point X, as claimed in the mean-value theorem. Then concoct a function g whose derivative is never zero in $I = (a, b)$, and end up with a contradiction.]

4. Let f be the function defined as follows:

$$f(x) = x^2 \sin(1/x), \qquad \text{if } x \neq 0;$$
$$f(0) = 0.$$

a) Find the formula for $f'(x)$ when $x \neq 0$, by the differentiation rules.

b) Show that $f'(x)$ does not have a limit as x approaches 0.

c) Show that the difference quotient at 0,

$$\frac{f(0 + h) - f(0)}{h} = h \sin(1/h),$$

does have a limit as $h \to 0$ (use the squeeze limit law), and, consequently, that $f'(0)$ exists and is zero.

Thus, f is an everywhere differentiable function whose derivative f' is not a continuous function.

5. Prove the zero-crossing property of f' from the extreme-value property of f. [*Hint:* Suppose f' does change sign on I, and choose two points c, d, in I, where f' has opposite signs, with $c < d$. Apply Theorem 2 from Chapter 6 to show that if $f'(c) > 0 > f'(d)$, then the maximum value of f on $[c, d]$ cannot be assumed at either c or d. Now apply Theorem 3 in Chapter 6, etc.]

6. Earlier we came to the shape principle by combining the zero-crossing property for f' with Theorems 3 and 3'. Prove it now directly from the extreme-value property of a continuous function and Theorem 3, Chapter 6. Here is what you are to prove:

If f is continuous on an interval I and has no critical points in the interior of I, then f is monotone on I.

[*Hint:* Suppose f is not monotone. Then there must be three points $a < b < c$ in I such that $f(b)$ is either the maximum or the minimum of the values of f at these three points. Then $f'(x) = 0$ at some point between a and c.]

We know from algebra that a polynomial $p(x)$ of degree n,

$$p(x) = x^n + a_{n-1}x^{n-1} + \cdots + a_0,$$

has at most n real roots. That is, there are *at most* n distinct real numbers that are solutions of the equation $p(x) = 0$. There may very well be *fewer* than n. Thus, $p(x) = x^2 + 1$ is a polynomial of degree 2 having *no* real roots.

7. Prove that a polynomial of *odd degree* has at least one real root.

8. Call a polynomial p of degree n *simple* if it does have n distinct real roots. Prove that if p is simple, then so is its derivative $q = p'$.

2. MACLAURIN POLYNOMIALS

If a function f is the sum of a power series,

$$f(x) = a_0 + a_1 x + \cdots + a_n x^n + \cdots = \sum_0^\infty a_j x^j$$

on $(-r, r)$, then $f(x)$ is *approximated* by the nth partial sum

$$s(x) = a_0 + a_1 x + \cdots + a_n x^n.$$

We can get an idea of the nature of this approximation from the infinite series. We can factor out x^{n+1} from the remainder series $r(x) = f(x) - s(x)$,

$$
\begin{aligned}
r(x) &= a_{n+1}x^{n+1} + \cdots + a_{n+j}x^{n+j} + \cdots \\
&= x^{n+1}[a_{n+1} + a_{n+2}x + \cdots + a_{n+j}x^{j-1} + \cdots] \\
&= x^{n+1}h(x),
\end{aligned}
$$

where h is the sum of a power series converging on $(-r, r)$. In particular, h is continuous, and on any *closed* subinterval I, say $I = [-r/2, r/2]$, $|h|$ has a maximum value K. Thus

$$
|f(x) - s(x)| \le K|x|^{n+1}
$$

on I. This shows the general nature of the approximation of f by its partial-sum polynomial $s(x)$; as x becomes small, the error becomes small like a multiple of x^{n+1}. But unless we know that K is small, we can only be sure that this approximation is good locally around the origin, i.e., for sufficiently small x. Moreover, K cannot be small unless $f^{(n+1)}(0)$ is small, because

$$
K \ge |h(0)| = |a_{n+1}| = \frac{f^{(n+1)}(0)}{(n+1)!}.
$$

The polynomial $s(x)$ above is very special, being the nth partial sum of the Maclaurin series for $f(x)$. We shall call it the nth *Maclaurin polynomial* of $f(x)$, or the *Maclaurin polynomial of degree n*. Its formula is

$$
p_n(x) = f(0) + f'(0)x + \frac{f''(0)}{2} + \cdots + \frac{f^{(n)}(0)}{n!}x^n.
$$

EXAMPLE 1   The fourth Maclaurin polynomial of e^x is

$$
p(x) = 1 + x + \frac{x^2}{2} + \frac{x^3}{6} + \frac{x^4}{24}.
$$

EXAMPLE 2   The fifth Maclaurin polynomial of $\cos x$ is

$$
p(x) = 1 - \frac{x^2}{2!} + \frac{x^4}{4!}.
$$

In a power series the nth partial sum is generally understood to be the sum through the term a_nx^n, regardless of how many preceding coefficients a_n are zero.

We can also form Maclaurin polynomials for functions that are not sums of power series. All that is required is that the function in question have derivatives up through some order n around the origin.

EXAMPLE 3   The function

$$
f(x) = x^{7/2} + x^3 - 2x^2 + x + 3
$$

has derivatives through the order 3 around the origin, and you can check that its second-degree Maclaurin polynomial is

$$f(0) + \frac{f'(0)}{1} x + \frac{f''(0)}{2!} x^2 = 3 + x - 2x^2.$$

We could also form its third-degree Maclaurin polynomial. Note, however, that $f^{(4)}(0)$ does not exist, because

$$\frac{d^4 x^{7/2}}{dx^4} = \frac{7}{2} \cdot \frac{5}{2} \cdot \frac{3}{2} \cdot \frac{1}{2} x^{-1/2},$$

which is undefined at $x = 0$. Therefore, f does not have any further Maclaurin polynomials.

EXAMPLE 4 The function $\tan x$ can be proved to be the sum of its Maclaurin series near the origin, but we don't know any way of obtaining the series coefficients except to compute them by the Maclaurin formula. We can obtain the first few coefficients in this manner, but the differentiations required become more and more cumbersome, and after a while we give up. Here we shall compute the third Maclaurin polynomial. We have

$$f(0) = 0;$$

$$\frac{d}{dx} \tan x = \sec^2 x = \tan^2 x + 1, \qquad f'(0) = 1;$$

$$\frac{d^2}{dx^2} \tan x = 2 \tan x \sec^2 x$$
$$= 2 \tan^3 x + 2 \tan x, \qquad f''(0) = 0;$$

$$\frac{d^3}{dx^3} \tan x = 6 \tan^2 x \sec^2 x + 2 \sec^2 x$$
$$= 6 \tan^4 x + 8 \tan^2 x + 2, \qquad f'''(0) = 2.$$

The third Maclaurin polynomial of $\tan x$ is thus

$$p(x) = x + \frac{x^3}{3}.$$

It is also the fourth Maclaurin polynomial. (This can be proved without calculation from simple facts about odd and even functions.)

The question now is whether a Maclaurin polynomial of f *always* approximates f in the local way that we discovered above. That is, if f has derivatives through order n at the origin, so that we can form its nth-degree Maclaurin polynomial, p, does it necessarily follow that

$$|f(x) - p(x)| \le K|x|^{n+1}$$

on some interval I about the origin?

We start by noting one feature that persists in the general situation: the remainder $r(x) = f(x) - p(x)$ always has zero derivatives at the origin through order n. This is because a polynomial is its own Maclaurin polynomial.

LEMMA 1   *If*

$$p(x) = c_0 + c_1 x + \cdots + c_n x^n,$$

then

$$c_k = \frac{p^{(k)}(0)}{k!}$$

for $k = 0, \ldots, n$.

Proof. This is a special case of Theorem 14 in the last chapter, but with only a finite sum involved the proof in this case is independent of infinite series. We just compute the derivatives of $p(x)$ and see that the leading term (constant term) of the polynomial $p^{(k)}(x)$ is $k!\, c_k$. So $p^{(k)}(0) = k!\, c_k$. ∎

This lemma has the following consequence.

LEMMA 2   *If*

$$p(x) = a_0 + a_1 x + \cdots + a_n x^n$$

is a Maclaurin polynomial for $f(x)$, *and if we set* $r(x) = f(x) - p(x)$, *then*

$$0 = r(0) = r'(0) = \cdots = r^{(n)}(0).$$

That is, the nth Maclaurin polynomial of $f - p$ *is zero.*

Proof. By assumption, the derivatives of f exist up through the order n at the origin and the coefficients a_k of the polynomial p have the values

$$a_k = \frac{f^{(k)}(0)}{k!}$$

for $k = 0, 1, \ldots, n$. By Lemma 1 the coefficients a_k also have the values

$$a_k = \frac{p^{(k)}(0)}{k!},$$

for $k = 0, 1, \ldots, n$. Therefore, $p^{(k)}(0) = f^{(k)}(0)$ for $k = 0, 1, \ldots, n$, and since $r^{(k)}(x) = f^{(k)}(x) - p^{(k)}(x)$ for all the derivatives that exist, it follows that $r^{(k)}(0) = 0$ for $k = 0, 1, \ldots, n$. ∎

Our approximation question can now be rephrased: If $g(x)$ is a function defined on an interval I about the origin, and if the derivatives $g'(0)$, $g''(0), \ldots, g^{(n)}(0)$ all exist and all have the value 0, that is, if the nth Maclaurin polynomial of g exists and is identically zero, must it necessarily follow that

$$|g(x)| \le K |x|^{n+1},$$

i.e., that

$$\left| \frac{g(x)}{x^{n+1}} \right| \le K,$$

on some interval about the origin?

For the function f considered earlier we saw that K had to be at least as big as $|f^{(n+1)}(0)|/(n+1)!$, and this suggests that we are going to need the existence of one more derivative.

LEMMA 3   *Suppose that $g(x)$ has derivatives up through the order $n + 1$ in some interval I about the origin, and that*

$$0 = g(0) = g'(0) = \cdots = g^{(n)}(0).$$

Then for each x in I, there is a number X lying between 0 and x, such that

$$g(x) = \frac{g^{(n+1)}(X)}{(n + 1)!} x^{n+1}.$$

Proof. We apply the parametric mean-value theorem repeatedly to the quotient $g(x)/x^{n+1}$. First

$$\frac{g(x)}{x^{n+1}} = \frac{g(x) - g(0)}{x^{n+1} - 0} = \frac{g'(\alpha)}{(n + 1)\alpha^n}$$

for some α between 0 and x. Then

$$\frac{g'(\alpha)}{(n + 1)\alpha^n} = \frac{g''(\beta)}{(n + 1)n\beta^{n-1}}$$

for some point β between 0 and α. We can continue in this way right up through the nth derivative, because all the derivatives $g^{(k)}(0)$ are 0. Thus,

$$\frac{g(x)}{x^{n+1}} = \frac{g'(\alpha)}{(n + 1)\alpha^n} = \frac{g''(\beta)}{(n + 1)n\beta^{n-1}} = \cdots = \frac{g^{(n)}(\xi)}{(n + 1)!\,\xi} = \frac{g^{(n+1)}(X)}{(n + 1)!},$$

and this final value for the quotient gives us the lemma. ∎

Lemmas 2 and 3 combine as follows:

THEOREM 5   *If f has derivatives up through the order $n + 1$ in some interval I about the origin, and if $s(x)$ is the nth Maclaurin polynomial of f, then*

$$f(x) - s(x) = \frac{f^{(n+1)}(X)}{(n + 1)!} x^{n+1},$$

where X is a number depending on x and lying between 0 and x.

Proof.   Lemmas 2 and 3 show that if $r(x) = f(x) - s(x)$, then

$$r(x) = \frac{r^{(n+1)}(X)}{(n + 1)!} x^{n+1}.$$

Moreover, $r^{(n+1)} = f^{(n+1)}$, because $s(x)$ is a polynomial of degree at most n and so its $(n + 1)$st derivative is zero. The theorem follows. ∎

The expression of $f(x)$ in the form $f(x) = s(x) + r(x)$ now becomes a formula for $f(x)$ made up of its nth Maclaurin polynomial plus an explicit formula for the remainder $r(x)$:

$$f(x) = f(0) + \frac{f'(0)}{1} x + \frac{f''(0)}{2!} x^2 + \cdots + \frac{f^{(n)}(0)}{n!} x^n + \frac{f^{(n+1)}(X)}{(n + 1)!} x^{n+1}.$$

This is called the *Maclaurin formula with remainder.*

The remainder formula is very useful. For example, if we can determine that $|f^{(n+1)}(x)|$ assumes the maximum value M on the interval I, then

$$|f(x) - s(x)| \le \frac{M}{(n + 1)!} |x|^{n+1}$$

for all x in I, and we have the error-estimating inequality we were looking for.

Beyond this, if f is infinitely differentiable, then we can fix x and vary n, and possibly show that the nth remainder $r_n(x) = f(x) - s_n(x)$ approaches 0. We would then have another way of proving that f is the sum of its Maclaurin series.

EXAMPLE 5   For $f(x) = e^x$, the Maclaurin formula with remainder is

$$e^x = s_n(x) + r_n(x) = 1 + \frac{x}{1} + \frac{x^2}{2!} + \cdots + \frac{x^n}{n!} + e^X \frac{x^{n+1}}{(n+1)!}$$

for some X between 0 and x. If x is positive then $e^X \le e^x$, and

$$r_n(x) \le \frac{e^x x^{n+1}}{(n+1)!}.$$

(If x is negative, then $e^X \le 1$ and we omit the e^x factor.) We know that $x^n/n! \to 0$ as $n \to \infty$. Therefore $r_n(x) \to 0$ for all x. That is

$$s_n(x) = e^x - r_n(x) \to e^x,$$

so

$$e^x = \lim_{n \to \infty} s_n(x) = \lim_{n \to \infty} \sum_{k=0}^{n} \frac{x^k}{k!} = \sum_{k=0}^{\infty} \frac{x^k}{k!}.$$

We thus have a second proof that e^x is the sum of its Maclaurin series. However when x is positive the above estimate of the remainder $r_n(x)$ is not as good as the one we obtained in Chapter 11 by comparing the remainder series with the geometric series.

PROBLEMS FOR SECTION 2

Find the nth Maclaurin polynomial $p_n(x)$ of the function $f(x)$, where:

1. $f(x) = \arcsin x, \quad n = 3$ 2. $f(x) = \log(1 + x), \quad n = 3$

3. $f(x) = \sec x, \quad n = 3$ 4. $f(x) = \log(\cos x), \quad n = 4$

5. $f(x) = \sin\left(x + \frac{\pi}{4}\right), \quad n = 4$ 6. $f(x) = (1 + x)^a, \quad n = 4$

7. $f(x) = \tan x, \quad n = 4$

Find the nth Maclaurin formula with remainder, $f(x) = p_n(x) + r_n(x)$, for each function f below. Also determine a constant K such that

$$|f(x) - p_n(x)| \le Kx^{n+1}$$

on the interval $I = \left[-\frac{1}{2}, \frac{1}{2}\right]$.

8. $f(x) = \arcsin x, \quad n = 2$ 9. $f(x) = \sec x, \quad n = 2$

10. $f(x) = \tan x, \quad n = 2$ 11. $f(x) = e^{-x}, \quad n = 4$

12. $f(x) = \sin\left(x - \frac{\pi}{4}\right), \quad n = 4$ 13. $f(x) = (1 + x)^{3/2}, \quad n = 3$

14. $f(x) = \int_0^x e^{-t^2}\,dt, \quad n = 3$

15. Prove that every Maclaurin polynomial of an odd (even) function is odd (even).

16. The second Maclaurin polynomial of $f(x)$ is $2 + (x/2) - (x^2/8)$. Show that the second Maclaurin polynomial of $1/f(x)$ is

$$\frac{1}{2} - \frac{x}{8} + \frac{x^2}{16},$$

by computing the derivatives of $g = 1/f$ and evaluating them at the origin.

17. Let the function $g(x)$ in the above problem have the second Maclaurin polynomial $A + Bx + Cx^2$. Supposing that f and g both have continuous third derivatives, we can write

$$f(x) = 2 + \frac{x}{2} - \frac{x^2}{8} + x^3 F(x),$$

$$g(x) = A + Bx + Cx^2 + x^3 G(x),$$

where F and G are continuous. Use the fact that $f(x)g(x) = 1$ to solve for the coefficients A, B, and C.

18. The second Maclaurin polynomial of $f(x)$ is $1 + x + (x^2/6)$. Find the second Maclaurin polynomial of $g = 1/f$:

a) by the method of Problem 16;

b) by the method of Problem 17.

19. Find the third Maclaurin polynomial for $\tan x$ by the method of Problem 17, using this time the equation $\cos x \tan x = \sin x$ and the known polynomials for sine and cosine.

20. Prove that $\sin x$ is the sum of its Maclaurin series by using the Maclaurin formula with remainder, as in Example 5 in the text.

21. Prove that $1/(1 - x)$ is the sum of its Maclaurin series over the interval $(-1, 1/2)$ by using the Maclaurin formula with remainder.

22. Suppose that f has derivatives up through order $n - 1$ on an interval I about the origin, with

$$f(0) = f'(0) = \cdots = f^{(n-1)}(0) = 0.$$

Suppose furthermore that $f^{(n-1)}(x)$ is differentiable at $x = 0$, so that $f^{(n)}(0)$ exists. Show that $f(x)/x^n \to f^{(n)}(0)/n!$ as $x \to 0$. (Apply l'Hôpital's rule.)

23. If f has derivatives of order $n - 1$ on an interval about the origin and if $f^{(n-1)}(x)$ is differentiable at $x = 0$, then

$$f(0), \qquad f'(0), \qquad \cdots, \qquad f^{(n)}(0)$$

all exist, and we can write down the nth Maclaurin polynomial $p_n(x)$ for f. Show that

$$\lim_{x\to 0} \frac{f(x) - p_n(x)}{x^n} = 0.$$

(Assume and apply the conclusion in the preceding problem.)

24. Let p and q be two polynomials of degree n at most, and suppose each of them is an nth-order approximation of f around the origin, in the sense that

$$\lim_{x\to 0} \frac{f(x) - p(x)}{x^n} = 0,$$

$$\lim_{x\to 0} \frac{f(x) - q(x)}{x^n} = 0.$$

Show that the two polynomials p and q must necessarily be the same. In view of the above problem, we see that this property uniquely characterizes the nth Maclaurin polynomial of f.

25. Use Problem 24 to prove that every Maclaurin polynomial of an even (odd) function is even (odd).

3. TAYLOR POLYNOMIALS

There are uniquely determined polynomials that approximate f locally about the point $x = a$ in exactly the manner that the Maclaurin polynomials approximate f about the origin. The polynomials are now written in the form

$$p(x) = c_0 + c_1(x - a) + \cdots + c_n(x - a)^n,$$

because then the coefficients c_k have the same derivative formulas as before, except that now they involve the derivative of f at $x = a$:

$$c_k = \frac{f^{(k)}(a)}{k!}.$$

These more general polynomials are called *Taylor polynomials*. Thus

$$s_n(x) = f(a) + \frac{f'(a)}{1}(x - a) + \cdots + \frac{f^{(n)}(a)}{n!}(x - a)^n,$$

is the nth *Taylor polynomial for f at $x = a$*. It can be formed for any function having n derivatives at $x = a$.

EXAMPLE 1   The third Taylor polynomial for $f(x) = \log x$ at $x = 1$ is

$$p(x) = f(1) + \frac{f'(1)}{1}(x - 1) + \frac{f''(1)}{2!}(x - 1)^2 + \frac{f'''(1)}{3!}(x - 1)^3$$

$$= 0 + \frac{1}{1}(1 - x) + \frac{(-1)}{2!}(x - 1)^2 + \frac{2}{3!}(x - 1)^3$$

$$= (x - 1) - \frac{1}{2}(x - 1)^2 + \frac{1}{3}(x - 1)^3.$$

EXAMPLE 2   The second Taylor polynomial for $f(x) = \sin x$ around $a = \pi/4$ is

$$f(\pi/4) + \frac{f'(\pi/4)}{1}(x - \pi/4) + \frac{f''(\pi/4)}{2!}(x - \pi/4)^2$$

$$= \sin(\pi/4) + [\cos(\pi/4)](x - \pi/4) - [\sin(\pi/4)](x - \pi/4)^2/2$$

$$= \frac{\sqrt{2}}{2}\left[1 + \left(x - \frac{\pi}{4}\right) - \frac{(x - \pi/4)^2}{2}\right].$$

The results and proofs for Taylor polynomials repeat the earlier treatment of Maclaurin polynomials practically verbatim. First is the fact that a polynomial is equal to its Taylor polynomial of the same degree. This requires the preliminary observation that a polynomial $p(x)$ can always be rewritten as a polynomial in $(x - a)$. We simply set $x = [(x - a) + a]$ and expand all powers of x. For example,

$$
\begin{aligned}
x^3 - x &= [(x - 2) + 2]^3 - [(x - 2) + 2] \\
&= [(x - 2)^3 + 6(x - 2)^2 + 12(x - 2) + 8] - [(x - 2) + 2] \\
&= (x - 2)^3 + 6(x - 2)^2 + 11(x - 2) + 6.
\end{aligned}
$$

This new form for $p(x)$ is called its *expansion about the point $x = a$*.

We now have the following results, with the same proofs as before.

LEMMA 4 *If*

$$
p(x) = c_0 + c_1(x - a) + \cdots + c_n(x - a)^n,
$$

then

$$
c_k = \frac{p^{(k)}(a)}{k!}
$$

for $k = 0, 1, \ldots, n$.

LEMMA 5 *If*

$$
p(x) = c_0 + c_1(x - a) + \cdots + c_n(x - a)^n
$$

is the Taylor polynomial for f at $x = a$, and if we set $r(x) = f(x) - p(x)$, then

$$
0 = r(a) = r'(a) = \cdots = r^{(n)}(a).
$$

That is, the nth Taylor polynomial of $f(x) - p(x)$ at a is 0.

LEMMA 6 *If $g(x)$ has derivatives through the order $n + 1$ in some interval I about $x = a$, and if*

$$
g(a) = g'(a) = \cdots = g^{(n)}(a) = 0,
$$

then for each x in I there is a number X lying between a and x such that

$$
g(x) = \frac{g^{(n+1)}(X)}{(n + 1)!} (x - a)^{n+1}.
$$

THEOREM 6 *If f has derivatives through the order $n + 1$ in some interval I about $x = a$, then*

$$
f(x) = \sum_{k=0}^{n} \frac{f^{(k)}(a)}{k!} (x - a)^k + \frac{f^{(n+1)}(X)}{(n + 1)!} (x - a)^{n+1},
$$

where X is some number depending on x and lying between a and x.

The identity in the theorem, expressing f as the sum

$$
f(x) = s_n(x) + r_n(x)
$$

of its nth Taylor polynomial at $x = a$ plus the associated remainder term, is called the *Taylor formula with remainder*.

EXAMPLE 3   The polynomial in Example 2 approximates sin x with the error (remainder)

$$\frac{-\cos(X)}{3!}\left(x - \frac{\pi}{4}\right)^3,$$

where X is some number between x and $\pi/4$. Since $|\cos X| \leq 1$, the error is at most

$$\frac{\left|x - \dfrac{\pi}{4}\right|^3}{3!}.$$

For values of x near $\pi/4$, this approximation to sin x will have an accuracy that could be obtained from the Maclaurin series only by using many more terms. However, this advantage is offset in part by the fact that here we have to contend with decimal approximations to $\pi/4$.

EXAMPLE 4   The Taylor remainder formula shows again that any polynomial $p(x)$ of degree n is equal to its own nth Taylor polynomial at any point $x = a$ because the remainder is then 0. Thus, if $p(x) = x^3 - x$, then

$$p'(x) = 3x^2 - 1, \qquad p''(x) = 6x, \qquad p'''(x) = 6;$$
$$p(2) = 6, \qquad p'(2) = 11, \qquad p''(2) = 12, \qquad p'''(2) = 6;$$
$$p(x) = 6 + \frac{11}{1}(x - 2) + \frac{12}{2!}(x - 2)^2 + \frac{6}{3!}(x - 2)^3.$$

Therefore,

$$x^3 - x = 6 + 11(x - 2) + 6(x - 2)^2 + (x - 2)^3.$$

The whole theory of Taylor polynomial approximations and Taylor series can be reduced to the corresponding Maclaurin theory by a simple device. We just define a new function g from the function f by setting

$$g(t) = f(t + a).$$

What this does in effect is to slide, or translate, the graph of f to a new position to obtain the graph of g. The equation $g(t) = f(t + a)$ says that the behavior of g around $t = 0$ is identical to the behavior of f around $t = a$.

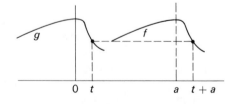

Here is how this works out for infinite series. Suppose that g is the sum of a power series,

$$g(t) = c_0 + c_1 t + c_2 t^2 + \cdots + c_n t^n + \cdots$$

on an interval $(-c, c)$, and suppose that we set $t = x - a$ and $f(x) = g(x - a)$. Then the above series turns into a series expansion of $f(x)$:

$$f(x) = c_0 + c_1(x - a) + c_2(x - a)^2 + \cdots + c_n(x - a)^n + \cdots.$$

As t runs across the interval $(-c, c)$, then $x = t + a$ runs across $(a - c, a + c)$. For any x in this interval, the series $\sum c_n(x - a)^n$ is the same as the series $\sum c_n t^n$ for $t = x - a$ in the interval $(-c, c)$, Thus, $f(x)$ is the sum of the x series on $(a - c, a + c)$.

Conversely, if we can express a function $f(x)$ as the sum of a convergent power series of the form

$$f(x) = c_0 + c_1(x - a) + \cdots + c_n(x - a)^n + \cdots$$

on an interval centered at $x = a$, say $J = (a - c, a + c)$, then we have

$$f(t + a) = c_0 + c_1 t + \cdots + c_n t^n + \cdots$$

for t in the interval $I = (-c, c)$, and this is the Maclaurin series for $g(t) = f(t + a)$ there.

The coefficients c_n are, of course, the Maclaurin coefficients for the function g,

$$c_n = \frac{g^{(n)}(0)}{n!},$$

but they now can be reevaluated in terms of the derivatives of f at the point $x = a$. This follows from the relationship $f(x) = g(x - a)$. Differentiating repeatedly, we see that $f'(x) = g'(x - a)$, $f''(x) = g''(x - a)$, etc., and therefore

$$f'(a) = g'(0), \qquad f''(a) = g''(0), \qquad \text{etc.}$$

Thus, $f^{(n)}(a) = g^{(n)}(0)$ for all n, and so

$$c_n = \frac{f^{(n)}(a)}{n!}$$

for all n. The series for f can thus be rewritten

$$f(x) = f(a) + \frac{f'(a)}{1!}(x - a) + \frac{f''(a)}{2!}(x - a)^2 + \cdots + \frac{f^{(n)}(a)}{n!}(x - a)^n + \cdots$$

This is called the *Taylor series for f around x = a*.

A function f that is infinitely differentiable on an interval $(a - c, a + c)$ may or may not be the sum of its Taylor series there, just as a function g that is infinitely differentiable on $(-c, c)$ may or may not be the sum of its Maclaurin series.

In order to prove that an infinitely differentiable function f is the sum of its Taylor series at $x = a$, we can try the "initial-value problem" approach that we used in Section 8 of Chapter 11, or we can try to show that the remainder approaches 0 in the Taylor formula with remainder. Nothing here is new, since the Taylor series for f is equivalent to the Maclaurin series for $g(t) = f(t + a)$.

PROBLEMS FOR SECTION 3

Find the nth Taylor polynomial of the function f around the point a, where:

1. $f(x) = \log x$, $n = 3$, $a = 1$
2. $f(x) = \sin x$, $n = 4$, $a = \pi/4$
3. $f(x) = \sin x$, $n = 4$, $a = \pi/6$
4. $f(x) = \sin x$, $n = 4$, $a = \pi/3$
5. $f(x) = e^x$, $n = 3$, $a = a$
6. $f(x) = \tan x$, $n = 3$, $a = \pi/4$

Find the nth Taylor formula with remainder, $f(x) = p_n(x) + r_n(x)$, about the point $x = a$, where:

7. $f(x) = \cos x$, $n = 4$, $a = \pi/4$
8. $f(x) = \cos x$, $n = 4$, $a = \pi/3$
9. $f(x) = \sin x$, $n = 4$, $a = \pi/6$
10. $f(x) = e^x$, $n = n$, $a = a$
11. $f(x) = x^{1/3}$, $n = 3$, $a = 1$
12. $f(x) = x^{-1/4}$, $n = 3$, $a = 1$
13. $f(x) = x^r$, $n = 3$, $a = 1$
14. $f(x) = \tan x$, $n = 2$, $a = \pi/4$
15. $f(x) = \log \cos x$, $n = 3$, $a = \pi/4$

Write out the Taylor series for:

16. e^x about $x = a$
17. $\sqrt{x}$ about $x = 1$
18. $1/x$ about $x = 1$
19. $\sin x$ about $x = \pi/2$
20. $\log x$ about $x = 1$
21. $\sin x$ about $x = \pi$
22. $\sin x$ about $x = \pi/4$

23. Carry out the details of the calculation of the Taylor remainder formula from the Maclaurin remainder formula.

24. Show that

$$\frac{\sqrt{2}}{2}\left[1 - \frac{\pi}{20} - \frac{\pi^2}{800}\right]$$

approximates $\sin \pi/5$ with an error less than 0.001.

25. Express $p(x) = x^3 + 2x^2 + 1$ as its Taylor polynomial about $a = 1$.

26. Express $p(x) = x^4 - 5x^3 + 10x$ as its Taylor polynomial about $a = 2$.

27. Prove that $\sin x$ is the sum of its Taylor series about $\pi/2$ by using the Taylor formula with remainder and proving that the remainder goes to zero.

28. Show that the Taylor series for $\sin x$ about $a = \pi/6$ expresses the identity

$$\sin x = \sin\left(x - \frac{\pi}{6}\right)\cos\frac{\pi}{6} + \cos\left(x - \frac{\pi}{6}\right)\sin\frac{\pi}{6}.$$

29. Prove that e^x is the sum of its Taylor series about $x = a$, by evaluating the remainder and showing that it goes to zero.

4. THE MODIFIED TRAPEZOIDAL RULE

In this section we shall go back and pick up the estimation of definite integrals where we abandoned it at the end of Chapter 9. The trapezoidal rule can be vastly improved by including a correction that takes into account the exact directions of the graph at the two endpoints. Here is the result.

The modified trapezoidal rule. *If the fourth derivative $f^{(4)}$ is continuous and bounded by K on the interval $[a, b]$, then*

$$\int_a^b f \approx \text{trapezoidal estimate} - (\Delta x)^2 \frac{f'(b) - f'(a)}{12},$$

with an error at most

$$\frac{K(b-a)}{720}(\Delta x)^4.$$

Let us see how the modified formula improves things with our test calculation of $\int_1^{3/2} dx/x = \log(3/2)$. We use the same four-fold subdivision of $[1, (3/2)]$, but now add the endpoint-derivative correction to the trapezoidal estimate that we looked at earlier. Then

$$\log\frac{3}{2} = \int_1^{3/2}\frac{dx}{x} \approx \frac{1}{8}\left[\frac{1}{2} + \frac{8}{9} + \frac{4}{5} + \frac{8}{11} + \frac{1}{3}\right] - \frac{1}{64}\left(\frac{1 - 4/9}{12}\right)$$

$$= \left\{\frac{1}{16} + \frac{1}{9} + \frac{1}{10} + \frac{1}{11} + \frac{1}{24}\right\} - \frac{5}{6912}$$

$$= +\begin{cases} 0.06250000\cdots \\ 0.11111111\cdots \\ 0.10000000\cdots \\ 0.09090909\cdots \\ 0.04166666\cdots \end{cases}$$

$$\begin{aligned} &\quad 0.40618686\cdots \\ &-0.00072338\cdots \end{aligned}$$

$$= \quad 0.4054634\cdots.$$

Here $f^{(4)}(x) = 24/x^5$, which is bounded by $K = 24$ over the interval $[1, 3/2]$. The error in the improved estimate is therefore at most

$$\frac{K(b-a)}{720}(\Delta x)^4 = \frac{24}{720}\cdot\frac{1}{2}\cdot\left(\frac{1}{8}\right)^4 = \frac{1}{245{,}760} < \frac{10^{-5}}{2} = 0.000005.$$

Thus, we have calculated that

$$\log 3/2 = 0.405463\cdots$$

with an error less than 0.000005 in magnitude, and have five-place accuracy. (However, we don't know whether 0.40546 or 0.40547 is the *closest* five-place decimal. Five-place tables give the latter.)

We give the proof in order to show an example of calculus in action in a more complex situation than we have met so far. The proof is long and intricate, and requires the combination of several techniques. Key roles are played by results given in problems, so these problems constitute part of the total proof.

LEMMA 7 *If q is a polynomial of degree at most two, then*

$$q(b) - q(a) = \frac{q'(b) + q'(a)}{2}(b - a)$$

for any two points a and b.

Proof. If $q(x) = Ax^2 + Bx + C$, then

$$
\begin{aligned}
q(b) - a(a) &= A(b^2 - a^2) + B(b - a) \\
&= [A(b + a) + B](b - a) \\
&= \frac{[(2Ab + B) + (2Aa + B)]}{2}(b - a) \\
&= \frac{q'(b) + q'(a)}{2}(b - a). \quad \blacksquare
\end{aligned}
$$

There are similar (but more complicated) identities for higher-degree polynomials. We shall need the next one. Note that it includes Lemma 7 as a special case.

LEMMA 8   *If p is a polynomial of degree at most four, then*

$$
p(b) - p(a) = \frac{p'(b) + p'(a)}{2}(b - a) - \frac{p''(b) - p''(a)}{12}(b - a)^2
$$

for any two points a and b.

Proof. We can write p as its own Taylor polynomial at a and also as its Taylor polynomial at b. Thus,

$$
\begin{aligned}
p(x) = p(a) + p'(a)(x - a) + \frac{p''(a)}{2}(x - a)^2 + \frac{p'''(a)}{6}(x - a)^3 \\
+ \frac{p^{(4)}(a)}{24}(x - a)^4,
\end{aligned}
$$

$$
\begin{aligned}
p(x) = p(b) + p'(b)(x - b) + \frac{p''(b)}{2}(x - b)^2 + \frac{p'''(b)}{6}(x - b)^3 \\
+ \frac{p^{(4)}(b)}{24}(x - b)^4.
\end{aligned}
$$

Now evaluate the first equation at $x = b$, the second at $x = a$, and subtract the second from the first. The resulting equation can be written

$$
\begin{aligned}
2[p(b) - p(a)] = [p'(b) + p'(a)](b - a) - \frac{p''(b) - p''(a)}{2}(b - a)^2 \\
+ \frac{p'''(b) + p'''(a)}{6}(b - a)^3.
\end{aligned}
$$

The fourth-degree terms cancel because the fourth derivative of p is a constant. We can now replace the third-order term by a second-order term, by applying Lemma 7 to the second-degree polynomial $q(x) = p''(x)$. The new numerical coefficient on the second-order term is

$$
-\left(\frac{1}{2}\right) + \left(\frac{1}{3}\right) = -\frac{1}{6},
$$

and the lemma now follows upon dividing by 2. $\blacksquare$

LEMMA 9   *If the fourth derivative of f is continuous and is bounded by K over the interval $[a, b]$, then the difference between $\int_a^b f$ and the number*

$$
\frac{1}{2}(b - a)(f(b) + f(a)) - \frac{1}{12}(b - a)^2(f'(b) - f'(a))
$$

is at most

$$\frac{K}{720}(b - a)^5$$

in magnitude.

Proof. Let $p(x)$ be the polynomial of degree at most 3 determined by the conditions

$$p(a) = f(a), \qquad p(b) = f(b), \qquad p'(a) = f'(a), \qquad p'(b) = f'(b).$$

Problems 4 and 5 develop the existence and uniqueness of such a polynomial p. Then the difference $g(x) = f(x) - p(x)$ is zero at a and b, as is its derivative $g'(x)$. Moreover, $g^{(4)}(x) = f^{(4)}(x)$, since $p^{(4)} = 0$, so $g^{(4)}$ is bounded by K on the interval $[a, b]$. It follows from Extra Problem 40 of Chapter 9 that

$$\left| \int_a^b g \right| \le \frac{K}{720}(b - a)^5.$$

But

$$\int_a^b g = \int_a^b f - \int_a^b p$$

$$= \int_a^b f - \left[\frac{1}{2}(b - a)(p(b) + p(a)) - \frac{1}{12}(b - a)^2(p'(b) - p'(a)) \right]$$

by Lemma 8. (Note that if $q(x)$ is an antiderivative of $p(x)$, then q is a polynomial of degree at most 4.) Since $p(b) = f(b), \ldots, p'(a) = f'(a)$, we end up with the assertion of Lemma 9. ∎

Proof of modified rule. Let the points $a = x_0, x_1, \ldots, x_n = b$ subdivide the interval $[a, b]$ into n equal subintervals of common length $\Delta x = (b - a)/n$. We apply Lemma 7 to each subinterval, and have the approximations:

$$\int_{x_{n-1}}^{x_n} f \approx \Delta x \frac{f(x_n) + f(x_{n-1})}{2} - (\Delta x)^2 \frac{f'(x_n) - f'(x_{n-1})}{12},$$

$$\vdots$$

$$\int_{x_1}^{x_2} f \approx \Delta x \frac{f(x_2) + f(x_1)}{2} - (\Delta x)^2 \frac{f'(x_2) - f'(x_1)}{12},$$

$$\int_{x_0}^{x_1} f \approx \Delta x \frac{f(x_1) + f(x_0)}{2} - (\Delta x)^2 \frac{f'(x_1) - f'(x_0)}{12},$$

where each approximation is in error by at most

$$\frac{K}{720}(\Delta x)^5.$$

Now we add these approximate equalities. The sum of the integrals is $\int_{x_0}^{x_n} f = \int_a^b f$. The sum of the first terms on the right is the trapezoidal approximation to $\int_a^b f$. Finally, the sum of the second terms is

$$- (\Delta x)^2 \frac{f'(x_n) - f'(x_0)}{12} = - (\Delta x)^2 \frac{f'(b) - f'(a)}{12},$$

since all the intermediate evaluations $f'(x_k)$ cancel out.

We thus end up with

$$\int_a^b f \approx \text{trapezoidal estimate} - (\Delta x)^2 \frac{f'(b) - f'(a)}{12}.$$

Since the error at each step is at most

$$\frac{K}{720}(\Delta x)^5,$$

the total error is at most

$$n\frac{K}{720}(\Delta x)^5 = \frac{K(b-a)}{720}(\Delta x)^4,$$

where we have again used the fact that $n\Delta x = b - a$. This proves the modified rule. ∎

PROBLEMS FOR SECTION 4

1. What would the error be in the estimate of $\log 2$ provided by the modified trapezoidal rule if $[1, 2]$ were divided into 10 subintervals? 20 subintervals?

2. Estimate $\log 2$ by the modified trapezoidal rule, using a six-fold subdivision of $[1, 2]$. Compute the error estimate as the first step, and use it as a guide in deciding how far to carry out the decimal expansions.

3. Carry out the details of the proof of Lemma 8.

4. a) Show that the most general polynomial p of degree at most 3, such that $p(a) = p'(a) = 0$, can be written

   $$p(x) = A(x - a)^3 + B(x - a)^2.$$

   [*Hint:* Write p as its own Taylor polynomial at a.]

   b) Show that if $b \neq a$, then we can arrange to give $p(b)$ and $p'(b)$ any values we wish by choosing the coefficients A and B suitably. Show, in fact, that if C and D are any numbers, then the equations

   $$p(b) = C, \qquad p'(b) = D$$

   can be *uniquely* solved for A and B in terms of C and D.

5. Assuming the conclusions of the preceding problem, show that we can find a unique polynomial $p(x)$, of degree at most 3, such that

   $$p(a), \quad p'(a), \quad p(b), \quad \text{and} \quad p'(b)$$

   have any four assigned values.

6. Strengthen the modified trapezoidal rule by showing that if the fourth derivative $f^{(4)}$ does not change sign on $[a, b]$, then the error has the sign of $f^{(4)}$. (Look back at Chapter 9, Extra Problem 40. It gives more information than we have used.)

   Now look again at the $\log(3/2)$ example in the text and show, furthermore, that the decimal expansion of $\log(3/2)$ begins $0.40546\cdots$.

7. Show that the modified trapezoidal estimate gives the exact answer if f is a polynomial of degree at most 3.

8. Show that the error is exactly $(\Delta x)^4(b - a)/720$ times the constant value of $f^{(4)}(x)$ if f is a fourth-degree polynomial. (This will require the more precise result in Extra Problem 40, Chapter 9.)

9. Assume the above problem, and apply the modified trapezoidal rule to a fourth-degree polynomial using the trivial subdivision of $[a, b]$ (no subdividing points). State the result as a lemma about fifth-degree polynomials, analogous to Lemma 8.

The identity

$$\pi = 4 \arctan 1 = 4 \int_0^1 \frac{dx}{1 + x^2}$$

can be used to calculate π. In the following problems we consider the ten-fold subdivision of $[0, 1]$ by the one-place decimals $0.0, 0.1, \ldots, 0.9, 1.0$. The calculation of the higher derivatives of $f(x) = 1/(1 + x^2)$ becomes rather complicated, and you may assume that $|f^{(4)}(x)| \leq 24$ on the interval $[0, 1]$. The correct beginning of the decimal expansion of π is

$$\pi = 3.141592653589 \cdots.$$

10. Show that the error in the approximation of $\pi = 4 \int_0^1 dx/(1 + x^2)$ is:

   a) less than 0.0067 when the unmodified trapezoidal sum is used (Chapter 9, Section 10);

   b) less than $0.000014 = 1.4(10)^{-5}$ when the modified trapezoidal rule is used.

A modern desk calculator gave the following values for $1/(1 + x^2)$ at the subdividing points:

1/1.01	0 . 9 9 0 0 9 9 0 0 9 9 0 0
1/1.04	0 . 9 6 1 5 3 8 4 6 1 5 3 8
1/1.09	0 . 9 1 7 4 3 1 1 9 2 6 6 0
1/1.16	0 . 8 6 2 0 6 8 9 6 5 5 1 7
1/1.25	0 . 8 0 0 0 0 0 0 0 0 0 0 0
1/1.36	0 . 7 3 5 2 9 4 1 1 7 6 4 7
1/1.49	0 . 6 7 1 1 4 0 9 3 9 5 9 7
1/1.64	0 . 6 0 9 7 5 6 0 9 7 5 6 0
1/1.81	0 . 5 5 2 4 8 6 1 8 7 8 4 5

It is clear from the top line that the calculator did not round off at the last place, and the above values are therefore the twelfth-place truncations in the decimal expansions. So each approximates its entry with an error lying between 0 and 10^{-12}. Note that one row is exact.

11. Compute the sum of the above twelfth-place decimals, and modify it to correspond to the trapezoidal sum. (Two entries are missing, as well as the factor Δx.) Then multiply by 4. The result is an approximation of π. Show that the error is approximately 0.0017. Show also that this error is due almost entirely to the trapezoidal sum, and not to the above approximation of this sum. That is, show that the error in your approximation of the exact trapezoidal sum is insignificant as compared to the error in the trapezoidal sum itself.

12. Now modify the approximation by the endpoint-derivative correction term, and compare with the expansion of π. Your answer should be correct through eight decimal places.

13. The actual estimate of π obtained above is much better than theoretically predicted error (Problem 10). Go over the proof of the modified trapezoidal rule with the present case in mind and try to see why this might be.

5. SIMPSON'S RULE

Simpson's rule is a good compromise between accuracy and simplicity. It is not as accurate as the modified trapezoidal rule, by a factor of 4, but it avoids the two endpoint-derivative evaluations. The basic estimate depends on a rather complicated integration by parts and an approximation of f by a polynomial p, similar to the derivations in the last section but less straightforward. This time p is the quadratic polynomial having the same values as f at three equally spaced points. We shall omit the proof.

The basic estimate. *The "weighted" Riemann-sum estimate*

$$\int_{a-h}^{a+h} f(x)\,dx \approx \frac{h}{3}[f(a-h) + 4f(a) + f(a+h)]$$

is accurate to within an error of

$$\frac{K}{90}h^5,$$

if the fourth derivative of f exists and is bounded by K over the interval $[a-h, a+h]$.

Simpson's rule. *If the points $a = x_0, x_1, \ldots, x_n = b$ subdivide the interval $[a, b]$ into an even number of equal subintervals of length $\Delta x = (b-a)/n$ each, then*

$$\int_a^b f \approx \frac{\Delta x}{3}[f(x_0) + 4f(x_1) + 2f(x_2) + 4f(x_3) + \cdots + 4f(x_{n-1}) + f(x_n)].$$

If $f^{(4)}(x)$ is bounded by K over the interval $[a, b]$, then the error in the approximation is at most

$$\frac{K(b-a)}{180}(\Delta x)^4.$$

Note that the coefficient pattern in this estimating sum, including the outside factor $(1/3)$, is

$$\frac{1}{3}, \frac{4}{3}, \frac{2}{3}, \frac{4}{3}, \frac{2}{3}, \frac{4}{3}, \ldots, \frac{2}{3}, \frac{4}{3}, \frac{1}{3}.$$

Since there are $n + 1$ terms, the sum of these coefficients is n, the number of intervals in the subdivision. The right side is a "weighted" average of the values $f(x_i)$, where $i = 0, 1, \ldots, n$.

The proof of Simpson's rule simply adds up the three-point basic estimates for the $n/2$ successive blocks of two subintervals each in the given subdivision.

EXAMPLE 1   We consider our canonical $\log(3/2)$ problem for the last time. Now the estimate is

$$\log\frac{3}{2} = \int_1^{3/2} \frac{dt}{t} \approx \frac{\Delta x}{3}\left[\frac{1}{x_0} + \frac{4}{x_1} + \frac{2}{x_2} + \frac{4}{x_3} + \frac{1}{x_4}\right]$$

$$= \frac{1}{24}\left[1 + \frac{32}{9} + \frac{8}{5} + \frac{32}{11} + \frac{2}{3}\right]$$

$$= \frac{1}{24} + \frac{4}{27} + \frac{1}{15} + \frac{4}{33} + \frac{1}{36}$$

$$= \begin{cases} 0.0416666 \cdots \\ 0.1481481 \cdots \\ 0.0666666 \cdots \\ 0.1212121 \cdots \\ \underline{0.0277777 \cdots} \\ 0.4054711 \cdots . \end{cases}$$

Since $f^{(4)}(x) = 24/x^5$ is bounded by $K = 24$ over the interval $\left[1, \frac{3}{2}\right]$, the error in the above estimate is at most

$$\frac{K(b-a)}{180}(\Delta x)^4 = \frac{24}{180} \cdot \frac{1}{2} \cdot \left(\frac{1}{8}\right)^4$$

$$= \frac{1}{61{,}440} < \frac{1}{6}(10)^{-4} < 0.000017.$$

Thus,

$$0.40545 < \log(3/2) < 0.40549$$

and

$$\log \frac{3}{2} \approx 0.4055$$

to the nearest four decimal places.

PROBLEMS FOR SECTION 5

1. Show that Simpson's rule is exact for a polynomial of degree 3 or less.

2. Apply Simpson's rule to $f(x) = x^4$ over the interval $[-a, a]$ with $\Delta x = a$. Show that the error is exactly

$$\frac{K(b-a)}{180}(\Delta x)^4$$

in magnitude, where K is the constant value of $f^{(4)}$. This shows that the error formula cannot be improved.

3. Estimate $\pi = 4 \int_0^1 dx/(1 + x^2)$ by applying Simpson's rule to the data provided for Problem 11, Section 4. This is just a heavy dose of arithmetic, but the use of calculus in applied problems generally ends this way. Your error should be approximately $2.6(10)^{-5}$.

4. Now compute the expected error from the statement of Simpson's rule and compare with the actual error obtained above.

13
Vectors in the Plane and in Space

The interaction between vectors and calculus leads to a huge increase in the range of problems to which calculus can be successfully applied, and such vector calculus is one of the principal topics of advanced courses in calculus. In this book we can only introduce the subject and give a few applications.

This chapter develops the algebra of vectors in the plane and in space, along with its geometric interpretations. Chapter 14 then treats the differential calculus for vector functions of a parameter t, and Chapter 16 takes up vector interpretations of differentiation that arise in the study of functions of two or three independent variables. Chapter 15 is an introduction to such functions, containing no vector material.

1. VECTORS IN THE PLANE

A *vector quantity* is any quantity having both a magnitude and a direction.

EXAMPLE 1   The vector velocity of a particle moving in the plane (or in space) is specified by the direction in which the particle is moving and by its speed (the magnitude of its velocity).

EXAMPLE 2   Forces are vector quantities. For example, the gravitational force exerted by the moon on a circling rocket is directed toward the center of the moon, and its magnitude is proportional to $1/d^2$, where d is the distance from the rocket to the center of the moon.

EXAMPLE 3   In the plane (or in space) a directed line segment, or *arrow*, is a vector quantity, since it has both a direction and a magnitude (length).

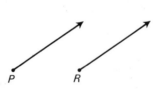

Any vector can be represented geometrically by an arrow. Thus a force applied to an object at a point P can be represented by an arrow stemming from P in the direction of the force, the magnitude of the force being given by the arrow length. The same force, applied to an object at a point R, would then be represented by an arrow having the same length and direction but stemming from R. In other words, two *different* arrows represent the *same* vector force if they have the same direction and length.

The velocity of a moving object can be represented by an arrow in the same way, and in the same way two different objects having the same vector velocity call for two different arrows having the same direction and length.

We use the notation $\overrightarrow{PQ}$ for the directed line segment (arrow) *from P to Q*. Two arrows that represent the same vector are said to be equal. Thus the equation

$$\overrightarrow{PQ} = \overrightarrow{RS}$$

means that the two arrows have the same length and direction; although different as directed line segments, they are the same as vectors.

The situation is exactly like the representation of rational numbers by fractions. We write

$$\frac{4}{6} = \frac{2}{3},$$

meaning that the two fractions represent the same rational numbers; although different as fractions they are the same as rational numbers. And just as we refer to the *number* 2/3, meaning the number represented by the fraction, so shall we refer to the *vector* $\overrightarrow{PQ}$, meaning the vector represented by the arrow.

Equality of arrows is preserved under various operations. For example, it follows from the geometry of parallelism that equal arrows have equal projections on a line l:

I) If $\overrightarrow{PQ} = \overrightarrow{AB}$, and if P', etc., are the points on a line l obtained by dropping perpendiculars from P, etc., then $\overrightarrow{P'Q'} = \overrightarrow{A'B'}$.

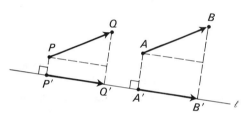

This projection principle is just as true in space as in the plane, the only difference being that in space we probably visualize the foot P' of the perpendicular dropped from P as obtained by passing the *plane* through P perpendicular to l.

Two vectors forces applied to a point P act in combination as though they were a single force, called their *resultant*, or *sum*. The sum vector is found by the parallelogram rule. Vector velocities also combine by the parallelogram rule; in fact, any two vectors of the same kind combine in this way.

The Parallelogram Rule. The sum (resultant) of the vectors $\overrightarrow{PQ}$ and $\overrightarrow{PR}$ is the vector $\overrightarrow{PS}$, where S is the fourth vertex of the parallelogram having PQ and PR as adjacent sides.

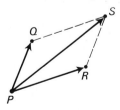

EXAMPLE 4 A bow exerts a force of twenty pounds on an arrow just before release. That force is the vector sum of the two forces exerted by the bow ends along the string of the bow.

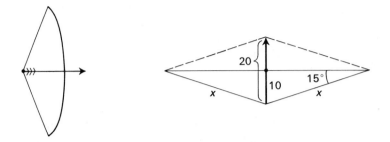

In the righthand figure above, this vector sum is the vertical diagonal of the parallelogram. If the pulled string makes an angle of 15° with the line through the bow tips, then the common magnitude x of these two string-tension forces is given by

$$\frac{10}{x} = \sin 15° \approx 0.26,$$

$$x \approx \frac{10}{0.26} \approx 40 \text{ pounds.}$$

EXAMPLE 5 A canoeist wishes to paddle across a river to a point directly opposite on the other bank. He can paddle at 4 miles per hour and the river current runs 2 miles per hour. In what direction must he aim his canoe?

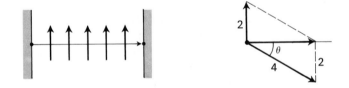

Solution. His actual velocity is the vector sum of the water velocity and his velocity relative to the water. He wants that vector sum to be perpendicular to the river bank. So he must aim upstream at such an angle θ with the perpendicular that $\sin \theta = 2/4$. Thus $\theta = 30°$.

No actual procedure was specified above for completing the parallelogram. One possibility is to "lay off" from Q the arrow $\overrightarrow{QS}$ that is equal to $\overrightarrow{PR}$.

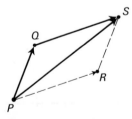

We start from Q, and proceed in the direction given by $\overrightarrow{PR}$ a distance equal to the length of $\overrightarrow{PR}$, thus determining the terminal point S. This procedure has the advantage of covering a special case that is missed by the parallelogram rule as first stated, namely, the case where $\overrightarrow{PQ}$ and $\overrightarrow{PR}$ are collinear and no genuine parallelogram exists. We shall therefore adopt the "laying off" characterization as the definition of vector addition.

DEFINITION   *The sum of the vectors $\overrightarrow{PQ}$ and $\overrightarrow{PR}$ is the vector $\overrightarrow{PS}$, where the point S is determined by laying off the vector $\overrightarrow{PR}$ from the point Q, so that*

$$\overrightarrow{QS} = \overrightarrow{PR}$$

When we want to refer to a vector without pinning ourselves down to one of its representing arrows, we shall generally use a lower case boldface letter such as **u**, **v**, or **x**. Such direct designation is useful and suggestive. For example, the sum of two vectors **u** and **v** was defined in terms of arrow representatives, and if its direct designation as **u** + **v** is to make any sense then it must be independent of which *particular* arrow was chosen to represent **u**. The following figure illustrates this needed fact. See Theorem 1.

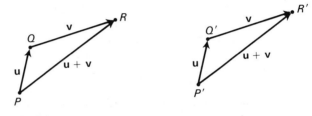

The equation

$$\mathbf{u} = \overrightarrow{PQ}$$

is read: "**u** is the vector (represented by the arrow) $\overrightarrow{PQ}$."

Another important vector operation is multiplying a vector by a number.

DEFINITION   *The product of a vector **u** and a positive number t, designated t**u**, is the vector **v** having the same direction as **u**, but t times the magnitude of **u**. If t is negative, then **v** = t**u** has the direction opposite to that of **u** and $|t|$ times the magnitude of **u**.*

EXAMPLE 6   If a wire carrying an electric current i is positioned near a magnet, then the wire experiences a (vector) force **F**. If the current i is doubled then so is **F**, and if i is reversed then so is **F**. That is, if i_0 and $\mathbf{F}_0$ are the original current and force, and if the current is changed to the new value

$$i = ci_0$$

where c can be positive or negative, then the new force **F** is the corresponding multiple of $\mathbf{F}_0$:

$$\mathbf{F} = c\mathbf{F}_0.$$

(This phenomenon could be called the electric-motor principle.)

In vector discussions, numbers and numerical quantities are called *scalars*. In the example above, the equation $i = ci_0$ is a scalar equation, while $\mathbf{F} = c\mathbf{F}_0$ is a vector equation. The product $c\mathbf{F}_0$ is the vector $\mathbf{F}_0$ multiplied by the scalar c.

It follows from the geometry of proportional figures that *the operation of multiplying by a scalar t is preserved when arrows are projected on a line:*

II) If $\overrightarrow{PR} = t(\overrightarrow{PQ})$, and if P', Q', and R' are the points on the line l obtained by dropping perpendiculars from P, Q, and R, then $\overrightarrow{P'R'} = t(\overrightarrow{P'Q'})$.

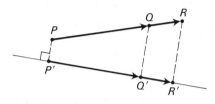

We shall see that all vector properties and laws are consequences of the two projection principles (I) and (II), coupled with elementary facts about signed distances along a number line given at the beginning of Chapter 1.

So far we have been discussing vectors without assuming them to be confined to a particular plane, and without making any use of a coordinate system. We turn now to the coordinate plane. Until further notice, all vectors will be assumed to lie in the standard x,y plane. Now each vector $\mathbf{u}$ determines and is determined by an ordered pair of numbers (u_1, u_2) called its *components*, defined as follows:

DEFINITION   *If* $\mathbf{u} = \overrightarrow{PQ}$, *where* $P = (a, b)$ *and* $Q = (x, y)$, *then the components of* $\mathbf{u}$ *are the coordinate differences*

$$(u_1, u_2) = (x - a, y - b).$$

Note that the stem coordinate is subtracted from the tip coordinate. Of course it must be checked that these numbers depend only on $\mathbf{u}$, and not on the particular arrow representing $\mathbf{u}$. Lemma 1 does this.

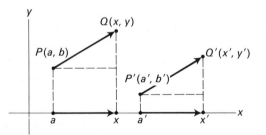

LEMMA 1   *If* $\overrightarrow{PQ} = \overrightarrow{P'Q'}$, *where* $P' = (a', b')$ *and* $Q' = (x', y')$, *P and Q being as above, then*

$$x' - a' = x - a, \qquad y' - b' = y - b.$$

Proof. This is a consequence of the projection principle (I). The equal arrows $\overrightarrow{PQ}$ and $\overrightarrow{P'Q'}$ have equal projections on the x-axis, so the arrows $\overrightarrow{ax}$ and $\overrightarrow{a'x'}$ are equal on this number line. That is, the signed distance from a to x equals the signed distance from a' to x':

$$x - a = x' - a'.$$

Similarly, $y - b = y' - b'$. ∎

A vector in the coordinate plane can thus be identified with an ordered pair of numbers, its components, just as earlier we identified a point in the coordinate plane with its pair of coordinates. It turns out that the vector operations have particularly simple characterizations as operations on number pairs.

THEOREM 1   *If* $\mathbf{u} = (u_1, u_2)$ *and* $\mathbf{v} = (v_1, v_2)$ *then*

$$\mathbf{u} + \mathbf{v} = (u_1 + v_1, u_2 + v_2).$$

That is, vector addition is given by ordinary arithmetic addition of corresponding vector components. In particular, the sum of two vectors is independent of which arrows represent them.

Proof. Consider any pair of head-to-tail arrows $\overrightarrow{PQ}$ and $\overrightarrow{QR}$ representing $\mathbf{u}$ and $\mathbf{v}$ respectively. Since the coordinate differences from P to Q are (u_1, u_2) and the coordinate differences Q to R are (v_1, v_2), it follows that the coordinate differences P to R are $(u_1 + v_1, u_2 + v_2)$. Since $\overrightarrow{PR}$ is the sum vector, by definition, the theorem follows. ∎

The calculation in proving Lemma 1 came down to the fact that on a coordinate line (say the x-axis) there are only two directions, positive and negative, so that a vector lying along the line is specified by a signed distance, i.e., by a number that may have either sign. In this situation the product of a vector and a number is simply the product of two numbers, but they are interpreted differently: the product of the scalar t and the vector (signed distance) x is the vector (signed distance) tx.

This observation amounts to the rule for the components of the product $t\mathbf{u}$: the projection on the x-axis of the vector $\mathbf{u} = (a, b)$ is the one dimensional vector (signed distance) a, so the projection of $t(a, b)$ is ta, by the projection law (II) and the above remark.

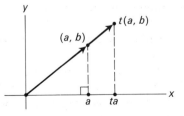

Similarly for the y-axis. The components of $t(a, b)$ are thus the numbers ta and tb, and we have proved:

THEOREM 2   *If* $\mathbf{u} = (a, b)$, *then* $t\mathbf{u} = (ta, tb)$.

EXAMPLE 7   If $\mathbf{a} = (1, 2)$ and $\mathbf{b} = (3, -1)$, then

$$\mathbf{a} + \mathbf{b} = (1, 2) + (3, -1) = (1 + 3, 2 - 1) = (4, 1)$$

$$2\mathbf{a} + 4\mathbf{b} = 2(1, 2) + 4(3, -1) = (2, 4) + (12, -4) = (14, 0)$$

$$(-3)\mathbf{a} + \mathbf{b} = (-3)(1, 2) + (3, -1) = (-3, -6) + (3, -1) = (0, -7).$$

The two vector operations satisfy certain laws of algebra. These laws permit problems to be set up and solved by vector manipulations similar to the procedures of ordinary algebra.

The Vector Laws

For any vectors $\mathbf{u}$, $\mathbf{v}$, and $\mathbf{w}$,

V1. $\mathbf{u} + \mathbf{v} = \mathbf{v} + \mathbf{u}$,

V2. $(\mathbf{u} + \mathbf{v}) + \mathbf{w} = \mathbf{u} + (\mathbf{v} + \mathbf{w})$.

There is a special vector, designated $\mathbf{0}$ and called the zero vector, such that

V3. $\mathbf{u} + \mathbf{0} = \mathbf{0} + \mathbf{u} = \mathbf{u}$

for every vector $\mathbf{u}$. Moreover, for each vector $\mathbf{u}$ there is a uniquely determined "opposite" vector, designated $-\mathbf{u}$ and called the *negative of* $\mathbf{u}$, such that

V4. $\mathbf{u} + (-\mathbf{u}) = \mathbf{0}$.

The above laws concern vector addition only. Multiplication by scalars satisfies the further laws:

V5. $t(\mathbf{u} + \mathbf{v}) = t\mathbf{u} + t\mathbf{v}$

V6. $(s + t)\mathbf{u} = s\mathbf{u} + t\mathbf{u}$

V7. $s(t\mathbf{u}) = (st)\mathbf{u}$

V8. $1\mathbf{u} = \mathbf{u}$.

The first two laws are the commutative and associative laws for vector addition. The fifth and sixth are distributive laws: they state that the operation of multiplying a vector by a scalar distributes over both addition of vectors and addition of numbers. The seventh law is a mixed associative law relating multiplication of numbers to the multiplication of a vector by a number.

The zero vector has the components $(0, 0)$ and is represented geometrically by "degenerate" arrows $\overrightarrow{PP}$ that go nowhere; they have zero length and no direction. The negative of $\mathbf{u} = (u_1, u_2) = \overrightarrow{PQ}$ is given algebraically by $-\mathbf{u} = (-u_1, -u_2)$, and geometrically by the reversed arrow: $-\mathbf{u} = \overrightarrow{QP}$. In terms of these arrows and the "laying off" definition of addition, the equation $\mathbf{u} + (-\mathbf{u}) = \mathbf{0}$ is

$$\overrightarrow{PQ} + \overrightarrow{QP} = \overrightarrow{PP}.$$

The following figures show how the commutative and associative laws for vector addition can be verified geometrically.

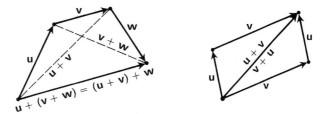

The trouble with such geometric proofs is that it is usually impossible to draw figures in sufficiently "general position" to represent all cases. It is much easier to give purely algebraic proofs on the basis of Theorems 1 and 2 and the corresponding laws for numbers. For example, if $\mathbf{u} = (u_1, u_2)$ and $\mathbf{v} = (v_1, v_2)$, then

$$
\begin{aligned}
t[\mathbf{u} + \mathbf{v}] &= t[(u_1, u_2) + (v_1, v_2)] \\
&= t(u_1 + v_1, u_2 + v_2) \quad \text{(Theorem 1)} \\
&= (t(u_1 + v_1), t(u_2 + v_2)) \text{ (Theorem 2)} \\
&= (tu_1 + tv_1, tu_2 + tv_2) \quad \text{(distributive law for numbers)} \\
&= (tu_1, tu_2) + (tv_1, tv_2) \quad \text{(Theorem 1)} \\
&= t(u_1, u_2) + t(v_1, v_2) \quad \text{(Theorem 2)} \\
&= t\mathbf{u} + t\mathbf{v}.
\end{aligned}
$$

Subtracting $\mathbf{u}$ is defined as adding $-\mathbf{u}$:

$$\mathbf{v} - \mathbf{u} = \mathbf{v} + (-\mathbf{u}).$$

Numerically, this amounts simply to subtracting corresponding components, since

$$
\begin{aligned}
(v_1, v_2) - (u_1, u_2) &= (v_1, v_2) + (-u_1, -u_2) \\
&= (v_1 - u_1, v_2 - u_2).
\end{aligned}
$$

We can now interpret the definition of vector components. Note first that if the point P has *coordinates* (a, b), then the vector $\overrightarrow{OP}$, called the *position vector* of P, has *components* (a, b). So if $P = (a, b)$ and $Q = (x, y)$ then the equation defining the components of $\mathbf{u} = \overrightarrow{PQ}$,

$$(u_1, u_2) = (x - a, y - b),$$

can now be written

$$(u_1, u_2) = (x, y) - (a, b)$$

and interpreted as the vector identity

$$\overrightarrow{PQ} = \overrightarrow{OQ} - \overrightarrow{OP}.$$

In other words, the coordinate differences in the formula for the components of $\overrightarrow{PQ}$ simply reflect the expression of $\overrightarrow{PQ}$ as the difference of the position vectors to the arrow endpoints Q and P.

The fact that the *coordinates* of a point P are the *components* of its position vector $\mathbf{u} = \overrightarrow{OP}$ means that a point and its position vector are algebraically indistinguishable.

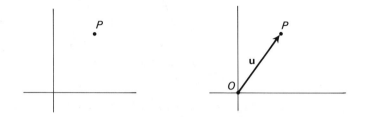

In order to avoid ambiguity, we shall now take the vector interpretation of a pair of numbers as the primary one, and when we want to refer to a point by its coordinates we shall normally do so by way of its position vector. Thus "the point $\mathbf{u}$" means "the point P such that $\mathbf{u} = \overrightarrow{OP}$," or "the point having the position vector $\mathbf{u}$." We can thus refer to the arrow from (the point) $\mathbf{u}$ to (the point) $\mathbf{v}$, and even write $\overrightarrow{\mathbf{uv}}$. It represents the vector $\mathbf{v} - \mathbf{u}$.

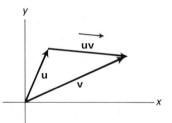

Component-coordination notation is traditionally inconsistent. The components of a vector $\mathbf{u}$ are normally indicated by subscripts on the corresponding lightface letter, as in

$$\mathbf{u} = (u_1, u_2).$$

On the other hand, the position vector of a variable point is usually indicated

$$\mathbf{x} = (x, y),$$

in order to keep the standard x,y usage and also in order to avoid subscripts, since a subscripted letter may suggest a constant. The position vector of a *fixed* point is treated in either manner, as in

$$\mathbf{x}_0 = (x_0, y_0), \quad \text{or} \quad \mathbf{a} = (a, b), \quad \text{or} \quad \mathbf{a} = (a_1, a_2).$$

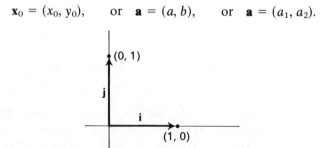

Finally, we introduce a traditional method for expressing a vector in terms of its components. The unit vectors along the coordinate axes are designated **i** and **j**,

$$\mathbf{i} = (1, 0), \qquad \mathbf{j} = (0, 1),$$

and are called the *standard basis vectors*. Then:

LEMMA 2 *Any vector* $\mathbf{u} = (u_1, u_2)$ *is given by*

$$\mathbf{u} = u_1\mathbf{i} + u_2\mathbf{j}.$$

That is, any vector **u** is a *linear combination* of the standard basis vectors **i** and **j**, with coefficients equal to the components of **u**.

Proof. By Theorems 1 and 2, we have

$$\mathbf{u} = (u_1, u_2) = (u_1, 0) + (0, u_2)$$
$$= u_1(1, 0) + u_2(0, 1)$$
$$= u_1\mathbf{i} + u_2\mathbf{j}. \quad \blacksquare$$

When vectors are written this way, their manipulation involves constant use of the laws of vector algebra. For example, the computation of the coefficients (components) of $\mathbf{u} + \mathbf{v}$,

$$\mathbf{u} + \mathbf{v} = (u_1\mathbf{i} + u_2\mathbf{j}) + (v_1\mathbf{i} + v_2\mathbf{j})$$
$$= (u_1 + v_1)\mathbf{i} + (u_2 + v_2)\mathbf{j},$$

would require, if written out in complete detail, several applications of the associative law for vector addition, one application of the commutative law, and two applications of the distributive law $(s + t)\mathbf{u} = s\mathbf{u} + t\mathbf{u}$.

PROBLEMS FOR SECTION 1

1. Solve the bow problem if the pulled bow string exerts a force of 20 pounds on the arrow when the angle between the bow tips and the deflected string is 10°.

2. Solve the bow problem when an angle of 30° produces a force of 30 pounds.

3. Solve the canoe problem when the canoe and stream speeds are $2.8 \approx 2\sqrt{2}$ and 2 miles per hour, respectively.

4. Solve the canoe problem when the two speeds are $2.3 \approx 4/\sqrt{3}$ and 2.

5. A small plane flies 200 mph. The pilot wishes to fly due north at a time when a westerly wind of 40 mph is blowing. How should he set his course? What will be his true speed (relative to ground)?

6. Do the same problem for plane and wind speeds of 130 and 50 mph, respectively.

7. Draw a figure for the geometric verification of the vector law **V**5.

8. For each pair of vectors **x** and **a** given below, find the requested algebraic combination.
   a) $\mathbf{x} = (1, 4)$, $\mathbf{a} = (2, 1)$; find $\mathbf{x} - \mathbf{a}$.
   b) $\mathbf{x} = (3/2, 1/4)$, $\mathbf{a} = (7/3, 1/6)$; find $\mathbf{x} + 6\mathbf{a}$.
   c) $\mathbf{x} = (3/2, 1/4)$, $\mathbf{a} = (7/3, 1/6)$; find $6(\mathbf{x} + \mathbf{a})$.
   d) $\mathbf{x} = (1, 4)$, $\mathbf{a} = (2, 7)$; find $\dfrac{1}{2}\mathbf{x} - 2\mathbf{a}$.

Prove the following laws from the component characterizations of the vector operations and the laws of algebra for numbers.

9. $\mathbf{a} + \mathbf{x} = \mathbf{a}$ if and only if $\mathbf{x} = \mathbf{0}$

10. $\mathbf{a} + \mathbf{u} = \mathbf{0}$ if and only if $\mathbf{u} = -\mathbf{a}$

11. $(\mathbf{u} + \mathbf{x}) + \mathbf{a} = \mathbf{u} + (\mathbf{x} + \mathbf{a})$

12. $(s + t)\mathbf{x} = s\mathbf{x} + t\mathbf{x}$

13. $0\mathbf{a} = \mathbf{0}$

14. $(-1)\mathbf{a} = -\mathbf{a}$

15. A theorem of geometry says that if two sides of a quadrilateral are parallel and equal, then the other two sides are also parallel and equal, so the figure is a parallelogram. Formulate and prove this as a theorem about arrows and the vector operations.

2. THE PARAMETRIC EQUATION OF A LINE

Remember that when we refer to "the point $\mathbf{x}$" we mean the point P having the position vector $\mathbf{x}$, so that $\mathbf{x} = \overrightarrow{OP}$. Algebraically, $\mathbf{x}$ and P are indistinguishable: the coordinates of P are the components of $\mathbf{x}$.

A line l is determined if we know a point $\mathbf{x}_0$ on l and a nonzero vector $\mathbf{u}$ having one of the two directions along l. For then a point $\mathbf{x}$ lies on l if and only if the arrow from $\mathbf{x}_0$ to $\mathbf{x}$ is parallel to $\mathbf{u}$, i.e., if and only if

$$\mathbf{x} - \mathbf{x}_0 = t\mathbf{u},$$

or

$$\mathbf{x} = \mathbf{x}_0 + t\mathbf{u},$$

from some scalar t.

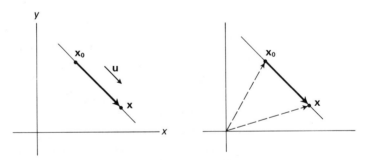

As the real number t runs from $-\infty$ to $+\infty$, the varying point $\mathbf{x}$ sweeps out the line. In this situation, t is called a *parameter*, and the above equation is a *parametric equation* for the line. This is consistent with the use of the word *parameter* in Chapter 10. The parameter t is not "part of" the point $\mathbf{x}$, but t determines $\mathbf{x}$.

The result above is important, and we restate it as a theorem.

THEOREM 3   *The line through the point $\mathbf{x}_0$ in the direction of the nonzero vector $\mathbf{u}$ has the parametric equation*

$$\mathbf{x} = \mathbf{x}_0 + t\mathbf{u}.$$

EXAMPLE 1   The line through the point $\mathbf{x}_0 = (1, 2)$ in the direction of $\mathbf{u} = (3, -1)$ has the parametric equation

$$\mathbf{x} = \mathbf{x}_0 + t\mathbf{u}$$
$$= (1, 2) + t(3, -1).$$

This is the equation

$$(x, y) = (1 + 3t, 2 - t),$$

and is equivalent to the pair of numerical (scalar) equations

$$x = 1 + 3t,$$
$$y = 2 - t.$$

Note that these are parametric equations for the line in exactly the sense of Section 1 in Chapter 10. The parameter can easily be eliminated to obtain an x,y-equation for the line. For instance, $t = 2 - y$ from the second equation, so

$$x = 1 + 3t = 1 + 3(2 - y) = 7 - 3y.$$

The line has the equation

$$x + 3y = 7.$$

Two points determine a line. If $\mathbf{a}$ and $\mathbf{b}$ are (the position vectors of) two distinct points on l, then $\mathbf{u} = \mathbf{b} - \mathbf{a}$ gives the direction of l, so its parametric equation $\mathbf{x} = \mathbf{a} + t\mathbf{u}$ can be written

$$x = \mathbf{a} + t(\mathbf{b} - \mathbf{a}).$$

EXAMPLE 2   The line through the points $\mathbf{a} = (1, 2)$ and $\mathbf{b} = (4, 1)$ has the parametric equation

$$\mathbf{x} = (1, 2) + t[(4, 1) - (1, 2)]$$
$$= (1, 2) + t(3, -1),$$

and is the line of Example 1.

The two-point equation $\mathbf{x} = \mathbf{a} + t(\mathbf{b} - \mathbf{a})$ can also be written

$$\mathbf{x} = (1 - t)\mathbf{a} + t\mathbf{b},$$

or

$$\mathbf{x} = s\mathbf{a} + t\mathbf{b},$$

where $s + t = 1$. In the form

$$\mathbf{x} - \mathbf{a} = t(\mathbf{b} - \mathbf{a}),$$

it says that $\mathbf{x}$ is t times as far from $\mathbf{a}$ as $\mathbf{b}$ is. For example, $\mathbf{x}$ will be the midpoint of the segment $\mathbf{ab}$ if $t = 1/2$, and solving for $\mathbf{x}$ with this value of t yields the midpoint formula

$$\mathbf{x} = \frac{\mathbf{a} + \mathbf{b}}{2}.$$

When $t = 2/3$, we get the point $2/3$ of the way from **a** to **b**, and its formula comes out to be

$$\mathbf{x} = \frac{1}{3}\mathbf{a} + \frac{2}{3}\mathbf{b}.$$

It divides the segment ab in the ratio 2 to 1. In general,

THEOREM 4   *If $s + t = 1$ and both numbers are positive, then*

$$\mathbf{x} = s\mathbf{a} + t\mathbf{b}$$

is the point on the segment **ab** *that divides it in the ratio t/s.*

Note that the dividing point is nearer the endpoint having the *larger coefficient.*

EXAMPLE 3

$$\mathbf{x} = \frac{\mathbf{a} + \mathbf{b}}{2} = \frac{(1, 2) + (4, 4)}{2}$$

$$= \frac{1}{2}(5, 6) = \left(\frac{5}{2}, 3\right).$$

EXAMPLE 4   The point $(3/4)$ of the way from **a** to **b** on the above segment is

$$\mathbf{x} = \frac{1}{4}\mathbf{a} + \frac{3}{4}\mathbf{b}$$

$$= \frac{(1, 2) + 3(4, 4)}{4}$$

$$= \frac{(13, 14)}{4} = \left(\frac{13}{4}, \frac{7}{2}\right).$$

We can already solve some problems in geometry by this new algebra.

EXAMPLE 5   Prove that the medians of a triangle are concurrent.

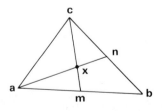

Solution. We start off by computing the point **x** where the two medians shown above intersect. Since **x** lies on the segment **mc**, it can be written

$$\mathbf{x} = t\mathbf{m} + (1 - t)\mathbf{c}$$

for some number t between 0 and 1. Moreover, $\mathbf{m} = (\mathbf{a} + \mathbf{b})/2$. Thus,

$$\mathbf{x} = \frac{t}{2}\mathbf{a} + \frac{t}{2}\mathbf{b} + (1 - t)\mathbf{c}.$$

But $\mathbf{x}$ also lies on the segment $\mathbf{am}$, so that, similarly,

$$\mathbf{x} = (1 - s)\mathbf{a} + \frac{s}{2}\mathbf{b} + \frac{s}{2}\mathbf{c}$$

for some number s between 0 and 1. These two representations must give the same point $\mathbf{x}$, and this will happen if $s = t$ and $(t/2) = (1 - t)$. This gives $s = t = (2/3)$, and

$$\mathbf{x} = \frac{\mathbf{a} + \mathbf{b} + \mathbf{c}}{3}.$$

Since this representation is symmetric in the letters $\mathbf{a}$, $\mathbf{b}$, and $\mathbf{c}$, the point $\mathbf{x}$ lies on all three medians. (In more detail, if we had started with another pair of medians, the calculation of their point of intersection would be the same as the above calculation except that the letters $\mathbf{a}$, $\mathbf{b}$ and $\mathbf{c}$ would be permuted. The resulting formula for the point of intersection would thus be the same as the above formula except for this permutation of the letters, and the two points of intersection are thus the same point.)

PROBLEMS FOR SECTION 2

Write out the following examples of the parametric equation of Theorem 3, and reduce each equation to the equivalent pair of scalar equations.

1. $\mathbf{x}_0 = (1, 2)$, $\mathbf{u} = (-3, 0)$ 2. $\mathbf{x}_0 = (-2, 3)$, $\mathbf{u} = (1, 1)$

3. $\mathbf{x}_0 = (0, 0)$, $\mathbf{u} = (3, 4)$ 4. $\mathbf{x}_0 = (0, 0)$, $\mathbf{u} = (a, b)$

5. $\mathbf{x}_0 = (x_0, y_0)$, $\mathbf{u} = (1, 1)$ 6. $\mathbf{x}_0 = (2, -1)$, $\mathbf{u} = (1, 4)$

7. Find the point of intersection of the lines $\mathbf{x} = (1, 2) + s(1, 1)$ and $\mathbf{x} = (2, 2) + t(3, -1)$.

In each of the following problems, write the parametric equation of the line through the given points. Also give the equivalent pair of scalar equations.

8. $\mathbf{x}_0 = (1, 4)$, $\mathbf{x}_1 = (2, -2)$ 9. $\mathbf{x}_0 = (1, 1)$, $\mathbf{x}_1 = (3, 0)$

10. $\mathbf{x}_0 = (1, 1)$, $\mathbf{x}_1 = (4, 1)$ 11. $\mathbf{x}_0 = (2, 0)$, $\mathbf{x}_1 = (2, 1)$

12. $\mathbf{x}_0 = (-1, 1/2)$, $\mathbf{x}_1 = (1/2, 3/4)$

In each of the following problems, find the point $\mathbf{x}$ which divides the line segment into the given ratio.

13. $\mathbf{x}_0 = (1, 1)$, $\mathbf{x}_1 = (-6, 0)$; $\mathbf{x}$ is 1/3 of the way from $\mathbf{x}_1$ to $\mathbf{x}_0$

14. $\mathbf{x}_0 = (0, 1)$, $\mathbf{x}_1 = (10, 4)$; $\mathbf{x}$ is 1/4 of the way from $\mathbf{x}_0$ to $\mathbf{x}_1$

15. $\mathbf{x}_0 = (1/2, 1/3)$, $\mathbf{x}_1 = (4/3, 1/2)$; $\mathbf{x}$ is 2/5 of the way from $\mathbf{x}_0$ to $\mathbf{x}_1$

16. $\mathbf{x}_0 = (-1, -4)$, $\mathbf{x}_1 = (-6, -2)$; $\mathbf{x}$ is 7/8 of the way from $\mathbf{x}_1$ to $\mathbf{x}_0$

17. $\mathbf{x}_0 = (4, -3)$, $\mathbf{x}_1 = (-1, 6)$; $\mathbf{x}$ is 5/6 of the way from $\mathbf{x}_0$ to $\mathbf{x}_1$

18. Prove that the diagonals of a parallelogram bisect one another.

19. Prove that the line segment determined by the midpoints of two sides of a triangle is parallel to the third side and half its length.

20. In Example 5 it was noted that if $s = t$ and $1 - t = t/2$, so that $s = t = 2/3$, then the two representations of $\mathbf{x}$ are the same. It was implicit that $s = t = 2/3$ was the *only* solution, but this was never proved. Show this now, by showing that if the two representations of $\mathbf{x}$ are subtracted, then the resulting single equation can be cast into the form

$$k_1(\mathbf{b} - \mathbf{a}) = k_2(\mathbf{c} - \mathbf{a}),$$

where the coefficients k_1 and k_2 involve s and t. Conclude that $k_1 = k_2 = 0$, etc.

21. Let $\mathbf{x}_1$, $\mathbf{x}_2$, $\mathbf{x}_3$ and $\mathbf{x}_4$ be the vertices of an arbitrary quadrilateral. Show that the midpoints of the sides are the vertices of a parallelogram.

22. A theorem of geometry says that if the opposite sides of a quadrilateral are parallel, then the opposite sides are also equal, and the figure is a parallelogram. However, this clearly can be false if all four points lie on a straight line.

Prove the following correct algebraic version of the theorem.

Theorem. *If* $(\mathbf{a}_2 - \mathbf{x}_2) = s(\mathbf{a}_1 - \mathbf{x}_1)$ *and* $(\mathbf{a}_2 - \mathbf{a}_1) = t(\mathbf{x}_2 - \mathbf{x}_1)$, *then either* $s = t = 1$ *or the four points are collinear.*

23. Prove the following geometric theorem.

Theorem. *A line parallel to one side of a triangle cuts the other two sides proportionately. (Take one vertex at the origin.)*

3. THE DOT PRODUCT

We shall use the ordinary absolute-value symbol $|\mathbf{x}|$ for the magnitude of the vector $\mathbf{x} = (x, y)$:

$$|\mathbf{x}| = \sqrt{x^2 + y^2}.$$

Geometrically, $|\mathbf{x}|$ is the *distance* from the origin to the point $\mathbf{x}$ and also the *length* of any arrow representing $\mathbf{x}$. According to, the definition of $t\mathbf{a}$,

$$|t\mathbf{a}| = |t| \cdot |\mathbf{a}|.$$

A *unit vector* is a vector of length one. Any nonzero vector $\mathbf{x}$ determines a unique unit vector $\mathbf{u}$ in the same direction, that is, a vector $\mathbf{u}$ such that

$$|\mathbf{u}| = 1,$$
$$\mathbf{u} = t\mathbf{x}, \qquad t > 0.$$

For these requirements lead to

$$1 = |t\mathbf{x}| = |t| \cdot |\mathbf{x}|, \qquad \text{and} \qquad |t| = \frac{1}{|\mathbf{x}|}.$$

But t is positive, so $t = 1/|\mathbf{x}|$. Thus $\mathbf{u}$ is uniquely determined as

$$\mathbf{u} = \frac{\mathbf{x}}{|\mathbf{x}|}.$$

The fact that **u** defined this way is a unit vector can of course be checked directly. It is called the *normalization* of **x**.

EXAMPLE 1   The unit vector in the direction of $\mathbf{x} = (1, 2)$ is $\mathbf{u} = \mathbf{x}/|\mathbf{x}| = (1, 2)/\sqrt{5} = (1/\sqrt{5}, 2/\sqrt{5})$.

Since there is exactly one unit vector in a given direction, we sometimes identify a direction with its unit vector.

The trigonometric identity

$$\cos^2\theta + \sin^2\theta = 1$$

can be interpreted as saying that

$$(\cos\,\theta, \sin\,\theta)$$

is a unit vector. Moreover, every unit vector **u** can be expressed this way: since **u** is a point (u, v) on the unit circle, it follows that

$$\mathbf{u} = (\cos\,\theta, \sin\,\theta),$$

where θ is the angle from the direction of the positive x-axis to the direction of **u**.

In the general situation, where $\mathbf{u} = \mathbf{x}/|\mathbf{x}|$ is the unit vector (direction) of **x**,

$$(\cos\,\theta, \sin\,\theta) = \mathbf{u} = \frac{\mathbf{x}}{|\mathbf{x}|} = \left(\frac{x}{r}, \frac{y}{r}\right),$$

where $r = |\mathbf{x}| = \sqrt{x^2 + y^2}$. The cross-multiplied forms

$$\mathbf{x} = |\mathbf{x}|\mathbf{u},$$

$$(x, y) = r(\cos\,\theta, \sin\,\theta),$$

can be called the polar-coordinate representation of the vector $\mathbf{x} = (x, y)$, that is, its representation as the product of its length times its direction.

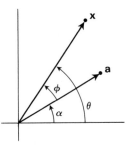

Now let **a** be a second nonzero vector, with unit vector

$$(\cos\,\alpha, \sin\,\alpha) = \frac{\mathbf{a}}{|\mathbf{a}|} = \left(\frac{a}{l}, \frac{b}{l}\right),$$

where $l = |\mathbf{a}| = \sqrt{a^2 + b^2}$. Then $\phi = \theta - \alpha$ is the angle from $\mathbf{a}$ to $\mathbf{x}$, and

$$\cos \phi = \cos(\theta - \alpha) = \cos\theta\cos\alpha + \sin\theta\sin\alpha$$

$$= \frac{x}{r}\cdot\frac{a}{l} + \frac{y}{r}\cdot\frac{b}{l} = \frac{ax + by}{rl}.$$

We could also compute $\sin\phi$. But it turns out that the $\sin\phi$ formula doesn't generalize to three dimensions, whereas the $\cos\phi$ formula does. This reflects the fact that there is no natural way to define a *signed* angle *from* one direction *to* another in three dimensions. We can only look at the *undirected* angle *between* two vectors, and such an angle is specified by its cosine (since $\cos\phi = \cos(-\phi)$).

The numerator in the above formula for $\cos\phi$ comes up so often that we give it a special name. It is called the *dot product* of the vectors $\mathbf{a}$ and $\mathbf{x}$, and is designated $\mathbf{a}\cdot\mathbf{x}$.

DEFINITION   *The dot product $\mathbf{a}\cdot\mathbf{x}$ of the vectors $\mathbf{a} = (a, b)$ and $\mathbf{x} = (x, y)$ is defined by*

$$\mathbf{a}\cdot\mathbf{x} = ax + by.$$

EXAMPLE 2   The dot product of the vectors $(1, 4)$ and $(-3, 2)$ is

$$(1, 4)\cdot(-3, 2) = 1(-3) + 4\cdot 2 = -3 + 8 = 5.$$

EXAMPLE 3   The length $|\mathbf{x}|$ of the vector $\mathbf{x}$ is

$$|\mathbf{x}| = (\mathbf{x}\cdot\mathbf{x})^{1/2},$$

since $|\mathbf{x}|^2 = x^2 + y^2 = x\,x + y\,y = \mathbf{x}\cdot\mathbf{x}$.

The formula for $\cos\phi$ that we computed above is an important property of the dot product.

THEOREM 5   *If ϕ is the angle between two nonzero vectors $\mathbf{a}$ and $\mathbf{x}$, then*

$$\cos\phi = \frac{\mathbf{a}\cdot\mathbf{x}}{|\mathbf{a}|\,|\mathbf{x}|}.$$

COROLLARY   *The vectors $\mathbf{a}$ and $\mathbf{x}$ are perpendicular if and only if $\mathbf{a}\cdot\mathbf{x} = 0$.*

Proof. The angle ϕ between the vectors is $90°$ $(\pi/2)$ if and only if $\cos\phi = 0$, and this is equivalent to $\mathbf{a}\cdot\mathbf{x} = 0$, by the formula of the theorem. ∎

EXAMPLE 4   In order to check the vectors $\mathbf{a} = (1, 2)$ and $\mathbf{x} = (4, -2)$ for perpendicularity, we compute

$$\mathbf{a}\cdot\mathbf{x} = ax + by = 1\cdot 4 + 2(-2) = 0.$$

Thus, they are perpendicular ($\mathbf{a} \perp \mathbf{x}$).

EXAMPLE 5 Find the angle ϕ between the vectors $(1, -1)$ and $(3, 2)$.

Solution

$$\cos \phi = \frac{\mathbf{a} \cdot \mathbf{x}}{|\mathbf{a}| \, |\mathbf{x}|} = \frac{1 \cdot 3 + (-1)2}{\sqrt{1 + 1}\sqrt{9 + 4}} = \frac{1}{\sqrt{26}},$$

and $\phi = \arccos 1/\sqrt{26}$.

EXAMPLE 6 Show that the straight line

$$ax + by + c = 0$$

is perpendicular to the vector $\mathbf{a} = (a, b)$.

Solution. Let $\mathbf{x}_1$ and $\mathbf{x}_0$ be any two distinct points on the line. Then

$$ax_1 + by_1 + c = 0, \qquad ax_0 + by_0 + c = 0;$$

and subtracting these two equations gives the equation

$$a(x_1 - x_0) + b(y_1 - y_0) = 0.$$

But this just says that

$$\mathbf{a} \cdot (\mathbf{x}_1 - \mathbf{x}_0) = 0$$

and hence that $\mathbf{a} \perp (\mathbf{x}_1 - \mathbf{x}_0)$. Since the direction of $\mathbf{x}_1 - \mathbf{x}_0$ is the direction of the arrow $\overrightarrow{\mathbf{x}_0 \mathbf{x}_1}$ along the line, we see that the vector $\mathbf{a}$ is perpendicular to the line.

Reversing the above steps gives a new derivation of the equation of a line. Given a line l in the coordinate plane, choose a fixed point $\mathbf{x}_0 = (x_0, y_0)$ on l and a fixed vector $\mathbf{a} = (a, b)$ perpendicular to the direction of l. Then a point $\mathbf{x}$ will be on l if and only if the arrow $\overrightarrow{\mathbf{x}_0 \mathbf{x}}$ is perpendicular to the vector $\mathbf{a}$. In view of Theorem 5, this is exactly the condition

$$\mathbf{a} \cdot (\mathbf{x} - \mathbf{x}_0) = 0.$$

So

$$a(x - x_0) + b(y - y_0) = 0,$$

or

$$ax + by + c = 0 \qquad (c = -ax_0 - by_0),$$

is an equation of the line l.

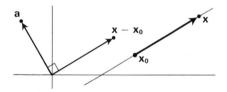

We now do for the dot product what we did in the beginning for vector addition and multiplication by scalars: We list the algebraic properties that let us use $\mathbf{x} \cdot \mathbf{a}$ directly, without recourse to its component formula. They are:

(s1) $$\mathbf{x} \cdot \mathbf{x} = |\mathbf{x}|^2,$$

(s2) $$\mathbf{x} \cdot \mathbf{a} = \mathbf{a} \cdot \mathbf{x},$$

(s3) $$(\mathbf{x}_1 + \mathbf{x}_2) \cdot \mathbf{a} = \mathbf{x}_1 \cdot \mathbf{a} + \mathbf{x}_2 \cdot \mathbf{a},$$

(s4) $$(t\mathbf{x}) \cdot \mathbf{a} = t(\mathbf{x} \cdot \mathbf{a}).$$

As before, the proofs depend on properties of numbers via components. Thus,

$$
\begin{aligned}
(\mathbf{x}_1 + \mathbf{x}_2) \cdot \mathbf{a} &= ((x_1, y_1) + (x_2, y_2)) \cdot (a, b) \\
&= (x_1 + x_2, y_2 + y_2) \cdot (a, b) \\
&= (x_1 + x_2)a + (y_1 + y_2)b \\
&= (x_1 a + y_1 b) + (x_2 a + y_2 b) \\
&= (x_1, y_1) \cdot (a, b) + (x_2, y_2) \cdot (a, b) \\
&= \mathbf{x}_1 \cdot \mathbf{a} + \mathbf{x}_2 \cdot \mathbf{a},
\end{aligned}
$$

which is (s3).

Note that, because of the commutative law (s2), both (s3) and (s4) can be turned around:

$$\mathbf{a} \cdot (\mathbf{x}_1 + \mathbf{x}_2) = \mathbf{a} \cdot \mathbf{x}_1 + \mathbf{a} \cdot \mathbf{x}_2,$$

$$\mathbf{a} \cdot (t\mathbf{x}) = t(\mathbf{a} \cdot \mathbf{x}).$$

These formal properties of the dot product will be basic in the discussion of the three-dimensional dot product in Section 5. Here we make only one application. It is an important theorem about resolving an arbitrary vector into "components" parallel and perpendicular to a given vector.

THEOREM 6   *Let* $\mathbf{a}$ *be a fixed nonzero vector. Then every vector* $\mathbf{x}$ *can be expressed in a unique way as a sum*

$$\mathbf{x} = c\mathbf{a} + \mathbf{n}$$

of a vector $c\mathbf{a}$ *parallel to* $\mathbf{a}$ *and a vector* $\mathbf{n}$ *perpendicular to* $\mathbf{a}$.

Proof. Supposing that $\mathbf{x}$ has an expression of the above form, we can find what the constant c must be by taking the dot product with $\mathbf{a}$:

$$
\begin{aligned}
\mathbf{x} \cdot \mathbf{a} &= c\mathbf{a} \cdot \mathbf{a} + \mathbf{n} \cdot \mathbf{a} \\
&= c(\mathbf{a} \cdot \mathbf{a}),
\end{aligned}
$$

where $\mathbf{n} \cdot \mathbf{a} = 0$ because $\mathbf{n} \perp \mathbf{a}$. Thus c must have the value

$$c = \frac{\mathbf{x} \cdot \mathbf{a}}{\mathbf{a} \cdot \mathbf{a}}.$$

Moreover, if c is defined this way then $\mathbf{x} - c\mathbf{a}$ is perpendicular to $\mathbf{a}$, for then

$$(\mathbf{x} - c\mathbf{a}) \cdot \mathbf{a} = \mathbf{x} \cdot \mathbf{a} - c(\mathbf{a} \cdot \mathbf{a})$$

is zero. So we set $\mathbf{n} = \mathbf{x} - c\mathbf{a}$ and have the decomposition of the theorem. ▮

The vector $c\mathbf{a}$ is called the *vector component of* $\mathbf{x}$ *parallel to* $\mathbf{a}$. If $\mathbf{a}$ is a unit vector then the coefficient c is just the dot product

$$c = \mathbf{x} \cdot \mathbf{a}.$$

Moreover, $|c| = |c\mathbf{a}|$, since $|\mathbf{a}|$ is 1. In this case the coefficient c is called the *scalar component of* $\mathbf{x}$ *in the direction* $\mathbf{a}$. The word *component* by itself is used, ambiguously, in both senses. It may be helpful to look back at the original context of the word. If

$$\mathbf{u} = (u_1, u_2) = u_1\mathbf{i} + u_2\mathbf{j}$$

then $u_1\mathbf{i}$ is the vector component of $\mathbf{u}$ along the x-axis, and u_1 is the scalar component of $\mathbf{u}$ in the direction $\mathbf{i}$, (i.e., in the direction of the positive x-axis).

If $\mathbf{b}_1$ and $\mathbf{b}_2$ are perpendicular unit vectors, we say that they form an *orthonormal basis* for the coordinate plane.

THEOREM 7   *If the pair* $\mathbf{b}_1$, $\mathbf{b}_2$ *is an orthonormal basis for the plane, then every vector* $\mathbf{x}$ *can be written*

$$\mathbf{x} = x_1\mathbf{b}_1 + x_2\mathbf{b}_2,$$

where x_1 *and* x_2 *are the scalar components of* $\mathbf{x}$ *in the directions* $\mathbf{b}_1$ *and* $\mathbf{b}_2$, *respectively.*

Proof. By Theorem 6,

$$\mathbf{x} = x_1\mathbf{b}_1 + \mathbf{n},$$

where $\mathbf{n}$ is perpendicular to $\mathbf{b}_1$ and hence is a scalar multiple of $\mathbf{b}_2$,

$$\mathbf{n} = c\mathbf{b}_2,$$

for some scalar c. Substituting this expression for $\mathbf{n}$ in the first equation, and then dotting with $\mathbf{b}_2$, shows that

$$c = \mathbf{x} \cdot \mathbf{b}_2 = x_2. \quad \blacksquare$$

EXAMPLE 7   The vectors $\mathbf{a}_1 = (3, 4)$ and $\mathbf{a}_2 = (8, -6)$ are perpendicular. Normalize them, so that they will form an orthonormal basis, and then compute the components of $(1, 0)$ with respect to this basis.

Solution

$$|\mathbf{a}_1| = \sqrt{9 + 16} = 5,$$

$$|\mathbf{a}_2| = \sqrt{64 + 36} = 10.$$

The basis is

$$\mathbf{b}_1 = \frac{\mathbf{a}_1}{|\mathbf{a}_1|} = \frac{1}{5}(3, 4) = \left(\frac{3}{5}, \frac{4}{5}\right),$$

$$\mathbf{b}_2 = \frac{\mathbf{a}_2}{|\mathbf{a}_2|} = \frac{1}{10}(8, -6) = \left(\frac{4}{5}, -\frac{3}{5}\right).$$

Then

$$(1, 0) = c_1\mathbf{b}_1 + c_2\mathbf{b}_2,$$

where

$$c_1 = (1, 0) \cdot \mathbf{b}_1 = (1, 0) \cdot \left(\frac{3}{5}, \frac{4}{5}\right) = \frac{3}{5}$$

and

$$c_2 = (1, 0) \cdot \mathbf{b}_2 = (1, 0) \cdot \left(\frac{4}{5}, \frac{3}{5} \right) = \frac{4}{5}.$$

EXAMPLE 8   Find the components of $(2, 1)$ with respect to the above basis.

Solution

$$c_1 = (2, 1) \cdot \left(\frac{3}{5}, \frac{4}{5} \right) = \frac{6}{5} + \frac{4}{5} = 2$$

$$c_2 = (2, 1) \cdot \left(\frac{4}{5}, -\frac{3}{5} \right) = \frac{8}{5} - \frac{3}{5} = 1.$$

Note that these are also the components of $(2, 1)$ with respect to the standard basis. This implies that there is some unusual symmetry in the configuration, and the diagram below shows what it is.

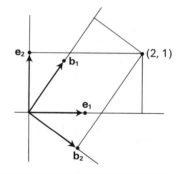

PROBLEMS FOR SECTION 3

Determine the length of each of the following vectors, and give the unit vector in the same direction. Draw each vector as an arrow from the origin and also as an arrow from the point $(-1, 2)$.

1. $\mathbf{v} = (3, 4)$ 2. $\mathbf{x} = (4, -1)$
3. $\mathbf{a} = (-3, 1)$ 4. $\mathbf{x} = (1/2, 1/2)$
5. $\mathbf{v} = (1, -2)$

Find $\mathbf{a} \cdot \mathbf{x}$ and the angle ϕ between $\mathbf{a}$ and $\mathbf{x}$ ($\phi = $ arc cos (something)). Draw $\mathbf{a}$ and $\mathbf{x}$ as arrows from the origin, and label ϕ.

6. $\mathbf{a} = (2, 1), \mathbf{x} = (1, 4)$ 7. $\mathbf{a} = (1/2, -2), \mathbf{x} = (1, -1)$
8. $\mathbf{a} = (3, -1/2), \mathbf{x} = (-1/2, 2/3)$ 9. $\mathbf{a} = (1/4, -1/3), \mathbf{x} = (1/3, 1/4)$
10. $\mathbf{a} = (2, -1), \mathbf{x} = (2, 2)$

11. $\mathbf{u}$ is a unit vector perpendicular to $\mathbf{a} = (1, 2)$. Determine $\mathbf{u}$ (two answers).
12. $\mathbf{u}$ is a unit vector making the angle $\pi/3$ with $\mathbf{a} = (1, 0)$. Determine $\mathbf{u}$ (two answers).
13. $\mathbf{u}$ is a unit vector making the angle $\pi/3$ with $\mathbf{a} = (1, 1)$. Determine $\mathbf{u}$.
14. The same question when $\mathbf{a} = (1, 2)$.
15. Determine the collection of all vectors $\mathbf{x}$ making the angle $\pi/3$ with $\mathbf{a} = (1, 0)$.

16. Find the equation of the line through $x_0 = (1, -5)$ perpendicular to the vector $a = (-2, 1)$.

17. Find the equation of the line through the point $x_0 = (3, 2)$ perpendicular to the vector $(-2, 3)$.

18. Let l be the line through the point x_0 perpendicular to the nonzero vector a. Show that l goes through the origin if and only if x_0 is perpendicular to a.

19. Let l be the line through the point x_0 perpendicular to the nonzero vector a. Show that its distance from the origin is $|a \cdot x_0|/|a|$.

20. Prove the law (s4): $t(x \cdot a) = (tx) \cdot a$.

The vectors $b_1 = (2/\sqrt{5}, 1/\sqrt{5})$ and $b_2 = (-1/\sqrt{5}, 2/\sqrt{5})$ form an orthonormal basis. For each of the following vectors x, compute its components x_1 and x_2 with respect to the basis $\{b_1, b_2\}$, by Theorem 7. Then directly verify the correctness of the equation

$$x = x_1 b_1 + x_2 b_2.$$

21. $x = (3, 4)$ 22. $x = (-1, 1)$

23. $x = (0, 1)$ 24. $x = (1/2, 1/2)$

25. $x = (1, 0)$

For each pair a and x given below, find the (scalar) component of x in the direction of a.

26. $a = (1, 1), x = (2, 1)$ 27. $a = (1, -2), x = (2, 1)$

28. $a = (1, -2), x = (1, 2)$ 29. $a = (10, 1), x = (1, 0)$

30. $a = (1, 0), x = (10, 1)$ 31. $a = (10, 1), x = (0, 1)$

32. If a is a given nonzero vector and k is a given constant, we know that $a \cdot x = k$ is the equation of a line perpendicular to a. Now reinterpret this locus as the collection of all vectors x having a certain component property.

4. COORDINATES IN SPACE

A vector in space has *three* components. They are determined in the same manner as in the plane, except that now a *three*-dimensional coordinate system is involved, having three mutually perpendicular lines in space through a common origin O as axes. We generally draw and label such space axes as shown below. This configuration is called a *right-handed* coordinate system, because if the thumb of the right hand points in the direction of the positive z-axis, then the curl of the fingers gives the positive direction of rotation in the x, y-plane, from the positive x-axis to the positive y-axis.

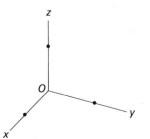

Each axis is given a coordinate system with common origin at O and a common unit of distance. The plane through the y- and z-axes is then a coordinate plane, called the yz-coordinate plane. It is viewed as a vertical "back wall" plane. The xy-plane is a horizontal "floor" plane, and it has the normal xy-plane appearance when viewed from a point on the positive z-axis. The xz-plane is a vertical "left wall" plane perpendicular to the yz-plane. Figures in these coordinate planes have to be distorted when they are drawn on the page. In particular, the fact that their axes are perpendicular cannot be shown and must be visualized.

The three coordinate planes divide space into 8 regions, called *octants*, coming together at O. Imagine a room corner inside a building. Then the floor extends into other rooms and the walls extend down to the floor below. We can indicate the extended edges that we can't see by dotted lines or light lines. There are eight rooms sharing the corner point O, four on our floor having O as a floor point, and four on the floor below having O as a ceiling corner point.

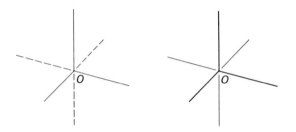

Consider now an arbitrary point p in space. Its coordinates can be obtained by dropping perpendiculars to the axes, just as in the plane. However, the resulting figure, shown at the left below, is hard to visualize. We therefore describe the process differently.

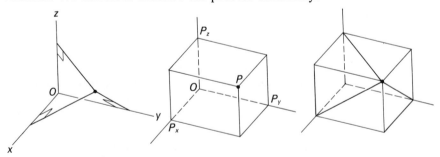

First, pass the plane through p perpendicular to the x-axis. Then the x-coordinate of p is determined by where this plane intersected the x-axis. If the other two coordinates of p are defined in the same way, then the three planes through p and the three coordinate planes altogether determine a rectangular "coordinate box" for p. It can be described as the rectangular box having three of its edges along the coordinate axes and having p as the vertex diagonally opposite to the origin O. The middle figure above shows this box, and it should appear to be three-dimensional. In the right figure the three original perpendicular lines are now seen as lying on faces of the box.

Just as we did in one and two dimensions, we identify the geometric point p with its coordinate triple (x, y, z) and speak of the point (x, y, z).

If the line segment joining the points

$$p_1 = (x_1, y_1, z_1) \qquad \text{and} \qquad p_2 = (x_2, y_2, z_2)$$

is not parallel to any coordinate plane, then p_1 and p_2 are diagonally opposite vertices of a certain rectangular box, just as were p and the origin O in the above coordinate box.

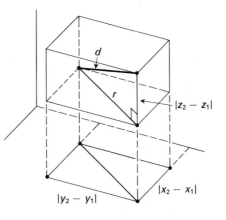

The edge lengths are now given by the coordinate differences, as shown, and the distance formula is again a consequence of the Pythagorean theorem:

$$d^2 = r^2 + (z_2 - z_1)^2 = (x_2 - x_1)^2 + (y_2 - y_1)^2 + (z_2 - z_1)^2.$$

Thus:

The distance d between the points (x_1, y_1, z_1) and (x_2, y_2, z_2) in coordinate three-space is given by

$$d = \sqrt{(x_2 - x_1)^2 + (y_2 - y_1)^2 + (z_2 - z_1)^2}.$$

There is still the problem that if we indicate a space point P by a dot in the usual way, then we don't really know where P is, because there is a whole line in space that looks "end on" like a single point. The simplest way to fix the location of P is to show also the point Q directly below it in the xy-plane. If you can visualize the point Q, shown below, as a point in the floor plane, and P as a point directly over it, they you should be able to "see" P in space.

We can also convey the space location of P by other auxiliary lines, the coordinate box giving probably the strongest impression of where P is.

PROBLEMS FOR SECTION 4

Draw a three-dimensional coordinate system and locate the following points on it.

1. $P_1(6, 2, 1)$
2. $P_2(-6, 3, 1)$
3. $P_3(1, -2, 5)$
4. $P_4(2, 3, 4)$
5. $P_5(2, 1, -3)$
6. $P_6(-1, -2, 3)$
7. $P_7(2, -1, -4)$
8. $P_8(-2, 3, -1)$
9. $P_9(-6, -2, -1)$
10. $P_{10}(2, 1, -4)$

11. Find the distance between P_1 and P_4.
12. Find the distance between P_4 and P_9.
13. Find the distance between P_{10} and P_1.
14. Find the distance between P_6 and P_5.
15. Find the distance between P_3 and P_2.

What is the locus of a point:

16. Whose x-coordinate is always 0?
17. Whose y-coordinate is always 3?
18. Whose x- and z-coordinates are always 0?

5. VECTORS IN SPACE

At this point the reader is asked to go over the discussion in Section 1 and note that everything said there is valid in space, the sole difference being that a space vector has three components instead of two. In particular, the component characterization of the sum $x + a$ and the product tx are now

$$(x, y, z) + (a, b, c) = (x + a, y + b, z + c),$$

$$t(x, y, z) = (tx, ty, tz).$$

The magnitude $|x|$ of the vector $x = (x, y, z)$ is the distance from the origin to x, and its formula is

$$|x| = \sqrt{x^2 + y^2 + z^2}.$$

Then the formula for the distance d between the space points x_1 and x_2 can be rewritten, in terms of vector subtraction, as

$$d = |x_1 - x_2|.$$

The *direction* of $a = (a, b, c)$ is again presented by its unit vector

$$u = \frac{a}{|a|} = \left(\frac{a}{r}, \frac{b}{r}, \frac{c}{r} \right),$$

where $r = |\mathbf{a}| = \sqrt{a^2 + b^2 + c^2}$. But the geometric interpretation of the components of $\mathbf{u}$ necessarily changes slightly from what it was in the two-dimensional case. There we could write

$$\mathbf{u} = (\cos \theta, \sin \theta),$$

where θ is the angle from the positive x-axis to the direction $\mathbf{u}$. This clearly does not generalize to three dimensions, but the new interpretation of $\mathbf{u}$ is nevertheless very close.

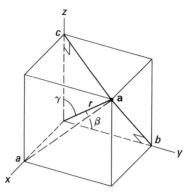

If a coordinate box is drawn with vertices at $\mathbf{a}$, and if the *face* diagonals are drawn from $\mathbf{a}$ (two are shown above), then it is apparent that

$$\frac{a}{r} = \cos \alpha, \qquad \frac{b}{r} = \cos \beta, \qquad \frac{c}{r} = \cos \gamma,$$

where α, β, and γ are the angles between the direction of $\mathbf{a}$ and the positive x-, y-, and z-axes. Thus,

$$\mathbf{u} = \frac{\mathbf{a}}{|\mathbf{a}|} = (\cos \alpha, \cos \beta, \cos \gamma),$$

and $\mathbf{u}$ is the triple of so-called "direction cosines" of $\mathbf{a}$.

If direction cosines are introduced in the plane, as shown below, then $\theta - \phi = \pi/2$, and

$$\cos \phi = \cos\left(\theta - \frac{\pi}{2}\right) = \sin \theta.$$

Thus,

$$(\cos \theta, \cos \phi) = (\cos \theta, \sin \theta),$$

and the direction cosines reduce to the cosine and sine of the single angle θ. Remember, however, that when the plane direction is presented as $(\cos \theta, \sin \theta)$, then θ must be the *signed* angle *from* the positive x-axis *to* the vector direction. On the other hand, if the direction is written $(\cos \theta, \cos \phi)$, then θ

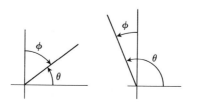

and ϕ can be taken to be the positive *unsigned* angles *between* the vector direction and the axes, because of the identity $\cos \alpha = \cos(-\alpha)$.

The parametric equation of the line through the distinct points $\mathbf{x}_0$ and $\mathbf{x}_1$ comes out to be

$$\mathbf{x} = \mathbf{x}_0 + t(\mathbf{x}_1 - \mathbf{x}_0),$$

just as before.

EXAMPLE 1   The line through $\mathbf{x}_o = (1, 2, 3)$ and $\mathbf{x}_1 = (3, -2, 1)$ has the parametric equation

$$\mathbf{x} = (1, 2, 3) + t(2, -4, -2).$$

This is the equation

$$(x, y, z) = (1 + 2t, 2 - 4t, 3 - 2t),$$

and is equivalent to the three scalar equations

$$x = 1 + 2t, \qquad y = 2 - 4t, \qquad z = 3 - 2t.$$

Here the parameter t cannot be eliminated, and the parametric equations are *not* equivalent to a single linear equation in x, y, and z. In fact, we shall see in the next section that such an equation has a *plane* as its graph.

EXAMPLE 2   The midpoint of the segment from $\mathbf{x}_1 = (1, 2, 3)$ to $\mathbf{x}_2 = (3, -4, 5)$ is

$$\frac{\mathbf{x}_1 + \mathbf{x}_2}{2} = \frac{(1, 2, 3) + (3, -4, 5)}{2} = \frac{(4, -2, 8)}{2} = (2, -1, 4).$$

EXAMPLE 3   We saw earlier that the three medians of the triangle $\mathbf{a}_1\mathbf{a}_2\mathbf{a}_3$ are concurrent in the point $(\mathbf{a}_1 + \mathbf{a}_2 + \mathbf{a}_3)/3$. This point is actually the *centroid* of the triangle. It is algebraically the average value of the three vectors $\mathbf{a}_1$, $\mathbf{a}_2$, and $\mathbf{a}_3$. The proof given for this median result is valid for a triangle in space. Now consider four space points $\mathbf{a}_1$, $\mathbf{a}_2$, $\mathbf{a}_3$, and $\mathbf{a}_4$ not lying in any one plane. They are the vertices of *tetrahedron*. Through each vertex we draw a line to the centroid of the opposite side, and we ask whether these four lines are concurrent.

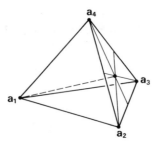

Here we shall just guess the answer and then verify that it is correct.

We guess that

$$\frac{\mathbf{a}_1 + \mathbf{a}_2 + \mathbf{a}_3 + \mathbf{a}_4}{4},$$

the *average* of the four vertex vectors, is the point we want. Is it on the segment joining the vertex $\mathbf{a}_1$ to the centroid $(\mathbf{a}_2 + \mathbf{a}_3 + \mathbf{a}_4)/3$? Yes, because the equation

$$\frac{\mathbf{a}_1 + \mathbf{a}_2 + \mathbf{a}_3 + \mathbf{a}_4}{4} = \frac{1}{4}\mathbf{a}_1 + \frac{3}{4}\left(\frac{\mathbf{a}_2 + \mathbf{a}_3 + \mathbf{a}_4}{3}\right)$$

shows it to be the point 3/4 of the way along the segment from the vertex to the centroid.

The dot product in three dimensions has essentially the same definition,

$$(x, y, z) \cdot (r, s, t) = xr + ys + zt,$$

and it has the same properties (s1) through (s4), for the same reasons. In view of this second example of a dot product, we can take these properties as the *axioms* for a dot product. Here they are again, for ready reference:

(s1) $\mathbf{x} \cdot \mathbf{x} = |\mathbf{x}|^2,$

(s2) $\mathbf{x} \cdot \mathbf{a} = \mathbf{a} \cdot \mathbf{x},$

(s3) $(\mathbf{x}_1 + \mathbf{x}_2) \cdot \mathbf{a} = \mathbf{x}_1 \cdot \mathbf{a} + \mathbf{x}_2 \cdot \mathbf{a},$

(s4) $(t\mathbf{x}) \cdot \mathbf{a} = t(\mathbf{x} \cdot \mathbf{a}).$

The formula relating the dot product to $\cos \phi$ cannot be proved as before. Instead, we shall see that it is a consequence of the dot-product axioms. Consider first the special case of perpendicularity.

THEOREM 8 *The vectors $\mathbf{a}$ and $\mathbf{x}$ are perpendicular if and only if $\mathbf{a} \cdot \mathbf{x} = 0$.*

Proof. By the Pythagorean theorem, $\mathbf{a}$ and $\mathbf{x}$ are perpendicular if and only if

$$|\mathbf{a}|^2 + |\mathbf{x}|^2 = |\mathbf{a} - \mathbf{x}|^2.$$

Using (s1) and then the two versions of (s3), this beomes

$$\mathbf{a} \cdot \mathbf{a} + \mathbf{x} \cdot \mathbf{x} = (\mathbf{a} - \mathbf{x}) \cdot (\mathbf{a} - \mathbf{x})$$
$$= \mathbf{a} \cdot (\mathbf{a} - \mathbf{x}) - \mathbf{x} \cdot (\mathbf{a} - \mathbf{x})$$
$$= \mathbf{a} \cdot \mathbf{a} - \mathbf{a} \cdot \mathbf{x} - \mathbf{x} \cdot \mathbf{a} + \mathbf{x} \cdot \mathbf{x}.$$

Since $\mathbf{x} \cdot \mathbf{a} = \mathbf{a} \cdot \mathbf{x}$, this equation reduces to $2(\mathbf{a} \cdot \mathbf{x}) = 0$, and hence to $\mathbf{a} \cdot \mathbf{x} = 0$, proving the theorem. ∎

EXAMPLE 4 We see that $\mathbf{a} = (-1, 2, 3)$ and $\mathbf{x} = (3, 3, -1)$ are perpendicular, by computing

$$\mathbf{a} \cdot \mathbf{x} = (-1)3 + 2 \cdot 3 + 3 \cdot (-1) = -3 + 6 - 3 = 0.$$

EXAMPLE 5   Find the most general vector $\mathbf{x}$ perpendicular to

$$\mathbf{a}_1 = (1, 2, 3) \qquad \text{and} \qquad \mathbf{a}_2 = (2, 3, 1).$$

Solution. $\mathbf{x} = (x, y, z)$ will be perpendicular to these two vectors if and only if $\mathbf{a}_1 \cdot \mathbf{x} = 0$ and $\mathbf{a}_2 \cdot \mathbf{x} = 0$, that is, if and only if

$$x + 2y + 3z = 0,$$
$$2x + 3y + z = 0.$$

Eliminating z (by multiplying the second equation by 3 and subtracting), we see that

$$5x + 7y = 0 \qquad \text{and} \qquad y = -(5/7)x.$$

Similarly, we find that $z = x/7$. Thus a vector $\mathbf{x}$ is perpendicular to both $\mathbf{a}_1 = (1, 2, 3)$ and $\mathbf{a}_2 = (2, 3, 1)$ if and only if it can be written in the form

$$\mathbf{x} = \left(x, -\frac{5}{7}x, \frac{1}{7}x \right) = x\left(1, -\frac{5}{7}, \frac{1}{7} \right).$$

Next we repeat an argument given in Section 3.

THEOREM 6   *If $\mathbf{x}$ and $\mathbf{a}$ are any two vectors, with $\mathbf{a} \neq 0$, then there is a uniquely determined scalar c such that $\mathbf{x} - c\mathbf{a}$ is perpendicular to $\mathbf{a}$.*

Proof. This is obvious geometrically from a diagram, but we want to deduce it just from the axioms. If $\mathbf{x} - c\mathbf{a} \perp \mathbf{a}$, then

$$0 = (\mathbf{x} - c\mathbf{a}) \cdot \mathbf{a} = \mathbf{x} \cdot \mathbf{a} - c(\mathbf{a} \cdot \mathbf{a})$$
$$= \mathbf{x} \cdot \mathbf{a} - c\,|\mathbf{a}|^2,$$

by (s3), (s4), and (s1). Therefore,

$$c = \frac{\mathbf{x} \cdot \mathbf{a}}{|\mathbf{a}|^2}.$$

Conversely, with this value of c, the calculation above can be reversed and $\mathbf{x} - c\mathbf{a}$ is perpendicular to $\mathbf{a}$. ∎

The $\cos \phi$ law is implicit in this result, for we now have a right triangle as shown below at the left if $c > 0$, and at the right if $c < 0$. In particular, the sign of c is the sign of $\cos \phi$, and since $\cos \phi = $ adjacent/hypotenuse except for sign, it follows that

$$\cos \phi = \frac{c\,|\mathbf{a}|}{|\mathbf{x}|} = \frac{\mathbf{x} \cdot \mathbf{a}}{|\mathbf{a}|^2} \cdot \frac{|\mathbf{a}|}{\mathbf{x}} = \frac{\mathbf{x} \cdot \mathbf{a}}{|\mathbf{x}|\,|\mathbf{a}|}.$$

We have proved:

THEOREM 9 *If ϕ is the angle between the nonzero vectors **a** and **x**, then*

$$\cos \phi = \frac{\mathbf{a} \cdot \mathbf{x}}{|\mathbf{a}|\,|\mathbf{x}|}.$$

REMARK This argument could have been used in the plane as well.

EXAMPLE 6 Find the angle between $(1, 2, 1)$ and $(1, -1, 2)$.

Solution

$$\cos \phi = \frac{\mathbf{a} \cdot \mathbf{x}}{|\mathbf{a}|\,|\mathbf{x}|} = \frac{1 \cdot 1 + 2(-1) + 1 \cdot 2}{\sqrt{6}\sqrt{6}} = \frac{1}{6},$$

so

$$\phi = \arccos\left(\frac{1}{6}\right).$$

Just as in the plane, the unit vectors along the coordinate axes are given special designations and are called the standard basic vectors. Now there are three,

$$\mathbf{i} = (1, 0, 0) \qquad \mathbf{j} = (0, 1, 0), \qquad \mathbf{k} = (0, 0, 1),$$

and, as before,

$$\mathbf{x} = (x, y, z) = x\mathbf{i} + y\mathbf{j} + z\mathbf{k}$$

for every vector **x**.

PROBLEMS FOR SECTION 5

In the following, find the unit vector in the same direction as the given vector.

1. $(2, 1, 2)$
2. $(-3, 6, -6)$
3. $(7, -6, -6)$
4. $(-3, 4, 12)$
5. $(1/2, 1/2, 1/4)$

Given **x** and **a**, find $\mathbf{x} \cdot \mathbf{a}$ and the angle ϕ between **x** and **a**.

6. $\mathbf{x} = (2, 1, 2)$, $\mathbf{a} = (1, 1, 0)$
7. $\mathbf{x} = (1/2, 0, -1/2)$, $\mathbf{a} = (2, 1, 2)$
8. $\mathbf{x} = (3, 1, \sqrt{2})$, $\mathbf{a} = (1, 0, 0)$
9. $\mathbf{x} = (1, 1/2, -1)$, $\mathbf{a} = (0, 1, 1/2)$
10. $\mathbf{x} = (1, 1, 1)$, $\mathbf{a} = (2, -2, 1)$ ($\phi = \arctan(\ )$)

Write the parametric equation of the line through the two given points.

11. $(1, -2, 1)$, $(2, 1, -1)$
12. $(1/2, 0, 1)$, $(1/2, -1, 2)$
13. $(-2, -1, 2)$, $(1, 3, 4)$
14. $(4, 1/4, -1)$, $(-1, 1/2, 1)$
15. $(-1, -2, 1)$, $(2, 0, 2)$

Write the scalar equations of the line through the two given points.

16. $(-3, 1, 2)$, $(2, -1, -1)$
17. $(1, -1, 4)$, $(1, 1/2, 0)$
18. $(-1, 1/2, -1/3)$, $(-2, -2, 1)$
19. $(-2, -1, 0)$, $(1/2, -1, -2)$
20. $(1/4, 1/2, 1/3)$, $(1, 0, 1)$

Write the parametric equation of the line through $\mathbf{x}_0$ in the direction of the vector $\mathbf{a}$, where:

21. $\mathbf{x}_0 + (0, 0, 0)$, $\mathbf{a} = (1, -2, -1)$.   22. $\mathbf{x}_0 = (4, -2, 2)$, $\mathbf{a} = (-2, 1, 1)$.

23. $\mathbf{x}_0 = (1, 1, 1)$, $\mathbf{a} = (2, 0, -1)$.   24. $\mathbf{x}_0 = (-3, 1, 2)$, $\mathbf{a} = (5, -2, -3)$.

25. Let $\mathbf{a}$ and $\mathbf{l}$ be two unit vectors, making angles, α, β, γ, and λ, μ, ν, respectively, with the coordinate axes. Interpret the formula

$$\cos \phi = \cos \alpha \cos \lambda + \cos \beta \cos \mu + \cos \gamma \cos \nu.$$

Find the most general vector $\mathbf{x}$ perpendicular to $\mathbf{a}_1$ and $\mathbf{a}_2$ in the following problems.

26. $\mathbf{a}_1 = (1, 2, 3)$, $\mathbf{a}_2 = (0, 1, 2)$ 27. $\mathbf{a}_1 = (1, 0, 1)$, $\mathbf{a}_2 = (1, 1, 0)$

28. $\mathbf{a}_1 = (2, -1, 1)$, $\mathbf{a}_2 = (0, 2, 2)$

6. PLANES

We saw earlier that the dot-product perpendicularity condition led to an easy derivation of the equation of a line. In three dimensions, the same proof leads to the equation of a *plane*.

Let $\mathbf{x}_0 = (x_0, y_0, z_0)$ be any fixed point and let $\mathbf{a} = (a, b, c)$ be a fixed nonzero vector. Let π be the plane that passes through the point $\mathbf{x}_0$ and is perpendicular to the direction of $\mathbf{a}$. A point $\mathbf{x}$ will lie on the plane π if and only if the arrow $\overrightarrow{\mathbf{x}_0\mathbf{x}}$ is perpendicular to $\mathbf{a}$. Thus, $\mathbf{x}$ is on π if and only if

$$\mathbf{a} \cdot (\mathbf{x} - \mathbf{x}_0) = 0,$$

or

$$a(x - x_0) + b(y - y_0) + c(z - z_0) = 0.$$

This equation is therefore the "point–direction" equation of the plane.

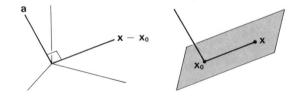

If we set

$$(ax_0 + by_0 + cz_0) = d,$$

the equation takes the general linear form

$$ax + by + cz = d.$$

Conversely, any such first-degree equation is the equation of a plane, provided that at least one of the coefficients a, b, and c is really there (nonzero). For, if $\mathbf{x}_0 = (x_0, y_0, z_0)$ is any fixed triple satisfying the equation

$$ax_0 + by_0 + cz_0 = d,$$

then subtracting the two equations gives the equivalent form

$$a(x - x_0) + b(y - y_0) + c(z - z_0) = 0,$$

or

$$\mathbf{a} \cdot (\mathbf{x} - \mathbf{x}_0) = 0.$$

Thus, the points $\mathbf{x} = (x, y, z)$ whose coordinates satisfy the equation $ax + by + cz - d = 0$ are exactly the points $\mathbf{x}$ such that $\mathbf{x} - \mathbf{x}_0$ is perpendicular to $\mathbf{a}$, that is, the points lying on the plane π through $\mathbf{x}_0$ and perpendicular to $\mathbf{a}$. Altogether we have proved:

THEOREM 10 *The plane through the point (x_0, y_0, z_0) and perpendicular to the nonzero vector $\mathbf{a} = (a, b, c)$ has the equation*

$$\mathbf{a} \cdot (\mathbf{x} - \mathbf{x}_0) = 0,$$

or

$$ax + by + cz = d,$$

where $d = \mathbf{a} \cdot \mathbf{x}_0$. Conversely, the graph of the equation

$$ax + by + cz = d$$

is a plane perpendicular to $\mathbf{a} = (a, b, c)$, provided that $\mathbf{a}$ is not the zero vector.

EXAMPLE 1 Find the plane through the point $\mathbf{x}_0 = (2, -1, 4)$ and perpendicular to the vector $\mathbf{n} = (-3, 5, 1)$.

Solution. Substituting in the equation

$$\mathbf{n} \cdot (\mathbf{x} - \mathbf{x}_0) = 0$$

gives

$$-3(x - 2) + 5(y + 1) + 1(z - 4) = 0,$$

or

$$-3x + 5y + z = -7$$

as the equation of the plane.

If the nonzero vectors $\mathbf{a} = (a_1, a_2, a_3)$ and $\mathbf{b} = (b_1, b_2, b_3)$ are not parallel then there is an essentially unique (unique up to a scalar multiple) nonzero vector $\mathbf{x}$ that is perpendicular to them both, and it is useful to have a scheme for writing $\mathbf{x}$. The conditions on $\mathbf{x}$, $\mathbf{a} \cdot \mathbf{x} = 0$ and $\mathbf{b} \cdot \mathbf{x} = 0$, are the following two equations in three unknowns:

$$a_1 x + a_2 y + a_3 z = 0,$$
$$b_1 x + b_2 y + b_3 z = 0.$$

Solving them by the usual method of elimination leads to a particular solution vector that is called the *cross product* of $\mathbf{a}$ and $\mathbf{b}$ and designated $\mathbf{a} \times \mathbf{b}$, namely:

DEFINITION $$\mathbf{a} \times \mathbf{b} = (a_2 b_3 - b_2 a_3, a_3 b_1 - b_3 a_1, a_1 b_2 - b_1 a_2).$$

At some point it must be shown that $\mathbf{a} \times \mathbf{b}$ is not zero. This can be checked directly from the formula, and will be left as an exercise. In summary:

THEOREM 11 *If $\mathbf{a}$ and $\mathbf{b}$ are two nonparallel (and nonzero) space vectors then a vector $\mathbf{x}$ is perpendicular to them both if and only if $\mathbf{x} = t(\mathbf{a} \times \mathbf{b})$ for some scalar t.*

EXAMPLE 2   Find the plane that contains the three points $\mathbf{x}_0 = (1, 2, 1)$, $\mathbf{x}_1 = (2, -2, 0)$, and $\mathbf{x}_2 = (1, 0, -3)$.

Solution. The most elementary procedure is to substitute the above values into the general equation of a plane and thus obtain three equations for the unknown coefficients.

A more efficient procedure is to note that the vectors $\mathbf{x}_1 - \mathbf{x}_0 = (1, -4, -1)$ and $\mathbf{x}_2 - \mathbf{x}_0 = (0, -2, -4)$ are parallel to the plane, so their cross product

$$(16 - 2, 0 + 4, -2 - 0) = (14, 4, -2) = 2(7, 2, -1)$$

must be normal to the plane. Therefore its equation is

$$7(x - 1) + 2(y - 2) - (z - 1) = 0,$$

or

$$7x + 2y - z = 10.$$

EXAMPLE 3   Find the direction of the line of intersection of the two planes

$$2x + 3y - z = 4 \quad \text{and} \quad x + y + z = 0.$$

Solution. The coefficient vectors $(2, 3, -1)$ and $(1, 1, 1)$ are normal to the two planes and hence to the line of intersection, so their cross product gives the desired direction. It is

$$(3 + 1, -1 - 2, 2 - 3) = (4, -3, -1).$$

The "standard basis expansion" of $\mathbf{x}$ can be helpful here: the cross product $\mathbf{a} \times \mathbf{b}$ is simply the formal expansion of the 3×3 determinant

$$\begin{vmatrix} \mathbf{i} & \mathbf{j} & \mathbf{k} \\ a_1 & a_2 & a_3 \\ b_1 & b_2 & b_3 \end{vmatrix}.$$

For by following the usual expansion procedure (using minors of the top row), we get

$$\mathbf{i} \begin{vmatrix} a_2 & a_3 \\ b_2 & b_3 \end{vmatrix} - \mathbf{j} \begin{vmatrix} a_1 & a_3 \\ b_1 & b_3 \end{vmatrix} + \mathbf{k} \begin{vmatrix} a_1 & a_2 \\ b_1 & b_2 \end{vmatrix}$$

$$= (a_2 b_3 - b_2 a_3)\mathbf{i} + (a_3 b_1 - b_3 a_1)\mathbf{j} + (a_1 b_2 - b_1 a_2)\mathbf{k}.$$

This may be the best way of computing cross products. Thus in Example 3 above we write down the determinant

$$\begin{vmatrix} \mathbf{i} & \mathbf{j} & \mathbf{k} \\ 2 & 3 & -1 \\ 1 & 1 & 1 \end{vmatrix}$$

and read the answer $4\mathbf{i} - 3\mathbf{j} - \mathbf{k} = (4, -3, -1)$.

PROBLEMS FOR SECTION 6

Write the equation of the plane through the given point and perpendicular to the given vector.

1. Point $(0, 0, 0)$; vector $(1, 2, 1)$ 2. Point $(-1, 1, 1/2)$; vector $(3, -1, -2)$

3. Point $(4, 1, 1)$; vector $(1/2, 1, 0)$ 4. Point $(1, -1, 4)$; vector $(-1, -2, 1)$

5. Point $(6, 4, -6)$; vector $(1/3, 1/4, -1/6)$

In Problems 6 through 9 find an independent pair of vectors **a** and **x** each perpendicular to the given vector.

6. $(1, 1, 0)$ 7. $(1, 2, -1)$ 8. $(2, 2, 2)$ 9. $(5, -1, 3)$

10. Find the point of intersection of the line
$$\mathbf{x} = (1, 2, -1) + t(2, -2, 3)$$
and the plane
$$3x + y - z = 2.$$

11. Let π be the plane through the origin and the two points $\mathbf{a}_1(2, -1, 1)$ and $\mathbf{a}_2(1, 3, 1)$. Prove that its intersection with the xy-coordinate plane is the line
$$\mathbf{x} = t(1, -4, 0).$$
Eliminate the parameter and find the xy-equation of the line.

12. Find the plane π through the two lines
$$\mathbf{x} = (1, 2, -1) + t(2, -2, 3),$$
$$\mathbf{x}' = (1, 2, -1) + s(5, 1, 0).$$

13. Find the line of intersection of the planes
$$x + 2y - z = 0 \quad \text{and} \quad 3x - y + 2z = 0.$$

14. Find the plane through the origin perpendicular to each of the planes
$$x + y + z = 3 \quad \text{and} \quad x - 2y - z = 1.$$

15. Find the plane through $(1, 2, -1)$ perpendicular to each of the planes
$$2x - y + z = 0 \quad \text{and} \quad x + 3y - 4z = 1.$$

EXTRA PROBLEMS FOR CHAPTER 13

The vectors $\mathbf{b}_1 = (2/\sqrt{5}, 1/\sqrt{5})$ and $\mathbf{b}_2 = (-1/\sqrt{5}, 2/\sqrt{5})$ form an orthonormal basis for the plane. For each of the following vectors **x**, compute its components x_1 and x_2 with respect to the basis $\{\mathbf{b}_1, \mathbf{b}_2\}$, by Theorem 7. Then directly verify the correctness of the equation
$$\mathbf{x} = x_1\mathbf{b}_1 + x_2\mathbf{b}_2.$$

1. $\mathbf{x} = (\sqrt{5}, \sqrt{5})$ 2. $\mathbf{x} = (6/\sqrt{5}, -2/\sqrt{5})$ 3. $\mathbf{x} = (-1/\sqrt{5}, 1/\sqrt{5})$

4. $\mathbf{x} = (2/\sqrt{5}, -1/\sqrt{5})$ 5. $\mathbf{x} = (\sqrt{5}, -2/\sqrt{5})$

6. Let $\{\mathbf{b}_1, \mathbf{b}_2\}$ be an orthonormal basis for the plane. Prove that if x_1 and x_2 are the components of **x** and u_1 and u_2 are the components of **u** (all with respect to the basis $\mathbf{b}_1, \mathbf{b}_2$), then
$$\mathbf{x} \cdot \mathbf{u} = x_1u_1 + x_2u_2.$$

7. Show that if ϕ is the angle from the direction of $\mathbf{a} = (a_1, a_2)$ to the direction of $\mathbf{b} = (b_1, b_2)$, then

$$\sin \phi = \frac{b_2 a_1 - b_1 a_2}{|\mathbf{a}|\,|\mathbf{b}|}.$$

Let $\mathbf{a}$ be any nonzero vector and set

$$\mathbf{n} = (-a_2, a_1).$$

Show that $\mathbf{n} \perp \mathbf{a}$ and that the direction of $\mathbf{n}$ *leads* the direction of $\mathbf{a}$ by $90°$.

8. The plane can be distinguished from space by the fact that two nonparallel lines in the plane *must* intersect. Assuming this, show that if $\mathbf{a}$ and $\mathbf{b}$ are two nonparallel (and nonzero) vectors in the plane, then any other vector $\mathbf{c}$ can be expressed in the form

$$\mathbf{c} = s\mathbf{a} + t\mathbf{b}$$

in one and only one way (i.e., for a uniquely determined pair of scalars s and t). [*Hint:* consider the two lines with parameters s and t given by $\mathbf{x} = \mathbf{c} - s\mathbf{a}$ and $\mathbf{x} = t\mathbf{b}$.]

The scalars s and t in the above representation of $\mathbf{c}$ are called the *components of $\mathbf{c}$ with respect to the basis $\{\mathbf{a}, \mathbf{b}\}$*. They can be found by solving a pair of simultaneous equations. For example, to find the components of $(2, 1)$ with respect to $\mathbf{a} = (1, 3)$ and $\mathbf{b} = (2, -1)$, we have to solve the vector equation

$$(2, 1) = s\mathbf{a} + t\mathbf{b} = s(1, 3) + t(2, -1)$$
$$= (s + 2t, 3s - t).$$

This is equivalent to the pair of scalar equations

$$s + 2t = 2$$
$$3s - t = 2,$$

and solving them simultaneously gives $s = 4/7$, $t = 5/7$. Thus

$$(s, 1) = \frac{4}{7}(1, 3) + \frac{5}{7}(2, -1).$$

Find the components of each of the following vectors with respect to the basis $(1, 2)$, $(-1, -1)$.

9. $\mathbf{x} = (7, 1)$ 10. $\mathbf{x} = (-3, 4)$

11. $\mathbf{x} = (6, 1)$ 12. $\mathbf{x} = (-1, -3)$

13. $\mathbf{x} = (0, 4)$ 14. $\mathbf{e}_1 = (1, 0)$

15. $\mathbf{e}_2 = (0, 1)$

16. Let $\mathbf{s}_1 = (s_1, t_1)$ and $\mathbf{s}_2 = (s_2, t_2)$ be the component pairs of

$$\mathbf{e}_1 = (1, 0) \quad \text{and} \quad \mathbf{e}_2 = (0, 1)$$

with respect to a basis $\{\mathbf{a}_1, \mathbf{a}_2\}$. Show that the component pair of any vector $\mathbf{x} = (x, y)$ is $x\mathbf{s}_1 + y\mathbf{s}_2$.

17. Now go back to Problems 9 through 15. Use the components of $\mathbf{e}_1$ and $\mathbf{e}_2$, as determined in Problems 14 and 15, to write out the answers to Problems 9 through 13 by the formula just proved in Problem 16.

18. Prove Problem 8 by algebra. That is, solve the equation

$$\mathbf{x} = s\mathbf{a}_1 + t\mathbf{a}_2$$

for s and t in terms of x and y, by the usual technique for solving two equations in two unknowns. You will find that this solution process works only if $\mathbf{a}_1$ and $\mathbf{a}_2$

satisfy a certain algebraic condition, and that this condition is equivalent to their not being parallel.

19. The component pair $\mathbf{s} = (s, t)$ of a vector $\mathbf{x}$ with respect to a basis $\{\mathbf{a}_1, \mathbf{a}_2\}$ is just another point in the coordinate plane. Show that the operation taking $\mathbf{x}$ to $\mathbf{s}$ (also referred to as the *mapping* $\mathbf{x} \to \mathbf{s}$) is linear. That is, show that if $\mathbf{s}_1$ and $\mathbf{s}_2$ are the component pairs of $\mathbf{x}_1$ and $\mathbf{x}_2$, then

$$a\mathbf{s}_1 + b\mathbf{s}_2$$

is the component pair of $a\mathbf{x}_1 + b\mathbf{x}_2$, for any scalars a and b.

   If we let ϕ be the function that assigns to each vector $\mathbf{x}$ its component pair $\mathbf{s}$, $\mathbf{s} = \phi(\mathbf{x})$, then the conclusion you are to prove can be stated as:

$$\phi(a\mathbf{x}_1 + b\mathbf{x}_2) = a\phi(\mathbf{x}_1) + b\phi(\mathbf{x}_2)$$

That is, ϕ "preserves" linear combinations, and for this reason we call ϕ a linear mapping.

Like the dot product, the cross product has a collection of algebraic properties that we can establish from its definition, and which let us then use the cross product without necessarily referring back to its components. Prove the following properties of the cross product.

20. $\mathbf{a}_1 \times \mathbf{a}_2 = 0$ if and only if $\mathbf{a}_1$ and $\mathbf{a}_2$ are dependent. [*Hint:* With $\mathbf{a}_1 = (a_1, b_1)$, etc., suppose $a_1 \neq 0$ and set $k = (a_2/a_1)$. Show that if $\mathbf{a}_1 \times \mathbf{a}_2 = 0$, then $\mathbf{a}_2 = k\mathbf{a}_1$.]

21. $\mathbf{a}_1 \times \mathbf{a}_2 = -(\mathbf{a}_2 \times \mathbf{a}_1)$

22. $(\mathbf{a}_1 \times \mathbf{a}_2) \cdot \mathbf{a}_3 = \mathbf{a}_1 \cdot (\mathbf{a}_2 \times \mathbf{a}_3)$

23. $\mathbf{a}_1 \times \mathbf{a}_2$ is perpendicular to both $\mathbf{a}_1$ and $\mathbf{a}_2$. (Use Problems 22 and 20.)

24. $\mathbf{a} \times (\mathbf{u}_1 + \mathbf{u}_2) = (\mathbf{a} \times \mathbf{u}_1) + (\mathbf{a} \times \mathbf{u}_2)$

25. $|\mathbf{a}_1 \times \mathbf{a}_2|^2 = |\mathbf{a}_1|^2 |\mathbf{a}_2|^2 - |\mathbf{a}_1 \cdot \mathbf{a}_2|^2$

26. $|\mathbf{a}_1 \times \mathbf{a}_2| = |\mathbf{a}_1| |\mathbf{a}_2| \sin \phi$, where ϕ is the angle between the directions of $\mathbf{a}_1$ and $\mathbf{a}_2$ (from Problem 25 and the dot-product cosine formula).

27. $|\mathbf{a}_1 \times \mathbf{a}_2|$ is the area of the parallellogram having $\mathbf{a}_1$ and $\mathbf{a}_2$ for sides.

28. $|(\mathbf{a}_1 \times \mathbf{a}_2) \cdot \mathbf{a}_3|$ is the volume of the parallelopiped having edges $\mathbf{a}_1$, $\mathbf{a}_2$, and $\mathbf{a}_3$.

29. Let the vectors $\mathbf{a}$ and $\mathbf{b}$ be independent (nonparallel and nonzero). Show that the triple $\mathbf{a}$, $\mathbf{b}$, $\mathbf{a} \times \mathbf{b}$ has right-handed orientation. (Let π be the plane through the origin containing $\mathbf{a}$ and $\mathbf{b}$, and hence perpendicular to $\mathbf{a} \times \mathbf{b}$. Imagine π rotating continuously to the position of the x,y-plane, carrying $\mathbf{a}$ and $\mathbf{b}$ along with it to a pair of vectors in the x,y-plane. Then $\mathbf{a} \times \mathbf{b}$ rotates continuously to a vector pointing along the z-axis, and the magnitude of $\mathbf{a} \times \mathbf{b}$ remains constant, by Problem 26, etc.)

30. We have seen, in the examples and problems, that two nonparallel vectors $\mathbf{a}_1$ and $\mathbf{a}_2$ determine a nonzero vector $\mathbf{a}$ perpendicular to them both, and uniquely up to scalar multiples. That is, if $\mathbf{x}$ is any other vector perpendicular to both $\mathbf{a}_1$ and $\mathbf{a}_2$, then $\mathbf{x} = t\mathbf{a}$ for some scalar t. Use this fact to prove the following result:

   If $\mathbf{x}$ is a vector perpendicular to $\mathbf{a}_1$, $\mathbf{a}_2$, and $\mathbf{a}$, then $\mathbf{x} = 0$.

   In particular, if $\mathbf{a}_1$, $\mathbf{a}_2$, and $\mathbf{a}_3$ are any three mutually perpendicular nonzero vectors, then the only vector perpendicular to them all is 0.

31. Use the result in Problem 30 to prove the following space generalization of Theorem 7:

   Theorem 12. *If $\mathbf{b}_1$, $\mathbf{b}_2$, and $\mathbf{b}_3$ are orthonormal vectors in space, then* *every vector* **x** *can be written*

$$\mathbf{x} = x_1\mathbf{b}_1 + x_2\mathbf{b}_2 + x_3\mathbf{b}_3,$$

where x_i is the scalar component of **x** *in the direction* $\mathbf{b}_i$, $x_i = \mathbf{x} \cdot \mathbf{b}_i$, *for* $i = 1, 2, 3$.

Such a triple $\{\mathbf{b}_1, \mathbf{b}_2, \mathbf{b}_3\}$ is called an *orthonormal basis* for space vectors.

32. Let $\{\mathbf{b}_i\}_1^3$ be an orthonormal basis, and let the vectors **x** and **a** have components $\{x_i\}_1^3$ and $\{a_i\}_1^3$ with respect to this basis. Prove that

$$\mathbf{a} \cdot \mathbf{x} = a_1 x_1 + a_2 x_2 + a_3 x_3 = \sum_{i=1}^{3} x_i.$$

33. The vectors $\mathbf{a}_1 = (2, 1, 1)$ and $\mathbf{a}_2 = (1, -1, -1)$ are orthogonal (perpendicular). Find an orthonormal basis $\{\mathbf{b}_1, \mathbf{b}_2, \mathbf{b}_3\}$ such that $\mathbf{b}_1$ and $\mathbf{b}_2$ are the directions of $\mathbf{a}_1$ and $\mathbf{a}_2$, respectively.

34. Same problem for the vectors $\mathbf{a}_1 = (1, 2, 3)$ and $\mathbf{a}_2 = (3, -3, 1)$.

35. Same problem for $\mathbf{a}_1 = (1, 1, 1)$ and $\mathbf{a}_2 = (1, -1, 0)$.

36. Check that

$$\mathbf{b}_1 = \frac{1}{3}(2, 1, 2), \qquad \mathbf{b}_2 = \frac{1}{3}(2, -2, -1), \qquad \text{and} \qquad \mathbf{b}_3 = \frac{1}{3}(1, 2, -2)$$

form an orthonormal basis.

Using the above basis and Theorem 12 (Problem 31), express each of the following vectors **x** in the form

$$\mathbf{x} = x_1\mathbf{b}_1 + x_2\mathbf{b}_2 + x_3\mathbf{b}_3.$$

Then verify directly the correctness of each expansion.

37. $\mathbf{x} = \mathbf{e}_1 = (1, 0, 0)$

38. $\mathbf{x} = \mathbf{e}_2 = (0, 1, 0)$

39. $\mathbf{x} = \mathbf{e}_3 = (0, 0, 1)$

40. $\mathbf{x} = (1, 2, 3)$

41. $\mathbf{x} = (4, 0, -2)$

42. $\mathbf{x} = (1, -1, 1)$

43. $\mathbf{x} = (-2, -2, 1)$

44. $\mathbf{x} = (3, 1, -2)$

14

Vector-Valued Functions

This chapter is concerned with vector aspects of the calculus of paths. The first four sections treat plane paths, Section 5 recapitulates this vector calculus for space paths and briefly discusses a new phenomenon (torsion) that the third dimension makes possible. Finally, Section 6 presents Newton's astonishing derivation of the universal law of gravitation from Kepler's observations about planetary motion.

1. THE TANGENT VECTOR TO A PATH

In the same way that a pair of numbers x and y combine to form a single vector

$$\mathbf{x} = (x, y),$$

a pair of functions

$$x = g(t),$$
$$y = h(t),$$

can be combined to form a single *vector-valued* function

$$(x, y) = (g(t), h(t))$$

or

$$\mathbf{x} = \mathbf{g}(t).$$

In order to discuss the continuity of a vector-valued function we fix a value,

$$\mathbf{x}_0 = (x_0, y_0) = (g(t_0), h(t_0)),$$

and then give t an increment Δt to get a new value

$$\mathbf{x} = (x_0 + \Delta x, y_0 + \Delta y) = (g(t_0 + \Delta t), h(t_0 + \Delta t)).$$

The *vector increment* is then

$$\Delta \mathbf{x} = \mathbf{x} - \mathbf{x}_0 = (\Delta x, \Delta y).$$

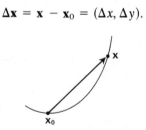

It is the vector of the arrow $\overrightarrow{\mathbf{x}_0\mathbf{x}}$ from the fixed point $\mathbf{x}_0$ to the varying point $\mathbf{x}$. Then the vector-valued function is *continuous* at $\mathbf{x}_0$ if $\Delta \mathbf{x} \to 0$ as $\Delta t \to 0$. This means that the magnitude of $\Delta \mathbf{x}$ approaches 0,

$$|\Delta \mathbf{x}| = \sqrt{(\Delta x)^2 + (\Delta y)^2} \to 0.$$

and this occurs just when both Δx and Δy approach 0 as Δt approaches 0. That is:

A vector-valued function $\mathbf{x} = (g(t), h(t))$ *is continuous at* $t = t_0$ *if and only if both component functions* g *and* h *are continuous there.*

In the same way we find that a vector function $\mathbf{x} = (g(t), h(t))$ has the limit $\mathbf{l} = (l, m)$ as t approaches t_0 if and only if $g(t) \to l$ and $h(t) \to m$ as $t \to t_0$.

The discussion so far has been preliminary; our main concern is the *derivative* of a vector-valued function **x** of a parameter *t*. There is nothing new in the definition except the use of vector limits. We form the difference quotient

$$\frac{\Delta \mathbf{x}}{\Delta t} = \frac{1}{\Delta t}(\Delta x, \Delta y) = \left(\frac{\Delta x}{\Delta t}, \frac{\Delta y}{\Delta t}\right),$$

and take the limit as $\Delta t \to 0$. Since

$$\frac{\Delta x}{\Delta t} \to \frac{dx}{dt} = g'(t)$$

and

$$\frac{\Delta y}{\Delta t} \to \frac{dy}{dt} = h'(t),$$

it follows that the vector

$$\frac{\Delta \mathbf{x}}{\Delta t} = \left(\frac{\Delta x}{\Delta t}, \frac{\Delta y}{\Delta t}\right)$$

approaches the vector

$$\left(\frac{dx}{dt}, \frac{dy}{dt}\right) = (g'(t), h'(t))$$

as Δt approaches 0. We state this conclusion as a theorem.

THEOREM 1 *The vector function*

$$\mathbf{x} = (x, y) = (g(t), h(t))$$

is differentiable if and only if its component functions g and h are both differentiable, and then

$$\frac{d\mathbf{x}}{dt} = \left(\frac{dx}{dt}, \frac{dy}{dt}\right) = (g'(t), h'(t)).$$

We can also write

$$\mathbf{x} = \mathbf{g}(t) \qquad \text{and} \qquad \frac{d\mathbf{x}}{dt} = \mathbf{g}'(t).$$

This gives us a new way to look at paths. A smooth path is simply a continuously differentiable vector-valued function **g** with **g**′ nowhere zero.

Now consider the geometric meaning of the above limit. If $\Delta t > 0$, then the vector $\Delta \mathbf{x}/\Delta t$ has the same direction as the increment vector $\Delta \mathbf{x} = \mathbf{x} - \mathbf{x}_0$. So if we locate the vector $\Delta \mathbf{x}/\Delta t$ at $\mathbf{x}_0$, then it will lie along the

arrow $\overrightarrow{\mathbf{x}_0\mathbf{x}}$. It will be longer than $\overrightarrow{\mathbf{x}_0\mathbf{x}}$ if $\Delta t < 1$, the multiplying scalar $1/\Delta t$ then being greater than 1. The direction of the limit vector

$$\frac{d\mathbf{x}}{dt} = \lim_{\Delta t \to 0} \frac{\Delta \mathbf{x}}{\Delta t}$$

is thus the limit of the direction of the secant arrow $\overrightarrow{\mathbf{x}_0\mathbf{x}_1}$ as $\mathbf{x}_1$ approaches $\mathbf{x}_0$ along the path. But the limit of the direction of the secant arrow is what we *mean* by the tangent direction. Therefore:

> *The path derivative $d\mathbf{x}/dt$ can be interpreted geometrically as a vector tangent to the path. It is called the path tangent vector.*

We normally draw it as the arrow from the path point $\mathbf{x}$ to $\mathbf{x} + d\mathbf{x}/dt$.

EXAMPLE 1   Find the tangent vector to the path

$$\mathbf{x} = (x, y) = (t^3 - 3t, t^2)$$

at the point corresponding to $t = -1$. The same for $t = 0$, $t = \dfrac{3}{2}$, and $t = 2$.

Solution. At the general point $\mathbf{x} = (x, y) = (t^3 - 3t, t^2)$, the path has the tangent vector

$$\frac{d\mathbf{x}}{dt} = \left(\frac{dx}{dt}, \frac{dy}{dt}\right) = (3t^2 - 3, 2t).$$

At $t = -1$, the path point $\mathbf{x}$ is $(2, 1)$ and the tangent vector is

$$\left.\frac{d\mathbf{x}}{dt}\right|_{t=-1} = (3(-1)^2 - 3, 2(-1)) = (0, -2).$$

When $t = 0$,

$$\mathbf{x} = (0, 0), \qquad \frac{d\mathbf{x}}{dt} = (-3, 0).$$

When $t = \dfrac{3}{2}$,

$$\mathbf{x} = \left(-\frac{9}{8}, \frac{9}{4}\right), \qquad \frac{d\mathbf{x}}{dt} = \left(\frac{15}{4}, 3\right).$$

When $t = 2$,

$$\mathbf{x} = (2, 4), \qquad \frac{d\mathbf{x}}{dt} = (9, 4).$$

This path is the same one we began our path discussion with in Chapter 10. If we locate the vector $d\mathbf{x}/dt$ at the path point $\mathbf{x}$ where it is tangent, i.e., if we consider the arrow from $\mathbf{x}$ to $\mathbf{x} + d\mathbf{x}/dt$, then the four tangent vectors computed above look like this.

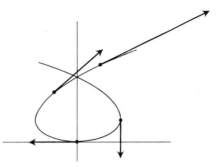

When the vector $\mathbf{a}$ is constant and the scalar t is variable, we shall feel free to write the product $t\mathbf{a}$ in the more traditional order, $\mathbf{a}t$.

EXAMPLE 2 Show that a path whose tangent vector is constant and not zero must necessarily be a straight line.

Solution. Let $\mathbf{x} = (g(t), h(t))$ be the path and let $\mathbf{a}$ be the nonzero constant tangent vector. Our hypothesis is that

$$\frac{d\mathbf{x}}{dt} = (g'(t), h'(t)) = \mathbf{a} = (a, b),$$

for all t. That is,

$$g'(t) = a,$$
$$h'(t) = b,$$

for all t. Integrating these two equations, we see that

$$x = g(t) = at + c,$$
$$y = h(t) = bt + d,$$

or

$$\mathbf{x} = \mathbf{a}t + \mathbf{c}.$$

This is the parametric equation for a line in vector form.

EXAMPLE 3 Find the line tangent to the curve $\mathbf{x} = (x, y) = (2 \cos t, \sin t)$ at the point corresponding to $t_0 = \pi/6$.

Solution. At $t_0 = \pi/6$,

$$\mathbf{x}_0 = (2 \cos \pi/6, \sin \pi/6) = (\sqrt{3}, 1/2),$$
$$\frac{d\mathbf{x}}{dt}\bigg|_{\pi/6} = (-2 \sin t, \cos t)_{\pi/6} = (-1, \sqrt{3}/2).$$

The line through $\mathbf{x}_0$ in the direction of the vector $\mathbf{a}$ has the parametric equation

$$\mathbf{x} = \mathbf{x}_0 + \mathbf{a}t.$$

Here the direction is given by the tangent vector, and the equation of the line is

$$\mathbf{x} = (\sqrt{3}, 1/2) + (-1, \sqrt{3}/2)t$$

$$= (\sqrt{3} - t, 1/2 + (\sqrt{3}/2)t).$$

The component equations are

$$x = \sqrt{3} - t,$$

$$y = \frac{1 + \sqrt{3}t}{2}.$$

EXAMPLE 4   Show that if the vector variable $\mathbf{x} = (x, y)$ is a differentiable function of the parameter t, then $|\mathbf{x}| = \sqrt{x^2 + y^2}$ is a constant if and only if $\mathbf{x}$ is always perpendicular to $d\mathbf{x}/dt$.

Solution. The assumption of constant magnitude can be written

$$x^2 + y^2 = |\mathbf{x}|^2 = \text{constant},$$

where x and y are differentiable functions of t. Differentiating with respect to t gives

$$2x\frac{dx}{dt} + 2y\frac{dy}{dt} = 0,$$

or

$$2\left(\mathbf{x} \cdot \frac{d\mathbf{x}}{dt}\right) = 0.$$

Thus $\mathbf{x}$ is perpendicular to $d\mathbf{x}/dt$. Conversely, if these vectors are perpendicular, then the above argument can be reversed, to give

$$\frac{d}{dt}(x^2 + y^2) = 0,$$

$$x^2 + y^2 = \text{constant}.$$

EXAMPLE 5   If $\mathbf{x}_0 = (g(t_0), h(t_0))$ is the point closest to the origin on the path $\mathbf{x} = (g(t), h(t))$, then the tangent vector to the path at $\mathbf{x}_0$ is necessarily perpendicular to the vector $\mathbf{x}_0$.

This is because $|\mathbf{x}|^2 = g^2(t) + h^2(t)$ has its minimum value at $\mathbf{x}_0$, so its derivative is zero there. But

$$\frac{d |\mathbf{x}|^2}{dt} = 2\left(\mathbf{x} \cdot \frac{d\mathbf{x}}{dt}\right),$$

as we saw in Example 4, so

$$\mathbf{x}_0 \cdot \frac{d\mathbf{x}}{dt}\bigg|_0 = 0,$$

as claimed.

A tangent vector

$$\frac{d\mathbf{x}}{dt} = (g'(t), h'(t))$$

can perfectly well be the zero vector $(0, 0)$, in which case it does not specify a unique direction. However, this is exactly what we ruled out when we defined a smooth path, so a smooth path has a continuously varying, nowhere-zero, tangent vector. In particular, a smooth path has a continuously changing direction.

PROBLEMS FOR SECTION 1

Find the derivative $d\mathbf{x}/dt$ for each of the following vector functions. Also, determine the points where $d\mathbf{x}/dt$ is perpendicular to $\mathbf{x}$.

1. $\mathbf{f}(t) = (t, 1/t)$
2. $\mathbf{x} = (e^{2t}, e^{-t})$
3. $\Phi(t) = (t^2, t + 3)$
4. $\mathbf{x} = (2t, t + 1)$
5. $\mathbf{x} = (\cos t, \sin t)$
6. $\mathbf{u} = (\sqrt{1 - t^2}, t)$
7. $\mathbf{g}(t) = (3 \cos t, 2 \sin t)$
8. $\mathbf{x} = (t^2, (1 + t)^2)$

Find the vector tangent to each of the paths below at the given point. Then write down a parametric equation for the tangent line at the point.

9. $\mathbf{x} = (40t, 30t - 16t^2)$; when $t = 1$
10. $\mathbf{x} = (4 \cos t, 3 \sin t)$; when $t = \pi/3$
11. $\mathbf{x} = (e^t \cos t, e^t \sin t)$; when $t = 0$
12. $\mathbf{x} = (t^2, t^3)$; when $t = 2$
13. $\mathbf{x} = (e^t + e^{-t}, e^t - e^{-t})$; when $t = 0$
14. $\mathbf{x} = (1/t, 1/(t^2 + 1))$; when $t = -1$
15. $\mathbf{x} = (\log t, 2/t)$; when $t = 1$
16. $\mathbf{x} = ((t^2 + 1)/(t^2 + 2), 2t)$; when $t = 0$
17. The ellipse

$$\frac{x^2}{a^2} + \frac{y^2}{b^2} = 1$$

has the parametric equation

$$\mathbf{x} = (a \cos \theta, b \sin \theta).$$

Use this representation to find a parametric equation of the tangent line to the ellipse at $(x_0, y_0) = (a \cos \theta_0, b \sin \theta_0)$.

18. The path $\mathbf{x} = (a \sec \theta, b \tan \theta)$ runs along the hyperbola $(x/a)^2 - (y/b)^2 = 1$. Use this parametric representation of the hyperbola to find the equation of its tangent line at $(x_0, y_0) = (a \sec \theta_0, b \tan \theta_0)$.

19. Redraw the figure showing the meaning of Theorem 1 for the case when Δt is negative.

20. Show that the line tangent to the path $\mathbf{x} = \mathbf{g}(t)$ at the point $\mathbf{x}_0 = \mathbf{g}(t_0)$ has the parametric equation

$$\mathbf{x} - \mathbf{x}_0 = \mathbf{g}'(t_0)(t - t_0).$$

21. Show that a path whose tangent vector has a constant *direction* must lie along a

straight line. (The hypothesis is that $d\mathbf{x}/dt = f(t)\mathbf{u}$, where $\mathbf{u}$ is a constant unit vector.)

22. Show that a path $\mathbf{x} = \mathbf{g}(t)$ is uniquely determined by its derivative $\mathbf{g}'(t)$ and an initial value $\mathbf{x}_0 = \mathbf{g}(t_0)$. That is, if

$$\mathbf{g}(t) = (g(t), h(t)) \qquad \text{and} \qquad \mathbf{l}(t) = (l(t), m(t))$$

are two paths such that $\mathbf{g}'(t) = \mathbf{l}'(t)$ for all t and $\mathbf{g}(t_0) = \mathbf{l}(t_0)$, then

$$\mathbf{g}(t) = \mathbf{l}(t) \qquad \text{for all } t.$$

23. Let $\mathbf{x} = \mathbf{g}(t) = (g(t), h(t))$ and $\mathbf{x} = \mathbf{p}(s) = (p(s), q(s))$ be two paths that do not intersect. Let d be the shortest distance between them, and suppose that we can find a closest pair of points,

$$\mathbf{x}_1 = \mathbf{g}(t_0) \qquad \text{and} \qquad \mathbf{x}_2 = \mathbf{l}(s_0),$$

i.e., a pair such that

$$|\mathbf{x}_2 - \mathbf{x}_1| = d.$$

Prove that the segment $\mathbf{x}_1\mathbf{x}_2$ is perpendicular to both curves.

24. If one path is a straight line l in the above problem, then the tangent to the other path must be parallel to l at the point closest to l. Use this principle to find the point on $\mathbf{x} = (t, t^2 - 1)$ closest to the line $\mathbf{x} = (-s - 1, 2s)$.

2. VECTOR FORMS OF THE DERIVATIVE RULES

In the beginning sections of Chapter 13 we were sometimes able to prove things about vectors without referring explicitly to their components. For example, we proved in this way that the medians of a triangle are concurrent. Instead of explicitly calculating with components we were able to use the vector laws.

In a similar way, derivative calculations involving vector functions of t become more transparent and frequently more efficient if we use vector forms of the derivative rules, as given below.

1. If $\mathbf{x}$ and $\mathbf{u}$ are differentiable vector functions of t, then so is $\mathbf{x} + \mathbf{u}$ and

$$\frac{d}{dt}(\mathbf{x} + \mathbf{u}) = \frac{d\mathbf{x}}{dt} + \frac{d\mathbf{u}}{dt}.$$

2. If w is a differentiable scalar function of t and $\mathbf{x}$ is a differentiable vector function of t, then $w\mathbf{x}$ is a differentiable vector function of t and

$$\frac{d}{dt}(w\mathbf{x}) = w\frac{d\mathbf{x}}{dt} + \frac{dw}{dt}\mathbf{x}.$$

3. If $\mathbf{x}$ and $\mathbf{u}$ are differentiable vector functions of t, then $\mathbf{x} \cdot \mathbf{u}$ is a differentiable scalar function of t and

$$\frac{d}{dt}(\mathbf{x} \cdot \mathbf{u}) = \mathbf{x} \cdot \frac{d\mathbf{u}}{dt} + \frac{d\mathbf{x}}{dt} \cdot \mathbf{u}.$$

Rules (2) and (3) are simply the product rule for the two products we have for plane vectors.

4. If $\mathbf{x}$ is a differentiable vector function of the parameter r, and if $r = f(t)$ is differentiable, then $\mathbf{x}$ is a differentiable vector function of t, and

$$\frac{d\mathbf{x}}{dt} = \frac{d\mathbf{x}}{dr}\frac{dr}{dt}.$$

This is the rule for changing the path parameter; it is another chain rule.

5. If $d\mathbf{x}/dt$ is identically zero, then $\mathbf{x}$ is a constant vector.

These reformulations all follow from the original derivative rules and Theorem 1. For example,

$$\frac{d}{dt}[\mathbf{x} \cdot \mathbf{u}] = \frac{d}{dt}[(x, y)\cdot(u, v)] = \frac{d}{dt}(xu + yv)$$

$$= x\frac{du}{dt} + \frac{dx}{dt}u + y\frac{dv}{dt} + \frac{dy}{dt}v$$

$$= (x, y)\cdot\left(\frac{du}{dt}, \frac{dv}{dt}\right) + \left(\frac{dx}{dt}, \frac{dy}{dt}\right)\cdot(u, v)$$

$$= \mathbf{x} \cdot \frac{d\mathbf{u}}{dt} + \frac{d\mathbf{x}}{dt} \cdot \mathbf{u}.$$

EXAMPLE We reconsider Examples 2 and 4 from the last section. In the second example we were given that $d\mathbf{x}/dt$ is the constant vector $\mathbf{a}$. One such vector function of t is $\mathbf{a}t$, since

$$\frac{d}{dt}(\mathbf{a}t) = \mathbf{a},$$

by rule (2) above. Thus

$$\frac{d}{dt}(\mathbf{x} - \mathbf{a}t) = 0$$

(rule (1)), and so

$$\mathbf{x} - \mathbf{a}t = \text{constant vector } \mathbf{c},$$

$$\mathbf{x} = \mathbf{a}t + \mathbf{c}$$

(rule (5)). This was the conclusion of Example 2.

In the fourth example we considered a path $\mathbf{x} = \mathbf{g}(t)$ for which $|\mathbf{x}|$ is constant. We can now compute

$$0 = \frac{d}{dt}|\mathbf{x}|^2 = \frac{d}{dt}(\mathbf{x} \cdot \mathbf{x}) = \mathbf{x} \cdot \frac{d\mathbf{x}}{dt} + \frac{d\mathbf{x}}{dt} \cdot \mathbf{x}$$

$$= 2\mathbf{x} \cdot \frac{d\mathbf{x}}{dt},$$

by rule (3), and so $\mathbf{x} \perp d\mathbf{x}/dt$, as before.

PROBLEMS FOR SECTION 2

1. Let $\mathbf{x} = \mathbf{f}(t)$ be a twice-differentiable path such that $d^2\mathbf{x}/dt^2 \equiv 0$. Prove that $\mathbf{x} = \mathbf{a}t + \mathbf{c}$, so the path is simply the parametric representation of a straight line.

2. Prove the rule (4) by a component calculation, the way (3) was proved in the text.

3. Prove the rule (5).

4. Let $\mathbf{x} = \mathbf{f}(t)$ be a path whose tangent vector has a constant direction. Prove that the path lies along a straight line (without resorting to component calculations).

5. The tangent vector of the path $\mathbf{x} = \mathbf{f}(t)$ is everywhere perpendicular to $\mathbf{x}$. Prove that the path runs along a circle about the origin.

6. The tangent vector of the path $\mathbf{x} = \mathbf{f}(t)$ is everywhere perpendicular to the constant vector $\mathbf{a}$ ($\mathbf{a} \neq 0$). Prove that the path runs along a line perpendicular to $\mathbf{a}$.

7. Let $\mathbf{x} = \mathbf{f}(t)$ be a path whose tangent vector $d\mathbf{x}/dt$ is always in the direction of $\mathbf{x}$. Prove that $\mathbf{x}$ runs along a line through the origin. (Write $\mathbf{x} = |\mathbf{x}|\,\mathbf{u}$, where $\mathbf{u}$ is unit vector, presumably varying with t. Apply rule (2) and remember that $d\mathbf{u}/dt$ is necessarily perpendicular to $\mathbf{u}$, as in the second part of the Example.)

8. If $\mathbf{x}$ is a differentiable vector function of t, prove from rule (3) that

$$\frac{d}{dt}|\mathbf{x}| = \frac{\mathbf{x} \cdot \dfrac{d\mathbf{x}}{dt}}{|\mathbf{x}|}.$$

9. If $0 < c < a$, and $\mathbf{f} = (c, 0)$ and $\mathbf{g} = (-c, 0)$, we know (Chapter 10, Extra Problem 29) that the set of points $\mathbf{x}$ such that

$$|\mathbf{x} - \mathbf{f}| + |\mathbf{x} - \mathbf{g}| = 2a$$

is the ellipse

$$\frac{x^2}{a^2} + \frac{y^2}{b^2} = 1,$$

where $b^2 = a^2 - c^2$. We can consider the above ellipse to be parametrized in some way, so that $\mathbf{x}$ traces the elliptical path with parameter t. The actual parametrization doesn't matter. Now differentiate the locus identity

$$|\mathbf{x} - \mathbf{f}| + |\mathbf{x} - \mathbf{g}| = 2a$$

with respect to t, using Problem 8, and so prove that the arrows $\overrightarrow{\mathbf{f}\mathbf{x}}$ and $\overrightarrow{\mathbf{g}\mathbf{x}}$ make equal angles with the tangent to the ellipse at $\mathbf{x}$.

This is the so-called *optical property* of the ellipse, because it shows that the rays from a point source of light placed at $\mathbf{g}$ reflect off the ellipse and all come together (are "focused") at $\mathbf{f}$ (and vice versa). Each of the points $\mathbf{f}$ and $\mathbf{g}$ is called a *focus* of the ellipse.

10. It was shown in Chapter 10, Section 4, that the polar graph of the equation $r(1 - \epsilon \cos \theta) = 1$ is an ellipse if $0 < \epsilon < 1$. An important additional fact is that this ellipse has one focus at the origin. Prove this now from the values of a and b given there. (This is just algebra.)

11. The vector forms of the derivative rules can be proved without resorting to component calculations by going back to the difference quotient definition of the derivative. Show that if $\mathbf{x}$ and $\mathbf{u}$ are differentiable functions of the parameter t and if $z = \mathbf{x} \cdot \mathbf{u}$ then

$$\Delta z = \mathbf{x} \cdot \Delta\mathbf{u} + \Delta\mathbf{x} \cdot \mathbf{u} + \Delta\mathbf{x} \cdot \Delta\mathbf{u}$$

just as was the case for the ordinary product of two real-valued functions considered in Chapter 5. Then finish the proof of rule (3) as in Chapter 5.

12. Prove the second rule in this direct component-free manner.

3. ARC LENGTH AS PARAMETER

We saw in Chapter 10 that the arc length s of a smooth plane path $\mathbf{x} = (x, y) = (g(t), h(t))$, measured *forward* from some fixed point, is determined by the equation

$$\frac{ds}{dt} = \sqrt{\left(\frac{dx}{dt}\right)^2 + \left(\frac{dy}{dt}\right)^2}.$$

The right side is now recognizable as the magnitude of the path tangent vector

$$\frac{d\mathbf{x}}{dt} = \left(\frac{dx}{dt}, \frac{dy}{dt}\right).$$

The equation can thus be rewritten

$$\left|\frac{d\mathbf{x}}{dt}\right| = \frac{ds}{dt},$$

and says:

The magnitude of the path tangent vector is the rate of change of the path arc length s with respect to the parameter t.

Note that if the parameter t happens to be equal to the arc length s, or more generally, if $s = t + c$, then

$$\left|\frac{d\mathbf{x}}{dt}\right| = \frac{ds}{dt} = 1$$

and the tangent vector is always a unit vector. The same argument works backward and shows that if the tangent vector is always a unit vector then $ds/dt = 1$ and $s = t +$ constant. Thus

THEOREM 2 *The parameter of a smooth plane path is equal to the path arc length (plus a constant) if and only if the tangent vector is always a unit vector.*

In general the parameter t will not be the arc length, but we can always reparameterize the path by its arc length. The argument goes as follows. Since the tangent vector to a smooth path is never zero (the definition of smoothness), $ds/dt = |d\mathbf{x}/dt|$ is always positive. Therefore s is an increasing function of t which has an everywhere-differentiable inverse, by the inverse function theorem. That is, t is a differentiable function of s, say

$$t = \phi(s).$$

Then

$$x = g(t) = g(\phi(s)) = l(s),$$
$$y = h(t) = h(\phi(s)) = m(s)$$

is a new smooth parameterization of the path, with the path length s as the new parameter. Strictly speaking, this change of parameters gives us a new path. In any event, we can now write the tangent vector $d\mathbf{x}/dt$ in the form

$$\frac{d\mathbf{x}}{dt} = \frac{d\mathbf{x}}{ds}\frac{ds}{dt},$$

by rule (4) from the last section. This is the "polar coordinate" representation for the tangent vector, ds/dt being its magnitude and $d\mathbf{x}/ds$ being its direction, i.e., the unit tangent vector $\mathbf{T}$ that we think of as its direction.

We have noted in several contexts that a vector function of constant magnitude is necessarily perpendicular to its derivative. When the vector function is the unit tangent vector $\mathbf{T}$ to a path we can say more.

THEOREM 3   *Let $\mathbf{x} = \mathbf{g}(s)$ be a path with arc length as parameter, and suppose that the unit tangent vector $\mathbf{T}$ is a differentiable function of s. Then the vector $d\mathbf{T}/ds$ is perpendicular to the path, and points in the direction in which the path is turning. Moreover, its magnitude is the absolute curvature of the path. Thus $d\mathbf{T}/ds$ has the "polar coordinate" representation*

$$\frac{d\mathbf{T}}{ds} = \kappa\mathbf{N},$$

where κ is the absolute path curvature and $\mathbf{N}$ is the unit vector normal to the path in the direction in which the path is turning.

Proof. The first step is "old hat." We differentiate the identity

$$|\mathbf{T}|^2 = \mathbf{T} \cdot \mathbf{T} = 1$$

by the product law for dot products:

$$\mathbf{T} \cdot \frac{d\mathbf{T}}{ds} + \frac{d\mathbf{T}}{ds} \cdot \mathbf{T} = 0.$$

Therefore,

$$2\left(\mathbf{T} \cdot \frac{d\mathbf{T}}{ds}\right) = 0$$

and the vectors $\mathbf{T}$ and $d\mathbf{T}/ds$ are perpendicular. Since $\mathbf{T}$ is tangent to the path, $d\mathbf{T}/ds$ is perpendicular to the path.

For the second assertion we can only give an intuitive argument. Consider the approximate equality

$$\frac{\Delta\mathbf{T}}{\Delta s} \approx \frac{d\mathbf{T}}{ds}.$$

Taking Δs positive, this says, in particular, that the direction of $d\mathbf{T}/ds$ is approximately the same as the direction of $\Delta\mathbf{T}$. But this is approximately the direction in which the curve is turning. Therefore the direction of $d\mathbf{T}/ds$ is approximately the direction in which the curve is turning, and since the approximation can presumably be made arbitrarily good by taking Δs sufficiently small, the two directions must in fact be equal. (It would be more appropriate to conclude from the above plausibility argument that it seems correct to *define* the direction in which the path is bending to be the direction of the vector $d\mathbf{T}/ds$.)

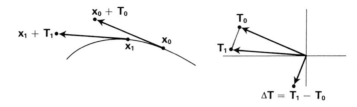

Now write the unit vector **T** in the form

$$\mathbf{T} = (\cos \phi, \sin \phi),$$

where ϕ is the angle from the positive x-axis to the direction of the path. Then

$$\frac{d\mathbf{T}}{ds} = \frac{d\mathbf{T}}{d\phi}\frac{d\phi}{ds} = (-\sin \phi, \cos \phi)\frac{d\phi}{ds}.$$

But $d\phi/ds$ was our definition of the *curvature* of a path. And $(-\sin \phi, \cos \phi)$ is another unit vector. Therefore

$$\left|\frac{d\mathbf{T}}{ds}\right| = |(-\sin \phi, \cos \phi)|\left|\frac{d\phi}{ds}\right| = 1 \cdot \kappa = \kappa,$$

where $\kappa = |d\phi/ds|$ is the absolute curvature of the path. The theorem is thus proved. ∎

EXAMPLE A path that doesn't curve should lie along a straight line. This conjecture is easy to prove using the above formula. For if $\kappa = 0$, then $d\mathbf{T}/ds = 0$ and **T** is a constant, say, $\mathbf{T} = \mathbf{a}$. Since $\mathbf{T} = d\mathbf{x}/ds$ we conclude, as before, that

$$\mathbf{x} = \mathbf{x}_0 + s\mathbf{a}.$$

Here s is arc length. If the path had been given in terms of some other parameter t, then the arc length s is a function of t, say $s = f(t)$, and then the path runs along the above line according to the equation

$$\mathbf{x} = \mathbf{x}_0 + f(t)\mathbf{a}.$$

PROBLEMS FOR SECTION 3

1. a) Reparameterize the path

$$(x, y) = (e^t \cos t, e^t \sin t)$$

   with the arc length s as parameter. (This involves finding s as a function of t, then t as a function of s, then substituting.)

   b) Then compute $\mathbf{T} = d\mathbf{x}/ds$ and $d\mathbf{T}/ds$, and verify that they are orthogonal.

   c) Compute $\kappa = |d\mathbf{T}/ds|$.

2. Do the same for the path $(x, y) = \dfrac{2}{3}(\cos^3 t, \sin^3 t)$.

3. The same for the path

$$(x, y) = \left(\frac{4}{3}t^{3/2}, \frac{1}{2}t^2 - t\right).$$

4. VELOCITY AND ACCELERATION

Now let the parameter t be time, so that the path $\mathbf{x} = \mathbf{g}(t)$ is the trajectory of a moving particle, and ds/dt is its speed. We could define the velocity of the particle as the time rate of change of its position $\mathbf{x}$, that is, as the tangent vector $d\mathbf{x}/dt$. But it is intuitively more satisfying to define the velocity as the vector $\mathbf{v}$ whose direction is the direction in which the particle is moving, and whose magnitude is the speed ds/dt at which the particle is moving. It then *follows* that

$$\mathbf{v} = \frac{d\mathbf{x}}{dt},$$

because $d\mathbf{x}/dt$ meets both of these requirements.

Acceleration is the *time rate of change of velocity*. It is thus the vector

$$\mathbf{a} = \frac{d\mathbf{v}}{dt} = \frac{d^2\mathbf{x}}{dt^2}.$$

What is interesting about vector acceleration is that it is due *both* to change in speed *and* to change in direction. If *either* the speed is changing *or* the direction is changing, then the particle is accelerating.

Think how one is affected by changes in velocity when riding in a car. If the car speeds up while going down a straight road the back of the seat presses harder against the rider, who attributes the extra pressure to the straight-ahead acceleration of the car. Now consider what happens when the car travels with *constant* speed along a winding road. When riding around a curve, one is thrown sideways. This is a consequence of the *sidewise* acceleration of the car, and the greater or lesser extent that one feels this sidewise pressure is a rough measure of the magnitude of the sidewise acceleration. On this basis we can reach some conclusions about this perpendicular acceleration. One is thrown more to the side when going at the same speed around a sharper curve, or travelling at a higher speed around the same curve. Thus the acceleration must increase in magnitude if *either* the speed or the curvature of the path is increased.

The exact formula for these different acceleration effects is contained in the following theorem.

THEOREM 4   *If $\mathbf{x} = \mathbf{g}(t)$ is the path of a moving particle, with the time t as parameter, then the acceleration $\mathbf{a}$ has the formula*

$$\mathbf{a} = \frac{dv}{dt}\mathbf{T} + \kappa v^2 \mathbf{N},$$

where v is the speed ds/dt, $\mathbf{T}$ and $\mathbf{N}$ are the unit tangent and normal vectors, and κ is the absolute curvature of the path.

Proof. We start with

$$\mathbf{v} = \frac{d\mathbf{x}}{dt} = \frac{d\mathbf{x}}{ds}\frac{ds}{dt} = v\mathbf{T}$$

and differentiate again:

$$\mathbf{a} = \frac{d\mathbf{v}}{dt} = v\frac{d\mathbf{T}}{dt} + \frac{dv}{dt}\mathbf{T}.$$

But

$$\frac{d\mathbf{T}}{dt} = \frac{d\mathbf{T}}{ds}\frac{ds}{dt} = \kappa\mathbf{N}v$$

from Theorem 3, and substituting this value into the formula for **a** gives the formula of the theorem. ∎

It follows from the theorem that the component of the acceleration vector **a** normal to the curve is

$$\mathbf{a} \cdot \mathbf{N} = \kappa v^2,$$

in the direction in which the curve is bending. This shows explicitly how the "sidewise" acceleration varies with curvature κ and speed v.

EXAMPLE   A circle of radius r has curvature

$$\kappa = \frac{1}{r}.$$

Thus a particle that moves along this circle with *constant* speed v is nevertheless accelerating, according to the formula

$$\mathbf{a} = \kappa v^2 \mathbf{N} = \left(\frac{v^2}{r}\right)\mathbf{N}.$$

That is, its acceleration is directed toward the center of the circle and is of magnitude v^2/r.

PROBLEMS FOR SECTION 4

Find the velocity and acceleration vectors, the speed v, and the tangential and normal components (dv/dt and κv^2) of the acceleration, for each of the following motions:

1. $(x, y) = (\cos 3t, \sin 3t)$ 2. $(x, y) = (3 \cos t, 3 \sin t)$

3. $(x, y) = (3 \cos t, 2 \sin t)$ 4. $(x, y) = (t, \sin t)$

5. $(x, y) = (t, t^2)$ 6. $(x, y) = (t^2, t^3)$

7. $(x, y) = (2 \cos 3t, 3 \sin 2t)$ 8. $(x, y) = (\cos t \sin 3t, \sin t \sin 3t)$

9. A road lies along the parabola $100y = x^2$. A truck moving along the road is so loaded that the normal component of its acceleration must not exceed 25. What limitation does this impose on its speed as it rounds the vertex of the parabola?

10. A particle moves along the ellipse $(x, y) = (3 \cos t, 2 \sin t)$ with constant speed 5. What are the maximum and minimum values of the magnitude of its acceleration?

11. Is it possible for a particle to move along a curving path without experiencing any sideways acceleration? Consult the formula of Theorem 4.

12. If you travel along a road on which the curvature is never zero, how must you adjust your speed so that your sidewise acceleration has a constant magnitude?

13. There are two circular tracks, the large one having twice the radius of the

smaller. If you are comfortable driving 50 mph on the smaller track, how much faster can you drive on the larger track without experiencing a greater radial (sidewise) acceleration?

14. Suppose that the parameter in the standard parametric equations of the ellipse

$$\frac{x^2}{a^2} + \frac{y^2}{b^2} = 1$$

is the time t, so that the parametric equations

$$x = a \cos t \quad \text{and} \quad y = b \sin t$$

describe the motion of a particle moving along the ellipse. Show that the acceleration vector

$$\mathbf{a} = \frac{d^2\mathbf{x}}{dt^2}$$

is always directed to the center of the ellipse (the origin).

5. PATHS IN SPACE

We shall not generally graph space paths, because drawing three-dimensional configurations is so complicated. However, three coordinate functions of a parameter t,

$$x = g(t), \qquad y = h(t), \qquad z = k(t),$$

can be viewed as defining a path in space exactly as two functions do in the plane. If t is interpreted as time, the path can be interpreted as the trajectory of a moving particle. The path will be called smooth if the three coordinate functions all have continuous derivatives and if the three derivatives are never simultaneously zero.

A space path can be viewed as a single vector-valued function of t,

$$\mathbf{x} = (x, y, z) = (g(t), h(t), k(t)),$$

and, exactly as in the plane, the derivative of this vector-valued function is the vector function

$$\frac{d\mathbf{x}}{dt} = \left(\frac{dx}{dt}, \frac{dy}{dt}, \frac{dz}{dt} \right) = (g'(t), h'(t), k'(t)).$$

As before, $d\mathbf{x}/dt$ can be interpreted as the tangent vector to the path at the point $\mathbf{x} = (g(t), h(t), k(t))$.

EXAMPLE 1 The path $\mathbf{x} = (\cos t, \sin t, t/2)$ is a spiral rising over the unit circle in the xy-plane. Its tangent vector is

$$\frac{d\mathbf{x}}{dt} = (-\sin t, \cos t, 1/2).$$

When $t = \pi/2$, the path point is $(0, 1, \pi/4)$ and the tangent vector is $(-1, 0, 1/2)$.

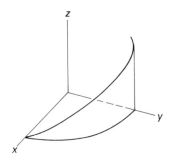

EXAMPLE 2   Show that a space path $\mathbf{x} = \mathbf{g}(t)$ lies entirely on the surface of a sphere about the origin if and only if

$$\mathbf{x} \perp d\mathbf{x}/dt \qquad \text{for all } t.$$

Solution. To say that the path lies on the surface of a sphere about O is to say that $|\mathbf{x}|^2 = x^2 + y^2 + z^2$ has a constant value, say

$$x^2 + y^2 + z^2 = r^2,$$

along the path. Differentiating with respect to t gives

$$2x\frac{dx}{dt} + 2y\frac{dy}{dt} + 2z\frac{dz}{dt} = 0,$$

or

$$2\mathbf{x} \cdot \frac{d\mathbf{x}}{dt} = 0.$$

Thus $\mathbf{x} \perp d\mathbf{x}/dt$ for such a path. Conversely, if $\mathbf{x}$ and $d\mathbf{x}/dt$ are always perpendicular, then this argument can be reversed to show that $|\mathbf{x}|$ is constant.

Just as in the plane, vector derivative calculations like the one above can often be better understood in terms of the direct vector forms of the derivative rules given in Section 2.

We add the product rule for the cross product.

6. If $\mathbf{a}$ and $\mathbf{b}$ are differentiable vector functions of t then so is $\mathbf{x} = \mathbf{a} \times \mathbf{b}$, and

$$\frac{d\mathbf{x}}{dt} = \mathbf{a} \times \frac{d\mathbf{b}}{dt} + \frac{d\mathbf{a}}{dt} \times \mathbf{b}.$$

The proof can be given in terms of components, using the rather complicated definition of the cross product. But it is better to apply the properties of the cross product listed in Extra Problems 20 through 27 in Chapter 13. (See the Extra Problems at the end of this chapter.)

So far we have discussed length for plane paths only. When we turn to space paths, a sin of omission that we committed in Chapter 10 catches up with us. The straightforward definition of path length, as the limit of the lengths of inscribed polygons, involves approximation by finite sums that are not quite Riemann sums. It can be shown, nevertheless, that these sums do converge to the path-length integral,

$$\int_a^b \sqrt{\left(\frac{dx}{dt}\right)^2 + \left(\frac{dy}{dt}\right)^2}\, dt$$

in the plane, and

$$\int_a^b \sqrt{\left(\frac{dx}{dt}\right)^2 + \left(\frac{dy}{dt}\right)^2 + \left(\frac{dx}{dt}\right)^2}\, dt$$

in space. We managed to avoid having to go into this in the plane, but the problem remains with us in space. So we shall simply assume without proof that a smooth space path has length, and that

$$\frac{\text{chord length}}{\text{arc length}} \to 1$$

as chord length approaches 0. Then the argument that we gave in Section 3 of Chapter 10 applies practically verbatim, and proves that

$$\left(\frac{ds}{dt}\right)^2 = \left(\frac{dx}{dt}\right)^2 + \left(\frac{dy}{dt}\right)^2 + \left(\frac{dz}{dt}\right)^2.$$

Thus the magnitude of the path tangent vector $d\mathbf{x}/dt$ is again

$$\left|\frac{d\mathbf{x}}{dt}\right| = \frac{ds}{dt}.$$

Theorem 2 therefore holds for a space path, with the same proof, as does Theorem 3 down to the final statement

$$\left|\frac{d\mathbf{T}}{ds}\right| = \kappa.$$

There we had to use a plane argument that does not generalize to space, because our definition of curvature is tied to the plane and does not generalize. However, the formula

$$\kappa = \left|\frac{d\mathbf{T}}{ds}\right|$$

would have been a reasonable alternative definition for the curvature of a path. That is, it is reasonable to consider the curvature of a path to be the magnitude of the rate of change of the unit tangent vector $\mathbf{T}$ with respect to the arc length s. Then the last part of Theorem 3 proves the two possible curvature definitions in the plane to be equivalent. But, for a space curve, the only possible definition of the absolute curvature κ is our new one,

$$\kappa = \left|\frac{d\mathbf{T}}{ds}\right|.$$

With this definition, Theorem 3 shows that a space path satisfies the same law

$$\frac{d\mathbf{T}}{ds} = \kappa\mathbf{N},$$

where the unit vector $\mathbf{N}$ is normal to the path.

The proof of Theorem 4 is also the same. Here are the statements of these three theorems again (always assuming s measured forward).

THEOREM 5 *The parameter of a smooth space path is equal to the path arc length (plus a constant) if and only if the tangent vector is always a unit vector.*

THEOREM 6 *Let $\mathbf{x} = \mathbf{g}(s)$ be a twice-differentiable space path with arc length as parameter and unit tangent vector $\mathbf{T}$. Then*

$$\frac{d\mathbf{T}}{ds} = \kappa\mathbf{N},$$

where $\mathbf{N}$ is the unit vector normal to the path in the direction in which the path is turning, and κ is (by definition) the absolute curvature of the path.

THEOREM 7 *If $\mathbf{x} = \mathbf{g}(t)$ is the path of a moving particle, with time t as parameter, then the tangential and normal components of the acceleration $\mathbf{a}$ are given by*

$$\mathbf{a} = \frac{dv}{dt}\mathbf{T} + \kappa v^2\mathbf{N},$$

where v is the speed $ds/dt = |d\mathbf{x}/dt|$, and $\mathbf{T}$, κ, and $\mathbf{N}$ are as in the above theorem.

EXAMPLE 3 Consider again the spiral

$$\mathbf{x} = (\cos t, \sin t, t/2)$$

of Example 1. We see that the tangent vector

$$\frac{d\mathbf{x}}{dt} = (-\sin t, \cos t, 1/2)$$

has constant magnitude

$$\left|\frac{d\mathbf{x}}{dt}\right| = \frac{\sqrt{5}}{2}.$$

Thus $ds/dt = \sqrt{5}/2$, and if we start measuring s at the point when $t = 0$, then

$$s = \frac{\sqrt{5}}{2}t.$$

Changing to the parameter s, the path is

$$\mathbf{x} = \left(\cos\frac{2}{\sqrt{5}}s, \sin\frac{2}{\sqrt{5}}s, \frac{s}{\sqrt{5}}\right).$$

Then the unit tangent vector is

$$\mathbf{T} = \frac{d\mathbf{x}}{ds} = \frac{2}{\sqrt{5}}\left(-\sin\frac{2}{\sqrt{5}}s, \cos\frac{2}{\sqrt{5}}s, \frac{1}{2}\right),$$

so

$$\frac{d\mathbf{T}}{ds} = \frac{4}{5}\left(-\cos\frac{2}{\sqrt{5}}s, -\sin\frac{2}{\sqrt{5}}s, 0\right) = \frac{4}{5}\mathbf{N},$$

where $\mathbf{N}$ is a unit vector. Comparing this with

$$\frac{d\mathbf{T}}{ds} = \kappa\mathbf{N},$$

we see that the absolute curvature has the constant value

$$\kappa = \frac{4}{5}.$$

This constantly rising spiral is called a *circular helix*. We shall see in the problems that it is also *twisting* at a constant rate.

We conclude with a brief description of this notion of a twisting space curve, leaving the details to the set of extra problems. Suppose that $\mathbf{x} = \mathbf{f}(t)$ is a smooth space curve with nonzero curvature κ, so that the unit tangent and normal vectors $\mathbf{T}$ and $\mathbf{N}$ are well defined. Then

$$\mathbf{B} = \mathbf{T} \times \mathbf{N}$$

is a unit vector perpendicular to both $\mathbf{T}$ and $\mathbf{N}$ called the unit *binomial* vector to the curve. The plane containing $\mathbf{T}$ and $\mathbf{N}$ (as arrows drawn from the path point) can be considered to be the plane that "momentarily" contains the curve, and it is called the *osculating* plane to the curve at that point. If π is the plane passing through three distinct points $\mathbf{x}_0$, $\mathbf{x}_1$, and $\mathbf{x}_2$ on the curve, then it can be shown that π approaches the osculating plane at $\mathbf{x}_0$ as $\mathbf{x}_1$ and $\mathbf{x}_2$ approach $\mathbf{x}_0$.

The osculating plane is determined by its unit normal vector $\mathbf{B}$, so the *derivative* of $\mathbf{B}$, $d\mathbf{B}/ds$, is the vector rate at which the osculating plane is changing its space orientation, or the (vector) rate at which the curve is *twisting*. It will be seen in the problems that $d\mathbf{B}/ds$ is parallel to $\mathbf{N}$. When the scalar multiple is written with a minus sign,

$$\frac{d\mathbf{B}}{ds} = -\tau\mathbf{N}$$

that τ (Greek tau) is called the *torsion* of the curve. The reason for the minus sign will appear in the problems.

PROBLEMS FOR SECTION 5

Find the tangent vector to each of the following paths at the given point.

1. $\mathbf{x} = (t, t^2, t^3)$; $t = -1$
2. $\mathbf{x} = (2\cos t, 3\sin t, t^2)$; $t = \pi/6$
3. $\mathbf{x} = (3t, t - 2/3, \sin t)$; $t = \pi/3$
4. $\mathbf{x} = (\sec t, \tan t, \sin t)$; $t = \pi/3$
5. $\mathbf{x} = (\log t, \sqrt{t}, 1/t)$; $t = 4$

Prove the following derivative rules for space paths.

6. If $\mathbf{u}$ and $\mathbf{x}$ are differentiable vector functions of t, then so is $a\mathbf{u} + b\mathbf{x}$ for any constants a and b, and

$$\frac{d}{dt}(a\mathbf{u} + b\mathbf{x}) = a\frac{d\mathbf{x}}{dt} + b\frac{d\mathbf{u}}{dt}.$$

7. If w is a differentiable scalar function of t, and $\mathbf{x}$ is a differentiable vector function of t, then $w\mathbf{x}$ is a differentiable vector function of t, and

$$\frac{d}{dt}w\mathbf{x} = w\frac{d\mathbf{x}}{dt} + \frac{dw}{dt}\mathbf{x}.$$

8. If $\mathbf{x}$ and $\mathbf{u}$ are differentiable vector functions of t, then $\mathbf{x} \cdot \mathbf{u}$ is a differentiable scalar function of t, and

$$\frac{d}{dt}\mathbf{x} \cdot \mathbf{u} = \mathbf{x} \cdot \frac{d\mathbf{u}}{dt} + \frac{d\mathbf{x}}{dt} \cdot \mathbf{u}.$$

9. If $\mathbf{x}$ is a differentiable vector function of t and if $t = \phi(r)$ is differentiable, then $\mathbf{x}$ is a differentiable vector function of r, and

$$\frac{d\mathbf{x}}{dr} = \frac{dt}{dr}\frac{d\mathbf{x}}{dt}.$$

Try to use the above rules in proving the remaining problems.

10. a) Find the smallest distance between the lines

$$\mathbf{x} = t(1, 2, -1) \qquad \text{and} \qquad \mathbf{u} = (1, 1, 1) + s(3, -2, 2).$$

   b) If $\mathbf{x}_0$ and $\mathbf{u}_0$ are the points on the two lines obtained in part (a), show that $\mathbf{x}_0 - \mathbf{u}_0$ is perpendicular to the directions of both lines.

11. If $\mathbf{x}_0$ and $\mathbf{u}_0$ are the points on any two space paths at which the paths are closest together, show that the vector $\mathbf{x}_0 - \mathbf{u}_0$ is perpendicular to both paths.

12. Let $\mathbf{x} = \mathbf{f}(t)$ be a space path that does not pass through the origin, and let $\mathbf{x}_0 = \mathbf{f}(t_0)$ be a point on the path that is closest to the origin. Show that the path tangent vector at $\mathbf{x}_0$ is perpendicular to $\mathbf{x}_0$.

13. Let $\mathbf{x} = \mathbf{f}(t)$ be a twice-differentiable path such that

$$\frac{d^2\mathbf{x}}{dt^2} \equiv 0.$$

Prove that $\mathbf{x} = \mathbf{a}t + \mathbf{k}$, so that the path is simply the standard parametric representation of a straight line.

14. Let $\mathbf{x} = \mathbf{f}(t)$ be a path whose tangent vector has a constant direction. Prove that the path lies along a straight line.

15. Let $\mathbf{x} = \mathbf{f}(t)$ be a path whose tangent vector is everywhere perpendicular to $\mathbf{x}$. Prove that the path lies on a sphere about the origin. Also prove the converse.

16. Let $\mathbf{x} = \mathbf{f}(t)$ be a path whose tangent vector is everywhere perpendicular to the constant vector $\mathbf{a}$ ($\mathbf{a} \neq 0$). Prove that the path lies in a plane perpendicular to $\mathbf{a}$.

17. Let $\mathbf{x}$ be a path whose tangent vector is always parallel to the vector $\mathbf{x}$. Prove that $\mathbf{x}$ runs along a line through the origin.

18. Let $\mathbf{x} = \mathbf{f}(t)$ be a differentiable vector function of t, and set $\dot{\mathbf{x}} = d\mathbf{x}/dt$ (Newton's dot notation). Prove the rule

$$\frac{d}{dt}|\mathbf{x}| = \frac{\mathbf{x} \cdot \dot{\mathbf{x}}}{|\mathbf{x}|}.$$

[*Hint.* See Problem 8.]

19. The absolute curvature $\kappa = |d\mathbf{T}/ds|$ has the following formula in terms of derivatives with respect to the parameter t.

$$\kappa^2 = \frac{|\dot{\mathbf{x}}|^2 \, |\ddot{\mathbf{x}}|^2 - (\dot{\mathbf{x}} \cdot \ddot{\mathbf{x}})^2}{|\dot{\mathbf{x}}|^6}.$$

Prove this formula, in the following steps:

a) $\dot{\mathbf{x}} = |\dot{\mathbf{x}}| \, \mathbf{T}$

b) $\ddot{\mathbf{x}} = |\dot{\mathbf{x}}|^2 \dfrac{d\mathbf{T}}{ds} + \dfrac{(\dot{\mathbf{x}} \cdot \ddot{\mathbf{x}})}{|\dot{\mathbf{x}}|} \mathbf{T}$

c) Compute $\ddot{\mathbf{x}} \cdot \ddot{\mathbf{x}}$ from (b) and solve for $\kappa^2 = \left(\dfrac{d\mathbf{T}}{ds} \cdot \dfrac{d\mathbf{T}}{ds} \right)$.

Compute the absolute curvature κ of the following space paths from the formula for κ in Problem 19.

20. $\mathbf{x} = (\cos t, \sin t, t/2)$. (Compare with Example 3.)

21. $\mathbf{x} = (t, t^2/2, t^3/3)$

22. $\mathbf{x} = (\cos t, \sin t, \sin t)$

23. The formula in Problem 19 applies to plane curves as well as to space curves. Show that, in the plane, it reduces to our earlier formula

$$\kappa = \frac{|\ddot{y}\dot{x} - \ddot{x}\dot{y}|}{[(\dot{x})^2 + (\dot{y})^2]^{3/2}}.$$

(This requires looking at components.)

24. Let $\mathbf{x} = \mathbf{f}(t)$ be a space path of zero curvature. Prove that it necessarily runs along a straight line. (Use the formula in Problem 19, and consider two cases. If $\ddot{\mathbf{x}}$ is identically zero, then the path is the standard parametric representation of a straight line. (This was an earlier problem, but work it out again.) If $\ddot{\mathbf{x}} \neq 0$, then show that $\cos \phi = \pm 1$, where ϕ is the angle between $\dot{\mathbf{x}}$ and $\ddot{\mathbf{x}}$. Finally, apply Problem 17.)

6. KEPLER'S LAWS AND NEWTON'S UNIVERSAL LAW OF GRAVITATION

In Section 4, the acceleration vector of a moving plane particle was resolved into its components parallel and perpendicular to the velocity vector $\mathbf{x}'(t)$. It is also possible to resolve the acceleration vector into components parallel and perpendicular to the path position vector $\mathbf{x}(t)$, and this resolution of the acceleration vector has a spectacular application. Using it, we shall be able to derive Newton's universal law of gravitation from Kepler's laws of planetary motion as an almost routine exercise in calculus. Newton did this in the seventeenth century, and it was one of his greatest achievements.

Kepler's laws

The first two laws concern the orbit of a single planet, while the third law relates the various planetary orbits to each other.

I. *A planet moves in a plane through the sun in such a way that the vector from the sun to the planet sweeps out equal areas in equal times. That is, the vector sweeps out area at a constant rate.*

II. *The path of a planet is an ellipse with the sun at one focus.*

III. *The **square** of the period T of a planetary orbit is proportional to the **cube** of its semimajor axis a. That is, the ratio T^2/a^3 is the same for all the planets.*

The new resolution of the acceleration vector $\ddot{\mathbf{x}}(t)$ that we need for Kepler's laws can be obtained most easily by using polar coordinates. We write $\mathbf{x}$ in its polar coordinate form,

$$\mathbf{x} = r(\cos\theta, \sin\theta) = r\mathbf{u},$$

where $\mathbf{u}$ is the unit vector of $\mathbf{x}$, $r = |\mathbf{x}|$, and r and θ are functions of t. Then

$$\dot{\mathbf{u}} = \frac{d}{dt}(\cos\theta, \sin\theta) = (-\sin\theta, \cos\theta)\dot{\theta} = \dot{\theta}\mathbf{p},$$

where $\mathbf{p} = (-\sin\theta, \cos\theta)$ is the unit vector perpendicular to $\mathbf{u}$ and leading $\mathbf{u}$ by $\pi/2$ (since $(-\sin\theta, \cos\theta) = (\cos(\theta + \pi/2), \sin(\theta + \pi/2))$). Similarly,

$$\dot{\mathbf{p}} = \frac{d}{dt}(-\sin\theta, \cos\theta) = -\dot{\theta}\mathbf{u}.$$

Therefore, the velocity vector $\mathbf{v} = \dot{\mathbf{x}}$ and acceleration vector $\mathbf{a} = \dot{\mathbf{v}}$ can be written

$$\mathbf{v} = \dot{\mathbf{x}} = \frac{d}{dt}(r\mathbf{u}) = \dot{r}\mathbf{u} + r\dot{\mathbf{u}} = \dot{r}\mathbf{u} + r\dot{\theta}\mathbf{p},$$

$$\mathbf{a} = \dot{\mathbf{v}} = (\ddot{r}\mathbf{u} + \dot{r}\dot{\theta}\mathbf{p}) + (\dot{r}\dot{\theta}\mathbf{p} + r\ddot{\theta}\mathbf{p} - r(\dot{\theta})^2\mathbf{u}).$$

Thus,

$$\mathbf{a} = (\ddot{r} - r\dot{\theta}^2)\mathbf{u} + (r\ddot{\theta} + 2\dot{r}\dot{\theta})\mathbf{p},$$

and this is the formula we have been looking for.

We can now analyze the consequences of Kepler's first law. If polar coordinates are introduced in the plane containing the planetary orbit, with the sun as the origin, then the area formula in Chapter 10, Section 5, says that the derivative of the area swept out with respect to the polar angle θ is $r^2/2$. Therefore,

$$\frac{d}{dt}(\text{area swept out}) = \frac{r^2}{2}\frac{d\theta}{dt} = \frac{\dot{\theta}r^2}{2}.$$

Kepler's law says that area is being swept out at a constant rate, so

$$\frac{\dot{\theta}r^2}{2} = \text{constant},$$

and

$$\frac{d}{dt}\left(\frac{\dot{\theta}r^2}{2}\right) = \frac{\ddot{\theta}r^2 + 2\dot{\theta}r\dot{r}}{2} = 0.$$

In particular, $\ddot{\theta}r + 2\dot{\theta}\dot{r} = 0$, so the new formula obtained above for the path acceleration vector reduces to

$$\mathbf{a} = (\ddot{r} - r\dot{\theta}^2)\mathbf{u},$$

which is a vector directed along the line through the origin. We have proved the following theorem.

THEOREM 8   *If a plane path* $\mathbf{x}(t)$ *has the property that the position vector* $\mathbf{x}(t)$ *sweeps out area at a constant rate, then the acceleration vector* $\ddot{\mathbf{x}}(t)$ *is always directed along the line through* $\mathbf{x}(t)$ *and* O *(i.e.,* $\ddot{\mathbf{x}}(t)$ *is always a scalar multiple of* $\mathbf{x}(t)$*).*

Note that this result is *independent of the shape of the path*, and that it includes, as a special case, the result in Section 4 about uniform motion around a circle.

Note also in this situation that the magnitude of the acceleration vector is just the magnitude of its radial component, which by the component formula is

$$|\mathbf{a}| = |\ddot{r} - r\dot{\theta}^2|.$$

Kepler's second law says that a planet moves in an elliptical path with one focus at the sun. If we take the origin at the sun, and the major axis of the ellipse along the x-axis, then the elliptical path has the polar equation

$$r = \frac{k}{1 - \epsilon \cos \theta}.$$

where $0 < \epsilon < 1$ and k is positive. (See Example 4 in Section 4 of Chapter 10 and Problem 10 in Section 2 of this chapter.) We can now take the major step toward Newton's inverse-square law of gravitational attraction.

THEOREM 9   *Kepler's first and second laws together imply that the acceleration of a planet is directed toward the sun, and is inversely proportional to the square of the distance from the sun in magnitude.*

Proof. We have already seen, in Theorem 5, that the acceleration of the planet is directed along the line to the sun, and that its magnitude is therefore the absolute value of its radial component, which is

$$|\mathbf{a}| = |\ddot{r} - r\dot{\theta}^2|,$$

by the new resolution formula for $\mathbf{a} = \ddot{\mathbf{x}}$. Here

$$r = \frac{k}{1 - \epsilon \cos \theta},$$

by Kepler's second law (and our discussion above). We differentiate this equation, and use the fact that $r^2 \dot{\theta}$ is a constant C, by Kepler's first law, obtaining

$$\dot{r} = \frac{-k\epsilon \sin \theta}{(1 - \epsilon \cos \theta)^2}\, \dot{\theta}.$$

$$= \frac{-\epsilon \sin \theta}{k} r^2 \dot{\theta}$$

$$= -\frac{C}{k} \epsilon \sin \theta.$$

Differentiating again yields the equation

$$\ddot{r} = -\frac{C}{k} \epsilon \cos \theta \cdot \dot{\theta}.$$

The radial component of the acceleration can now be computed:

$$\ddot{r} - r\dot{\theta}^2 = -\frac{C}{k} \epsilon(\cos \theta)\dot{\theta} - \frac{C\dot{\theta}}{r}$$

$$= -C\dot{\theta}\left[\frac{1}{r} + \frac{\epsilon \cos \theta}{k}\right]$$

$$= -C\dot{\theta}\left[\frac{1 - \epsilon \cos \theta}{k} + \frac{\epsilon \cos \theta}{k}\right] = -\frac{C}{k}\dot{\theta}$$

$$= -\left(\frac{C^2}{k}\right)\frac{1}{r^2}.$$

This shows the acceleration to be inversely proportional to r^2 in magnitude, and to be directed toward the origin, completing the proof of the theorem. ∎

THEOREM 10 *Kepler's third law implies that the constant of proportionality C^2/k is the same for all the planets.*

Proof. This calculation will be left as an exercise. It depends on the following facts.

1. The semimajor and semiminor axes of the ellipse $r = k/(1 - \epsilon \cos \theta)$ are

$$a = \frac{k}{1 - \epsilon^2} \quad \text{and} \quad b = \sqrt{1 - \epsilon^2}\, a$$

(See Chapter 10, Section 4).

2. The area of an ellipse is πab (Chapter 8, Problem 68 in the problem set at the end of the chapter).

3. If T is the period of the planet, that is, the time it takes the planet to make one complete revolution, then the area enclosed by its elliptical orbit is $CT/2$. This is because $C/2$ is the constant rate at which the position vector **x** sweeps out area, by the definition of C.

According to Newton's second law, the (vector) acceleration **a** of a moving body is related to the (vector) force **f** (which causes the acceleration) by the equation

$$\mathbf{f} = m\mathbf{a},$$

where m is the mass of the body. Therefore, our three theorems above can be summarized as follows:

The motion of each planet is governed by a force directed toward the sun, of magnitude

$$k\frac{m}{r^2},$$

where r is the distance from the planet to the sun, m is the mass of the planet, and the constant k is the same for all planets.

It is easy to perform a mental experiment showing that the force is also directly proportional to the mass of the attracting body (the sun); for if the sun were divided into two equal parts, then the total force exerted by the sun on the planet would surely be the sum of the two equal forces exerted by its two equal parts, so each part would exert half of the total force. Reasoning in this way, we are led at once to Newton's universal law of gravitation:

If two bodies of mass m and M are at a distance r apart, then they exert on each other an attractive (gravitional) force of magnitude

$$\gamma \frac{mM}{r^2},$$

where γ is a universal constant (independent of m, M, and r).

PROBLEMS FOR SECTION 6

1. Prove Theorem 10.
2. Prove the converse of Theorem 8 (by reversing the chain of reasoning used there). That is, prove that if the acceleration vector $\mathbf{a} = \ddot{\mathbf{x}}(t)$ is always parallel to $\mathbf{x}(t)$, then the position vector $\mathbf{x}$ sweeps out area at a constant rate.

The next three problems are aimed at proving the following theorem.

Theorem 11. If a particle moves in the plane in such a way that its acceleration is directed toward the origin and is proportional to $1/r^2$ in magnitude, then the particle moves along a conic section (ellipse, parabola, or hyperbola) with the origin as a focus.

We already know from the second problem that then $r^2\dot{\theta}$ is a nonzero constant C, which we shall assume to be positive. In particular, $\dot{\theta}$ is positive, so θ is an increasing function of t, and therefore t is a function of θ. That is, there is a uniquely determined time t_0 at which θ reaches any given value θ_0, this means, in particular, that r is a function of θ along the path. Our problem is to determine the nature of this function.

3. Use the chain rule $\dot{u} = (du/d\theta)\dot{\theta}$ to show that

$$\frac{d\left(\frac{1}{r}\right)}{d\theta} = -\frac{\dot{r}}{C}$$

(where C is the constant value of $r^2\dot{\theta}$), and hence that

$$\frac{d^2\left(\frac{1}{r}\right)}{d\theta^2} = \frac{-\ddot{r}}{C\dot{\theta}}.$$

4. The theorem hypothesis amounts to the equation

$$\ddot{r} - r(\dot{\theta})^2 = -\frac{k}{r^2},$$

where k is a positive constant. Why is this? Use this equation and the results of the problem above to show that

$$\frac{d^2\left(\frac{1}{r}\right)}{d\theta^2} + \frac{1}{r} = \frac{k}{C^2}.$$

5. In Chapter 19 it will be shown that the differential equation $\ddot{y} + y = C$ has the general solution $y = A \cos t + B \sin t + C$. Assuming this, show that the general solution of the above differential equation can be written in the form

$$kr = \frac{C^2}{1 - \epsilon \cos(\theta - \theta_0)},$$

where ϵ is a positive constant. According to Chapter 10, Section 4, the graph of this equation is an ellipse if $\epsilon < 1$, a parabola if $\epsilon = 1$, and a hyperbola if $\epsilon > 1$.

6. Suppose that $0 < \epsilon < 1$, and that the ellipse

$$kr = \frac{C^2}{1 - \epsilon \cos \theta}$$

is the path of a particle moving in such a way that $r^2 \dot{\theta} = C$. Show that

$$\frac{T^2}{a^3} = \frac{4\pi^2}{k},$$

where T is the period of the motion (the time required for one complete revolution) and a is the semimajor axis of the ellipse. (This is Kepler's third law.)

EXTRA PROBLEMS FOR CHAPTER 14

The following problems lean heavily on the vector cross product and its properties given in Extra Problems 20 through 27 at the end of the Chapter 13.

It will be shown below that the absolute curvature κ and the torsion τ of a space curve $\mathbf{x} = \mathbf{f}(t)$ have the following cross-product formulas.

Let $\mathbf{p}$ be the cross product of the first and second derivatives of $\mathbf{x} = \mathbf{f}(t)$:

$$\mathbf{p} = \mathbf{f}'(t) \times \mathbf{f}''(t) = \dot{\mathbf{x}} \times \ddot{\mathbf{x}}.$$

Then

$$\kappa = \frac{|\mathbf{p}|}{|\dot{\mathbf{x}}|^3} \qquad \text{and} \qquad \tau = \frac{\mathbf{p} \cdot \dddot{\mathbf{x}}}{|\mathbf{p}|^2}.$$

Use these formulas to compute the curvature and torsion of the following space curves.

1. $\mathbf{x} = (t, t^2/2, t^3/3)$ 2. $\mathbf{x} = (t^2 + 1, t^2 - 2t, 2t - 1)$

3. $\mathbf{x} = (\sin t, \cos t, bt)$ 4. $\mathbf{x} = (\sin at, \cos at, bt)$

5. $\mathbf{x} = (c \sin at, c \cos at, bt)$ 6. $\mathbf{x} = (\cos t, \sin t, \sin t)$

7. $\mathbf{x} = (e^t, e^{2t}, e^{-t})$ 8. $\mathbf{x} = (\log t, t^3, t^4)$

9. $\mathbf{x} = (\sin t, \tan t, \sec t)$

We turn now to the proofs of the curvature and torsion formulas.

10. Show that if $\mathbf{a}$ and $\mathbf{b}$ are perpendicular unit vectors, then $\mathbf{c} = \mathbf{a} \times \mathbf{b}$ is a unit vector. Thus the binormal to a space curve, $\mathbf{B} = \mathbf{T} \times \mathbf{N}$, is a unit vector.

11. The acceleration formula of Theorem 7 is valid whether t is considered to be time or not, and has the neutral form

(*)
$$\ddot{\mathbf{x}} = \left(\frac{d}{dt}|\dot{\mathbf{x}}|\right)\mathbf{T} + \kappa\,|\dot{\mathbf{x}}|^2\,\mathbf{N}.$$

Show therefore that

(**)
$$\mathbf{p} = \dot{\mathbf{x}} \times \ddot{\mathbf{x}} = \kappa\,|\dot{\mathbf{x}}|^3\,\mathbf{B},$$

and therefore, since $\mathbf{B}$ is a unit vector, that

$$\kappa = \frac{|\mathbf{p}|}{|\dot{\mathbf{x}}|^3}.$$

12. Show, by differentiating the identities

$$\mathbf{B}\cdot\mathbf{T} = 0, \qquad \mathbf{B}\cdot\mathbf{B} = 1, \qquad \mathbf{B}\cdot\ddot{\mathbf{x}} = 0$$

a) That $d\mathbf{B}/ds$ is perpendicular to both $\mathbf{T}$ and $\mathbf{B}$ and hence can be written in the form

$$\frac{d\mathbf{B}}{ds} = -\tau\mathbf{N};$$

b) that

$$\mathbf{B}\cdot\dddot{\mathbf{x}} = -\dot{\mathbf{B}}\cdot\ddot{\mathbf{x}}.$$

13. Use the results in Problems 11 and 12 to show that

$$\mathbf{p}\cdot\dddot{\mathbf{x}} = \kappa^2\,|\dot{\mathbf{x}}|^6\,\tau$$

and thus derive the formula for τ.

14. Show that the present formula for κ equals the formula given in Problem 19 of Section 5.

15. Let $\mathbf{x} = \mathbf{f}(t)$ be a space path having curvature everywhere nonzero (so that the torsion τ is everywhere defined). Show that if $\tau \equiv 0$, then the path lies in a plane.

16. Prove the derivative rule for the cross product of two vector functions, using properties of the cross product from the list given in Extra Problems of Chapter 13. [*Hints:* Note first that $(\mathbf{a}_1 + \mathbf{a}_2) \times \mathbf{b} = \mathbf{a}_1 \times \mathbf{b} + \mathbf{a}_2 \times \mathbf{b}$ (from Problems 24 and 21) and that $|\mathbf{a} \times \mathbf{b}| \le |\mathbf{a}|\cdot|\mathbf{b}|$ (from Problem 25). Now show that if $\mathbf{x} = \mathbf{a} \times \mathbf{b}$ then

$$\Delta\mathbf{x} = (\mathbf{a} + \Delta\mathbf{a}) \times \mathbf{b} + \mathbf{a} \times (\mathbf{b} + \Delta\mathbf{b}) + \Delta\mathbf{a} \times \Delta\mathbf{b},$$

and finish as usual.]

15
Functions of Two Variables

More often than not the dependency relationships that arise in mathematics and its applications involve several variables. The volume of a pyramid with a square base depends on its base edge b and altitude h, according to the formula

$$V = \frac{1}{3} b^2 h.$$

Thus, V is a function of the two independent variables b and h.

The temperature T indicated on a weather map varies with position, and is thus a function of latitude and longitude. If we add the variation of T with time also, then T is a function of three variables, say, x, y, and t.

The state of a small homogeneous body of gas is determined by its pressure p, volume V, and temperature T. According to Boyle's law, these state variables are related by

$$pV = k(T + C),$$

where k and C are constants. This equation determines any one of the state variables as a function of the other two.

We investigate the variation of such a function by a "divide-and-conquer" strategy. We first study the variation that occurs when only one of the independent variables is permitted to change. The tool here is the *partial derivative*, which is just the ordinary derivative with respect to the one changing variable. Then we learn how to combine the information provided by the several partial derivatives in order to understand the way in which the function varies. This "multidimensional" calculus is an enormous subject that we will barely touch upon in this book.

In this chapter we shall consider only the separate partial derivatives, and two relatively simple applications, to maximum-minimum problems and to the equation of a tangent plane to a graph in three dimensions. In Chapter 16 we consider the combined action of the partial derivatives, and investigate the simplest consequences of their vector interpretation.

1.   GRAPHS IN COORDINATE SPACE

The graph of an equation in x, y, and z is the space configuration made up of all the points (x, y, z) satisfying the equation, and it will generally be some kind of two-dimensional surface.

EXAMPLE 1   The graph of

$$x^2 + y^2 + z^2 = 25$$

is the sphere about the origin of radius 5. For, taking square roots and replacing x by $x - 0$, etc., we get the equivalent equation

$$\sqrt{(x - 0)^2 + (y - 0)^2 + (z - 0)^2} = 5,$$

which says that the distance from $(0, 0, 0)$ to (x, y, z) is 5, by the distance formula. The graph consists of all such points (x, y, z) and is thus the sphere. Here is a drawing of the part of the sphere in the first octant.

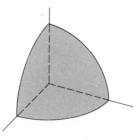

It was shown by a vector argument in Chapter 13, Section 6 that the graph of the equation

$$ax + by + cz + d = 0$$

is always a plane, provided that the coefficients a, b, and c are not all zero. In the simplest cases this can be seen directly, without the intervention of vectors, and these direct considerations have their own value.

EXAMPLE 2   A horizontal plane is specified by the constant z-coordinate common to all of its points. For example, the graph of the equation

$$z = 2$$

is the horizontal plane two units above the xy-coordinate plane. More generally, $z = c$ is the equation of the horizontal plane lying at a distance c above (a distance $|c|$ below if c is negative) the xy-coordinate plane. In particular, $z = 0$ is the equation of the xy-coordinate plane itself. Every horizontal plane is an xy-plane in its own right. For example, the equations

$$x^2 + y^2 = 1, \qquad z = 2$$

specify the unit circle in the plane $z = 2$.

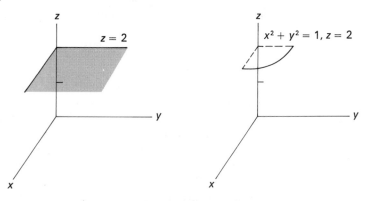

Similarly, $x = a$ is the equation of the vertical plane parallel to the yz-coordinate plane and lying at a distance a from it, as measured in the direction of the positive x-axis. And $y = b$ is a vertical plane parallel to the xz-coordinate plane.

The intersection of the latter two planes is the *vertical line* consisting of

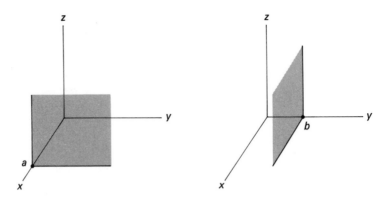

all points of the form (a, b, z), i.e., the vertical line passing through the point $(a, b, 0)$ in the xy-coordinate plane.

These simple planes can be useful in the study of a more complicated space figure F. For example, if we can determine how F intersects each horizontal plane $z = c$, then by visualizing several of these horizontal plane sections of F, we can build up a picture of F.

The word *section* can be used whenever two configurations intersect: the A section of B is simply the intersection of A and B, but with the implication that the intersection is a piece of B that may reveal structural information about B.

EXAMPLE 3   Plane sections can even be used to study other planes. Consider, for example, the space graph of the equation

$$2x + y = 1.$$

Its horizontal plane sections are all the same, being in each case the line in the plane $z = c$ having the equation $2x + y = 1$. This means that the *space graph of $2x + y = 1$ is the *vertical plane* passing through the *line*

$$2x + y = 1$$

in the xy-coordinate plane.

Alternatively, we can argue that since the equation $2x + y = 1$ does not contain z, a point (x_0, y_0, z_0) lies on its graph if and only if (x_0, y_0, z) also does, for *any* value of z. The graph thus contains the whole vertical line through each of its points. In order to specify *which* vertical lines, we need only specify how the graph intersects any given horizontal plane. Since its intersection with the xy-coordinate plane is the line l in that plane having the equation $2x + y = 1$, the space graph consists of the collection of all vertical lines through l, and hence is the vertical plane through l.

EXAMPLE 4   The same reasoning applies to any other equation involving only x and y. Thus the equation

$$x^2 + y^2 = 1$$

has a circle for its graph in the xy-coordinate plane, but its space graph is the doubly infinite circular *cylinder* made up of all vertical lines through this circle. These lines are called the *rulings* on the cylinder, and their common direction is the direction of the cylinder. Since this is the direction of the z-axis, we say the cylinder is parallel to the z-axis.

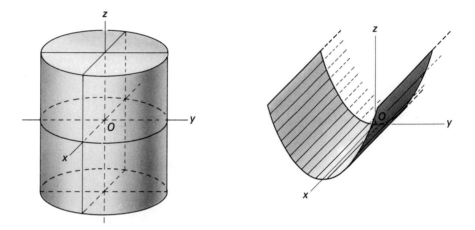

The graph of $z = y^2$ is a parabolic cylinder parallel to the x-axis, and the graph of $x^2 + 2z^2 = 1$ is an elliptical cylinder parallel to the y-axis.

EXAMPLE 5 The two parabolic cylinders $2z = 4 - y^2$ and $2z = 4 - x^2$, together with the coordinate planes, bound a solid figure in the first octant, as sketched at the right. It has square cross sections parallel to the xy-plane.

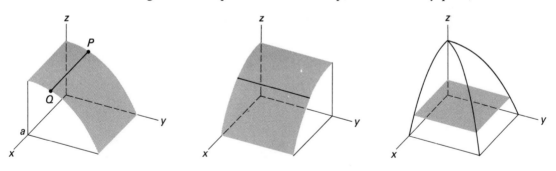

EXAMPLE 6 So far we have examined only the simplest examples of first- and second-degree equations. The general first-degree equation always has a plane for its graph, a vertical plane if the equation does not contain z, and is thus of the form

$$ax + by + c = 0,$$

and a nonvertical plane otherwise. In the second case the equation can always be solved for z and thus written in the form

$$z = ax + by + c.$$

That its graph is always a plane was shown in Chapter 13, by vector considerations, but it can also be verified directly by noting that all the vertical and horizontal plane sections of the graph are straight lines.

The graph of any second-degree equation is called a *quadric* surface. It can be shown that *every* plane section of such a surface is a conic or a degenerate conic, no matter how the plane is oriented. Here are a few examples.

EXAMPLE 7 The graph of

$$z = \text{(quadratic in both } x \text{ and } y)$$

is called a *paraboloid*. Its vertical sections are parabolas. There are two types, with standard forms:

$$z = \frac{x^2}{a^2} + \frac{y^2}{b^2} \qquad \text{(elliptic paraboloid)},$$

$$z = \frac{x^2}{a^2} - \frac{y^2}{b^2} \qquad \text{(hyperbolic paraboloid)}.$$

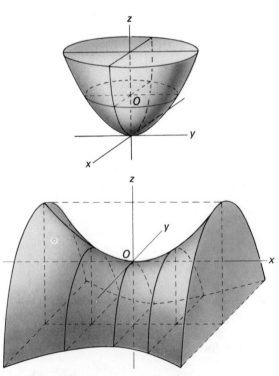

Their horizontal sections are ellipses and hyperbolas, respectively. If $a = b$ in the first case, so that the horizontal sections are circles, the surface is a *paraboloid* of *revolution*. It can be generated by rotating about the z-axis its section in any plane through the z-axis.

EXAMPLE 8 The graph of

$$z^2 = x^2 + y^2$$

is a *cone*. It is a *circular cone* about the z-axis, meaning that its sections perpendicular to the z-axis are circles centered on the z-axis. In the same way

$$z^2 = \frac{x^2}{a^2} + \frac{y^2}{b^2}$$

is an *elliptical cone*. In each case, the origin is the *vertex* of the cone, and the cone consists of all straight lines (its *rulings*, or *generators*) through the vertex and the points of a suitably chosen plane section. A circular cone is a cone of revolution.

EXAMPLE 9 The graph of

$$\frac{x^2}{a^2} + \frac{y^2}{b^2} + \frac{z^2}{c^2} = 1$$

is called an *ellipsoid*. It is a bounded surface, and its intersection with *any* plane (that intersects it) is an ellipse in that plane. This is pretty easy to see for a plane parallel to a coordinate plane, but it is by no means obvious in general.

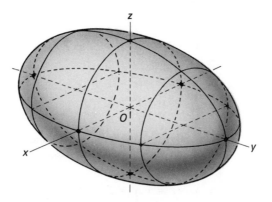

If two of the three constants a, b, and c are equal then we have an *ellipsoid* of *revolution*.

EXAMPLE 10   Finally, the graph of

$$\frac{x^2}{a^2} + \frac{y^2}{b^2} - \frac{z^2}{c^2} = k$$

is a *hyperboloid*. It has one sheet if $k > 0$ and two sheets if $k < 0$. Its horizontal sections are ellipses and its vertical sections are hyperbolas. If $a = b$ the surface in each case is a hyperboloid of revolution (about the z-axis).

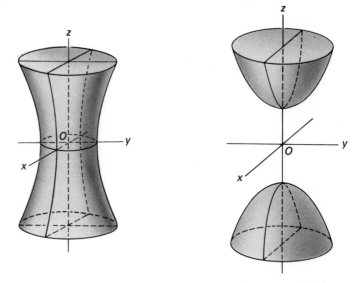

PROBLEMS FOR SECTION 1

Identify each of the following surfaces, including the nature of the section perpendicular to each axis. (For example: a hyperboloid of two sheets, with elliptical sections perpendicular to the x-axis and hyperbolic sections perpendicular to the y- and z-axes.) If it is a surface of revolution, give the axis.

1. $x^2 + 2y^2 + z^2 = 1$ 2. $x^2 + 2y^2 - z^2 = 1$

3. $x^2 + 2y^2 - z^2 = -1$ 4. $x^2 - 2y^2 - z^2 = 1$

5. $x^2 - 2y^2 + z^2 = 1$ 6. $x^2 + 2y^2 + z^2 = 0$

7. $x^2 + 2y^2 - z^2 = 0$ 8. $x^2 - 2y^2 + z^2 = 0$

9. $x^2 + 2y^2 - z = 0$ 10. $x^2 - 2y^2 - z = 0$

11. $x^2 + 2y + z^2 = 0$ 12. $x^2 + z^2 = 0$

13. $x^2 - z^2 = 0$

Draw the figure representing the solid in the first octant bounded by the coordinate planes and the following surfaces.

14. $x + y + z = 2$ 15. $y + z = 1$, $x = 2$

16. $x + z = 1$, $y = 2$ 17. $x + y = 1$, $z = 2$

18. $x^2 + y^2 = 1$, $z = 2$

19. $x^2 + y^2 = 1$, $x + z = 1$

20. $x^2 + y^2 + z^2 = 25$, $y = 4$

21. $y^2 + z^2 = 16$, $x^2 + y^2 = 16$

22. $x^2 + y^2 = 16$, $x + y + z = 8$

23. $4x^2 + 4y^2 + z^2 = 64$, $y + z = 4$

24. $x^2 + y^2 = z^2 + 9$, $z = 4$

2. FUNCTIONS OF TWO VARIABLES; PARTIAL DERIVATIVES

Functions of several variables involve the same language and the same conventions that are used for functions of one variable. Thus, each of the expressions $x^2 + y^2$ and $x \sin y$ defines a function $f(x, y)$ of the two independent variables x and y, since each of them has a uniquely determined value for each pair of numbers x and y. Similarly,

$$g(x, y, z, w) = (x^2 + y^2)zw$$

is a function of four variables. Except for an occasional remark, we shall consider only functions of two variables in this chapter.

In general, a function $f(x, y)$ will not be defined for all pairs of numbers (x, y). For example,

$$f(x, y) = y \log x$$

is defined only when $x > 0$. Thus:

DEFINITION   *A function of two variables is an operation that determines a unique output number z when applied to any given input pair of numbers (x, y) chosen from a certain domain.*

Each of the equations

$$z = x^2 + y,$$

$$xyz = 1,$$

$$z^3 = x/y,$$

determines z as a function of x and y. Thus a function of *two* variables can be defined by an equation in *three* variables, just as a function of one variable can be defined by an equation in two variables. Similarly,

$$e^w = xyz$$

determines w as a function of the *three* variables x, y, and z.

The domain of a function of two variables $f(x, y)$ will generally be some sort of region in the plane bounded by one or more curves. For example, $f(x, y) = \sqrt{1 - x^2 - y^2}$ is defined when $1 - x^2 - y^2 \geq 0$, that is, when

$$x^2 + y^2 \leq 1.$$

The domain of f is the "unit circular disk," and it is bounded by the unit circle $x^2 + y^2 = 1$.

The domain of

$$g(x, y) = y \log x$$

is the right half-plane, and its boundary is the y-axis.

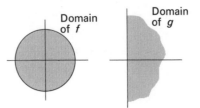

A point (a, b) is *in the interior* of a domain D if some small circular disk about (a, b) lies entirely in D. Any other point of D is a *boundary point*. Thus a point is a boundary point of D if every circle about it contains both points in D and points not in D. As stated above, the boundary points of a domain normally lie along one or more boundary curves.

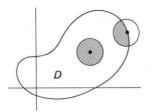

If $z = x^2y^3 + xy + y^2$, and if we hold y fixed, then z becomes a function of x alone and we can calculate its derivative. This derivative is called the *partial derivative of z with respect to x*. In Leibniz notation it is written

$$\frac{\partial z}{\partial x},$$

where ∂x is read 'del x'. Thus, treating y as a constant and computing the derivative of z with respect to x, we have

$$\frac{\partial z}{\partial x} = 2xy^3 + y.$$

Similarly, we obtain $\partial z/\partial y$ by regarding x as a constant and computing the derivative of z with respect to y:

$$\frac{\partial z}{\partial y} = 3x^2y^2 + x + 2y.$$

This procedure is followed for any number of independent variables. For example, if

$$w = x^2 - xy + y^2 + 2yz + 2z^2 + z,$$

then

$$\frac{\partial w}{\partial x} = 2x - y,$$

$$\frac{\partial w}{\partial y} = -x + 2y + 2z,$$

$$\frac{\partial w}{\partial z} = 2y + 4z + 1.$$

In addition to the Leibniz notation for partial derivatives, we shall need function notation, analogous to the f' notation for functions of one variable. We shall use

$$D_1 f \quad \text{and} \quad D_2 f$$

for the partial derivative of f with respect to its first and second variables, respectively. Thus, if $z = f(x, y)$, then

$$\frac{\partial z}{\partial x} = D_1 f(x, y),$$

$$\frac{\partial z}{\partial y} = D_2 f(x, y).$$

As in the case of functions of one variable, the function notation is superior to the Leibniz notation when substitutions have to be shown. For example, the value of $\partial z/\partial x$ at $(x, y) = (a, b)$ is given in the two notations by

$$D_1 f(a, b) = \frac{\partial z}{\partial x}\bigg|_{(a, b)}.$$

EXAMPLE   If $f(x, y) = xy + x^2 - y$, then

$$D_1 f(x, y) = y + 2x, \qquad D_2 f(x, y) = x - 1,$$
$$D_1 f(2, 3) = 3 + 2 \cdot 2 = 7, \qquad D_2 f(2, 3) = 2 - 1 = 1.$$

At a couple of points we shall have to mention continuity, in order to state things correctly. This notion extends naturally to a function of two variables $f(x, y)$:

The function f is continuous at the point (a, b) if its value $f(x, y)$ can be made to be as close to $f(a, b)$ as we wish by taking the point (x, y) suitably close to (a, b).

In the plane, the closeness of two points is measured by the distance between them, so being suitably close to (a, b) means lying within a suitably small circle about (a, b). This is a simultaneous condition on x and y, and the notion of continuity thus involves simultaneous variation in x and y.

Since this chapter is only a preview of functions of more than one variable, we shall not look into whether there are reasonable ways in which such a function can be discontinuous. Every function that will come up will be continuous wherever it is defined. The reason lies along the lines of the Chapter 3 theorem, that a differentiable function is automatically continuous. But when two independent variables are involved, the argument from derivatives to continuity is more subtle, and we shall just state the appropriate theorem from Appendix 3.

THEOREM   *Suppose that both partial derivatives of $f(x, y)$ exist everywhere inside a circle C, and are everywhere less in magnitude than a fixed constant k. Then f is continuous inside of C.*

PROBLEMS FOR SECTION 2

1. Write out the definitions of $\partial z/\partial x$ and $\partial z/\partial y$ (where $z = f(x, y)$) as limits of difference quotients, using the increment notation.

For each of the following functions, find $\partial z/\partial x$ and $\partial z/\partial y$.

2. $z = 2x - 3y$ 3. $z = 3x^2 - 2xy + y$

4. $z = \sqrt{x^2 + y^2}$ 5. $z = e^{ax+by}$

6. $z = e^{xy}$ 7. $z = \sin(2x + 3y)$

8. $z = \cos((x^2 - y^2)/2)$ 9. $z = xye^{-x^2/2}$

10. $z = y \log(x + y)$ 11. $z = y \log(xy)$

12. $z = e^{-x} \cos y$ 13. $z = \sin ax \cos by$

14. $z = \dfrac{x}{y} - \dfrac{y}{x}$ 15. $z = \arctan \dfrac{x}{y}$

16. $z = x^y$ 17. $z = \dfrac{x + y}{x - y}$

In the problems below, compute the partial derivatives using function notation, and determine the indicated values.

18. $f(u, v) = 3u^2 + 2uv + v^3$, $D_1 f(2, -3)$, $D_2 f(1, 1)$

19. $f(s, t) = se^{st}$, $D_1 f(2, 3)$, $D_2 f(2, -1)$

20. $f(x, y) = \sin(3x + y^2)$, $D_1 f(\pi, 0)$, $D_2 f(0, \sqrt{\pi})$

21. $f(x, y, z) = xy^2 z^3$, $D_1 f(0, 2, -2)$, $D_2 f(0, 2, -2)$, $D_3 f(1, 2, -2)$

22. $f(x, u, t) = x^2 + ut^3$, all partials at $(2, 0, 1)$

23. $f(s, t) = s^2 \sin t$, $D_1 f(a, b)$, $D_2 f(a, 0)$

A function can have a property called "homogeneity of degree n" that we won't define here, but which can be shown to be equivalent to the condition

(*) $$xD_1 f(x, y) + yD_2 f(x, y) = nf(x, y).$$

24. Show that $z = \arctan(y/x)$ is homogeneous of degree 0. That is, show that

$$0 = x\frac{\partial z}{\partial x} + y\frac{\partial z}{\partial y}.$$

Each of the following functions is homogeneous. In each case verify (*) and determine the degree of homogeneity.

25. $z = x^4 + y^4 - x^2 y^2$ 26. $z = (x + y)/(x - y)$

27. $z = x/(x^2 + y^2)$

28. Polar coordinates and rectangular coordinates are related by the equations

$$x = r \cos \theta, \qquad r = \sqrt{x^2 + y^2},$$

$$y = r \sin \theta, \qquad \theta = \arctan \frac{y}{x}.$$

Find all the partial derivatives of these functions.

29. The boundary of a plane region G need not consist of one or more curves. What is the domain of the function $f(x, y) = 1/(x^2 + y^2)$? What is its boundary?

30. Find a function whose domain boundary consists of just the two points $(1, 0)$ and $(-1, 0)$.

31. Find a function whose domain boundary consists of the unit circle $x^2 + y^2 = 1$ plus the origin $(0, 0)$.

3. FUNCTION GRAPHS; TANGENT PLANES

The graph of a function of two variables $f(x, y)$ is the graph of the equation

$$z = f(x, y).$$

If we interpret the domain D of f as a subset of the xy-coordinate plane, then the graph of f is a surface S that is *spread out* over D, in the sense that each vertical line through a point $(x, y, 0)$ of D intersects S exactly once, in the point $(x, y, z) = (x, y, f(x, y))$.

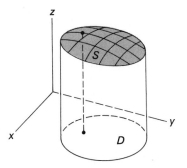

It is very helpful in visualizing such a surface S to consider the *curves* in which it intersects several vertical planes

$$x = a \qquad \text{and} \qquad y = b,$$

that is, the section of S in these planes. Its section in the plane $x = a$ is the graph in this yz-plane of the function of one variable $f(a, y)$, that is, the graph of the equation

$$z = f(a, y).$$

It is the *slice* in which the plane $x = a$ cuts S.

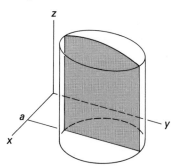

The slope of this curve is the derivative of z with respect to y, while x is held fixed at the value $x = a$, and this is exactly the partial derivative

$$\left. \frac{\partial z}{\partial y} \right|_{x=a} = D_2 f(a, y).$$

Thus we can interpret $\partial z / \partial y$ as *the slope of the graph of $z = f(x, y)$ in the y-direction.*

Similarly, the section of S in the plane $y = b$ is the graph in that xz-plane of the equation $z = f(x, b)$.

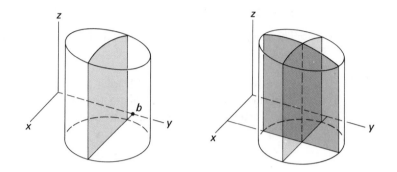

Its slope is $\partial z / \partial x$, evaluated along $y = b$, that is, $D_1 f(x, b)$. Thus, $\partial z / \partial x$ can be interpreted as *the slope of the graph $z = f(x, y)$ in the x-direction.*

If (a, b) is in the domain of f, then the two sections above necessarily intersect over $(a, b, 0)$, the point of intersection being

$$(a, b, z) = (a, b, f(a, b)).$$

If we sketch two or three curves of each type, even very roughly, we get an impression of how the surface is curving. We did this in the figure on page 549.

Finally, the section of S in the horizontal plane $z = c$ is the graph in that xy-plane of the equation

$$c = f(x, y).$$

This horizontal curve may or may not intersect a given vertical slice, say the one in the plane $y = b$. It will just in case the function of x

$$z = f(x, b)$$

assumes the value c. An example where the curves do not intersect is shown below for the parabolic cap $z = 4 - x^2 - y^2$.

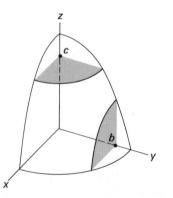

When we draw the graphs

$$c = f(x, y)$$

in the xy-coordinate plane, that is, in the domain D of f, it is natural to call them the *level* curves of f. An xy-figure showing several such level curves is an alternate way of picturing the behavior of $f(x, y)$.

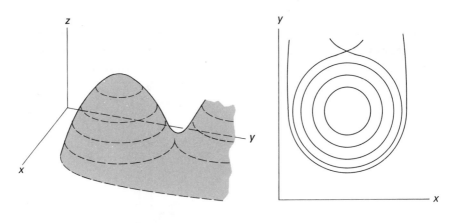

EXAMPLE 4   The level curves of the parabolic cap

$$5z = 25 - (x^2 + y^2)$$

are just circles. Nevertheless, their spacing can indicate the shape of the space graph of the equation. This is the principle of a contour map.

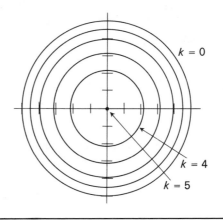

The tangent plane to the graph of $z = f(x, y)$ at a point (x_0, y_0, z_0) is the plane through that point having slopes in the x and y directions equal to the slopes of the surfaces at that point. The following example works this out in detail. In such problems it is convenient to replace the general equation of a nonvertical plane

$$z = ax + by + c$$

by its *point–slope* form

$$z - z_0 = a(x - x_0) + b(y - y_0).$$

The latter is obtained by subtracting the identity

$$z_0 = ax_0 + by_0 + c$$

from the general equation.

EXAMPLE 5   We shall find the equation of the plane π that is tangent to the surface $4z = 25 - 3x^2 - y^2$ at the point $(x_0, y_0, z_0) = (2, 1, 3)$.

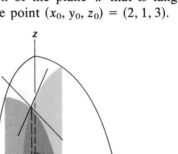

(It should be checked that this point is on the surface.) We assume that there *is* a tangent plane and that our only problem is to find its equation. In Chapter 16 we will convince ourselves that the plane that we get meets every criterion of tangency.

Since the tangent plane contains the point $(2, 1, 3)$, its point–slope equation is

$$z - 3 = a(x - 2) + b(y - 1).$$

The surface is the graph of the function

$$f(x, y) = \frac{25 - 3x^2 - y^2}{4}.$$

At the point of tangency the surface and plane must have the same slope in the x-direction. That is, we must have $D_1 f(2, 1) = a$. But

$$D_1 f(x, y) = -\frac{3x}{2},$$

and

$$D_1 f(2, 1) = -3.$$

Therefore, $a = -3$.

Similarly, the surface and tangent plane must have the same slope in the y-direction at the point of tangency, so that

$$b = D_2 f(2, 1) = -\left.\frac{2y}{4}\right|_{(2,1)} = -\frac{1}{2}.$$

The equation of the tangent plane is therefore

$$z - 3 = -3(x - 2) - \frac{1}{2}(y - 1)$$

or
$$2z = -6x - y + 19.$$

The same argument shows that the plane
$$z - z_0 = a(x - x_0) + b(y - y_0)$$
will be tangent to the surface
$$z = f(x, y)$$
at the point $(x_0, y_0, z_0) = (x_0, y_0, f(x_0, y_0))$ if
$$a = D_1 f(x_0, y_0),$$
and
$$b = D_2 f(x_0, y_0).$$

Thus the equation of the tangent plane to the graph of
$$z = f(x, y)$$
at the point over (x_0, y_0) is
$$z - z_0 = [D_1 f(x_0, y_0)](x - x_0) + [D_2 f(x_0, y_0)](y - y_0),$$
where $z_0 = f(x_0, y_0)$.

EXAMPLE 6   Find the equation of the plane tangent to the surface $z = xy^2$ at the point over $(x_0, y_0) = (-3, 2)$.

Solution. We compute the remaining coefficients for the above form of the tangent plane equation:
$$z_0 = f(-3, 2) = (-3)(2)^2 = -12,$$
$$D_1 f(-3, 2) = y^2 \big|_{(-3, 2)} = 2^2 = 4,$$
$$D_2 f(-3, 2) = 2xy \big|_{(-3, 2)} = 2(-3)(2) = -12.$$

Substituting these values in the general form above gives
$$z + 12 = 4(x + 3) - 12(y - 2),$$
or
$$z = 4x - 12y + 24,$$
as the equation of the tangent plane.

Tangent planes to surfaces that are not explicitly given as function graphs will be discussed in the next chapter (Section 7). We can get some idea of what to expect by considering a quadric surface, such as the sphere
$$x^2 + y^2 + z^2 = 14.$$
Although a sphere is not a function graph, we can consider it to be a

combination of *two* function graphs, forming its upper and lower hemispheres. The functions in question can be found by solving for z in the equation above, and are

$$f(x, y) = \sqrt{14 - (x^2 + y^2)},$$
$$g(x, y) = -f(x, y).$$

We can find the *nonvertical* tangent planes to the sphere by finding the tangent planes to these two function graphs.

EXAMPLE 7   The point $(3, 1, -2)$ lies on the above sphere, on the lower function graph. In order to find the tangent plane at this point, we compute the partials of this function at $(x, y) = (3, 1)$:

$$D_1g(x, y) = -\frac{-x}{\sqrt{14 - (x^2 + y^2)}}; \qquad D_1g(3, 1) = \frac{3}{\sqrt{14 - 10}} = \frac{3}{2};$$

$$D_2g(x, y) = -\frac{-y}{\sqrt{14 - (x^2 + y^2)}}; \qquad D_2g(3, 1) = \frac{1}{\sqrt{14 - 10}} = \frac{1}{2}.$$

The tangent plane thus has the equation

$$z - z_0 = \frac{3}{2}(x - x_0) + \frac{1}{2}(y - y_0),$$

or

$$z + 2 = \frac{3}{2}(x - 3) + \frac{1}{2}(y - 1),$$

or

$$3x + y - 2z = 14.$$

EXAMPLE 8   Instead of solving explicitly for the upper and lower functions in the above example, it is easier simply to assume that the equation has function solutions and to calculate their partial derivatives by implicit differentiation. The situation is exactly like that in Chapter 5, except for notation. We assume that the equation

$$x^2 + y^2 + z^2 = 14$$

defines z implicitly as a function of x and y. If y is held constant then z is a function of x alone, and we can compute its derivative by the chain rule, as in Chapter 5. The only difference is that we are computing the partial derivative $\partial z/\partial x$ instead of the ordinary derivative. We get

$$2x + 2z\frac{\partial z}{\partial x} = 0,$$

so $\partial z/\partial x = -x/z$. Similarly, $\partial z/\partial y = -y/z$. The tangent plane at x_0, y_0, z_0 thus has the equation

$$z - z_0 = -\frac{x_0}{z_0}(x - x_0) - \frac{y_0}{z_0}(y - y_0)$$

and substituting $(x_0, y_0, z_0) = (3, 1, -2)$ gives the same answer as before.

REMARK   The two function-graph hemispheres in these examples are joined along the equatorial seam, where the tangent planes are all vertical. However, from a purely geometric viewpoint a sphere doesn't have a seam and all tangent planes are on equal footing. The vector approach to tangent planes given in the next chapter doesn't require such a function graph decomposition and treats all tangent planes equally.

PROBLEMS FOR SECTION 3

Sketch the level curves for each of the following functions at the given values of z.

1. $z = y/x$; $z = 0, 1, 2, 3, 4$
2. $z = y/x^2$; $z = 0, 1, 2, 3$
3. $z = x + y$; $z = -2, -1, 0, 1, 2$
4. $z = x - y$; $z = -2, -1, 0, 1, 2$
5. $z = xy$; $z = -2, -1, 0, 1, 2$
6. $z = x^2 + y^2$; $z = 1, 2, 3, 4$
7. $z = x^2 + 4y^2$; $z = 1, 2, 3, 4$
8. $z = x^2 - y^2$; $z = -2, -1, 0, 1, 2$
9. $z^2 = x^2 + y^2 + 1$; $z = 1, 2, 3, 4$
10. $z^2 = x^2 + y^2 - 1$; $z = 0, 1, 2, 3$

Assuming that each of the following equations defines z implicitly as a function of x and y, find the partial derivative $\partial z/\partial x$ and $\partial z/\partial y$ by implicit differentiation.

11. $\dfrac{x^2}{a^2} + \dfrac{y^2}{b^2} + \dfrac{z^2}{c^2} = 1$
12. $Ax^2 + Bxy + Cy^2 + Dz^2 + E = 0$
13. $x^3 + y^3 + z^3 = 1$
14. $z^3 + xyz = C$
15. $ze^{xyz} = 1$
16. $\sin xy + \sin yz + \sin xz = 0$
17. $e^{xz} \sin yz = 1$
18. $xy \arctan(z/x) = 1$

In the following problems, determine the equation of the tangent plane to the surface at the point indicated.

19. $xy + yz + zx = 11$, $(1, 2, 3)$
20. $xyz = x + y + z$, $(1/2, 1/3, -1)$
21. $z = \log \sqrt{x^2 + y^2}$, $(-3, 4, \log 5)$
22. $z = e^x \sin y$, $(1, \pi/2, e)$
23. $z = xy$, $(2, -1, -2)$
24. $z = 3x^2 + 2y^2 - 11$, $(2, 1, 3)$

Use implicit differentiation to find the tangent planes to the following surfaces at the given points.

25. $x^2 + y^2 + z^2 = 6$; $(-1, 2, 1)$
26. $x^2 + 2y^2 + 5z^2 = 6$; $\left(2, \dfrac{1}{2}, -\dfrac{1}{2}\right)$
27. $x^2 + 2y^2 - z^2 = 2$; $(1, 1, 1)$, $(1, 1, -1)$
28. $x^2 + 2y^2 - z^2 = -1$; $(1, 1, 2)$
29. Using implicit differentiation, show that the equation of the tangent plane to the sphere

$$x^2 + y^2 + z^2 = r^2$$

at (x_0, y_0, z_0) can be written

$$xx_0 + yy_0 + zz_0 = r^2.$$

30. Obtain a similar result for the ellipsoid

$$\frac{x^2}{a^2} + \frac{y^2}{b^2} + \frac{z^2}{c^2} = 1.$$

4. MAXIMUM — MINIMUM PROBLEMS

Suppose that the function $z = f(x, y)$ assumes a maximum (or minimum) value at a point (x_0, y_0) in the interior of its domain. If we hold y constant at the value y_0, then $f(x, y_0)$ is a function of the one variable x having its maximum value at $x = x_0$, and so its derivative must be zero there, as in Chapter 6. That is, $\partial z/\partial x = 0$ at the point $(x, y) = (x_0, y_0)$. Similarly, $\partial z/\partial y = 0$ at this point. The equations

$$\frac{\partial z}{\partial x} = 0, \qquad \frac{\partial z}{\partial y} = 0$$

are thus two equations in two unknowns that are satisfied by the maximum point (x_0, y_0), and we may be able to solve them simultaneously and so determine (x_0, y_0).

In analogy with the earlier definition for a function of one variable, we call such a point (x_0, y_0) where both partial derivatives are zero a *critical point* of f.

EXAMPLE 1   Find the dimensions of the rectangular box with open top and one-cubic-foot capacity that has the smallest surface area.

Solution. The box is shown in the figure.

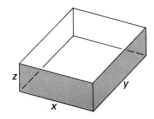

Its volume is xyz and this is required to be 1:

$$xyz = 1.$$

The total surface area is

$$A = xy + 2xz + 2yz.$$

We can eliminate one variable, say z, by solving for z in the first equation and substituting in the second, giving $z = 1/xy$ and

$$A = xy + \frac{2}{y} + \frac{2}{x}.$$

Now A is a function of x and y. Its domain is the first quadrant, because x and y have to be positive, but no other limitation is placed on their values. If A has a minimum value, it has to be at a critical point, that is, a point where

$$\frac{\partial A}{\partial x} = y - \frac{2}{x^2} = 0,$$

$$\frac{\partial A}{\partial y} = x - \frac{2}{y^2} = 0.$$

So we solve these equations simultaneously. First,

$$yx^2 = 2,$$
$$xy^2 = 2.$$

Dividing gives $x/y = 1$ and $x = y$. Therefore, $x^3 = 2$ and $x = y = 2^{1/3}$. Then

$$z = \frac{1}{xy} = 2^{-2/3}.$$

The *proportions* of the box are given by

$$x = y = 2z.$$

The box has a square bottom, with edge length twice the height of the box.

REMARK In Chapter 6 we had a theorem guaranteeing that in a certain situation a function assumes a maximum value, and telling how to find it. Here we have no such theoretical guarantee that a maximum exists. We have only argued that *if f* has a maximum value at an interior point, *then* its partial derivatives must both be zero there. This means that, in any given problem, you should convince yourself that it is reasonable to expect a maximum or minimum value somewhere. For example, in the problem above you probably feel that among all rectangular boxes of a fixed capacity, there has to be one with minimal surface area. But this is tricky business, and we ought to nail down the existence of extreme values in a manner analogous to (but of necessity more complicated than) the argument in Chapter 6. (See Examples 3 and 4.)

EXAMPLE 2 If we restrict the domain of

$$z = f(x, y) = y^3 + xy - x^3$$

to the unit circular disk $x^2 + y^2 \leq 1$, does f have a maximum value at an interior point?

Solution. An interior maximum point (x, y) has to be a critical point of f. That is,

$$\frac{\partial z}{\partial x} = y - 3x^2 = 0,$$

$$\frac{\partial z}{\partial y} = 3y^2 + x = 0.$$

at such a point. We can solve these equations simultaneously by substituting from the first into the second, to get

$$27x^4 + x = 0.$$

Factoring this equation,

$$x(27x^3 + 1) = 0,$$

shows its solutions to be $x = 0$, $x = -1/3$. The corresponding y values, obtained from the critical point equations, are $y = 0$ and $y = 1/3$, respectively. The critical points are thus $(0, 0)$ and $(-1/3, 1/3)$, and

$$f(0, 0) = 0,$$

$$f\left(-\frac{1}{3}, \frac{1}{3}\right) = -\frac{1}{27}$$

However,

$$f(0, 1) = 1,$$

so neither critical point yields a maximum value for f. Therefore f does not assume a maximum value at an interior point of its restricted domain.

This example can be pursued further. We call a domain *closed* if it includes its boundary and *finite* if it lies entirely inside some circle. The restricted domain in the above example is both closed and finite. We shall now assume the following theoretical maximum principle:

If a function f is continuous on a finite, closed domain D, then f assumes a maximum value and a minimum value on D.

EXAMPLE 3   It follows from this maximum principle that the function f of Example 2 has a maximum value somewhere on its domain $x^2 + y^2 \le 1$. We saw there that the maximum value could not be at an interior point, and it therefore must be somewhere on the boundary circle $x^2 + y^2 = 1$. We thus want the maximum of $z = y^3 + xy - x^3$ subject to the auxiliary condition $x^2 + y^2 = 1$. But this is now a one-variable problem. We can either solve for y in $x^2 + y^2 = 1$ and substitute in the equation for z, or we can use the method of auxiliary variables. This remaining one-variable problem is rather complicated, and we shall not go any further with it. Instead, we shall work out a simpler problem of the same type.

EXAMPLE 4   Find the maximum value of

$$z = x^2 + 4x^2y^2 - y^2$$

on the closed circular disk $x^2 + y^2 \le 1$.

Solution. The maximum value occurs either at an interior point or at a boundary point. The only possible interior location is a critical point, determined by the critical point equations

$$\frac{\partial z}{\partial x} = 2x + 8xy^2 = 0,$$

$$\frac{\partial z}{\partial y} = 8x^2y - 2y = 0.$$

These equations reduce to

$$x(1 + 4y^2) = 0,$$
$$y(4x^2 - 1) = 0.$$

The first shows that $x = 0$, and then the second says that $y = 0$. Thus the only critical point is the origin $(x, y) = (0, 0)$. The value of z there is zero, and this is not the maximum value of z since we see by inspection that $z = 1$ at $(1, 0)$.

The maximum therefore occurs on the boundary curve $x^2 + y^2 = 1$, where the problem reduces to a one-variable problem. In this example, the reduction is very simple. On the boundary $y^2 = 1 - x^2$, so

$$z = x^2 + 4x^2y^2 - y^2 = x^2 + (1 - x^2)(4x^2 - 1)$$
$$= 6x^2 - 4x^4 - 1.$$

This is over the x-interval $[-1, 1]$. We saw in Chapter 6 that the maximum value of z on this interval must occur at an endpoint or at a point where $dz/dx = 0$. At the two endpoints $z = 1$. On the other hand, the equation

$$\frac{dz}{dx} = 12x - 16x^3 = 0$$

has the roots $x = 0, \pm\sqrt{3}/2$, and the corresponding values of z are $z = -1$ and $z = 6(3/4) - 4(3/4)^2 - 1 = 5/4$. This second value is larger than the common endpoint value, so z has its maximum value 5/4 at the boundary points determined by $x = \pm\sqrt{3}/2$. At these points $y^2 = 1 - x^2 = 1/4$ and $y = \pm1/2$. There are thus four maximum points, symmetric with respect to the two axes, and the one in the first quadrant is $(\sqrt{3}/2, 1/2)$.

PROBLEMS FOR SECTION 4

Each of the following functions is defined on the whole plane and has a minimum or maximum value. Find this extreme value and the point (or points) where it occurs.

1. $z = x^2 + 2xy + 2y^2 - 6y$
2. $z = 4x + 6y - x^2 - y^2$
3. $z = x^4 - x^2y^2 + y^4 + 4x^2 - 6y^2$
4. $z = x^4 + 2xy^2 + y^4$
5. $z = 3x^4 - 4x^3y + y^6$

Let D be the first quadrant. Each of the following functions has a maximum or a minimum on D. Find this extreme value and the point where it occurs.

6. $z = xy + \dfrac{1}{x} + \dfrac{1}{y}$
7. $z = 8y^3 + x^3 - 3xy$

8. $z = y + 2x - \log xy$

9. Find the minimum distance between the parabola $y = x^2$ and the line $y = x - 1$.

10. A rectangular box with capacity one cubic foot is to be made with an open top. The material for the bottom costs half again as much as the material for the sides. Find the dimensions of the least expensive box.

11. Find the maximum and minimum values of

$$f(x, y) = x^2 + y^2 - xy - x$$

on the unit square $-1 \le x \le 1$, $-1 \le y \le 1$.

12. Find the maximum and minimum values of

$$f(x, y) = 2x^4 - 3x^2y^2 + 2y^4 - x^2$$

on the closed unit disk $x^2 + y^2 \le 1$.

13. Find the minimum distance between the path $(x, y) = (t, t^2 + 1)$ and the path $(x, y) = (-s - 1, 2s)$.

14. The temperature in degrees Fahrenheit at each point (x, y) in the region $0 \le x$, $y \le 1$ is given by $T = 48xy - 32x^3 - 24y^2$. Find the points of maximum and minimum temperature and the temperature at each of these points.

15. Find positive numbers x, y, and z such that $x + 3y + 2z = 18$ and xyz is a maximum. [*Hint.* Use the technique of implicit differentiation.]

5. WHAT ABOUT HIGHER DIMENSIONS?

By now it is clear that visualizing and representing equation graphs in three dimensions is much more complicated than in two dimensions, and the corresponding problems for equations in four or more variables are hopeless. It is natural to wonder whether the effort spent on the three-dimensional situation was worth it, and to ask how one can go about trying to understand the behavior of functions of three or more variables.

What we do is to proceed more or less by analogy with the two- and three-dimensional cases, keeping in mind both their similarities and their differences. For example, we consider the set of all quadruples of numbers such as $(1, 3, -4, 2)$ as forming a four-dimensional coordinate space, and we consider the graph of an equation in four variables as forming a three-dimensional "surface" in this 4-space. The graph of a function of three variables $f(x, y, z)$ is the graph of the equation

$$w = f(x, y, z).$$

We view the three-dimensional domain D of f as lying in the three-dimensional coordinate "plane" consisting of all points of the form $(x, y, z, 0)$. Then the graph of f is a three-dimensional "surface" spread out "over" the domain D. If (a, b, c) is in the domain of f, we picture it as the point

$$(a, b, c, 0) \qquad \text{in 4 space}$$

and the graph point over it is

$$(a, b, c, f(a, b, c)).$$

In the same way we view the graph of a function of n variables as forming an n-dimensional "surface" in $(n + 1)$-dimensional coordinate space. We thus discuss higher-dimensional situations in geometric language suggested by our experience in two and three dimensions, and we still reason geometrically, but in a looser, less concrete way.

As our geometric grip loosens, and we reason more by geometric analogy than by direct geometric visualization, we have to bolster our weaker intuition by paying more attention to its algebraic and analytic counterparts. It is here that the theory of vector spaces comes to our aid. The algebraic study of vector spaces gives us very sharp and clear notions of planes and lines and perpendicularity in higher dimensions, and we can then use these flat objects to approximate curving configurations near points of tangency in almost exactly the same way that a tangent line approximates a curve and a tangent plane approximates a surface.

We took a first look at vectors in Chapter 13, but a whole course in linear algebra is needed before one can feel confident about the geometry of $(n + 1)$-dimensional space and the behavior of functions of n variables.

EXTRA PROBLEMS FOR CHAPTER 15

If a second-degree equation has no mixed-product terms (xy, yz, xz), then its graph can be identified by completing squares just as in the plane, and it always turns out to be a standard quadric surface of one of the types listed in Section 1, except for being centered at a point other than the origin.

1. By completing squares show that
$$x^2 + y^2 + z^2 + 2x - 4z = 0$$
   is a sphere centered at $(-1, 0, 2)$.

2. By completing squares show that
$$x^2 + y^2 - z^2 + 2x + 2z = 0$$
   is a cone with vertex at $(-1, 0, 1)$ and with axis parallel to the z-axis.

Identify the graphs of the following equations.

3. $x^2 + y^2 + z + 4 = 0$ 4. $x^2 + y^2 + 2x + z = 0$

5. $x^2 + y^2 + 2y - z = 0$ 6. $x^2 - y^2 + 2x + 4y + z = 0$

7. $x^2 + 2y^2 + z^2 + 4y + 2z = 0$ 8. $x^2 + 2y^2 - z^2 + 4x - 4y = 0$

9. $x^2 + 2y^2 - z^2 + 4x - 4y = -7$ 10. $x^2 - 2y^2 - z^2 + 4x - 4y = 0$

11. $x^2 - 2y^2 - z^2 + 4y - 4z = 0$

12. Prove that the graph of $z^2 = x^2 + y^2$ is a cone with origin as vertex. (Use the fact that the straight line through the origin and the point (a, b, c) has parametric equations
$$x = at, \qquad y = bt, \qquad z = ct.)$$

13. Find the straight lines passing through the origin that lie on the paraboloid
$$z = \frac{x^2}{a^2} - \frac{y^2}{b^2}.$$

14. Find the point on the plane $x + 2y + 3z = 4$ closest to the origin.

15. Find the point on the surface $xyz = 4$ closest to $(1, 0, 0)$.

16. Find the maximum value of xyz when the point (x, y, z) lies on the plane $2x + y + 3z = 10$.

17. Find the maximum volume of a rectangular box that is inscribed in the ellipsoid
$$x^2 + \frac{y^2}{4} + \frac{z^2}{9} = 1.$$

18. Find the maximum volume of a rectangular box inscribed in the ellipsoid
$$\frac{x^2}{a^2} + \frac{y^2}{b^2} + \frac{z^2}{c^2} = 1.$$

19. Find the maximum volume of a rectangular box having three sides on the coordinate planes and one vertex on the plane
$$\frac{x}{a} + \frac{y}{b} + \frac{z}{c} = 1,$$
   where a, b, and c are positive.

20. Given positive numbers $a_1, a_2, \ldots, a_n$, find the numbers $x_1, \ldots, x_n$ such that

$$\sum_1^n a_i x_i = 1$$

and $\sum_1^n x_i^2$ is as small as possible.

21. Given positive numbers $a_1, a_2, \ldots, a_n$, find the numbers $x_1, x_2, \ldots, x_n$ such that

$$\sum_1^n x_i^2 = 1$$

and $\sum_1^n a_i x_i$ is as large as possible.

16
The Chain Rule in Several Variables

The chain rule for functions of several variables has an importance beyond anything suggested by the one-variable case. One could say that the chain rule takes on new dimensions. The present chapter explores some of this new territory in two and three dimensions. More advanced courses pursue the same ideas in their full generality.

The first five sections are restricted to functions of two variables. The next three sections then go over the same material when there are three independent variables.

The last two sections of the chapter take up higher-order partial derivatives.

1. THE CHAIN RULE

Since a partial derivative of a function of several variables is nothing but the ordinary derivative of a suitable restriction of that function, it follows that, by and large, the rules for computing partial derivatives are the same as for ordinary derivatives. Thus the sum, product, and quotient rules are all the same.

The chain rule acquires a new form, however, since now it concerns the composition of an outside function of *two or more variables* with *two or more inside functions.*

EXAMPLE 1   If $z = x^2 + y^2$ and if $x = 2t$ and $y = t^3$, then

$$z = (2t)^2 + (t^3)^2 = t^2(4 + t^4).$$

Here is the Leibniz form of the chain rule for a function of two variables. We call $f(x, y)$ *smooth* if it has continuous partial derivatives.

CHAIN RULE   *If z is a smooth function of x and y, and if x and y are differentiable functions of t, then z is a differentiable function of t, and*

$$\frac{dz}{dt} = \frac{\partial z}{\partial x}\frac{dx}{dt} + \frac{\partial z}{\partial y}\frac{dy}{dt}.$$

Note that the formula is "dimensionally" correct in the same way the old formula was. The positions of x cancel symbolically, one being up and one down, as do the positions of y, so that symbolically the righthand side reduces to z up and t down, which agrees with the left.

Of course, z will be a function of t only if the domains and ranges overlap. We can picture this overlapping in the xy-plane by considering the two functions

$$x = g(t) \qquad \text{and} \qquad y = h(t)$$

as together defining a path

$$(x, y) = (g(t), h(t)).$$

Then this path must run, for a while at least, in the region D that is the domain of $z = f(x, y)$. The path could very well meander in and out of D, and in this case the domain of the composite function

$$z = f(x, y) = f(g(t), h(t))$$

is made up of two or more separate t-intervals, corresponding to the separate pieces of the path lying in D. Such diagrams lead us to interpret the composite-function derivative dz/dt as the derivative of $f(x, y)$ *along the path.*

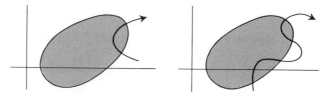

The chain rule will be proved in the next section, and its meaning will become clearer in terms of vectors in Section 4. Here we shall merely perform some computations with it.

EXAMPLE 2 We verify the correctness of the chain rule in the situation of Example 1. We had

$$z = x^2 + y^2, \qquad x = 2t, \qquad \text{and} \qquad y = t^3.$$

By the chain rule,

$$\frac{\partial z}{\partial t} = \frac{\partial z}{\partial x}\frac{dx}{dt} + \frac{\partial z}{\partial y}\frac{dy}{dt}$$

$$= 2x \cdot 2 + 2y \cdot 3t^2$$

$$= 2(2t) \cdot 2 + 2t^3 \cdot 3t^2 = 8t + 6t^5.$$

On the other hand,

$$z = (x^2 + y^2) = (2t)^2 + (t^3)^2 = 4t^2 + t^6,$$

from which the same answer,

$$dz/dt = 8t + 6t^5,$$

follows by direct computation.

EXAMPLE 3 What is the formula for $\partial z/\partial u$ if

$$z = f(x, y), \qquad x = g(u, v), \qquad y = h(u, v)?$$

Solution. Here z has become a function of u and v, and we find $\partial z/\partial u$ by holding v constant and taking the ordinary derivative with respect to u. This reduces the present situation to the chain rule, and we have

$$\frac{\partial z}{\partial u} = \frac{\partial z}{\partial x}\frac{\partial x}{\partial u} + \frac{\partial z}{\partial y}\frac{\partial y}{\partial u}.$$

This formula is the chain rule for partial derivatives.

EXAMPLE 4   Calculate $\partial z/\partial u$ in two ways when

$$z = x^2 + y^2, \qquad x = uv^2, \qquad y = u^2v.$$

Solution 1. Substitute the expressions for x and y:

$$z = x^2 + y^2 = (uv^2)^2 + (u^2v)^2$$
$$= u^2v^4 + u^4v^2.$$

Then compute $\partial z/\partial u$:

$$\frac{\partial z}{\partial u} = 2uv^4 + 4u^3v^2.$$

Solution 2. Use the chain rule for partial derivatives:

$$\frac{\partial z}{\partial u} = \frac{\partial z}{\partial x}\frac{\partial x}{\partial u} + \frac{\partial z}{\partial y}\frac{\partial y}{\partial u}$$
$$= 2xv^2 + 2y2uv$$
$$= 2(uv^2)v^2 + 2(u^2v)2uv$$
$$= 2uv^4 + 4u^3v^2.$$

These Leibniz expressions treat z ambiguously. Two quite different functions were being differentiated when we wrote $\partial z/\partial x$ and $\partial z/\partial u$.

EXAMPLE 5   The following very common situation is a good example of how bad things can be. If z is a function of x and y, and if y is a function of x, then z is a function of x and

$$\frac{dz}{dx} = \frac{\partial z}{\partial x} + \frac{\partial z}{\partial y}\frac{dy}{dx}.$$

Correctly interpreted, this is a valid and important special case of the chain rule. But the correct interpretation requires us to keep in mind that dz/dx and $\partial z/\partial x$ are derivatives of two entirely different functions! In simple situations we can put up with such loose talk, but we must always be ready to come to the rescue with more precise formulations of the chain rule involving function notation.

Good enough for most purposes is the intermediate form, part Leibniz notation and part function notation, as follows:

CHAIN RULE   *If $z = f(x, y)$ with f smooth, and if x and y are differentiable functions of t, then z is a differentiable function of t, and*

$$\frac{dz}{dt} = D_1f(x, y)\frac{dx}{dt} + D_2f(x, y)\frac{dy}{dt}.$$

In this form the Leibniz notation is permitted for the derivatives with respect to the ultimate independent variable or variables, but the function notation is used for the partials with respect to the intermediate variables. Consider again the important situation of Example 5.

EXAMPLE 6   If $z = f(x, y)$ and y is a function of x, then we have the special case $t = x$ in the above formula: z becomes a function of x alone, and the formula becomes

$$\frac{dz}{dx} = D_1 f(x, y) + D_2 f(x, y) \frac{dy}{dx}.$$

Note how this fomulation avoids the ambiguity of the pure Leibniz formula.

Most precise, but also most ungainly, is the pure function form. Leaving out the differentiability hypotheses, it is:
   If $\phi(t) = f(g(t), h(t))$, *then*

$$\phi'(t) = D_1 f(g(t), h(t)) g'(t) + D_2 f(g(t), h(t)) h'(t).$$

PROBLEMS FOR SECTION 1

In Problems 1 through 5 find dz/dt by the chain rule.

1. $z = x^2 - 2xy + y^2$, $x = (t + 1)^2$, $y = (t - 1)^2$
2. $z = x \sin y$, $x = 1/t$, $y = \arctan t$
3. $z = \log \sqrt{(x - y)/(x + y)}$, $x = \sec t$, $y = \tan t$
4. $z = x^2 y^3$, $x = 2t^3$, $y = 3t^2$
5. $z = \log(x^2 + y^2)$, $x = e^{-t}$, $y = e^t$
6. The altitude of a right circular cone is 8 inches and is increasing at 0.5 in/min. The radius of the base is 10 inches and is decreasing at the rate of 0.2 in/min. How fast is the volume changing?

Calculate $\partial z / \partial v$ and $\partial z / \partial u$ in two ways in the following problems.

7. $z = x^2 + y^2$, $x = u + v$, $y = u - v$
8. $z = x^2 + 3xy + y^2$, $x = \sin u + \cos v$, $y = \sin u - \cos v$
9. $z = e^{xy}$, $x = u^2 + 2uv$, $y = 2uv + v^2$
10. $z = \sin(4x + 5y)$, $x = u + v$, $y = u - v$
11. Show that $z = f(y/x)$ is homogeneous of degree 0; that is, show that

$$x \frac{\partial z}{\partial x} + y \frac{\partial z}{\partial y} = 0.$$

12. Recall from the last chapter that we call $z = f(x, y)$ *homogeneous of degree n if*

$$(*) \qquad\qquad x \frac{\partial z}{\partial x} + y \frac{\partial z}{\partial y} = nz.$$

Suppose that

$$z = f(u, v), \qquad u = g(x, y), \qquad \text{and} \qquad v = h(x, y),$$

where f is homogeneous of degree m and g and h are each homogeneous of degree n. Show that z is homogeneous of degree mn as a function of x and y.

That is,

$$z = f(g(x, y), h(x, y))$$

is homogeneous of degree mn.

13. The condition $(*)$ above is not the usual definition of homogeneity. Show that a function $z = f(x, y)$ will satisfy $(*)$ if it has the property

$$(**) \qquad\qquad f(tx, ty) = t^n f(x, y).$$

(Differentiate with respect to t.)

14. Show, conversely, that if a function satisfies $(*)$, then it satisfies $(**)$. You will need the following fact:

If $du/dt = n(u/t)$, then $u = ct^n$.

(See Problem 34 in the Extra Problems for Chapter 5.)

2. PROOF OF THE CHAIN RULE

The proof of this theorem will depend on the fact that if we hold one variable fixed in $f(x, y)$, then the mean-value theorem can be applied to the remaining function of the other variable. For example, if we set $g(x) = f(x, b)$, then the mean-value property

$$g(x + \Delta x) - g(x) = g'(X)\Delta x$$

becomes

$$f(x + \Delta x, b) - f(x, b) = D_1 f(X, b)\Delta x.$$

With this observation we can prove the theorem.

THEOREM 1 *If the functions $x = g(t)$ and $y = h(t)$ are differentiable at t_0, and if $z = f(x, y)$ is smooth inside some small circle about the point*

$$(x_0, y_0) = (g(t_0), h(t_0)),$$

then the composite function

$$k(t) = f(g(t), h(t))$$

is differentiable at t_0, and

$$k'(t_0) = D_1 f(x_0, y_0)g'(t_0) + D_2 f(x_0, y_0)h'(t_0).$$

Proof. The proof is almost the same as the earlier proof for functions of one variable (Chapter 5, Theorem 6'). If we give t an increment Δt, then x and y acquire increments Δx and Δy, and $z = f(x, y)$ acquires the increment

$$\Delta z = f(x + \Delta x, y + \Delta y) - f(x, y).$$

(We have dropped the zero subscripts to simplify the notation.) What is new in this situation is that the intermediate variables x and y together form a *variable point* (x, y) with new value $(x + \Delta x, y + \Delta y)$. In order to bring in the partial derivatives of f, we make this point change in two steps, first changing x by Δx and *then* y by Δy. That is, we take one step horizontally off the path and then a step vertically back onto the path. Then Δz is the sum of its two partial changes:

$$\Delta z = [f(x + \Delta x, y + \Delta y) - f(x + \Delta x, y)] + [f(x + \Delta x, y) - f(x, y)].$$

Each bracket involves a change in only one variable, and can be rewritten by the mean-value theorem for functions of that variable. Thus,

$$\Delta z = D_2 f(x + \Delta x, Y) \Delta y + D_1 f(X, y) \Delta x,$$

where Y lies between y and $y + \Delta y$, and X lies between x and $x + \Delta x$. Now divide by Δt,

$$\frac{\Delta z}{\Delta t} = D_2 f(x + \Delta x, Y) \frac{\Delta y}{\Delta t} + D_1 f(X, y) \frac{\Delta x}{\Delta t},$$

and let Δt approach 0. The limit on the right exists and has the value on the right below. Therefore, dz/dt exists and

$$\frac{dz}{dt} = D_2 f(x, y) \frac{dy}{dt} + D_1 f(x, y) \frac{dx}{dt},$$

which is the chain rule. Note that we have needed the continuity of the partial derivative $D_2 f$ in order to conclude that

$$D_2 f(x + \Delta x, Y) \rightarrow D_2 f(x, y)$$

as $\Delta t \rightarrow 0$. The continuity of D_1 was used, too. ∎

PROBLEMS FOR SECTION 2

The Leibniz form for the chain rule for a function of three variables is:

If w is a smooth function of x, y, and z, and if x, y, and z are differentiable functions of t, then w is a differentiable function of t, and

$$\frac{dw}{dt} = \frac{\partial w}{\partial x}\frac{dx}{dt} + \frac{\partial w}{\partial y}\frac{dy}{dt} + \frac{\partial w}{\partial z}\frac{dz}{dt}.$$

1. State the pure function form of this rule.
2. Prove this three-variable chain rule, following the plan of the two-variable proof in the text. You will have to express Δw as the sum of *three* "partial" increments.
3. State the chain rule for a function of four variables.
4. State the chain rule for a function of n variables $x_1, x_2, \ldots, x_n$.

3. IMPLICIT FUNCTIONS

The chain rule for functions of two variables lets us discuss implicit functions in a general context that was unavailable in Chapter 5.

The graph of an equation of the form

$$f(x, y) = k$$

is called a *level set* of the function f, as we saw in the last chapter. For example, the level sets of the function

$$f(x, y) = x^2 + y^2$$

are the circles about the origin:

$$x^2 + y^2 = k.$$

Generally, as in this example, a level set of a smooth function will be some sort of a smooth curve, although probably not a function graph. But it may be made up of two or more separate function graphs $y = \phi(x)$, just as the above level set $x^2 + y^2 = k$ consists of the graphs of the two functions

$$y = \phi_1(x) = \sqrt{k - x^2} \qquad \text{and} \qquad y = \phi_2(x) = -\sqrt{k - x^2}.$$

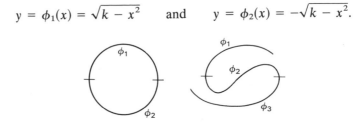

By hypothesis, $z = f(x, y)$ has the constant value k along the graph of any such function $y = \phi(x)$,

$$z = f(x, \phi(x)) = k,$$

so if f and ϕ are smooth functions then

(*) $$\frac{dz}{dx} = D_1 f + D_2 f \frac{dy}{dx} = 0.$$

The middle term is just the chain-rule evaluation of dz/dx when $z = f(x, y)$ and y is a function of x. The result is zero because z is constant as a function of x. If $D_2 f$ is not zero, the above equation can then be solved for dy/dx:

$$\frac{dy}{dx} = \frac{-D_1 f(x, y)}{D_2 f(x, y)},$$

or

$$\phi'(x) = \frac{-D_1 f(x, \phi(x))}{D_2 f(x, \phi(x))}.$$

In the language of Chapter 5, a function $y = \phi(x)$, such that

$$f(x, \phi(x)) = k$$

is a *function solution* of the equation

$$f(x, y) = k,$$

and is defined *implicitly* by this equation. So we have obtained above a general formula for the derivative of an implicitly defined function.

Note that if $D_2f = 0$ in $(*)$, then also $D_1f = 0$, and the point $(x, y) = (x, \phi(x))$ is a critical point of f. So if the graph of ϕ contains no critical point of f, then $D_2f \neq 0$ along the graph and we can solve for $\phi'(x)$ as above.

Altogether we have shown that:

THEOREM 2   *If $f(x, y)$ is a smooth function and $y = \phi(x)$ is a differential function defined implicitly by the equation $f(x, y) = k$, and if there are no critical points of f on the graph of ϕ, then*

$$\phi'(x) = -\frac{D_1f(x, \phi(x))}{D_2f(x, \phi(x))}.$$

This whole analysis works in the same way when we consider a piece of the level set $f(x, y) = k$, along which x is a smooth function of y,

$$x = \psi(y).$$

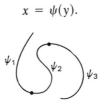

Such a piece is not a standard function graph, because its independent variable runs vertically, but we would still call ψ an implicitly defined function. This time we have

$$z = f(\psi(y), y) = k,$$

$$\frac{dz}{dy} = D_1f\frac{dx}{dy} + D_2f = 0,$$

$$\psi'(y) = \frac{dx}{dy} = -\frac{D_2f(x, y)}{D_1f(x, y)} = -\frac{D_2f(\psi(y), y)}{D_1f(\psi(y), y)},$$

again provided that $(x, y) = (\psi(y), y)$ is not a critical point of f.

So far, the condition that there be no critical points of f along the piece of the level set we are looking at has been a requirement tacked on at the last minute to permit dividing by D_2f or D_1f in the derivative formula. If (x_0, y_0) is a critical point on the level set $f = k$, then we don't get a derivative formula for an implicitly defined function through that point. This raises the possibility that near a critical point the level set may not even *be* a smooth function graph, and simple examples confirm this suspicion.

EXAMPLE 1   The function

$$f(x, y) = y^2 - x^2$$

has a critical point at the origin. Its level set through the origin, the graph of $y^2 - x^2 = 0$, is the pair of straight lines $y = x$ and $y = -x$, so in the neighborhood of the origin the level set is not a function graph at all.

EXAMPLE 2   The function

$$f(x, y) = y^3 - x^2$$

has a critical point at the origin. Its level set through the origin is the graph of the function $y = \phi(x) = x^{2/3}$, but $\phi'(0)$ does not exist. The level set has a cusp at the origin.

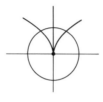

EXAMPLE 3   The level set $y^2 - x^3 = 0$ is the graph of $x = \psi(y) = y^{2/3}$. It is a nonstandard function graph, and is the image of the graph in Example 2 under the rotation of the plane over the 45° line $y = x$. Again, the origin is a bad point: the graph has a cusp there, ψ' fails to exist there, and

$$f(x, y) = y^2 - x^3$$

has a critical point there.

Such examples suggest the true significance of the "no critical point" condition: It is what guarantees that, in the short run, a level set necessarily lies along a smooth function graph, either standard or nonstandard. Near a noncritical point (x_0, y_0), the equation $f(x, y) = k$ does define y as a smooth function of x, or x as a smooth function of y, depending on whether

$$D_2 f(x_0, y_0) \neq 0 \qquad \text{or} \qquad D_1 f(x_0, y_0) \neq 0.$$

This theorem is called the *implicit-function theorem*. Here it is again, stated somewhat more exactly.

THEOREM 3   *Suppose that $f(x, y)$ is smooth, and that $D_2 f(x_0, y_0) \neq 0$, where (x_0, y_0) is a point on the level set $f(x, y) = k$. Then the level set runs along a smooth function graph as it goes through (x_0, y_0). That is, there is a rectangle about (x_0, y_0) within which the equation*

$$f(x, y) = k$$

determines y as a smooth function of x. Similarly, if $D_1(x_0, y_0) \neq 0$, then $f(x, y) = k$ defines x as a smooth function of y in a suitably small rectangle about (x_0, y_0).

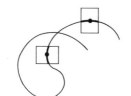

EXAMPLE 4 Use the implicit-function theorem to determine whether or not the equation

$$x^3 + y - y^3 = 0$$

determines y as a function of x locally about each point (x_0, y_0) on the graph.

Solution. The above equation is of the form

$$f(x, y) = 0$$

and we have to determine whether $D_2 f(x, y)$ is ever zero along the level set. We have

$$D_2 f(x, y) = 1 - 3y^2,$$

which is zero at $y = \pm 1/\sqrt{3}$. Substituting these values in the level set equation gives

$$x^3 \pm \left(\frac{1}{\sqrt{3}} - \frac{1}{3\sqrt{3}} \right) = 0$$

or

$$x = \mp \left(\frac{2}{3\sqrt{3}} \right)^{1/3} = \mp \frac{2^{1/3}}{\sqrt{3}}.$$

So there are two points on the level set at which $D_2 f = 0$,

$$\pm \left(\frac{2^{1/3}}{\sqrt{3}}, -\frac{1}{\sqrt{3}} \right).$$

At every other point (x_0, y_0) on the level set the second partial $D_2 f$ is different from zero, so the level set is a function graph near (x_0, y_0).

Suppose, now, that the level set $f(x, y) = k$ contains no critical point of f. Then locally the level set is always a smooth function graph, standard or nonstandard. That is, along a sufficiently short stretch of the level set through any given point P_0, either y is a smooth function of x,

$$y = \phi(x),$$

or x is a smooth function of y,

$$x = \psi(y).$$

Now this is exactly the local characterization of the locus of a smooth path, as we saw in Chapter 10. Doesn't this mean that the whole level set is (the locus of) a smooth path? Not quite, because of what the following example shows.

EXAMPLE 5   The level set

$$5(x^2 + y^2) - (x^2 + y^2)^2 = 4$$

is the pair of concentric circles

$$x^2 + y^2 = 1, \qquad x^2 + y^2 = 4.$$

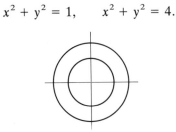

So what we *can* say is that such a smooth level set is made up of one or more smooth-path loci. In any case, we can discuss its *local* properties in terms of a smooth path

$$(x, y) = (g(t), h(t))$$

that runs along it. The advantage of this "path" language is that it permits us to discuss a situation in which locally one variable is a function of the other without having to say which is which.

The proof of the implicit-function theorem is frequently omitted from a first course. In later years, a much more general form of the theorem is proved by much more powerful methods than are available to us now. However, Theorem 3 has an elementary and very instructive proof. We shall give it in Section 9.

PROBLEMS FOR SECTION 3

Use Theorem 3 to determine for each of the following cubic equations $f(x, y) = 0$, whether or not the equation determines y as a function of x locally about each point (x_0, y_0) on the graph. Determine any exceptional points (x_0, y_0) near which the equation may *not* determine y as a function of x. (According to the theorem these are the simultaneous solutions of the two equations $f(x, y) = 0$ and $D_2 f(x, y) = 0$.)

1. $x^3 + y + y^3 = 0$ 2. $1 - xy + y^3 = 0$
3. $1 + xy^2 + y^3 = 0$ 4. $x^3 + xy + y^3 = 0$
5. $x^3 + xy^2 + y^3 = 0$.

6. Study the local solvability of the equation

$$1 - xy + x^2 \log y = 0$$

for y in terms of x.

4. THE GRADIENT OF A FUNCTION AND THE VECTOR FORM OF THE CHAIN RULE

The two partial derivatives of $f(x, y)$ can be viewed as the components of a vector. This vector is called the *gradient* of f and is designated grad f. Thus,

$$\operatorname{grad} f = \left(\frac{\partial z}{\partial x}, \frac{\partial z}{\partial y}\right) = (D_1 f, D_2 f),$$

and

$$[\operatorname{grad} f](a, b) = \left(\frac{\partial z}{\partial x}, \frac{\partial z}{\partial y}\right)\Bigg|_{(x,y) = (a,b)} = (D_1 f(a, b), D_2 f(a, b)).$$

Note that a critical point of f is just a point where the vector grad f is zero.

EXAMPLE 1 If $z = f(x, y) = (x^2 + 2y)/4$, then

$$(\operatorname{grad} f)(x, y) = (\partial z/\partial x, \partial z/\partial y) = (x/2, 1/2).$$

Thus,

$$(\operatorname{grad} f)(-3, 0) = (x/2, 1/2)_{(-3,0)} = (-3/2, 1/2).$$

When we view the point (x, y) in the coordinate plane as a vector

$$\mathbf{x} = (x, y),$$

then a function of two variables $z = f(x, y)$ can be viewed as a function of one vector variable $\mathbf{x}$, and we shall often write

$$f(x, y) = f(\mathbf{x}).$$

For example, grad $f(a, b) = \operatorname{grad} f(\mathbf{a})$.

Also, we view grad $f(\mathbf{a})$ as a vector located at the point $\mathbf{a}$, i.e., as an arrow drawn from $\mathbf{a}$. Then, for each point $\mathbf{x}$, we have associated the located vector grad $f(\mathbf{x})$, and grad f is a *field of vectors*, or a *vector field*. We can sketch the vector field grad f by drawing some of its arrows.

EXAMPLE 2 If $(x, y) = (x^2 + y^2)/4$, then

$$(\operatorname{grad} f)(x, y) = \left(\frac{\partial}{\partial x}\left(\frac{x^2 + y^2}{4}\right), \frac{\partial}{\partial y}\left(\frac{x^2 + y^2}{4}\right)\right) = \left(\frac{x}{2}, \frac{y}{2}\right) = \frac{1}{2}(x, y).$$

Thus

$$\operatorname{grad} f(1, 2) = \frac{1}{2}(1, 2) = \left(\frac{1}{2}, 1\right),$$

$$\operatorname{grad} f(0, -1) = \frac{1}{2}(0, -1) = \left(0, -\frac{1}{2}\right), \qquad \text{etc.}$$

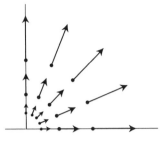

EXAMPLE 3   In Example 1 we had

$$(\text{grad } f)(x, y) = \left(\frac{x}{2}, \frac{1}{2}\right),$$

and we can directly draw some of these vectors as arrows based at the corresponding points (x, y).

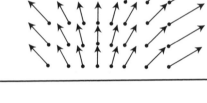

The chain rule

$$\frac{dz}{dt} = \frac{\partial z}{\partial x}\frac{dx}{dt} + \frac{\partial z}{\partial y}\frac{dy}{dt}$$

$$= (D_1 f)\frac{dx}{dt} + (D_2 f)\frac{dy}{dt}$$

can now be given the following very important reformulation:

CHAIN RULE   *If $z = f(x, y)$ is smooth, and if x and y are differentiable functions of t, then z is a differentiable function of t, and*

$$\frac{dz}{dt} = \text{grad } f \cdot \frac{d\mathbf{x}}{dt}$$

$$= \text{grad } f \cdot \text{path tangent vector.}$$

This dot-product formula for the derivative of a function along a path has many consequences. To begin with, consider the level curve

$$f(x, y) = k,$$

where k is an arbitrary constant. We saw in the last section that such a curve is locally a smooth path wherever grad f is not zero. More explicitly, if (x_0, y_0) is a point of the level set $f(x, y) = k$ at which grad f is not zero, then there is a small rectangle R about (x_0, y_0) whose intersection with the level set is the locus of a smooth path.

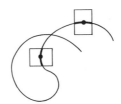

Suppose, then, that $\mathbf{x} = \mathbf{g}(t)$ is a smooth path running along the level curve $f(x, y) = k$. Then $z = f(x, y)$ has the constant value k along the path, and $dz/dt = 0$. Therefore

$$0 = \frac{dz}{dt} = \text{grad } f \cdot \frac{d\mathbf{x}}{dt},$$

so the vector grad f is perpendicular to the path tangent vector. Since the path traces the level curve, its tangent vector is tangent to the curve. We have thus proved the following theorem.

THEOREM 4 *If the function $f(x, y)$ is smooth and $\mathbf{x}_0 = (x_0, y_0)$ is not a critical point of f, then* grad $f(\mathbf{x}_0)$ *is perpendicular to the level curve of f through $\mathbf{x}_0$.*

A vector perpendicular to a curve is said to be *normal* to the curve.

EXAMPLE 4 According to the theorem, a vector normal to the ellipse

$$\frac{x^2}{6} + \frac{4y^2}{6} = \frac{25}{6}$$

at the point $(x_0, y_0) = (3, 2)$ is given by

$$\text{grad}\left(\frac{x^2}{6} + \frac{4y^2}{6}\right)_{(3,2)} = \left(\frac{2}{6}x, \frac{8}{6}y\right)\Big|_{(3,2)} = \left(1, \frac{8}{3}\right).$$

In the figure below, we draw this normal vector located at $\mathbf{x}_0$.

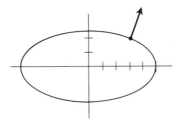

The theorem gives us an easy way to write down the equation of the tangent line to a level curve at a point $\mathbf{x}_0$ where grad $f \neq 0$. Since the tangent line is perpendicular to grad $f(\mathbf{x}_0)$, its equation is

$$\text{grad } f(\mathbf{x}_0) \cdot (\mathbf{x} - \mathbf{x}_0) = 0,$$

as we saw in Section 3 of Chapter 13.

If the gradient of f is zero at $\mathbf{x}_0$, this equation will not give a line because both coefficients will be zero. In fact, we know from examples in the last section that the level curve $f(x, y) = k$ may not *have* a tangent line at a critical point.

EXAMPLE 5   Find the line tangent to the curve

$$x^3 + xy + y^3 = 5$$

at the point $\mathbf{x}_0 = (2, -1)$.

Solution. First verify that $(2, -1)$ is on the graph:

$$2^3 + 2(-1) + (-1)^3 = 8 - 2 - 1 = 5.$$

Then compute grad $f(2, -1)$, where $f(x, y) = x^3 + xy + y^3$:

$$\text{grad } f(2, -1) = (3x^2 + y, x + 3y^2)_{(2,-1)} = (11, 5).$$

Substituting these values in the above formula for the tangent line gives

$$\begin{aligned} \text{grad } f(\mathbf{x}_0) \cdot (\mathbf{x} - \mathbf{x}_0) &= (11, 5) \cdot (x - 2, y + 1) \\ &= 11(x - 2) + 5(y + 1) = 0 \end{aligned}$$

or

$$11x + 5y - 17 = 0.$$

This is the equation of the line tangent to the curve $x^3 + xy + y^5 = 5$ at the point $(2, -1)$.

EXAMPLE 6   Show that the general tangent-line formula above agrees with our earlier formula for the line tangent to the graph of a function $y = g(x)$.

Solution. This graph is the level curve

$$g(x) - y = 0,$$

and the gradient of this special function $f(x, y) = g(x) - y$ is

$$\text{grad } f(x, y) = (g'(x), -1).$$

The dot-product formula

$$\text{grad } f(\mathbf{x}_0) \cdot (\mathbf{x} - \mathbf{x}_0) = 0$$

reduces in this case to

$$g'(x_0)(x - x_0) - 1(y - y_0) = 0,$$

or

$$y - y_0 = g'(x_0)(x - x_0),$$

which is the original tangent-line formula.

There is a converse to Theorem 4, as follows.

THEOREM 5 *Suppose the path* $\mathbf{x} = (g(t), h(t))$ *and the function* $z = f(x, y)$ *are so related that at every point of the path the path direction is perpendicular to* grad f. *Then the path runs along a level curve of* f.

Proof. The derivative of $z = f(x, y)$, along the path

$$\frac{dz}{dt} = \text{grad } f \cdot \frac{d\mathbf{x}}{dt},$$

is everywhere zero, by hypothesis, so z, as a function of t, must therefore be a constant k. That is,

$$z = f(x, y) = k$$

along the path $(x, y) = (g(t), h(t))$, which is another way of saying that the path runs along the level curve $f(x, y) = k$. ∎

PROBLEMS FOR SECTION 4

For each of the following functions determine grad f at the given point.

1. $f(x, y) = x^2 + y^2$; $(3, 4)$

2. $f(x, y) = \arctan \dfrac{x}{y}$; $(1, 2)$

3. $f(x, y) = \cos(xy)$; $(1, \pi/6)$

4. $f(x, y) = e^x \tan y$; $(0, \pi/4)$

5. $f(x, y) = x^3 + 2x^2y - 3xy + 4xy^2 - y^3$; $(1, 1)$

6. $f(x, y) = x^{2y}$; $(2, 1)$

7. $f(x, y) = 2x - y$; (a, b)

In the following problems the functions are divided by a constant, just to make the vectors short enough so they won't clutter up the page. You can use a different constant if you wish. Sketch the gradient field of $f(x, y)$ when:

8. $4f(x, y) = \dfrac{x + y}{4}$.

9. $f(x, y) = \dfrac{(y - x^2)}{4}$

10. $f(x, y) = -\dfrac{x}{4y}$.

Find a vector, other than the zero vector, that is normal to each curve at the given point.

11. $\sqrt{x^2 - y^2} = 1$; $(\sqrt{2}, 1)$

12. $xe^{xy} = 2e$; $(2, 1/2)$

13. $\arcsin \sqrt{x^2 + y^2} = \pi/6$; $(1/4, \sqrt{3}/4)$

14. $\arctan xe^y = \pi/4$; $(1/e, 1)$

Using gradients, find the tangent line to each curve below at the given point.

15. $xy = 2$; $(1, 2)$

16. $xy^2 = 1$; $(1/4, 2)$

17. $x^2 + y^2 = 25$; $(3, 4)$

18. $x^2/9 + y^2/4 = 1$; $(2, (2/3)\sqrt{5})$

19. $x^2 - 2xy - 3y^2 = 5$; $(2, -1)$

20. $x^2 + xy + y^2 - 2x + y = 0$; $(2, -3)$

21. $x^2 - 3xy^2 + 2y^3 = 16$; $(0, 2)$

22. $xe^{xy} = 2$; $(2, 0)$

23. $(x + y)\arctan xy = (5/8)\pi$; $(2, 1/2)$

24. $xy \log x = 2e$; $(e, 2)$

25. Show that the level curves of $f(x, y) = x^2 + y^2$ are everywhere perpendicular to the level curves of $g(x, y) = y/x$. Sketch a few curves from each family.

26. Show that the level curves of $f(x, y) = x^2 - y^2$ are everywhere perpendicular to the level curves of $g(x, y) = xy$. Sketch a few curves from each family.

27. Let $f(x, y) = k$ be a level curve not containing the origin, and let (x_0, y_0) be a point on this level set that is closest to the origin. Prove that

$$\mathbf{x}_0 = (x_0, y_0) \qquad \text{and} \qquad \text{grad } f(\mathbf{x}_0)$$

are parallel vectors.

28. Show that the tangent line to the ellipse

$$\frac{x^2}{a^2} + \frac{y^2}{b^2} = 1$$

at the point (x_0, y_0) has the equation

$$\frac{xx_0}{a^2} + \frac{yy_0}{b^2} = 1.$$

29. Prove the corresponding result for the hyperbola

$$\frac{x^2}{a^2} - \frac{y^2}{b^2} = 1.$$

30. Show that the equation of the tangent line to the parabola

$$y = kx^2$$

at the point (x_0, y_0) can be written

$$\tfrac{1}{2}(y + y_0) = kxx_0.$$

31. Show that the tangent line to the conic

$$ax^2 + 2bxy + cy^2 + d = 0$$

at the point (x_0, y_0) can be written

$$ax_0x + b(x_0y + y_0x) + cy_0y + d = 0.$$

5. DIRECTIONAL DERIVATIVES

We are now in a position to compute the rate at which $f(x, y)$ changes as the point (x, y) moves through (x_0, y_0) in the direction (of the unit vector) $\mathbf{u}$. This rate of change will be called *the derivative of f in the direction* $\mathbf{u}$ *at the point* (x_0, y_0), and will be designated

$$D_{\mathbf{u}}f(x_0, y_0).$$

If we go a distance t from the point $\mathbf{x}_0$ in the direction $\mathbf{u}$, we end up at the point $\mathbf{x} = \mathbf{x}_0 + t\mathbf{u}$. This is true because $\mathbf{u}$ is a unit vector: the distance from $\mathbf{x}_0$ to $\mathbf{x}$ is

$$|\mathbf{x} - \mathbf{x}_0| = |(\mathbf{x}_0 + t\mathbf{u}) - \mathbf{x}_0| = |t\mathbf{u}| = |t|\,|\mathbf{u}|,$$

and this is t because $|\mathbf{u}| = 1$ and t is positive. The average rate of change of f from $\mathbf{x}_0$ to $\mathbf{x}_1$ is thus

$$\frac{f(\mathbf{x}_0 + t\mathbf{u}) - f(\mathbf{x}_0)}{t}.$$

As usual, the true, or "instantaneous," rate of change is defined as the limit of the average rate of change, so:

DEFINITION
$$D_{\mathbf{u}}f(\mathbf{x}_0) = \lim_{t \to 0} \frac{f(\mathbf{x}_0 + t\mathbf{u}) - f(\mathbf{x}_0)}{t}.$$

In order to evaluate this *directional derivative*, note that it is the ordinary derivative of a certain function of t: if

$$\phi(t) = f(\mathbf{x}_0 + t\mathbf{u}) = f(x_0 + tu, y_0 + tv),$$

then

$$D_{\mathbf{u}}f(\mathbf{x}_0) = \lim_{t \to 0} \frac{\phi(t) - \phi(0)}{t} = \phi'(0).$$

But ϕ is a composite function. It is the restriction of $f(x, y) = f(\mathbf{x})$ to the straight line path $\mathbf{x} = \mathbf{x}_0 + t\mathbf{u}$. So the derivative $\phi'(t)$ is calculated by the chain rule, and has the value

$$\phi'(t) = \text{grad } f(\mathbf{x}_0 + t\mathbf{u}) \cdot \frac{d\mathbf{x}}{dt}$$

$$= \text{grad } f(\mathbf{x}_0 + t\mathbf{u}) \cdot \mathbf{u}.$$

We only want $\phi'(0)$. Thus:

THEOREM 6 *The directional derivative $D_{\mathbf{u}}f(\mathbf{x}_0)$ is the value at $\mathbf{x}_0$ of the derivative of f along the straight-line path through $\mathbf{x}_0$ in the direction $\mathbf{u}$. Its chain-rule evaluation is*

$$D_{\mathbf{u}}f(\mathbf{x}_0) = \text{grad } f(\mathbf{x}_0) \cdot \mathbf{u}.$$

EXAMPLE 1 Find the derivative of $z = x^2 y$ at the point $\mathbf{x}_0 = (2, -3)$ in the direction of the vector $\mathbf{a} = (3, 4)$.

Solution. Since $\mathbf{a} = (3, 4)$ is not a unit vector, we first have to find its direction (the unit vector having the direction of $\mathbf{a}$). This is

$$\mathbf{u} = \frac{\mathbf{a}}{|\mathbf{a}|} = \frac{(3, 4)}{\sqrt{3^2 + 4^2}} = \left(\frac{3}{5}, \frac{4}{5}\right).$$

Next, we compute

$$(\text{grad } f)(2, -3) = \left(\frac{\partial z}{\partial x}, \frac{\partial z}{\partial y}\right)_{(2,-3)}$$

$$= (2xy, x^2)_{(2,-3)} = (-12, 4).$$

Then,

$$D_{\mathbf{u}}f(\mathbf{x}_0) = \text{grad } f(\mathbf{x}_0) \cdot \mathbf{u}$$

$$= (-12, 4) \cdot \left(\frac{3}{5}, \frac{4}{5}\right)$$

$$= \frac{-36 + 16}{5} = -4.$$

If $\mathbf{u} = (1, 0)$, then $D_{\mathbf{u}}f$ is simply the first partial derivative of f:

$$D_{(1,0)}f = D_1 f = \frac{\partial z}{\partial x}$$

(where $z = f(x, y)$). Similarly $D_{(0,1)}f$ is the second partial derivative $D_2f = \partial z/\partial y$. These facts can be read off from the chain-rule formula for $D_{\mathbf{u}}f$ or they can be seen directly from the original limit definition of $D_{\mathbf{u}}f$. The symbol D has been used ambiguously in the contexts D_1 and $D_{\mathbf{u}}$, but no harm results.

Now recall from Section 3 of Chapter 13 that if $\mathbf{u}$ is a unit vector, then the dot product $\mathbf{a} \cdot \mathbf{u}$ can be interpreted as the component of the vector $\mathbf{a}$ in the direction $\mathbf{u}$. Also, $\mathbf{a} \cdot \mathbf{u} = |\mathbf{a}| \cos \phi$, where ϕ is the angle between the direction of $\mathbf{a}$ and $\mathbf{u}$. The formula

$$D_{\mathbf{u}}f(\mathbf{x}) = \operatorname{grad} f(\mathbf{x}) \cdot \mathbf{u}$$

thus shows that:

The directional derivative $D_{\mathbf{u}}f(\mathbf{x})$ is the component of the vector $\operatorname{grad} f(\mathbf{x})$ in the direction $\mathbf{u}$, and can be expressed

$$D_{\mathbf{u}}f(\mathbf{x}) = |\operatorname{grad} f(\mathbf{x})| \cos \phi,$$

where ϕ is the angle between $\mathbf{u}$ and the vector $\operatorname{grad} f(\mathbf{x})$.

If $\mathbf{x}$ is held fixed, but the direction $\mathbf{u}$ of the derivative is allowed to vary, then $\cos \phi$ varies between -1 and $+1$ in this formula. In particular, $D_{\mathbf{u}}f(\mathbf{x})$ has its maximum value $|(\operatorname{grad} f)(\mathbf{x})|$, when $\mathbf{u}$ is in the direction of $(\operatorname{grad} f)(\mathbf{x})$, for then $\cos \phi$ has its maximum value 1. Also

$$D_{\mathbf{u}}f(\mathbf{x}) = 0$$

whenever the direction $\mathbf{u}$ is perpendicular to $\operatorname{grad} f(\mathbf{x})$. This can be thought of as the "local" expression of the fact that f is constant along any path that runs perpendicular to the vector field $\operatorname{grad} f$. (See Theorem 5.)

EXAMPLE 2   If $f(x, y) = xy^2$, find the maximum value of the directional derivatives of f at $\mathbf{x}_0 = (1, 1)$ and the direction $\mathbf{u}$ in which f has this maximum rate of change.

Solution. First,

$$\operatorname{grad} f(1, 1) = (y^2, 2xy)_{(1,1)} = (1, 2).$$

Then, as we saw above, $D_{\mathbf{u}}f(1, 1)$ will have the maximum value

$$|\operatorname{grad} f(1, 1)| = |(1, 2)| = \sqrt{5},$$

when $\mathbf{u}$ is the direction of $\operatorname{grad} f(1, 1)$,

$$\mathbf{u} = \frac{\operatorname{grad} f(1, 1)}{|\operatorname{grad} f(1, 1)|} = \frac{(1, 2)}{\sqrt{5}} = \left(\frac{1}{\sqrt{5}}, \frac{2}{\sqrt{5}}\right).$$

We can interpret the relation between $\operatorname{grad} f(\mathbf{x})$ and the directions in which the directional derivative $D_{\mathbf{u}}f(\mathbf{x})$ has its maximum, zero, and minimum values in terms of our experiences on a hilly countryside.

Two straight highways intersecting at right angles can be used as xy-axis for the neighboring countryside, and then the terrain itself is the graph of a certain function $z = f(x, y)$. If we start at a point on a hillside and walk levelly along the side of the hill, we are walking along a level curve of f. (Strictly speaking we are walking *over* a level curve. The level curve itself is normally drawn in the xy-base plane.) If we now turn and start walking upward at right angles to the level curve we are necessarily moving in the direction of *greatest steepness* of the hill, for our new direction, specified as a direction in the base plane, is the direction of the *gradient of f*, and we have just seen that this is the direction in which $z = f(x, y)$ has the maximum rate of change. A stream of water flows down the hillside along the shortest path to the bottom, and this means that it is always running in the direction of greatest steepness downward, which is the direction of $-$grad f.

The countryside $z = f(x, y)$ can be pictured in the xy-base plane by a series of level curves. These are the contour lines of a geodetic map. We get from one contour line to the next most quickly along the perpendicular direction. This is pretty clear geometrically, and is a geometric interpretation of the fact that the maximum rate of change of $z = f(x, y)$ is in the direction of the gradient of f. As we move perpendicularly to the contour lines, the number of lines we cross per unit distance is a rough measure of the magnitude of grad f. When the contour lines bunch together they indicate a steep hillside and a large rate of change of $z = f(x, y)$. With a little experience interpreting geodetic contour lines, one gets a direct three-dimensional image from looking at the two-dimensional map.

PROBLEMS FOR SECTION 5

1. Establish the formula

$$\frac{\text{grad } f(\mathbf{x}_0) \cdot \mathbf{a}}{|\mathbf{a}|}$$

for the derivative of f in the direction of the vector $\mathbf{a}$ at the point $\mathbf{x}_0$.

For each of the functions below, find the derivative at the given point and in the direction of the given vector $\mathbf{a}$. (Either use the formula of Problem 1, or work directly from Theorem 6 as in Example 1 in the text.)

2. $f(x, y) = x^2 y$; $(1, 2)$; $\mathbf{a} = (3/5, 4/5)$

3. $f(x, y) = x^2 - y^2$; $(1/2, -1)$; $\mathbf{a} = (\sqrt{3}/2, 1/2)$

4. $f(x, y) = xe^{xy}$; $(2, 1)$; $\mathbf{a} = (12/13, 5/13)$

5. $f(x, y) = \log[(x + y)/(x - y)]$; $(1/2, 2)$; $\mathbf{a} = (1/4, 3/4)$

6. $f(x, y) = \cos x \sin y$; $(\pi/3, \pi/3)$; $\mathbf{a} = (1/2, 1/3)$

7. $f(x, y) = e^{x \tan y}$; $(1, \pi/4)$; $\mathbf{a} = (1/e, 2/e)$

8. $f(x, y) = \sqrt{x^2 + y^2}$; $(3, 4)$; $\mathbf{a} = (1/4, 1/3)$

9. $f(x, y) = \arctan x/y$; $(1/2, 1/3)$; $\mathbf{a} = (1/5, 1/12)$

10. $f(x, y) = x^y$; $(e, 0)$; $\mathbf{a} = (0, 1)$

11. $f(x, y) = x^{xy}$; $(1, 2)$; $\mathbf{a} = (2, 0)$

Find the maximum value of the directional derivative of the given function at the point indicated, and the direction of the maximum rate of change.

12. $g(x, y) = xy^2 + yx^2$; $(1/2, 1)$

13. $g(x, y) = x^2 y^2 - 2xy^3 + x$; $(1/2, 2)$

14. $g(x, y) = \tan x \sec y$; $(\pi/4, \pi/3)$

15. $g(x, y) = \log \dfrac{x^2 - y^2}{x^2 + y^2}$; $(1/2, 1/4)$

16. $g(x, y) = ye^{x/y}$; $(0, 2)$

17. $g(x, y) = \dfrac{x - y}{x + y}$; $(1/4, 3/4)$

18. $g(x, y) = \sqrt{xy}$; $(2, 8)$

19. $g(x, y) = \arctan x/y$; $(1, 1/2)$

20. Write out the definition of the first partial derivative $D_1 f$ as the limit of a difference quotient, and show that it agrees with the definition of the directional derivative $D_{\mathbf{u}} f$ when $\mathbf{u} = \mathbf{e}_1 = (1, 0)$.

21. Suppose that $z = f(x, y)$ and $(x, y) = (g(t), h(t))$. In Section 1 we interpreted the chain-rule derivative dz/dt as the derivative of f along the path $\mathbf{x} = \mathbf{g}(t)$. Show now that:

    The derivative of f along the path $\mathbf{x} = \mathbf{g}(t)$ is equal to the directional derivative of f in the direction of the path, multiplied by the magnitude $|d\mathbf{x}/dt|$ of the path tangent vector.

22. Given $f(x, y) = x^3 y$, find the directions $\mathbf{u}$ for which

$$D_{\mathbf{u}} f(2, 1) = 10.$$

23. If we know the derivative of f at $\mathbf{x}_0$ in two independent directions, then we can solve for grad $f(\mathbf{x}_0)$. Suppose that $D_{\mathbf{u}} f(\mathbf{x}_0)$ has the value 6 in the direction of the vector $(3, 4)$ and the value $-\sqrt{5}$ in the direction of the vector $(2, -4)$. Find grad $f(\mathbf{x}_0)$.

24. Find grad $f(\mathbf{x}_0)$ if $D_{\mathbf{u}} f(\mathbf{x}_0) = 1$ in the direction of $(1, -\sqrt{3})$ and $D_{\mathbf{u}} f(\mathbf{x}_0) = -1$ in the direction of $(1, \sqrt{3})$.

Write down the derivative of f at $\mathbf{a} = (a, b)$ in the direction of the unit vector $\mathbf{u} = (u, v)$, for each of the following functions.

25. $f(x, y) = x + y$

26. $f(x, y) = (x^2 + y^2)/2$

27. $f(x, y) = xy$

6. GRADIENTS AND DIRECTIONAL DERIVATIVES IN SPACE

If f is a function of three variables each of which is a function of t, then there is a correspondinding chain rule that you probably can guess.

THEOREM 7   *If $w = f(\mathbf{x}) = f(x, y, z)$ is smooth, and if*

$$\mathbf{x} = (x, y, z) = (g(t), h(t), k(t))$$

is a differentiable path running through the domain of f, then along the path w is a differentiable function of t and

$$\frac{dw}{dt} = \frac{\partial w}{\partial x}\frac{dx}{dt} + \frac{\partial w}{\partial y}\frac{dy}{dt} + \frac{\partial w}{\partial z}\frac{dz}{dt}$$

$$= D_1 f \frac{dx}{dt} + D_2 f \frac{dy}{dt} + D_3 f \frac{dz}{dt}.$$

The proof is just like the two-dimensional proof, the only change being that Δw is now broken down into the sum of three partial increments corresponding to separate changes in x, y, and z.

The gradient of a function of three variables $f(x, y, z)$ is the three-dimensional vector

$$\text{grad } f = (D_1 f, D_2 f, D_3 f),$$

and the three-dimensional chain rule has the same dot-product formula as before,

$$\frac{dw}{dt} = \text{grad } f \cdot \frac{d\mathbf{x}}{dt},$$

where, of course, the three-dimensional dot product is involved.

The directional derivatives of a function of three variables $f(x, y, z)$ are defined the same way as in the plane, with the same final formulation:

For any space direction $\mathbf{u}$, *that is, any three-dimensional unit vector* $\mathbf{u}$, *the derivative of* f *in the direction* $\mathbf{u}$ *at the point* $\mathbf{x}$ *is*

$$D_{\mathbf{u}}f(\mathbf{x}) = \text{grad } f(\mathbf{x}) \cdot \mathbf{u}.$$

Thus $D_{\mathbf{u}}f(\mathbf{x}_0)$ is the value at $\mathbf{x}_0$ of the derivative of f along the straight line $\mathbf{x} = \mathbf{u}t + \mathbf{x}_0$ through $\mathbf{x}_0$ in the direction $\mathbf{u}$.

EXAMPLE Find the derivative of $f(x, y, z) = xy^2z^3$ at the point $(1, 1, 1)$ in the direction of the vector $(2, 1, -2)$.

Solution. The gradient of f at $(1, 1, 1)$ is

$$\text{grad } f(1, 1, 1) = (y^2z^3, 2xyz^3, 3xy^2z^2)|_{(1,1,1)} = (1, 2, 3).$$

The unit vector $\mathbf{u}$ in the direction of $\mathbf{a} = (2, 1, -2)$ is

$$\mathbf{u} = \frac{\mathbf{a}}{|\mathbf{a}|} = \frac{(2, 1 - 2)}{\sqrt{4 + 1 + 4}} = \left(\frac{2}{3}, \frac{1}{3}, -\frac{2}{3}\right).$$

Then

$$D_{\mathbf{u}}f(1, 1, 1) = \text{grad } f(1, 1, 1) \cdot \mathbf{u}$$

$$= (1, 2, 3) \cdot \left(\frac{2}{3}, \frac{1}{3}, -\frac{2}{3}\right)$$

$$= \left(\frac{2}{3} + \frac{2}{3} - \frac{6}{3}\right) = -\frac{2}{3}.$$

PROBLEMS FOR SECTION 6

Determine the gradient vector of each of the following functions at the given point.

1. $f(x, y, z) = \cos xy + \sin xz$; $P = (0, 2, -1)$
2. $f(x, y, z) = ax + by + cz$; $P = (x_0, y_0, z_0)$
3. $f(x, y, z) = x^2 + 3xy + y^2 + z^2$; $P = (1, 0, 2)$
4. $f(x, y, z) = x \cos y + y \cos z + z \cos x$; $P = (\pi/2, \pi/2, 0)$
5. $f(x, y, z) = \log(x^2 + y^2) + e^{yz}$; $P = (3, 4, 0)$

Find the derivative of each of the following functions at the given point in the direction of the given vector.

6. $g(x, y, z) = xyz$; $P = (1, 1/2, 2)$; $\mathbf{a} = (3, -1, 2)$
7. $g(x, y, z) = x^2 + 2xy - y^2 + xz + z^2$; $P = (2, 1, 1)$; $\mathbf{a} = (1, 2, 2)$
8. $g(x, y, z) = x^2y + xye^z - 2xze^y$; $P = (1, 2, 0)$; $\mathbf{a} = (2, -3, 6)$

9. $g(x, y, z) = \log \sqrt{x^2 + y^2 + z^2}$; $P = (1, 2, -1)$; $\mathbf{a} = (1, -2, 2)$

10. $g(x, y, z) = xe^{yz} + yze^x$; $P = (-4, 2, -2)$; $\mathbf{a} = (2, 2, -1)$

11. The directional derivative $D_{\mathbf{u}}f(\mathbf{x}_0)$ has the value 3, 1, and -1 in the directions of the positive coordinate axes. Find grad $f(\mathbf{x}_0)$.

12. The directional derivative $D_{\mathbf{u}}f(\mathbf{x}_0)$ has the value 3, 1, and -1 in the directions of the vectors

$$(0, 1, 1), \qquad (1, 0, 1), \qquad \text{and} \qquad (1, 1, 0),$$

respectively. Find grad $f(\mathbf{x}_0)$.

13. Find the direction of *maximum rate of change* of

$$f(x, y, z) = x^2 + xy - 2y^2 + 4xz + z^2$$

at the point $(2, 3, -2)$. What is this maximum rate of change? In what direction is the rate of change at f at this point *minimum?*

14. Find the maximum rate of change and the direction in which it occurs for

$$f(x, y, z) = \log[(x + y)/(x + z)]$$

at the point $(1, 2, 1)$.

7. LEVEL SURFACES AND TANGENT PLANES

We don't try to visualize the graph of a function of three variables

$$w = f(x, y, z),$$

because it would involve trying to picture a four-dimensional configuration. However, the equation

$$f(x, y, z) = c$$

defines (in general) a two-dimensional surface in three-dimensional space, called a *level surface* of f, and such surfaces *can* be visualized. For example, the level surfaces of

$$f(x, y, z) = x^2 + y^2 + z^2$$

are simply spheres about the origin, such as

$$x^2 + y^2 + z^2 = 4.$$

Although it is possible to draw pictures of such surfaces, it is hard to do, and we shall continue to argue without using pictures.

Now suppose that

$$\mathbf{x} = (x, y, z) = (g(t), h(t), k(t))$$

is any differentiable path lying in the level surface

$$f(x, y, z) = c.$$

Then $w = f(x, y, z)$ has the constant value c along this path, and its path derivative dw/dt is zero. That is,

$$\text{grad } f \cdot \frac{d\mathbf{x}}{dt} = 0,$$

by the chain rule. We therefore have the space analogue of Theorem 4.

THEOREM 8 *At every point* $\mathbf{x}_0 = (x_0, y_0, z_0)$ *on the level surface* $f(x, y, z) = k$, *the vector* grad $f(\mathbf{x}_0)$ *is perpendicular to the surface, in the sense that* grad $f(\mathbf{x}_0)$ *is perpendicular to every differentiable curve lying in the surface and passing through* $\mathbf{x}_0$.

A direction perpendicular to a surface is said to be *normal* to the surface.

EXAMPLE 1 Find a vector normal to the surface $xyz = 1$ at the point $(1, 2, 1/2)$.

Solution. This is a level surface of $f(x, y, z) = xyz$ and grad f is a normal vector, by the theorem. Therefore,

$$\text{grad } f(1, 2, 1/2) = (yz, xz, xy)_{(1,2,1/2)}$$
$$= (1, 1/2, 2)$$

is a normal vector.

EXAMPLE 2 Find a parametric equation for the line normal to the surface

$$x^2 + y^2 + z^2 = 14$$

at the point $\mathbf{x}_0 = (1, 2, 3)$.

Solution. This is the level surface

$$f(x, y, z) = 14,$$

where $f(x, y, z) = x^2 + y^2 + z^2$. Since grad $f(\mathbf{x}_0)$ is normal to the surface, it is parallel to the line we want. A vector parametric equation of the line is thus

$$\mathbf{x} - \mathbf{x}_0 = t \text{ grad } f(\mathbf{x}_0),$$

or

$$(x, y, z) - (1, 2, 3) = t(2x_0, 2y_0, 2z_0) = t(2, 4, 6).$$

This can be written in the form

$$(x, y, z) = (2t + 1)(1, 2, 3)$$

or

$$(x, y, z) = s(2, 4, 6),$$

where $s = 2t + 1$.

EXAMPLE 3 Show that the line normal to the surface

$$x^2 + y^2 + z^2 = k$$

at any point $\mathbf{x}_0 = (x_0, y_0, z_0)$ passes through the origin. (This is the general form of the special result in Example 2. We expect it to be true because this level surface is a sphere about the origin, and any line perpendicular to a sphere should be a diameter.)

Solution. If $f(x, y, z) = x^2 + y^2 + z^2$, then

$$\text{grad } f(\mathbf{x}_0) = (2x, 2y, 2z)_{\mathbf{x}_0} = (2x_0, 2y_0, 2z_0) = 2\mathbf{x}_0.$$

The parametric equation for the line is thus

$$\mathbf{x} - \mathbf{x}_0 = t \operatorname{grad} f(\mathbf{x}_0) = 2t\mathbf{x}_0,$$

or

$$\mathbf{x} = (2t + 1)\mathbf{x}_0 = s\mathbf{x}_0,$$

where $s = 2t + 1$. The points $\mathbf{x}$ on the line are the scalar multiples of $\mathbf{x}_0$, and $\mathbf{x} = 0$ when $s = 0$.

EXAMPLE 4 Suppose that the space path

$$\mathbf{x} = (x, y, z) = (g(t), h(t), k(t))$$

is perpendicular to the gradient of the function $F(x, y, z)$ at each path point. Show that the path lies entirely on a level surface of F.

Solution. We simply have to show that F is constant along the curve, that is, that the composite function

$$w = F(g(t), h(t), k(t))$$

is a constant function of t. By the chain rule,

$$\frac{dw}{dt} = \operatorname{grad} F \cdot \frac{d\mathbf{x}}{dt},$$

and this is everywhere 0, by hypothesis. Thus $dw/dt = 0$, w is constant, say

$$w = F(g(t), h(t), k(t)) = k,$$

and the path lies on the level surface

$$F(x, y, z) = k.$$

We can now give a satisfactory account of the tangent plane to a surface.

DEFINITION *Let $f(x, y, z)$ be smooth, and let $\mathbf{x}_0$ be a point on the level surface $f(x, y, z) = k$ at which $\operatorname{grad} f$ is not zero. Then the tangent plane to the surface at $\mathbf{x}_0$ is the plane through $\mathbf{x}_0$ that contains all the tangent vectors at $\mathbf{x}_0$ to all the smooth paths lying in the surface and passing through $\mathbf{x}_0$.*

According to Theorem 8, the tangent plane is perpendicular to $\operatorname{grad} f(\mathbf{x}_0)$, and its equation is therefore

$$\operatorname{grad} f(\mathbf{x}_0) \cdot (\mathbf{x} - \mathbf{x}_0) = 0,$$

or

$$[D_1 f(\mathbf{x}_0)](x - x_0) + [D_2 f(\mathbf{x}_0)](y - y_0) + [D_3 f(\mathbf{x}_0)](z - z_0) = 0.$$

EXAMPLE 5 Find the plane tangent to the surface $xe^y \sin z = 1$ at the point $\mathbf{x}_0 = (2, 0, \pi/6)$.

Solution. First verify that $(2, 0, \pi/6)$ is on the surface:

$$2e^0 \sin\left(\frac{\pi}{6}\right) = 2 \cdot 1 \cdot \frac{1}{2} = 1.$$

Then compute grad $f(2, 0, \pi/6)$, where $f(x, y, z) = xe^y \sin z$:

$$\text{grad } f(2, 0, \pi/6) = (e^y \sin z, xe^y \sin z, xe^y \cos z)_{(2,0,\pi/6)}$$
$$= (1/2, 1, \sqrt{3}).$$

Substituting these values in the above formula for the tangent plane gives

$$\text{grad } f(\mathbf{x}_0) \cdot (\mathbf{x} - \mathbf{x}_0) = (1/2, 1, \sqrt{3}) \cdot (x - 2, y, z - \pi/6)$$

$$= 1/2(x - 2) + y + \sqrt{3}(z - \pi/2) = 0,$$

or

$$x + 2y + 2\sqrt{3}z - (2 + \sqrt{3}\pi) = 0.$$

This is the equation of the plane tangent to the surface $xe^y \sin z = 1$ at the point $(2, 0, \pi/6)$.

EXAMPLE 6 Show that the above formula for a tangent plane agrees with the formula in Chapter 15 for the plane tangent to the graph of the function $z = g(x, y)$.

Solution. This graph is the level surface $g(x, y) - z = 0$, and the gradient of this special function

$$f(x, y, z) = g(x, y) - z$$

is

$$\text{grad } f(x, y, z) = (D_1 g(x, y), D_2 g(x, y), -1).$$

The dot-product formula

$$\text{grad } f(\mathbf{x}_0) \cdot (\mathbf{x} - \mathbf{x}_0) = 0$$

thus reduces in this case to

$$[D_1 g(x_0, y_0)](x - x_0) + [D_2 g(x_0, y_0)](y - y_0) - (z - z_0) = 0,$$

or

$$z - z_0 = [D_1 g(x_0, y_0)](x - x_0) + [D_2 g(x_0, y_0)](y - y_0).$$

This is the Chapter 15 formula.

PROBLEMS FOR SECTION 7

In Problems 1 through 5, find a vector normal to the level surface at the given point.

1. $-2x^2 - 6y^2 + 3z^2 = 4$; $(1, -1, -2)$ 2. $x^2y - y^2z + z^2x = 33$; $(2, 1/2, 4)$
3. $xe^{xy} - ze^{yz} = 0$; $(-1, 2, -1)$ 4. $\sin(x^2yz) = 1/2$; $(-1, 1/2, \pi/2)$
5. a) $\log xyz = 0$; $(1/2, 1/3, 6)$
   b) $xyz = 1$; $(1/2, 1/3, 6)$

Find a parametric equation for the line normal to the surface at the given point.

6. $z = xy$; $(2, -1, -2)$ 7. $z = e^x \sin y$; $(1, \pi/2, e)$

8. $z = x^2 y^2$; $(-2, 2, 16)$ 9. $xy + yz + xz = 1$; $(2, 3, -1)$

10. $\log \sqrt{x^2 + y^2 - z^2} = 0$; $(1, -1, 1)$

Find the plane tangent to the given surface at the given point.

11. $-2x^2 - 6y^2 + 3z^2 = 4$; $(1, -1, 2)$

12. $x^2 y - y^2 z + z^2 x = 33$; $(2, 1/2, 4)$

13. $xe^{xy} - ze^{yz} = 0$; $(-1, 2, -1)$

14. $z = e^x \sin y$; $(1, \pi/2, e)$

15. $\log \sqrt{x^2 + y^2 - z^2} = 0$; $(1, -1, 1)$

16. Find the point(s) on the level surface

$$\frac{x^2}{9} + \frac{y^2}{4} + z^2 = 1$$

at which the tangent plane is perpendicular to $(1, 1, \sqrt{3})$.

17. The surface

$$z = -(x^{2/3} + y^{2/3})$$

has a "spike" at the origin and has no tangent plane there. Given any nonvertical plane

$$z = ax + by$$

through the origin; show that there is a circle $x^2 + y^2 = r^2$ within which the plane lies above the surface. (That is, any nonvertical plane can be supported on the point of a vertical pin.)

8. THE LAGRANGE-MULTIPLIER METHOD

We often have to find the maximum value of a function $f(\mathbf{x}) = f(x, y, z)$ on a level surface S defined by $g(x, y, z) = 0$. That is, we want to maximize $f(\mathbf{x})$ subject to the "constraint" $g(\mathbf{x}) = 0$. Suppose that f attains its maximum value on S at a point $\mathbf{x}_0$ not at the boundary of S. We claim that then grad $f(\mathbf{x}_0)$ and grad $g(\mathbf{x}_0)$ are necessarily parallel:

$$\text{grad } f(\mathbf{x}_0) = k \text{ grad } g(\mathbf{x}_0),$$

for some constant k. In order to see this, let $\mathbf{x} = \mathbf{p}(t)$ be any path lying in S and passing through $\mathbf{x}_0$: $\mathbf{x}_0 = \mathbf{p}(t_0)$. Then $f(\mathbf{p}(t))$ has its maximum value at t_0 and its derivative must be zero there. So, for every such path,

$$\text{grad } f(\mathbf{x}_0) \cdot \mathbf{p}'(t_0) = 0.$$

Thus grad $f(\mathbf{x}_0)$ is normal to S at $\mathbf{x}_0$, and hence of the form k grad $g(\mathbf{x}_0)$.

According to the above analysis, the point $\mathbf{x}_0$ where $f(\mathbf{x})$ has its maximum value on the level surface $g(\mathbf{x}) = 0$ satisfies the equations

$$g(\mathbf{x}_0) = 0,$$
$$\text{grad}(f - kg)(\mathbf{x}_0) = 0,$$

for some constant k. The second equation is a vector equation, so altogether we have four equations in the four unknowns

$$x_0, \quad y_0, \quad z_0, \quad k,$$

and we may be able to solve them to determine the maximum point $\mathbf{x}_0$. This is the method of *Lagrange multipliers*.

EXAMPLE 1 Suppose we want to find the maximum value of $f(x) = x + 2y - 2z$ on the sphere $x^2 + y^2 + z^2 = 4$, using the Lagrange multiplier k. This means solving the equations

$$x^2 + y^2 + z^2 = 4,$$
$$(1, 2, -2) - k(2x, 2y, 2z) = 0,$$

simultaneously. The vector equation is equivalent to the three scalar equations

$$2kx = 1, \quad 2ky = 2, \quad 2kz = -2,$$

and substituting from them into the first equation gives

$$\frac{1}{4k^2} + \frac{1}{k^2} + \frac{1}{k^2} = 4,$$

and $k = \pm\dfrac{3}{4}$. The multiplier equation then determines $\mathbf{x}$:

$$\mathbf{x} = (x, y, z) = \pm\frac{4}{3}\left(\frac{1}{2}, 1, -1\right).$$

These are the *critical points* for the constrained function. We can see that the plus sign given f is its maximum value and the minus sign its minimum value.

The same method works for functions of any number of variables, and it generalizes to situations involving two or more constraints. Some linear algebra is now required, so the more general treatment has to be put off to a later course. For a function of two variables the Lagrange-multiplier method is equivalent to our earlier method of auxiliary variables.

EXAMPLE 2 Consider again the problem of finding the most efficient proportions for a container in the shape of a circular cylinder (Example 5, Section 5, Chapter 6). We want to minimize the total surface area

$$A = 2\pi r^2 + 2\pi rh,$$

subject to the constraint of constant volume, say

$$V = \pi r^2 h = 1.$$

The Lagrange-multiplier equation $\operatorname{grad} f = k \operatorname{grad} g$ is then

$$2\pi(2r + h, r) = k\pi(2rh, r^2),$$

or

$$2(2r + h, r) = kr(2h, r).$$

This vector equation by itself determines the proportions of the container, for it requires that $2h = 2r + h$ and hence $h = 2r$. The constraint equation $V = 1$ can then be used to obtain the particular values of r and h.

PROBLEMS FOR SECTION 8

1. Find the minimum value of $f(x, y, z) = x^2 + 3y^2 + 2z^2$ on the plane $x + y - z = 1$.

2. Find the maximum value of $f(x, y, z) = x + y + z$ on the ellipsoid

$$\frac{x^2}{4} + y^2 + \frac{z^2}{9} = 1.$$

3. Find the maximum and minimum values of $f(x, y) = x^2 + x + 2y^2$ on the unit circle $x^2 + y^2 = 1$.

4. Find the maximum and minimum values of $f(x, y) = x^2 + xy - y^2$ on the unit circle $x^2 + y^2 = 1$.

5. Find the point on the ellipse

$$\frac{x^2}{4} + y^2 = 1$$

that is closest to the line $y - x = 4$.

6. Find the point on the hyperbola $x^2 - y^2 = 1$ that is closest to the line $2x + y = 0$.

7. Find the ellipse

$$\frac{x^2}{a^2} + \frac{y^2}{b^2} = 1$$

that passes through the point $(1, 4)$ and has minimum area. (The area of the above ellipse is πab.)

8. Find the ellipsoid

$$\frac{x^2}{a^2} + \frac{y^2}{b^2} + \frac{z^2}{c^2} = 1$$

that passes through the point $(2, 1, 3)$ and has minimum volume. (The volume of the above ellipsoid is $\frac{4}{3}\pi abc$.)

9. Find the rectangle of maximum perimeter (and with sides parallel to the coordinate axes) that can be inscribed in the ellipse $x^2 + 2y^2 = 1$.

10. Find the rectangle of maximum area (and having sides parallel to the coordinate axes) that can be inscribed in the ellipse $x^2 + 2y^2 = 1$.

9. IMPLICIT FUNCTIONS AGAIN

In the examples we have looked at, the graph of an equation of the form

$$f(x, y, z) = k$$

has always turned out to be a surface in space. The question is whether this necessarily has to happen. The situation turns out to be exactly analogous to our two-variable discussion in Section 3. The surface $f(x, y, z) = k$ is a level set of the function of three variables $w = f(x, y, z)$, and it can be shown, exactly as in the case of two variables, that if the point (x_0, y_0, z_0) on this level set is not a critical point of f, then near (x_0, y_0, z_0) the equation

$$f(x, y, z) = k$$

either determines z as a smooth function of x and y,

$$z = \phi(x, y),$$

or it determines y as a smooth function of x and z,

$$y = \psi(x, z),$$

or it determines x as a smooth function of y and z.

$$x = \chi(y, z).$$

That is, we can view the level set

$$f(x, y, z) = k$$

near (x_0, y_0, z_0) either as a surface spread out over a region in the xy-coordinate plane, *or* as a surface spread out over a region in the yz-coordinate plane. These possibilities are not mutually exclusive. Two, or even all three, can occur simultaneously. For example, the linear equation

$$x + 2y - 3z = 5$$

can be solved for each of the variables in terms of the other two, so we can view it in all three ways.

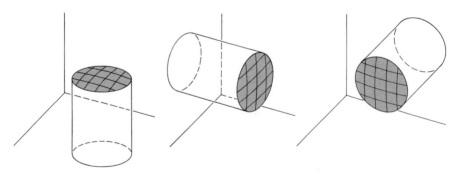

The implicit-function theorem lets us fill a logical gap in the derivation of the tangent-plane formula in the last section. We assumed there that grad $f(\mathbf{x}_0)$ has the *unique* direction normal to the level surface $f(\mathbf{x}) = k$, in the sense of Theorem 8. It is intuitively obvious that a smooth surface has a unique normal direction at each of its points, but we ought to be able to prove this analytically, and it is the implicit-function theorem that lets us do it. It tells us that if grad $f(\mathbf{x}_0) \neq 0$, then the level surface $f(\mathbf{x}) = k$ is a smooth function graph near $\mathbf{x}_0$. Which variable is the dependent variable only affects the figure we draw, and doesn't matter analytically. So suppose that the level surface S near $\mathbf{x}_0$ is the graph of

$$z = g(x, y),$$

where g is smooth and $z_0 = g(x_0, y_0)$. The xz-plane $y = y_0$ intersects S in the curve $z = g(x, y_0)$. This is the path

$$x = x, \qquad y = y_0, \qquad z = g(x, y_0),$$

with parameter x, and its tangent vector at $\mathbf{x}_0$ is

$$(1, 0, D_1 g(x_0, y_0)).$$

Similarly, the yz-plane $x = x_0$ intersects S in a path with tangent vector

$$(0, 1, D_2 g(x_0, y_0))$$

at $\mathbf{x}_0$. These two vectors are not parallel, so there is a unique direction perpendicular to them both. Thus the direction normal to S at $\mathbf{x}_0$ is uniquely determined.

The above discussion is based on the hypothesis that grad $f(\mathbf{x}_0) \neq 0$. If $\mathbf{x}_0$ is a critical point of f, then the level surface may not have a tangent plane at that point.

EXAMPLE 1   The level surface

$$f(x, y, z) = x^2 + y^2 - z^2 = 0$$

is a cone with vertex at the origin. A cone does not have a tangent plane at its vertex, so the origin must be a critical point of f. We see that it is:

$$\operatorname{grad} f(0) = (2x, 2y, -2x)_{(0,0,0)} = \mathbf{0}.$$

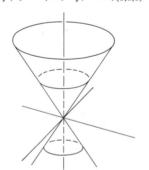

We turn now to the proof of the implicit-function theorem. The details will be given for the two-dimensional case, mainly because the illustrative figure is easier to draw in two dimensions. However, the proof for a function of three variables is practically the same.

If $D_2 f(x_0, y_0)$ is not zero, we can suppose it is positive (otherwise, look at $-f$), and then by continuity we can choose a square about (x_0, y_0) on which $D_2 f$ has a positive lower bound m. Also, by continuity, we can suppose that $|D_1 f|$ has an upper bound M on the square. The crux of the matter is then the following theorem.

THEOREM 9   *Suppose that $z = f(x, y)$ is a smooth function defined throughout a square S of side $2b$ about a point (x_0, y_0), and suppose that on S the partial derivatives* *satisfy the inequalities*

$$\frac{\partial z}{\partial y} > m, \qquad \left|\frac{\partial z}{\partial x}\right| < M,$$

where m and M are positive constants. We trim the square down to a rectangle R about (x_0, y_0) of width $2a$, where

$$a = \frac{bm}{M},$$

and we take k as the value $f(x_0, y_0)$ at the center of R. Then in R the level set

$$f(x, y) = k$$

is the graph of a smooth function $y = \phi(x)$, with domain the closed interval $I = [x_0 - a, x_0 + a]$ forming the base of R.

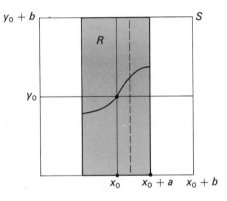

Proof. We shall prove that $f(x, y) > k$ along the top edge of the rectangle R. It will then follow, in exactly the same way, that $f(x, y) < k$ along the bottom edge of R. Therefore, along any vertical segment $x = x_1$ crossing R, $f(x_1, y)$ is an increasing function of y that starts with a value less than k and ends with a value greater than k. So $f(x_1, y)$ assumes the value k exactly once, by the intermediate-value theorem. That is, the level set $f(x, y) = k$ intersects each vertical segment of R exactly once. That is, in R, the level set $f = k$ is the graph of a function ϕ with domain

$$I = [x_0 - a, x_0 + a].$$

So the fact that the level set in R is a function graph follows from the fact that $f(x, y) > k$ along the top edge of the rectangle R. To prove this we go vertically from the center (x_0, y_0). Along this vertical segment $z = f(x_0, y)$ is an increasing function of y, and its total increase is at least mb since its derivative is everywhere larger than m. So, at the center of the top edge, we have

$$f(x_0, y_0 + b) > k + mb.$$

Along the top edge we have a function of x,

$$f(x, y_0 + b),$$

with derivative bounded by M. In particular, its derivative is bounded below by $-M$, so for the same reason it cannot decrease from its central value by more than Ma:

$$f(x, y_0 + b) \geq f(x_0, y_0 + b) - Ma.$$

Combining this with the preceding inequality, and remembering that

$$mb - Ma = 0,$$

we see that

$$f(x, y_0 + b) > k.$$

The proof that $f(x, y_0 - b) < k$ goes in exactly the same way and we omit it. So we have our implicitly defined function ϕ.

It remains to be shown that ϕ is smooth. To do this, we go back to the proof of the chain rule and copy out the formula

$$\Delta z = f(x + \Delta x, y + \Delta y) - f(x, y)$$
$$= \Delta x D_1 f(X, y) + \Delta y D_2 f(x + \Delta x, Y),$$

where X is some number between x and $(x + \Delta x)$, and Y is some number between y and $(y + \Delta y)$. Here the two points (x, y) and $(x + \Delta x, y + \Delta y)$ are on the level set $f = k$, so $\Delta z = k - k = 0$ and

$$\frac{\Delta y}{\Delta x} = -\frac{D_1 f(X, y)}{D_2 f(x + \Delta x, Y)}.$$

It then follows from the derivative bounds that

$$\left| \frac{\Delta y}{\Delta x} \right| < \frac{M}{m},$$

or

$$|\Delta y| < \frac{M}{m} |\Delta x|,$$

so ϕ is continuous. But now, as $\Delta x \to 0$, we have

$$\Delta y \to 0 \qquad \text{and} \qquad Y \to y, \qquad \text{as well as } X \to x.$$

Since the partial derivatives $D_1 f$ and $D_2 f$ are continuous we can take limits in the above expression for $\Delta y / \Delta x$ and conclude that $y = \phi(x)$ is differentiable and

$$\frac{dy}{dx} = -\frac{D_1 f(x, y)}{D_2 f(x, y)}.$$

In terms of ϕ, this is the formula

$$\phi'(x) = -\frac{D_1 f(x, \phi(x))}{D_2 f(x, \phi(x))}. \quad \blacksquare$$

The basic theorem for a function of three variables has exactly the structure of Theorem 9.

THEOREM 10   *Suppose that $w = f(x, y, z)$ is a smooth function defined throughout a cube C of side $2b$ about a point (x_0, y_0, z_0), and suppose that in C the partial derivatives satisfy the inequalities*

$$\frac{\partial w}{\partial z} > m, \qquad \left| \frac{\partial w}{\partial x} \right| < M, \qquad \left| \frac{\partial w}{\partial y} \right| < M,$$

where m and M are positive constants. We thin the cube down to a rectangular box R about (x_0, y_0, z_0) with square base S of side $2a$, where

$$a = \frac{bm}{2M},$$

and we take k as the value $f(x_0, y_0, z_0)$ at the center of R. Then in R the level set

$$f(x, y, z) = k$$

is the graph of a smooth function $z = \phi(x, y)$, with domain the square S forming the base of R.

The proof is identical to that of the earlier theorem except for a couple of points, and we shall mention only the changes.

In the proof that $f > k$ on the top square of R, that is, that $f(x, y, z_0 + b) > k$, we go from the center point $(x_0, y_0, z_0 + b)$ to the general point $(x, y, z_0 + b)$ in *two* steps, an x-change step and then a y-change step. Each of these steps permits a maximum decrease in the value of f of amount Ma, so the total decrease is less than $2Ma$. This explains the new factor 2 in the definition of a.

The other change is in notation. In the proof that the implicitly defined function ϕ is smooth, we hold either x or y fixed and show that the *partial* derivative of ϕ with respect to the other variable exists, concluding, say, that

$$\frac{\partial z}{\partial y} = -\frac{D_2 f(x, y, z)}{D_3 f(x, y, z)}. \quad \blacksquare$$

PROBLEMS FOR SECTION 9

1. In the proof of the implicit-function theorem, we used the fact that if $f'(t) > m$ on an interval $[c, d]$, then

$$f(d) - f(c) > m(d - c).$$

   Give a careful proof of this fact.

2. In the notation of the theorem, prove that

$$f(x, y_0 - b) < k$$

   for every x in the interval $[-a, a]$. (Imitate the proof given for the inequality $f(x, y_0 + b) > k$, with suitable modifications.)

3. What inequality law (or laws) is needed for the conclusion $|\Delta y/\Delta x| < M/m$ in the proof of the differentiability of the implicitly defined function ϕ?

4. Give all the details of the final limit evaluation that yields the formula for dy/dx.

5. Write out the proof of Theorem 10.

10. HIGHER-ORDER PARTIAL DERIVATIVES

Another new situation that arises in the study of partial derivatives is the occurrence of *mixed* higher-order derivatives, like $\partial^2 z/(\partial x\, \partial y)$. The notation means what one would expect:

$$\frac{\partial^2 z}{\partial y\, \partial x} = \frac{\partial}{\partial y}\left(\frac{\partial z}{\partial x}\right),$$

$$\frac{\partial^2 z}{\partial x\, \partial y} = \frac{\partial}{\partial x}\left(\frac{\partial z}{\partial y}\right).$$

For example, if

$$z = x^2 y^3 + x^4 y,$$

then

$$\frac{\partial z}{\partial x} = 2xy^3 + 4x^3 y,$$

$$\frac{\partial^2 z}{\partial y\, \partial x} = \frac{\partial}{\partial y}\left(\frac{\partial z}{\partial x}\right) = 6xy^2 + 4x^3,$$

$$\frac{\partial z}{\partial y} = 3x^2 y^2 + x^4,$$

$$\frac{\partial^2 z}{\partial x\, \partial y} = \frac{\partial}{\partial x}\left(\frac{\partial z}{\partial y}\right) = 6xy^2 + 4x^3.$$

Note in the above example that

$$\frac{\partial^2 z}{\partial x\, \partial y} = \frac{\partial^2 z}{\partial y\, \partial x}.$$

Such equality is not accidental, for this "commutative" law is a theorem.

THEOREM 11 *If $f(x, y)$ has continuous partial derivatives of the first and second order, then*

$$D_1 D_2 f(x, y) = D_2 D_1 f(x, y).$$

Proof. We shall show that both mixed partials are approximately equal to the "second-order difference quotient"

(*) $$\frac{f(x + h, y + k) - f(x + h, y) - f(x, y + k) + f(x, y)}{hk}.$$

Consider first the function

$$\phi(x) = f(x, y + k) - f(x, y),$$

where y and $(y + k)$ are both held fixed. The numerator in the above expression (*) is then $\phi(x + h) - \phi(x)$, and hence can be written

$$\phi(x + h) - \phi(x) = h\phi'(X) = h[D_1 f(X, y + k) - D_1 f(X, y)]$$

by the mean-value theorem. Here, X is some number lying between x and $(x + h)$. Since $D_1 f(X, y)$ is differentiable as a function of y, we can apply the mean-value theorem once more:

$$h[D_1 f(X, y + k) - D_1 f(X, y)] = hk D_2 D_1 f(X, Y),$$

where Y lies between y and $y + k$. Thus the double difference quotient (*) has the values $D_2 D_1 f(X, Y)$.

We now start over again with the function

$$\psi(y) = f(x + h, y) - f(x, y),$$

where now x and $x + h$ are held fixed. The numerator in $(*)$ is then $\psi(y + k) - \psi(y)$, and proceeding exactly as before, through two applications of the mean-value theorem, we find this time that the double difference quotient $(*)$ has the value $D_1 D_2 f(X', Y')$, where Y' lies between y and $y + k$ and X' lies between x and $x + h$.

Altogether we have evaluated $(*)$ in two different ways. In particular, we have proved that there are numbers X and X' both lying between x and $x + h$, and numbers Y and Y' both lying between y and $y + k$, such that

$$D_2 D_1 f(X, Y) = D_1 D_2 f(X', Y').$$

Now, let h and k both approach zero. Then X and X' both approach x, and Y and Y' both approach y, and since the partial derivatives $D_2 D_1 f$ and $D_1 D_2 f$ are continuous, the above equation becomes

$$D_2 D_1 f(x, y) = D_1 D_2 f(x, y)$$

in the limit. This proves the theorem. ∎

REMARK   Theorem 11 holds equally well for functions of more than two variables. For example if $z = f(u, v, w, x, y)$, and f has continuous partial derivatives of first and second order, then

$$\frac{\partial^2 z}{\partial u\, \partial w} = \frac{\partial^2 z}{\partial w\, \partial u}, \qquad \frac{\partial^2 z}{\partial v\, \partial y} = \frac{\partial^2 z}{\partial y\, \partial v}, \qquad \text{etc.}$$

The reason is that all variables but two are held fixed when any one of these mixed partials is investigated, so Theorem 11 applies to the remaining function of two variables.

PROBLEMS FOR SECTION 10

Find *all* of the second partial derivatives of the following functions.

1. $z = 3x^2 - 2xy + y$ 2. $z = x \cos y - y \cos x$

3. $z = e^{x^2 y}$ 4. $z = \sin(2x^2 + 3y)$

5. $z = x^2 + 3xy + y^2$ 6. $z = y \log x$

7. $z = xy$ 8. $z = \sin 3x \cos 4y$

9. $z = \dfrac{x}{y^2} - \dfrac{y}{x^2}$ 10. $z = \arctan y/x$

11. Let f be a function of one variable and set

$$z = f(x - ct).$$

Show that

$$\frac{\partial^2 z}{\partial t^2} = c^2 \frac{\partial^2 z}{\partial x^2}.$$

12. Show, more generally, that

$$z = f(x - ct) + g(x + ct)$$

satisfies the above partial differential equation.

13. If $z = \log(x^2 + y^2)$, show that

$$\frac{\partial^2 z}{\partial x^2} + \frac{\partial^2 z}{\partial y^2} = 0.$$

14. If $z = \arctan \dfrac{y}{x}$, show that

$$\frac{\partial^2 z}{\partial x^2} + \frac{\partial^2 z}{\partial y^2} = 0.$$

15. If $z = \dfrac{e^{-x^2/4y}}{\sqrt{y}}$, show that

$$\frac{\partial z}{\partial y} = \frac{\partial^2 z}{\partial x^2}.$$

11. THE CLASSIFICATION OF CRITICAL POINTS

We return now to functions of two variables. In Chapter 6 we saw that if x_0 is a critical point of f at which the second derivative $f''(x_0)$ exists and is nonzero, then f has a relative maximum or minimum at x_0, depending on the sign of $f''(x_0)$. On the other hand, if $f''(x_0) = 0$, then nothing can be concluded.

A similar result holds for a function $z = f(x, y)$ of two variables, the requirement that $f''(x_0) \neq 0$ being replaced by $D(x_0, y_0) \neq 0$, where the "discriminant" function $D(x, y)$ is defined by

$$D(x, y) = \left(\frac{\partial^2 z}{\partial x\, \partial y}\right)^2 - \frac{\partial^2 z}{\partial x^2}\frac{\partial^2 z}{\partial y^2}.$$

However, a new possibility occurs now, namely, that f may have a maximum at (x_0, y_0) along one line through this point, and a minimum at (x_0, y_0) along a different line. The resulting configuration of the graph of f around (x_0, y_0) is called a *saddle point*, a term that the figure below shows to be appropriate.

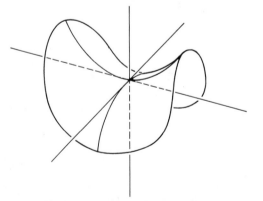

In this situation, the level set through $\mathbf{x}_0 = (x_0, y_0)$ is a pair of intersecting curves, dividing the neighborhood of $\mathbf{x}_0$ into four quadrant-like pieces on which $f(x, y) - f(x_0, y_0)$ is alternately positive and negative.

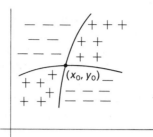

This time we shall start with the proofs, although the theory could be postponed without affecting the examples. We consider first the case $D(x_0, y_0) < 0$. Since $D(x, y)$ is continuous, we can restrict our attention to a small circular disk R about $\mathbf{x}_0$ on which $D(x, y)$ is everywhere negative. Inspection of the formula for $D(x, y)$ shows that then $\partial^2 z/\partial x^2$ and $\partial^2 z/\partial y^2$ can never be zero and must have the same sign throughout R. We shall suppose that both are positive.

So our total hypotheses are that $D(x, y)$ is negative throughout R, $\partial^2 z/\partial x^2$ and $\partial^2 z/\partial y^2$ are both positive throughout R, while the center point (x_0, y_0) in R is a critical point of f. Under these assumptions we shall show that $f(x_0, y_0)$ is the minimum value of f on R.

It is sufficient to show that f is concave up along every diameter of R. This could be handled by second-order directional derivatives, but since we haven't done anything with higher order directional derivatives we must work things out from scratch by way of the chain rule. So let $\mathbf{u} = (u, v)$ be any unit vector and let ϕ be the restriction of f to the diameter $\mathbf{x} = \mathbf{x}_0 + t\mathbf{u}$,

$$\phi(t) = f(x_0 + tu, y_0 + tv),$$

where $|t|$ is less than the radius of R. Then

$$\phi'(t) = u\,D_1f + v\,D_2f,$$
$$\phi''(t) = u^2\,D_1D_1f + uv\,D_1D_2f + vu\,D_2D_1f + v^2\,D_2D_2f$$
$$= v^2\,D_{22}f + 2uv\,D_{12}f + u^2\,D_{11}f,$$

where the doubly indexed symbols indicate the second partial derivatives from the line above (in reversed order), and where all the partial derivatives are evaluated at the point $(x_0 + tu, y_0 + tv)$.

Now remember from algebra that if

$$b^2 - ac < 0,$$

then the equation

$$av^2 + 2bvu + cu^2 = 0$$

has no solution pair (u, v) except $(0, 0)$. This is just the usual quadratic equation condition, restated in a form that treats a and c symmetrically. We apply this to the above quadratic expression for $\phi''(t)$, recalling that

$$D(x, y) = (D_{12})^2 - D_{11}D_{22} < 0$$

by hypothesis, and conclude that $\phi''(t)$ is never zero. Moreover, the expression for $\phi''(t)$ is a continuous function of $\mathbf{u} = (u, v)$ as well as t, and since it is positive when $u = 0$ (because $D_{22}f > 0$ by hypothesis) it must always be positive. In particular, $\phi(t)$ is concave up as a function of t for each fixed unit vector $\mathbf{u}$, and we have proved that $f(x, y)$ is concave up on every diameter of R. That is, f has a relative minimum at (x_0, y_0).

Going back to the beginning, if $Df = (D_{12}f)^2 - D_{11}fD_{22}f$ is negative at (x_0, y_0) but the common sign of D_{11} and D_{22} is negative, then the above argument repeats almost verbatim to show that ϕ'' is always negative; hence ϕ is concave down on every diameter of R and f has a relative maximum at (x_0, y_0). The following theorem has been proved.

THEOREM 12   *Suppose that $f(x, y)$ has continuous partial derivatives of the first and second order, and suppose that $\mathbf{x}_0 = (x_0, y_0)$ is a critical point of f at which the discriminant $D(x, y)$ is negative. Then f has a relative minimum at $\mathbf{x}_0$ or a relative maximum at $\mathbf{x}_0$, depending on whether $D_{11}f(x_0, y_0)$ and $D_{22}f(x_0, y_0)$ are both positive or both negative.*

There remains the case where (x_0, y_0) is a critical point of $f(x, y)$ at which the discriminant is positive. We could continue in terms of the unit vector $\mathbf{u} = (u, v)$ and the corresponding homogeneous quadratic polynomial, but instead we shall retreat to the more familiar inhomogeneous polynomial in m, where m is the slope of the vector $\mathbf{u}$, $m = v/u$. This introduces an exceptional case, the direction $(0, 1)$ for which $m = \infty$, but we are not aiming at a complete proof. So suppose that $D_{22}f(x_0, y_0) \neq 0$, and consider the quadratic polynomial

$$q(m) = am^2 + 2bm + c,$$

where

$$a = D_{22}f(x_0, y_0),$$
$$b = D_{12}f(x_0, y_0),$$
$$c = D_{11}f(x_0, y_0).$$

The quadratic formula for the roots of $q(m)$ is

$$m = \frac{-b \pm \sqrt{b^2 - ac}}{a},$$

where $b^2 - ac$ is the discriminant $Df(x_0, y_0)$, which we are now assuming to be positive. The quadratic formula therefore gives two roots,

$$m_1 = (-b - \sqrt{D})/a, \qquad m_2 = (-b + \sqrt{D})/a,$$

and $q(m)$ has the corresponding factorization:

$$q(m) = a(m - m_1)(m - m_2).$$

In particular, $q(m)$ changes sign as m crosses m_1, say from $-$ to $+$. Then $q(m)$ is positive for any m between m_1 and m_2. Now $v^2 q(m)$ is the value we found for $\phi''(0)$, where $\phi(t)$ is the restriction of $f(x, y)$ to the line through (x_0, y_0) of slope m. So, for any m between m_1 and m_2, $f(x, y)$ is concave up as $\mathbf{x} = (x, y)$ moves through $\mathbf{x}_0 = (x_0, y_0)$ along the line of slope m. Similarly, $f(x, y)$ is concave down as $\mathbf{x}$ moves through $\mathbf{x}_0$ along a line whose slope m is either less than m_1 or greater than m_2. This means that $\mathbf{x}_0$ is a saddle point for f (although the proof is not complete).

We have given above a partial proof of the following theorem.

THEOREM 13   *If $f(x, y)$ has continuous partial derivatives of the first and second order, and if (x_0, y_0) is a critical point of f at which the discriminant of f is positive, then (x_0, y_0) is a saddle point for f.*

EXAMPLE 1   It is clear on the face of it that $z = x^2 + y^2$ is minimum at the origin, but let us disregard the obvious and see what the machinery of critical-point analysis has to say. The first partial derivatives $\partial z/\partial x = 2x$ and $\partial z/\partial y = 2y$ vanish simultaneously only at the origin, which is therefore the only critical point. The second partial derivatives are all constants, with the values

$$\frac{\partial^2 z}{\partial x^2} = 2, \qquad \frac{\partial^2 z}{\partial y^2} = 2, \qquad \frac{\partial^2 z}{\partial x \, \partial y} = 0.$$

The discriminant thus has the constant value

$$0^2 - 2 \cdot 2 = -4.$$

So Theorem 12 tells us that $f(x, y) = x^2 + y^2$ has a relative minimum at the origin. The graph of f is a paraboloid of revolution.

EXAMPLE 2   Classify the critical point or points of $z = x^2 - y^2$.

Solution. We find that $\partial z/\partial x = 2x$ and $\partial z/\partial y = -2y$ are simultaneously zero only at the origin, which is thus the only critical point. The second partial derivatives are again all constant,

$$\frac{\partial^2 z}{\partial x^2} = 2, \qquad \frac{\partial^2 z}{\partial y^2} = -2, \qquad \frac{\partial^2 z}{\partial x \, \partial y} = 0,$$

and the discriminant $D(x, y)$ has the constant value

$$\left(\frac{\partial^2 z}{\partial x \, \partial y} \right)^2 - \frac{\partial^2 z}{\partial x^2} \frac{\partial^2 z}{\partial y^2} = 0^2 - 2(-2) = 4.$$

This time the origin is a saddle point by Theorem 13. The graph is a hyperbolic paraboloid.

EXAMPLE 3   Study the critical points of the function $z = x^3 y + 3x + y$.

Solution. The equations

$$\frac{\partial z}{\partial x} = 3x^2 y + 3 = 0,$$

$$\frac{\partial z}{\partial y} = x^3 + 1 = 0$$

determine the solutions: $x = -1$, from the second equation, and then $y = -1$ from the first. Therefore $(-1, -1)$ is the only critical point. The discriminant

$$\left(\frac{\partial^2 z}{\partial x \, \partial y} \right)^2 - \frac{\partial^2 z}{\partial x^2} \frac{\partial^2 z}{\partial y^2} = (3x^2)^2 - (6xy) \cdot 0 = 9x^4$$

is positive at $(-1, -1)$, so this point is a saddle point, by Theorem 13.

PROBLEMS FOR SECTION 11

Classify the critical points of each of the following functions of x and y.

1. $x^2 + 2y^2 + 2x - y$
2. $x - y + xy$
3. $x^3 - xy + y^2$
4. $x^2 + 3xy + y^2$
5. xy (Draw a few level curves.)
6. $x^2y^2 + 2x - 2y$
7. $x^2y + y^2 + x$
8. $x^3 - 3xy + y^3$
9. $x^2 - xy + y^3 - x$
10. $x^4 + y^4$
11. $x^4 + y^4 + 4x - 4y$
12. $x^4 - y^4 + 4x - 4y$
13. $x^2 + y^2 + \dfrac{2}{xy}$
14. $\dfrac{1}{x^2} + \dfrac{4}{y^2} + xy$
15. $e^x + e^y - e^{x+y}$
16. $x \sin y$

17. Prove the following corollary of (part of) Theorem 12.

> **Theorem.** Suppose that the discriminant of $f(x, y)$ is negative at the point (x_0, y_0), and that the second partial derivatives $\partial^2 z/\partial x^2$ and $\partial^2 z/\partial y^2$ are both positive there. Prove that, for points (x, y) sufficiently close to (x_0, y_0), the graph of $z = f(x, y)$ lies entirely above its tangent plane at (x_0, y_0).

EXTRA PROBLEMS FOR CHAPTER 16

1. If x and y are independent variables, then their differentials dx and dy are new independent variables, wholly independent of x and y in their values, but used along with x and y. The definition of the differential of a *dependent* variable is analogous to that given earlier for a function of one independent variable (at the end of Chapter 5).

> **Definition.** If $z = f(x, y)$, then
> $$dz = D_1 f(x, y)\, dx + D_2 f(x, y)\, dy.$$

Show that this relationship between dx, dy, and dz remains true even if x and y are functions of other variables, say $x = g(s, t)$ and $y = h(s, t)$, so that the underlying independent variables are s and t throughout.

If u and v are both functions of x and y, show that:

2. $d(u + v) = du + dv$
3. $d(uv) = u\, dv + v\, du$
4. $d\left(\dfrac{u}{v}\right) = \dfrac{v\, du - u\, dv}{v^2}$
5. If $w = f(u)$, then $dw = f'(u)\, du$
6. If $w = f(xz, yz)$, show that
$$x\frac{\partial w}{\partial x} + y\frac{\partial w}{\partial y} = z\frac{\partial w}{\partial z}.$$

7. Let f be a smooth function, and let the domain D of $f(x, y)$ be in one piece, in the sense that a smooth path can be drawn in D between any two points of D. Show that f is a constant function if and only if $\operatorname{grad} f$ is identically 0. (This depends on, and generalizes, Theorem 7 in Chapter 3 (Section 8).)

8. Show that
$$\operatorname{grad}(f + g)(X) = \operatorname{grad} f(X) + \operatorname{grad} g(X)$$
$$\operatorname{grad} cf(X) = c \operatorname{grad} f(X).$$

9. Let D be a domain in one piece, as in Problem 7, and suppose f and g are two functions with domain D such that
$$\operatorname{grad} f = \operatorname{grad} g$$

on D. Use Problems 7 and 8 to show that there is a constant C such that

$$f(x, y) = g(x, y) + C$$

for all (x, y) in D.

10. Find a function f such that

$$\text{grad } f(X) \equiv X.$$

What is the most general such function?

11. Show that

$$\text{grad } fg(X) = f(X) \text{ grad } g(X) + g(X) \text{ grad } f(X).$$

12. A smooth path of length L runs from the point X_0 to the point X_1. A function f, defined on a region containing the path, has both partial derivatives bounded by a constant K. Show that

$$|f(X_1) - f(X_0)| \leq 2KL.$$

13. Show that every ellipse of the form

$$x^2 + 2y^2 = C$$

is orthogonal to every parabola of the form

$$y = kx^2.$$

14. Show that the sphere

$$x^2 + y^2 + z^2 = r^2$$

and the cone

$$z^2 = x^2 + y^2$$

intersect orthogonally.

15. Show that the ellipsoid

$$x^2 + y^2 + 2z^2 = C$$

and the paraboloid

$$z = k(x^2 \times y^2)$$

intersect orthogonally.

16. Write out the proof of Theorem 7.

17. Show that the equation of the tangent line to the conic

$$ax^2 + by^2 + cx + dy + e = 0$$

at the point (x_0, y_0) can be written

$$ax_0 x + by_0 y + \frac{c}{2}(x + x_0) + \frac{d}{2}(y + y_0) + e = 0.$$

18. Show that the equation of the tangent plane to the ellipsoid

$$\frac{x^2}{a^2} + \frac{y^2}{b^2} + \frac{z^2}{c^2} = 1$$

at the point (x_0, y_0, z_0) can be written

$$\frac{xx_0}{a^2} + \frac{yy_0}{b^2} + \frac{zz_0}{c^2} = 1.$$

19. Prove the analogous result for the hyperboloid

$$\frac{x^2}{a^2} + \frac{y^2}{b^2} - \frac{z^2}{c^2} = k,$$

    $(k \neq 0)$.

20. Prove a similar result for the paraboloid

$$z = \frac{x^2}{a^2} + \frac{y^2}{b^2}.$$

21. A tangent plane to the surface

$$x^{1/2} + y^{1/2} + z^{1/2} = 1$$

    has intercepts a, b, and c on the axes. Show that $a + b + c$ is independent of the point of tangency.

22. A tangent plane to the surface

$$x^{2/3} + y^{2/3} + z^{2/3} = 1$$

    has intercepts a, b, and c on the axes. Show that $a + b + c$ is independent of the the point of tangency.

23. If $z = e^x \sin y$, show that

$$\frac{\partial^2 z}{\partial x^2} + \frac{\partial^2 z}{\partial y^2} = 0.$$

24. If $w = (x^2 + y^2 + z^2)^{-1/2}$, prove that

$$\frac{\partial^2 w}{\partial x^2} + \frac{\partial^2 w}{\partial y^2} + \frac{\partial^2 w}{\partial z^2} = 0.$$

25. Show that if $z = f(x, y)$, then

$$\frac{\partial^2 z}{\partial x \, \partial y \, \partial x} = \frac{\partial^3 z}{\partial y \, \partial x^2}.$$

26. Show that if $z = e^{x^2 - y^2} \sin(2xy)$, then

$$\frac{\partial^2 z}{\partial x^2} + \frac{\partial^2 z}{\partial y^2} = 0.$$

27. Show that the rectangular box of maximum total surface area that can be inscribed in a sphere is a cube.

28. Find the maximum and minimum values of

$$f(x, y, z) = x - 2y + 5z$$

    on the sphere $x^2 + y^2 + z^2 = 30$.

29. Find the minimum value of $x^2 + y^2 + z^2$ on the plane $x - 2y + 5z = 1$.

30. Find the point of the plane $2x - 3y - z = 10$ closest to the origin.

31. Find the volume of the largest rectangular solid with sides parallel to the coordinate planes that can be inscribed in the ellipsoid

$$\frac{x^2}{9} + \frac{y^2}{4} + z^2 = 1.$$

32. Same problem for the ellipsoid

$$\frac{x^2}{a^2} + \frac{y^2}{b^2} + \frac{z^2}{c^2} = 1.$$

33. Find the maximum and minimum values of $f(x, y, z) = (8x - y + 27z)/2$ on the surface

$$x^4 + y^4 + z^4 = 1.$$

17
Double and Triple Integrals

A continuous function of two variables $f(x, y)$ can be integrated over a plane region G in much the same way that a continuous function of one variable is integrated over an interval. The number we get is called the *double integral* of f over G, and is designated

$$\iint_G f(x, y)\, dx\, dy.$$

Similarly, a continuous function of three variables $f(x, y, z)$, defined over a finite region G in space, has a *triple integral*

$$\iiint_G f(x, y, z)\, dx\, dy\, dz,$$

and a continuous function of n variables $f(x_1, \ldots, x_n)$ has an n-tuple integral over an analogous domain in "n-dimensional space." Such multiple integrals are as important as partial derivatives for the more advanced theory and applications of calculus. The present chapter is only an introduction to this large subject, and is restricted to some properties and applications of double and triple integrals.

Like the one-variable integral $\int_a^b f$, the double integral $\iint_G f$ has an extensive array of interpretations. Here are a few.

1. If f is positive, then $\iint_G f(x, y)\, dx\, dy$ is the volume of the region that lies between the surface $z = f(x, y)$ and the xy-plane, with the plane region G as its base.

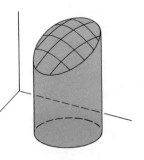

2. For any continuous f,

$$\frac{\displaystyle\iint_G f(xy)\, dx\, dy}{\text{area of } G}$$
is the average value of f over G.

3. If G is a plane lamina with variable mass density $\rho(x, y)$, then

$$\iint_G \rho(x, y)\, dx\, dy$$

is the mass of G.

4. If $\rho(x, y)$ is a probability density function defined over the plane, then $\iint_G \rho$ is the probability that a randomly chosen point will land in the region G.

5. The surface area of the piece of the graph of $z = f(x, y)$ that lies over G is given by

$$\iint_G \sqrt{1 + \left(\frac{\partial z}{\partial x}\right)^2 + \left(\frac{\partial z}{\partial y}\right)^2}\, dx\, dy.$$

This is only a partial list. In general, if Q is any quantity that can be interpreted as being spread out over the plane, and if the distribution of Q has a continuous density $\rho(x, y)$, as discussed at the end of Chapter 4, then the amount of Q lying over the plane region G is

$$\iint_G \rho(x, y)\, dx\, dy.$$

Like the definite integral $\int_a^b f$, the double integral $\iint_G f$ is ultimately defined as the limit of certain Riemann sums. But the purely analytic theory will be postponed, and we shall proceed provisionally on the basis of geometric reasoning. We shall assume certain geometrically obvious properties of the volume of a solid body in space, and shall develop the double integral on the basis of these assumptions. This is analogous to our earlier treatment of the integral $\int_a^b f$, which was based on geometrically obvious properties of the areas of plane regions.

1. VOLUMES BY ITERATED INTEGRATION

Suppose that the region G in the xy-plane is bounded by the graphs of continuous functions g and h, and lies between the x values $x = a$ and $x = b$, as in either figure below.

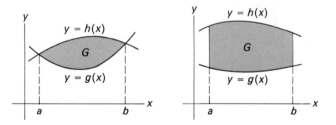

We consider the solid based on G and bounded above by the piece of the surface $z = f(x, y)$ lying over G, where f is a positive continuous function.

This solid is the intersection of the region under the surface $z = f(x, y)$ and the cylinder having G as its base. We shall often refer to it as the solid *based* on G and *capped* by the surface $z = f(x, y)$.

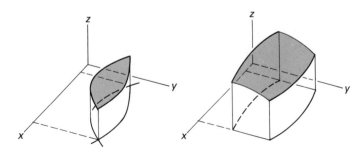

The cross section of this solid in the plane $x = x_0$ is the plane region under the graph of $z = f(x_0, y)$, from $y_1 = g(x_0)$ to $y_2 = h(x_0)$. Its area is

$$A(x_0) = \int_{y_1}^{y_2} f(x_0, y)\, dy = \int_{g(x_0)}^{h(x_0)} f(x_0, y)\, dy,$$

by the area considerations in Chapter 4. We assume that the cross section changes continuously as the slicing plane is varied, in the manner discussed in Section 5 of Chapter 9.

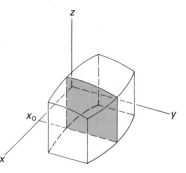

(The general proof of this fact for an arbitrary positive continuous function $z = f(x, y)$ is beyond our present capabilities, but it is easy enough to prove for the types of functions we actually meet in practice. See Problems 36 and 37 in the extra problems at the end of this chapter.) Therefore,

$$V = \int_a^b A(x)\, dx,$$

by Theorem 8 of Chapter 9. Substituting the above value for $A(x)$, we end up with a new volume formula:

$$V = \int_a^b \left[\int_{g(x)}^{h(x)} f(x, y)\, dy \right] dx.$$

This evaluation of V is called an *iterated* integral. We first integrate $f(x, y)$ with respect to y, holding x fixed. The limits of integration will depend on the fixed value of x, and so will the resulting value of the integral. It was called $A(x)$ above. Then we integrate this function of x between fixed limits a and b, and obtain the number that is called the iterated integral. All in all, we start with a function of two variables $f(x, y)$, and first "integrate y out," leaving a function of the one variable x, and then "integrate x out," ending up with a number.

Here is the theorem we have proved (supposing that the x cross section varies continuously with x):

THEOREM 1   *Let S be the solid based on the region G in the xy-plane and capped by the surface $z = f(x, y)$, where f is continuous and nonnegative on G. Suppose, furthermore, that in the xy-plane G is bounded below and above by the graphs of the continuous functions g and h, and left and right by the straight lines $x = a$ and $x = b$. Then the volume V of S is given by the iterated integral*

$$V = \int_a^b \left[\int_{g(x)}^{h(x)} f(x, y)\, dy \right] dx.$$

EXAMPLE 1 Evaluate the iterated integral

$$\int_0^1 \left(\int_{x^2}^x xy^2 \, dy \right) dx.$$

Sketch the "domain of integration."

Solution. We first compute the inside y integral, treating x as a constant:

$$\int_{x^2}^x xy^2 \, dy = \left. \frac{xy^3}{3} \right]_{x^2}^x = \frac{1}{3}(x^4 - x^7).$$

Then we compute the outside x integral:

$$\frac{1}{3} \int_0^1 (x^4 - x^7) \, dx = \frac{1}{3} \left[\frac{x^5}{5} - \frac{x^8}{8} \right]_0^1 = \frac{1}{3} \left(\frac{1}{5} - \frac{1}{8} \right) = \frac{1}{40}.$$

Thus,

$$\int_0^1 \left(\int_{x^2}^x xy^2 \, dy \right) dx = \frac{1}{40}.$$

For each fixed x_0 between $x = 0$ and $x = 1$ the inside function of y is integrated from $y = x_0^2$ to $y = x_0$. The domain of integration is thus the plane region between the graphs $y = x^2$ and $y = x$, as shown below.

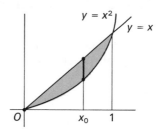

EXAMPLE 2 Find the volume of the space region under the graph of $z = x + y$ based on the triangle in the xy-plane that is cut off from the first quadrant by the line $x + y = 1$.

Solution. The figure is shown below. Such a figure should usually be drawn. We don't really need it here, because when the domain of integration is provided explicitly, as it is here, the volume can be computed by just plugging into the iterated integration formula. The triangular domain is bounded above and below by the graphs $y = 1 - x$ and $y = 0$, and runs left to right from $x = 0$ to $x = 1$. The volume integral is therefore

$$V = \int_0^1 \left[\int_0^{1-x} (x + y) \, dy \right] dx.$$

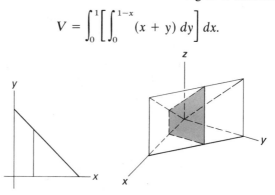

For the inside integral we have

$$\int_0^{1-x} (x + y)\, dy = \left[xy + \frac{y^2}{2} \right]_0^{(1-x)} = x(1 - x) + \frac{(1 - x)^2}{2}$$

$$= \frac{1}{2}(1 - x^2).$$

Therefore,

$$V = \int_0^1 \frac{1 - x^2}{2}\, dx = \frac{1}{2}\left[x - \frac{x^3}{3} \right]_0^1 = \frac{1}{3}.$$

REMARK   The brackets or parentheses setting off the inside integral of an iterated integral are generally omitted, it being understood that the inside integral is computed first. We would thus write the above iterated integral as follows:

$$V = \int_0^1 \int_0^{1-x} (x + y)\, dy\, dx.$$

EXAMPLE 3   Find the volume of the region in the first octant under the graph of $z = f(x, y) = 1 - y - x^2$.

Solution.   The domain is the region in the xy-plane bounded by the coordinate axes and the level curve $f(x, y) = 0$, that is, the parabola $y = 1 - x^2$.

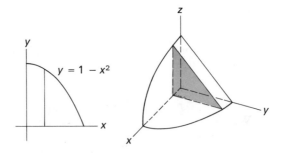

So the inside y integral runs from $y = 0$ to $y = 1 - x^2$, and the volume formula is

$$V = \int_0^1 \int_0^{1-x^2} (1 - y - x^2)\, dy\, dx.$$

The figure for the solid at the right above also shows the edge curves in the other two coordinate planes. The edge curve in the xz-coordinate plane is obtained by setting $y = 0$ in the equation

$$z = 1 - y - x^2,$$

and is the parabola $z = 1 - x^2$. In the zy-coordinate plane it is the straight line $z = 1 - y$. These edge curves, together with the cross section at x, are enough to give a good visual image of the solid.

Here is the volume computation:

$$V = \int_0^1 \int_0^{1-x^2} (1 - y - x^2) \, dy \, dx$$

$$= \int_0^1 \left[(1 - x^2)y - \frac{y^2}{2} \right]_0^{1-x^2} dx$$

$$= \int_0^1 \frac{(1 - x^2)^2}{2} \, dx = \frac{1}{2}\left[x - \frac{2x^3}{3} + \frac{x^5}{5} \right]_0^1 = \frac{4}{15}.$$

PROBLEMS FOR SECTION 1

Compute the following iterated integrals. In each case sketch the region that is the domain of integration in the xy-plane.

1. $\displaystyle\int_0^1 \int_0^1 xy \, dy \, dx$

2. $\displaystyle\int_0^1 \int_0^1 dy \, dx$

3. $\displaystyle\int_0^{\pi/2} \int_0^x \cos y \, dy \, dx$

4. $\displaystyle\int_0^{\pi/2} \int_0^{\sin x} dy \, dx$

5. $\displaystyle\int_{-1}^1 \int_x^1 xy \, dy \, dx$

6. $\displaystyle\int_0^1 \int_0^{\sqrt{1-x^2}} x \, dy \, dx$

7. $\displaystyle\int_{-1}^1 \int_{-\sqrt{1-x^2}}^{\sqrt{1-x^2}} y \, dy \, dx$

8. $\displaystyle\int_1^2 \int_x^{2x} y \, dy \, dx$

9. $\displaystyle\int_0^1 \int_{x^2}^x (x + y) \, dy \, dx$

10. $\displaystyle\int_{1/4}^{3/2} \int_{x^2}^x (x + y) \, dy \, dx$

Use iterated integration to find the volume of each of the space regions described in Problems 11 through 21 below. Sketch each configuration.

11. The unit cube $0 \le x \le 1, 0 \le y \le 1, 0 \le z \le 1$.

12. The solid based on the rectangle $0 \le x \le a, 0 \le y \le b$, in the xy-plane, and capped by the plane $z = c$.

13. The solid based on the unit square and capped by the plane $z = x + y$.

14. The solid based on the first quadrant of the unit circle $x^2 + y^2 \le 1$ and capped by the plane $z = y$.

15. The piece of the first octant cut out by the plane $x + y = 1$ and the parabolic cylinder $z = 1 - y^2$.

16. The piece of the first octant cut out by the plane $x + y = 1$ and the parabolic cylinder $z = 1 - x^2$.

17. The piece of the first octant cut out by the plane $z = x$ and the cylinder $x^2 + y^2 = 1$.

18. The piece of the first octant cut out by the plane

$$\frac{x}{a} + \frac{y}{b} + \frac{z}{c} = 1,$$

where a, b, and c are all positive.

19. The interior of the ellipsoid

$$\frac{x^2}{a^2} + \frac{y^2}{b^2} + \frac{z^2}{c^2} = 1.$$

20. The solid bounded by the parabolic cylinders $z = 1 - x^2$, $x = 1 - y^2$, $x = y^2 - 1$, and the plane $z = 0$. (Divide the base region into two pieces and compute the volume as the sum of two integrals.)

21. The solid bounded by the parabolic cylinder $x = 2 - y^2$ and the planes $y = x$, $z = 0$, and $z = 1 - y$. (Here also the base region will have to be divided into two pieces.)

22. If $f_1(x, y) \le f_2(x, y)$ over the region G, then the solid cut out from the cylinder through G by the graphs of f_2 and f_1 has the volume

$$V = \iint_G (f_2(x, y) - f_1(x, y)) \, dy \, dx.$$

Justify this.

23. Find the volume of the solid lying over the unit square $0 \le x \le 1$, $0 \le y \le 1$, and bounded above and below by the surface $z = y$ and $z = y^3$. (Apply the formula in Problem 22.)

24. Find the volume of the solid lying over the quarter of the unit circle in the first quadrant and bounded above and below by the surface $z = y$ and $z = y^3$.

25. Find the volume of the piece of the first octant lying between the surfaces $z = y$ and $z = 1 - x^2$.

26. For some base regions G, it may be convenient and even necessary to interchange the role of x and y and apply the volumes-by-slicing formula to varying y-slices. Draw the base region G for which the following iterated integral is a correct volume formula, and verify its correctness by giving an argument based on volumes-by-slicing, as we did in the text:

$$V = \int_a^b \left(\int_{g(y)}^{h(y)} f(x, y) \, dx \right) dy.$$

27. Redo Problem 17 by this method.

28. Redo Problem 20 by this method. Now the base region can be left in one piece.

29. Redo Problem 21 by this method. This time the base region can be left intact.

2. RIEMANN SUMS AND THE DOUBLE INTEGRAL

Let $f(x, y)$ be a continuous function defined on a plane region G. Then we can define what we mean by a Riemann sum for f over G, and we shall see that the relationship between the Riemann sums for f and the definite integral of f is the same for such a function of two variables as it is for a function of one variable.

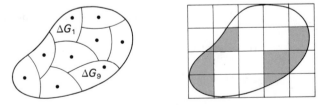

In order to define a Riemann sum for $f(x, y)$ over G, we first subdivide G in any manner into a finite number of small subregions ΔG_i where $i = 1, \ldots, n$, as indicated in the figures above. The simplest way to do this is to impose a lattice of rectangles over G, as at the right, but it is desirable for flexibility to allow more general subdivisions. Next, we choose an arbitrary "evaluation point" (x_i, y_i) in the subregion ΔG_i, for each i. Then the Riemann sum for f associated with this subdivision and evaluation set is

$$\sum_{i=1}^{n} f(x_i, y_i)\, \Delta A_i,$$

where ΔA_i is the area of the subregion ΔG_i.

EXAMPLE 1   We consider $f(x, y) = y/x$ over the unit square G having its lower left corner at $(1, 0)$. We subdivide the square into four congruent subsquares, and use the lower left corner of each as its evaluation point. Then the Riemann sum for y/x over G associated with this subdivision and this evaluation set has the value

$$f(1, 0)\,\Delta A_1 + f\!\left(\frac{3}{2}, 0\right)\!\Delta A_2 + f\!\left(1, \frac{1}{2}\right)\!\Delta A_3 + f\!\left(\frac{3}{2}, \frac{1}{2}\right)\!\Delta A_4$$

$$= \frac{0}{1}\cdot\frac{1}{4} + \frac{0}{3/2}\cdot\frac{1}{4} + \frac{1/2}{1}\cdot\frac{1}{4} + \frac{1/2}{3/2}\cdot\frac{1}{4}$$

$$= \frac{1}{4}\!\left[0 + 0 + \frac{1}{2} + \frac{1}{3}\right]$$

$$= \frac{1}{4}\cdot\frac{5}{6} = \frac{5}{24}.$$

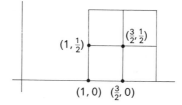

EXAMPLE 2   If we use the *upper right* corners as evaluation points in the above situation, then the Riemann sum has the value

$$f\!\left(\frac{3}{2}, \frac{1}{2}\right)\!\Delta A_1 + f\!\left(2, \frac{1}{2}\right)\!\Delta A_2 + f\!\left(\frac{3}{2}, 1\right)\!\Delta A_3 + f(2, 1)\,\Delta A_4$$

$$= \frac{1/2}{3/2}\cdot\frac{1}{4} + \frac{1/2}{2}\cdot\frac{1}{4} + \frac{1}{3/2}\cdot\frac{1}{4} + \frac{1}{2}\cdot\frac{1}{4}$$

$$= \frac{1}{4}\!\left[\frac{1}{3} + \frac{1}{4} + \frac{2}{3} + \frac{1}{2}\right] = \frac{7}{16}.$$

EXAMPLE 3   Find the Riemann sum for the above function f and Region G if G is divided into nine equal squares and the lower left corners are taken as evaluation points.

Solution. The points on the x-axis again contribute zero terms, and the remaining 6 terms are

$$f\left(1,\frac{1}{3}\right)\Delta A_4 + f\left(\frac{4}{3},\frac{1}{3}\right)\Delta A_5 + f\left(\frac{5}{3},\frac{1}{3}\right)\Delta A_6 + f\left(1,\frac{2}{3}\right)\Delta A_7$$

$$+ f\left(\frac{4}{3},\frac{2}{3}\right)\Delta A_8 + f\left(\frac{5}{3},\frac{2}{3}\right)\Delta A_9$$

$$= \frac{1/3}{1}\cdot\frac{1}{9} + \frac{1/3}{4/3}\cdot\frac{1}{9} + \frac{1/3}{5/3}\cdot\frac{1}{9} + \frac{2/3}{1}\cdot\frac{1}{9} + \frac{2/3}{4/3}\cdot\frac{1}{9} + \frac{2/3}{5/3}\cdot\frac{1}{9}$$

$$= \frac{1}{9}\left[\frac{1}{3} + \frac{1}{4} + \frac{1}{5} + \frac{2}{3} + \frac{2}{4} + \frac{2}{5}\right] = \frac{47}{180}.$$

When we vary the subdivision and evaluation set, we get a different value for the Riemann sum. We shall see that if the function f and region G are reasonably well-behaved, then this varying Riemann sum $\sum$ approaches a limit I as the subdivision is taken finer and finer. More exactly, we can make the error in the approximation $\sum \approx I$ as small as we wish simply by using a subdivision all of whose subregions ΔG_i have suitably small size. This limit I is called the *double integral* of f over G. Thus, when we say that the double integral exists and is a certain number, we mean that the varying Riemann sum approaches this number as its limit as the maximum size of its subregions approaches zero.

DEFINITION   *The double integral of f over G, designated*

$$\iint\limits_{G} f(x, y)\, dA,$$

is the limit (if it exists) of the Riemann sums for f over G in the above sense.

The basic theorem is then:

THEOREM 2   *If $f(x, y)$ is continuous on a bounded closed region G, then the double integral*

$$\iint\limits_{G} f(x, y)\, dA$$

exists.

Moreover, when the region G is simple enough, the double integral of f over G can be evaluated by iterated integration.

THEOREM 3   *Let $f(x, y)$ be a continuous function defined over the closed plane region G, and suppose that G is bounded above and below in the xy-plane by the graphs of $y = h(x)$ and $y = g(x)$, and runs from $x = a$ to $x = b$. Then*

$$\iint\limits_{G} f(x, y)\, dA = \int_a^b \int_{g(x)}^{h(x)} f(x, y)\, dy\, dx.$$

If, instead, the left and right parts of the boundary of G lie along the graphs $x = p(y)$ and $x = q(y)$, respectively, and G extends from $y = c$ to $y = d$, then

$$\iint\limits_{G} f(x, y) \, dA = \int_{c}^{d} \int_{p(y)}^{q(y)} f(x, y) \, dx \, dy.$$

In the long run Theorems 2 and 3 have to be given purely analytic proofs. Such proofs are too theoretical to be included here, but we can prove Theorem 2 subject to natural assumptions about volume, and using one final property of a continuous function. On this basis we shall show in the next section that if V is the volume of the solid S based on G and capped by the graph of f, then V is approximated arbitrarily closely by the Riemann sums for f over G. That is, the Riemann sums approach V as their limit. So we have a geometric proof of the existence of the double integral $\iint_{G} f(x, y) \, dA$. Theorem 3 is then simply the statement that this same volume can be computed by iterated integration, by Theorem 1 and its alternate form with variables reversed. The geometric proof of Theorem 2 will be given in the next section.

Theorems 2 and 3 have important applications, most of which come about as follows. We want to compute some quantity Q from geometry or physics, or from some other discipline. We first show that Q can be estimated by a certain type of finite sum. We then observe that this finite sum can also be interpreted as a Riemann sum for a certain function $\rho(x, y)$ over a region G. But this means that

$$Q = \iint\limits_{G} \rho(x, y) \, dA,$$

since both Q and the double integral are approximated arbitrarily closely by the same estimating number. Finally, we *compute* Q by iterated integration, by virtue of Theorem 3.

In more detail, a quantity Q can be calculated in this way if it can be interpreted as a "distribution" over the plane with a continuous "density" $\rho(x, y)$, as discussed at the end of Chapter 4. We first show that if ΔQ is the amount of the quantity lying over an incremental region ΔG, and if (x_0, y_0) is an arbitrary "evaluation point" in ΔG, then

$$\Delta Q \approx \rho(x_0, y_0) \, \Delta A,$$

with an error that is small in comparison with the area ΔA of ΔG. When we sum these estimates over a subdivision of a region G, we see that the amount Q of the quantity over G is given approximately by

$$Q \approx \sum \rho(x_i, y_i) \, \Delta A_i,$$

the error being small in comparison with the area A of G. Since the sum on the right is also a Riemann sum for ρ over G, we conclude that Q is given exactly by the double integral

$$Q = \iint\limits_{G} \rho(x, y) \, dA.$$

There are several applications of this procedure in the remaining sections of this chapter.

PROBLEMS FOR SECTION 2

1. Evaluate the Riemann sum for $f(x, y) = x/(1 + y)^2$ over the unit square $0 \le x \le 1, 0 \le y \le 1$, using the subdivision obtained by dividing the square into nine congruent subsquares and choosing the *upper right* vertex of each square as evaluation point.

2. Evaluate the Riemann sum for the same function and subdivision but with the *lower left* vertices as evaluation points.

3. Prove that if the double integrals of f and g exist over a region G, then so does the double integral of $af(x, y) + bg(x, y)$ and

$$\iint_G [af(x, y) + bg(x, y)] \, dA = a \iint_G f(x, y) \, dA + b \iint_G g(x, y) \, dA.$$

Assume that the usual limit laws hold here: The limit of a sum is the sum of the limits, etc.

4. Prove that if $f(x, y)$ has the constant value k over the region G, then its double integral over G exists and

$$\iint_G f = kA,$$

where A is the area of G.

5. The proof to be given shortly that a continuous function has a double integral over a bounded closed region G works only for positive functions. Prove from this that it is nevertheless true that an arbitrary continuous function $f(x, y)$ has a double integral over G. [Hint: Let k be any number less than the minimum value of f on G. Then $g(x, y) = f(x, y) - k$ is positive on G. Now apply the above two problems.]

6. Prove that if the base region G is divided into two nonoverlapping subregions G_1 and G_2, then

$$\iint_G f \, dA = \iint_{G_1} f \, dA + \iint_{G_2} f \, dA.$$

(This should again reduce to the assumption that a certain limit law remains valid for this type of limit.)

A corollary of Theorem 3 is that if the base region G can be described in *both* ways specified there, then the two iterated integrals are necessarily equal:

$$(*) \qquad \int_a^b \int_{g(x)}^{h(x)} f(x, y) \, dy \, dx = \int_c^d \int_{p(y)}^{q(y)} f(x, y) \, dx \, dy.$$

In Problems 7 through 12, an iterated integral is given that specifies a base region G in one of the above ways, and you are to work out the correct limits of integration for the other integral. (This may require that the region G be divided into two or more parts, with the given integral expressed as the sum of two or more integrals.) In each case *draw* the region G from the given limits of integration and then describe G the other way.

7. $\displaystyle\int_0^1 \int_0^x f(x, y)\, dy\, dx$ 8. $\displaystyle\int_0^1 \int_0^{x^2} \cdots\, dy\, dx$

9. $\displaystyle\int_0^1 \int_{x^2}^x \cdots\, dy\, dx$ 10. $\displaystyle\int_0^1 \int_{1-x}^{1+x} \cdots\, dy\, dx$

11. $\displaystyle\int_{-1}^1 \int_0^{y^2} \cdots\, dx\, dy$ 12. $\displaystyle\int_0^1 \int_{y^2}^1 \cdots\, dx\, dy$

The iterated integral identity ($*$) can be used to evaluate either integral in terms of the other. This is called *reversing the order of integration*. Sometimes a difficult integration can be made easier by reversing the order of integration. Compute the following iterated integrals by reversing the order of integration.

13. $\displaystyle\int_0^1 \int_y^1 e^{-x^2}\, dx\, dy$ 14. $\displaystyle\int_0^1 \int_0^{y^2} y \cos (1 - x)^2\, dx\, dy$

15. $\displaystyle\int_0^1 \int_x^1 \sqrt{1 - y^2}\, dy\, dx$

3. PROOF OF THEOREM 2

We assume, to begin with, that any solid body has a uniquely determined volume. By a solid body we mean a region in space bounded by (pieces of) a finite number of surfaces. For example, a circular cylinder is a solid body bounded by three surfaces, its two bases and its curved lateral surface. A cube is also a cylinder, but it is a solid body bounded by six surfaces, its six faces.

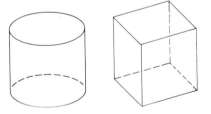

Furthermore, we assume:
1. The volume of a cylinder is given by

$$V = Ah$$

where A is the area of its base and h is its altitude.

2. If a solid body is divided into two nonoverlapping pieces, then its volume is the sum of the volumes of its two constituents

$$V = V_1 + V_2$$

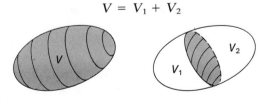

Note that:

2'. If a solid S includes a solid S_1, then the volume of S is greater than the volume of S_1.

This follows from principle (2), because

$$V = V_1 + V_2 > V_1,$$

where V is the volume of S, V_1 is the volume of S_1, and V_2 is the volume of the solid S_2 obtained by removing S_1 from S.

We are interested in the volume of the solid S based on a region G in the xy-plane and capped by the graph $z = f(x, y)$, where f is a positive continuous function. We consider, to begin with, a small incremental closed subregion ΔG of G, and the piece ΔS of the solid S lying over ΔG. Let m and M be the minimum and maximum values of f on the subregion ΔG, and let C_m and C_M be the cylinders based on ΔG with altitudes m and M respectively. (These will not generally be circular cylinders. They are shown circular in the figure for simplicity and clarity.) Then C_m is included in ΔS, which is included in C_M, so

$$\text{volume } (C_m) \leq \text{volume } (\Delta S) \leq \text{volume } (C_M),$$

by the volume principle (2'). By (1), this is the inequality

$$m \, \Delta A \leq \Delta V \leq M \, \Delta A,$$

where ΔA is the area of ΔG.

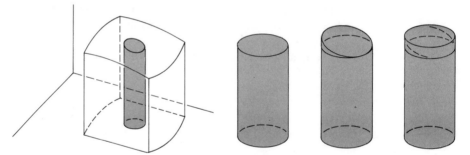

Now let (x_0, y_0) be an arbitrary "evaluation point" in ΔG. Then

$$m \leq f(x_0, y_0) \leq M,$$

so

$$m \, \Delta A \leq f(x_0, y_0) \, \Delta A \leq M \, \Delta A.$$

Thus the numbers ΔV and $f(x_0, y_0) \, \Delta A$ both lie between $m \, \Delta A$ and $M \, \Delta A$, so they differ by less than $M \, \Delta A - m \, \Delta A$. That is,

$$\Delta V \approx f(x_0, y_0) \, \Delta A,$$

with an error less than $(M - m) \, \Delta A$ in magnitude. If the values of f on the incremental subregion ΔG vary by less than a number e, that is, if $M - m < e$, then the error in the above estimate is less than $e \, \Delta A$.

Now suppose that we are given a positive number e, and suppose that we can then subdivide the region G into a finite number of subregions ΔG_i, where $i = 1, \ldots, n$, that are small enough so that the values of f vary by less than e over each subregion ΔG_i. (That is, if M_i and m_i are the maximum and minimum values of f on ΔG_i, then $M_i - m_i < e$, for each subregion ΔG_i.) Let (x_i, y_i) be an arbitrary evaluation point in the subregion ΔG_i, for each i. Then, for each i, we know from the above reasoning that the volume ΔV_i

based on ΔG_i can be estimated as the product $f(x_i, y_i)$ times the area ΔA_i of ΔG_i,

$$\Delta V_i \approx f(x_i, y_i) \Delta A_i,$$

with an error less than $e \Delta A_i$. The sum of these errors is less than

$$\sum e \Delta A_i = e \sum \Delta A_i = eA,$$

where A is the area of G. Also $V = \sum \Delta V_i$ is the volume of the solid S based on G. So when we sum the above estimates, we end up with the estimate

$$V \approx \sum_{i=1}^{n} f(x_i, y_i) \Delta A_i,$$

with an error less than eA.

We have thus shown that the volume V that we are interested in can be estimated by a Riemann sum for f over G. Moreover, we can make the error bound eA as small as we wish by first taking e small and then choosing a corresponding sufficiently fine subdivision. That is, V is the limit of the Riemann sums of f over G. In particular, the double integral

$$\iint\limits_{G} f(x, y) \, dA$$

exists, proving Theorem 2, and its value is the volume V of the space region based on G and capped by the graph of f.

The only thing that remains to be examined is our assumption that if we are given any positive number e, then any suitably "fine" subdivision will consist of subregions ΔG_i on each of which the values of f vary by less than e.

We define the *diameter* of a closed region to be the maximum distance apart of any two of its points. Thus the diameter of a square of side s is $\sqrt{2}s$, the length of its diagonal, and the diameter of a triangle is the longest of its side lengths. When we say that a subdivision is "suitably fine," we just mean that the diameters of its subregions ΔG_i are all less than a suitably small number d. The fact that we can find such a number d for a given e is a new property of a continuous function on a closed domain. Here is a direct statement of this property without any reference to subdivisions.

THEOREM 4   *If $f(x, y)$ is continuous on a bounded closed region G, and if we are given any positive number e, then we can find a positive number d that is suitably small in the following sense: Whenever two points (x, y) and (x', y') in G are closer together than the distance d, then*

$$|f(x, y) - f(x', y')| < e.$$

This property says that we have *uniform* control over the continuous variation of $f(x, y)$ over the whole of G. The theorem states that any function $f(x, y)$ that is continuous on a bounded closed region G is necessarily *uniformly* continuous in the above sense. The proof of this theorem is beyond the level of a first course, but we can verify it for the situations that usually arise in practice. (See Problems 1 and 2.) In any event, if we are given e we can now proceed as follows. First we find the above uniform control number d. Then we subdivide G into subregions ΔG_i each of which has diameter less than d. Then the variation of f is less than e on each subregion ΔG_i and the proof of Theorem 2 is complete.

PROBLEMS FOR SECTION 3

We have been using the diameter of a closed region G as a measure of how big G is, in connection with Riemann sum subdivisions. There is a refinement of this measure that we might call the *length* of G. We say that

$$\text{length of } G \leq l$$

if any two points of G can be connected by a path that lies in G and has length less than or equal to l. If, in addition, there is a pair of points in G for which the shortest connecting path has length l, then the length of G is exactly l.

1. Now let G be a region whose length is less than l, and let $f(x, y)$ be continuously differentiable on G, with both partial derivatives bounded by a constant K. Prove that

$$|f(x_1, y_1) - f(x_2, y_2)| \leq \sqrt{2}Kl$$

   for any two points (x_1, y_1) and (x_2, y_2) in G.

2. Let f be as in the problem above and let G be a convex region. This means that if $\mathbf{x}_1$ and $\mathbf{x}_2$ are any two points of G, then the line segment joining them lies entirely in G. Prove that f is uniformly continuous on G.

3. Suppose that G is a region that can be subdivided into subregions ΔG_i having arbitrarily small length. Let $f(x, y)$ be a positive continuously differentiable function on G, with both partial derivatives bounded by K.

   a) Show first that if G is divided into subregions ΔG_i all having length less than e, and if (x_i, y_i) is an arbitrary evaluation point in ΔG_i for each i, then the volume V of the solid based on G and capped by the surface $z = f(x, y)$ is given approximately by

$$V \approx \sum f(x_i, y_i) \, \Delta A_i,$$

   with an error less than $\sqrt{2}KAe$. (A is the area of G and ΔA_i is the area of ΔG_i.)

   b) Conclude that the double integral $\iint_G f$ exists, where the limit of the Riemann sum is taken in a sense slightly different from before.

4. THE MASS OF A PLANE LAMINA

In Chapter 4 we defined the density of a lamina at the point (x_0, y_0) as the limit

$$\rho(x_0, y_0) = \lim \frac{\Delta m}{\Delta A},$$

where Δm and ΔA are, respectively, the mass and area of an incremental region ΔG containing the point (x_0, y_0), and where the limit is taken as ΔG shrinks down on this point, that is, as the diameter of ΔG approaches 0.

Thus

$$\rho(x_0, y_0) \approx \frac{\Delta m}{\Delta A},$$

where the error can be made as small as desired by taking the diameter of ΔG suitably small.

Density is thus a sort of derivative of mass with respect to area. We shall now see that the mass m can be recovered from its "two-dimensional derivative" $\rho(x, y)$ by two-dimensional integration. That is, the mass of the piece of the lamina occupying a region G is given by the double integral of ρ over G,

$$\iint_G \rho(x, y)\, dA.$$

To begin with, we subdivide the lamina G into small sublaminas ΔG_i, and choose an evaluation point (x_i, y_i) in ΔG_i for each i. Then,

$$\frac{\Delta m_i}{\Delta A_i} \approx \rho(x_i, y_i)$$

for each i. We assume that we can make all the errors simultaneously less than a preassigned positive number e by choosing the subdivision fine enough (i.e., by taking the diameters of the subregions ΔG_i all less than some suitably small number). Then, for each index i,

$$\Delta m_i \approx \rho(x_i, y_i)\, \Delta A_i,$$

with an error less than $e\, \Delta A_i$. The sum of the errors is thus less than eA, where A is the area of G. Therefore, the total mass m of the region G is given approximately by

$$m \approx \sum_{i=1}^{n} \rho(x_i, y_i)\, \Delta A_i,$$

with an error at most eA. And we can make this approximation as good as we wish, since in the beginning we can take e to be *any* positive number, however small.

So far we have been looking at the sum $\sum \rho(x_i, y_i)\, \Delta A_i$ as an estimate of the total mass m of the lamina. But this sum is also a Riemann sum approximating the double integral $\iint_G \rho(x, y)\, dA$. So the double integral and the mass m are both estimated with arbitrarily small error by the same number, and it follows that m is equal to the double integral. Thus:

THEOREM 5   *If G is a plane lamina with variable density $\rho(x, y)$, then the total mass m of G is*

$$m = \iint_G \rho(x, y)\, dA.$$

REMARK   We assumed without proof that the errors in the estimates $\Delta m_i/\Delta A_i \approx \rho(x_i, y_i)$ can be made small simultaneously, and to this extent the proof of the theorem is incomplete.

EXAMPLE   The density of a square lamina of side length $2a$ is proportional to the square of the distance from the center. Find the mass of the lamina.

Solution. If we center the square at the origin then the density $\rho(x, y)$ is given by

$$\rho(x, y) = k(x^2 + y^2),$$

where k is the constant of proportionality. This was our hypothesis. The mass of the lamina is therefore

$$m = k \int_{-a}^{a} \int_{-a}^{a} (x^2 + y^2)\, dx\, dy = k \int_{-a}^{a} \left[\frac{x^3}{3} + xy^2 \right]_{-a}^{a} dy$$

$$= 2k \int_{-a}^{a} \left(\frac{a^3}{3} + ay^2 \right) dy = 2k \left[\frac{a^3 y}{3} + \frac{ay^3}{3} \right]_{-a}^{a}$$

$$= \frac{8}{3} ka^4.$$

PROBLEMS FOR SECTION 4

In each of the following problems find the mass of the plane lamina G having the given shape and density.

1. G is the unit square with lower left vertex at the origin, and $\rho(x, y) = xy$.

2. G is the piece of the unit circle $x^2 + y^2 \le 1$ in the first quadrant and $\rho(x, y) = xy$.

3. G is the lamina between the graphs of $y = x$ and $y = x^2$ and $\rho(x, y) = \sqrt{x}$.

4. G is the parabolic cap bounded by the parabola $y = 4 - x^2$ and the x-axis. The density is proportional to the distance from the x-axis.

5. G is the same as in Problem 4, but the density is proportional to the distance from the y-axis.

6. G is the unit circle $x^2 + y^2 \le 1$ and the density is proportional to the distance from the y-axis.

7. G is the unit circle and the density is proportional to the distance from the x-axis.

8. G is the unit circle and the density is proportional to the sum of the distances from the two coordinate axes.

9. G is the square symmetric about the origin with one vertex at (a, a). Its density is proportional to the distance from the line $x + y = 0$ (that is, $\rho(x, y) = k|x + y|$.)

10. G is the piece of the first quadrant under the line $x + y = 1$ and the density is e^{x+y}.

11. G is the lamina between the graph of $y = \sin x$ and the x-axis, from $x = 0$ to $x = \pi$. The density is proportional to the distance from the y-axis.

12. G is the above lamina and the density is proportional to the distance from the x-axis.

13. G is the lamina between the graph $y = e^{-x}$ and the x-axis, from $x = 0$ to infinity; $\rho(x, y) = x$.

5. MOMENTS AND CENTER OF MASS OF A PLANE LAMINA

The proof of the following theorem is similar in spirit to the proof of the mass formula in the last section. It will be given below.

THEOREM 6 *If a plane lamina with variable density $\rho(x, y)$ occupies the region G, then the moment M of the lamina about the axis $x = r$ is given by*

$$M = \iint_G (x - r)\rho(x, y)\, dA.$$

By definition, the center of mass $(\bar{x}, \bar{y})$ of the lamina is its balancing point. In particular, the lamina balances on the axis $x = \bar{x}$, so the moment of the lamina about the axis $x = \bar{x}$ is zero. We therefore find $\bar{x}$ from the equation $M = 0$, which is

$$0 = \iint_G (x - \bar{x})\rho(x, y)\, dA = \iint_G x\rho(x, y)\, dA - \bar{x}\iint_G \rho(x, y)\, dA.$$

Solving for $\bar{x}$ in this equation, and for $\bar{y}$ in the corresponding moment equation about the axis $y = \bar{y}$, we have the corollary:

COROLLARY *The center of mass $(\bar{x}, \bar{y})$ of the above lamina has the coordinates*

$$\bar{x} = \frac{\displaystyle\iint_G x\rho(x, y)\, dA}{\displaystyle\iint_G \rho(x, y)\, dA} = \frac{\text{moment about the } y\text{-axis}}{\text{total mass } m},$$

$$\bar{y} = \frac{\displaystyle\iint_G y\rho(x, y)\, dA}{\displaystyle\iint_G \rho(x, y)\, dA} = \frac{\text{moment about the } x\text{-axis}}{\text{total mass } m}.$$

EXAMPLE Find the center of mass of the square G having lower left vertex at the origin and upper right vertex at (a, a), if its density is proportional to the square of the distance from the origin.

Solution. If k is the constant of proportionality for the density,

$$\rho(x, y) = k(x^2 + y^2),$$

then the calculation in the example in the last section shows that the mass of the square lamina is $2ka^4/3$.

For the moment about the y-axis, we have

$$\int_0^a \int_0^a x \cdot k(x^2 + y^2) \, dx \, dy = k \int_0^a \left[\frac{x^4}{4} + \frac{x^2 y^2}{2} \right]_0^a dy$$

$$= k \int_0^a \left(\frac{a^4}{4} + \frac{a^2 y^2}{2} \right) dy = k \left[\frac{a^4 y}{4} + \frac{a^2 y^3}{6} \right]_0^a$$

$$= k \left(\frac{a^5}{4} + \frac{a^5}{6} \right) = \frac{5}{12} ka^5.$$

Thus

$$\bar{x} = \frac{\text{moment about y-axis}}{\text{total mass}} = \frac{\frac{5}{12} ka^5}{\frac{2}{3} ka^4} = \frac{5}{8} a.$$

Finally, since the whole configuration is symmetric in x and y, we also have $\bar{y} = 5a/8$.

We turn now to the proof of Theorem 6.

According to our earlier discussion of the general nature of moments (Section 8 of Chapter 9), the moment about a line parallel to the y-axis will decrease if the mass of the lamina is shifted to the left, and will increase if the mass is moved to the right. Therefore, if ΔG is an incremental piece of the lamina lying over an interval $[x', x'']$ on the x-axis, and if the mass Δm on ΔG is slid left and accumulated at a point on the line $x = x'$, then the moment of ΔG about the axis $x = r$ is decreased to the moment of this point mass, which is $(x' - r) \Delta m$. That is

$$(x' - r) \Delta m \le \Delta M,$$

where ΔM is the moment of ΔG.

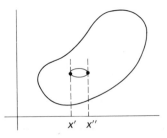

Similarly, the moment of ΔG is increased by sliding the mass of ΔG to the right and accumulating it at a point on the line $x = x''$. So, altogether, we have the moment ΔM of ΔG squeezed between these two point mass moments

$$(x' - r) \Delta m \le \Delta M \le (x'' - r) \Delta m.$$

Now let (x_0, y_0) be any "evaluation point" in ΔG. Then $x' \le x_0 \le x''$, so

$$(x' - r) \Delta m \le (x_0 - r) \Delta m \le (x'' - r) \Delta m.$$

Comparing these two inequalities shows that

$$\Delta M \approx (x_0 - r)\,\Delta m,$$

with an error at most $(x'' - x')\,\Delta m$ in magnitude. Since $x'' - x'$ is the width of the incremental lamina ΔG, we can restate the above estimate as follows:

If the width of the incremental lamina ΔG is less than d, then

$$\Delta M \approx (x_0 - r)\,\Delta m$$

with an error less than $d\,\Delta m$.

We now proceed just as we did when studying the mass distribution over G. We subdivide G into small incremental sublaminas ΔG_i, and choose an arbitrary evaluation point (x_i, y_i) in each piece ΔG_i. We suppose that the widths of the sublaminas ΔG_i are all less than some number d. Then we can apply the above estimate of the moment ΔM_i of ΔG_i in terms of its mass Δm_i and its evaluation point (x_i, y_i), and have

$$\Delta M_i \approx (x_i - r)\,\Delta m_i,$$

with an error less than $d\,\Delta m_i$, for each index i. The total moment M of the lamina is the sum $\sum \Delta M_i$ of its incremental moments, and its total mass m is the sum $\sum \Delta m_i$ of its incremental masses. Finally, the sum of the errors in the above approximations is less than

$$\sum d\,\Delta m_i = d \sum \Delta m_i = dm.$$

So when we add up these estimates of the incremental moments we end up with the approximation

$$M \approx \sum (x_i - r)\,\Delta m_i,$$

with an error less than dm. Note that this error can be made as small as we wish by taking the maximum width d of the subregions ΔG_i suitably small.

We now recall that $\Delta m_i \approx \rho(x_i, y_i)\,\Delta A_i$, with an error less than $e\,\Delta A_i$ (when the diameter of ΔG_i is suitably small). So

$$(x_i - r)\,\Delta m_i \approx (x_i - r)\rho(x_i, y_i)\,\Delta A_i,$$

with an error less than $|x_i - r|e\,\Delta A_i$. Let D be the maximum value $|x_i - r|$ can have. Thus, D is the maximum distance from the axis $x = r$ to points of the lamina G. Then the error in the last estimate is less than $De\,\Delta A_i$, and the sum of these errors is less than $\sum De\,\Delta A_i = De \sum \Delta A_i = DeA$. Thus

$$\sum (x_i - r)\,\Delta m_i \approx \sum (x_i - r)\rho(x_i, y_i)\,\Delta A_i,$$

with an error at most DeA. Here again, the error can be made as small as we wish by choosing e suitably small. And when we combine this estimate with the one for M above, we see that

$$M \approx \sum (x_i - r)\rho(x_i, y_i)\,\Delta A_i,$$

with an error that can be made arbitrarily small (by choosing the subdividing regions ΔG_i small enough).

Since the sum on the right is also a Riemann sum for the double integral $\iint_G (x - r)\rho(x, y)\,dA$, we conclude as before that M is given exactly by this double integral. (The double integral and the moment M are each approximated arbitrarily closely by the same estimating number, so they must be equal.) This completes the proof of the theorem. ∎

PROBLEMS FOR SECTION 5

In each of the following problems find the center of mass of the lamina G having the given shape and density. Assume the given value of the mass.

1. G is the unit square with lower left vertex at the origin, and $\rho(x, y) = xy$. Its mass is 1/4.

2. G is the piece of the unit circle $x^2 + y^2 \le 1$ in the first quadrant and $\rho(x, y) = xy$. Its mass is 1/8.

3. G is the lamina between the graphs of $y = x$ and $y = x^2$ and $\rho(x, y) = \sqrt{x}$. The mass is 4/35.

4. G is the parabolic cap bounded by the parabola $y = 4 - x^2$ and the x-axis. The density is proportional to the distance from the x-axis. The mass is $256k/15$, where k is the constant of proportionality for the density.

5. G is the same as in Problem 4, but the density is proportional to the distance from the y-axis. The mass is $8k$.

6. G is the right half of the unit circle $x^2 + y^2 \le 1$ and the density is proportional to the distance from the y-axis. The mass is $2k/3$.

7. G is the right half of the unit circle and the density is proportional to the distance from the x-axis. The mass is $2k/3$.

8. G is the first quadrant of the unit circle and the density is proportional to the sum of the distances from the two coordinate axes. The mass is $2k/3$.

9. G is the square symmetric about the origin with one vertex at (a, a). Its density is proportional to the distance from the line $x + y = 0$ (that is, $\rho(x, y) = k|x + y|$. The mass is $8ka^3/3$.

10. G is the piece of the first quadrant under the line $x + y = 1$ and the density is e^{x+y}. The mass is 1.

11. G is the lamina between the graph of $y = \sin x$ and the x-axis, from $x = 0$ to $x = \pi$. The density is proportional to the distance from the y-axis. The mass is πk.

12. G is the above lamina and the density is proportional to the distance from the x-axis. The mass is $\pi k/4$.

13. G is the lamina between the graph $y = e^{-x}$ and the x-axis, from $x = 0$ to infinity; $\rho(x, y) = x$. The mass is 1.

6. THE DOUBLE INTEGRAL IN POLAR COORDINATES

Let G be a region in the xy-plane lying between the polar graphs

$$r = g(\theta) \quad \text{and} \quad r = h(\theta),$$

from $\theta = \alpha$ to $\theta = \beta$, as at the left above. With θ restricted to lie between α

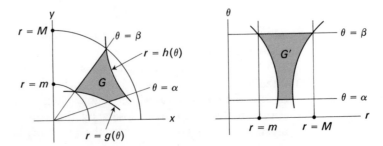

and β, each point (x, y) in G determines unique polar coordinates r and θ, where $r = (x^2 + y^2)^{1/2}$ and θ is the unique angle between α and β such that $(\sin \theta, \cos \theta) = (x/r, y/r)$. Thus each point (x, y) in G determines a unique point (r, θ) in the Cartesian plane (with axes labeled r and θ). These (r, θ)-points fill out the region G' bounded by the graphs of $r = g(\theta)$ and $r = h(\theta)$, and the horizontal lines $\theta = \alpha$ and $\theta = \beta$, as shown in the righthand figure above. We say that G' is the image of G under the mapping $(x, y) \rightarrow (r, \theta)$. The areas of G and G' are different, but we know how to compute them both (Section 6 in Chapter 4 and Section 5 in Chapter 10). They are

$$A = \int_\alpha^\beta \frac{[h^2(\theta) - g^2(\theta)]}{2} \, d\theta$$

and

$$A' = \int_\alpha^\beta [h(\theta) - g(\theta)] \, d\theta.$$

There is a simple comparison between these two areas that seems very crude, but it becomes more precise when applied to the incremental subregions of a subdivision.

LEMMA 1 *There is a point (r_0, θ_0) in the region G' having the ratio A/A' as its first coordinate. That is,*

$$A = r_0 A'.$$

where r_0 is the r-coordinate of some point (r_0, θ_0) in the region G'.

Proof. If m and M are the minimum and maximum values of the function $(h + g)/2$ over the integration interval $[\alpha, \beta]$, then the A integrand above lies between $m(h - g)$ and $M(h - g)$. This inequality becomes

$$mA' \leq A \leq MA'$$

upon integrating. Thus A/A' lies between m and M, so it is a value of the function $(h + g)/2$ at some point θ_0, by the intermediate-value principle. (This argument is a special case of the proof of the weighted mean-value theorem for integrals—see page 362, Problem 41.) We now have

$$A = \frac{h(\theta_0) + g(\theta_0)}{2} A',$$

or

$$A = r_0 A',$$

where $r_0 = [h(\theta_0) + g(\theta_0)]/2$. Moreover, the point

$$(r_0, \theta_0) = ([h(\theta_0) + g(\theta_0)]/2, \theta_0)$$

is in G'; in fact, it is the midpoint of the segment crossing G' from the left boundary point $(g(\theta_0), \theta_0)$ to the right boundary point $(h(\theta_0), \theta_0)$. The lemma is thus proved. ∎

The role played by the conversion factor r_0 in passing from A' to A becomes clear when a small region near the origin is compared with one further away (see figure).

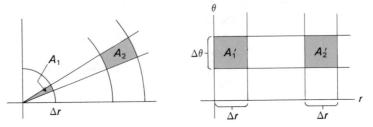

The above lemma is the heart of the following theorem.

THEOREM 7   *If $f(x, y)$ is continuous on the bounded closed region G, and if we define $F(r, \theta)$ by*

$$F(r, \theta) = f(r \cos \theta, r \sin \theta),$$

then

$$\iint_G f(x, y) \, dA = \iint_{G'} F(r, \theta) r \, dA',$$

where G' is the image of G under the mapping $(x, y) \to (r, \theta)$.

Proof. If we subdivide G into small subregions ΔG_i, then their images $\Delta G_i'$ under the mapping $(x, y) \to (r, \theta)$ form a corresponding subdivision of G'. For each i, we choose an evaluation point (r_i, θ_i) in $\Delta G_i'$ such that $\Delta A_i = r_i \Delta A_i'$ (by the lemma), and we take (x_i, y_i) as the corresponding point in ΔG_i,

$$(x_i, y_i) = (r_i \cos \theta_i, r_i \sin \theta_i).$$

Then,

$$\iint_G f(x, y) \, dA \approx \sum f(x_i, y_i) \, \Delta A_i,$$

$$\iint_{G'} F(r, \theta) r \, dA' \approx \sum F(r_i, \theta_i) r_i \, \Delta A_i',$$

where each Riemann-sum approximation can be made as good as desired by choosing the subdivision fine enough. But the two Riemann sums are equal, because

$$f(x_i, y_i) = f(r_i \cos \theta_i, r_i \sin \theta_i) = F(r_i, \theta_i),$$

and

$$\Delta A_i = r_i \, \Delta A_i'$$

by the choice of r_i. The two double integrals are thus both approximated as closely as desired by the same number, so they must be equal. ∎

The theorem above is the simplest case of a "change of variables" in a multiple integral, analogous to the substitution law for single integrals. The general theory of such transformations of multiple integrals is studied in advanced calculus. It is an important subject, with many applications. For example, the present theorem provides us with a new way to evaluate a double integral by iterated integration.

THEOREM 8 *Let $f(x, y)$ be continuous on a bounded closed region G, and suppose that the boundary of G lies along the polar graphs $r = g(\theta)$, and $r = h(\theta)$, and the rays $\theta = \alpha$ and $\theta = \beta$, where $\alpha < \beta$ and $g(\theta) < h(\theta)$ for all θ between α and β. Then*

$$\iint\limits_{G} f(x, y)\, dA = \int_{\alpha}^{\beta}\left[\int_{g(\theta)}^{h(\theta)} F(r, \theta) r\, dr\right] d\theta,$$

where $F(r, \theta) = f(r \cos \theta, r \sin \theta)$.

Proof. This is an immediate corollary of Theorems 3 and 7. ∎

EXAMPLE 1 Compute the double integral

$$\int_{G} e^{-(x^2+y^2)}\, dA,$$

where G is the circular disk $x^2 + y^2 \le a^2$.

Solution. In polar coordinates G can be described by the conditions $0 \le r \le a$, $0 \le \theta \le 2\pi$. Moreover,

$$e^{-(x^2+y^2)} = e^{-r^2}.$$

Thus

$$\int_{G} e^{-(x^2+y^2)}\, dA = \int_{0}^{2\pi}\left[\int_{0}^{a} e^{-r^2} r\, dr\right] d\theta$$

$$= \int_{0}^{2\pi} [-e^{-r^2}/2]_0^a\, d\theta$$

$$= \int_{0}^{2\pi} \frac{1 - e^{-a^2}}{2}\, d\theta = \pi(1 - e^{-a^2}).$$

This is an example of a double integral that is easier to compute than its one-dimensional analogue. In fact, the integral

$$\int_{0}^{a} e^{-x^2}\, dx$$

cannot be computed at all in terms of elementary functions.

EXAMPLE 2 Compute the volume of a ball of radius a.

Solution. The ball is bounded by the spherical surface

$$x^2 + y^2 + z^2 = a^2.$$

Moreover, the total volume of the ball is eight times the volume of the piece of the ball in the first octant. Thus

$$V = 8\iint\limits_{G} \sqrt{a^2 - (x^2 + y^2)}\, dA,$$

where G is the part of the circular disk $x^2 + y^2 \le a^2$ lying in the first quadrant. By Theorem 8, we can compute this double integral as the following iterated integral in polar coordinates:

$$V = 8 \int_0^{\pi/2} \left[\int_0^a \sqrt{a^2 - r^2}\, r \, dr \right] d\theta$$

$$= 8 \int_0^{\pi/2} \left[-\frac{1}{3}(a^2 - r^2)^{3/2} \right]_0^a d\theta$$

$$= \frac{8a^3}{3} \int_0^{\pi/2} d\theta = \frac{8a^3}{3} \cdot \frac{\pi}{2} = \frac{4}{3}\pi a^3.$$

Here again we are helped by the extra r in the integrand of the polar coordinate integral.

EXAMPLE 3   Find the center of mass of a homogeneous semicircular disk of radius R.

Solution. Assuming that the lamina occupies the right half of the circle $x^2 + y^2 \le R^2$ and has constant density 1, its moment M about the y-axis is $\iint_G x \, dA$, where G can be described in terms of polar coordinates by the conditions that r runs from 0 to R and θ runs from $-\pi/2$ to $\pi/2$. Also $x = r \cos \theta$, so by Theorem 8,

$$M = \iint_G x \, dA = \int_{-\pi/2}^{\pi/2} \left[\int_0^R (r \cos \theta) r \, dr \right] d\theta$$

$$= \int_{-\pi/2}^{\pi/2} \left[\frac{r^3 \cos \theta}{3} \right]_{r=0}^{r=R} d\theta = \frac{R^3}{3} \int_{-\pi/2}^{\pi/2} \cos \theta \, d\theta$$

$$= \frac{R^3}{3} [\sin \theta]_{-\pi/2}^{\pi/2} = \frac{2}{3} R^3.$$

The x-coordinate of the center of mass is the moment about the y-axis divided by the area (= the mass):

$$\bar{x} = \frac{M}{A} = \frac{\dfrac{2}{3} R^3}{\dfrac{\pi R^2}{2}} = \frac{4}{3} \frac{R}{\pi}.$$

The disk will balance on the x-axis, by symmetry, so $\bar{y} = 0$.

PROBLEMS FOR SECTION 6

1. Find the volume of the solid in the half-space $z \ge 0$ based on the circular disk $x^2 + y^2 \le 9$ and capped by the sphere $x^2 + y^2 + z^2 = 25$.

2. Find the volume of the solid cut out of the spherical ball $x^2 + y^2 + z^2 \le b^2$ by the cylinder $x^2 + y^2 = a^2$, where $0 < a \le b$.

3. Find the volume of the cone $0 \le z \le 1 - r$.

4. Find the volume of the cone $0 \le z \le (a - r)h/a$.

5. Find the volume of the solid based on the interior of the cardioid $r = 1 + \cos \theta$ and capped by the cone $z = 2 - r$.

6. Find the volume of the solid based on the interior of the circle $r = \cos \theta$ and capped by the cone $z = 1 - r$.

7. Find the volume of the solid based on the interior of the circle $r = \cos \theta$ and capped by the sphere $r^2 + z^2 = 1$.

8. Find the volume of the solid based on the interior of the circle $r = \cos \theta$ and capped by the plane $z = x$.

9. Show that the mass of a circular lamina of radius R whose density is proportional to the distance from the center is $2\pi k R^3/3$, where k is the constant of proportionality.

10. Compute the center of mass of the right half of the above lamina.

11. Compute the mass and center of mass of the semicircular lamina $0 \le r \le 1$, $0 \le \theta \le \pi$ if the density is proportional to $r(1 - r)$.

12. Compute the mass and center of mass of the semicircular lamina $0 \le r \le 1$, $0 \le \theta \le \pi$, if its density is proportional to $\sqrt{1 - r^2}$.

13. Find the volume of the solid obtained by intersecting the cylinders

$$x^2 + y^2 = a^2 \qquad \text{and} \qquad x^2 + z^2 = a^2.$$

7. SURFACE AREA

Let π and P be two planes that intersect at an angle θ. Their line of intersection will be called their axis.

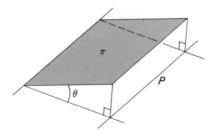

When we project from π vertically down onto P, lengths parallel to the axis remain unchanged, but lengths perpendicular to the axis are multiplied by $\cos \theta$, and hence are decreased by this constant factor.

It follows that if a region R on π is projected vertically downward to P, then it is compressed by the factor $\cos \theta$ in the direction perpendicular to the axis, and its area is therefore decreased by the factor $\cos \theta$:

$$\text{area of image region} = \cos \theta(\text{area of } R).$$

We assume this as self-evident. However, a proof could be constructed by starting with rectangles having one side parallel to the axis, and then approximating more general regions by finite collections of such rectangles, somewhat in the manner of our Riemann sum approximations. We shall use this relationship in reverse.

LEMMA 2   *If G is a region in the xy-plane, and if π is the plane $z = ax + by + c$, then the area of the region on π lying over G is*

$$\sqrt{1 + a^2 + b^2}(\text{area of } G).$$

Proof. We know that the vector $(-a, -b, 1)$ is normal to π, and that if θ is the angle between this vector and the z-axis, then

$$\cos \theta = \frac{1}{\sqrt{1 + a^2 + b^2}}.$$

(See Section 5 in Chapter 13.) Since θ is also the angle between π and the xy-plane, a region G' on π has its area decreased by the factor $\cos \theta = 1/\sqrt{1 + a^2 + b^2}$ when it is projected vertically downward to the xy-plane, becoming the region G on the xy-plane. That is,

$$\text{area of } G = \frac{1}{\sqrt{1 + a^2 + b^2}} \, (\text{area of } G'),$$

and the lemma is just the cross-multiplied form of this equation. ∎

Now consider the piece of the surface $z = f(x, y)$ lying over the region G in the xy-plane. Its tangent plane at $z_0 = f(x_0, y_0)$,

$$z - z_0 = \frac{\partial z}{\partial x}\bigg|_{(x_0, y_0)} (x - x_0) + \frac{\partial z}{\partial y}\bigg|_{(x_0, y_0)} (y - y_0),$$

remains very nearly parallel to the surface near the point of tangency. So if ΔG is an incremental subregion containing the evaluation point (x_0, y_0), and

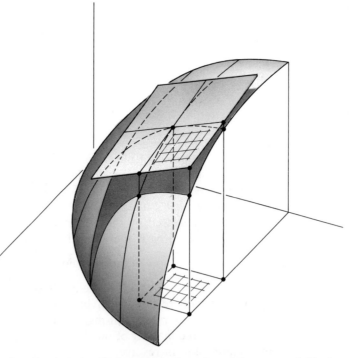

if ΔS is the area of the little piece of surface lying over ΔG, then ΔS is approximately equal to the corresponding area on the tangent plane, which is

$$\sqrt{1 + \left(\frac{\partial z}{\partial x}\right)_0^2 + \left(\frac{\partial z}{\partial y}\right)_0^2} \; \Delta A,$$

by the lemma. We have used the abbreviation $(\partial z/\partial x)_0$ for the value of the derivative at (x_0, y_0). It seems intuitively clear that this approximation improves as ΔA approaches 0, in the sense that the ratio of the two areas approaches 1. This is the same thing as saying that

$$\Delta S \approx \sqrt{1 + \left(\frac{\partial z}{\partial x}\right)_0^2 + \left(\frac{\partial z}{\partial y}\right)_0} \; \Delta A,$$

with an error that is small compared to ΔA. Therefore if we subdivide G into suitably small subregions ΔG_i and add up the above estimates, we see that the area S of the surface over G is given approximately by

$$S \approx \sum \sqrt{1 + \left(\frac{\partial z}{\partial x}\right)_i + \left(\frac{\partial z}{\partial y}\right)_i} \; \Delta A_i,$$

with an error that is small compared to the area A of G. The partial derivatives are evaluated at arbitrarily chosen points (x_i, y_i) in ΔG_i.

The discussion above is not a real proof, but it should seem plausible and it is the best we can do for surface area.

Since the sum on the right above is a Riemann sum for the function $\sqrt{1 + (\partial z/\partial x)^2 + (\partial z/\partial y)^2}$, we conclude as usual that S is given exactly by the double integral of this function over G.

We have thus shown that the following theorem is a consequence of plausible geometric assumptions.

THEOREM 9 *If $f(x, y)$ is a continuously differentiable function whose domain includes the region G, then the area of the piece of the surface $z = f(x, y)$ lying over G is*

$$\iint\limits_{G} \sqrt{1 + \left(\frac{\partial z}{\partial x}\right)^2 + \left(\frac{\partial z}{\partial y}\right)^2} \; dA.$$

EXAMPLE Find the formula for the area of the surface of a sphere.

Solution. The sphere of radius R about the origin has the equation

$$x^2 + y^2 + z^2 = R^2.$$

In order to compute the partial derivatives $\partial z/\partial x$ and $\partial z/\partial y$, we can solve for z and differentiate, or we can simply differentiate implicitly in the equation as it stands. The latter method gives

$$2x + 2z \frac{\partial z}{\partial x} = 0,$$

so $\partial z/\partial x = -x/z$. Similarly, $\partial z/\partial y = -y/z$. Since the total surface area S is eight times the area in the first octant, we have

$$S = 8 \iint\limits_{G} \sqrt{1 + \left(\frac{\partial z}{\partial x}\right)^2 + \left(\frac{\partial z}{\partial y}\right)^2} \; dA$$

$$= 8 \iint\limits_{G} \sqrt{1 + \left(\frac{x}{z}\right)^2 + \left(\frac{y}{z}\right)^2} \; dA,$$

where G is the quarter circle in the first quadrant. If we write 1 as z/z in the integrand, we see that

$$S = 8 \iint_G \frac{R}{z} \, dA.$$

In polar coordinates, $z = \sqrt{R^2 - (x^2 + y^2)} = \sqrt{R^2 - r^2}$, so

$$\iint_G \frac{R}{z} \, dA = R \int_0^{\pi/2} \left[\int_0^R \frac{r \, dr}{\sqrt{R^2 - r^2}} \right] d\theta$$

$$= R \int_0^{\pi/2} [-\sqrt{R^2 - r^2}]_0^R \, d\theta = R^2 \int_0^{\pi/2} d\theta$$

$$= \frac{\pi}{2} R^2.$$

The total area S is eight times this, so the formula for the area of the surface of a sphere of radius R is

$$S = 4\pi R^2.$$

PROBLEMS FOR SECTION 7

Find the area of each of the following surfaces.

1. The piece of the paraboloid $z = 4 - x^2 - y^2$ that lies above the xy-plane. (Polar coordinates are simplest.)

2. The surface cut out of the cone $z = 1 - r$ by the cylinder $r = \cos \theta$.

3. The piece of the sphere $x^2 + y^2 + z^2 = 25$ lying over the circular disk $x^2 + y^2 \le 16$.

4. The piece of the sphere $x^2 + y^2 + z^2 = R^2$ cut out by the cylinder $x^2 + y^2 = a^2$ and lying above the xy-plane.

5. The piece of the sphere $x^2 + y^2 + z^2 = 1$ lying over the interior of the circle $r = \cos \theta$.

6. The piece of the plane $z = 6 - 2x - 3y$ lying over the unit square with lower left corner at the origin.

7. The piece of the above plane that is cut off by the first octant.

8. The piece of the plane $z = ax + by + c$ lying over the region G. (That is, prove that the Lemma is a special case of Theorem 9.)

9. The cone of altitude h and base radius a.

10. When $r \le a$, the equation

$$az = h(a - r)$$

describes the cone whose altitude is h and whose base is the circular disk $r \le a$. Let G be any region lying in this disk. Show that the area of the part of the cone lying over G is

$$\sqrt{1 + \left(\frac{h}{a}\right)^2} \text{ (area of } G\text{)}.$$

8. ITERATED TRIPLE INTEGRATION

Suppose that a solid S, occupying a region G in space, has a variable continuous density $\rho(x, y, z)$. When we consider how to estimate the total mass m of S in terms of its density function ρ, we are led to an approximation of m by a finite sum $\sum \rho(x_i, y_i, z_i) \Delta V_i$ having all the earmarks of a Riemann sum, except for involving incremental volumes ΔV_i rather than the incremental areas ΔA_i (or incremental lengths Δx_i) of our earlier discussions. Then m is given exactly by the limit of this three-dimensional Riemann sum as the incremental volumes ΔV_i approach zero in a suitable manner. This limit is called the *triple integral* of ρ *over* G, and is designated $\iiint_G \rho(x, y, z)\, dV$. Thus, the mass m of S is the triple integral of its density function ρ:

$$ m = \iiint_G \rho(x, y, z)\, dV. $$

The development outlined above will be given in the final section, and will lead to the notion of the triple integral $\iiint_G f(x, y, z)\, dV$ for any continuous function f. The reasoning will be similar to that used earlier for the double integral. Again, it will turn out that the multiple integral can be evaluated by iterated integration, that is by "integrating out" the variables one at a time. We shall assume this for the moment, and turn to the practical problem of determining the limits of integration. They occur at the boundary of the integration domain G, and a sketch of G is almost essential in order to get them right.

EXAMPLE 1 Use iterated integration to evaluate the triple integral

$$ \iiint_G x\, dV, $$

where G is the region cut off from the first octant by the plane $x + y + z = 2$.

Solution. The region G is shown in the figure below. Suppose that we decide to integrate first with respect to z. This means holding x and y fixed and integrating the resulting function of z. Geometrically, the domain of this integration is a vertical line segment crossing G from the xy-plane to the plane $x + y + z = 2$. Along this segment, z varies from $z = 0$ to $z = 2 - x - y$.

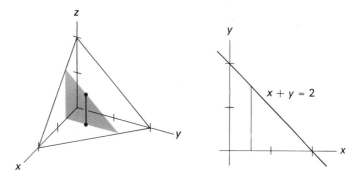

So the first integral is

$$\int_0^{2-x-y} x\,dz = xz \Big]_{z=0}^{z=2-x-y}$$

$$= x(2 - x - y).$$

With z integrated out, we have left a double integral of a function of x and y, over the region R in the xy-plane that is the base of the original space region G. Here R is the triangle cut off in the first quadrant by the line $x + y = 2$. If we choose to integrate next with respect to y, the calculation finishes off as follows:

$$\int_0^2 \int_0^{2-x} x(2 - x - y)\,dy\,dx = \int_0^2 \left[-x\frac{(2 - x - y)^2}{2} \right]_0^{2-x} dx$$

$$= \frac{1}{2} \int_0^2 x(2 - x)^2\,dx$$

$$= \frac{1}{2} \int_0^2 (4x - 4x^2 + x^3)\,dx$$

$$= \frac{1}{2} \left[2x^2 - \frac{4x^3}{3} + \frac{x^4}{4} \right]_0^2$$

$$= 4 - \frac{16}{3} + 2 = \frac{2}{3}.$$

In proving that a triple integral can be evaluated by iterated integration, the key step is a *piercing* formula, expressing a triple integral as a double integral of a varying single integral, as follows:

We choose any axis, and call it, say, the t-axis. Let π be the base plane $t = 0$, and let R be the base of the solid G in π. That is, R is obtained by projecting all the points of G perpendicularly onto π. For each point p in R, we suppose that the line through p perpendicular to π (and therefore parallel to the t-axis) intersects the space region G in an interval $[T_1, T_2]$. This interval will vary with p, so what we really are saying is that G runs from a lower surface $t = T_1(p)$ to an upper surface $t = T_2(p)$. The piercing formula evaluates the triple integral of f over G by first forming the one-dimensional integral of f over the interval $[T_1(p), T_2(p)]$ for each p, $\int_{T_1(p)}^{T_2(p)} f\,dt$, and then computing the double integral over R of the resulting function of p:

$$\iiint_G f\,dV = \iint_R \left[\int_{T_1(p)}^{T_2(p)} f\,dt \right] dA.$$

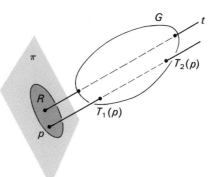

We shall prove this formula in Section 10. By taking the t-axis in the piercing formula as the z-axis, we obtain the iteration evaluation procedure we were using in Example 1. In the same way, we can justify integrating out the variables x, y, z in any other iteration order.

Here is a formal statement of this evaluation identity. It is proved simply by inserting the double-integral evaluation formula of Theorem 3 into the piercing formula.

THEOREM 10 *Let G be the space region based on the region R in the xy-plane and capped by the graph $z = \phi(x, y)$. Suppose also that R is bounded in the xy-plane by the lower graph $y = g(x)$, the upper graph $y = h(x)$, and the lines $x = a$ and $x = b$. Then*

$$\iiint\limits_{G} f(x, y, z) \, dV = \int_a^b \int_{g(x)}^{h(x)} \int_0^{\phi(x,y)} f(x, y, z) \, dz \, dy \, dx.$$

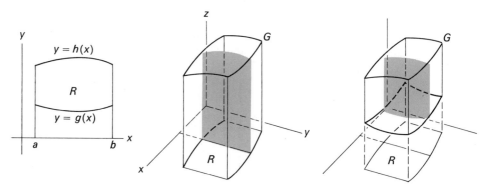

A slightly more general domain is shown in the righthand figure. Here, G lies over the same base region R in the xy-plane, but runs from the lower surface $z = \psi(x, y)$ to the upper surface $z = \phi(x, y)$. Now,

$$\iiint\limits_{G} f \, dV = \int_a^b \int_{g(x)}^{h(x)} \int_{\psi(x,y)}^{\phi(x,y)} f \, dz \, dy \, dx.$$

EXAMPLE 2 Suppose G is the region between the parabolic surfaces $z = x^2 + y^2$ and $z = 2 - (x^2 + y^2)$. The intersection curve of these surfaces is the circle

$x^2 + y^2 = 1$ in the plane $z = 1$. The part of G in the first octant is shown below. The base region R in the xy-plane is the interior of the unit circle, with y ranging therefore from $y = -\sqrt{1 - x^2}$ to $y = \sqrt{1 - x^2}$. Thus

$$\iiint\limits_{G} f\, dV = \int_{-1}^{1} \int_{-\sqrt{1-x^2}}^{\sqrt{1-x^2}} \int_{x^2+y^2}^{2-(x^2+y^2)} f\, dz\, dy\, dx.$$

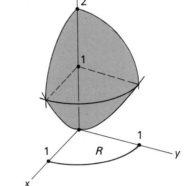

EXAMPLE 3   Use iterated triple integration to find the volume of the tetrahedral region of Example 1.

Solution. The volume of a space region G is simply the triple integral over G of the constant 1,

$$V = \iiint\limits_{G} 1\, dV$$

(because the volume equals the total mass of a mass distribution having constant density one). Thus

$$V = \int_{0}^{2} \int_{0}^{2-x} \int_{0}^{2-x-y} dz\, dy\, dx$$

$$= \int_{0}^{2} \int_{0}^{2-x} (2 - x - y)\, dy\, dx$$

$$= \int_{0}^{2} \left[-\frac{(2 - x - y)^2}{2} \right]_{0}^{2-x} dx = \frac{1}{2} \int_{0}^{2} (2 - x)^2\, dx$$

$$= -\frac{1}{6} (2 - x)^3 \bigg]_{0}^{2} = \frac{4}{3}.$$

There is nothing sacred about the order in which the variables are integrated when we evaluate a multiple integral by iterated integration, and sometimes a different order will make for an easier calculation, as we have already seen in the double-integral sections. Here is a triple-integral example.

EXAMPLE 4 Compute the volume of the wedge cut from the cylinder $y^2 + z^2 = 1$ in the first octant by the planes $y = x$ and $x = 0$.

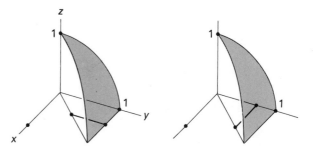

Solution. In the order we have been using so far, this sets up as the iterated integral

$$V = \int_0^1 \int_x^1 \int_0^{\sqrt{1-y^2}} dz \, dy \, dx = \int_0^1 \int_x^1 \sqrt{1 - y^2} \, dy \, dx,$$

and now we have to compute the rather nasty integral $\int \sqrt{1 - y^2} \, dy$. But if we interchange the order of the x and y integrations, we have

$$V = \int_0^1 \int_0^y \int_0^{\sqrt{1-y^2}} dz \, dx \, dy = \int_0^1 \int_0^y \sqrt{1 - y^2} \, dx \, dy$$

$$= \int_0^1 y\sqrt{1 - y^2} \, dy = -\frac{1}{2} \int_1^0 u^{1/2} \, du \qquad (u = 1 - y^2)$$

$$= \frac{1}{3} u^{3/2} \Big]_0^1 = \frac{1}{3}.$$

After Section 10 we would be in a position to make a Riemann-sum estimate of turning moments. We shall skip this development, however, since the reasoning proceeds exactly as in the two-dimensional case, with analogous conclusions:

The moment M about the origin of a mass distribution with density $\rho(x, y, z)$ is the vector

$$M = (M_x, M_y, M_z) = \left(\iiint x\rho \, dV, \iiint y\rho \, dV, \iiint z\rho \, dV \right).$$

The center of mass of the distribution is given by

$$(\bar{x}, \bar{y}, \bar{z}) = \frac{M}{m} = \left(\frac{M_x}{m}, \frac{M_y}{m}, \frac{M_z}{m} \right),$$

where $m = \iiint \rho \, dV$ is the total mass of the distribution.

EXAMPLE 5 Find the centroid of the tetrahedron in Example 1.

Solution. The centroid of a space region G is the center of mass of a distribution in G having a constant density k, and k may as well be taken to have the value $k = 1$, since it factors out and cancels in the quotients defining the center of mass.

The moment of the above tetrahedron about the origin thus has the x-component $M_x = \iiint_G x \, dV$, which we computed in Example 1 to be 2/3. The mass of the tetrahedron is its volume, which we found in Example 3 to be 4/3. Thus,

$$\bar{x} = \frac{M_x}{m} = \frac{2/3}{4/3} = \frac{1}{2}.$$

Moreover, the configuration is symmetric with respect to the variables x, y, and z, so $\bar{y}$ and $\bar{z}$ must also have the value 1/2. Thus

$$(\bar{x}, \bar{y}, \bar{z}) = \left(\frac{1}{2}, \frac{1}{2}, \frac{1}{2}\right) = \frac{1}{2}(1, 1, 1).$$

EXAMPLE 6 Find the centroid of the piece of the unit ball lying in the first octant.

Solution. The unit sphere has the equation $x^2 + y^2 + z^2 = 1$, so

$$M_z = \iiint_G z \, dV = \int_0^1 \int_0^{\sqrt{1-x^2}} \int_0^{\sqrt{1-x^2-y^2}} z \, dz \, dy \, dx$$

$$= \int_0^1 \int_0^{\sqrt{1-x^2}} \left[\frac{z^2}{2}\right]_0^{\sqrt{1-x^2-y^2}} dy \, dz = \frac{1}{2} \int_0^1 \int_0^{\sqrt{1-x^2}} (1 - x^2 - y^2) \, dy \, dz$$

$$= \frac{1}{2} \int_0^1 \left[(1 - x^2)y - \frac{y^3}{3}\right]_0^{\sqrt{1-x^2}} dx = \frac{1}{3} \int_0^1 (1 - x^2)^{3/2} \, dx$$

$$= \frac{1}{3} \int_0^{\pi/2} \cos^4 u \, du \qquad (\text{from } x = \sin u)$$

$$= \frac{1}{3} \int_0^{\pi/2} \left(\frac{1 + \cos 2u}{2}\right)^2 du$$

$$= \frac{1}{12} \int_0^{\pi/2} \left(1 + 2\cos 2u + \left(\frac{1 + \cos 4u}{2}\right)\right) du$$

$$= \frac{1}{12} \int_0^{\pi/2} \frac{3}{2} \, du \qquad \left(\text{since } \int_0^{\pi/2} \cos 2u \, du = 0, \text{ etc.}\right)$$

$$= \frac{\pi}{16}.$$

Since the volume of one eighth of the unit ball is $\pi/6$, we have

$$\bar{z} = \frac{M_z}{V} = \frac{\pi/16}{\pi/6} = \frac{3}{8}.$$

By symmetry, $\bar{x}$ and $\bar{y}$ have the same value, so the centroid is

$$(\bar{x}, \bar{y}, \bar{z}) = \frac{3}{8}(1, 1, 1).$$

PROBLEMS FOR SECTION 8

Find the volume of each of the following solids by triple (iterated) integration.

1. The tetrahedron cut from the first octant by the plane $3x + 2y + z = 6$.

2. The rectangular solid cut off from the first octant by the planes $x = a$, $y = b$, $z = c$.

3. The tetrahedron cut off from the first octant by the plane

$$\frac{x}{a} + \frac{y}{b} + \frac{z}{c} = 1.$$

4. The solid cut off from the first octant by the parabolic surfaces $z = 1 - x^2$ and $y = 1 - x^2$.

5. The solid cut off from the first octant by the parabolic surfaces $z = 1 - x^2$ and $x = 1 - y^2$.

6. The solid cut off from the first octant by the plane $z = x$ and the cylinder $x^2 + y^2 = 1$.

7. The solid cut off from the first octant by the plane $z = y$ and the cylinder $x^2 + y^2 = 1$.

8. The solid cut off from the first octant by the cylinders $x^2 + y^2 = 1$ and $x^2 + z^2 = 1$.

9. The solid bounded by the cylinders $x^2 + y^2 = 1$ and $y^2 + z^2 = 1$.

10. The solid bounded above by $z = 1 - x^2$ and below by $z = y^2$.

11. Compute the triple integral of $f(x, y, z) = y$ over the region G lying in the first octant between the xy-plane and the surface $z = 4 - (x^2 + y^2)$.

12. Compute the triple integral of $f(x, y, z) = xy$ over the region G cut off from the first octant by the ellipsoid

$$\frac{x^2}{a^2} + \frac{y^2}{b^2} + \frac{z^2}{c^2} = 1.$$

13. Compute the triple integral of $f(x, y, z) = xyz$ over the solid cut off from the first octant by the cylinders $x^2 + z^2 = 4$ and $x^2 + y^2 = 4$.

14. Calculate the centroid of the wedge cut out of the cylinder $x^2 + y^2 = 1$ by the plane $z = y$ above and the plane $z = 0$ below.

15. Calculate the centroid of the solid cut from the first octant by the ellipsoid

$$\frac{x^2}{a^2} + \frac{y^2}{b^2} + \frac{z^2}{c^2} = 1.$$

(Assume that the volume of the ellipsoid is $(4/3)\pi abc$.)

9. CYLINDRICAL AND SPHERICAL COORDINATES

There is another possibility for evaluating the double integral in the piercing formula. If there is some element of symmetry about the t-axis in the integrand f or the domain G, then it may be easier to compute the double integral as an iterated integral in polar coordinates, using Theorems 7 and 8 from Section 6.

For example, suppose we want to integrate the function $f(x, y, z)$ over the space region G that is based on the region R in the xy-plane and capped by the graph of $z = \phi(x, y)$. Just as in the last section, we first integrate z from 0 to $\phi(x, y)$, and have left the double integral over R

$$\int\int_R \left[\int\int_0^{\phi(x,y)} f(x, y, z)\, dz \right] dx\, dy.$$

Now suppose that R is most easily described in polar coordinates, as in Section 6. Then this remaining double integral can be computed in polar coordinates, as in that section.

EXAMPLE 1 Find the mass of the solid S described by the inequalities $x^2 + y^2 \leq 1$, $0 \leq z \leq x$, if the density of S is given by

$$\rho(x, y, z) = z\sqrt{x^2 + y^2}.$$

Solution.

$$m = \int\int\int_S \rho\, dV = \int\int\int_S z\sqrt{x^2 + y^2}\, dV$$

$$= \int\int_R \int_0^x \sqrt{x^2 + y^2}\, z\, dz\, dx\, dy$$

$$= \int\int_R \frac{x^2}{2} \sqrt{x^2 + y^2}\, dx\, dy.$$

Here the integration limits for R are most easily described in polar coordinates, as $-\pi/2 \leq \theta \leq \pi/2$, $0 \leq r \leq 1$. So we evaluate this double integral as an iterated integral in polar coordinates, by Theorem 8. This means replacing $dx\, dy$ by $r\, dr\, d\theta$, and setting $x = r\cos\theta$, $x^2 + y^2 = r^2$. We now have

$$\int\int_{R'} \frac{(r\cos\theta)^2}{2} \cdot r \cdot r\, dr\, d\theta = \int_0^1 \frac{r^4}{2}\, dr \int_{-\pi/2}^{\pi/2} \cos^2\theta\, d\theta = \frac{1}{20}\pi.$$

EXAMPLE 2 In practice, we would notice on reading the above problem that we want to replace x and y by polar coordinates and would do so from the beginning:

$$m = \int\int\int_S \rho\, dV = \int_{-\pi/2}^{\pi/2} \int_0^1 \left[\int\int_0^{r\cos\theta} rz\, dz \right] r\, dr\, d\theta \text{ etc.}$$

Here is the general statement of this procedure.

THEOREM 11 *Suppose that the space region G is based on the region R in the xy-plane and is capped by the surface $z = \phi(x, y)$. Suppose that R is bounded by the polar graphs $r = g(\theta)$ and $r = h(\theta)$ and the rays $\theta = \alpha$ and $\theta = \beta$, where $\alpha < \beta$,* *and $g(\theta) < h(\theta)$ on $[\alpha, \beta]$. Then*

$$\iiint\limits_{G} f(x,\, y,\, z)\, dV = \int_{\alpha}^{\beta} \int_{g(\theta)}^{h(\theta)} \left[\int_{0}^{\psi(r,\theta)} F(r,\, \theta,\, z)\, dz \right] r\, dr\, d\theta$$

where $F(r,\, \theta,\, z) = f(r\cos\theta,\, r\sin\theta,\, z)$, and $\psi(r,\, \theta) = \phi(r\cos\theta,\, r\sin\theta)$.

When we set $x = r\cos\theta$, $y = r\sin\theta$, and interpret r and θ as polar coordinates of the point (x, y), then we call the triple (r, θ, z) the "cylindrical" coordinates of the point $(x, y, z) = (r\cos\theta, r\sin\theta, z)$. There is a simple reason for the terminology; in this interpretation the graph of $r = a$ is the cylinder of radius a about the z-axis.

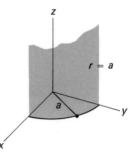

In general, the *cylindrical graph* of an equation such as $r = f(\theta, z)$, or $z = g(r, \theta)$, or $h(r, \theta, z) = 0$ consists of the points in xyz-space whose cylindrical coordinates satisfy the equation.

Also, we can interpret the triple (r, θ, z) as a new point in Cartesian $r\theta z$-space. Then Theorem 11 can be reformulated as follows.

THEOREM 12 *If G' is the region in the Cartesian $r\theta z$-space corresponding to the region G in xyz-space, and if we define F from f by $F(r,\, \theta,\, z) = f(r\cos\theta,\, r\sin\theta,\, z)$, then*

$$\iiint\limits_{G} f(x,\, y,\, z)\, dV = \iiint\limits_{G'} F(r,\, \theta,\, z) r\, dV'.$$

Proof. The image region G' consists of those points (r, θ, z) for which the point $(x, y, z) = (r\cos\theta, r\sin\theta, z)$ is in G. (Remember that only one value of θ is chosen for each pair (x, y), normally, the unique θ in $[0, 2\pi)$ such that $(\cos\theta, \sin\theta) = (x/r, y/r)$.) In the situation of Theorem 11, G' is described by the limits of integration in the iterated integral on the right. Therefore, by Theorem 10, this iterated integral is the value of the triple integral

$$\iiint\limits_{G'} F(r,\, \theta,\, z) r\, dV'.$$

We thus have two triple integrals equal to the same iterated integral, and hence to each other. ∎

The *spherical* coordinates of a point are shown in the accompanying figure. Essentially, they combine the polar angle θ in the xy-plane with polar coordinates ρ, ϕ in the *half-plane* determined by θ and the z-axis. Note that ϕ ranges from 0 to π only. Referring to the figure, we see first that $z = \rho \cos \phi$ and $r = \rho \sin \phi$.

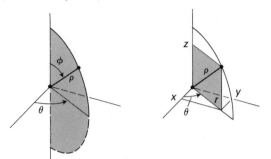

Then

$$x = r \cos \theta = \rho \sin \phi \cos \theta,$$
$$y = r \sin \theta = \rho \sin \phi \sin \theta.$$

The change-of-coordinate formulas are thus

$$x = \rho \sin \phi \cos \theta,$$
$$y = \rho \sin \phi \sin \theta,$$
$$z = \rho \cos \phi.$$

REMARK   Note that we are using ρ here for the distance to the origin; earlier ρ was our standard symbol for density.

Triple integrals in which the integrand f or the domain of integration G exhibit some elements of spherical symmetry about the origin may be easier to compute by iterated integration in spherical coordinates. Here is the basic theorem.

THEOREM 13   *Let G'' be the region in ρ, ϕ, θ-space that is the image of G under the spherical-coordinate mapping $(x, y, z) \to (\rho, \phi, \theta)$. That is, G'' consists of the triples (ρ, ϕ, θ) for which the corresponding points (x, y, z) lie in G. Let $f(x, y, z)$ be a continuous function on G, and define the function K by*

$$K(\rho, \phi, \theta) = f(\rho \sin \phi \cos \theta, \rho \sin \phi \sin \theta, \rho \cos \phi).$$

Then

$$\iiint\limits_{G} f(x, y, z) \, dV = \iiint\limits_{G''} K(\rho, \phi, \theta) \rho^2 \sin \phi \, dV''.$$

Sketch of Proof.   This is just the "square" of Theorem 12. We start with the formula at the end of that theorem,

$$\iiint\limits_{G'} F(r, \theta, z) r \, dV',$$

and apply Theorem 12 to it, this time making the change to polar coordinates ρ, ϕ in the z,r-plane: $z = \rho \cos \phi$, $r = \rho \sin \phi$. This converts $F(r, \theta, z)$ to $K(\rho, \phi, \theta)$, r to $\rho \sin \phi$, and dV' to $\rho \, dV''$, and Theorem 13 then follows. ∎

We can now evaluate the integral on the right in the theorem above as an iterated integral in the variables ρ, ϕ, θ (by Theorem 10), where the limits of integration are determined by describing G in terms of spherical coordinates in xyz-space.

EXAMPLE 2 Compute the volume of the sphere of radius a by using spherical coordinates.

Solution. The volume of a body is the same as its mass if its density is constant and equal to one. That is, the volume of G is obtained as the triple integral over G of the constant function 1. Also, the sphere of radius a is described in terms of the spherical-coordinate variables by the inequalities $0 \le \rho \le a$, $0 \le \phi \le \pi$, $0 \le \theta \le 2\pi$. Therefore,

$$V = \iiint_G 1 \, dV = \iiint_{G''} \rho^2 \sin \phi \, dV''$$

$$= \int_0^{2\pi} \int_0^{\pi} \int_0^a \rho^2 \sin \phi \, d\rho \, d\phi \, d\theta$$

$$= \int_0^{2\pi} \int_0^{\pi} \frac{\rho^3}{3} \sin \phi \Big]_{\rho=0}^{a} d\phi \, d\theta$$

$$= \frac{a^3}{3} \int_0^{2\pi} \int_0^{\pi} \sin \phi \, d\phi \, d\theta = \frac{a^3}{3} \int_0^{2\pi} [-\cos \phi]_0^{\pi} \, d\theta$$

$$= \frac{2a^3}{3} \int_0^{2\pi} d\theta = \frac{2a^3}{3} \cdot 2\pi = \frac{4}{3} \pi a^3.$$

PROBLEMS FOR SECTION 9

In Problems 1 through 7 find the volume of the given solid by iterated integration in cylindrical or spherical coordinates.

1. The region between the paraboloid $z = x^2 + y^2$ and the plane $z = 4$.

2. The region cut from the sphere $x^2 + y^2 + z^2 = b^2$ by the cylinder $x^2 + y^2 = a^2$ (where $a \le b$).

3. The "ice cream cone" cut from the sphere $x^2 + y^2 + z^2 = 4$ by the vertical cone with vertex at the origin and semi-vertex angle equal to $\pi/6$.

4. The region cut from the sphere $x^2 + y^2 + z^2 = a^2$ by the upper half of the cone generated by rotating the line $z = mx$ about the z-axis. Use cylindrical coordinates.

5. The region cut from the sphere $\rho = a$ by the cone $\phi = \alpha$ (spherical coordinates). Interpret your answer for the case $\alpha > \pi/2$, and, in particular for $\alpha = \pi$.

6. The region between the paraboloid $z = x^2 + y^2$ and the plane $z = 2x$.

7. The region bounded by the sphere $x^2 + y^2 + z^2 = 6$ and the paraboloid $z = x^2 + y^2$.

8. Find the mass of a ball of radius a if its density is proportional to the distance from the center.

9. Find the mass of a ball of radius a if its density is inversely proportional to the distance from the center.

10. The density of a ball is $k\rho^{\alpha}$, where ρ is the distance to the origin and α is a constant. Find what limitation is placed on the exponent α by the requirement that the mass of the ball be finite.

11. The density of a ball centered at the origin is proportional to the distance from the xy-plane. Find the mass of the ball.

12. The density of a ball centered at the origin is proportional to the distance from the z-axis. Find the mass of the ball.

13. A cylinder is bounded by the cylindrical surface $x^2 + y^2 = a^2$ and the planes $z = 0$, $z = b$. Its density is zr. Find its mass.

14. Find the centroid of the region between the xy-plane and the paraboloid $z = 4 - (x^2 + y^2)$.

15. Find the centroid of the "ice cream cone" region of Problem 3.

16. Find the centroid of the region of Problem 7.

17. Find the centroid of the region cut off from the first octant by the sphere $x^2 + y^2 + z^2 = a^2$.

18. Write out the proof of Theorem 11, using the piercing formula.

10. THE TRIPLE INTEGRAL

The definition of the triple integral is exactly analogous to our earlier definition of the double integral. Let $f(x, y, z)$ be any continuous function defined over a bounded closed region G in space. A Riemann sum for f over G is defined in the following way: We subdivide G in any manner into a finite number of small subregions ΔG_i, and in each of the subregions ΔG_i we choose an arbitrary evaluation point (x_i, y_i, z_i). Then the Riemann sum for f defined by this subdivision of G and the evaluation set is

$$\sum f(x_i, y_i, z_i) \, \Delta V_i,$$

where ΔV_i is the volume of G_i, for each i.

If we change to a different subdivision and evaluation set, then the Riemann sum will probably change in value. Suppose that this varying Riemann sum $\sum$ approaches a limit L as the subdivision is taken finer and finer. More precisely, suppose that we can make $\sum - L$ as small as we wish simply by taking a subdivision whose subregions ΔG_i all have diameters less than a suitably small positive number d. If this happens, then the limit L of the varying Riemann sum $\sum$ is called the *triple integral of f over G*, and is designated

$$\iiint\limits_{G} f(x, y, z) \, dV.$$

That is,

DEFINITION
$$\iiint\limits_{G} f(x, y, z) \, dV = \lim \sum_{i=1}^{n} f(x_i, y_i, z_i) \, \Delta V_i,$$

provided that the limit exists in the above sense.

The basic theorem is that this limit does exist for any continuous function f.

THEOREM 14 *If $f(x, y, z)$ is continuous on the bounded closed region G, then its triple integral over G exists.*

So far we have parroted our earlier double-integral discussion almost verbatim. However, the proof of Theorem 14 has to be different, because the graph of a function of three variables lies in four-dimensional space, and we have no secure intuitions about four-dimensional volume. Instead, we shall prove this theorem by interpreting a positive continuous function as the density $\rho(x, y, z)$ of a mass distribution, and then applying certain intuitive principles that govern the relationship between a mass distribution and its density ρ. We shall then have a "physical" proof of the theorem.

Suppose, then, that the continuous function $\rho(x, y, z)$ is the density of a distribution of mass through a space region G. Let m be the total mass of G and let V be the volume of G. We assume:

I. If the density $\rho(x, y, z)$ has the constant value k throughout G, then $m = kV$.

II. If $\rho(x, y, z) \le k$ throughout G, then $m \le kV$. If $\rho(x, y, z) \ge k$ throughout G, then $m \ge kV$.

These principles let us analyze the relationship between m and ρ. Starting with some given small positive number e, we subdivide G into a finite number of subregions ΔG_i on each of which the variation of ρ is less then e. That is, if ρ'_i and ρ''_i are the minimum and maximum values of $\rho(x, y, z)$ on the subregion ΔG_i, then $\rho''_i - \rho'_i < e$, for each i. This restriction on the variation of ρ will be met automatically if we subdivide G into subregions whose diameters are all less than a suitably small number d. (See Section 3.)

According to principle (II) above, the mass Δm_i occupying the subregion ΔG_i satisfies the inequality

$$\rho'_i \Delta V_i \le \Delta m_i \le \rho''_i \Delta V_i,$$

for each i. Now let (x_i, y_i, z_i) be an arbitrary evaluation point in ΔG_i, for each i. Then,

$$\rho'_i \le \rho(x_i, y_i, z_i) \le \rho''_i,$$

by the definition of ρ'_i and ρ''_i, so we also have the inequality

$$\rho'_i \Delta V_i \le \rho(x_i, y_i, z_i) \Delta V_i \le \rho''_i \Delta V_i.$$

Since two numbers lying in the same interval differ by at most the length of the interval, it follows, for each i, that the numbers Δm_i and $\rho(x_i, y_i, z_i) \Delta V_i$ differ by at most $(\rho''_i - \rho'_i) \Delta V_i$ in magnitude. That is,

$$\Delta m_i \approx \rho(x_i, y_i, z_i) \Delta V_i,$$

with an error at most $(\rho''_i - \rho'_i) \Delta V_i$. Since $\rho''_i - \rho'_i \le e$ for each i, the sum of these errors is at most

$$\sum e \, \Delta V_i = e \sum \Delta V_i = eV,$$

and since $m = \sum m_i$ is the total mass of G, we conclude that

$$m \approx \sum \rho(x_i, y_i, z_i) \, \Delta V_i$$

with an error at most eV.

Note that we can make this error as small as we wish by choosing e suitably small to begin with and then using any subdivision for which the diameters of the subregions ΔG_i are all less than the associated small number d. This shows that m is the limit of the varying Riemann sum as the subdivision of G is taken finer and finer. Therefore the triple integral $\iiint_G \rho(x, y, z) \, dV$ exists (and has the value m).

Since any positive continuous function $f(x, y, z)$ can be interpreted as the density of a mass distribution, we thus have a "physical" proof of Theorem 14, for the case that f is positive. Moreover, any function that is not positive can be reduced to a positive f by a simple device (Problem 6), so we are done. ∎

The above proof of Theorem 14 is based on physical intuitions about density and mass. However, with a little more mathematical sophistication, the reasoning we used can be converted into a sound analytic proof.

We turn now to the piercing theorem.

THEOREM 15   *Let G be the region in space based on the bounded closed plane region R in the xy-plane and capped by the surface $z = \phi(x, y)$, where ϕ is continuous on R. Let f be continuous on G. Then*

$$\iiint_G f \, dV = \iint_R \left(\int_0^{\phi(x,y)} f(x, y, z) \, dz \right) dA.$$

Sketch of proof. Let $g(x, y)$ be the function defined by the inner integral on the right:

$$g(x, y) = \int_0^{\phi(x,y)} f(x, y, z) \, dz.$$

It can be shown that g is continuous, and we simply assume this. Then g is integrable over the plane region R, and

$$\iint_R g \, dA \approx \sum g(x_i, y_i) \, \Delta A_i,$$

where the sum on the right is any Riemann sum for g over R.

The evaluation point (x_i, y_i) can be chosen anywhere in the ith incremental subregion ΔR_i, and we choose it at a point where the surface function ϕ assumes its average value over ΔR_i, so that

$$\phi(x_i, y_i) = \frac{1}{\Delta A_i} \iint_{\Delta R_i} \phi \, dA.$$

Now fix i and subdivide the z-interval from 0 to $\phi(x_i, y_i)$:

$$0 = z_0 < z_1 < \cdots < z_n = \phi(x_i, y_i).$$

We then have the corresponding Riemann sum approximation

$$g(x_i, y_i) = \int_0^{\phi(x_i, y_i)} f(x_i, y_i, z) \, dz \approx \sum_{m=1}^{n} f(x_i, y_i, z_m) \, \Delta z_m.$$

The planes $z = z_m$ slice the cylinder based on ΔR_i into cylindrical blocks, except that we replace the top plane by a piece of the graph of $z = \phi(x, y)$. The mth block has volume $\Delta V_{i,m} = \Delta z_m \, \Delta A_i$, and this holds even for the top block, because of the way we chose (x_i, y_i):

$$\Delta V_{i,n} = \iint_{\Delta R_i} [\phi(x, y) - z_{n-1}] \, dA = [\phi(x_i, y_i) - z_{n-1}] \Delta A_i = \Delta z_n \, \Delta A_i.$$

The Riemann sum estimate above can therefore be rewritten

$$g(x_i, y_i) \, \Delta A_i \approx \sum_{m=1}^{n} f(x_i, y_i, z_m) \, \Delta V_{i,m},$$

after multiplying by ΔA_i.

Finally, we add up these approximations for the different incremental bases ΔR_i, and end up with

$$\sum_i g(x_i, y_i) \, \Delta A_i \approx \sum_{i,m} f(x_i, y_i, z_m) \, \Delta V_{i,m},$$

where the sum on the right is now a Riemann sum for $\iiint_G f \, dV$. All of these approximations can be made as close as desired by choosing sufficiently fine subdivisions, so in the limit we have

$$\iint_R g \, dA = \iiint_G f \, dV,$$

which is what we wanted. ∎

PROBLEMS FOR SECTION 10

Prove the following four properties of the triple integral from its definition as the limit of Riemann sums, assuming that the standard limit laws are valid for the type of limit used in this definition.

1. If f is the constant function k and if V is the volume of the space region G, then

$$\iiint_G f \, dV = kV.$$

2. $\iiint_G (af + bg) \, dV = a \iiint_G f \, dV + b \iiint_G g \, dV.$

3. If G is divided into two nonoverlapping subregions G_1 and G_2, then

$$\iiint_G f \, dV = \iiint_{G_1} f \, dV + \iiint_{G_2} f \, dV.$$

4. If $f \geq 0$ on G, then $\iiint_G f \, dV \geq 0$.

5. Now prove from Problems 1, 2, and 4 that, if f is bounded above by the

constant k on G, then

$$\iiint_G f\,dV \le kV,$$

where V is the volume of G.

6. Assuming that the triple integral exists for a positive continuous function (Theorem 14, as proved in the text), prove that it exists for any continuous function (Theorem 14 as stated). (*Hint:* Let k be the maximum value of f on G, write $f = k - (k - f)$ and apply various results already established.)

7. Prove Theorem 15 for an arbitrary continuous function by reducing to a positive function, in the manner suggested in the above problem.

EXTRA PROBLEMS FOR CHAPTER 17

Find the volume of each of the space regions described in Problems 1 through 5 below.

1. The solid based on the rectangle $0 \le x \le a$, $0 \le y \le b$ in the xy-plane and capped by the surface $z = b^2 - y^2$.

2. The solid cut from the first octant by the plane $x + z = 1$ and the cylinder $x^2 + y^2 = 1$.

3. The solid bounded by the parabolic cylinders $y = 1 - x^2$, $y = x^2 - 1$, $z = 1 - y^2$, and the plane $z = 0$.

4. The solid cut from the first octant by the plane $y + z = 1$ and the ellipse $y^2 + 2x^2 = 1$.

5. The solid bounded by the sphere $x^2 + y^2 + z^2 = 1$ and the square cylinder $|x| + |y| = 1$.

6. Suppose we choose N points $(x_1, y_1), \ldots, (x_N, y_N)$ that are scattered in a fairly uniform way throughout a plane region G. That is, there must be roughly the same number of chosen points "per unit area" wherever we look in G. If the area of G is A, give some kind of an argument to support the claim that

$$\frac{A}{N} \sum_{i=1}^{N} f(x_i, y_i)$$

is approximately equal to the double integral of f over G.

7. Let G be the unit square $0 \le x \le 1$, $0 \le y \le 1$.

   a) Compute $\iint_G xy\,dA$ by iterated integration.

   b) Compute the above estimate of this double integral for the nine vertices obtained by dividing the unit square into four equal subsquares.

8. Work out the above problem for $f(x, y) = x + y$.

9. The same for $f(x, y) = x^2 + y^2$.

10. The same for $f(x, y) = x^2y^2$.

11. The estimates in Problems 9 and 10 were not very good. They would improve with a finer subdivision, of course, but there is something else wrong. In a sense a vertex is surrounded by only 1/4 as much area as an interior point and an edge point that is not a vertex is surrounded by 1/2 as much area as an interior point. We are therefore not "weighting" the effect of these boundary points correctly. In order to adjust for this edge effect, suppose we count each of the four corner values $f(x_i, y_i)$ once, each of the four edge midpoint values twice,

and the middle value $f(1/2, 1/2)$ four times. Then we have effectively increased the number of terms in the sum from 9 to 16, so we now divide by 16 rather than by 9. Recompute the estimate for $f(x, y) = x^2 + y^2$, using this weighted sum.

12. Do the above problem for $x^2 y^2$.

13. a) If the unit square is divided into 9 equal subsquares and if their 16 vertices are weighted as in Problem 10, show that we must now divide by $N = 36$.

    b) Recompute the estimate for the double integral of $x^2 y^2$ over the unit square using this new weighted sum.

    c) Compute the double integral again and compare answers.

14. Find the moments of the plane region bounded by the polar graph of $r = (\cos \theta)^{1/3}$.

15. Find the center of mass of the plane lamina bounded by the cardioid $r = 1 + \cos \theta$ if its density is proportional to $|y|$.

16. Find the centroid of the right-hand leaf of the region bounded by polar graph of $r = \cos 2\theta$.

17. Find the centroid of the cardioid $r = a(1 + \cos \theta)$.

18. Show that the graphs of the functions $f(x, y) = x^2 + y^2$, $g(x, y) = 2xy$ have the same area over any region R in the xy-plane.

19. The same for $f(x, y) = h(x) + k(y)$, $g(x, y) = h(x) - k(y)$.

20. A function $v(x, y)$ is said to be conjugate to a function $u(x, y)$ if

$$\frac{\partial u}{\partial x} = \frac{\partial v}{\partial y}, \qquad \frac{\partial u}{\partial y} = -\frac{\partial v}{\partial x}.$$

Show that if $v(x, y)$ is conjugate to $u(x, y)$ over a region R in the xy-plane, then the graph of u over R has the same surface area as the graph of v over R.

Referring to the above result, show that each of the following pairs of functions have graphs of equal surface area over any region R over which they both are defined.

21. $x^2 - y^2$, $2xy$ 22. $x^3 - 3xy^2$, $3x^2 y - y^3$

23. $e^x \cos y$, $e^x \sin y$ 24. $\frac{1}{2} \log(x^2 + y^2)$, $\arctan \dfrac{y}{x}$

25. If the graph of a positive function $y = f(x)$ is revolved about the x-axis, the area of the surface of revolution it sweeps out from $x = a$ to $x = b$ is given by

$$A = 2\pi \int_a^b f(x)\sqrt{1 + [f'(x)]^2} \, dx.$$

Prove this formula from the general area formula for the graph of a function $z = g(x, y)$.

26. Suppose that $z = f(r)$, where $r = \sqrt{x^2 + y^2}$. Show that the area of the graph of $z = f(r)$ over the circle $x^2 + y^2 \le a^2$ equals

$$2\pi \int_0^a r\sqrt{1 + [f'(r)]^2} \, dr.$$

27. When the circle $(x - b)^2 + z^2 = a^2$, $0 < a < b$ is revolved about the z-axis, the resulting surface is a torus. Its equation can be written

$$(r - b)^2 + z^2 = a^2, \quad r = \sqrt{x^2 + y^2}.$$

Find its total surface area.

28. Find the volume of the "cored apple" obtained by removing from the ball $x^2 + y^2 + z^2 \le b^2$ its intersection with the cylinder $x^2 + y^2 \le a^2$, where $a < b$.

29. Find the moments about the coordinate axes of the piece of the above solid cut off by the first octant.

30. Find the centroid of the intersection of the first octant with the half-space

$$\frac{x}{a} + \frac{y}{b} + \frac{z}{c} \le 1,$$

where a, b, and c are positive constants.

31. Given $0 < a < b$, let S be the spherical shell with inner radius a and outer radius b. Find the mass of the piece of S lying in the first octant.
    Show that the limit of the centroid as $a \to b$ is the point $(b/2, b/2, b/2)$.

32. Find the volume of the solid bounded by the hyperboloid $z^2 - x^2 - y^2 = 1$ and the plane $z = 2$.

33. Find the volume of the solid bounded by the paraboloid $z = 2 - (x^2 + y^2)$ and the upper sheet of the hyperboloid $z^2 - x^2 - y^2 = 2$.
    Same problem, but with the lower sheet of the hyperboloid.

34. Find the volume of the region based on the circular disc $x^2 + y^2 \le r^2$ and capped by the surface $z = 1/\sqrt{x^2 + y^2}$. (This will involve an improper integral. First calculate the volume over the ring $a^2 \le x^2 + y^2 \le r^2$ and then let $a \to 0$.)

35. A ball centered at the origin is made of a magical material of density

$$\rho(x, y, z) = \frac{1}{x^2 + y^2 + z^2}.$$

Show that its mass is nevertheless finite. Calculate the mass of the ball if its radius is r. (Another improper integral. First find the mass of the shell $a^2 \le x^2 + y^2 + z^2 \le r^2$.)

36. Let S be the solid based on the rectangle $a \le x \le b$, $c \le y \le d$, and capped by the surface $z = f(x, y)$. Suppose that the partial derivative $\partial z/\partial x$ exists and is bounded by $B(|\partial z/\partial x| \le B)$ on this rectangle. Let R_{x_0} be the cross section of S in the plane $x = x_0$. Thus, R_{x_0} is the region in this plane under the graph $z = f(x_0, y)$, from $y = c$ to $y = d$. We want to show that R_{x_0} varies continuously as x_0 varies near a fixed $x_0 = k$, and for this purpose we draw all these regions R_{x_0} in a single yx-plane. (See left below.)

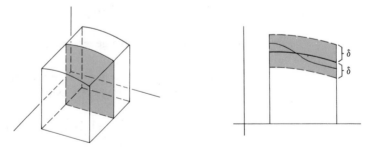

We choose any positive number δ and draw the band of vertical width 2δ centered on the curve $z = f(k, y)$, as shown in the righthand figure. (The band runs from the bottom curve $(z = f(k, y) - \delta)$ up to the top curve $z = (f(k, y) + \delta)$.)

Show, then, that the graph $z = f(x_0, y)$ lies in this band for every x_0 in the interval $(k - \delta/B, k + \delta/B)$. Since the band width 2δ can be taken as small as desired, this shows that R_{x_0} varies continuously with x_0.

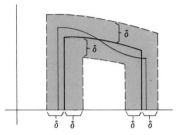

37. Consider the situation of the above problem, but with the base rectangle replaced by the base region G lying between the graphs $y = g(x)$ and $y = h(x)$, from $x = a$ to $x = b$. Assume that the derivatives g' and h' are bounded by the same B. Now the cross section R_{x_0} has varying width. Prove, however, that R_{x_0} is squeezed between the minimum and maximum regions of the type shown above, for every x_0 in the interval $(k - \delta/B, k + \delta/B)$.

18
Green's Theorem

This chapter brings together several topics that earlier were developed independently—mixed partial derivatives, paths, double integrals—and shows them to be related in a surprising and powerful way. The culminating theorem, Green's theorem, is important in applications and in other branches of mathematics, especially in the theory of functions of a complex variable. And it forms a bridge to one of the principal subjects of advanced calculus: the combination of multidimensional integral and differential calculus in multidimensional versions of the Fundamental Theorem of Calculus.

1. WORK AND ENERGY IN ONE-DIMENSIONAL MOTION

The acceleration of any moving body is known to be proportional to the total force F acting on it. Thus one-dimensional motion along the x-axis always satisfies the equation $d^2x/dt^2 = cF$, where c is the constant of proportionality. It is customary to put the constant on the other side of the equation, and thus write

$$F = k\frac{d^2x}{dt^2}.$$

A different body will have a different constant of proportionality. In fact, experiment shows that if the mass of the body is doubled, then it requires double the force in order to produce the same acceleration. So k represents the mass of the body. Thus (in suitable units)

$$F = m\frac{d^2x}{dt^2} = ma,$$

where F is the total force acting on the body (directed along the x-axis), m is the mass of the body, and $a = d^2x/dt^2$ is its acceleration. This is Newton's second law of motion.

The complete history of a particle moving under a given force F is obtained by solving this differential equation, as was done for the special case of harmonic motion in Chapter 6. But for many purposes all we need to know is the accumulated effect of the force, in terms either of the length of time over which F acts or of the distance over which F acts, i.e.,

$$\int_{t_1}^{t_2} F\,dt \qquad \text{or} \qquad \int_{x_1}^{x_2} F\,dx.$$

Here x_1 is the initial position of the particle at the initial time t_1, and x_2 and t_2 are the terminal position and time.

In order to see the meaning of the first integral we rewrite Newton's equation in terms of the velocity $v = dx/dt$,

$$F = m\frac{dv}{dt},$$

and have

$$\int_{t_1}^{t_2} F\,dt = \int_{t_1}^{t_2} m\frac{dv}{dt}\,dt = \int_{v_1}^{v_2} m\,dv = mv_2 - mv_1,$$

where v_1 and v_2 are the initial and terminal velocities of the particle. The quantity mv is called the *momentum* of the particle. Note that it is a signed quantity, being positive (negative) as the velocity is positive (negative). This corresponds to the fact that in two or more dimensions momentum is a vector quantity. In words, the calculation above shows that:

The time integral of F over the time interval $[t_1, t_2]$ equals the change in momentum from $t = t_1$ to $t = t_2$.

Since time is the underlying variable in the motion, we can compute the other F integral, $\int_{x_1}^{x_2} F\, dx$, by making the change of variable

$$dx = \left(\frac{dx}{dt}\right) dt = v\, dt,$$

and have

$$\int_{x_1}^{x_2} F\, dx = \int_{t_1}^{t_2} Fv\, dt = \int_{t_1}^{t_2} m\frac{dv}{dt} v\, dt = \int_{t_1}^{t_2} mv\frac{dv}{dt}\, dt$$

$$= \int_{v_1}^{v_2} mv\, dv = \tfrac{1}{2}mv^2 \bigg]_{v_1}^{v_2} = \tfrac{1}{2}m(v_2)^2 - \tfrac{1}{2}m(v_1)^2.$$

The integral $\int_{x_1}^{x_2} F\, dx$ is called the *work* done by the force F during the motion, and the quantity $mv^2/2$ is called the *kinetic energy* of the particle. In these terms, the calculation above shows that:

The work done by the total applied force F during a motion is equal to the change in kinetic energy from the beginning to the end of the motion.

This work-energy law lets us answer some questions about a solution to Newton's equation without having to find the solution.

EXAMPLE 1 The harmonic motion discussion in Section 8 of Chapter 6 concerned a spring of stiffness k exerting a negative ("restoring") force F of amount $F = -kx$. Use the work-energy law to find the amplitude of the motion when a spring of stiffness 5 is attached to a particle of mass 3 and given an initial velocity $v_0 = 2$ at $x_0 = 0$.

Solution. The amplitude A is the maximum value of x during the motion, and hence is the value of x for which v first becomes zero. The work-energy equation

$$\int_{x_0}^{x} F\, dx = \tfrac{1}{2}mv^2 - \tfrac{1}{2}mv_0^2$$

thus becomes

$$\int_0^A (-5x)\, dx = 0 - \tfrac{1}{2}3(2)^2.$$

So

$$-\frac{5}{2}A^2 = -6,$$

and

$$A = 2\sqrt{\frac{3}{5}}.$$

The momentum and energy laws obtained above were the results of purely formal calculations. We still have to consider whether the various integrals actually make sense. The dt integrals are "along the motion," as follows. We are considering a solution $x(t)$ of Newton's equation with initial values (t_1, x_1) and terminal value (t_2, x_2). Then v_1 and v_2 are the initial and terminal values of $v = x'(t)$. In theory the force F is permitted to be a smooth function of the three variables t, x, and v, $F = F(t, x, v)$, but along the motion F reduces to a function of t alone, $F = F(t, x(t), x'(t))$. So there is no problem about the meaning of the integrations with respect to t.

The changes to the dv and dx integrals appear to be applications of the direct substitution rule for definite integrals, but it will be found that the change to dx doesn't fit the rule in the general case where $F = F(x, t, v)$. Fortunately, there are many important situations in which F is a function of x alone, $F = F(x)$, and then the change-of-variable equation

$$\int_{t_1}^{t_2} F(x(t))x'(t)\, dt = \int_{x_1}^{x_2} F(x)\, dx$$

is correct.

However, there is a hidden ramification even here. Suppose that F is everywhere defined and positive, and that the initial velocity v_1 is negative. Since

$$\frac{dv}{dt} = \frac{F}{m} > 0,$$

v is an increasing function of t, and we suppose that it eventually becomes positive, so $v_2 > 0$. The motion thus starts out with x decreasing, then passes through a critical point (t_0, x_0) where $v = 0$ and the direction of motion changes, and finishes with x increasing. In other words, x does not just run straight across the interval $[x_1, x_2]$, as the work integral $\int_{x_1}^{x_2} F(x)\, dx$ might suggest, but first backs up to x_0 and then moves forward from x_0 to x_2. The t integral encompasses this whole motion, but the x integral does not, unless we think of $\int_{x_1}^{x_2}$ as standing for $\int_{x_1}^{x_0} + \int_{x_0}^{x_2}$.

So our viewpoint should be that the t integral is the basic integral definition of the work done by F along the motion:

DEFINITION   *The work done by the force F along a motion from $t = t_1$ to $t = t_2$ is $\int_{t_1}^{t_2} Fv\, dt$.*

Then the calculation above shows that:

If the force F is a function of x alone, then the work done by F over any motion starting at $x = x_1$ and ending at $x = x_2$ is given by

$$\int_{x_1}^{x_2} F(x)\, dx.$$

REMARK 1   In the general case, where F varies with t and/or v as well as with x, and the integral $\int_{x_1}^{x_2} F\, dx$ therefore doesn't make any direct sense, the equation

$$\int_{x_1}^{x_2} F\, dx = \int_{t_1}^{t_2} Fv\, dt$$

is taken as the *definition* of $\int_{x_1}^{x_2} F\, dx$.

REMARK 2 If the total force F is the sum of two or more forces, say $F = F_1 + F_2$, then the work done by F is the sum

$$\int_{x_1}^{x_2} F\, dx = \int_{x_1}^{x_2} F_1\, dx + \int_{x_1}^{x_2} F_2\, dx,$$

and we naturally consider the separate F_1 and F_2 integrals to be the parts of the total work attributable to these constituent forces. Thus $\int_{x_1}^{x_2} F_1\, dx$ is the work done by F_1, etc.

REMARK 3 If the force F is constant then

$$\int_{t_1}^{t_2} F\, dt = F(t_2 - t_1), \qquad \text{and} \qquad \int_{x_1}^{x_2} F\, dx = F(x_2 - x_1).$$

So for a constant force F, the change in momentum is given by F times how *long* the force acts, and the change in kinetic energy is given by F times how *far* the force acts.

Now let F depend only on x and let $\phi(x)$ be any antiderivative of $F(x)$. Then $V = -\phi$ is called the *potential energy* of the force field $F(x)$; it is uniquely determined once we have decided what point x_0 is to be assigned zero energy. (Note that the equations $V'(x) = -F(x)$, $V(x_0) = 0$ form an initial value problem for V.) Since

$$\int_{x_1}^{x_2} F(x)\, dx = V(x_1) - V(x_2),$$

the work-energy identity can now be rewritten

$$V(x_1) + \tfrac{1}{2}mv_1^2 = V(x_2) + \tfrac{1}{2}mv_2^2.$$

The sum $E = V(x) + \tfrac{1}{2}mv^2$ of the potential and kinetic energy is called the total energy of the particle, and the above equation says that the total energy remains constant as the particle moves under the force F. This is the law of *conservation of energy*. The work done by F from x_1 to x_2 can be viewed as so much potential energy converted into kinetic energy.

EXAMPLE 2 The restoring force exerted by a compressed or extended spring is

$$F = -kx.$$

The potential energy of the spring is thus

$$V(x) = \frac{kx^2}{2} + C.$$

When a particle of mass m is attached and the system is set into oscillatory motion, the total energy remains constant:

$$\tfrac{1}{2}mv^2 + \tfrac{1}{2}kx^2 \equiv \text{constant}.$$

In Example 1 we had $m = 3$ and $k = 5$, so

$$3v^2 + 5x^2 = C$$

in that motion. The initial condition $v_0 = 2$, $x_0 = 0$ shows that $C = 12$, and since the amplitude A is the value of $|x|$ for which $v = 0$, we end up with the same equation as before,

$$5A^2 = 12.$$

In view of the second remark above, about the work attributable to one of several constituent forces, we can compute the work done by a force F in situations where F is obviously not the total force and no connection with initial and terminal velocity is implied.

EXAMPLE 3   When a quantity of gas is held at a constant temperature, its pressure p and total volume V are related by the equation

$$pV = \text{constant}.$$

A circular cylinder 1 ft in diameter has a variable length determined by a piston at one end. It is filled with gas at a pressure of 50 lb/in^2 when its length is 10 ft, and the gas is then further compressed by moving the piston until the length is reduced to 5 ft. Supposing that the gas is held at a constant temperature, how much work is done on the gas by the moving piston?

Solution. The volume of the enclosed gas is proportional to the length x of the cylinder, and the force F with which the piston pushes against the gas is proportional to the gas pressure p, so the equation $pV = \text{constant}$ can be replaced by

$$Fx = C.$$

Initially $x_0 = 10$ ft. The force is negative, being in the direction of decreasing x, and

$$F_0 = -pA \ (= -\text{pressure times piston area, in square inches})$$
$$= -50 \times \pi 6^2 \approx -5652 \text{ lb.},$$

so $C = F_0 x_0 = -56,520$. The work done is thus

$$W = \int_{10}^{5} F \, dx = -\int_{10}^{5} \frac{56,520}{x} \, dx$$
$$= 56,520 \, (\log 10 - \log 5)$$
$$= 56,520 \log 2 \approx 56,520 \times 0.693$$
$$\approx 39,168 \text{ foot-pounds.}$$

EXAMPLE 4   A tank is in the shape of an inverted cone 10 feet high and 10 feet in diameter at the top. It is filled with water from an inlet at the bottom. How much work is done in pumping the water in?

Solution. Water weighs 62.4 pounds per cubic foot, so y feet of water presses on the bottom with a pressure of 62.4y pounds per square foot.

Imagine the water being pushed into the conical tank by a pump having a piston of area A square feet. Neglecting the recovery strokes of the piston, the total distance d that the piston travels in filling the tank is given by

$$dA = \text{volume of tank,}$$

$$d = \frac{1}{A}\left(\frac{1}{3}\pi r^2 h\right) = \frac{250\pi}{3A}.$$

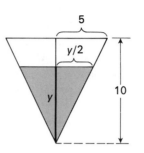

At the point when the piston has travelled x feet, the tank contains Ax cubic feet of water, and its depth y is related to x by

$$\frac{1}{3}\pi(y/2)^2 y = Ax.$$

So

$$y = \left(\frac{12Ax}{\pi}\right)^{1/3}.$$

Since the force exerted by the piston is 62.4 Ay,

$$W = \int_0^d 62.4\, Ay\, dx$$

$$= \int_0^d 62.4\, A\left(\frac{12Ax}{\pi}\right)^{1/3} dx$$

$$= 62.4\, A^{4/3}\left(\frac{12}{\pi}\right)^{1/3}\left(\frac{3}{4}x^{4/3}\right)\Big|_0^d$$

$$= 62.4\, A^{4/3}\left(\frac{12}{\pi}\right)^{1/3}\frac{3}{4}\left(\frac{250\pi}{3A}\right)^{4/3}$$

$$= 62.4\pi \cdot 625 \approx 122{,}460 \text{ ft-lbs.}$$

PROBLEMS FOR SECTION 1

1. How much work is done by a force of magnitude $F(x) = 5x$ on a particle that moves from $x = 2$ to $x = 4$?

2. Same question if the particle moves from $x = -2$ to $x = 2$.

3. How much work is done by a force of magnitude $2/x$ in a motion from $x = 1$ to $x = 2$?

In Problems 4 through 10, first write down the law of conservation of energy for the given system, and then use it to solve the problem.

4. A particle of mass 2 moves along the positive x-axis under a (total) force $F(x) = 1/x^2$. Its velocity at $x = 1$ is -1. (a) At what point will it turn around and start moving in the positive x-direction? (b) What will its subsequent velocity be at $x = 2$? (c) What is its limiting speed as $x \to \infty$?

5. A particle of mass 3 is attached to a spring of stiffness 5. Find the relationship between the amplitude of its motion and its maximum speed.

6. Solve the above problem for a particle of mass m attached to a spring of stiffness k.

7. A particle of mass 3 is attached to a spring of stiffness 2. Find its amplitude A if its speed at $x = A/2$ is 5.

8. A particle is repelled from the origin by a force of magnitude $1/x^{3/2}$. A particle of mass 1 starts at $x = 4$ with an initial velocity of $+1$. What is its limiting velocity as $x \to \infty$?

9. In the same situation, the initial velocity is -1. How close will the particle come to the origin?

10. A particle of mass m is repelled from the origin with a force of magnitude $1/x^{3/2}$. If it starts from rest at $x = 4$, what is its limiting velocity as $x \to \infty$?

11. A tank has a square cross section, 10 ft on a side, and is 20 ft tall. Find the work done in filling the tank with water from a bottom inlet, as in Example 4.

12. a) An irregularly shaped tank is being filled with water from the bottom, in the manner of Example 4. Suppose that the water depth y (in feet) is a known function

$$y = f(v)$$

of the volume v (in cubic feet) of water in the tank. Follow the scheme of Example 4 to show that the work done in filling the tank to a volume V_0 is

$$W = \int_0^{V_0/A} 62.4 \, Af(Ax) \, dx$$

where A is the area of the pump piston in square feet.

b) Show that the answer above is independent of A.

13. A tank h feet high has a capacity of V cubic feet. It requires W foot-pounds of work to fill the tank with water from the bottom. Show that it requires $62.4 \, Vh - W$ foot-pounds of work to pump out the tank from the top. (*Hint:* the combined act of filling and pumping out the tank amounts to raising V cubic feet of water through h feet.)

14. A cylinder with a movable piston at one end contains 300 cubic feet of gas at a pressure of 100 lb/in^2. The gas is compressed to 1/3 of its original volume (at a constant temperature) by moving the piston. How much work does the piston do during this motion?

15. It can be proved that the work done by gas in an expanding container is always given by

$$W = \int_{V_1}^{V_2} p \, dV \quad \text{foot-pounds,}$$

where p is the pressure in *pounds per square foot*, and V is the volume in cubic feet. This proof is independent of the formula relating p and V. Verify the above formula when the gas is confined to a cylinder with variable length determined by a piston at one end and is held at a constant temperature.

16. Verify the formula in the above problem for gas expanding a spherical balloon (at a constant temperature).

17. The pressure and volume of a certain body of gas are related by

$$pV^\alpha = k,$$

where α and k are positive constants. Find the work done if the gas expands from $V = V_0$ to $V = V_1$.

18. The formula in Problem 15 also holds for the tank-filling problems, with p a pressure at the pump. Show that this is so, by starting with the general formula of Problem 12.

2. WORK ALONG PLANE PATHS: LINE INTEGRALS

In the plane, force and acceleration are vector quantities, and Newton's law is the vector law

$$\mathbf{F} = m\mathbf{a}.$$

Here $\mathbf{F}$ is the total (vector) force acting on a particle, m is the mass of the particle, and $\mathbf{a}$ is the (vector) acceleration of the particle. This law governs all plane motion.

We keep the t-integral definition of work that we decided on in the last section, except that the product of the vector $\mathbf{F}$ with the velocity vector $\mathbf{v}$ has to be the dot product to permit the analogous calculations to go through. The particle trajectory is now a plane path $\mathbf{x} = \boldsymbol{\gamma}(t)$.

DEFINITION *The work W done by the force $\mathbf{F}$ along the motion $\mathbf{x} = \boldsymbol{\gamma}(t)$ over the time interval $[t_1, t_2]$ is*

$$W = \int_{t_1}^{t_2} (\mathbf{F} \cdot \mathbf{v})\, dt = \int_{t_1}^{t_2} [\mathbf{F}(\boldsymbol{\gamma}(t)) \cdot \boldsymbol{\gamma}'(t)]\, dt.$$

The work-energy calculation is now the same as before. If $\mathbf{F}$ is the total force acting on the particle, so that $\mathbf{F} = m\, d\mathbf{v}/dt$ by Newton's law, and if $v = |\mathbf{v}| = (\mathbf{v} \cdot \mathbf{v})^{1/2}$ is the speed of the particle, then

$$W = \int_{t_1}^{t_2} (\mathbf{F} \cdot \mathbf{v})\, dt = \int_{t_1}^{t_2} \left(m\frac{d\mathbf{v}}{dt} \cdot \mathbf{v} \right) dt$$

$$= \int_{t_1}^{t_2} \frac{1}{2} m \frac{d}{dt} (\mathbf{v} \cdot \mathbf{v})\, dt = \frac{1}{2} m \int_{t_1}^{t_2} \frac{d}{dt} (v^2)\, dt$$

$$= \frac{1}{2} mv^2 \bigg]_{t_1}^{t_2} = \frac{1}{2} m(v_2)^2 - \frac{1}{2} m(v_1)^2.$$

So again:

The work done by the total force $\mathbf{F}$ acting on a particle along its motion is equal to the change in kinetic energy of the particle from the beginning to the end of the motion.

Now suppose that the force $\mathbf{F}$ governing the motion arises from a *field of force*, i.e., from a varying force $\mathbf{F} = \mathbf{F}(\mathbf{x})$ that depends only on the position $\mathbf{x}$. In the one-dimensional case we noted a very simple integral formula for the work done by this kind of force, and as a consequence we were led to the notion of potential energy and the law of conservation of energy. It is natural to ask whether similar things happen in the plane. To begin with:

Does the relationship $d\mathbf{x} = \mathbf{v}\,dt$ convert the work integral to a pure $\mathbf{x}$-integral,

$$\int_{t_i}^{t_1} (\mathbf{F} \cdot \mathbf{v})\,dt = \int_{x_1}^{x_2} \mathbf{F}(\mathbf{x}) \cdot d\mathbf{x}$$

as it did on the line? This turns out to be a more complicated question in the plane, and it leads to some new aspects of integration. Unlike the one-dimensional case, the $\mathbf{x}$-integral here has no meaning by itself; it can be viewed only as a new notation for the work integral, suggested by the relationship $d\mathbf{x} = \mathbf{v}\,dt$ and defined in the equation above. It remains to be established that there is some point to doing this. This definition is stated formally below, with two changes. First, the notation for the $\mathbf{x}$-integral is modified so as to indicate the path of integration rather than just its endpoints. This is necessary because we can go from $\mathbf{x}_0$ to $\mathbf{x}_1$ along many different paths, and the value of the integral may very well change from path to path.

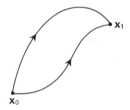

Also, since the same situation arises in other applications, the work-energy context will be discarded at this point. We shall state the definition analytically (i.e., independent of interpretation) as the definition of a *line integral*, and proceed in these neutral terms from now on.

DEFINITION   *If $\mathbf{x} = \boldsymbol{\gamma}(t)$ is any smooth path in the plane, defined over the parameter interval $[t_1, t_2]$, and if $\mathbf{p}(\mathbf{x}) = (p(x, y), q(x, y))$ is any vector field defined on a region containing the path $\boldsymbol{\gamma}$, then the* line integral *of the vector field $\mathbf{p}(\mathbf{x})$ along the path $\boldsymbol{\gamma}$, designated*

$$\int_{\gamma} \mathbf{p} \cdot d\mathbf{x} \qquad or \qquad \int_{\gamma} p\,dx + q\,dy$$

is defined to be

$$\int_{t_1}^{t_2}\left[\mathbf{p} \cdot \frac{d\mathbf{x}}{dt} \right] dt = \int_{t_1}^{t_2}\left(p\frac{dx}{dt} + q\frac{dy}{dt} \right) dt,$$

where for each t the vector field $\mathbf{p}(\mathbf{x})$ is evaluated at the path point $\mathbf{x} = \boldsymbol{\gamma}(t)$.

The precise definition is thus

$$\int_{\gamma} \mathbf{p} \cdot d\mathbf{x} = \int_{t_1}^{t_2} [\mathbf{p}(\boldsymbol{\gamma}(t)) \cdot \boldsymbol{\gamma}'(t)]\,dt.$$

REMARK   If a path $\boldsymbol{\gamma}$ has a finite number of corners, but is smooth between each adjacent pair of corners, then we form the line integral over $\boldsymbol{\gamma}$ by simply adding up the line integrals over its smooth pieces.

EXAMPLE 1 The line integral $\int_\gamma y\,dx + xy\,dy$ over the path γ given by $(x, y) = (t + t^3, t^2)$ on the parameter interval $[0, 1]$ has the value

$$\int_\gamma y\,dx + xy\,dy = \int_0^1 \left(y\frac{dx}{dt} + xy\frac{dy}{dt} \right) dt$$

$$= \int_0^1 [t^2(1 + 3t^2) + (t + t^3)t^2 2t]\,dt$$

$$= \int_0^1 (2t^6 + 5t^4 + t^2)\,dt$$

$$= \frac{2}{7} + 1 + \frac{1}{3}$$

$$= 1\frac{13}{21}.$$

EXAMPLE 2 Evaluate the line integral of the vector field $\mathbf{p}(\mathbf{x}) = (x, y^2)$ over the quarter of the unit circle in the first quadrant, parameterized by $(x, y) = (\cos\theta, \sin\theta)$.

Solution

$$\int_\gamma \mathbf{p} \cdot d\mathbf{x} = \int_\gamma x\,dx + y^2\,dy$$

$$= \int_0^{\pi/2} \left(x\frac{dx}{d\theta} + y^2\frac{dy}{d\theta} \right) d\theta$$

$$= \int_0^{\pi/2} [\cos\theta(-\sin\theta) + \sin^2\theta(\cos\theta)]\,d\theta$$

$$= \left[\frac{\cos^2\theta}{2} + \frac{\sin^3\theta}{3} \right]_0^{\pi/2}$$

$$= \frac{1}{3} - \frac{1}{2}$$

$$= -\frac{1}{6}.$$

Calculations involving line integrals depend on properties that a line integral inherits from its definition as a definite integral. Some properties carry over directly. For example, it follows directly from the definition of a line integral that if $\mathbf{p}(\mathbf{x})$ and $\mathbf{r}(\mathbf{x})$ are any two vector fields, then

$$\int_\gamma (\mathbf{p} + \mathbf{r}) \cdot d\mathbf{x} = \int_\gamma \mathbf{p} \cdot d\mathbf{x} + \int_\gamma \mathbf{r} \cdot d\mathbf{x}.$$

Also, if a path γ is broken up into a succession of "partial paths" γ_i by subdividing the parameter interval $[a, b]$ of γ,

$$a = t_0 < t_1 < \cdots < t_n = b,$$

γ_i being the restriction of γ to the parameter interval $[t_{i-1}, t_i]$, then

$$\int_{\gamma} \mathbf{p} \cdot d\mathbf{x} = \int_{\gamma_1} \mathbf{p} \cdot d\mathbf{x} + \int_{\gamma_2} \mathbf{p} \cdot d\mathbf{x} + \cdots + \int_{\gamma_n} \mathbf{p} \cdot d\mathbf{x}.$$

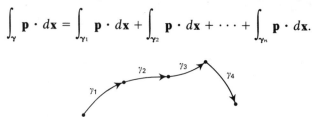

In this situation γ is sometimes called the sum of the partial paths γ_i, and is written

$$\gamma = \gamma_1 + \gamma_2 + \cdots + \gamma_n.$$

Other properties show up in new forms in the line-integral setting. In the remainder of this section we shall be concerned with ramifications of the substitution rule. We start by showing that a line integral over a path γ is independent of the particular way in which γ happens to be parameterized. This is the first indication that the line-integral concept is fruitful.

THEOREM 1   *A line integral is invariant under any change of path parameter that preserves path endpoints. That is, if*

1. $\mathbf{x} = \boldsymbol{\lambda}(s)$ *is a path with parameter interval* $[s_1, s_2]$,
2. $s = h(t)$ *is a differentiable change of path parameter defined on* $[t_1, t_2]$, *with* $s_1 = h(t_1)$ *and* $s_2 = h(t_2)$, *and*
3. $\mathbf{x} = \boldsymbol{\gamma}(t)$ *is the reparameterized path* $\boldsymbol{\gamma}(t) = \boldsymbol{\lambda}(h(t))$ *on the parameter interval* $[t_1, t_2]$, *then*

$$\int_{\gamma} \mathbf{p} \cdot d\mathbf{x} = \int_{\lambda} \mathbf{p} \cdot d\mathbf{x}.$$

Proof. This calculation combines the chain rule for functions of one variable with the direct substitution rule for the definite integral:

$$\int_{\gamma} \mathbf{p} \cdot d\mathbf{x} = \int_{t_1}^{t_2} [\mathbf{p}(\boldsymbol{\gamma}(t)) \cdot \boldsymbol{\gamma}'(t)] \, dt$$

$$= \int_{t_1}^{t_2} [\mathbf{p}(\boldsymbol{\lambda}(h(t))) \cdot \boldsymbol{\lambda}'(h(t))h'(t)] \, dt$$

$$= \int_{t_1}^{t_2} [\mathbf{p}(\boldsymbol{\lambda}(h(t))) \cdot \boldsymbol{\lambda}'(h(t))]h'(t) \, dt$$

$$= \int_{s_1}^{s_2} [\mathbf{p}(\boldsymbol{\lambda}(s)) \cdot \boldsymbol{\lambda}'(s)] \, ds$$

$$= \int_{\lambda} \mathbf{p} \cdot d\mathbf{x}. \quad \blacksquare$$

REMARK   If a change of parameter $s = h(t)$ *reverses* the path endpoints, then the line integral changes sign.

Proof. The calculation is the same down to the next to the last line, which is now

$$\int_{s_2}^{s_1} [\quad] \, ds = -\int_{s_1}^{s_2} = -\int_{\lambda} \mathbf{p} \cdot d\mathbf{x}. \quad \blacksquare$$

Now suppose that a path $\mathbf{x} = \boldsymbol{\gamma}(s)$ runs along a curve C from the initial point x_0 to the terminal point x_1. We can draw the curve C, mark the initial and terminal points, and indicate the direction of the path by an arrow. For example, the path might run along the parabola $y = x^2$ from the point $(-1, 1)$ to the point $(2, 4)$.

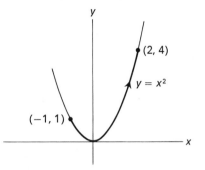

It is reasonable to call such a marked portion of a curve a *directed curve segment*. Theorem 1 says that all paths tracing the same directed curve segment give the same answer in line-integral calculations. For example, the parabolic arc marked as above contains all the essential information about the path of a line integral: any two paths that run along the parabola $y = x^2$, starting at $(-1, 1)$ and ending at $(2, 4)$, will determine the same line integral (of a given vector field). We can therefore relax the line-integral notation, replacing the reference to a particular parameterization $\boldsymbol{\gamma}$ by the directed curve segment C that it traces, as in

$$\int_C \mathbf{p} \cdot d\mathbf{x}.$$

EXAMPLE 3 Evaluate the line integral

$$\int_C y^2 \, dx - x^3 \, dy$$

where C is the above directed curve segment.

Solution. By the theorem, we can parameterize the curve segment in any manner consistent with the given initial and terminal points, and a simple choice is to take x as the parameter. The path is thus

$$(x, y) = (x, x^2)$$

over the x-interval $[-1, 2]$, and the integral is

$$\int_C y^2 \, dx - x^3 \, dy = \int_{-1}^{2} [(x^2)^2 - x^3(2x)] \, dx$$

$$= \int_{-1}^{2} (-x^4) \, dx = -\frac{x^5}{5}\Bigg]_{-1}^{2} = -\frac{33}{5}.$$

EXAMPLE 4   Evaluate the line integral of Example 3, where C this time is the unit square traced counterclockwise, as shown below.

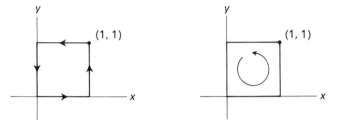

Solution. Here it is natural to use x as parameter along the horizontal segments and y along the vertical segments. Note that the integral along the top segment is then $-\int_0^1 y^2\,dx$, because it is traced in the opposite direction. The complete line integral is then

$$\int_C y^2\,dx - x^3\,dy = \int_0^1 0^2\,dx - \int_0^1 1^3\,dy - \int_0^1 1^2\,dx - \int_0^1 0^3\,dy$$

$$= -2.$$

EXAMPLE 5   A path along a directed curve segment could even reverse directions for a while and then change back. For example, let λ be the path along the circular arc $y = 1 - \sqrt{1 - x^2}$, from $x = -a$ to $x = a$, with x as parameter on this interval. Thus λ traces the arc once, from left to right.

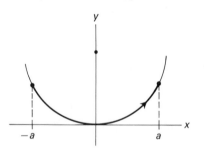

Now let γ be a path swept out by three beats of a pendulum swinging along this arc. Thus γ might be obtained from λ by the change of parameter

$$x = h(t) = a \cos t$$

on the t-interval $[-\pi, 2\pi]$. Note that this parameter change preserves the initial and terminal points of the path, since

$$-a = h(-\pi)$$
$$a = h(2\pi),$$

and Theorem 1 therefore applies. But γ traces the arc three times: across, back, and across again.

It is convenient to use the word *path* ambiguously for either a directed curve segment or (as formerly) for a parametric representation of a directed curve segment. Since the essence of a path is the directed curve segment it traces (according to Theorem 1), this ambiguity causes no trouble.

A path is called *closed* if it ends where it begins (i.e., if its terminal point is its initial point). The unit square path in Example 4 was closed. If C' and C'' are two paths running from $\mathbf{x}_0$ to $\mathbf{x}_1$, then $C'-C''$ is a closed path starting and ending at $\mathbf{x}_0$.

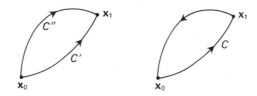

Conversely, a closed path C starting and ending at $\mathbf{x}_0$ can always be decomposed into two pieces and then written $C = C'-C''$ by cutting it at any intermediate point x_1 and letting C'' be the second piece traced backwards. A path may cross itself one or more times.

A path that does not cross itself is called *simple*. Thus a *simple closed* path $\boldsymbol{\gamma}$ on the parameter interval $[a, b]$ has the property that $\boldsymbol{\gamma}(t) = \boldsymbol{\gamma}(t')$ for $t < t'$ if and only if $t = a$ and $t' = b$.

A path along a curve that crosses itself has to be marked more carefully, because there are two possibilities for where the path goes at a crossing point. An example is shown below.

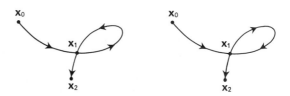

Since the loop is traced in opposite directions in the two paths, the line integral along that piece of the path will change signs. In more detail, suppose that we are given a vector field $\mathbf{p}(\mathbf{x})$, and suppose that the line integral of $\mathbf{p}$ along the first path, call it I, is broken up into a sum of three terms

$$I = I_0 + I_1 + I_2,$$

where I_0 is the integral along the piece of the path from $\mathbf{x}_0$ to $\mathbf{x}_1$, I_2 from $\mathbf{x}_1$ to $\mathbf{x}_1$ (i.e., around the loop as first marked), and I_2 from $\mathbf{x}_1$ to $\mathbf{x}_2$. Then the line integral of $\mathbf{p}$ along the *second* path is

$$I_0 - I_1 + I_2.$$

These two line integrals differ by $2I_1$.

PROBLEMS FOR SECTION 2

1. Compute the integral of the vector field $(-y, x)$ around the unit circle $x^2 + y^2 = 1$ traced counterclockwise.

2. Compute the integral of the vector field in the above problem around the unit square (with opposite corners $(0, 0)$ and $(1, 1)$), again traced "counterclockwise."

In Problems 3 through 7, compute $\int_C \mathbf{p} \cdot d\mathbf{x}$ for the given vector field $\mathbf{p}$ and directed curve segment C.

3. $\mathbf{p}(x, y) = (xy, x^2)$; C is the line segment from $(0, 1)$ to $(2, -3)$.

4. $\mathbf{p}(x, y) = (xy, x^2)$; C is the broken line going from $(0, 1)$ to $(2, -3)$ in a vertical step followed by a horizontal step.

5. $\mathbf{p}(x, y) = xy\mathbf{i} + x^2\mathbf{j}$; C is the graph of $y = 1 - x^2$ from $(0, 1)$ to $(2, -3)$.

6. $\mathbf{p}(x, y) = xy\mathbf{i} + x^2\mathbf{j}$; C is given parametrically by $\mathbf{x} = (2t^2, 1 - 4t)$ on $[0, 1]$.

7. a) $\mathbf{p}(\mathbf{x}) = (y^2, 2xy)$; C is the curve of Problem 1.
   b) C is the curve of Problem 3.

8. a) Compute $\dfrac{d}{dt}(xy^2)$ supposing that both x and y are functions of t.
   b) With the above answer in mind, show that

$$\int_C y^2\, dx + 2xy\, dy = x_1 y_1^2 - x_0 y_0^2$$

   for *any* path C from (x_0, y_0) to (x_1, y_1).

9. Show that

$$\int_C \frac{x\, dx + y\, dy}{\sqrt{x^2 + y^2}} = \sqrt{x_1^2 + y_1^2} - \sqrt{x_0^2 + y_0^2}$$

   where C is any path from (x_0, y_0) to (x_1, y_1) and avoiding the origin. (*Hint:* Compute $d\sqrt{x^2 + y^2}/dt$.)

10. Show that

$$\int_C \frac{dx}{y} - \frac{x\, dy}{y^2} = \frac{x_1}{y_1} - \frac{x_0}{y_0}$$

   along any path from $\mathbf{x}_0$ to $\mathbf{x}_1$ that does not touch the x-axis.

11. Compute the work done by the force $\mathbf{F}(\mathbf{x}) = (x + y, x - y)$ along the path $xy^2 = 4$ from $(1, 2)$ to $(4, 1)$.

12. Compute the work done by the force

$$\mathbf{F}(\mathbf{x}) = \frac{\mathbf{i}}{y} - \frac{x\mathbf{j}}{y^2}$$

   along the line segment from $(1, 2)$ to $(2, 4)$.

13. Show that if

$$\mathbf{F}(\mathbf{x}) = \left(\frac{x}{r^3}, \frac{y}{r^3}\right), \qquad \text{where} \qquad r = \sqrt{x^2 + y^2},$$

   then the work done by $\mathbf{F}$ is the same for all paths from $\mathbf{x}_0$ to $\mathbf{x}_1$. (*Hint:* Compute $d(1/r)/dt$.)

The line integral definition

$$\int_C \mathbf{p} \cdot d\mathbf{x} = \int_{t_1}^{t_2} [\mathbf{p}(\boldsymbol{\gamma}(t)) \cdot \boldsymbol{\gamma}'(t)]\, dt$$

works just as well in three dimensions as in two, with the same proof of independence of parameter. The space integral is conventionally expressed as

$$\int_C p\,dx + q\,dy + r\,dz \qquad \text{or} \qquad \int_C p_1\,dx + p_2\,dx + p_2\,dz.$$

Compute $\int_C \mathbf{p} \cdot d\mathbf{x}$ for the given vector field and path, where:

14. $\mathbf{p}(\mathbf{x}) = (x + y, y + z, z + x)$, and C is the line segment from $(1, 2, 1)$ to $(2, 0, 3)$.

15. $\mathbf{p}(\mathbf{x}) = x\mathbf{i} + y\mathbf{j} + z\mathbf{k}$, and C is the curve $(x, y, z) = (t, t^2, t^3)$, from $(0, 0, 0)$ to $(1, 1, 1)$.

16. Show that

$$\int_C yz\,dx + zx\,dy + xy\,dz = x_1 y_1 z_1 - x_0 y_0 z_0$$

along any path from $\mathbf{x}_0$ to $\mathbf{x}_1$.

17. Find the work done by the force

$$\mathbf{F}(\mathbf{x}) = (x, y, z)$$

along the path $y = 2x$, $z = x^2$ from $x = 0$ to $x = 2$.

3. GRADIENT VECTOR FIELDS

In our earlier one-dimensional discussion we defined the *potential energy* $V(x)$ of a varying force $F(x)$ by

$$V'(x) = -F(x).$$

That is, V is the negative of an antiderivative of F, and is uniquely determined up to an additive constant of integration. Since every continuous function has an antiderivative, potential functions always exist.

The analogue for a vector force field $\mathbf{F}(\mathbf{x})$ in the plane would be a function $V(x, y)$ such that

$$\text{grad } V(x, y) = -\mathbf{F}(x, y).$$

In the purely mathematical setting, the minus sign associated with the potential energy is generally discarded:

DEFINITION *A potential function for the vector field* $\mathbf{p}(\mathbf{x})$ *is any function* ϕ *such that*

$$\text{grad } \phi(\mathbf{x}) = \mathbf{p}(\mathbf{x}).$$

Then the potential energy of the field would be $-\phi$ *plus a suitable constant.*

But here is where the big difference between the line and higher dimensional spaces shows itself: in general, a plane vector field $\mathbf{p}(\mathbf{x})$ cannot be expressed as the gradient of a function. The plane situation is thus richer in possibilities than the line, and requires deeper investigation. For example, are there conditions on a vector field $\mathbf{p}(\mathbf{x})$ that will guarantee it to be a gradient field? If it is not, can we find a way of measuring by how much it fails to be a gradient field, and interpret this discrepancy in some meaningful way?

We first note that if $\mathbf{p}(\mathbf{x})$ *is* a gradient field, then the results of Section 1 generalize completely.

THEOREM 2   *Let* $\mathbf{p}(\mathbf{x}) = \mathbf{p}(x, y)$ *be a gradient field on a region G in the plane, with potential function* $\phi(x, y)$ *on G, and let* $\mathbf{x}_0$ *and* $\mathbf{x}_1$ *be any two points of G. Then*

$$\int_C \mathbf{p}\, d\mathbf{x} = \phi(\mathbf{x}_1) - \phi(\mathbf{x}_0)$$

for any path C in G that runs from $\mathbf{x}_0$ *to* $\mathbf{x}_1$.

Proof. If C is given parametrically by $\mathbf{x} = \boldsymbol{\gamma}(t)$ on the interval $[a, b]$, then

$$\int_C \mathbf{p} \cdot d\mathbf{x} = \int_a^b [\mathbf{p}(\boldsymbol{\gamma}(t)) \cdot \boldsymbol{\gamma}'(t)]\, dt$$

$$= \int_a^b [\mathrm{grad}\ \phi(\boldsymbol{\gamma}(t)) \cdot \boldsymbol{\gamma}'(t)]\, dt$$

$$= \int_a^b \frac{d}{dt}(\phi(\boldsymbol{\gamma}(t)))\, dt$$

$$= \phi(\boldsymbol{\gamma}(b)) - \phi(\boldsymbol{\gamma}(a))$$

$$= \phi(\mathbf{x}_1) - \phi(\mathbf{x}_0). \quad \blacksquare$$

EXAMPLE 1   Compute the line integral of the field $\mathbf{p}(\mathbf{x}) = (2xy, x^2)$ along the path $(x, y) = (t^3 - t, t^2)$ over the t-interval $[-1, 2]$.

Solution. The given field is the gradient of $\phi(x, y) = x^2 y$, so the line integral equals

$$\phi(x_1, y_1) - \phi(x_0, y_0) = \phi(6, 4) - \phi(0, 1)$$
$$= 6^2 \cdot 4 - 0^2 \cdot 1 = 144.$$

COROLLARY   *Let* $\mathbf{F} = \mathbf{F}(\mathbf{x})$ *be a gradient force field, with potential energy* $V(\mathbf{x}) = -\phi(\mathbf{x})$, *and suppose that* $\mathbf{F}$ *is the only force acting on a moving particle of mass m. Then the total energy of the particle*

$$E = \frac{1}{2} mv^2 + V(\mathbf{x})$$

remains constant during the motion.

Proof. The work-energy law from Section 2 combines with Theorem 2 above to give

$$\frac{1}{2} mv^2 - \frac{1}{2} mv_0^2 = \int_C \mathbf{F} \cdot d\mathbf{x} = \phi(\mathbf{x}) - \phi(\mathbf{x}_0) = V(\mathbf{x}_0) - V(\mathbf{x}),$$

where C is the path the particle has followed from $\mathbf{x}_0$ to $\mathbf{x}$. Thus $\frac{1}{2}mv^2 + V(\mathbf{x})$ has the constant value $\frac{1}{2}mv_0^2 + V(\mathbf{x}_0)$. $\quad \blacksquare$

A particle that is subject only to a gradient force field thus obeys the law of conservation of energy, and such a field is said to be *conservative.*

EXAMPLE 2 A particle of mass 3 moves from the point $(2, 0)$ to the point $(0, \frac{1}{2})$ under the action of the force field $\mathbf{F}(x, y) = (-x/r^3, -y/r^3)$, where $r = \sqrt{x^2 + y^2}$. Its initial speed is 1. What is its speed at the point $(0, \frac{1}{2})$?

Solution. The given field is the gradient field of $1/r$, so $V(\mathbf{x}) = -1/r = -1/\sqrt{x^2 + y^2}$. At $(2, 0)$ the total energy of the particle is

$$\frac{1}{2} mv^2 + V(\mathbf{x}) = \frac{3}{2}(1)^2 - \frac{1}{2} = 1.$$

Since the total energy remains constant during its motion, at the point $(0, \frac{1}{2})$ we have

$$1 = \frac{3}{2}v^2 - \frac{1}{1/2}, \qquad v = \sqrt{2}.$$

Theorem 2 shows that the line integral of a gradient field is *independent of path:* it depends on the path endpoints $\mathbf{x}_0$ and $\mathbf{x}_1$, but has the same value for all paths from $\mathbf{x}_0$ to $\mathbf{x}_1$.

The converse is also true.

THEOREM 3 *Let $\mathbf{p}(\mathbf{x}) = (p_1(x), p_2(x))$ be a continuous vector field on a plane region G having line integrals in G that are independent of path in the above sense. Then $\mathbf{p}(\mathbf{x})$ is a gradient field.*

Proof. Choose any fixed point $\mathbf{x}_0 = (x_0, y_0)$ in G and define $\phi(\mathbf{x})$ by

$$\phi(\mathbf{x}) = \int_{\mathbf{x}_0}^{\mathbf{x}} \mathbf{p}(\mathbf{x}) \cdot d\mathbf{x}$$

where the integral refers to the common value of all line integrals $\int_C \mathbf{p} \cdot d\mathbf{x}$ along paths C running from $\mathbf{x}_0$ to $\mathbf{x}_1$. We claim that then ϕ is a smooth function, and that

$$\operatorname{grad} \phi = \mathbf{p}.$$

It will be sufficient to show that $D_1\phi = p_1$. To do this, note that

$$\phi(x + \Delta x, y) - \phi(x, y) = \int_x^{x+\Delta x} p_1(s, y) \, ds,$$

because the ϕ difference is given by the line integral along *any* path from (x, y) to $(x + \Delta x, y)$, and the horizontal line segment with length s as parameter is the simplest. By the integral form of the mean-value theorem, the integral on the right above equals

$$p_1(X, y) \, \Delta x$$

for some number X between x and $x + \Delta x$. Thus

$$\lim_{\Delta x \to 0} \frac{\phi(x + x, y) - \phi(x, y)}{\Delta x} = \lim_{\Delta x \to 0} p_1(X, y) = p_1(x, y),$$

so

$$D_1\phi(x, y) = p_1(x, y)$$

as claimed. ∎

Here is another way of saying "independence of path".

THEOREM 4   *The vector field* $\mathbf{p}(\mathbf{x})$ *is a gradient field if and only if*

$$\int_C p(\mathbf{x}) \cdot d\mathbf{x} = 0$$

for all closed paths C *in* R.

Proof. If C is a closed path, then its initial and terminal points are the same, so if $\phi(\mathbf{x})$ is a potential function for $\mathbf{p}(\mathbf{x})$ then

$$\int_C \mathbf{p} \cdot d\mathbf{x} = \phi(\mathbf{x}_1) - \phi(\mathbf{x}_0)$$
$$= \phi(\mathbf{x}_0) - \phi(\mathbf{x}_0) = 0.$$

Conversely, if the integral of $\mathbf{p}$ around every closed path is 0, then its integral is independent of path, because if C' and C'' are any two paths from x_0 to x_1 then $C = C'-C''$ is a closed path, so

$$0 = \int_C \mathbf{p} \cdot d\mathbf{x} = \int_{C'} \mathbf{p} \cdot d\mathbf{x} - \int_{C''} \mathbf{p} \cdot d\mathbf{x}$$

and

$$\int_{C'} = \int_{C''}. \quad \blacksquare$$

There are now two questions: How do we recognize when a field is a gradient field? And the practical one: How then do we find a potential function ϕ?

The first question has a deceptively easy answer. If $p_1 = \partial\phi/\partial x$ and $p_2 = \partial\phi/\partial y$, then

$$\frac{\partial p_1}{\partial y} = \frac{\partial p_2}{\partial x},$$

because this is just the equality of the mixed partials $\partial^2\phi/\partial y\, \partial x = \partial^2\phi/\partial x\, \partial y$. In other words, supposing that the component functions p_1 and p_2 have continuous partial derivatives:

LEMMA 1   *A necessary condition for the vector field* $\mathbf{p}(\mathbf{x})$ *to be a gradient field is the identity*

$$\frac{\partial p_1}{\partial y} = \frac{\partial p_2}{\partial x}.$$

A field having this property is said to be *closed*. So now the question is whether a closed field is necessarily a gradient field. The answer is tricky: *it depends on whether or not the domain region G has holes in it*. If it does not (and is therefore what is called a *simply connected* domain) then everything works out. We shall consider only the case of a region that is *rectangular*, in the sense that its boundary is a rectangle with sides parallel to the coordinate axes. The crux of the matter is Green's theorem, which will be taken up in the next section. It has the following theorem as a corollary.

THEOREM 5   *Let* $\mathbf{p}(\mathbf{x})$ *be a closed vector field on a region G and let C be the boundary of a rectangular subregion of G. Then*

$$\int_C p(\mathbf{x}) \cdot d\mathbf{x} = 0.$$

From this we can prove:

THEOREM 6   *If* $\mathbf{p}(\mathbf{x})$ *is a closed vector field defined on a rectangular region G then* $\mathbf{p}(\mathbf{x})$ *is a gradient field.*

Proof. Fix any point (x_0, y_0) in G. Then we can go to any other point (x, y) along a path C_1 consisting of a vertical segment followed by a horizontal segment, and also along a path C_2 consisting of a horizontal segment followed by a vertical segment.

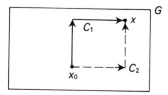

(If $\mathbf{x}$ is directly above or below $\mathbf{x}_0$, then the horizontal leg is missing from each path—or is considered to be present as a trivial segment of length zero. And the vertical leg is missing when $\mathbf{x}$ is directly to the right or left of $\mathbf{x}_0$.)

By Green's theorem

$$\int_{C_1} \mathbf{p} \cdot d\mathbf{x} = \int_{C_2} \mathbf{p} \cdot d\mathbf{x}$$

since $\int_{C_2 - C_1} = 0$. We define $\phi(x, y)$ to be the common value of these two line integrals, and can then prove $D_1\phi = p_1$ and $D_2\phi = p_2$ just as in Theorem 3. ∎

EXAMPLE 3   We can follow the scheme of Theorem 6 to compute a potential function ϕ. For example, we see by inspection that

$$\mathbf{p}(x, y) = (2x + y, x + 2y)$$

is closed, since $\partial p_1/\partial y = 1 = \partial p_2/\partial x$. The theorem then guarantees that we can obtain ϕ by integrating $\mathbf{p}$ "up and over" from (x_0, y_0) to (x, y):

$$
\begin{aligned}
\phi(x, y) &= \int_{y_0}^{y} (x_0 + 2y)\, dy + \int_{x_0}^{x} (2x + y)\, dx \\
&= (x_0 y + y^2)_{y_0}^{y} + (x^2 + xy)_{x_0}^{x} \\
&= (x_0 y + y^2) - (x_0 y_0 + y_0^2) + (x^2 + xy) - (x_0^2 + x_0 y) \\
&= x^2 + xy + y^2 - (x_0^2 + x_0 y_0 + y_0^2).
\end{aligned}
$$

In a sense, half of the computation in the above example is wasted, because all we need is the fact that

$$\phi(x, y) = x^2 + xy + y^2 + C.$$

The line-integral computation of Theorem 6 and Example 3 is needed theoretically, in order to guarantee the existence of a potential function ϕ, but in practice there is a simpler procedure that eliminates the wasted computation.

EXAMPLE 4 We recompute the potential function ϕ of the above example. Since $\partial\phi/\partial x = 2x + y$, we can try to find ϕ by "partially integrating" $2x + y$ with respect to x. That is, we treat y as a constant and find an antiderivative with respect to x. Then

$$\phi(x, y) = x^2 + xy + C(y),$$

where the constant of this x integration is a function of y that still must be found. By hypothesis, $\partial\phi/\partial y = x + 2y$, so

$$x + 2y = \frac{\partial}{\partial y}\phi(x, y) = x + C'(y).$$

Thus $C'(y) = 2y, C(y) = y^2 + C,$ and

$$\phi(x, y) = x^2 + xy + y^2 + C.$$

This new procedure is guaranteed to work. Here it is again, in general terms. We first find a function $\alpha(x, y)$ by partially integrating $p_1(x, y)$ with respect to x. In the above example $\alpha(x, y) = x^2 + xy$. Since

$$\frac{\partial}{\partial x}(\phi(x, y) - \alpha(x, y)) = p_1(x, y) - p_1(x, y) = 0,$$

the difference $\phi - \alpha$ is a function of y only, call it $C(y)$. We are thus *guaranteed* that the computation

$$C'(y) = \frac{\partial}{\partial y}(\phi - \alpha) = p_2(x, y) - \frac{\partial}{\partial y}\alpha(x, y)$$

will turn out to involve only y. In the above example, it was

$$C'(y) = (x + 2y) - x = 2y.$$

We then integrate to find $C(y)$ and are done.

EXAMPLE 5 The field $(2xy^3, 3x^2y^2)$ is closed, since $\partial p_1/\partial y = 6xy^2 = \partial p_2/\partial x$. We can therefore find its potential function as above:

$$\phi(x, y) = \int (2xy^3)\, dx = x^2y^3 + C(y),$$

$$C'(y) = \frac{\partial}{\partial y}[\phi(x, y) - x^2y^3] = p_2(x, y) - \frac{\partial}{\partial y}x^2y^3$$

$$= 3x^2y^2 - 3x^2y^2 = 0,$$

so

$$C(y) = C \quad \text{and} \quad \phi(x, y) = x^2y^3 + C.$$

Gradient fields in space have essentially the same theory as in the plane, but the calculations are more complicated because of the added dimension. We first note:

LEMMA 2 *A necessary condition for*

$$\mathbf{p}(\mathbf{x}) = (p_1(x, y, z), p_2(x, y, z), p_3(x, y, z))$$

to be a gradient field is the set of three equations:

$$\frac{\partial p_1}{\partial y} = \frac{\partial p_2}{\partial x},$$

$$\frac{\partial p_2}{\partial z} = \frac{\partial p_3}{\partial y},$$

$$\frac{\partial p_3}{\partial x} = \frac{\partial p_1}{\partial z}.$$

Proof. If grad $\phi = \mathbf{p}$, then the above equations are the identities for mixed partials:

$$\frac{\partial^2 \phi}{\partial y\, \partial x} = \frac{\partial^2 \phi}{\partial x\, \partial y},$$

$$\frac{\partial^2 \phi}{\partial z\, \partial y} = \frac{\partial^2 \phi}{\partial y\, \partial z},$$

$$\frac{\partial^2 \phi}{\partial x\, \partial z} = \frac{\partial^2 \phi}{\partial z\, \partial x}.$$

A field satisfying the above conditions is called *closed*. ∎

THEOREM 7 *If G is a rectangular box region in space and if* $\mathbf{p}(\mathbf{x})$ *is a closed vector field on G, then* $\mathbf{p}$ *is a gradient field.*

Sketch of Proof. Fix any point (x_0, y_0, z_0) in G. Then we can go to any other point (x, y, z) in G along a path B_1 consisting of three line segments that are parallel to the x, y, and z axes in that order, or along a path B_2 consisting of line segments parallel to the y, z, and x axes in order, or, finally, along a path B_3 consisting of line segments parallel to the z, x, and y axes in order. We claim that

$$\int_{B_1} \mathbf{p} \cdot d\mathbf{x} = \int_{B_2} \mathbf{p} \cdot d\mathbf{x} = \int_{B_3} \mathbf{p} \cdot d\mathbf{x}.$$

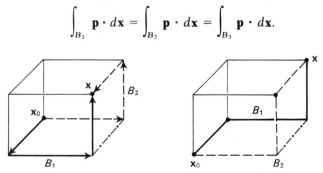

The figure shows that $B_1 - B_2$ can be rewritten as $C_1 + C_2$, where C_1 is a horizontal square traced counterclockwise when viewed from above and C_2

is a vertical square parallel to the xz-plane traced counterclockwise when viewed from the right. But $\int_{C_2} = 0$ by the earlier two-dimensional theory for the variables y and z, and similarly $\int_{C_1} = 0$. Thus

$$\int_{B_1 - B_2} \mathbf{p} \cdot d\mathbf{x} = \int_{C_1 + C_2} \mathbf{p} \cdot d\mathbf{x} = 0,$$

so

$$\int_{B_1} \mathbf{p} \cdot d\mathbf{x} = \int_{B_2} \mathbf{p} \cdot d\mathbf{x}.$$

We can now define $\phi(x, y, z)$ as the common value of the above three line integrals, and the proof that $D_i \phi = p_i, i = 1, 2, 3$, is the same as before. ∎

In practice we find the potential function ϕ by the second of the two earlier methods.

EXAMPLE 6   Show that

$$\mathbf{p}(\mathbf{x}) = (2x + y, x + 2y + z, y + 2z)$$

is a gradient field and find its potential function ϕ.

Solution. Since

$$\frac{\partial p_1}{\partial y} = 1 = \frac{\partial p_2}{\partial x},$$

$$\frac{\partial p_2}{\partial z} = 1 = \frac{\partial p_3}{\partial y},$$

$$\frac{\partial p_3}{\partial x} = 0 = \frac{\partial p_1}{\partial z},$$

the field is closed. In order to find ϕ we can partially integrate p_1 with respect to x, and get

$$\phi(x, y, z) = x^2 + xy + C(y, z),$$

where the constant of integration must now be taken to be a function of the two remaining variables. Then

$$C(y, z) = \phi - (x^2 + xy),$$

so

$$\frac{\partial C}{\partial y} = \frac{\partial \phi}{\partial y} - x = p_2(x, y, z) - x$$

$$= (2y + x + z) - x = 2y + z.$$

Therefore

$$C(y, z) = \int (2y + z)\, dy = y^2 + yz + D(z).$$

Finally,

$$D(z) = C(y, z) - (y^2 + yz)$$
$$= \phi(x, y, z) - (x^2 + xy) - (y^2 + yz),$$

so

$$D'(z) = \frac{\partial \phi}{\partial z} - 0 - y = p_3 - y$$
$$= (2z + y) - y = 2z,$$

and

$$D(z) = z^2 + C.$$

Altogether

$$\phi(x, y, z) = x^2 + xy + C(y, z)$$
$$= x^2 + xy + y^2 + yz + D(z)$$
$$= x^2 + xy + y^2 + yz + z^2 + C.$$

Note again how the variables dropped out when they were supposed to, first x in the calculation of $\partial C(y, z)/\partial y$, and then y in the calculation of $C'(z)$. It can be proved, by following the above steps in general language, that this has to happen.

If $r = \sqrt{x^2 + y^2}$, then a "central" potential $\phi(\mathbf{x}) = \psi(r)$ has the gradient

$$\text{grad } \phi(\mathbf{x}) = \left(\frac{x\psi'(r)}{r}, \frac{y\psi'(r)}{r} \right) = \frac{\psi'(r)}{r} (x, y).$$

Vector fields in this form are easily recognizable, and then ψ can be found by a single integration.

EXAMPLE 7   The field

$$\left(\frac{x}{x^2 + y^2}, \frac{y}{x^2 + y^2} \right) = \left(\frac{x}{r^2}, \frac{y}{r^2} \right) = \frac{1}{r^2} (x, y)$$

is of the above form with $\psi'(r) = 1/r$. Therefore

$$\psi(r) = \log r, \quad \text{and} \quad \phi(\mathbf{x}) = \tfrac{1}{2} \log(x^2 + y^2).$$

The same thing works in space.

EXAMPLE 8   The field

$$\left(\frac{x}{r}, \frac{y}{r}, \frac{z}{r} \right) = \frac{1}{r} (x, y, z) = \frac{1}{r} \mathbf{x}$$

is of the form $\dfrac{\psi'(r)}{r} (x, y, z) = \dfrac{\psi'(r)}{r} \mathbf{x}$ with $\psi'(r) = 1$. So $\psi(r) = r$, and $\phi(\mathbf{x}) = \sqrt{x^2 + y^2 + z^2}$.

PROBLEMS FOR SECTION 3

In each of the following problems verify that the given vector field is closed and find a potential function by the line-integral method of Theorem 6 and Example 3.

1. $\mathbf{p}(x, y) = (x, y)$ 2. $\mathbf{p}(x, y) = (y, x)$

3. $\mathbf{p}(x, y) = (2xy, x^2)$

4. $\mathbf{p}(x, y) = (x^2 + 2xy, x^2 + y^2)$

5. $\mathbf{p}(x, y) = (2xy + y^2, 2xy + x^2)$

6. $\mathbf{p}(x, y) = (y/x^2, -1/x)$

In each of the following problems verify that the given vector field is closed, and find a potential function by the method of Examples 4 through 6.

7. $\mathbf{p}(x, y) = (xy^2, x^2y)$

8. $\mathbf{p}(x, y) = (x^2 + 2xy, x^2 + y^2)$

9. $\mathbf{p}(x, y) = (2xy + y^2, 2xy + x^2)$

10. $\mathbf{p}(x, y) = (1/y, -x/y^2)$

11. $\mathbf{p}(x, y, z) = (z, y, x)$

12. $\mathbf{p}(x, y, z) = (y + z, x + y, x + z)$

13. $\mathbf{p}(x, y, z) = \left(\dfrac{-x}{(x^2 + y^2 + z^2)^3}, \dfrac{-y}{(x^2 + y^2 + z^2)^3}, \dfrac{-z}{(x^2 + y^2 + z^2)^3} \right)$

Each of the following integrals is of the form

$$\int_{x_0}^{x_2} \mathbf{p} \cdot d\mathbf{x},$$

implying independence of path. Check this in each case, and then compute the integral. (In some cases it may be easier to compute the integral along a simple path than to find a potential function.)

14. $\displaystyle\int_{(1,1)}^{(2,-2)} (x + y) \, dx + (x - y) \, dy$

15. $\displaystyle\int_{(0,0)}^{(1,3)} (2x - y) \, dx + (2y - x) \, dy$

16. $\displaystyle\int_{(1,0)}^{(0,1)} (x^2 + y^2) \, dx + 2xy \, dy$

17. $\displaystyle\int_{(-1,-1)}^{(1,1)} x^2y^3 \, dx + x^3y^2 \, dy$

18. $\displaystyle\int_{(1,1)}^{(e,e)} x \log y^2 \, dx + \dfrac{x^2}{y} \, dy$

19. $\displaystyle\int_{(1,2)}^{(2,1)} \dfrac{x \, dx + y \, dy}{(x^2 + y^2)} = \int_{(1,2)}^{(2,1)} \dfrac{x \, dx + y \, dy}{r^2}$

20. $\displaystyle\int_{(1,2)}^{(2,3)} \dfrac{x \, dx + y \, dy}{r^3}$

21. $\displaystyle\int_{(0,0)}^{(-1,2)} x(x^2 + y^2)^{3/2} \, dx + y(x^2 + y^2)^{3/2} \, dy$

22. $\displaystyle\int_{(1,0)}^{(0,e)} \dfrac{\log r}{r} (x \, dx + y \, dy), \quad r = \sqrt{x^2 + y^2}$

23. Show that the field

$$\mathbf{p}(\mathbf{x}) = \left(\dfrac{-y}{x^2 + y^2}, \dfrac{x}{x^2 + y^2} \right)$$

is closed where defined. Show however that it is not a gradient field by computing its integral around the unit circle. (Use $(x, y) = (\cos \theta, \sin \theta)$.)

4. GREEN'S THEOREM

In this section we consider an *arbitrary* vector field $\mathbf{p}(\mathbf{x})$, presumably not a gradient field. The line integral of $\mathbf{p}(\mathbf{x})$ about a closed path C,

$$\int_C \mathbf{p}(\mathbf{x}) \cdot d\mathbf{x},$$

is then different from zero in general, and we shall see that it can be viewed as the *circulation* of the field about C. To this end we introduce a second interpretation of a vector field.

The flow of a fluid can be described by its *velocity* field, which assigns to each point $\mathbf{x}$ in space and each time t the vector velocity

$$\mathbf{v} = \mathbf{f}(\mathbf{x}, t)$$

of the fluid at that point in space and time. For example, the flow of air in the atmosphere is roughly described by a succession of charts of wind velocities. A better example, for the simplified situation we wish to consider, is provided by a shallow, quiet river. We suppose, first, that the water is in steady-state motion, i.e., that its velocity field is independent of time. Second, we confine our attention to the flow near the surface, and suppose that the motion is (essentially) two-dimensional. This means that the water moves in horizontal planes, and the motion is the same in all planes to the given depth. So the flow is described by a two-dimensional velocity vector field $\mathbf{v}(\mathbf{x})$ defined on the plane region determined by the river bed.

We now raise the following question: Can we tell, by working with the velocity field, whether there is any *circulation* in the flow of the stream? For example, can we recognize a whirlpool?

The first step is to compute the circulation around a closed path. Consider any closed path C and let $\mathbf{x} = \boldsymbol{\gamma}(s)$ be the parametric representation of C by its arc length measured from some field point. Then $\mathbf{v} \cdot \boldsymbol{\gamma}' = \mathbf{v}(\boldsymbol{\gamma}(s)) \cdot \boldsymbol{\gamma}'(s)$ is the component of the velocity vector $\mathbf{v}$ in the direction of the path tangent vector at the point $\boldsymbol{\gamma}(s)$. The integral

$$\int_C \mathbf{v} \cdot d\mathbf{x} = \int_{s_0}^{s_1} [\mathbf{v}(\boldsymbol{\gamma}(s)) \cdot \boldsymbol{\gamma}'(s)]\, ds$$

is therefore the accumulation over C of the tendency of the velocity field to be in the direction of C. If the integral is positive then there is more fluid velocity along the positive direction of C than along its negative direction, and the fluid shows net positive rotation about C. Thus the above line integral can be interpreted as the rate at which the fluid is circulating about C. Again:

If a vector field $\mathbf{p}(\mathbf{x})$ is interpreted as the velocity field of a fluid flow, then its line integral around a closed path C can be interpreted as the rate at which the fluid circulates about C. We shall therefore call $\int_C \mathbf{p} \cdot d\mathbf{x}$ the *circulation* of the vector field $\mathbf{p}$ about the closed path C.

REMARK Although we used arc length as parameter in order to obtain this interpretation, the line integral can be computed with any convenient parameterization, since it is independent of the choice of parameters.

EXAMPLE 1 Suppose we consider a circular disc of radius R rotating at a constant angular velocity of ω radians per second. (A rigid body doesn't seem much like a fluid, but this might seem to be a reasonable model of pure whirlpool motion.)

We consider the circulation of the "fluid" around the concentric circle of radius r. In the motion around this circle the velocity vector has the direction of the tangent vector and the constant magnitude ωr. So its circulation about the circle is $\omega r (2\pi r) = 2\omega \pi r^2$. In more detail, the motion

is given by

$$\mathbf{x} = (x, y) = (r \cos \omega t, r \sin \omega t),$$

with t running from 0 to $2\pi/\omega$. It has the velocity

$$\mathbf{v} = \frac{dx}{dt} = r\omega(-\sin \omega t, \cos \omega t).$$

The circulation integral, with t as parameter, is

$$\int_C \mathbf{v} \cdot d\mathbf{x} = \int_0^{2\pi/\omega} \left(\frac{dx}{dt} \cdot \frac{dx}{dt}\right) dt = (r\omega)^2 \int_0^{2\pi/\omega} dt = 2\omega\pi r^2.$$

With s as parameter, we get

$$\int_C \mathbf{v} \cdot d\mathbf{x} = \int_0^{2\pi r} \left(\frac{dx}{dt} \cdot \frac{dx}{ds}\right) ds = \omega r \int_0^{2\pi r} ds = 2\omega\pi r^2.$$

EXAMPLE 2   Continuing the above example, we compute the circulation about a sector of radius r and angular opening θ. The path C now consists of a circular arc C_0 of length $r\theta$ and two radii as shown below.

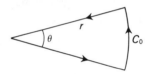

But the flow is perpendicular to the radii, so the integrand $\mathbf{v}(\mathbf{x}) \cdot d\mathbf{x}/ds$ is zero along these two pieces of the path. The circulation rate is thus just

$$\int_{C_0} \mathbf{v} \cdot d\mathbf{x} = \omega r \int_0^{\theta r} ds = \omega r \cdot \theta r$$

$$= 2\omega(\tfrac{1}{2}\theta r^2) = 2\omega A,$$

where A is the area of the sector.

EXAMPLE 3   Finally, we consider the circulation of the same flow about the "polar rectangle" bounded by circles of radii r_1 and r_2 ($r_1 < r_2$) and with an angular opening θ.

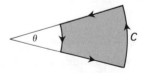

This path C is the difference between two sector paths, $C = C_2 - C_1$, as shown below, so

$$\int_C \mathbf{v} \cdot d\mathbf{x} = \int_{C_2} \mathbf{v} \cdot d\mathbf{x} - \int_{C_1} \mathbf{v} \cdot d\mathbf{x} = 2\omega(A_2 - A_1) = 2\omega A$$

where A is again the area of the region bounded by the path.

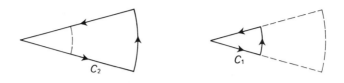

In these simple examples the circulation around the boundary of a region turned out to be proportional to the area of the region. Such examples raise the possibility of a two-dimensional aspect to circulation. The following figure shows that the circulation of a vector field around the boundary of a rectangle R can always be expressed as a finite sum of incremental circulations around the subregions of a subdivision of R.

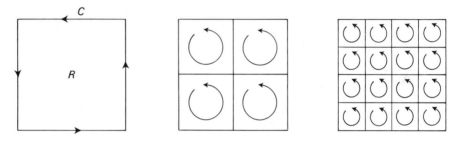

In the middle figure we have $\int_C = \sum_1^4 \int_{C_i}$ because the parts of the paths C_i interior to R consist of straight line segments that are traced once in each direction and hence cancel in the sum of the four line integrals. The circulation around the whole region is thus the sum of the circulations around its four pieces. The same remains true for any subdivision.

This procedure looks like a two-dimensional Riemann sum approximation, and suggests that the circulation of a vector field $\mathbf{p}(\mathbf{x})$ around the boundary of a plane region R can be viewed as the double integral *over R* of a "circulation density" for $\mathbf{p}(\mathbf{x})$.

To see what the density is, consider a single incremental rectangle ΔR, as shown below. The circulation around ΔR can be written

$$\int_y^{y+\Delta y} [q(x + \Delta x, t) - q(x, t)]\, dt + \int_x^{x+\Delta x} [p(t, y) - p(t, y + \Delta y)]\, dt.$$

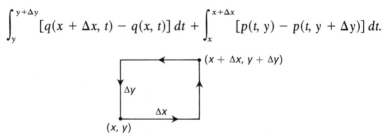

By applying the mean-value theorem twice we can rewrite the first integral, first as

$$\Delta y[q(x + \Delta x, Y) - q(x, Y)],$$

where Y is some number between y and $y + \Delta y$, and then as

$$\Delta y\, \Delta x D_1 q(X, Y),$$

where X is some number between x and $x + \Delta x$. The second integral can be evaluated in the same way, and comes out to be

$$-\Delta x\, \Delta y D_2 p(X', Y'),$$

where X' lies between x and $x + \Delta x$, and Y' between y and $y + \Delta y$.

When the incremental evaluations are added up we have the total circulation of the vector field $\mathbf{p}(\mathbf{x})$ around R equal to a Riemann sum for the *double integral over R* of the *function* $\partial q/\partial x - \partial p/\partial y$. (Actually it is the sum of two Riemann sums, for the functions $\partial q/\partial x$ and $-\partial p/\partial y$ separately.) In any case, this means that

$$\int_C \mathbf{p} \cdot d\mathbf{x} = \iint_R \left(\frac{\partial q}{\partial x} - \frac{\partial p}{\partial y}\right) dx\, dy,$$

and this is Greens's theorem for the rectangle R. It says that the circulation *around R* can be expressed as the double integral *over R* of the *circulation density* $\partial q/\partial x - \partial p/\partial y$. In particular, Green's theorem answers our original question: there is circulation in the vector field $\mathbf{p}(\mathbf{x})$ wherever the function $\partial q/\partial x - \partial p/\partial y$ is not zero.

The general form of Green's theorem requires a different approach, and one that is not so directly related to the velocity–circulation interpretation.

We shall call a region G *regular* if its boundary consists of a finite number of smooth arcs, each having only a finite number of critical points (vertical and horizontal tangents) and also, possibly, a finite number of vertical and horizontal line segments.

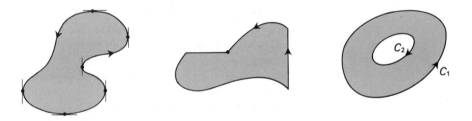

Let C be the boundary of G traced always in the direction that puts G to the left (i.e. traced *counterclockwise* with respect to nearby points of G). In the right figure above C is the union of two separate closed curves $C = C_1 + C_2$ that are traced in opposite directions, as shown. Then:

THEOREM 8   **Green's Theorem.** *If $\mathbf{p}(\mathbf{x}) = (p(\mathbf{x}), q(\mathbf{x}))$ is any smooth vector field defined on a regular region G and its boundary C, then*

$$\int_C \mathbf{p} \cdot d\mathbf{x} = \iint_G \left(\frac{\partial q}{\partial x} - \frac{\partial p}{\partial y}\right) dx\, dy.$$

Proof. We write $\mathbf{p}(\mathbf{x})$ as the sum of the two vector fields $(p(\mathbf{x}), 0)$ and $(0, q(\mathbf{x}))$ and prove the theorem for them separately. Consider the first of these two "component" fields. By drawing vertical lines through points of vertical tangency (and, possibly, through corners), we can subdivide G into a finite number of simpler regions $G_1, \ldots, G_n$ as shown.

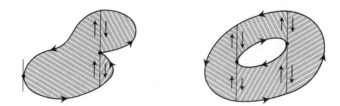

Note that the boundary line integral for G is then just the sum of the boundary line integrals for the subregions G_i, because the extra vertical boundary segments introduced in the subdividing process contribute line integrals that cancel in pairs, each being traced once upward and once downward. So if we can prove Green's theorem for each of these subregions G_i, $i = 1, \ldots, n$, then the sum of these n equations is Green's theorem for G.

We are thus reduced to proving Green's theorem for the field $(p(\mathbf{x}), 0)$ over a region G of the following simpler type: G is bounded above and below by function graphs, and left and right (possibly) by one or two vertical sides.

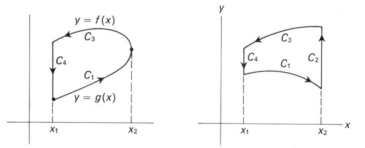

Having reduced the problem this far it is suddenly an easy consequence of the Fundamental Theorem of Calculus. We just evaluate the double integral as an iterated integral starting with y. In the calculation below, C_1 is the bottom curve, C_2 the vertical right-hand edge (if any), C_3 the top curve, *traced right to left*, and C_4 the left vertical edge (if any).

$$\iint_G \left(-\frac{\partial p}{\partial y} \right) dx\, dy = \int_{x_1}^{x_2} \left[\int_{g(x)}^{f(x)} -\frac{\partial p}{\partial y}\, dy \right] dx$$

$$= \int_{x_1}^{x_2} [p(x,\, g(x)) - p(x,\, f(x))]\, dx$$

$$= \int_{x_1}^{x_2} p(x,\, g(x))\, dx + 0 - \int_{x_1}^{x_2} p(x, f(x)\, dx + 0$$

$$= \int_{C_1} p\, dx + \int_{C_2} p\, dx + \int_{C_3} p\, dx + \int_{C_4} p\, dx$$

$$= \int_C p\, dx.$$

The C_2 and C_4 integrals are zero because x is constant along these edges. In detail, if t is any parameter for C_2, then

$$\int_{C_2} p\, dx = \int_{t_1}^{t_2} p \frac{dx}{dt}\, dt = \int_{t_1}^{t_2} p \cdot 0 \cdot dt = 0.$$

This completes the proof of Green's theorem for the first "component field" $(p(\mathbf{x}), 0)$.

By essentially the same proof we find that

$$\int_G \frac{\partial q}{\partial x}\, dx\, dy = \int_C q\, dy.$$

The principle difference is that this time we use *horizontal* subdividing lines for the preliminary decomposition of G into simpler regions, thus reducing the proof to the special case of a region that is bounded to the *left* and *right* by (nonstandard) function graphs, and, possibly, by horizontal line segments top and/or bottom.

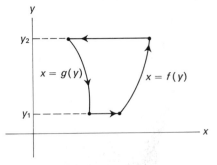

The sum of these two "component" Green's theorems is the general Green's theorem, as stated. ∎

EXAMPLE 4   Verify Green's theorem for the vector field $\mathbf{p}(\mathbf{x}) = (xy, y^2)$ over the unit circle and its interior.

Solution. a) We use $(x, y) = (\cos\theta, \sin\theta)$ on the unit circle. Then

$$\int_C \mathbf{p}\cdot d\mathbf{x} = \int_0^{2\pi} (\cos\theta\,\sin\theta(-\sin\theta) + \sin^2\theta\,\cos\theta)\, d\theta$$

$$= \int_0^{2\pi} (\sin^2\cos\theta - \sin^2\cos\theta)\, d\theta = 0.$$

b)
$$\iint_G \left(\frac{\partial q}{\partial x} - \frac{\partial p}{\partial y}\right) dx\, dy = \iint_G (0 - x)\, dx\, dy.$$

This integral can be calculated by iterated integration, but it must come out to be zero, because x is an odd function (of x) and the region G is symmetric about the y-axis, so the integrals of x over the parts of G to the left and right of the y-axis cancel.

EXAMPLE 5   Show that $\int_C - y\, dx$ is the area A of the region G whose boundary curve is C.

Solution. The line integral is of the form $\int p\,dx + q\,dy$ with $p(x, y) = -y$ and $q(x, y) = 0$. Then $\partial p/\partial y = -1$ and $\partial q/\partial x = 0$, so, by Green's theorem,

$$\int_C -y\,dx = \iint_G 1\,dx\,dy = A.$$

EXAMPLE 6 Use the above fact to compute the area inside the ellipse

$$\frac{x^2}{a^2} + \frac{y^2}{b^2} = 1.$$

Solution. The ellipse is given parametrically by $(x, y) = (a\cos\theta, b\sin\theta)$ on the parameter interval $[0, 2\pi]$. So

$$A = \int_C -y\,dx = \int_0^{2\pi} -(b\sin\theta)(-a\sin\theta)\,d\theta$$

$$= ab\int_0^{2\pi} \sin^2\theta\,d\theta = ab\int_0^{2\pi} \left(\frac{1 - \cos 2\theta}{2}\right) d\theta$$

$$= \frac{ab}{2}2\pi = \pi ab.$$

EXAMPLE 7 Use Green's theorem to show that the circulation of the rotating flow of Example 1 around the boundary of *any* regular region is equal to $2A$, where A is the area of the region.

Solution. The velocity field of Example 1 is the vector field

$$\mathbf{v}(\mathbf{x}) = r\omega(-\sin\omega t, \cos\omega t)$$
$$= (-\omega y, \omega x).$$

So

$$\int_C \mathbf{v} \cdot d\mathbf{x} = \iint_G \left(\frac{\partial v_2}{\partial x} - \frac{\partial v_1}{\partial y}\right) dx\,dy = \iint_G (\omega - (-\omega))\,dx\,dy$$

$$= 2\omega\iint_G 1\,dx\,dy = 2\omega A.$$

PROBLEMS FOR SECTION 4

1. Fluid flowing near a boundary is slowed down by viscosity. A simple model of such a flow is given by the field

$$\mathbf{v}(\mathbf{x}) = (0, x)$$

in the right half plane.

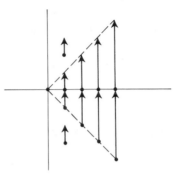

Compute the circulation of this flow about the unit square (with opposite corners at $(0, 0)$ and $(1, 1)$).

2. Compute the circulation of the above flow about the semicircular circumference bounding the right half of the unit circle.

Verify Green's theorem by directly computing the integrals involved when:

3. G is the unit square $0 \le x \le 1$, $0 \le y \le 1$ and the vector field is (xy, y).

4. G is the unit disc $x^2 + y^2 \le 1$ and the vector field is $(-y, x)$.

5. $\int_C xy(dx + dy)$, when C is the unit square.

6. Show that the area of a region G can be computed as the line integral $\int_C x \, dy$ where C is the boundary of G.

7. Use the above formula to compute the area of the ellipse $(x^2/a^2) + (y^2/b^2) = 1$, in the manner of Example 6.

8. Problem 4 and Example 5 combine to give the formula

$$A = \frac{1}{2} \int_C x \, dy - y \, dx$$

for the area of a region G in terms of a line integral around the boundary of G. This combined formula seems more complicated than its two ingredient formulas, but in some situations it actually results in an easier computation. Use the combined formula to compute the area inside the ellipse

$$\frac{x^2}{a^2} + \frac{y^2}{b^2} = 1.$$

9. Show that the area formula in Problem 8 can be rewritten

$$A = \frac{1}{2} \int_C x^2 \, d\left(\frac{y}{x}\right) = -\frac{1}{2} \int_C y^2 \, d\left(\frac{x}{y}\right).$$

Then use this reformulation to recompute the area inside the ellipse

$$\frac{x^2}{a^2} + \frac{y^2}{b^2} = 1.$$

10. Use the formula in Problem 9 to compute the area of the region between the upper branch of the hyperbola

$$\frac{y^2}{b^2} - \frac{x^2}{a^2} = 2$$

and the line $y = 2b$ $(b > 0)$. (Use $y = \sqrt{2}b \sec \theta$, $x = \sqrt{2}a \tan \theta$.)

11. As t runs from 0 to $+\infty$, the curve

$$(x, y) = \left(\frac{t}{1 + t^3}, \frac{t^2}{1 + t^3}\right)$$

describes a loop beginning and ending at the origin. Use the formula in Problem 9 to compute the area of this loop.

In Problems 12 through 21 compute the given line integral by Green's theorem. Here C_1 is the unit circle, C_2 is the unit square (with opposite vertices at $(0, 0)$ and $(1, 1)$), and C_3 is the triangle with vertices $(0, 0)$, $(1, 1)$, $(2, 0)$. In each case state the answer also for a more general contour C if possible.

12. $\displaystyle\int_{C_1} y\,dx + x\,dy$

13. $\displaystyle\int_{C_2} 2y\,dx + x\,dy$

14. $\displaystyle\int_{C_3} 2xy\,dx + x^2\,dy$

15. $\displaystyle\int_{C_1} (y^2 + 1)\,dx + 2x(y + 1)\,dy$

16. $\displaystyle\int_{C_2} e^x \sin y\,dx + e^x \cos y\,dy$

17. $\displaystyle\int_{C_1} y\tan^2 x\,dx + \tan x\,dy$

18. $\displaystyle\int_{C_1} \arctan y\,dx - \left(\frac{xy^2}{1 + y^2}\right)dy$

19. $\displaystyle\int_{C_2} \cos x \sin y\,dx + \sin x \cos y\,dy$

20. $\displaystyle\int_{C_3} \frac{xy\,dx}{1 + x} - \log(1 + x)\,dy$

21. $\displaystyle\int_{C_1} 2(2 + x)y\,dx + (2x + x^2)\,dy$

22. Let C be any regular simple closed curve that contains the origin its interior, traced counterclockwise, and let $\mathbf{p}(\mathbf{x})$ be the field

$$\left(\frac{-y}{x^2 + y^2}, \frac{x}{x^2 + y^2}\right).$$

Green's theorem cannot be applied because $\mathbf{p}(\mathbf{x})$ has a singularity at the origin. However, let C' be a small circle about the origin inside of C, also traced counterclockwise. Show that

$$\int_C \mathbf{p}\cdot d\mathbf{x} = \int_{C'} \mathbf{p}\cdot d\mathbf{x}.$$

(Apply Green's theorem to the region G bounded on the outside by C and on the inside by C'). Show therefore that

$$\int_C \mathbf{p}\cdot d\mathbf{x} = 2\pi.$$

Some Elementary Differential Equations

The laws governing natural phenomena are often expressed by differential equations. For example, we saw in Chapter 6 that the differential equation

$$\frac{dy}{dt} = ky$$

can be interpreted as the law of *normal population growth*. In the absence of inhibiting or stimulating factors, a population reproduces itself at a rate proportional to its size, which is exactly what the above equation says. Then, solving the equation, we are startled to learn that a population grows exponentially: The solution of the population equation is

$$y = y_0 e^{kt},$$

where the constant y_0 is the size of the initial population, i.e., the size at time $t = 0$. Moreover, we can now answer such quantitative questions as: How frequently will the population double in size? And when will it reach a magnitude that we have chosen, perhaps arbitrarily, as being critical?

Thus the differential equation states our understanding of the growth process as a *rate-of-change* phenomenon, and its solution shows us the growth pattern and predicts the future.

Elastic vibration is another example. We studied a simple idealized elastic system at the end of Chapter 6. It was instructive about frequency, amplitude, and phase, but it was not a very good model of reality because it did not take resistance into account. This topic will come up again at the end of the present chapter. By that time we will be able to handle the effects due to resistance and to an external driving force; and we will be in a position to analyze a wide variety of vibratory phenomena.

As these examples suggest, differential equations lead to important applications of calculus; many mathematicians feel that they are the most important applications. However, we shall only carry the subject far enough here to finish up our discussion of the simple elastic system. With that in mind we turn to a few beginning topics of the theory.

1. SEPARATION OF VARIABLES

The equation

$$3\frac{dy}{dx} + \frac{x}{y^2} = 0$$

states a relationship between an unknown function $y = f(x)$ and its derivative $dy/dx = f'(x)$, and is thus a differential equation for $y = f(x)$. It is a differential equation of the *first order* because it involves only the first derivative of f. A *solution* of the differential equation is a function $y = f(x)$ making the equation true, i.e., a function f such that

$$3f'(x) + \frac{x}{[f(x)]^2} \equiv 0.$$

Now imagine that y represents a solution in the original equation and let us see what we can learn about this solution by manipulating the equation into different but equivalent forms. If we multiply by y^2, the equation becomes

$$3y^2 \frac{dy}{dx} + x = 0.$$

We now recognize the first term as dy^3/dx, and the equation can be rewritten

$$\frac{d}{dx}\left[y^3 + \frac{x^2}{2}\right] = 0.$$

Since $du/dx = 0$ only if $u = $ constant, we conclude that

$$y^3 + \frac{x^2}{2} = C,$$

so

$$y = (C - x^2/2)^{1/3}.$$

Alternatively, after multiplying by y^2, we could write the equation as

$$\frac{d}{dx} y^3 = -x.$$

Therefore,

$$y^3 = \int - x\, dx = \frac{-x^2}{2} + C,$$

as before. We have thus "integrated" the equation and have found the most general solution function, $y = f(x)$. Note that the general solution contains a constant of integration.

The only tricky step above was recognizing $3y^2(dy/dx)$ as dy^3/dx. This step can be made more obvious by writing the differential equation in *differential* form,

$$3y^2\, dy + x\, dx = 0.$$

We then just integrate across, obtaining

$$\int 3y^2\, dy + \int x\, dx = C,$$

or

$$y^3 + \frac{x^2}{2} = C,$$

which we know to be the right answer.

In general:

THEOREM 1 *A differentiable function $y = f(x)$ satisfies the differential equation*

$$g(y)\, dy + h(x)\, dx = 0$$

if and only if it satisfies the equation

$$\int g(y)\, dy + \int h(x)\, dx = C$$

for some constant C. That is, if G(y) and H(x) are antiderivatives of g(y) and h(x), respectively, then the solutions of the differential equation are exactly the differentiable functions defined implicitly by the equation

$$G(y) + H(x) = C,$$

for all values of the constant C.

Proof. We argue backward. The function

$$G(f(x)) + H(x)$$

is constant if and only if its derivative is 0, i.e., if and only if

$$G'(f(x))f'(x) + H'(x) = 0.$$

That is, a differentiable function $y = f(x)$ satisfies the equation

$$G(y) + H(x) = \text{constant}$$

if and only if

$$G'(y)\frac{dy}{dx} + H'(x) = 0$$

or

$$g(y)\frac{dy}{dx} + h(x) = 0,$$

$$(\text{or} \quad g(y)\,dy + h(x)\,dx = 0).$$

Thus the solutions of the differential equation are exactly the differentiable functions defined implicitly by the equations $G(y) + H(x) = C$ for various constants C. ∎

A differential equation of the form

$$g(y)\,dy + h(x)\,dx = 0$$

is said to have *separated variables*; putting an equation into this form is *separating the variables*.

EXAMPLE 1   The equation

$$\frac{dy}{dx} + \frac{x}{y} = 0$$

can be put into this form by multiplying by $y\,dx$, the equation then becoming

$$y\,dy + x\,dx = 0.$$

By Theorem 1, this differential equation has the general solution

$$\int y\,dy + \int x\,dx = C,$$

or

$$\frac{y^2}{2} + \frac{x^2}{2} = C,$$

or

$$x^2 + y^2 = C$$

(replacing $2C$ by C). As curves in the coordinate plane, the solutions are the *circles about the origin.* A *function* is a solution if and only if its graph lies along one of these curves. The function solutions divide up into the upper half-circle functions

$$y = \sqrt{C - x^2},$$

and the lower half-circle functions

$$y = -\sqrt{C - x^2}.$$

Viewing a circle as the union of upper and lower half circles is geometrically unappealing, and in this sense it may be more satisfying to view the solutions of

$$h(x)\, dx + g(y)\, dy = 0$$

as the curves

$$H(x) + G(y) = C,$$

rather than as the functions whose graphs lie along these curves.

EXAMPLE 2   If we solve the above equation using the derivative form of the separated equation, then the steps could be shown like this:

$$\frac{dy}{dx} + \frac{x}{y} = 0,$$

$$y\frac{dy}{dx} + x = 0,$$

$$\frac{d}{dx}\left[\frac{y^2 + x^2}{2}\right] = 0,$$

$$y^2 + x^2 = \text{constant}$$

EXAMPLE 3   Solve the differential equation

$$\frac{dy}{dx} + xy = 0.$$

Solution. First separate the variables,

$$\frac{dy}{y} + x\, dx = 0,$$

and then integrate, obtaining

$$\log|y| + \frac{x^2}{2} = c.$$

Therefore,

$$\log |y| = c - \frac{x^2}{2},$$

$$|y| = e^{c-x^2/2}.$$

The right side is never 0, so y can never change sign. Therefore,

$$y = Ce^{-x^2/2},$$

where $C > 0$ if y is always positive and $C < 0$ if y is always negative. Note that the constant of integration appears here as the multiplicative constant C.

It may happen that the process of separating the variables loses a solution of the original equation. For example, we divided by y to separate variables in the above differential equation. This introduces the new requirement in the separated equation that $y \neq 0$, and warns us to check back to see whether the added restriction loses a solution. We see by inspection that the constant function $y = 0$ is indeed such a lost solution.

In this example, the lost solution is recovered when we change the form of the constant of integration from c to $C = e^c$. The solutions of the separated equation correspond to values of C different from 0, but we can try putting $C = 0$, and we find that we then recover the lost solution of the original equation.

Lost solutions may or may not be recovered in this way in the final form of the general solution. There can also be other types of "extraneous" solutions that are not covered by the "general" solution. However, we shall see in Section 4 that *linear* differential equations have no extraneous solutions; the general solution contains every solution.

EXAMPLE 4   A more realistic view (model) of population growth will take into account other factors besides the size of the population. For example, there may be some compelling reason why the population x can never grow beyond a limiting size L, due perhaps to a limitation on food or living space. In that case we would expect the growth rate to lessen as the population size increases, and to approach 0 as x approaches L. This hypothesis does not by itself determine the new growth law. The simplest way for such an inhibition to express itself is for the growth rate dx/dt to be proportional both to the population size x and to its remaining possible room for growth $L - x$. Then

$$\frac{dx}{dt} = kx(L - x).$$

There might, on the other hand, be a reason why the limiting size L is "doubly depressing" on the growth rate, so that

$$\frac{dx}{dt} = kx(L - x)^2.$$

In order to solve the first equation we separate variables and integrate:

$$\int \frac{dx}{x(L-x)} = \int k\,dt,$$

$$\frac{1}{L}\log\frac{x}{L-x} = kt + c,$$

$$\frac{x}{L-x} = Ce^{Lkt},$$

where the integration constant $C = e^c$ is positive. This equation can be solved for x, and the solutions are the functions

$$x = \frac{CLe^{Lkt}}{1 + Ce^{Lkt}}.$$

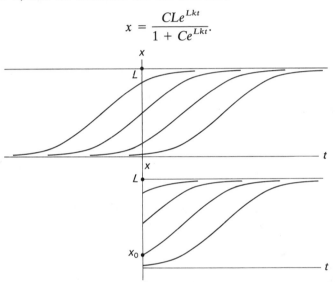

The solution curves fill up the strip $0 < x < L$, as indicated above. If we don't project these growth curves into the past, i.e., if we don't use *negative time*, then the solution curves are confined to the half strip $0 < x < L$, where $t \geq 0$. For any x_0 between 0 and L, there is a growth curve emanating from $(0, x_0)$, showing the evolution of a population that starts at size x_0. That is, $x = x_0$ at $t = 0$ is an *initial condition* that picks out a unique solution curve.

We lose the solutions $x = 0$ and $x = L$ when we separate variables in the above equation. The first lost solution is recovered in the general solution by giving the integration constant C the value of 0, but the solution $x = L$ is not part of the general solution. (In a sense, it corresponds to setting $C = \infty$.)

Note also that the lost solutions form the boundary of the region that is filled up by the solution curves of the general solution.

As the above example illustrates, differential equations frequently arise from word problems. Such problems have the added complication—and this may be the harder part of the solution—that first the appropriate differential equation has to be extracted from the statement of the problem. Here is a geometric example.

EXAMPLE 5   A curve has the property that its normal at a point (x_0, y_0) intersects the x-axis in a point (x_1, y_1) in such a way that $x_1 - x_0$ has the constant value a. Find all such curves.

Solution. The normal has the slope $-1/(dy/dx)_0$ and equation

$$y - y_0 = \frac{-1}{(dy/dx)_0}(x - x_0).$$

We obtain its x-intercept by setting $y = 0$, so

$$x_1 = x_0 + y_0\left(\frac{dy}{dx}\right)_0,$$

and

$$a = x_1 - x_0 = y_0\left(\frac{dy}{dx}\right)_0.$$

The differential equation of the curve is thus

$$y\frac{dy}{dx} = a.$$

Separating variables and integrating gives the solution

$$y^2 = 2ax + C.$$

The curves having the given property thus form a family of parabolas opening to the right.

EXAMPLE 6   If x_1 gallons of water at a temperature T_1 are mixed with x_2 gallons of water at a temperature T_2, the resulting uniform temperature T is

$$T = \frac{x_1 T_1 + x_2 T_2}{x_1 + x_2}.$$

An insulated 1000 gallon tank of water initially at the temperature 200°F is being cooled by running in water at 40°F, at the rate of 30 gallons per minute. The water is stirred, so that its temperature is approximately uniform, and the excess runs off. What is the water temperature after 1 hour?

Solution. Let T be the temperature at time t in minutes, so that $T_0 = 200$. In Δt minutes, $30\Delta t$ gallons of water are added at the temperature 40°, so

$$T + \Delta T = \frac{40(30\Delta t) + T(1000 - 30\Delta t)}{1000},$$

$$\Delta T = \frac{1200\Delta t - 30T\Delta t}{1000},$$

and

$$\frac{dT}{dt} = \frac{1}{100}(120 - 3T).$$

Solving this equation by separating the variables gives

$$T = 40 + 160e^{-3t/100}.$$

So

$$T_{60} = 40 + 160e^{-1.8} \approx 40 + 160(0.165)$$
$$\approx 66.4°.$$

PROBLEMS FOR SECTION 1

Solve the following differential equations.

1. $\dfrac{dy}{dx} = 2y$ 2. $y\dfrac{dy}{dx} = 1$

3. $\dfrac{dy}{dx} = ky$ 4. $\dfrac{dy}{dx} = y^2$

5. $\dfrac{dy}{dx} = xe^y$ 6. $\dfrac{dy}{dx} = ye^x$

7. $\dfrac{dy}{dx} = \sqrt{1 - y^2}$ 8. $\dfrac{dy}{dx} = x^2$

9. $\dfrac{dy}{dx} = \dfrac{2x}{y}$ 10. $\dfrac{dy}{dx} = \dfrac{2y}{x}$

11. $x^3\dfrac{dy}{dx} = y^2(x - 4)$ 12. $4xy\,dx + (x^2 + 1)\,dy = 0$

13. $dy/dx = f(x)$.   (Show that separation of variables amounts merely to integrating $f(x)$ for this type of equation.)

14. $\dfrac{dy}{dx} = y^2 \cos x$ 15. $x\,dy + \cos^2 y\,dx = 0$

16. A population grows at a rate proportional to its size, but, because of a steadily deteriorating environment, its growth rate is also *inversely proportional to time*. Find the law of population growth.

17. The growth rate of a population of size P is proportional to P, but is also inversely proportional to $P^{1/2}$, because of difficulties in maintaining the food supply. Find the law of population growth.

18. The velocity of water flowing from a pipe at the bottom of a cylindrical tank is proportional to the square root of the depth of the water in the tank. If the water level drops from 10 feet to 9 feet in the first hour, how long will it take to reach the level of x feet?

19. A chemical reaction converts a substance A into a substance B at a rate that is proportional to the amount of A present and inversely proportional to the amount of B. Find a formula relating the amount of B to the time t. (If x is the amount of A and y is the amount of B, assume that x and y have the initial values $x = c$, $y = 0$.)

20. Show directly from the differential equation of Example 4 that the solution curves have points of inflection at $x = L/2$.

21. An insulated 500 gallon tank is full of water at 210°F. Water at 60°F is running in at the rate of 10 gallons per minute, the excess running off through an overflow pipe. The water is stirred, so that its temperature can be assumed to be uniform. What is its temperature after a half hour?

22. The tangent to a curve at a point P intersects the x-axis at Q. The normal at P intersects the x-axis at R. The curve is such that the segments RP and PQ always have equal lengths. Find all such curves.

2. ONE-PARAMETER FAMILIES OF CURVES

We have considered a number of examples of differential equations, all in the general form

$$\frac{dy}{dx} = F(x, y).$$

In each case, the general solution contained a *constant* of *integration C*. For example, the general solution of

$$x\frac{dy}{dx} = 2y$$

is

$$y = Cx^2.$$

Each value of the constant C determines a solution curve. As C varies the solution curves vary, and in their totality they fill up some region G in the plane. In the above example the solution curves fill up the whole plane except for the y-axis.

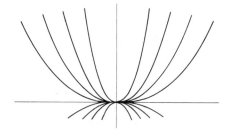

In the fourth example in the preceding section they fill up a strip or a half-strip. In such situations we call the constant of integration C a *parameter*, and we say that the solutions of the differential equation form a *one-parameter family* of curves filling up G. This use of the word "parameter" is different from, but related to, its use in connection with paths.

We can often recapture a differential equation from its general solution by differentiating and eliminating the parameter.

EXAMPLE 1   Along each curve of the one-parameter family

$$y = Cx^2,$$

we have

$$\frac{dy}{dx} = 2Cx.$$

Since y/x^2 has the constant value C along the curve, we can set $C = y/x^2$ in the second equation and thus recapture the differential equation

$$\frac{dy}{dx} = \frac{2y}{x}.$$

EXAMPLE 2   The parameter C in

$$y = x^2 + C$$

disappears upon differentiation. The differential equation of this one-parameter family is thus

$$\frac{dy}{dx} = 2x.$$

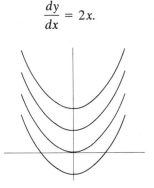

EXAMPLE 3   Sketch the one-parameter family

$$y = \sin(x + C),$$

where $x + C$ is restricted to lie in the interval $[-\pi/2, \pi/2]$. Find its differential equation.

Solution. For each value of C, x lies between $-\pi/2 - C$ and $\pi/2 - C$, and $y = \sin(x + C)$ runs from -1 to $+1$. The curves fill up the strip $-1 \le y \le 1$, as shown below.

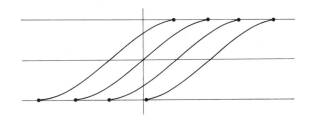

In principle, we can eliminate the parameter C by solving for C either in the given equation for y or in the differential equation for dy/dx, and then substituting its value in the other equation. We can describe the procedure as eliminating C *between* the two equations, ending up with a single equation relating y to dy/dx. Here we can first replace $y = \sin(x + C)$ by the equivalent equation

$$\arcsin y = x + C,$$

and then C drops out upon differentiating, giving the differential equation of the family at once:

$$\frac{1}{\sqrt{1-y^2}}\frac{dy}{dx} = 1,$$

or

$$\frac{dy}{dx} = \sqrt{1-y^2}.$$

Another procedure is to differentiate the given equation $y = \sin(x + C)$,

$$\frac{dy}{dx} = \cos(x + C),$$

and then use the fact that $\cos u = \sqrt{1 - \sin^2 u}$ for all u in the interval $[-\pi/2, \pi/2]$. Thus,

$$\frac{dy}{dx} = \cos(x + C) = \sqrt{1 - \sin^2(x + C)}$$

$$= \sqrt{1-y^2},$$

which is the same equation as before.

An *orthogonal trajectory* of a family of curves is a curve that crosses each curve of the family at right angles. The orthogonal trajectories of a one-parameter family of curves will themselves form a second one-parameter family, and they can be found by solving a differential equation.

EXAMPLE 4   Find the orthogonal trajectories of the family of parabolas $y = Cx^2$.

Solution. We have seen that the differential equation of this family of curves is

$$\frac{dy}{dx} = \frac{2y}{x}.$$

A curve cutting each of these parabolas orthogonally will have a slope that is everywhere the *negative reciprocal* of the above slope, so it will be a solution of the equation

$$\frac{dy}{dx} = -\frac{x}{2y},$$

or

$$2y\, dy + x\, dx = 0.$$

These orthogonal trajectories are thus the curves

$$y^2 + \frac{x^2}{2} = C,$$

a one parameter family of *ellipses*.

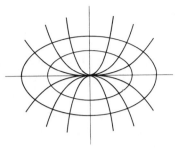

PROBLEMS FOR SECTION 2

In each of the following problems, find the differential equation of the given one-parameter family of curves. Sketch a few curves of the family, and indicate the region of the plane that they cover.

1. $y = x + C$

2. $y = Cx$

3. $x^2 + y^2 = C$

4. $y = \sqrt{x} + C$

5. $y = \sqrt{1 - x^2} + C$

6. $y = Ce^x/(1 + Ce^x)$

7. The one-parameter family of curves

$$y = (x + C)^2$$

has the added complication that *two* curves of the family pass through each point of its region G. Show that this is so, and sketch a few of the curves. Also find a differential equation of the family, and show from the differential equation why one would expect two solutions through each point.

8. Do the same for the one-parameter family

$$(y + C)^2 + x^2 = 1.$$

Find the orthogonal trajectories of each of the following families of curves. Sketch a few curves of the given family and a few of the orthogonal trajectories.

9. The straight lines $y = Cx$

10. The hyperbolas $y = \dfrac{1}{x} + C$

11. The parabolas $y^2 = x + C$

12. The hyperbolas $y = C/x$

13. The circles $x^2 + y^2 = C$

14. The curves $y = C - \arctan x$

15. The circles $x^2 + Cx + y^2 = 0$

16. The ellipses $a^2 y^2 + x^2 = C$

17. If $y = (x + C)/(1 - Cx)$, show that

$$\frac{dy}{dx} = \frac{1 + y^2}{1 + x^2}.$$

18. Solve the above differential equation, and show that its general solution can be expressed in the above form.

3. DIRECTION FIELDS

The differential equation

$$\frac{dy}{dx} = F(x, y)$$

itself has a geometric interpretation that is vivid and very useful. It assigns the slope $dy/dx = F(x, y)$ to the point (x, y) and hence determines a *direction* at each point. We say that the equation defines a *direction field* or a *field of directions* in the plane. We can sketch the direction field by drawing a short line segment (having the given slope) through each of a few points.

EXAMPLE 1   Sketch the direction field for the equation

$$\frac{dy}{dx} + \frac{x}{y} = 0.$$

Solution. Here the slope at (x, y) is given by

$$\frac{dy}{dx} = -\frac{x}{y}.$$

Since $-x/y$ is the negative reciprocal of y/x, which is the slope of the line segment from the origin to (x, y), the direction field consists of the directions perpendicular to these segments.

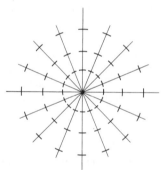

A solution of the differential equation is a curve having the direction $dy/dx = F(x, y)$ at each point (x, y) on the curve, i.e., a curve tangent at each of its points to the little segment showing the field directions at that point. With this in mind, it is practically obvious from the above figure that the solutions of

$$\frac{dy}{dx} = -\frac{x}{y}$$

are the circles about the origin.

EXAMPLE 2   The equation

$$\frac{dy}{dx} = y$$

has a feature similar to the first example, in that the slope dy/dx has the same value at each point along a line, this time a horizontal straight line.

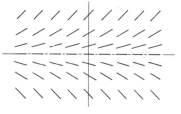

In practice we may meet differential equations of the form $dy/dx = F(x, y)$ where the function $F(x, y)$ is so complicated that we cannot explicitly solve the equation. However, we can still draw the direction field to see what the solutions look like, and this suggests a procedure for constructing approximate solutions. We start at some initial point (x_0, y_0) and proceed along the direction segment at that point to a new point (x_1, y_1). We then change directions and go along the direction segment at (x_1, y_1) to a third point (x_2, y_2), etc. We thus construct a "polygonal" function p, whose graph is linear over each of the intervals

$$[x_0, x_1], \quad [x_1, x_2], \quad [x_2, x_3], \quad \cdots,$$

and such that

$$p'(x_0) = F(x_0, p(x_0)), \qquad p'(x_1) = F(x_1, p(x_1)), \qquad \text{etc.}$$

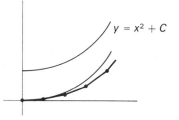

The figure suggests that if the points $x_0, x_1, x_2, \ldots$ are taken close together, then the polygonal function $p(x)$ computed as above will be a good approximation to an actual solution $f(x)$, at least in the short run. The figure also suggests that the error $|f(x) - p(x)|$ does grow as x moves away from x_0. However, there is a computation that one can make that shows how close together the points $x_0, x_1, x_2, \ldots$ should be taken in order to ensure that the error remains less than some preassigned tolerance over a given interval I.

PROBLEMS FOR SECTION 3

1. Sketch the direction field of the differential equation $(dy/dx) = x$. Use the direction field to sketch the solution curve through the point $(1, 1)$. Now find the equation of the solution curve passing through this point by solving the differential equation and evaluating the constant of integration.

2. Do the same for the equation $dy/dx = y/x$.

3. Do the same for the equation $dy/dx = xy$.

4. THE FIRST-ORDER LINEAR EQUATION

A first order linear differential equation is an equation of the form

$$\frac{dy}{dx} + a(x)y + b(x) = 0,$$

where the coefficients $a(x)$ and $b(x)$ are any functions of x. Superficially, the reason for the word *linear* is that we call $ay + b$ a linear expression in y. The deeper meaning of linearity will be discussed a little later.

A linear equation cannot generally be solved by separating the variables. However, if $b(x)$ is the zero function we have the special case

$$\frac{dy}{dx} + a(x)y = 0,$$

called the linear *homogeneous* equation, in which the variables can be separated. Proceeding as usual, we obtain the separated equation

$$\frac{dy}{y} + a(x)\,dx = 0,$$

which has the general solution

$$\int \frac{dy}{y} + \int a(x)\,dx = c,$$

or

$$\log|y| + A(x) = c,$$

where $A(x)$ is any particular antiderivative of $a(x)$. Then

$$|y| = e^c e^{-A(x)}.$$

And since the right side is never zero, y can never change sign and the general solution can be written

$$y = Ce^{-A(x)}.$$

Note that $y = 0$ is a solution of the homogeneous equation $dy/dx + a(x)y = 0$ that is lost in the separated equation $dy/y + a(x)\,dx = 0$, but is recovered in the general solution by giving the parameter C the value 0.

Note also that the growth equation $dy/dx - ky = 0$ is a special case of the linear homogeneous equation.

The solution of the inhomogeneous linear equation requires a new method, associated with a device that is very important in the theory and practice of differential equations.

What we do is to look for a solution y in the form $y = uv$, where v is a function that we already have in hand for some reason or other, and u is an unknown function to be determined. When uv is substituted for y in the given differential equation, the result is a new differential equation for u, and we hope that the new equation for u will be easier to solve than the old equation for y. It often will be if the multiplier function v is somehow related to the beginning equation.

As a first application of this procedure, we take v to be a nonzero solution of the linear homogeneous equation

$$\frac{dy}{dx} + a(x)y = 0.$$

(We know a formula for v, but there is no need to write it down. We do need the fact that v is never zero, so that we can write down $1/v$.) Then we try to find a solution y of the *inhomogeneous* equation

$$\frac{dy}{dx} + a(x)y = r(x)$$

in the form $y = vu$, where u is some function of x to be determined. Setting $y = vu$, we see that the inhomogeneous equation becomes

$$v\frac{du}{dx} + u\frac{dv}{dx} + a(x)uv = r(x)$$

or

$$u\left[\frac{dv}{dx} + a(x)v\right] + v\frac{du}{dx} = r(x)$$

or

$$v\frac{du}{dx} = r(x),$$

since, by assumption, the bracket is zero. Thus

$$\frac{du}{dx} = \frac{r(x)}{v}$$

and

$$u = \int\left(\frac{r(x)}{v}\right)dx + C.$$

Therefore,

$$y = uv = v\int\left(\frac{r(x)}{v}\right)dx + Cv.$$

Moreover, if y is *any* solution of the inhomogeneous equation, then we can set $u = y/v$, proceed as above, and conclude that y is given by the above formula. So there are no extraneous solutions. We have thus found a formula for all the solutions of the inhomogeneous linear equation, without having to know explicitly what the homogeneous solution v is. Of course, $v = e^{-A(x)}$, from the separation-of-variables calculation given earlier, so all in all we have proved the following theorem.

THEOREM 2   *The general solution of the linear differential equation*

$$\frac{dy}{dx} + a(x)y = r(x)$$

is given by

$$y = Ce^{-A(x)} + e^{-A(x)} \int e^{A(x)} r(x)\, dx,$$

where $A(x)$ is any particular antiderivative of the coefficient function $a(x)$, and C is the constant of integration. Moreover, every solution is of this form; there are no extraneous solutions.

The solution formula in Theorem 2 seems complicated and awkward at the moment, but its general form can be understood once we have looked into the nature of linearity. This will be the topic of the next section. Meanwhile, the formula should not be memorized. It is better to solve any particular inhomogeneous equation by retracing the steps in the proof of the theorem. This is called the *variation-of-parameter* method, for the following reason. The general solution of the homogeneous equation is a one-parameter family of solutions of the form $y = Cv(x)$. Here we replace the parameter C by an unknown function $u(x)$—hence the phrase "variation of parameter"—and try for a solution of the inhomogeneous equation in the form $y = u(x)v(x)$.

EXAMPLE 1 Solve the linear inhomogeneous equation

$$\frac{dy}{dx} - y = \cos x,$$

by the variation-of-parameter method.

Solution. We first find one solution of the homogeneous equation

$$\frac{dv}{dx} - v = 0$$

by separating variables (or by inspection). It is $v = e^x$. We then set $y = e^x u$ in the given equation, and get

$$ue^x + e^x \frac{du}{dx} - ue^x = \cos x,$$

$$e^x \frac{du}{dx} = \cos x,$$

$$\frac{du}{dx} = e^{-x} \cos x.$$

The integral of $e^{-x} \cos x$ is worked out by integrating by parts, and is

$$u = \int e^{-x} \cos x = \frac{1}{2}[e^{-x} \sin x - e^{-x} \cos x] + C.$$

Therefore,

$$y = e^x u = \frac{1}{2}[\sin x - \cos x] + Ce^x$$

is the general solution of the equation

$$\frac{dy}{dx} - y = \cos x.$$

PROBLEMS FOR SECTION 4

Solve the following equations by the variation-of-parameter method.

1. $\dfrac{dy}{dx} - 2y = e^x$

2. $\dfrac{dy}{dx} - 2y = e^{2x}$

3. $\dfrac{dy}{dx} + ay = e^x$

4. $\dfrac{dy}{dx} + ay = e^{bx}$

5. $\dfrac{dy}{dx} + y = x$

6. $\dfrac{dy}{dx} - y = x^2$

7. $\dfrac{dy}{dx} + \dfrac{y}{x} = x$

8. $\dfrac{dy}{dx} - \dfrac{y}{x} = x$

9. $\dfrac{dy}{dx} + \dfrac{y}{x} = \dfrac{1}{x}$

10. $\dfrac{dy}{dx} - \dfrac{y}{x} = \dfrac{1}{x}$

11. $\dfrac{dy}{dx} + ay = \sin x$

12. $\dfrac{dy}{dx} + y \sin x = \sin x$

13. $\dfrac{dy}{dx} + 2xy = x^3$

14. $\dfrac{dy}{dx} - \dfrac{y}{x} = \log x$

15. $\dfrac{dy}{dx} - y \tan x = 1$

16. $\dfrac{dy}{dx} + y \tan x = 1$

5. LINEARITY

We shall now examine some important properties of a linear differential equation that were implicit in the results of the last section. These properties will explain the general form of the rather complicated solution formula we obtained there, and they will provide a shortcut to the solutions of an important type of inhomogeneous equation. Finally, with the general nature of linearity in mind, we will be in a better position to understand the more complicated second-order equations that come up next.

A new notation is helpful here: We set

$$L(y) = \frac{dy}{dx} + a(x)y,$$

so that the linear equations

$$\frac{dy}{dx} + a(x)y = 0, \qquad \frac{dy}{dx} + a(x)y = r(x)$$

abbreviate to

$$L(y) = 0, \qquad L(y) = r(x).$$

EXAMPLE 1   If the equation is

$$\frac{dy}{dx} + xy = 0,$$

then $L(y) = \frac{dy}{dx} + xy$. In particular,

$$L(x^3) = \frac{d}{dx}(x^3) + x(x^3) = 3x^2 + x^4.$$

Similarly,

$$L(-x) = -1 - x^2,$$
$$L(x^3 - x) = -1 + 2x^2 + x^4$$
$$= L(x^3) - L(x);$$
$$L(e^x) = e^x(x + 1),$$
$$L(e^{-x^2}) = -xe^{-x^2},$$
$$L(e^{-x^2/2}) = 0.$$

THEOREM 3   *For any two differentiable functions $y_1 = f_1(x)$ and $y_2 = f_2(x)$, and any two constants c_1 and c_2,*

$$L(c_1y_1 + c_2y_2) = c_1L(y_1) + c_2L(y_2).$$

Proof.   We just compute.

$$L(c_1y_1 + c_2y_2) = \frac{d}{dx}(c_1y_1 + c_2y_2) + a(x)(c_1y_1 + c_2y_2)$$

$$= c_1\left[\frac{dy_1}{dx} + a(x)y_1\right] + c_2\left[\frac{dy_2}{dx} + a(x)y_2\right]$$

$$= c_1L(y_1) + c_2L(y_2). \quad \blacksquare$$

The theorem says that if

$$z = c_1y_1 + c_2y_2,$$

then

$$L(z) = c_1L(y_1) + c_2L(y_2).$$

That is, when we apply L to a linear combination of two functions y_1 and y_2, we get *that same* linear combination of the functions $L(y_1)$ and $L(y_2)$: applying L *preserves linear combination relationships*. For this reason we call L a *linear* differential operator, and we call the homogeneous equation

$$L(y) = 0,$$

or

$$\frac{dy}{dx} + a(x)y = 0,$$

a homogeneous *linear* differential equation. Similarly,

$$\frac{dy}{dx} + a(x)y = r(x)$$

or

$$L(y) = r(x),$$

is an *inhomogeneous linear* differential equation.

We already have explicit formulas for the solutions of these equations, but now we can see that the form these formulas take is due to the linearity of the differential operator L. That is, we can show that the solutions of $L(y) = 0$ and $L(y) = r(x)$ must have a certain general form simply because L is linear, and independently of any prior knowledge of these solutions.

THEOREM 4   *If $v = \phi(x)$ is one nonzero solution of the linear homogeneous equation*

$$L(y) = 0,$$

then the constant multiples of ϕ, $Cv = C\phi(x)$, form the family of all solutions.

   If $u = \psi(x)$ is any particular solution of the linear inhomogeneous equation

$$L(y) = r(x),$$

then its general solution is

$$y = Cv + u = C\phi(x) + \psi(x).$$

That is, the general solution of the inhomogeneous equation is the sum of the general solution of the homogeneous equation and any particular solution of the inhomogeneous equation.

Proof.   The linearity of L shows that if $L(y) = 0$, then $L(Cy) = CL(y) = 0$. Thus if $v = \phi(x)$ is one solution, then its constant multiples $Cv = C\phi(x)$ are all solutions. However, we do need Theorem 2 to tell us that there are no other solutions.

   Now suppose that we have found, somehow, one solution $u = \psi(x)$ of the inhomogeneous equation

$$L(u) = r(x).$$

Let y be any other solution. Then

$$L(y - u) = L(y) - L(u) = r(x) - r(x) = 0,$$

by Theorem 3, so $y - u$ is a solution of the homogeneous equation, and hence of the form

$$y - u = Cv = C\phi(x),$$

by the first part of the theorem above. Thus, every solution y of the inhomogeneous equation is of the form

$$y = u + Cv = \psi(x) + C\phi(x).$$

Finally, these functions are all solutions, for

$$L(y) = L(u + Cv) = L(u) + CL(v)$$
$$= r(x) + C \cdot 0$$
$$= r(x).$$

This completes the proof of the theorem. ∎

Note that the above proof depends almost entirely on the linearity of L as stated in Theorem 3.

If we now reexamine the explicit general solution that we obtained earlier,

$$y = Ce^{-A(x)} + e^{-A(x)} \int e^{A(x)} r(x) \, dx,$$

where $A(x)$ in an integral of $a(x)$, we can see the structure promised by the above theorem. It is the sum

$$y = Cv + u,$$

where

$$Cv = Ce^{-A(x)}$$

is the general solution of the homogeneous equation $L(y) = 0$, and

$$u = e^{-A(x)} \int e^{A(x)} r(x) \, dx$$

is a particular solution of the inhomogeneous equation $L(u) = r(x)$.

If a is a constant, we see by inspection that $u = e^{-ax}$ is a solution of the homogeneous equation

$$\frac{dy}{dx} + ay = 0.$$

The general solution of the inhomogeneous equation

$$\frac{dy}{dx} + ay = r(x)$$

then reduces, by virtue of Theorem 4, to finding a particular solution. The variation-of-parameter method is always available, but it can lead to a difficult integration, as we saw in the examples and problems in the last section. There is another, very simple, procedure, called the method of *undetermined coefficients*, that often works here. We illustrate it by some examples.

EXAMPLE 2   Solve the differential equation

$$\frac{dy}{dx} - 2y = 3 \sin x + \cos x.$$

Solution. Here the function $r(x) = 3 \sin x + \cos x$, and each of its derivatives, is a linear combination,

$$c_1 \sin x + c_2 \cos x,$$

of the two functions sin x and cos x. In this situation, with one exception to be mentioned later, the differential equation has a uniquely determined particular solution of the same form. In order to find it, we note that

$$L(\sin x) = \cos x - 2\sin x,$$
$$L(\cos x) = -\sin x - 2\cos x,$$

so, if $u = c_1 \sin x + c_2 \cos x$, then

$$\begin{aligned}
L(u) &= L(c_1 \sin x + c_2 \cos x)\\
&= c_1 L(\sin x) + c_2 L(\cos x)\\
&= c_1(\cos x - 2\sin x) + c_2(-\sin x - 2\cos x)\\
&= (c_1 - 2c_2)\cos x + (-2c_1 - c_2)\sin x.
\end{aligned}$$

We want $L(u)$ to be equal to $3\sin x + \cos x$. That is, we want

$$-2c_1 - c_2 = 3 \qquad \text{and} \qquad c_1 - 2c_2 = 1.$$

Solving these equations simultaneously, we find that $c_1 = c_2 = -1$. Therefore, $-\sin x - \cos x$ is a particular solution. Theorem 4 now tells us that the general solution of the inhomogeneous equation is the sum of this particular solution and the general solution of the homogeneous equation $dy/dx - 2y = 0$, which is Ce^{2x}. Thus

$$y = Ce^{2x} - \sin x - \cos x$$

is the general solution of the inhomogeneous equation

$$\frac{dy}{dx} - 2y = 3\sin x + \cos x.$$

EXAMPLE 3 Find the general solution of the differential equation

$$\frac{dy}{dx} - 2y = e^{-x}.$$

Solution. Here $r(x)$ and each of its derivatives is of the form ce^{-x}, so we look for a particular solution u of the same form: $u = ce^{-x}$. We see by inspection that $L(e^{-x}) = -3e^{-x}$, so

$$L(u) = L(ce^{-x}) = cL(e^{-x}) = -3ce^{-x},$$

and since this is required to be the function e^{-x}, the coefficient c must be $-1/3$. Thus, $-e^{-x}/3$ is a particular solution of the differential equation

$$\frac{dy}{dx} - 2y = e^{-x},$$

and its general solution is then

$$y = Ce^{2x} - \frac{1}{3}e^{-x},$$

by Theorem 4.

EXAMPLE 4 Find a general solution of the differential equation

$$\frac{dy}{dx} - 2y = e^{2x}.$$

Solution. Here we run into the difficulty alluded to earlier. We note that $L(e^{2x}) = 0$, so if $u = ce^{2x}$, then

$$L(u) = L(ce^{2x}) = cL(e^{2x}) = 0,$$

and it is impossible to find c so that $L(u) = e^{2x}$.

In general, if the function $r(x)$ on the right in the inhomogeneous equation

$$L(y) = r(x)$$

happens to be a solution of the homogeneous equation $L(y) = 0$, then the undetermined-coefficient method cannot be directly applied.

But, just in this case, it will be found that the variation-of-parameter method reduces to the completely trivial integration $v = \int dx = x + C$, and the general solution is

$$(x + C)r(x) = Cr(x) + xr(x).$$

So $xr(x)$ is the particular solution we were looking for.

Returning to Example 4, the particular solution is xe^{2x} and the general solution is

$$Ce^{2x} + xe^{2x} = (C + x)e^{2x}.$$

Problems 11 and 13 will show a couple of other ways of looking at this special case.

PROBLEMS FOR SECTION 5

In each of the following problems, find the general solution by using the method of undetermined coefficients.

1. $\dfrac{dy}{dx} - 2y = e^x$

2. $\dfrac{dy}{dx} - ay = e^{bx}$ $\qquad (a \neq b)$

3. $\dfrac{dy}{dx} + y = x$

4. $\dfrac{dy}{dx} + ay = x^2$

5. $\dfrac{dy}{dx} + ay = 1$

6. $\dfrac{dy}{dx} + 3y = \sin x$

7. $\dfrac{dy}{dx} + 2y = xe^x$

8. $\dfrac{dy}{dx} - y = xe^x$

9. $\dfrac{dy}{dx} + y = x \sin x$

10. $\dfrac{dy}{dx} + ay = e^x \sin x$

11. Show that the differential equation

$$\frac{dy}{dx} + ay = r(x)$$

is equivalent to

$$\frac{d}{dx}(e^{ax}y) = e^{ax}r(x).$$

Then find a solution by inspection in the case $r(x) = e^{-ax}$.

12. Show that the solution formula in Theorem 2 is an immediate consequence of the reformulated differential equation in Problem 11.

13. Show that if $L(y) = dy/dx + a(x)y$, then

$$L(xy) = xL(y) + y,$$

so

$$L(xy) = y \quad \text{if} \quad L(y) = 0.$$

14. Find the general solution of

$$\frac{dy}{dx} - y = e^x.$$

(Use Problem 11 or Problem 13 for the particular solution.)

15. Find the general solution of

$$\frac{dy}{dx} + 4y = e^{-4x}.$$

6. THE SECOND-ORDER LINEAR EQUATION

We now turn our attention to certain important *second-order* differential equations, i.e., equations involving the derivatives of f up through the order two. In the prime notation,

$$y' = \frac{dy}{dx}, \qquad y'' = \frac{d^2y}{dx^2}, \qquad \text{etc.,}$$

the equations we shall study are

$$y'' + a(x)y' + b(x)y = 0,$$

$$y'' + a(x)y' + b(x)y = r(x).$$

These equations are *linear*, as we shall see. This property leads to a relatively simple classification for the family of solution functions, much like the first-order situation that we summarized in Theorem 4. On the other hand, these equations have very important applications to natural phenomena associated with oscillation and vibration, resonance and damping, growth and decay. So our mathematical understanding of linearity helps us to describe and predict these natural phenomena.

Given the coefficient functions $a(x)$ and $b(x)$, we set

$$L(y) = y'' + a(x)y' + b(x)y$$

$$= f''(x) + a(x)f'(x) + b(x)f(x),$$

for any function $y = f(x)$ having two derivatives. Thus, if $a(x) = 2$ and

$b(x) = 3$, then

$$L(y) = y'' + 2y' + 3y,$$
$$L(f(x)) = f''(x) + 2f'(x) + 3f(x),$$
$$L(\sin x) = -\sin x + 2 \cos x + 3 \sin x$$
$$= 2(\sin x + \cos x),$$
$$L(e^x) = e^x + 2e^x + 3e^x = 6e^x,$$
$$L(e^{-x}) = e^{-x} - 2e^{-x} + 3e^{-x} = 2e^{-x}, \quad \text{etc.}$$

THEOREM 5 L is a linear differential operator. That is, if f_1 and f_2 are any functions having at least two derivatives, and if c_1 and c_2 are any two constants, then

$$L(c_1 f_1 + c_2 f_2) = c_1 L(f_1) + c_2 L(f_2).$$

Proof.   The proof is just a computational verification, like the proof of Theorem 3. We leave it as an exercise. ∎

The two differential equations we want to study are the linear homogeneous equation

$$L(f(x)) = 0$$

and the linear inhomogeneous equation

$$L(f(x)) = r(x).$$

The general properties of their solution families are stated in the following theorem.

THEOREM 6   If y_1 and y_2 are solutions of the homogeneous equation $L(y) = 0$, then so is any linear combination $y = c_1 y_1 + c_2 y_2$. Moreover, if y_1 and y_2 are independent, in the sense that neither is a constant multiple of the other, then

$$y = c_1 y_1 + c_2 y_2$$

is the general solution of $L(y) = 0$.
    If u is any particular solution of the inhomogeneous equation $L(u) = r(x)$, then

$$y = (c_1 y_1 + c_2 y_2) + u$$

is the general solution of the inhomogeneous equation (supposing y_1 and y_2 to be independent). That is, the general solution of the inhomogeneous equation is the sum of a particular solution and the general solution of the homogeneous equation.

Partial proof.   The proof of this theorem has an easy part and a hard part, and we are ready now only for the easy part.
    If $L(y_1) = 0$ and $L(y_2) = 0$, then

$$L(c_1 y_1 + c_2 y_2) = c_1 L(y_1) + c_2 L(y_2) = 0,$$

by Theorem 5. Thus the family of solutions of the homogeneous equation $L(y) = 0$ is closed under forming linear combinations. What we cannot prove here is that if y_1 and y_2 are independent solutions, then the functions $c_1 y_1 + c_2 y_2$ form the whole solution family.

Now suppose that u is some particular solution of the inhomogeneous equation $L(u) = r(x)$. If y is any other solution, then

$$L(y - u) = L(y) - L(u) = r - r = 0,$$

so $y - u$ is a solution of the homogeneous equation. Because $y = (y - u) + u$, we see that if we have one solution u of the inhomogeneous equation, then any other solution y can be written as u plus a solution of the homogeneous equation. ∎

According to Theorem 5, the general solution of the inhomogeneous equation depends on finding

1. the general solution of the homogeneous equation, and
2. a particular solution of the inhomogeneous equation.

The variation-of-parameter method is the basic theoretical tool in reducing (2) to (1). It now starts off like this: First, we find the general solution of the homogeneous equation in the form

$$y = c_1 v_1 + c_2 v_2.$$

Then we try for a particular solution of the inhomogeneous equation $L(y) = r(x)$, in the form

$$y = u_1 v_1 + u_2 v_2,$$

where u_1 and u_2 are unknown functions that have to be determined. Since we have replaced the parameters c_1 and c_2 by unknown functions u_1 and u_2, as in the first-order case, this method is called *variation of parameters*. But it is now more complicated than in our first-order examples, and it is best understood in the context of linear algebra. Moreover, we may not even be able to get started, because solving the homogeneous equation can be much harder in the second-order case. We shall therefore break off at this point, and leave variation of parameters to a future course.

The situation is much simpler if the coefficient functions $a(x)$ and $b(x)$ are both constants, and from now on this will be assumed. In this case, we can solve the homogeneous equation completely, and will do so in the next section. Moreover, independently of the homogeneous solution, we can find a particular solution of the inhomogeneous equation if the function $r(x)$ on the right lets us use the method of undetermined coefficients. This will be taken up now for the simplest context of the problem: to find a particular solution of the equation

$$L(y) = r(x),$$

when $r(x)$ is *not* itself a solution of the homogeneous equation.

EXAMPLE 1   Find a particular solution of the equation

$$L(y) = \sin x,$$

where

$$L(y) = y'' + 3y' + 2y.$$

Solution.   We try for a solution in the form

$$u = a \sin x + b \cos x.$$

That is, we try for a solution u in the form of a linear combination of $r(x)$ and its derivatives. Remember that, for this to succeed, it is essential that $r(x)$ and its derivatives involve only a finite number of functions. So the device cannot work for a function like $r(x) = 1/x$. Anyway, we see by inspection that

$$L(\sin x) = \sin x + 3 \cos x,$$
$$L(\cos x) = \cos x - 3 \sin x;$$

so

$$\begin{aligned}
L(u) &= L(a \sin x + b \cos x) \\
&= aL(\sin x) + bL(\cos x) \\
&= a(\sin x + 3 \cos x) + b(\cos x - 3 \sin x) \\
&= (a - 3b) \sin x + (3a + b) \cos x.
\end{aligned}$$

Thus, $L(u) = \sin x$ if and only if

$$a - 3b = 1, \qquad 3a + b = 0.$$

The solutions of these equations are $a = 1/10$, $b = -3/10$; so

$$u = \frac{1}{10} \sin x - \frac{3}{10} \cos x$$

is a particular solution of the inhomogeneous equation.

EXAMPLE 2   In order to obtain a particular solution of

$$L(y) = y'' + 3y' + 2y = e^t,$$

we try a linear combination of e^t and its derivative. But they are all the same function, so we just try

$$u = ce^t.$$

Since $L(e^t) = 6e^t$, we then have

$$L(u) = L(ce^t) = 6ce^t,$$

and the equation $L(u) = e^t$ reduces to $6ce^t = e^t$. Thus, $6c = 1$, and a particular solution is $u = e^t/6$.

EXAMPLE 3   It is always possible to find a solution of

$$L(y) = r_1(x) + r_2(x),$$

by finding a solution u_1 of $L(y) = r_1(x)$, a solution u_2 of $L(y) = r_2(x)$, and then adding:

$$L(u_1 + u_2) = L(u_1) + L(u_2) = r_1(x) + r_2(x).$$

For an equation like

$$y'' - 2y' + y = \sin x - \cos x,$$

this would be inefficient, since each separate solution would have to be of the form $a \sin x + b \cos x$, and there would be duplication of effort.

In the case of

$$y'' + 2y' - y = e^x + x,$$

however, x and its derivatives remain completely distinct from e^x and its derivatives, and in such a situation the divide-and-conquer strategy may pay off. We first calculate (mentally) that

$$L(e^x) = 2e^x,$$

so we will have $L(ce^x) = e^x$ if $c = 1/2$. That is, $u_1 = e^x/2$ is a particular solution of $L(y) = e^x$.

Next, we calculate what L does to x and its derivatives,

$$L(x) = 2 - x, \qquad L(1) = -1,$$

so $L(ax + b) = a(2 - x) + b(-1) = -ax + (2a - b)$. This will be the function x if $a = -1$ and $b = 2a = -2$. That is,

$$u_2 = -x - 2$$

is a particular solution of $L(y) = x$. Then

$$u = \frac{e^x}{2} - x - 2$$

is a particular solution of $L(y) = e^x + x$.

PROBLEMS FOR SECTION 6

In each of the following problems find a particular solution by the method of undetermined coefficients.

1. $y'' + y = x$
2. $y'' + 4y = e^x$
3. $y'' - 4y = e^x$
4. $y'' - 4y' = e^{3x}$
5. $y'' + 5y = \cos x$
6. $y'' - y = x^2$
7. $y'' + y' = x^2$
8. $y'' - y = Ax^2 + Bx + C$
9. $y'' + 2y = e^x + 2e^{-x}$
10. $y'' - y' = \cos x + 2x$
11. $y'' + y = xe^x$
12. $y'' + y' - 2y = x$
13. $y'' + y' - 2y = e^{-x}$
14. $y'' + y' - 2y = e^{2x}$
15. $y'' + y' - 2y = \sin x$
16. $y'' + y' = x^3$
17. $y'' + y' = x + \sin x$
18. $y'' + y' = x^2 e^{-x}$
19. $y'' - 2y' + y = x^2 e^x$
20. $y'' - 2y = xe^x + 1$
21. $y'' - 2y = e^x \sin x$
22. $y'' + 4y = x \sin x$
23. $y'' - 2y' + 3y = e^x \sin x$

24. Supposing that the differential operator L has constant coefficients, show that it commutes with differentiation. That is, show that

$$\frac{d}{dx} L(f(x)) = L(f'(x)).$$

7. THE HOMOGENEOUS EQUATION WITH CONSTANT COEFFICIENTS

In view of our first-order experience, it seems reasonable to check whether $y = e^{kx}$ can ever be a solution of the homogeneous equation

$$L(y) = y'' + ay' + by = 0.$$

EXAMPLE 1   Can $y = e^{kx}$ be a solution of

$$L(y) = y'' + 2y' - 3y = 0?$$

Solution.   We have

$$L(e^{kx}) = k^2 e^{kx} + 2k e^{kx} - 3 e^{kx}$$
$$= (k^2 + 2k - 3) e^{kx}.$$

This will be the zero function if and only if

$$k^2 + 2k - 3 = 0.$$

Since

$$k^2 + 2k - 3 = (k + 3)(k - 1),$$

the roots of the equation are

$$k = -3 \quad \text{and} \quad k = 1.$$

Thus, $y = e^{-3x}$ and $y = e^{x}$ are both solutions of $L(y) = 0$, and they are the only pure exponential solutions.

EXAMPLE 2   Continuing the above example,

$$y = c_1 e^{-3x} + c_2 e^{x}$$

is a solution for any constants c_1 and c_2, by Theorem 6. Moreover, since we cannot express e^{-3x} as a constant times e^{x}, Theorem 6 says, furthermore, that every solution of $L(y) = 0$ is of the above form. However, we haven't proved this part of the theorem, so let us see if we can find a direct proof for this example, using the product device again.

Let u be any solution of the equation and write $u = vy$, where v is to be determined and y is a solution we already know. Here we can take y as e^{-3x} or e^{x}, and the latter looks simpler. So we set $u = ve^{x}$. Then

$$u' = ve^{x} + v'e^{x},$$
$$u'' = ve^{x} + 2v'e^{x} + v''e^{x},$$

and

$$0 = L(u) = u'' + 2u' - 3u$$
$$= v(e^{x} + 2e^{x} - 3e^{x}) + 4v'e^{x} + v''e^{x}$$
$$= (v'' + 4v')e^{x}.$$

Therefore,

$$v'' + 4v' = 0,$$

which is a first-order equation in v'. Its general solution is

$$v' = c_1 e^{-4x},$$

so

$$v = c_1 e^{-4x} + c_2,$$

(with a different constant c_1) and finally

$$u = v e^x = c_1 e^{-3x} + c_2 e^x.$$

This proves that there are no other solutions than the ones we already had.

Now let us consider the general constant coefficient equation, which we shall write in the form

$$L(y) = y'' + 2py' + qy = 0.$$

Note that we are writing $2p$ for the coefficient of y'. For example, if $L(y) = y'' + 2y' - 3y$, then

$$p = 1 \quad \text{and} \quad q = -3.$$

This artificial notation simplifies the algebra. Then

$$L(e^{kx}) = k^2 e^{kx} + 2pk e^{kx} + q e^{kx}$$
$$= (k^2 + 2pk + q)e^{kx}.$$

This will be the zero function if and only if

$$k^2 + 2pk + q = 0,$$

which is called the *characteristic equation* of L. The solutions of the characteristic equation are given by the quadratic formula

$$k = \frac{-2p \pm \sqrt{4p^2 - 4q}}{2} = -p \pm \sqrt{p^2 - q}.$$

There are three cases.

Case I. If $p^2 > q$ and if we set $\Delta = \sqrt{p^2 - q}$, then there are two distinct real roots

$$k_1 = -p + \Delta \quad \text{and} \quad k_2 = -p - \Delta,$$

and the situation is exactly like Example 1. In fact, if we replace -3 and 1 by k_1 and k_2, the argument in Example 2 turns into a proof of the following theorem.

THEOREM 7 *If the characteristic equation*

$$k^2 + 2pk + q = 0$$

has distinct real solutions $k = k_1$, k_2, then the solutions of the linear homogeneous equation

$$y'' + 2py' + qy = 0$$

are exactly the functions of the form

$$y = c_1 e^{k_1 x} + c_2 e^{k_2 x}.$$

Case II. If $p^2 - q = 0$, then there is only one exponential solution e^{kx} given by $k = -p$. We illustrate this situation with an example.

EXAMPLE 3   For
$$y'' + 2y' + y = 0,$$
the characteristic equation
$$k^2 + 2k + 1 = (k + 1)^2 = 0$$
has only one root, $k = -1$. Thus, e^{-x} is the only pure exponential solution. But this leads to the general solution by the same device as in Example 2. Any other solution u can be written $u = ve^{-x}$, with v to be determined. Then,
$$u' = -ve^{-x} + v'e^{-x},$$
$$u'' = ve^{-x} - 2v'e^{-x} + v''e^{-x},$$
and
$$0 = L(u) = u'' + 2u' + u$$
$$= v(e^{-x} - 2e^{-x} + e^{-x}) + v''e^{-x}$$
$$= v''e^{-x}.$$
Therefore,
$$v'' = 0,$$
$$v' = c_1$$
$$v = c_1 x + c_2,$$
and
$$u = ve^{-x} = c_1 x e^{-x} + c_2 e^{-x}.$$

With general coefficients, the above argument proves the following theorem.

THEOREM 8   *If the characteristic equation*
$$k^2 + 2pk + q = 0$$
has exactly one real solution, i.e., if $q = p^2$ and the equation is $(k + p)^2 = 0$, then the solutions of
$$y'' + 2py' + qy = 0$$
are exactly the functions
$$y = c_1 x e^{-px} + c_2 e^{-px}.$$

Case III. If $p^2 < q$, then neither of the roots
$$k = -p \pm \sqrt{p^2 - q}$$
is real, and there are no pure exponential solutions. Let us, nevertheless, proceed as in Case II, and write an arbitrary solution u in the form
$$u = ve^{-px},$$

with v to be determined. Then,

$$u' = -pve^{-px} + v'e^{-px},$$
$$u'' = p^2ve^{-px} - 2pv'e^{-px} + v''e^{-px},$$

and

$$0 = L(u) = u'' + 2pu' + qu$$
$$= v(p^2 - 2p^2 + q)e^{-px} + v''e^{-px}.$$

The equation for u thus reduces to a simpler equation for v,

$$v'' = -\beta^2 v,$$

where $\beta = \sqrt{q - p^2}$. We know two solutions of this equation, namely

$$v = \sin \beta x \qquad \text{and} \qquad v = \cos \beta x.$$

Therefore, every linear combination

$$v = c_1 \sin \beta x + c_2 \cos \beta x$$

is a solution. We must show that there are no other solutions. The consistent thing to do would be to try the same device once more: set $v = w \cos \beta x$, and then find w by solving a first-order equation. This would work, but not as well as before, because we have to divide by $\cos \beta x$ in the process and then worry about the points where $\cos \beta x = 0$. Instead, we shall resort to the end of Section 4 in Chapter 4, where the proof for the case $\beta = 1$ was sketched. That proof works just as well for arbitrary β, and takes care of our present dilemma.

THEOREM 9 *If the characteristic equation*

$$k^2 + 2pk + q = 0$$

has no real solutions, i.e., if $p^2 - q < 0$, then the solutions of the equation

$$y'' + 2py' + qy = 0$$

are exactly the functions

$$y = e^{-p(x)}[c_1 \cos \beta x + c_2 \sin \beta x]$$

where $\beta = \sqrt{q - p^2}$.

Theorems 7 through 9 show that the general solution of the constant coefficient homogeneous equation

$$y'' + 2py' + qy = 0$$

is one of three types, depending on whether q is less than, equal to, or greater than p^2. If $q < p^2$, then we can write $q = p^2 - m^2$, where $m = \sqrt{p^2 - q}$. Similarly, if $q > p^2$, then q is of the form $q = p^2 + m^2$. We can then summarize the three theorems in the following form.

THEOREM 10   *The general solution of the constant coefficient homogeneous equation*

$$y'' + 2py' + qy = 0$$

   is:

$$e^{-px}[c_1 e^{mx} + c_2 e^{-mx}] \qquad \text{if } q = p^2 - m^2,$$

$$e^{-px}[c_1 x + c_2] \qquad \text{if } q = p^2,$$

$$e^{-px}[c_1 \cos mx + c_2 \sin mx] \qquad \text{if } q = p^2 + m^2.$$

Finally, Theorem 6 tells us that the general solution of the inhomogeneous equation

$$y'' + ay' + by = r(x)$$

is the sum of the general solution of the homogeneous equation

$$y'' + ay' + by = 0,$$

as determined in the above theorems, and a particular solution of the inhomogeneous equation, determined possibly by the method of undetermined coefficients.

EXAMPLE 4   Find the general solution of

$$y'' + 3y' + 2y = \sin t.$$

Solution.   The characteristic equation

$$k^2 + 3k + 2 = 0$$

factors into

$$(k + 2)(k + 1) = 0,$$

and hence has the two roots $k = -2, -1$. The general solution of the homogeneous equation $y'' + 3y' + 2y = 0$ is therefore

$$c_1 e^{-x} + c_2 e^{-2x}.$$

On the other hand, we found a particular solution of the inhomogeneous equation to be

$$\frac{1}{10} \sin x - \frac{3}{10} \cos x$$

in the first example of the last section. The general solution is therefore

$$c_1 e^{-x} + c_2 e^{-2x} + \frac{1}{10} \sin x - \frac{3}{10} \cos x,$$

by Theorem 6.

It may happen, when we write down the general solution for the homogeneous equation associated with

$$y'' + ay' + by = r(x),$$

that we discover the right-hand function $r(x)$ to be already one of the homogeneous solutions. In that case, the undetermined-coefficient procedure has to be modified. In view of our experience with the first-order equations, it is natural to try a linear combination of $xr(x)$ and its derivatives. We then have to calculate $L(xr(x))$, and it will shorten our work if we know ahead of time what the result of this calculation is.

LEMMA 1 *If $L(y) = y'' + ay' + by$, then*

$$L(xy) = xL(y) + (2y' + ay).$$

Therefore, if $y = f(x)$ is a solution of the homogeneous equation, $L(f) = 0$, then

$$L(xf) = 2f' + af.$$

The proof will be left as an exercise.

The peculiar form of the above result will make more sense after some of the problems. (See Extra Problem 47.)

EXAMPLE 5 Find the general solution of

$$y'' - 4y' + 4y = e^{2x}.$$

Solution. The characteristic equation is

$$k^2 - 4k + 4 = (k - 2)^2 = 0,$$

which has the single (repeated) root $k = 2$. The general solution of the homogeneous equation is therefore

$$e^{2x}(c_1 x + c_2),$$

by Theorem 8 or Theorem 10.

In this case, since $r(x) = e^{2x}$ and $xr(x) = xe^{2x}$ are both solutions of the homogeneous equation, we apply Lemma 1 to $x(xe^{2x}) = x^2 e^{2x}$:

$$L(x^2 e^{2x}) = xL(xe^{2x}) + 2(xe^{2x})' - 4(xe^{2x}) = 2e^{2x}.$$

So if $u = cx^2 e^{2x}$, then

$$L(u) = L(cx^2 e^{2x}) = cL(x^2 e^{2x}) = 2ce^{2x}.$$

This is required to be equal to $r(x) = e^{2x}$, so $c = 1/2$ and the particular solution is $x^2 e^{2x}/2$. The general solution is therefore

$$y = e^{2x}\left[\frac{x^2}{2} + c_1 x + c_2\right],$$

by Theorem 6.

EXAMPLE 6   Find a solution of

$$y'' + y = \sin x.$$

Solution. We would normally try a linear combination of $\sin x$ and its derivatives, $u = a \sin x + b \cos x$. But we find here that

$$L(\sin x) = 0 = L(\cos x).$$

So we try

$$u = ax \sin x + bx \cos x,$$

instead. By Lemma 1,

$$L(x \sin x) = xL(\sin x) + 2 \cos x = 0 + 2 \cos x$$
$$= 2 \cos x,$$
$$L(x \cos x) = -2 \sin x;$$

so

$$L(u) = L(ax \sin x + bx \cos x) = aL(x \sin x) + bL(x \cos x)$$
$$= 2a \cos x - 2b \sin x.$$

We want $L(u) = \sin x$, so $a = 0$ and $b = -1/2$. Thus the particular solution is

$$u = -\frac{1}{2} x \cos x,$$

and the general solution is

$$c_1 \sin x + \left(c_2 - \frac{x}{2}\right) \cos x.$$

PROBLEMS FOR SECTION 7

Find the general solution of each of the following equations.

1. $y'' + 2y = 0$ 2. $y'' + 2y = 1$ 3. $y'' + 2y = x^2$
4. $y'' + 2y = \sin x$ 5. $y'' + 4y = \sin 2x$ 6. $y'' - 2y = x + 1$
7. $y'' - y = e^x$ 8. $y'' - y' = 1$ 9. $y'' - y = e^x \sin x$
10. $y'' - y = xe^x$ 11. $y'' + y' - 2y = 0$ 12. $y'' + y' - 2y = x$
13. $y'' + y' - 2y = \sin x$ 14. $y'' + y' - 2y = e^x$
15. $y'' - 2y' + y = 0$ 16. $y'' - 2y' + y = x$
17. $y'' - 2y' + y = \sin x$ 18. $y'' - 2y' + y = e^x$
19. $y'' + 2y' + 2y = 0$ 20. $y'' + 2y' + 2y = x$
21. $y'' + 2y' + 2y = e^x$ 22. $y'' + 2y' + 2y = \sin x$

8. ELASTIC OSCILLATION

Although all sorts of elastic objects vibrate when they are struck or otherwise disturbed, the vibration normally dies away, as when a note is struck on a piano. We ascribe this attenuation to *resistance* to the motion, such as air resistance in the case of the piano string.

If the object is continuously "excited," then the vibration can continue indefinitely, despite resistance. Violin tones and organ tones are such sustained vibrations. A continuously excited vibration can even *increase* in intensity. This is the phenomenon of *resonance,* and it may be desirable or catastrophic. A suspension bridge is an elastic object, and a group of people marching in step over a suspension bridge at the natural frequency of the bridge can cause the bridge to vibrate with an increasing amplitude that may eventually destroy it. This is why soldiers were trained to break cadence when crossing a bridge. On the other hand, the current in electric circuits oscillates according to the same principles of vibration, and there resonance can be a desirable phenomenon.

The ideal mass–spring system of Chapter 6 becomes realistic when we add a resistance force proportional to the velocity, $-r\,dx/dt$. Then its motion is governed by Newton's law $ma - F = 0$, in the form

$$m\frac{d^2x}{dt^2} + \left[r\frac{dx}{dx} + sx\right] = 0,$$

or

$$\ddot{x} + \frac{r}{m}\dot{x} + \frac{s}{m}x = 0.$$

If, in addition, the system is continuously excited by a periodic external force of the form $c\sin\beta t$, then its motion is governed by Newton's law in the form

$$m\frac{d^2x}{dt^2} + r\frac{dx}{dt} + kx = c\sin\beta t.$$

(This is also the equation for the motion of electricity in a circuit to which a generator has been attached. The earlier equation governs the flow of electricity in a "free" circuit. We shall not analyze those electrical applications, however.)

Here again we have differential equations that state our understanding of a natural phenomenon, and their solutions will show the possible patterns of elastic behavior and permit quantitative predictions.

The analysis of this situation is broken down into the following problems.

PROBLEMS FOR SECTION 8

1. Consider first the case where the resistance r dominates the combined effect of the mass m and spring stiffness s, in the sense that

$$r^2 > 4ms.$$

Supposing that the initial velocity (at time $t = 0$) is zero, and the initial position is positive, show that the motion is of the form

$$x = c[ke^{-lt} - le^{-kt}],$$

where $0 < l < k$, and c is positive.

2. Assuming the above formula, show that the graph of x as a function of t has the form shown in the following diagram, with a single point of inflection at $t = [\log(k/l)]/(k - l)$.

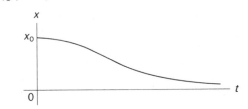

3. In the same situation of a dominant resistance, suppose that the initial position is at $x = 0$ but that the initial velocity is positive. Show that then the motion is of the form

$$x = c[e^{-lt} - e^{-kt}],$$

where $c > 0$ and $k > l > 0$.

4. Show that the graph of the above motion has the general features shown below.

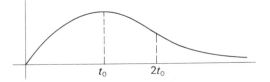

In particular, show that there is one critical point $t = t_0$ at which x has its maximum value, and one point of inflection, located at $2t_0$.

The next two problems analyze in more detail the same situation of a dominant resistance, when the initial position x_0 is positive, say $x_0 = 1$, and the initial velocity v_0, has various values.

5. Show first that if $x_0 = 1$, the motion has the form

$$x = ae^{-lt} + (1 - a)e^{-kt},$$

where $0 < l < k$. Then show that the initial velocity v_0 is less that $-k$ if and only if $a < 0$, and that the graph of the motion in this case has the general form shown below. (One change in sign, one critical point, one point of inflection.)

6. Show from the equation for x in the problem above:
   a) If $-k \le v_0 \le -k/(k + l)$, then the graph of the motion has the general form shown below (no change of sign, no critical point, no point of inflection).

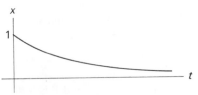

b) If $v_0 > - k/(k + l)$, then the graph is always positive and has one point of inflection, as below:

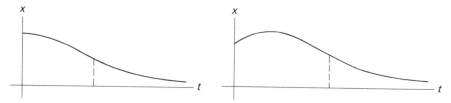

In the next two problems we assume that the resistance r balances the combined effect of the mass m and stiffness s, in the sense that $r^2 = 4ms$.

7. Show that if $r^2 = 4ms$ and if the initial conditions are $(x_0, v_0) = (1, 0)$, then the motion is given by

$$x = (1 + kt)e^{-kt},$$

where $k = r/2m$. Show also that the graph of this motion has the same features as in Problem 2 (but with a different point of inflection).

8. If the initial conditions are $(x_0, v_0) = (0, 1)$, show that the motion is given by

$$x = te^{-kt},$$

and that its graph has the same features as in Problem 4.

9. Consider finally the case where the resistance r is small in comparison with the combined effect of the mass m and spring stiffness s, in the sense that

$$r^2 < 4ms.$$

Show that then the motion is given by

$$x = ae^{-kt} \sin (\omega t + \theta),$$

where the exponential damping rate k has the value $k = r/2m$, and the frequency $\omega/2\pi$ is determined by

$$\omega = \frac{1}{2m} \sqrt{4ms - r^2}.$$

(See Sect. 9 in Ch. 6 for the source of the amplitude a and phase angle θ.)

10. Show that the graph of x, as a function of t in the above problem, has the general features shown below.

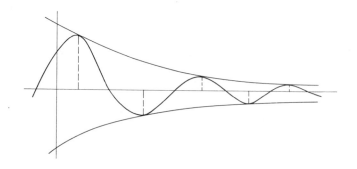

How is this graph affected by a decrease in the resistance r (m and s remaining unchanged)? What is the limiting configuration as r approaches zero?

We consider one more situation, in which an extra periodic "driving force" is impressed on the system, say a "sinusoidal" force, proportional to $\sin \alpha t$. The modified differential equation can be taken to be

$$(*) \qquad \ddot{x} + \frac{r}{m}\dot{x} + \frac{s}{m}x = \sin \alpha t,$$

where $r^2 < 4ms$, and where we have written $\sin \alpha t$ instead of the more general $c \sin \alpha t$, for simplicity. The motion is now a superposition (sum) of a transient vibration that dies out exponentially as discussed above (a solution of the homogeneous equation) and a "steady-state" sinusoidal vibration

$$A \sin (\alpha t + \beta)$$

(the particular solution of the inhomogeneous equation obtained by undetermined coefficients). Note that the total motion settles down to this steady-state simple harmonic motion as the transient component wears off. If in the beginning the transient and steady-state components tend to cancel one another, then the total effect is a vibration with an amplitude that builds up as time passes, and if the transient and steady-state frequencies are nearly the same, the limiting amplitude can be very large. This is the phenomenon of resonance.

11. Show that the particular solution of the above equation obtained by the method of undetermined coefficients can be written in the form

$$A \sin (\alpha t + \beta),$$

where the amplitude A is given by

$$A = \frac{1}{\sqrt{\left(\dfrac{s}{m} - \alpha^2\right)^2 + \left(\dfrac{\alpha r}{m}\right)^2}}.$$

Conclude that A is proportional to $1/r$ if the driving frequency $\alpha/2\pi$ is equal to the natural frequency $\sqrt{s/m}/2\pi$ of the undamped and undriven system.

12. With m, s, and r fixed, but α variable, prove that the maximum value of the amplitude A in the above problem is

$$A_{max} = \frac{2m^2}{r\sqrt{4ms - r^2}},$$

when $\alpha = \sqrt{ms - r^2/2}/m$.

13. If a simple electric circuit contains a total capacitance of amount $1/s$, a total resistance r, and a total inductance m, and if x is the amount of electricity that has moved past a fixed point in a displacement from the equilibrium position, then x fluctuates with time according to the same homogeneous equation

$$m\ddot{x} + r\dot{x} + sx = 0.$$

This is for a "free" circuit, unaffected by any outside influence. If a sinusoidal generator is attached, then the fluctuation of x is determined by the inhomogeneous equation $(*)$, with $\sin \alpha t$ possibly replaced by $c \sin \alpha t$ for some constant c.

But here we are more interested in the behavior of the *current* $i = \dot{x}$. Show that if x satisfies $(*)$, then $i = \dot{x}$ satisfies

$$(**) \qquad m\frac{d^2 i}{dt^2} + r\frac{di}{dt} + si = \alpha \cos (\alpha t).$$

14. Show without calculation, by applying a general result from Section 6 (see Problem 24), that if $A \sin(\alpha t + \beta)$ is a particular solution of $(*)$, then $\alpha A \cos(\alpha t + \beta)$ is a particular solution of $(**)$. Then show, from the result in Problem 11, that the amplitude $A' = \alpha A$ of this oscillation is given by

$$A' = \frac{1}{\sqrt{\left(\dfrac{s}{\alpha m} - \alpha\right)^2 + \left(\dfrac{r}{m}\right)^2}},$$

and hence that A' is maximum (for fixed s, m, and r) when

$$\alpha = \sqrt{\frac{s}{m}}.$$

Conclude that the resonant frequency is equal to the frequency of the undamped ($r = 0$) freely oscillating circuit.

9. SERIES SOLUTIONS OF DIFFERENTIAL EQUATIONS

Infinite series apply naturally to the solution of linear differential equations, because such a differential equation automatically determines the Maclaurin series of a solution function.

EXAMPLE 1 Consider the differential equation

$$y' - y = 0.$$

If we assume that a solution $y = f(x)$ is the sum of a power series

$$y = a_0 + a_1 x + a_2 x^2 + \cdots + a_n x^2 + \cdots,$$

then

$$y' = a_1 + 2a_2 x + 3a_3 x^2 + \cdots + (n+1)a_{n+1}x^n + \cdots$$

and

$$y' - y = (a_1 - a_0) + (2a_2 - a_1)x + (3a_3 - a_2)x^2 + \cdots$$
$$+ [(n+1)a_{n+1} - a_n]x^n + \cdots.$$

Since $y' - y$ is the zero function, its Maclaurin series coefficients must all be zero, so

$$a_1 = a_0,$$

$$a_2 = \frac{a_1}{2} = \frac{a_0}{2},$$

$$a_3 = \frac{a_2}{3} = \frac{a_0}{3!},$$

$$\vdots \qquad \vdots$$

$$a_{n+1} = \frac{a_n}{n+1} = \frac{a_0}{(n+1)!},$$

and

$$y = a_0 \left[1 + x + \frac{x^2}{2!} + \cdots + \frac{x^n}{n!} + \cdots \right],$$

where a_0 is an arbitrary constant, the "constant of integration." Of course, we recognize this solution to be $a_0 e^x$.

What we are doing here is essentially the reverse of our procedure in Section 8 of Chapter 11. There we noted that the function

$$y = 1 + x + \frac{x^2}{2} + \cdots + \frac{x^n}{n!} + \cdots$$

satisfies the differential equation $y' - y = 0$ (and we used this fact to conclude that $y = e^x$). Here we directly compute that, if the sum of a power series satisfies the differential equation $y' - y = 0$, then the series must be $a \sum_0^\infty x^n / n!$. Note that this calculation demands no prior knowledge of the exponential function, and if we have never seen e^x, we could still solve the differential equation $y' - y = 0$ as the sum of a power series in the above way and then use the series to compute the solution function. In this situation it would be necessary to check that the solution series does in fact converge.

EXAMPLE 2   If $y = f(x)$ is the sum of a power series

$$y = a_0 + a_1 x + \cdots + a_n x^n + \cdots,$$

and satisfies the differential equation

$$y'' + y = 0,$$

then the coefficients are determined by the same kind of recursive procedure as in the first example, except that this time the even coefficients go back to a_0 and the odd coefficients go back to a_1. Thus,

$$y = a_0 + a_1 x + a_2 x^2 + a_3 x^3 + a_4 x^4 + a_5 x^5 + \cdots,$$
$$y'' = 2a_2 + 3 \cdot 2a_3 x + 4 \cdot 3a_4 x^2 + 5 \cdot 4a_5 x^3 + 6 \cdot 5a_6 x^4 + 7 \cdot 6a_7 x^5 + \cdots,$$

and, since $y'' + y = 0$, we see that

$$a_2 = -\frac{a_0}{2}, \qquad\qquad a_3 = -\frac{a_1}{3!},$$

$$a_4 = -\frac{a_2}{4 \cdot 3} = \frac{a_0}{4!}, \qquad a_5 = -\frac{a_3}{5 \cdot 4} = \frac{a_1}{5!},$$

$$a_6 = -\frac{a_4}{6 \cdot 5} = -\frac{a_0}{6!}, \qquad a_7 = -\frac{a_5}{7 \cdot 6} = -\frac{a_1}{7!},$$

and so on. Thus

$$y = a_0 \left[1 - \frac{x^2}{2} + \frac{x^4}{4!} - \frac{x^6}{6!} + \cdots \right] + a_1 \left[x - \frac{x^3}{3!} + \frac{x^5}{5!} - \frac{x^7}{7!} + \cdots \right],$$

which we recognize to be $a_0 \cos x + a_2 \sin x$.

Again, if we knew nothing about the trigonometric functions, we could check that both series above converge everywhere to some functions $C(x)$ and $S(x)$ respectively, and we could compute these functions by the partial sums of the series and make up tables for them. Then the general solution of the differential equation $y'' + y = 0$ is expressed in terms of these two new functions by

$$y = a_0 C(x) + a_1 S(x),$$

as we saw above. This is only an academic exercise, but it has the following point: We can proceed in exactly this way with more complicated differential equations, computing the new functions that are their solutions as the sums of power series determined by the differential equation, and then studying these new solution functions on the basis of these power-series definitions.

The solutions of a second-order linear homogeneous equation with *variable* coefficients,

$$y'' + a(x)y' + b(x)y,$$

are generally not elementary functions, which means that there cannot be any general procedure for turning out solutions as finite combinations of familiar functions. Yet in some circumstances it is easy to compute the solutions as sums of power series in the manner of the above examples.

EXAMPLE 3 Solve the equation

$$y'' - xy = 0$$

by power series.

Solution. If

$$y = c_0 + c_1 x + c_2 x^3 + \cdots + c_n x^n + \cdots,$$

then

$$y'' = 2c_2 + 3 \cdot 2c_3 x + 4 \cdot 3c_4 x^2 + \cdots + (n + 2)(n + 1)c_{n+2}x^n + \cdots,$$
$$xy = c_0 x + c_1 x^2 + \cdots + c_{n-1}x^n \cdots,$$

and

$$y'' - xy = 2c_2 + (3 \cdot 2c_3 - c_0)x + (4 \cdot 3c_4 - c_1)x^2 + \cdots$$
$$+ ((n + 2)(n + 1)c_{n+2} - c_{n-1})x^n + \cdots.$$

Therefore $y'' - xy = 0$ if and only if

$$2c_2 = 0,$$
$$3 \cdot 2c_3 = c_0,$$
$$4 \cdot 3c_4 = c_1,$$
$$\cdot \quad \cdot$$
$$\cdot \quad \cdot$$
$$\cdot \quad \cdot$$
$$(n + 2)(n + 1)c_{n+2} = c_{n-1}.$$

These equations define the coefficients c_n recursively in terms of c_0 and c_1 (and c_2):

$$c_2 = 0, \qquad c_3 = \frac{c_0}{3 \cdot 2}, \qquad c_4 = \frac{c_1}{4 \cdot 3}, \qquad c_5 = \frac{c_2}{5 \cdot 4} = 0,$$

$$c_6 = \frac{c_3}{6 \cdot 5} = \frac{c_0}{6 \cdot 5 \cdot 3 \cdot 2}, \qquad c_7 = \frac{c_4}{7 \cdot 6} = \frac{c_1}{7 \cdot 6 \cdot 4 \cdot 3},$$

and so on. Thus

$$y = c_0 f(x) + c_1 g(x)$$

where

$$f(x) = 1 + \frac{x^3}{3 \cdot 2} + \frac{x^6}{6 \cdot 5 \cdot 3 \cdot 2} + \frac{x^9}{9 \cdot 8 \cdot 6 \cdot 5 \cdot 3 \cdot 2} + \cdots,$$

and

$$g(x) = x + \frac{x^4}{4 \cdot 3} + \frac{x'}{7 \cdot 6 \cdot 4 \cdot 3} + \frac{x^{10}}{10 \cdot 9 \cdot 7 \cdot 6 \cdot 4 \cdot 3} + \cdots.$$

There is no easy way to write down a formula for the nth coefficient in either of these series, but the pattern is clear. Moreover, it is easy to see, from the ratio test, that both series converge for all x, because the recursive relations give us simple explicit formulas for the ratios of successive terms. For example, $f(x)$ is of the form

$$f(x) = \sum_0^\infty a_n x^{3n},$$

where $a_n / a_{n-1} = 1/(3n(3n - 1))$. The ratio of the corresponding series terms is thus $x^3/(3n(3n - 1))$, which approaches 0 as $n \to \infty$, no matter what x is. Therefore, f and g are everywhere defined functions, and each satisfies the differential equation $y'' - xy = 0$, because the coefficients in its Maclaurin expansion satisfy the recursive relations that were equivalent to this differential equation. That $y = c_0 f(x) + c_1 g(x)$ is the general solution of this differential equation follows from Theorem 6.

PROBLEMS FOR SECTION 9

Show that the general solution of each of the following differential equations can be expressed as the sum of convergent power series. In each case, write down the recursion formula for the coefficients, and determine the convergence interval by the ratio test. Write down the first few terms of each series, and, if possible, the general term. (It is also instructive, where possible, to find the solutions by our earlier methods and compare answers.)

1. $y' - xy = 0$
2. $y'' + xy' = 0$
3. $y' - x^2 y = x$
4. $(1 - x)y' + y = 0$
5. $(1 - x^2)y' - xy = 0$
6. $y'' - 2y = 1$
7. $y'' - x^2 y = 0$
8. $y'' + 2x^2 y - xy = 0$
9. $(1 - x^2)y'' - y = 0$

10. Solve the equation

$$xy'' - y = 0$$

by the power series method. What conclusion can you draw about the possibility of expressing the general solution

$$y = af(x) + bg(x)$$

as a power series?

11. Supposing that $y = f(x)$ is one solution of

$$xy'' - y = 0$$

(the one found above), show that $z = u(x)y$ is another solution if and only if

$$u(x) = C_1 \int \frac{dx}{y^2} + C_2.$$

12. Show that no solution of the equation

$$x^2 y'' - y = 0$$

can be represented by a convergent Maclaurin series.

EXTRA PROBLEMS FOR CHAPTER 19

Solve the following differential equations.

1. $\dfrac{dy}{dx} = y^2 x$

2. $\dfrac{dy}{dx} = e^y$

3. $\sqrt{1 - x^2}\, dy + \sqrt{1 - y^2}\, dx = 0$

4. $\dfrac{dy}{dx} = \cos^2 y \sin x$

5. $dy/dx = (y/x) + (y/x)^2$. Here the variables cannot be separated. Show, however, that the substitution $y = vx$ reduces the given equation to one in x and v, in which the variable *can* be separated, and hence solve the equation. (Note that dy/dx becomes $d(xv)dx = v + x\, dv/dx$.)

6. Show that the equation

$$\frac{dy}{dx} = f\!\left(\frac{y}{x}\right)$$

reduces to a variables-separable equation in v and x under the substitution $y = vx$.

7. $(x^2 + y^2)\, dx - 2xy\, dy = 0$

8. A plane curve has the property that each of its tangents intersects the axis in two points whose midpoint is the point of tangency. Find all such curves.

9. The tangent to a plane curve at the point P on the curve intersects the x-axis at Q. The curve has the property that for every point of tangency P, the segment PQ is bisected by the y-axis. Find all such curves.

10. The tangent to a plane curve at the point P on the curve intersects the x-axis at the point Q. The curve has the property that the segments OP and PQ are always the same length. Find all such curves.

11. The growth rate of a population is the difference between the birth rate and the death rate. The birth rate is proportional to the population size, but the death rate is proportional to the square of its size. Find the law of population growth.

12. Solve the above problem if the death rate is proportional to $P^{3/2}$, where P is the population size.

13. Fifty pounds of salt are dissolved in a tank holding 100 gallons of water. Pure water is added at the rate of 1 gallon per minute and the solution is drained off at same rate. Assuming that the solution is agitated sufficiently so that mixture can be considered to be always uniform, how much salt is left after two hours?

14. A chemical dissolves in water at a rate that is jointly proportional to the amount undissolved and to the further amount that can be dissolved before saturation occurs.

    Suppose that x_0 pounds of the chemical are introduced into G gallons of pure water, and that the solution will become saturated after P pounds have dissolved. Find a formula for the amount of dissolved chemical as a function of time.

15. A mixture consists initially of equal parts of chemicals x and y, which combine in a reaction in the ratio of two parts of x to one part of y to produce a chemical z. The reaction rate is jointly proportional to the amount of x and y present. Find the amount of z that has been produced, as a function of time, supposing there is none to begin with.

16. It is experimentally determined that water flows out from a tank container through a hole in the bottom with a velocity v given by

$$v = 8\sqrt{h}\,\text{ft/sec},$$

    where h is the water depth in feet (assuming the hole to be large enough so that "edge effects" are negligible).

    Suppose that the tank is a vertical circular cylinder 10 feet in diameter and 15 feet high, and that the hole is 6 inches in diameter. How long will it take to empty a full tank?

17. Suppose that water is being added to the tank in the above problem at the constant rate of 2 cubic feet per second, the excess normally flowing off the top. How long will it take for the water level to fall 5 feet when the outlet valve at the bottom is opened?

18. A tank is a solid of revolution, with axis vertical, of such a shape that its water level falls at a constant rate when the valve at the bottom is opened. What is its shape? (Use the formula in Problem 16.)

When a one-parameter family is given by an equation of the form

$$f(x, y) = C,$$

so that its curves are level curves of the function f, the orthogonal trajectories have the direction of the gradient field of f. Their differential equation is thus

$$\frac{dy}{dx} = \frac{\partial f/\partial y}{\partial f/\partial x}.$$

For example, the orthogonal trajectories of the ellipses

$$x^2 + 2y^2 = C$$

have the direction $(2x, 4y)$, so their differential equation is

$$\frac{dy}{dx} = \frac{2y}{x}.$$

Such orthogonal trajectories are thus the curves of *steepest descent* for the function f. If $x = f(x, y)$ is a hillside on which rain falls, then the water will run off along paths on the hillside that lie over the one-parameter family of curves of steepest descent in the xy-plane.

Find the curves of steepest descent when:

19. $f(x, y) = x^2 - y^2$

20. $f(x, y) = x^2 y^3$

21. $f(x, y) = 1 - (x^2 + y^2)$

22. $f(x, y) = xy$

23. Rain falls on the ellipsoid

$$\frac{x^2}{a^2} + \frac{y^2}{b^2} + \frac{z^2}{c^2} = 1.$$

Along what paths will the water run off?

24. Show that the general solution of the differential equation

$$L(y) = b_1(x) + b_2(x)$$

is the sum of a particular solution of $L(y) = b_1(x)$, a particular solution of $L(y) = b_2(x)$, and the general solution of $L(y) = 0$.

25. Find the general solution of

$$\frac{dy}{dx} + y = 1 + e^x$$

by following the above prescription.

26. Solve

$$\frac{dy}{dx} - y = x + \sin x.$$

27. Find the general solution of

$$\frac{dy}{dx} - y = x + e^x.$$

28. Find the general solution of

$$\frac{dy}{dx} + y = e^x + e^{-x}.$$

29. If $L(y) = dy/dx - y/x$, compute

a) $L(x^4)$ b) $L(x^2)$ c) $L(x)$ d) $L(xe^x)$.

30. If $L(y) = (dy/dx) + a(x)y$, show that

$$uL(v) - vL(u) = u^2 \frac{d}{dx}\left(\frac{v}{u}\right).$$

31. If $L(y) = (dy/dx) + a(x)y$, show that

a) $L(uv) = uL(v) + v\dfrac{du}{dx}$,

b) $2L(uv) = uL(v) + vL(u) + \dfrac{d}{dx}(uv).$

32. Let L and M be any two first-order linear differential operators with constant coefficients, say

$$L(y) = y' + ay, \qquad M(y) = y' + by.$$

Show that L and M commute:

$$L[M(f(x))] = M[L(f(x))]$$

for any twice-differentiable function f.

In each of the following problems find a particular solution by the method of undetermined coefficients.

33. $y'' + y' = x + \sin x$ 34. $y'' + y' = x^2 e^{-x}$

35. $y'' - 2y' + y = x^2 e^x$ 36. $y'' - 2y = xe^x + 1$

37. $y'' - 2y = e^x \sin x$ 38. $y'' + 4y = x \sin x$

39. $y'' - 2y' + 3y = e^x \sin x$

40. Supposing that the differential operator L has constant coefficients, show that it commutes with differentiation. That is, show that

$$\frac{d}{dx} L(f(x)) = L(f'(x)).$$

41. Let D be the operation of differentiation, so that

$$Df = f', \qquad Df(x) = f'(x).$$

Then $L(f) = f'' + af' + bf$ can be thought of as a quadratic polynomial in D applied to f,

$$L(f) = D^2 f + aDf + bf = (D^2 + aD + b)f = p(D)f,$$

where p is the quadratic polynomial

$$p(x) = x^2 + ax + b.$$

Show that if $p(x)$ can be factored,

$$p(x) = (x - k_1)(x - k_2),$$

then

$$p(D)f = (D - k_1)(D - k_2)f,$$

where the right side is interpreted as

$$(D - k_1)[(D - k_2)f].$$

The above formula gives a systematic way of obtaining the solutions of the homogeneous equation. However, the third case, where the roots k_1 and k_2, obtained from the quadratic formula, involve the imaginary number $i = \sqrt{-1}$, carries us beyond the scope of this course.

42. Supposing that

$$D^2 + aD + b = (D - k_1)(D - k_2)$$

in the above sense, with $k_1 \neq k_2$, show that the general solution of the homogeneous equation

$$L(y) = (D^2 + aD + b)y = 0$$

is

$$y = C_1 e^{k_1 x} + C_2 e^{k_2 x}.$$

[*Hint:* Set $u = (D - k_2)y$. Then the homogeneous second-order equation $L(y) = 0$ reduces to the *first*-order linear equation

$$(D - k_1)u = 0.$$

Solving this gives u, and then $(D - k_2)y = u$ is a first-order inhomogeneous equation for y, that can be solved by variation of parameters, or undetermined coefficients.]

43. Now solve the homogeneous equation by the above scheme in the case where

$$D^2 + aD + b = (D - k)^2$$

(i.e., where $k_1 = k_2 = k$). In this case, use variation of parameters for the last step.

Use the method developed in Problem 42 to find the general solution of each of the following problems, by two applications of the first-order variation-of-parameter calculation.

44. $y'' - y' - 2y = e^x$ 45. $y'' - y' - 2y = e^{2x}$

46. $y'' - y = 1$

47. Show that

$$p(D)(xf) = xp(D)f + p'(D)f,$$

so

$$\text{if} \quad p(D)f = 0, \quad \text{then} \quad p(D)(xf) = p'(D)f.$$

(Here $p(x) = x^2 + ax + b, p'(x) = 2x + a$.)

48. Show that, if $p(D) = D^2 + aD + b$, then

$$p(D)(fg) = [p(D)f]g + [p'(D)f]g' + \frac{1}{2}[p''(D))f]g''.$$

Solve the following differential equations by power series. (There may be only one series solution.)

49. $y'' + xy + y = 0$ 50. $y'' + xy + 2y = 0$

51. $(1 + x^2)y' - 2xy = 0$ 52. $(1 + x^2)y'' + 2xy' - 2y = 0$

53. $xy'' + y' + xy = 0$

54. Suppose that y is one solution of the linear homogeneous equation

$$y'' + a(x)y' + b(x)y = 0,$$

and suppose that $y \neq 0$ on an interval I. Use the variation-of-parameter method (as in Example 2 and 3 of Section 7) to show that the general solution over I is given by

$$u = c_1 y \int \frac{e^{-A(x)}}{y^2} \, dx + c_2 y,$$

where $A(x)$ is any antiderivative of $a(x)$. This result can be used to complete the proof of Theorem 6 (for the interval I).

Appendix 1
Basic Inequality
Laws

We begin with the following two facts about positive numbers:

P1. If x and y are positive, then so are $x + y$ and xy.

P2. Zero is not positive. If $x \neq 0$, then x is positive or $-x$ is positive.

Starting from P1 and P2 as axioms, it is possible to prove all of the algebraic properties of inequalities. We shall sketch this development below, giving complete details for the proofs of some properties, partial proofs for others, and leaving some as exercises.

We call a number x *negative* if $-x$ is positive. A number x cannot be both positive and negative, for this would imply that $0 = x + (-x)$ is positive (by P1), which contradicts P2. Thus, P2 has the following reformulation:

P2′. A number x is either positive, or zero, or negative, and these possibilities are mutually exclusive.

1. Law of signs. *If x and y have the same sign (i.e., if both are positive or both are negative), then xy is positive. If x and y have opposite signs, then xy is negative.*

Proof. By hypothesis, x and y are either both positive or both negative. If x and y are both positive, then xy is positive, by P1. If x and y are both negative, then $-x$ and $-y$ are both positive, by P2, so $xy = (-x)(-y)$ is positive, by P1. Thus xy is positive in either case.

The other assertion will be left as an exercise.

2. *If $x \neq 0$, then x^2 is positive. In particular, $1 = 1 \cdot 1$ is positive.*

Proof. This is a corollary of (1), since if $x \neq 0$, then x^2 is a product in which both factors have the same sign.

3. *If $x \neq 0$, then x and $1/x$ have the same sign.*

Proof, by contradiction. Suppose x and $1/x$ have opposite signs. Then $1 = x(1/x)$ is negative, by (1). But 1 is positive, by (2). We therefore have a contradiction, so x and $1/x$ must have the same sign.

DEFINITION   *We say that x is less than y, and write $x < y$, if $y - x$ is positive. Similarly $x > y$ (x is greater than y) if $y < x$, that is, if $x - y$ is positive. In particular, $x > 0$ if and only if x is positive.*

4. *If $x < y$ and $y < z$, then $x < z$.*

Proof. By hypothesis, $x < y$ and $y < z$. Therefore $y - x$ and $z - y$ are both positive, by the definition of inequality. Therefore $z - x = (z - y) + (y - z)$ is positive, by P1. Therefore $x < z$, by the definition of inequality.

5. *If $x < y$, then $x + c < y + c$, no matter what number c is.*

Proof. If $x < y$, then $y - x$ is positive, by the definition of inequality. Then $(y + c) - (x + c) = y - x$ is positive. Then $x + c < y + c$, by the definition of inequality.

6. *If $x < y$ and if c is positive, then $cx < cy$.*

Proof. By hypothesis and the definition of inequality, $y - x$ and c are both positive. Therefore $cy - cx = c(y - x)$ is positive, by P1, and hence $cx < cy$, by the definition of inequality.

7. *If $x < y$ and $a < b$, then $x + a < y + b$.*
8. *If $0 < x < y$ and $0 < a < b$, then $ax < by$.*
9. *If x and y are positive and if $x < y$, then $1/x > 1/y$.*

That is, reciprocals of positive numbers are in opposite order.

Partial proof. We note that

$$\frac{1}{x} - \frac{1}{y} = \frac{y - x}{xy} = (y - x) \cdot \frac{1}{x} \cdot \frac{1}{y}$$

and that all three factors on the right are positive, for various reasons. It follows that $1/x > 1/y$.

10. *If $x < y$, then $-y < -x$.* Changing signs reverses an inequality.

11. *If $x < y$ and c is negative then $cx > cy$.* Multiplying by a negative number reverses an inequality.

DEFINITION *We say that x is less than or equal to y, and write $x \le y$ or $x \leqq y$, if either $x < y$ or $x = y$.* This so-called *weak* inequality satisfies a corresponding list of laws that we won't bother to give completely. Here is an example, to show how the two logical alternatives enter in.

12. *If $x \le y$, then $c + x \le c + y$, for any number c.*

Proof. If $x \le y$, then either $x < y$ or $x = y$. If $x < y$, then $x + c < y + c$, by (5). If $x = y$, then $x + c = y + c$ (the sum of a pair of numbers is uniquely defined). Therefore, in either case, $x + c \le y + c$.

13. *$|x| \le a$ if and only if $-a \le x \le a$.*

Proof. It follows from the definition of absolute value that $|x|$ is the larger of x and $-x$. The inequality $|x| \le a$ is thus equivalent to the pair of inequalities $-x \le a$, $x \le a$, and hence to the pair $-a \le x$, $x \le a$. But the latter pair is what we mean by the "continued" inequality $-a \le x \le a$.

14. *$|x + y| \le |x| + |y|$.*

Proof. Since

$$-|x| \le x \le |x|,$$
$$-|y| \le y \le |y|,$$

it follows that

$$-(|x|+|y|) \le x + y \le |x| + |y|,$$

by the weak inequality analogue of (8). But this is equivalent to $|x + y| \le |x| + |y|$ by (13).

Appendix 2
The Conic Sections

Properties of the parabola, ellipse, and hyperbola are treated in the problem sets at various places in the text, mostly on the basis of parametric and polar-coordinate representations involving the trigonometric functions. Here we shall give a straightforward algebraic treatment of some of these questions, starting with traditional locus definitions of the curves.

1. THE PARABOLA

DEFINITION   *A parabola is the locus traced by a point whose distance from a fixed point (called the focus) is equal to its distance from a fixed line (called the directrix).*

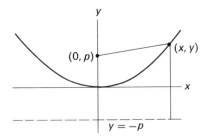

In order to obtain the equation of the parabola in standard form, we take the focus at $(0, p)$ on the y-axis and the directrix parallel to the x-axis with intercept $-p$. Then a point satisfies the locus condition if and only if

$$\sqrt{x^2 + (y - p)^2} = |y + p|.$$

Squaring and simplifying, we obtain first

$$x^2 + y^2 - 2py + p^2 = y^2 + 2py + p^2,$$

and then

$$x^2 = 4py.$$

This is the standard equation of the parabola.

EXAMPLE 1   Identify the graph of the equation $y - 2x^2 = 0$.

Solution.   We solve for x^2,

$$x^2 = (1/2)y,$$

and interpret the result as the standard form. Thus, $1/2 = 4p$, so $p = 1/8$. The graph is the parabola with focus at $(0, 1/8)$ and directrix $y = -1/8$.

Note that we never did write the given equation explicitly in the standard form $x^2 = 4py$. This would be

$$x^2 = 4\left(\frac{1}{8}\right)y.$$

Instead, we found the right value of p by setting $1/2 = 4p$, and that was all we needed to identify the graph.

The parabola $x^2 = 4py$ is symmetric about the y-axis. In terms of the locus definition, the axis of symmetry is the line through the focus perpendicular to the directrix. The *vertex* of a parabola is the point where it intersects its axis of symmetry; i.e., the point halfway from the focus to the directrix. The vertex of the parabola $y^2 = 4px$ is at the origin.

In the derivation above we assumed that p is positive. If p is negative, the algebra is the same, but the picture is different. The focus at $(0, p)$ and directrix $y = -p$ now give us a parabola opening *downward*. Thus, $x^2 = -4y$ is the equation $x^2 = 4py$ with $p = -1$; so its graph is the parabola with focus at $(0, -1)$ and directrix $y = 1$, a downward opening parabola with vertex at the origin.

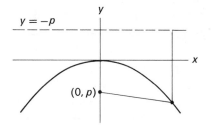

In the same way, the parabola with focus $(p, 0)$ on the x-axis and directrix $x = -p$ parallel to the y-axis has the equation

$$y^2 = 4px.$$

It is symmetric about the x-axis, with vertex at the origin, and opens to the right or to the left depending on whether p is positive or negative.

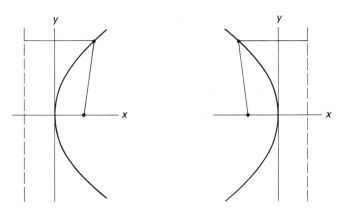

The proof does not have to be repeated. If we interchange the roles of x and y throughout an argument, then we automatically prove a new result, the statement of which is obtained from the old statement by simply interchanging the roles of x and y.

In summary:

THEOREM 1 *The equation of a parabola with vertex at the origin is*

$$x^2 = 4py$$

if its axis is the y-axis, and

$$y^2 = 4px$$

if its axis is the x-axis. In each case, p is the coordinate of the focus on the axis.

EXAMPLE 2   Identify the graph of $x + y^2 = 0$.

Solution. Solving for y^2,

$$y^2 = -x,$$

we have the form $y^2 = 4px$ with $p = -1/4$. So the graph is the parabola with focus at $(-1/4, 0)$ and directrix at $x = 1/4$. It opens to the left.

Now consider a parabola with vertex at (h, k). If we choose a new XY-axis system, with new origin at the point having old coordinates (h, k), then the vertex of the parabola is at the new origin; so the equation of the parabola in this new system will be

$$X^2 = 4pY$$

if its axis is vertical, and

$$Y^2 = 4pX$$

if its axis is horizontal, by Theorem 1. Since $X = x - h$ and $Y = y - k$ (Chapter 1, Section 8), we have the following generalization of Theorem 1.

THEOREM 2   *The equation of a parabola with vertex at the point (h, k) is*

$$(x - h)^2 = 4p(y - k)$$

if its axis is vertical, and

$$(y - k)^2 = 4p(x - h)$$

if its axis is horizontal. In each case p is the signed distance from the vertex to the focus along the axis of the parabola. Thus, the focus is at $(h, k + p)$ in the first case and $(h + p, k)$ in the second case.

EXAMPLE 3   Find the equation of the parabola with focus at $(4, 1)$ and directrix $x = 2$.

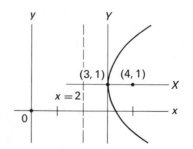

Solution. The vertex is halfway from the focus to the directrix, and hence is at $(h, k) = (3, 1)$. The focus at $(4, 1)$ is one unit to the right of the vertex at $(3, 1)$, so $p = 1$. And the parabola opens in the x-direction, so it has the y-squared equation. Its equation is thus:

$$(y - k)^2 = 4p(x - h)$$

with $(h, k) = (3, 1)$ and $p = 1$; that is

$$(y - 1)^2 = 4(x - 3),$$

or

$$y^2 - 2y - 4x + 13 = 0.$$

EXAMPLE 4 Find the equation of the parabola with focus at $(-2, 0)$ and with the line $y = 2$ as directrix.

Solution. The vertex is $(-2, 1)$, the point halfway between focus and directrix. The focus is one unit *below* the vertex, so $p = -1$, and the axis is vertical. So the equation is

$$(x - h)^2 = 4p(y - k)$$

with $(h, k) = (-2, 1)$ and $p = -1$, which works out to

$$x^2 + 4x + 4y = 0.$$

The equation $(x - h)^2 = 4p(y - k)$ can be solved for y, leading to an equivalent equation of the form

$$y = ax^2 + bx + c,$$

with $a \neq 0$. Thus;

Each vertical parabola is the graph of a quadratic function $f(x) = ax^2 + bx + c$.

Conversely, every quadratic function $f(x) = ax^2 + bx + c$, with $a \neq 0$, is the graph of a vertical parabola. To identify the parabola we transform the equation $y = ax^2 + bx + c$ into the standard form $(x - h)^2 = 4p(y - k)$ by completing the square on the x-terms, just as in Section 8 of Chapter 1.

EXAMPLE 5 Identify the graph of the equation $y = 2x^2 - 4x$.

Solution. We start by completing the square:

$$y + 2 = 2(x^2 - 2x + 1) = 2(x - 1)^2.$$

This equation can be written

$$(x - 1)^2 = (1/2)(y + 2),$$

which is in the standard form

$$(x - h)^2 = 4p(y - k),$$

with $(h, k) = (1, -2)$ and $p = 1/8$. The graph is thus the parabola with vertex at $(1, -2)$ and focus 1/8 unit above the vertex, at $(9/8, -2)$.

More generally, any equation in x and y that is quadratic in one variable and linear in the other has a parabolic graph that can be identified by completing a square.

EXAMPLE 6   Identify the graph of $y^2 - 6y - 2x + 7 = 0$.

Solution. We convert the equation to standard form, starting by completing the square on the y-terms, as follows:

$$y^2 - 6y + 9 - 2x + 7 = 9,$$
$$(y - 3)^2 = 2x + 2,$$
$$(y - 3)^2 = 2(x + 1).$$

This is of the form

$$(y - k)^2 = 4p(x - h)$$

with $(h, k) = (-1, 3)$ and $p = 1/2$. The graph is thus the parabola with vertex at $(-1, 3)$, opening to the right, with focus 1/2 unit to the right of the vertex at $(-1/2, 3)$ and with directrix $x = -3/2$.

A unique "vertical" parabola, with equation of the form

$$y = ax^2 + bx + c,$$

can be passed through any given set of three points having distinct x-coordinates, provided they are not collinear, for the three sets of numerical coordinates give us three equations in the three unknowns a, b, und c, and can be solved for these coefficients in the standard manner.

EXAMPLE 7   Find the vertical parabola that contains the points $(-1, 2)$, $(1, 1)$, $(2, 3)$.

Solution. The equation of the parabola, of the form

$$y = ax^2 + bx + c,$$

must be satisfied by each of the above three pairs of values. That is,

$$2 = a - b + c,$$
$$1 = a + b + c,$$
$$3 = 4a + 2b + c.$$

The first step in solving three equations in three unknowns is to eliminate one unknown, leaving two equations in two unknowns. For example, we can solve the first equation above for c, substitute in the other two equations and so reduce to two equations in a and b. A simpler way of accomplishing the same result is to subtract the first equation from each of the other two, since c will cancel in the subtraction. We then have

$$-1 = 2b,$$

$$1 = 3a + 3b.$$

Finally, $b = -1/2$, $a = 5/6$, $c = 2/3$, so the parabola has the equation

$$y = \frac{5}{6}x^2 - \frac{1}{2}x + \frac{2}{3}.$$

The geometry of the parabola is concerned largely with properties of the tangent line. We have to know how to differentiate in order to compute the slope of the tangent line, but from then on it is pure analytic geometry.

THEOREM 3 *The tangent line to the parabola*

$$x^2 = 4py$$

at the point (x_0, y_0) on the parabola has the equation

$$x_0 x = 2p(y + y_0).$$

Proof. The parabola $x^2 = 4py$ has the slope

$$\frac{dy}{dx} = \frac{2x}{4p} = \frac{x}{2p}.$$

The tangent line at the point (x_0, y_0) on the parabola thus has the slope $x_0/2p$, and its point–slope equation is

$$y - y_0 = \frac{x_0}{2p}(x - x_0).$$

We multiply out the right side and use the fact that $x_0^2 = 4py_0$ to get the form given in the theorem. ∎

The tangent-line equation and the slope formula $m_0 = x_0/2p$ will be used in the problems. Here we shall check only the optical property of the parabola.

THEOREM 4 *The tangent line to the parabola $x^2 = 4py$ at (x_0, y_0) makes equal angles with the focal radius to (x_0, y_0) and the vertical line at that point.*

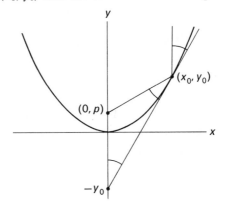

Proof. The tangent line $xx_0 = 2p(y + y_0)$ has y-intercept $y = -y_0$. The distance from the y-intercept to the focus $(0, p)$ is thus $p + y_0$. But this is also the distance from the point (x_0, y_0) to the directrix $y = -p$, and therefore is equal to the distance from (x_0, y_0) to the focus, by the locus definition of the parabola. The triangle in the figure above is thus isosceles and has equal base angles. Since the tangent meets all vertical lines at the same angle, this proves the theorem. $\blacksquare$

The theorem shows that the light rays emitted by a point source of light placed at the focus will all reflect off the parabola along vertical lines, forming a beam of light shining vertically without any weakening caused by spreading out. This is why a searchlight (or automobile headlight) has a reflector that is parabolic in cross section, and has a small intense light source at the focus. The ideal parallel beam will not quite be realized because the light source can only approximate a point source of light.

Telescopes use the same principle in reverse. The light rays arriving from a distant star form a parallel beam and are therefore focused onto a single point by a parabolic mirror. (Hence the word *focus*.) A camera placed (approximately) at the focus thus gathers all the light intercepted by the mirror, and this huge magnification of the light-gathering power of the camera lens makes it possible to take pictures of objects too faint to be recorded otherwise.

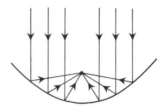

PROBLEMS FOR SECTION 1

Find the equations for each of the following parabolas.

1. Vertex at the origin; focus at $(2, 0)$
2. Vertex at the origin; focus at $(0, -1)$
3. Directrix $x = 1$; focus $(-1, 0)$
4. Vertex at the origin; directrix $y = 1/4$
5. Directrix the y-axis; focus $(1, 0)$
6. Directrix the y-axis; focus $(-2, 0)$
7. Directrix the x-axis; vertex $(0, 2)$
8. Directrix $x = 1$; focus at $(3, -1)$
9. Directrix $x = 1$; focus at the origin
10. Directrix $y = -2$; focus at the origin
11. Directrix $y = -2$; vertex at $(-2, 0)$

Identify each of the following parabolas by finding its vertex and focus.

12. $y = 2x^2$
13. $y^2 = 2x$
14. $y^2 + 2x = 0$
15. $y = 2x^2 + 1$

16. $y^2 + 2x + 1 = 0$

17. $3y + x^2 - 6 = 0$

18. $y = x^2 + 2x + 1$

19. $y = x^2 + 2x$

20. $y^2 + x + y = 0$

21. $2y^2 - 4y - x + 2 = 0$

22. $2y^2 - 4y - x = 0$

23. A chord of the parabola $x^2 = 4py$ has endpoints with x-coordinates x_1 and x_2. Show that the slope of the chord is $(x_1 + x_2)/4p$.

24. A focal chord (chord through the focus) of the parabola $x^2 = 4py$ runs from (x_1, y_1) to (x_2, y_2). Show that its length is $y_1 + y_2 + 2p$.

In each of the following problems find the equation of and identify the vertical parabola passing through the three given points.

25. $(1, 0), (2, -1), (3, 0)$

26. $(1, 0), (2, 1), (4, 0)$

27. $(-1, 1), (0, 0), (2, -1)$

28. There can be no parabola having a horizontal axis and passing through the three points in Problem 25. Why?

29. Find the parabola having a horizontal axis and passing through the three points in Problem 27.

30. Find the equation of the parabola with vertical axis, vertex at the origin, and tangent to the line $y = x - 2$.

31. Find the lines tangent to the parabola $y = x^2$ and passing through the point $(2, 1)$.

32. Show that the tangent line to the parabola $y^2 = 4px$ at the point (x_0, y_0) has slope $2p/y_0$.

33. The normal to a parabola at a point P on the parabola intersects the axis of the parabola at the point N. Show that P and N are at the same distance from the focus.

34. The tangent to a parabola at a point P intersects the directrix at the point T. Show that the segment PT subtends a right angle at the focus.

35. The tangent to a parabola at a point P intersects the tangent at the vertex at the point Q. If F is the focus, show that the segments FQ and PQ are perpendicular.

36. Prove that the midpoints of all chords of a parabola parallel to a given chord form a straight line that is parallel to the axis of the parabola.

37. Show that the lines tangent to a parabola at the ends of a focal chord (a chord through the focus) meet at right angles.

38. Show that the lines tangent to a parabola at the ends of a focal chord meet on the directrix.

39. Prove that the vertical line $x = x_0$ bisects every chord of the parabola $x^2 = 4py$ that is drawn parallel to the tangent line at (x_0, y_0).

40. Show that $x^2 = 4p(y + p)$ is the standard equation for a parabola with vertical axis and focus at the origin.

41. Show that every upward-opening parabola with focus at the origin intersects at right angles every downward-opening parabola with focus at the origin. (Use the equation in the above problem with two different values of p, one positive and one negative.)

2. THE ELLIPSE

Ellipses and hyperbolas also have focus–directrix locus definitions, but we shall start instead from another locus problem.

DEFINITION   *An ellipse is the locus traced by a point P whose distances from two fixed points F and F' have a constant sum, k,*

$$\overline{PF} + \overline{PF'} = k,$$

where k is greater than the distance $\overline{FF'}$. Each of the fixed points F and F' is called a focus of the ellipse (for reasons having to do with the optical property, which will be discussed later).

In order to obtain the simplest equation for this locus we take the x-axis along the segment FF' and the y-axis as its perpendicular bisector.

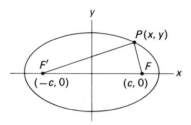

Then $F = (c, 0)$ and $F' = (-c, 0)$, where c is one half the segment length. Also, if a is the intercept of the locus on the positive x-axis, then the locus condition says that

$$(a - c) + (a + c) = k,$$

so $k = 2a$. The locus condition for $P = (x, y)$ can thus be rewritten

$$\sqrt{(x - c)^2 + y^2} + \sqrt{(x + c)^2 + y^2} = 2a,$$

where a and c are positive constants such that $a > c$. This, then, is the basic equation of the locus. In order to simplify it, we follow the usual procedure for eliminating radicals: solve for a radical, and square. If we solve for the second radical, square everything, and then simplify, we end up with

$$\sqrt{(x - c)^2 + y^2} = a - \left(\frac{c}{a}\right)x.$$

The ratio c/a occurring here is called the *eccentricity* of the ellipse and is designated e; note that $e = c/a$ is less than 1. The first simplifying step thus results in a formula for the length of the righthand "focal radius" PF. This is an important formula, and we record it as a theorem.

THEOREM 5   *Let E be the ellipse having foci $F = (c, 0)$ and $F' = (-c, 0)$, and x-intercepts at $\pm a$. Then for any point $P = (x, y)$ on E, the focal radii PF and PF' have the lengths*

$$\overline{PF} = a - ex, \qquad \overline{PF'} = a + ex,$$

where $e = c/a$.

Note that the second formula follows from the first, because $\overline{PF} + \overline{PF'} = 2a$. Continuing, we square the equation

$$\sqrt{(x - c)^2 + y^2} = a - \left(\frac{c}{a}\right)x$$

and simplify, this time ending up with

$$\left(\frac{a^2 - c^2}{a^2}\right)x^2 + y^2 = a^2 - c^2.$$

Setting $x = 0$ determines the y-intercepts as the points $\pm\sqrt{a^2 - c^2}$. We call the y-intercepts $\pm b$, as usual. That is, we set

$$b^2 = a^2 - c^2.$$

Then, after dividing by b^2, the above equation turns into the standard equation for the ellipse:

THEOREM 6 *An ellipse with foci located on the x-axis and centered at the origin has the equation*

$$\frac{x^2}{a^2} + \frac{y^2}{b^2} = 1.$$

Actually, the reasoning above does only half the job. We have shown that if a point (x, y) is on the locus, then its coordinates satisfy the standard equation. It must also be shown that the argument can be traced backward, so that if a point (x, y) satisfies the standard equation, then it is on the locus. This backwards process involves taking square roots twice, and it must be checked that no sign ambiguity arises. For example, we must check that if (x, y) is on the graph of

$$\frac{x^2}{a^2} + \frac{y^2}{b^2} = 1,$$

then $(a - ex)$ is positive. This will be left as an exercise.

Remember that $a > b$ in the standard equation above, the foci being located by $c^2 = a^2 - b^2$. If we start instead with foci at the points $(0, c)$ and $(0, -c)$ on the y-axis, then we end up with the same standard equation in terms of the intercepts a and b, but this time $b > a$ and $c^2 = b^2 - a^2$. The graph of the standard equation is thus:

1. An ellipse with foci on the x-axis if $a > b$;
2. A circle of radius $r = a = b$ if $a = b$;
3. An ellipse with foci on the y-axis if $a < b$.

We consider that we have identified an ellipse if we have found its standard equation, and hence know a and b. The larger of a and b is (the length of) the *semimajor axis* of the ellipse, and the smaller is (the length of) its *semiminor axis*. We then know the location of the foci, as we noted above, and everything else follows.

EXAMPLE 1 The graph of

$$\frac{x^2}{25} + \frac{y^2}{9} = 1$$

is the ellipse with foci at $\pm\sqrt{25 - 9} = \pm 4$ on the x-axis, and intercepts $\pm a = \pm 5$ on the x-axis and $\pm b = \pm 3$ on the y-axis. However, in view of the symmetry about the axes, we normally record only the positive intercepts $a = 5$, $b = 3$.

EXAMPLE 2 Identify the graph of

$$x^2 + 4y^2 = 9.$$

Solution 1. We throw the equation into the standard form

$$\frac{x^2}{9} + \frac{y^2}{9/4} = 1,$$

and then read off $a^2 = 9$, $b^2 = 9/4$. The graph is an ellipse with foci at

$$\pm c = \pm\sqrt{9 - \frac{9}{4}} = \pm\frac{3}{2}\sqrt{3}$$

on the x-axis, and positive intercepts $a = 3$, $b = 3/2$.

Solution 2. We recognize that the graph is an ellipse symmetric in the coordinate axes, so we compute the intercepts $a = 3$, $b = 3/2$ directly from the equation. Since $a > b$, we conclude that the foci are on the x-axis with $c = \sqrt{a^2 - b^2} = 3\sqrt{3}/2$.

EXAMPLE 3 The graph of $4x^2 + y^2 = 1$ is clearly going to be a central ellipse. Its positive intercepts are $a = 1/2$, $b = 1$; so its foci are on the y-axis and $c = \sqrt{1-(1/2)^2} = \sqrt{3}/2$.

EXAMPLE 4 Compute the lengths of the focal radii to the point(s) on the ellipse

$$\frac{x^2}{4} + \frac{y^2}{2} = 1$$

having x-coordinate 1.

Solution. We could solve for the y-coordinates and then use the distance formula, but the formulas for the focal radii are much simpler. We see that $a = 2$, $b = \sqrt{2}$, $c = \sqrt{4 - 2} = \sqrt{2}$, $e = c/a = \sqrt{2}/2$, and

$$\overline{PF} = a - ex = 2 - \frac{\sqrt{2}}{2} \cdot 1 = \frac{4 - \sqrt{2}}{2},$$

$$\overline{PF'} = a + ex = 2 + \frac{\sqrt{2}}{2} \cdot 1 = \frac{4 + \sqrt{2}}{2}.$$

The standard equation for an ellipse centered at (h, k) and having axes parallel to the coordinate axes is

$$\frac{(x - h)^2}{a^2} + \frac{(y - k)^2}{b^2} = 1.$$

The reasoning goes exactly as it did for the parabola.

EXAMPLE 5   Find the equation of the ellipse whose foci are at $(1, -2)$ and $(5, -2)$, and with $2a = 6$.

Solution. The center is midway between the foci, so $(h, k) = (3, -2)$ and $c = 2$. Since we are given that $a = 3$, it follows that $b = \sqrt{a^2 - c^2} = \sqrt{5}$. The standard equation of the ellipse is therefore

$$\frac{(x - 3)^2}{9} + \frac{(y + 2)^2}{5} = 1.$$

If we expand the squares and collect coefficients, the equation becomes

$$5x^2 + 9y^2 - 30x + 36y + 36 = 0.$$

EXAMPLE 6   Identify the graph of the equation

$$x^2 + 2y^2 + 4x - 4y = 0.$$

Solution. After completing the squares the equation is

$$(x + 2)^2 + 2(y - 1)^2 = 6,$$

from which we get standard equation

$$\frac{(x + 2)^2}{6} + \frac{(y - 1)^2}{3} = 1.$$

Thus, the graph is an ellipse centered at $(-2, 1)$ with $a = \sqrt{6}$ and $b = \sqrt{3}$. Then $c = \sqrt{6 - 3} = \sqrt{3}$, and since the major axis is parallel to the x-axis, the foci are at $(-2 - \sqrt{3}, 1)$ and $(-2 + \sqrt{3}, 1)$.

Consider again the formula for the right focal radius

$$PF = a - ex.$$

If we factor out e, the resulting expression $(a/e) - x$ can be interpreted as the distance from the point (x, y) to the vertical line $x = a/e$. That is, the distance from P to F is e times its distance from the line $x = a/e$. So:

THEOREM 7   *An ellipse can be characterized as the locus traced by a point whose distance from a fixed point F (the focus) is a constant e times its distance from a fixed line (the directrix), the constant e being less than 1.*

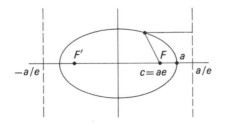

The ellipse $(x^2/a^2) + (y^2/b^2) = 1$ is also the focus–directrix locus defined by the left focus F' and a left directrix at $x = -a/e$. Many geometric properties of an ellipse involve its directrices and foci.

We shall now prove:

THEOREM 8 *The tangent line to the ellipse*

$$\frac{x^2}{a^2} + \frac{y^2}{b^2} = 1$$

at the point (x_0, y_0) on the ellipse has the standard equation

$$\frac{x_0 x}{a^2} + \frac{y_0 y}{b^2} = 1.$$

Proof. Calculus is needed to compute the slope of the tangent line. Differentiating the equation of the ellipse implicitly (Chapter 5, Section 6), we get

$$\frac{2x}{a^2} + \frac{2y}{b^2}\frac{dy}{dx} = 0$$

and

$$\frac{dy}{dx} = -\frac{b^2 x}{a^2 y}.$$

The tangent line at (x_0, y_0) thus has the slope $-b^2 x_0/a^2 y_0$, and its point–slope equation is

$$y - y_0 = -\frac{b^2 x_0}{a^2 y_0}(x - x_0).$$

This simplifies to the standard equation when we make use of the fact that

$$\frac{x_0^2}{a^2} + \frac{y_0^2}{b^2} = 1. \quad \blacksquare$$

The optical property of the ellipse is stated in terms of the tangent line.

THEOREM 9 *The focal radii to the point P on the ellipse make equal angles with the tangent line at P.*

Thus, if a point source of light is placed at one focus, say F', then all of its reflected rays will pass through the other focus F. The ellipse "focuses" the light on the point F, which is thus one of the two focal points of the optical system.

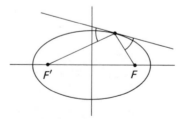

This focusing property of the ellipse explains a startling acoustical phenomenon that has been observed in some buildings. If a hall with walls made of stone, or some other acoustically reflecting material, is in the shape of an ellipse, then a person standing at one focus and speaking quietly can be heard with great clarity at the other focus, but not at points in between.

Proof of theorem. In order to prove the optical property, we make use of the focal radius formulas and the formula

$$d = \frac{|Ax_1 + By_1 + C|}{\sqrt{A^2 + B^2}}$$

for the distance from the point (x_1, y_1) to the line $Ax + By + C = 0$ (Extra Problem 32, in Chapter 1).

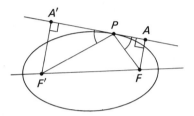

According to this formula, the distances from the foci to the tangent line $xx_0/a^2 + yy_0/b^2 = 1$ have the ratio

$$\frac{F'A'}{FA} = \frac{\left|\dfrac{-cx_0}{a^2} - 1\right|}{\left|\dfrac{cx_0}{a^2} - 1\right|}$$

$$= \frac{a^2 + cx_0}{a^2 - cx_0}$$

$$= \frac{a + ex_0}{a - ex_0},$$

which is the ratio $F'P/FP$ of the focal radii. The two right triangles are thus similar, and hence have equal angles at P. ▮

PROBLEMS FOR SECTION 2

In Problems 1 through 9, find the equation of the ellipse determined by the given conditions.

1. The foci are at $(-1, 0)$ and $(1, 0)$, and $2a = 3$.

2. The foci are at $(0, -1)$ and $(0, 1)$, and $2b = 3$.

3. The foci are at $(-1, 0)$ and $(1, 0)$, and $b = 1$.

4. The foci are at $(-1, 0)$ and $(1, 0)$, and the eccentricity ($e = c/a$) is 1/2.

5. The foci are at $(0, 1)$ and $(0, 3)$, and $e = 1/2$.

6. The foci are at $(0, 1)$ and $(4, 1)$, and $e = 3/4$.

7. The intercepts on the major and minor axes are $(-1, 1)$, $(3, 1)$, $(1, 2)$, and $(1, 0)$.

8. The foci are $(-1, 0)$, and $(1, 0)$, and the line $x = 2$ is a directrix.

9. The center is the origin, one focus is at $(0, 2)$, and the line $y = 3$ is a directrix.

10. Find the foci of the ellipse $x^2 + 4x + 2y^2 = 0$.

11. Same question for $2x^2 + 4x + y^2 = 0$.

12. Find the equation of the ellipse having the y-axis as a directrix, the point $(1, 0)$ as the corresponding focus, and eccentricity $1/2$.

13. Same question for the point $(1, 0)$, line $x = 3$, and $e = 2/3$.

14. Let P be a point on the ellipse $(x^2/a^2) + (y^2/b^2) = 1$ that is not an intercept. Supposing that $b < a$, show that the distance from P to the origin is greater than b and less than a. (This does not require calculus.)

15. Show that if $a^2 = b^2 + c^2$ and if the point (x, y) satisfies the equation $(x^2/a^2) + (y^2/b^2) = 1$, then the distance from (x, y) to $(c, 0)$ is $(a - ex)$, and the distance from (x, y) to $(-c, 0)$ is $(a + ex)$, where $e = c/a$. That is, work out the algebra for Theorem 5.

16. Complete the proof of Theorem 8 by supplying the missing algebraic calculation at the end of the proof.

17. Let k be a fixed positive constant. For each value of a greater than k, the equation

$$\frac{x^2}{a^2} + \frac{y^2}{a^2 - k^2} = 1$$

represents an ellipse; so, regarding a as a parameter, we thus have a one-parameter family of ellipses. Show that this is a *confocal* one-parameter family; that is, show that all ellipses of this form have the same foci.

18. The tangent to an ellipse at a point P intersects a directrix at the point D. Prove that the segment PD subtends a right angle at the corresponding focus.

19. Show that the product of the distances from the foci of an ellipse to a tangent is a constant, independent of the tangent.

20. The *vertices* of an ellipse are its intersections with its major axis. The lines from the vertices A' and A of an ellipse, through a point P on the ellipse, meet a directrix at D' and D. Show that the segment $D'D$ subtends a right angle at the corresponding focus.

21. A normal to an ellipse at a point P has intercepts Q_1 and Q_2 on the axes. Show that the product of the lengths of the segments PQ_1 and PQ_2 equals the product of the focal radii to P.

22. The tangent to an ellipse at a point P intersects the tangent at a vertex at the point Q. Show that the line through Q and the other vertex is parallel to the line through P and the center.

23. Show that every ellipse of the form

$$x^2 + 2y^2 = C$$

is orthogonal to every parabola of the form

$$y = kx^2.$$

(Let (x_0, y_0) be a point of intersection. Compute the two slopes at this point, and show that their product is -1.)

3. THE HYPERBOLA

DEFINITION   *A hyperbola is the locus traced by a point P whose distances from two fixed points F and F' have a constant positive difference k, where k is less than the distance between F' and F.*

The locus condition is thus

$$PF' - PF = k \qquad \text{or} \qquad PF - PF' = k,$$

that is,

$$PF' = PF \pm k.$$

The requirement $k < \overline{F'F}$ is necessary if there are to be locus points P that are not on the line through F and F', because any such point P satisfies triangle inequalities like $\overline{PF'} < \overline{PF} + \overline{FF'}$, so

$$k = PF' - PF < \overline{FF'}.$$

A little thought (or experimentation) will show that the locus has the general features shown in the following figure.

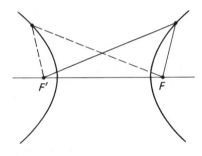

It has two branches, each of which cuts between F' and F. The right branch, bending around F, is the locus of the equation $PF' - PF = k$, while the left branch, bending around F', is the locus of the equation $PF - PF' = k$.

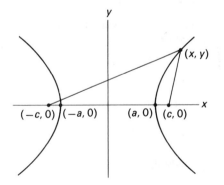

If we take the x-axis along the segment $F'F$ and the origin at its midpoint, then the coordinates of F and F' are $F = (c, 0)$ and $F' = (-c, 0)$, where c is one half the segment length. If a is the positive x-intercept of the locus, then $a < c$, and for this point the locus condition becomes

$$(a + c) - (c - a) = k.$$

So $k = 2a$, and the locus equation is

$$\sqrt{(x + c)^2 + y^2} = \sqrt{(x - c)^2 + y^2} \pm 2a.$$

The procedure now is exactly as it was in the case of the ellipse. The radicals are eliminated by twice squaring the equation and simplifying.

With $e = c/a$ as before, the first of these two steps yields the focal radius formulas

$$PF = \pm(ex - a),$$

$$PF' = \pm(ex + a),$$

where the plus signs are used for the right branch $(x > 0)$ and the minus sign for the left branch $(x < 0)$. Here the eccentricity e is greater than 1, since $e = c/a$ and $c > a$.

The second step reduces the equation

$$\sqrt{(x - c)^2 + y^2} = \pm\left(\frac{c}{a}x - a\right)$$

to

$$\frac{x^2}{a^2} - \frac{y^2}{b^2} = 1,$$

where $b^2 = c^2 - a^2$. This is the standard equation of the hyperbola.

It can be checked, conversely, that the graph of any equation of this form is a hyperbola with foci on the x-axis. Setting $c = \sqrt{a^2 + b^2}$, the graph is the hyperbola with foci at $F = (c, 0)$ and $F' = (-c, 0)$, and with $k = 2a$.

Certain aspects of the graph can be read directly from the standard equation $(x^2/a^2) - (y^2/b^2) = 1$. It is symmetric with respect to both coordinate axes. It has x-intercepts at $\pm a$, but no y-intercepts. Moreover, $x^2 \geq a^2$ on the graph. This is because the y^2 term is negative, so x^2/a^2 is at least 1. The graph thus consists of two separate pieces (branches), a part to the right of $x = a$ and a part to the left of $x = -a$.

EXAMPLE 1   The hyperbola with foci at $(2, 0)$ and $(-2, 0)$ and with $k = 2a = 2$ has the equation

$$x^2 - \frac{y^2}{3} = 1,$$

because the given data are $a = 1$, $c = 2$, and hence $b = \sqrt{c^2 - a^2} = \sqrt{4 - 1} = \sqrt{3}$.

EXAMPLE 2   For the hyperbola

$$\frac{x^2}{16} - \frac{y^2}{9} = 1$$

we have $a^2 = 16$ and $b^2 = 9$, so $c^2 = a^2 + b^2 = 25$, and the foci are at $(c, 0) = (5, 0)$ and $(-c, 0) = (-5, 0)$. Its eccentricity e is 5/4 (since $e = c/a$).

EXAMPLE 3   Find the equation of the hyperbola with foci on the x-axis, x-intercepts at ± 2, and eccentricity 3/2.

Solution. We are given that $a = 2$ and $e = c/a = 3/2$. Therefore, $c = 3$ and $b^2 = c^2 - a^2 = 9 - 4 = 5$. The equation is, therefore,

$$\frac{x^2}{4} - \frac{y^2}{5} = 1.$$

The branching of the hyperbola

$$\frac{x^2}{a^2} - \frac{y^2}{b^2} = 1$$

is related to another distinctive feature of the graph; it is asymptotic to each of the straight lines

$$y = \left(\frac{b}{a}\right)x, \qquad y = -\left(\frac{b}{a}\right)x,$$

in the manner shown in the figure.

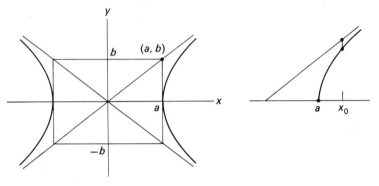

For example, the top of the right branch, whose equation is

$$y = \frac{b}{a}\sqrt{x^2 - a^2},$$

is asymptotic to the line

$$y = \left(\frac{b}{a}\right)x;$$

the two graphs come arbitrarily close together as they run out "to infinity." In order to see this, consider the vertical distance from the hyperbola to the line at a given value of x. It is

$$\frac{b}{a}x - \frac{b}{a}\sqrt{x^2 - a^2} = \frac{b}{a}(x - \sqrt{x^2 - a^2})$$

$$= \frac{b}{a}\frac{(x - \sqrt{x^2 - a^2})(x + \sqrt{x^2 - a^2})}{x + \sqrt{x^2 - a^2}}$$

$$= \frac{ab}{x + \sqrt{x^2 - a^2}},$$

and this is less than ab/x for all x larger than a. The vertical distance between the curves thus approaches zero as x tends to infinity. That is, the curves are asymptotic to each other as $x \to \infty$.

If we interchange the roles of x and y in the preceding discussion, we see that a hyperbola with foci at $(0, c)$ and $(0, -c)$ on the y-axis has the standard equation

$$\frac{y^2}{b^2} - \frac{x^2}{a^2} = 1$$

with $c^2 = a^2 + b^2$.

Note that *the axis containing the foci is not determined by the relative size of a and b*, as it was in the case of the ellipse, *but rather by where the minus sign occurs.* So a and b can be any relative size, and in particular they can be equal. In this case, the hyperbola is called *rectangular* or *equilateral.* Thus, an equilateral hyperbola has the equation

$$x^2 - y^2 = a^2 \qquad \text{or} \qquad y^2 - x^2 = a^2.$$

For given a and b, the equation

$$\frac{x^2}{a^2} - \frac{y^2}{b^2} = \pm 1$$

represents a pair of hyperbolas, one with foci on the x-axis and one with foci on the y-axis. Each of these hyperbolas is said to be the *conjugate* of the other. A pair of conjugate hyperbolas shares the same asymptotes; and they fit together to enclose the origin almost as though they formed a single closed curve.

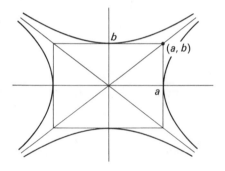

EXAMPLE 4   The hyperbola

$$x^2 - \frac{y^2}{4} = 1$$

has the asymptotes $y = 2x$ and $y = -2x$, because from the standard equation above we see that $a^2 = 1$ and $b^2 = 4$, so the asymptotes $y = \pm (b/a)x$ are the lines $y = \pm 2x$. The conjugate of the above hyperbola is the hyperbola

$$-x^2 + \frac{y^2}{4} = 1.$$

EXAMPLE 5   A hyperbola has asymptotes $y = \pm x$ and passes through the point $(2, 1)$. Find its equation.

Solution. Since the asymptote $y = (b/a)x$ is the line $y = x$, we have $b/a = 1$ and $a = b$, so the hyperbola is equilateral. Since the hyperbola contains a point (namely $(2, 1)$) "to the right of" the lines $y = \pm x$, its branches are right and left rather than up and down, and its equation has the form

$$\frac{x^2}{a^2} - \frac{y^2}{a^2} = 1.$$

And, since it contains $(2, 1)$,

$$\frac{4}{a^2} - \frac{1}{a^2} = 1.$$

So $a = \sqrt{3}$, and the hyperbola has the equation

$$\frac{x^2}{3} - \frac{y^2}{3} = 1,$$

or

$$x^2 - y^2 = 3.$$

PROBLEMS FOR SECTION 3

Find the foci, eccentricity, and asymptotes for each of the following hyperbolas.

1. $x^2 - y^2 = 1$ 2. $y^2 - x^2 = 1$ 3. $x^2 - y^2 = 2$

4. $x^2 - \dfrac{y^2}{3} = 3$ 5. $3y^2 - x^2 = 3$ 6. $y^2 - 4x^2 = 1$

7. $2x^2 - 3y^2 = 6$

Find the equation of the hyperbola where:

8. The foci are at $(\pm 3, 0)$ and the eccentricity is $3/2$.

9. It is symmetric about the coordinate axes; its x-intercepts are $(\pm 1, 0)$ and its eccentricity is 2.

10. Its foci are at $(\pm 3, 0)$ and its x-intercepts are $(\pm 2, 0)$.

11. Its foci are at $(0, \pm 3)$ and $a = 2$.

12. Its x-intercepts are $(\pm 1, 0)$ and it contains the two points $(3, \pm 1)$.

Find the equation of the hyperbola with asymptotes $y = \pm 2x$ if:

13. It passes through $(2, 0)$.

14. It has y-intercepts $(0, \pm 2)$.

15. Its foci are $(\pm 5, 0)$.

16. It passes through the point $(2, 2)$.

17. It passes through the point $(1, 1)$.

18. It passes through the point $(1, 3)$.

19. Two hyperbolas have the same asymptotes, one with foci on the x-axis and the other with foci on the y-axis. If their eccentrities are e and f, show that

$$\frac{1}{e^2} + \frac{1}{f^2} = 1.$$

20. Carry out the algebra (completing the square twice and simplifying each time) leading from the locus definition of the hyperbola to its standard equation.

21. Show by algebraic computation that if

$$\frac{x^2}{a^2} - \frac{y^2}{b^2} = 1$$

and if $c^2 = a^2 + b^2$, then the distance from (x, y) to $(c, 0)$ is $|ex - a|$, where $e = c/a$.

22. Show that the hyperbola

$$\frac{x^2}{2} - \frac{y^2}{2} = 1$$

is the locus of a point whose distance from $(2, 0)$ is $\sqrt{2}$ times its distance from line $x = 1$.

23. Show that the hyperbola

$$\frac{x^2}{a^2} - \frac{y^2}{b^2} = 1$$

is the locus of a point whose distance from the point $(c, 0)$ is e times its distance from the line $x = a/e$, where $c = \sqrt{a^2 + b^2}$ and $e = c/a$. [*Hint:* Use the focal radius formulas from the text discussion.]

24. Show that the hyperbola

$$\frac{x^2}{a^2} - \frac{y^2}{b^2} = 1$$

is the locus of a point whose distance from the point $(-c, 0)$ is e times its distance from the line $x = -a/e$.

25. Show that a line through a focus perpendicular to an asymptote intersects the asymptote at a point on a directrix.

26. A line through a point P on a hyperbola parallel to an asymptote intersects a directrix at Q. If F is the corresponding focus, show that the triangle FPQ is isosceles.

27. Prove that the tangent line to the hyperbola

$$\frac{x^2}{a^2} - \frac{y^2}{b^2} = 1$$

at the point (x_0, y_0) has the (standard) equation

$$\frac{xx_0}{a^2} - \frac{yy_0}{b^2} = 1.$$

28. *Optical property of the hyperbola.* Prove that the focal radii to a point P on a hyperbola make equal angles with the tangent line at P. (Follow the proof of the optical property of the ellipse, using the above equation of the tangent line.)

29. Prove that each hyperbola having foci F' and F intersects every ellipse having the same foci at right angles. [*Hint:* Combine the optical properties of the ellipse and hyperbola.]

30. Illustrate the above problem by taking the foci at $(\pm 3, 0)$ and sketching the hyperbolas whose eccentricities are 3 and 3/2 and the ellipse whose eccentricities are 3/4 and 1/2.

31. Show that the product of the distances from the foci of a hyperbola to the tangent at a point P on the hyperbola is constant; i.e., independent of P.

32. Show that the product of the focal radii to a point P on an equilateral hyperbola is equal to the square of the distance from P to the center of the hyperbola.

33. Show that the product of the distances from a point P on a hyperbola to the asymptotes is constant; i.e., independent of P.

34. The tangent to a hyperbola at a point P intersects the asymptotes at Q_1 and Q_2. Show that P is the midpoint of the segment $Q_1 Q_2$.

Appendix 3
Continuity

This appendix contains a brief introduction to the ε, δ characterizations of continuity and limits that are needed for advanced technical work.

Suppose we want to calculate π^3 accurately to six decimal places. Our problem is to determine how many places we have to take in the decimal expansion of π. That is, if a_n is the n-place truncation (or round off) of the decimal expansion of π, then how large must n be to ensure that

$$|\pi^3 - a_n^3| \le 10^{-6}/2?$$

In somewhat more general terms, the question is: given the tolerable error ε, how close must a number x be to π in order to ensure that

$$|\pi^3 - x^3| \le \varepsilon?$$

We can ask this question in neutral terms, and for a general function f, as follows:

Given a positive number ε, how close do we have to take x to a in order to ensure that

$$|f(a) - f(x)| < \varepsilon?$$

Or

How small should we take Δx in order to ensure that

$$|\Delta y| < \varepsilon,$$

where $\Delta y = f(x + \Delta x) - f(x)$?

This question concerns the quantitative aspect of continuity. We said, in Chapter 2, that a function f is continuous at $x = a$ if $f(x)$ can be made to be as close to $f(a)$ as we wish by taking x suitably close to a. This was a purely qualitative description. But in making computations we have to find out *how close* is "suitably close."

The smooth functions of calculus give us the simplest answer to this question, because of the following criterion. (It is the general statement of the procedure we were following in the examples in Section 9 of Chapter 6.)

The derivative criterion for continuity. *If $|f'|$ is bounded by the constant k on an interval I, that is, if $|f'(x)| \le k$ for every x in I, then*

$$|\Delta y| \le k\,|\Delta x|$$

for any two points x and $(x + \Delta x)$ in I.

Proof. By the mean-value principle,

$$\Delta y = f(x + \Delta x) - f(x) = f'(X) \cdot \Delta x,$$

where X is some number between x and $x + \Delta x$. But then $|f'(X)| \le k$ by hypothesis, so

$$|\Delta y| = |f'X| \cdot |\Delta x| \le k\,|\Delta x|,$$

as asserted. ∎

By virtue of this inequality, if now we want to ensure that

$$|\Delta y| < \varepsilon,$$

it is sufficient to see to it that

$$k\,|\Delta x| < \varepsilon,$$

i.e., to take

$$|\Delta x| < \frac{\varepsilon}{k}.$$

This is the simplest conceivable solution to the continuity problem.

EXAMPLE 1 If we consider $y = f(x) = x^3$ on the interval $I = [-4, 4]$, then

$$f'(x) = 3x^2 \leq 3 \cdot 4^2 = 48,$$

and hence

$$|\Delta y| \leq 48 \, |\Delta x|,$$

by the criterion above. Therefore, in order to ensure that $|\Delta y| < \varepsilon$, it is sufficient to take

$$48 \, |\Delta x| < \varepsilon, \qquad \text{or} \qquad |\Delta x| < \frac{\varepsilon}{48}.$$

In other words, if x and a are both on the interval $I = [-4, 4]$, then

$$|a^3 - x^3| < \varepsilon \qquad \text{whenever } |a - x| < \frac{\varepsilon}{48}.$$

Although the above criterion furnishes a very simple way to make the continuity calculation, it is important to find ways that are independent of the derivative, for two reasons.

First, we have to know that some functions are continuous before we can prove that they have derivatives.

EXAMPLE 2 Look back at the calculation of $d\sqrt{x}/dx = 1/2\sqrt{x}$. The difference quotient was

$$\frac{\Delta y}{\Delta x} = \frac{\sqrt{(x + \Delta x)} - \sqrt{x}}{\Delta x} = \frac{1}{\sqrt{x + \Delta x} + \sqrt{x}}.$$

In order to get the answer $1/2\sqrt{x}$, we have to know that $\sqrt{x + \Delta x}$ approaches $\sqrt{x}$ as $\Delta x \to 0$, i.e., that $\sqrt{x}$ is continuous. A prior continuity calculation is thus required, and to avoid circularity in our logic, we cannot base this calculation on the derivative of $\sqrt{x}$ and so can't use the derivative criterion.

So what do we do? Generally speaking, we make the same calculation and algebraic simplification of Δy that we did for the difference quotient in Chapter 3, but now we derive an inequality instead of taking a limit. In the above case, where

$$\frac{\Delta y}{\Delta x} = \frac{1}{\sqrt{x + \Delta x} + \sqrt{x}},$$

we note that since $\sqrt{x + \Delta x}$ is of necessity nonnegative, we make the right side larger (if anything) by omitting $\sqrt{x + \Delta x}$. That is,

$$\frac{\Delta y}{\Delta x} \leq \frac{1}{\sqrt{x}}.$$

Multiplying by Δx, and allowing for the possibility that Δx is negative, we get

$$|\Delta y| \le \frac{1}{\sqrt{x}}|\Delta x|.$$

Now we fix a positive number a and restrict x to the interval $[a, \infty)$. Then $1/\sqrt{x} \le 1/\sqrt{a}$, so

$$|\Delta y| \le k\,|\Delta x|,$$

where $k = 1/\sqrt{a}$. We have thus obtained an inequality like that of the derivative criterion by a direct calculation.

EXAMPLE 3   Let us see how this direct continuity calculation turns out for our original function $y = f(x) = x^3$ on the interval $[-4, 4]$. Here, the factoring procedure works most smoothly because it uses the two numbers r and x that we are restricting to $[-4, 4]$. We have

$$f(r) - f(x) = r^3 - x^3 = (r - x)(r^2 + rx + x^2),$$

so

$$|f(r) - f(x)| \le |r - x|\,(16 + 16 + 16) = 48\,|r - x|.$$

Thus, $|\Delta y| \le 48\,|\Delta x|$, as before.

If we try the increment approach, we get

$$\begin{aligned}
\Delta y = f(x + \Delta x) - f(x) &= (x + \Delta x)^3 - x^3 \\
&= x^3 + 3x^2\,\Delta x + 3x(\Delta x)^2 + (\Delta x)^3 - x^3 \\
&= \Delta x(3x^2 + 3x\,\Delta x + \Delta x^2).
\end{aligned}$$

In this form, it isn't useful to assume that both x and $(x + \Delta x)$ lie in $[-4, 4]$. Let us assume that x lies there, and that $|\Delta x|$ is at most 1; then $|3x^2 + 3x\,\Delta x + \Delta x^2| \le 48 + 12 + 1 = 61$, so

$$|\Delta y| \le 61\,|\Delta x|,$$

a different inequality stemming from different assumptions as to where the two points x and $(x + \Delta x)$ lie. In this case, in order to ensure that

$$|\Delta y| < \varepsilon,$$

it is sufficient to see to it that

$$61\,|\Delta x| < \varepsilon,$$

i.e., to take

$$|\Delta x| < \frac{\varepsilon}{61}.$$

The second reason for making the direct continuity calculation is that we frequently need to establish the continuity of f at points where f' does not exist.

EXAMPLE 4   Consider, again, $y = x^{1/3}$ around the origin. We know that $f'(x) = x^{-2/3}$ blows up at $x = 0$. However, since

$$\Delta y = f(0 + \Delta x) - f(0) = (0 + \Delta x)^{1/3} - 0^{1/3} = (\Delta x)^{1/3},$$

the inequality

$$|\Delta y| < \varepsilon$$

will hold if

$$|\Delta x|^{1/3} < \varepsilon, \quad \text{or} \quad |\Delta x| < \varepsilon^3.$$

Thus Δy does go to 0 with Δx, but much more slowly. For example, in order to make $|\Delta y| < 1/100$, we have to take

$$|\Delta x| < \left(\frac{1}{100}\right)^3 = \frac{1}{1,000,000}.$$

EXAMPLE 5 The upper semicircle $y = f(x) = \sqrt{1 - x^2}$ is another example. It is clear from geometry that this graph has vertical tangents at $x = 1$ and $x = -1$, and that f' fails to exist at these two points. Nevertheless, we can work out by simple algebra that

$$|\Delta y| \le \sqrt{2}\,|\Delta x|^{1/2}$$

at $x_0 = \pm 1$. Therefore,

$$|\Delta y| < \varepsilon$$

if

$$\sqrt{2}\,|\Delta x|^{1/2} < \varepsilon,$$

that is, if

$$|\Delta x| < \frac{\varepsilon^2}{2}.$$

Here again, $|\Delta y|$ approaches 0 much more slowly than Δx. For example, in order to ensure that $|\Delta y| < .01$, using the above inequalities, we have to take $|\Delta x| < (0.01)^2/2 = 0.00005$.

Let us summarize our various continuity calculations. In each case we wanted to know how small to take $|\Delta x|$ in order to ensure that $|\Delta y| < \varepsilon$, where ε is some arbitrary preassigned positive number which can be thought of as the tolerable error. In each case, we found a number δ (Greek delta) depending on ε which served as an adequate control for the size of Δx. Our various control numbers δ were

$$\delta = \frac{\varepsilon}{k}, \qquad \delta = \varepsilon^3, \qquad \delta = \frac{\varepsilon^2}{2}.$$

A general description of the continuity calculation would therefore run as follows: We first obtain an inequality which limits $|\Delta y|$ in terms of $|\Delta x|$. From this inequality, we see that in order to ensure that $|\Delta y| < \varepsilon$, it is sufficient to take $|\Delta x|$ less than a certain control number δ depending on ε.

This description of "effective" continuity is the modern *definition* of continuity.

DEFINITION *The function $y = f(x)$ is continuous at the point $x = x_0$ in its domain if for each positive "tolerable error" ε we can find an adequate "control" number δ. By this we mean that we can ensure that $|\Delta y| < \varepsilon$ by taking $|\Delta x| < \delta$:*

$$|\Delta x| < \delta \Rightarrow |\Delta y| < \varepsilon,$$

($\Rightarrow$ *means* implies), *or*

$$|x - x_0| < \delta \Rightarrow |f(x) - f(x_0)| < \varepsilon.$$

Let us consider next the "effective" continuity of the multiplication operation $y = uv$. In the examples above, the direct proof of continuity always started off as though its objective were a derivative, by setting up a formula for the dependent increment Δy, and manipulating it so that its dependence on Δx becomes apparent. The same scheme works here for the product function $y = uv$. If we start with fixed values $y_0 = u_0 v_0$ and give u and v increments Δu and Δv, then the new value of y is

$$y_0 + \Delta y = (u_0 + \Delta u)(v_0 + \Delta v).$$

Multiplying this out and subtracting $y = uv$, we see that

$$\Delta y = u_0 \, \Delta v + v_0 \, \Delta u + \Delta u \, \Delta v.$$

This increment formula is not new. It was the basis for our proof of the product rule for derivatives in Chapter 5. Here we use it differently, to get an inequality limiting $|\Delta y|$. First,

$$|\Delta y| \le |u_0| \cdot |\Delta v| + |v_0| \cdot |\Delta u| + |\Delta u| \cdot |\Delta v|.$$

Then suppose we choose some positive number δ less than 1, and restrict Δu and Δv to be at most δ in magnitude. Substituting the inequalities $|\Delta u| \le \delta$, $|\Delta v| \le \delta$, in the right side above, and noting that $\delta^2 < \delta$ because $\delta < 1$, we find

$$|\Delta y| \le |u_0| \, \delta + |v_0| \, \delta + \delta^2$$
$$< [|u_0| + |v_0| + 1] \, \delta.$$

If we want $|\Delta y| < \varepsilon$, it is thus sufficient to make

$$[|u_0| + |v_0| + 1] \, \delta = \varepsilon,$$

that is, to take

$$\delta = \frac{\varepsilon}{|u_0| + |v_0| + 1}.$$

We have shown that the increment in the product function $y = uv$ will be less than ε in magnitude if the increments Δu and Δv in the two factors are each bounded by

$$\delta = \frac{\varepsilon}{|u_0| + |v_0| + 1}.$$

We have thus effectively demonstrated the continuity of multiplication.

Now consider how the continuity calculation would go for the product of two continuous functions. Suppose that $u = f(x)$ and $v = g(x)$ are both continuous at $x = x_0$. Given ε, we want to find how close to take x to x_0 in order to guarantee that

$$|f(x)g(x) - f(x_0)g(x_0)| < \varepsilon,$$

or, more simply,

$$|uv - u_0 v_0| < \varepsilon.$$

We saw above that this inequality will hold if

$$\Delta u = u - u_0 \qquad \text{and} \qquad \Delta v = v - v_0$$

are each less than

$$\varepsilon' = \frac{\varepsilon}{|u_0| + |v_0| + 1}$$

in magnitude. Since $u = f(x)$ is continuous at x_0, we can find a positive control number δ_1 for the inequality

$$|\Delta u| < \varepsilon'.$$

That is, the inequality will hold whenever we restrict $\Delta x = x - x_0$ to be less than δ_1 in magnitude. Similarly, we can find a positive control number δ_2 for the continuous function $v = g(x)$, so that

$$|\Delta v| < \varepsilon'$$

whenever $|\Delta x| < \delta_2$. We want *both* $|\Delta u| < \varepsilon'$ and $|\Delta v| < \varepsilon'$, and this double requirement will be met if $|\Delta x|$ meets both control requirements,

$$|\Delta x| < \delta_1 \qquad \text{and} \qquad |\Delta x| < \delta_2.$$

We can state this final condition more simply by setting δ equal to the *smaller* of the two control numbers δ_1 and δ_2. So if $|\Delta x| < \delta$, then Δx meets both control requirements. Backtracking, we see that then both $|\Delta x| < \varepsilon'$ and $|\Delta v| < \varepsilon'$ hold, and therefore that $|uv - u_0v_0| < \varepsilon$. That is,

if $|x - x_0| < \delta$, then $|f(x)g(x) - f(x_0)g(x_0)| < \varepsilon$.

We have thus found a suitable control number δ for the product function $f(x)g(x)$ at $x = x_0$.

In this way, by explicitly tracking down how the continuity calculation goes, we have proved.

if f and g are continuous at $x = x_0$, then so is the product function fg.

This is the way all the limit and continuity laws are established. In every case we prove that a function is continuous, or that a limit exists and has a certain value, by exhibiting an explicit scheme that will furnish a control number δ for any given positive number ε. This means, of course, that we use the corresponding "effective" definition of a function limit:

DEFINITION *The function f has the limit l at $x = x_0$ if for every positive number ε, we can find a positive number δ such that*

$$0 < |x - x_0| < \delta \Rightarrow |f(x) - l| < \varepsilon.$$

Note that we don't allow x to take on the value x_0 when we consider whether or not $f(x)$ has the limit l as x approaches x_0.

The notion of sequential convergence is made precise in the same quantitative way.

DEFINITION *The sequence $\{a_n\}$ converges to the number l as its limit as n tends to infinity if, given any positive number ε, we can find an integer N such that*

$$n > N \Rightarrow |a_n - l| < \varepsilon.$$

Sooner or later one should work through all of this material in a systematic way. It is probably not appropriate to take the time to do this in a first calculus course, and here we shall just spotlight a few additional results in the exercises.

Finally, we consider the counterpart of the derivative continuity criterion for functions of two variables.

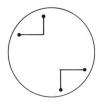

Suppose that on a circular disk R both partial derivatives of $z = f(x, y)$ exist and are bounded by k in magnitude. That is,

$$|D_1 f(x, y)| \leq k,$$
$$|D_2 f(x, y)| \leq k,$$

for all points (x, y) in R. Then

$$|\Delta z| \leq k(|\Delta x| + |\Delta y|)$$

for any two points (x, y) and $(x + \Delta x, y + \Delta y)$ in R.

Proof. Up to a point we just repeat the proof of the chain rule from Chapter 16 (pages 568–9). We go from the first point (x, y) to the second point $(x + \Delta x, y + \Delta y)$ in two steps, first a horizontal step to $(x + \Delta x, y)$ and then a vertical step. (It may be necessary to make the first step vertical and the second horizontal, as the figure shows.) Then Δz can be written as the sum of the two corresponding partial increments,

$$\Delta z = f(x + \Delta x, y + \Delta y) - f(x, y)$$
$$= [f(x + \Delta x, y + \Delta y) - f(x + \Delta x, y)] + [f(x + \Delta x, y) - f(x, y)],$$

and we can apply the mean-value principle for functions of one variable to each term on the right, obtaining

$$\Delta z = D_2 f(x + \Delta x, Y) \cdot \Delta y + D_1 f(X, y) \cdot \Delta x.$$

All this is exactly as before. But now we use the assumed inequalities on the partial derivatives and have at once the inequality we want:

$$|\Delta z| \leq k |\Delta y| + k |\Delta x| = k(|\Delta x| + |\Delta y|).$$

It was because of this derivative criterion that we could be sure that the functions we met in Chapter 15 were all continuous. ∎

REMARK   We can restate this conclusion in terms of the length $|\Delta \mathbf{x}|$ of the vector $\Delta \mathbf{x} = (\Delta x, \Delta y)$, because of the fact that

$$a + b \leq \sqrt{2}\sqrt{a^2 + b^2}$$

for any two numbers. (Square both sides and see what happens.) The inequality

$$|\Delta z| \leq k(|\Delta x| + |\Delta y|)$$

thus implies that

$$|\Delta z| \le \sqrt{2}\, k\, |\Delta \mathbf{x}|.$$

EXAMPLE 6   The two-variable criterion provides another way to establish the continuity inequality for the product operation $z = f(x, y) = xy$. The partial derivatives

$$\frac{\partial z}{\partial x} = y, \qquad \frac{\partial z}{\partial y} = x,$$

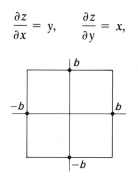

are both less than b in magnitude on the square $|x| \le b$, $|y| \le b$. Therefore,

$$|\Delta z| \le b(|\Delta x| + |\Delta y|)$$

for any two points (x, y) and $(x + \Delta x, y + \Delta y)$ in the square. If we set $(r, s) = (x + \Delta x, y + \Delta y)$, then we can rewrite this new version of the continuity inequality as

$$|rs - xy| \le b(|r - x| + |s - y|).$$

It is closely related to the earlier version, the differences being due to different assumptions about the magnitudes of x, y, and their increments.

PROBLEMS FOR APPENDIX 3

1. How many places in the decimal expansion

$$\log 4 = 1.3 \cdots$$

will be needed to approximate $(\log 4)^2$ with an error less than 0.01? (Use the derivative criterion.)

2. How many places in the decimal expansion

$$\log_{10} \pi = 0.4 \cdots$$

are needed to approximate $(\log_{10} \pi)^3$ with an error less than 0.005?

3. How many places in the expansion

$$\log 100 = 4.6 \cdots$$

are needed to approximate $(\log 100)^4$ with an error less than 0.005?

4. How many places in the decimal expansion

$$e = 2.7 \cdots$$

are needed to approximate e^5 with an error less than 0.002?

5. Prove the following counterpart of the derivative criterion.

If

$$|f'(x)| \geq b > 0$$

for every x in an interval I, then

$$|\Delta y| \geq b\,|\Delta x|$$

for any two points x and $(x + \Delta x)$ in I.

6. Show that if $e = 2.7\cdots$, then $e^3 < 23$, *without* computing $(2.8)^3$. (Use Problem 5 and the obvious inequalities $e > 8/3$, $3 - e > 0.2$.)

Prove the inequalities in Problems 7 through 11 by direct arguments (that is, without using calculus).

7. If $0 < x < a$, then

$$4x^3(a - x) < a^4 - x^4 < 4a^3(a - x).$$

8. If x and y are both greater than the positive number a, then

$$\left|\frac{1}{x} - \frac{1}{y}\right| < \frac{1}{a^2}|x - y|.$$

9. If $0 < x < a$ then $a^n - x^n < na^{n-1}(a - x)$.

10. Assuming that $|\sin x| \leq |x|$ for all x, show that

$$|\cos x - \cos y| \leq |x - y|.$$

11. If $0 < b < y$ then $y^{1/4} - b^{1/4} < (1/4)b^{-3/4}(y - b)$. (This can be proved directly from Problem 1 by relabeling.)

12. If $y = 1/x$, show by a direct estimation of Δy, as in Examples 2 and 3, that if $|x|$ and $|x + \Delta x|$ are both larger than a positive number b, then

$$|\Delta y| \leq \varepsilon \qquad \text{whenever } |\Delta x| \leq b^2\varepsilon.$$

13. If $y = x^4$, show by a direct algebraic estimate of Δy that if $|x|$ and $|x + \Delta x|$ are both less than a, then

$$|\Delta y| \leq \varepsilon \qquad \text{whenever } |\Delta x| \leq \frac{\varepsilon}{4a^3}.$$

14. Show that if $e = 2.7182\cdots$, then

$$e^4 - (2.718)^4 < 0.05.$$

(Use the preceding problem or Problem 7.)

15. Show that if $y = \sqrt{1 - x^2}$, then $y \leq \sqrt{2}(1 - x)^{1/2}$. Similarly, show that $y \leq \sqrt{2}(1 + x)^{1/2}$. This completes Example 5 in the text.

In each of the following four problems, find how small to take $\Delta x = x - a$ in order to ensure that $\Delta y = f(x) - f(a)$ will be less than ε in magnitude.

16. $f(x) = (x - 2)^{2/3}$; $a = 2$ 17. $f(x) = (4 - x^2)^{4/5}$; $a = 2$

18. $f(x) = \sqrt{x - 1}$; $a = 1$ 19. $f(x) = \sqrt{x^2 - 1}$; $a = 1$

20. Prove that if f is continuous and $f(a) \neq 0$, then $1/f$ is continuous at $x = a$.

21. Show by a direct algebraic argument that $f(x, y) = x + y$ is a continuous function of two variables. (This is much easier than the proof for the product in the text.)

22. Prove that if f and g are continuous at $x = a$, then so is $f + g$.

23. Give a rigorous ε, N proof of the law: If $a_n \to a$ and $b_n \to b$ as $n \to \infty$, then $a_n b_n \to ab$ as $n \to \infty$. (Imitate the proof that fg is continuous, replacing the control numbers δ_1 and δ_2 by control integers N_1 and N_2, etc.)

24. Prove the sign-preserving law: If $a_n \to a$ as $n \to \infty$, and if $a_n \geq 0$ for all n, then $a \geq 0$. [*Hint*: Suppose, on the contrary, that $a < 0$, and take $\varepsilon = |a| = -a$, in the definition of convergence. You should be able to show that, then,

$$a_n < 0 \qquad \text{for all } n \text{ larger than } N,$$

which is a contradiction.]

Appendix 4
The Principles
Underlying Calculus

In this appendix we shall complete the chain of proofs showing that all of calculus rests on the nested-interval property of the real numbers. There are many possible routes through this material. Here we shall repeatedly construct a nested sequence of intervals by the "bisection procedure." This device is not always feasible, but we use it whenever it is efficient. On other occasions we need variants of the completeness property, and several are developed in Section 2.

1. THE INTERMEDIATE-VALUE THEOREM

This is the theorem:

THEOREM *If f is continuous on the closed interval* [a, b] *and if l is any number between* f(a) *and* f(b), *then there is at least one point X in* [a, b] *for which* f(X) = l.

If we set $g(x) = f(x) - l$, and possibly change sign, then the intermediate-value theorem reduces to the zero-crossing theorem for a continuous function.

THEOREM 1 *If f is continuous on the closed interval* [a, b], *and if* f(a) < 0 < f(b), *then* f(X) = 0 *for some number X in* [a, b].

Let us see how we might go about computing such a number X. Suppose, to be specific, that $[a, b] = [1, 2]$. As a first step we divide $[1, 2]$ into 10 equal subintervals with endpoints

$$1.0, \quad 1.1, \quad 1.2, \quad \cdots, \quad 1.9, \quad 2.0.$$

Then we run through the corresponding values of f:

$$f(1.0), \quad f(1.1), \quad \cdots, \quad f(1.9), \quad f(2.0).$$

We shall suppose that we are never lucky enough to land exactly on a number where f is zero, so each value is either positive or negative. Since we start with the negative value $f(1.0)$ and end with the positive value $f(2.0)$, there will be a first sign change in this sequence of eleven values, say, between 1.2 and 1.3. Then

$$f(1.2) < 0 < f(1.3).$$

Our second step is to divide $[1.2, 1.3]$ into ten equal subintervals, with endpoints

$$1.20, \quad 1.21, \quad \cdots, \quad 1.29, \quad 1.30$$

and again select the first pair where f changes sign, say,

$$f(1.26) < 0 < f(1.27).$$

Continuing in this way, we construct an infinite decimal, beginning, say, $1.26745\ldots$, such that

$$f(1) < 0 < f(2)$$
$$f(1.2) < 0 < f(1.3)$$
$$f(1.26) < 0 < f(1.27)$$
$$\vdots$$
$$f(1.26745) < 0 < f(1.26746)$$

Now let X be the number having this infinite decimal expansion. Then $f(X)$ has to be 0. To see why, we note that the left-hand numbers 1, 1.2, 1.26, ... form an infinite sequence $\{a_n\}$ converging to X, and since f is continuous at X, we have

$$f(X) = \lim_{n \to \infty} f(a_n).$$

But $f(a_n) < 0$ for all n. Therefore $f(X) \le 0$ by the sign-preserving property of a convergent sequence. Similarly, since the sequence $\{b_n\}$ of right-hand numbers converges to X from above, and since $f(b_n) > 0$ for all n, it follows that

$$f(X) = \lim_{n \to \infty} f(b_n) \ge 0.$$

That is, $f(X)$ must be simultaneously ≤ 0 and ≥ 0, and hence must be 0.

The above process actually constitutes a proof of the zero-crossing theorem. Of course, to be called a proof it has to be described in more general terms. Moreover, it is somewhat more efficient and also more suitable for computing machines to use successive bisections rather than successive division into ten parts. This means that the number being computed is not now described by an infinite decimal, but instead is specified by an infinite sequence of closed intervals each of which is one half of its predecessor.

Proof of Theorem. Set $I_0 = [a, b]$ and let $I_1 = [a_1, b_1]$ be the left or right half of I_0 depending on whether the value of f at the midpoint $c = (a + b)/2$ is positive or negative. That is, if $f(c) > 0$, we set $a_1 = a$ and $b_1 = c$, and if $f(c) < 0$, we set $a_1 = c$ and $b_1 = b$. In either case,

$$f(a_1) < 0 < f(b_1).$$

(We suppose we are not lucky enough to find that $f(c) = 0$.) We continue bisecting in this way, at the nth step choosing $I_n = [a_n, b_n]$ as that half of I_{n-1} for which

$$f(a_n) < 0 < f(b_n).$$

Since $b_n - a_n = (b - a)/2^n$, the conditions for the nested-interval form of the completeness property are met, and the sequences $\{a_n\}$ and $\{b_n\}$ therefore converge to a common limit x. Then

$$f(x) = \lim_{n \to \infty} f(a_n) \le 0,$$
$$f(x) = \lim_{n \to \infty} f(b_n) \ge 0,$$

and so $f(x) = 0$.

2. COMPLETENESS

The nested-interval principle is only one of several mutually equivalent versions of the completeness property of the real numbers, and other versions are sometimes easier to use. Three of these alternative forms are proved here.

A set of numbers A is *bounded above* by a number b if $a \leq b$ for every a in A. For example, the interval $[0, 1]$ is bounded above by 1 (and also by 2, and by any other number greater than 1). So is the open interval $(0, 1)$. The set of all positive integers doesn't have an upper bound; it is *unbounded from above*. Note that if b is an upper bound to the set A, then so is any number b' larger than b.

In the examples above, the number 1 is not only an upper bound to both the closed interval $[0, 1]$ and the open interval $(0, 1)$, but in each case it is the *smallest possible upper bound*—any smaller number fails to be an upper bound. It is also the smallest upper bound to the sequence $\{(n - 1)/n\}$, where n runs through all positive integers.

The least upper bound principle says that such a smallest upper bound always exists.

THE LEAST UPPER BOUND PRINCIPLE   *If A is a nonempty set of numbers that is bounded above, then A has a least upper bound. That is, the collection B of all of the upper bounds of A contains a smallest number.*

Proof. Let b_0 be an upper bound of A and let a_0 be any number that is *not* an upper bound. For example, we could choose some number a from A and set $a_0 = a - 1$. Let I_0 be the closed interval $[a_0, b_0]$. Let $I_1 = [a_1, b_1]$ be the left or right half of I_0, depending on whether the midpoint $c = (a_0 + b_0)/2$ is or is not an upper bound of A. That is, if c is an upper bound, then we set $a_1 = a_0$ and $b_1 = c$, and if c is not an upper bound, then we set $a_1 = c$ and $b_1 = b_0$. In either case we have a new interval $I_1 = [a_1, b_1]$ for which the right endpoint b_1 is an upper bound of A and the left endpoint a_1 is not. We continue bisecting in this way, at the nth step choosing $I_n = [a_n, b_n]$ as that half of I_{n-1} for which b_n is an upper bound of A and a_n is not. Since $b_n - a_n = (b_0 - a_0)/2^n$, the nested-interval principle guarantees that the two endpoint sequences $\{a_n\}$ and $\{b_n\}$ converge to a common limit X. We now show that X is the least upper bound of A.

Consider any number a in A. Then $a \leq b_n$ for every n, since every b_n is an upper bound to A, so $a \leq \lim b_n = X$. Thus X *is* an upper bound to A. On the other hand, if c is any number smaller than X, then $c < a_n$ for all sufficiently large n (because $a_n \to X$ as $n \to \infty$), and since a_n is never an upper bound to A, then neither is c. So X is the smallest upper bound to A and we are done. ∎

Notation. We write

$$\text{lub } A$$

for the least upper bound to A.

Suppose now that A is a nonempty set that is bounded *below*, and let B be the (nonempty) collection of all its *lower* bounds. Then every a in A is an upper bound to B, so if

$$X = \text{lub } B,$$

then $X \leq a$ for every a in A. That is, X is also a lower bound to A. It is therefore the maximum element of B. Thus:

> *If A is a nonempty set that is bounded below, then A has a greatest lower bound* (glb A).

This complementary principle can be proved in other ways. We can imitate the proof of the lub property, with inequalities reversed. Or we can verify that

$$-\text{lub}(-A) = \text{glb } A.$$

Another form of completeness was touched on in Chapter 5, in connection with the definition of irrational exponents, such as 2^π. If $x < \pi < y$, then the number 2^π must satisfy the inequality

$$2^x < 2^\pi < 2^y.$$

So if we let A be the collection of all such numbers 2^x and B the collection of all such numbers 2^y, then 2^π must be a number separating A from B. The question is whether such a separating number necessarily exists.

THE SEPARATION PRINCIPLE *If A and B are (nonempty) sets of numbers, and if A is to the left of B, in the sense that $a \leq b$ for every number a in A and every number b in B, then there is at least one number X separating A from B. That is, there is a number X such that*

$$a \leq X \leq b$$

for every a in A and every b in B.

Proof. Let X be the least upper bound of the lower set A. Since every number b in B is an upper bound of A by hypothesis, and since X is the smallest of all the upper bounds to A, we have $X \leq b$ for every b in B. Thus

$$a \leq X \leq b$$

for every a in A and every b in B. ∎

A third useful and obvious variant of completeness is:

MONOTONE-LIMIT PRINCIPLE *If the sequence $\{a_n\}$ is increasing and bounded above, then it necessarily converges.*

If the numbers a_n increase with n, and are blocked from above by a number b, then it seems obvious that they must pile up at some point l short of b, in which case l must be the limit of the sequence. So this version of completeness is very intuitive.

For the technical proof we define l as the least upper bound of the sequence terms a_n and then prove that we can cause a_n to be as close to l as we wish simply by taking n large enough. So suppose we want a_n to be closer to l than the distance ε, where ε is some given positive number. Since

l is the smallest upper bound to the sequence, it follows that $l - \varepsilon$ is *not* an upper bound, i.e., that

$$l - \varepsilon < a_N$$

for at least one integer N. But then

$$l - \varepsilon < a_n$$

for all n beyond N, since $\{a_n\}$ is increasing. So we have

$$l - \varepsilon < a_n \le l,$$

or

$$0 \le l - a_n < \varepsilon,$$

for all such n, and this is exactly what we wanted to show. Thus $a_n \to l$ as $n \to \infty$.

3. THE RANGE OF A CONTINUOUS FUNCTION

If f is a continuous function on an interval I, then, by the intermediate-value theorem, the range of f over I is a set A having the following property:

($*$) *If x and y are any two points of A, with $x < y$, then A includes the whole closed interval $[x, y]$.*

We can now show that such a set A is necessarily an interval, at which point we will have proved that the range of a continuous function over an interval is always itself an interval (Theorem 2 in Chapter 2).

To start with, choose any fixed point x_0 in A, and let B be the intersection of A with the semi-infinite interval $[x_0, \infty)$. Note that B also satisfies ($*$).

Suppose first that B is bounded above and let b be its least upper bound. We claim that B is then either the closed interval $[x_0, b]$ or the semiopen interval $[x_0, b)$. To see this, let x be any number in $[x_0, b)$:

$$x_0 \le x < b.$$

Then x is not an upper bound to B, so B contains a number y greater than x. Then B includes the interval $[x_0, y]$, by ($*$), and hence contains x. So every point of $[x_0, b)$ is in B. On the other hand, since $b = \text{lub } B$, the only possibility for a point of B not being in $[x_0, b)$ is the point b itself. So

$$B = [x_0, b) \qquad \text{or} \qquad B = [x_0, b].$$

If B is not bounded above, the same argument shows that every point of $[x_0, \infty)$ is in B, so $B = [x_0, \infty)$ in this case.

We have thus shown that B is either a closed interval $[x_0, b]$ or a semiopen interval $[x_0, b)$, where b may be $+\infty$ in the second case. In exactly the same way, the intersection of A with the semi-infinite interval $(-\infty, x_0]$ is of the form $[a, x_0]$ or $(a, x_0]$, where a is permitted to be $-\infty$ in the second case. So A itself is one of the intervals

$$[a, b], \qquad [a, b), \qquad (a, b], \qquad (a, b),$$

where a may be $-\infty$ and/or b may be $+\infty$ whenever these possibilities make sense.

4. THE EXTREME-VALUE THEOREM

LEMMA *If f is continuous on a closed interval $[a, b]$, then f is bounded there.*

Proof. We show that f has an upper bound. The crucial observation is that if $a < c < b$ and if f is bounded above (by B_1) on $[a, c]$ and (by B_2) on $[c, b]$, then f is bounded above (by the larger of B_1 and B_2) on $[a, b]$. So if we assume that f is *not* bounded above on $[a, b]$ then f must be unbounded from above on at least one of the two halves of $[a, b]$. We can then repeatedly bisect, in the manner of Sections 1 and 2, and thus determine a nested sequence of closed intervals $I_n = [a_n, b_n]$, each being one half of its predecessor, such that f is unbounded from above on every I_n. Let X be the common limit of the endpoint sequences $\{a_n\}$ and $\{b_n\}$ (by the nested-interval principle), and choose any number B larger than $f(X)$. Since f is continuous at X, there is an interval I about the point X on which $f(x) < B$. But I contains all the left endpoints a_n from some point on (since $a_n \to X$ as $n \to \infty$), and similarly for the right endpoints, so I includes all the intervals I_n from some point on. In particular, f is bounded above by B on any such I_n, contradicting the fact that f is unbounded from above on I_n. So the assumption that f is unbounded from above on $[a, b]$ leads to a contradiction, and we are done. ∎

The proof that f has a lower bound on $[a, b]$ is essentially the same.

We now know that the values of f on $[a, b]$ form a nonempty set that is bounded. Let B be the least upper bound of this set. We claim that the number B is itself a value, i.e., that $B = f(X)$ for some X in $[a, b]$, in which case B is the *maximum* value of f on $[a, b]$. In order to see this, note first that B must be the least upper bound of f on one or the other (or perhaps both) of the two halves of $[a, b]$. It is left to the reader to check this. We can then repeatedly bisect as before, always choosing $I_n = [a_n, b_n]$ as one of the halves of I_{n-1} on which f has B as its least upper bound. If X is the unique number common to all the intervals I_n (by the nested-interval principle), then $f(X) = B$, and we are done. (The details are left to the reader. Show by an argument like that used in the Lemma that if $f(X) < B$, then f has an upper bound smaller than B on some of the intervals I_n, contradicting the fact that $B = \text{lub } f$ on each I_n.)

Alternative trick proof. If f never assumes the value B, then $B - f(x)$ is never zero and

$$g(x) = \frac{1}{B - f(x)}$$

is everywhere defined and continuous on $[a, b]$. Then g is bounded above, say by C, so we have

$$\frac{1}{B - f(x)} \le C,$$

$$B - f(x) \ge \frac{1}{C},$$

and hence

$$B - \frac{1}{C} \ge f(x).$$

Thus $B - (1/C)$ is an upper bound to f, contradicting the fact that B is the *smallest* upper bound. Therefore f must assume the value B.

We have thus proved, incompletely:

THE EXTREME-VALUE THEOREM   *If f is continuous on the closed interval $[a, b]$, then f assumes maximum and minimum values there.*

5. THE EXISTENCE OF ANTIDERIVATIVES

We now prove that every continuous function has an antiderivative. We shall give the proof for an increasing function f, using its left Riemann sum s_Δ. However, exactly the same proof works for an arbitrary continuous function if we replace s_Δ by the corresponding "lower" Riemann sum

$$l_\Delta = \sum_1^n m_i \, \Delta x_i,$$

where $m_i = f(\xi_i)$ is the *minimum* value of f on the ith subinterval $[x_{i-1}, x_i]$ for each i. Note that l_Δ is the left Riemann sum s_Δ if f is increasing.

For each x in $[a, b]$, let $A(x)$ be the least upper bound of all left Riemann sums l for f over the interval $[a, x]$. Note that every such Riemann sum $l = \sum f(x_{i-1}) \, \Delta x_i$ over $[a, x]$ is less than $f(x)(x - a)$, because its function values $f(x_{i-1})$ are all less than $f(x)$, f being increasing, while its interval widths Δx_i add exactly to $(x - a)$. So the collection of these Riemann sums l is bounded above by $f(x)(x - a)$ and hence *has* a least upper bound, by the completeness property of the real numbers.

We shall now show that $A(x)$ is an antiderivative of $f(x)$ on $[a, b]$.

THEOREM   *$A(x)$ is differentiable on $[a, b]$ and $A'(x) = f(x)$.*

Proof. Fix x and h, with $h > 0$. (We shall suppose that x is an interior point of $[a, b]$. At an endpoint the proof gives a one-sided derivative.) Let l be a left Riemann sum for f over $[a, x + h]$. We break the sum l into two parts at x, using the fact that the associated subdivision has a subdividing point x_m such that $x_m \leq x \leq x_{m+1}$.

The left part

$$\sum_{i=1}^m f(x_{i-1}) \, \Delta x_i + f(x_m)(x - x_m)$$

is a left Riemann sum for f over the interval $[a, x]$ and hence is $\leq A(x)$. The right part

$$f(x_m)(x_{m+1} - x) + \sum_{i=m+1}^n f(x_{i-1}) \, \Delta x_i$$

is less than $f(x + h) \cdot h$, because its function values are all less than $f(x + h)$, while its interval lengths add to h. Thus

$$l \leq A(x) + f(x + h) \cdot h.$$

This holds for every l over $[a, x + h]$ and hence for the least upper bound of these l's, so

$$A(x + h) \leq A(x) + f(x + h) \cdot h.$$

In other words, the right side is an upper bound to all the l's, and hence is $\geq$ their least upper bound $A(x + h)$.

On the other hand, if l is a left Riemann sum for f over $[a, x]$, then $l + f(x)h$ is a left Riemann sum for f over $[a, x + h]$, so

$$l + f(x)h \leq A(x + h).$$

This holds for all left Riemann sums l over $[a, x]$, and hence also for their least upper bound $A(x)$, so

$$A(x) + f(x)h \leq A(x + h).$$

We now have two inequalities relating $A(x + h)$ to $A(x)$, and together they show that

$$f(x) \leq \frac{A(x + h) - A(x)}{h} \leq f(x + h).$$

This was for positive h. If h is negative, then the roles of $A(x)$ and $A(x + h)$ are interchanged and the final inequality is reversed. In every case, the difference quotient $[A(x + h) - A(x)]/h$ is squeezed between $f(x)$ and $f(x + h)$. We can now let $h \to 0$ and conclude from the squeeze limit law that the derivative $A'(x)$ exists and $A'(x) = f(x)$. ∎

The proof of the existence of an antiderivative F is the crucial step needed to free the earlier development of the definite integral (Section 6 of Chapter 9) from all dependence on geometry. The only thing else that has to be done is to replace the area demonstrations given on pages 336–37 for the inequalities

$$s_\Delta \leq R \leq S_\Delta,$$
$$S_\Delta - s_\Delta \leq \delta[f(b) - f(a)]$$

by simple algebraic proofs that consist essentially of verifying that the various rectangular areas really are related as shown in the figures. Here R is an arbitrary Riemann sum for f associated with the subdivision Δ, s_Δ and S_Δ are the left and right Riemann sums for f over Δ, and δ is the mesh of Δ. The function f is continuous, positive, and increasing on $[a, b]$.

We can then complete the theory of Chapter 9. Since the bracketing inequality $s_\Delta \leq R \leq S_\Delta$ holds in particular for the special Riemann sum $R_F = F(b) - F(a)$ of Lemma 1 (page 341), it follows that the two numbers R and $R_F = F(b) - F(a)$ can differ at most by $S_\Delta - s_\Delta$, and we have

$$|R - [F(b) - F(a)]| \leq S_\Delta - s_\Delta \leq \delta[f(b) - f(a)].$$

Therefore, $R \to F(b) - F(a)$ as $\delta \to 0$, proving simultaneously that $\int_a^b f$ exists and that

$$\int_a^b f = F(b) - F(a).$$

This was for an increasing function. The integrability of any piecewise monotone, continuous function then follows from Theorem 4 (page 339).

Tables

Natural Logarithms of Numbers

n	$\log_e n$	n	$\log_e n$	n	$\log_e n$
0.0	*	4.5	1.5041	9.0	2.1972
0.1	7.6974	4.6	1.5261	9.1	2.2083
0.2	8.3906	4.7	1.5476	9.2	2.2192
0.3	8.7960	4.8	1.5686	9.3	2.2300
0.4	9.0837	4.9	1.5892	9.4	2.2407
0.5	9.3069	5.0	1.6094	9.5	2.2513
0.6	9.4892	5.1	1.6292	9.6	2.2618
0.7	9.6433	5.2	1.6487	9.7	2.2721
0.8	9.7769	5.3	1.6677	9.8	2.2824
0.9	9.8946	5.4	1.6864	9.9	2.2925
1.0	0.0000	5.5	1.7047	10	2.3026
1.1	0.0953	5.6	1.7228	11	2.3979
1.2	0.1823	5.7	1.7405	12	2.4849
1.3	0.2624	5.8	1.7579	13	2.5649
1.4	0.3365	5.9	1.7750	14	2.6391
1.5	0.4055	6.0	1.7918	15	2.7081
1.6	0.4700	6.1	1.8083	16	2.7726
1.7	0.5306	6.2	1.8245	17	2.8332
1.8	0.5878	6.3	1.8405	18	2.8904
1.9	0.6419	6.4	1.8563	19	2.9444
2.0	0.6931	6.5	1.8718	20	2.9957
2.1	0.7419	6.6	1.8871	25	3.2189
2.2	0.7885	6.7	1.9021	30	3.4012
2.3	0.8329	6.8	1.9169	35	3.5553
2.4	0.8755	6.9	1.9315	40	3.6889
2.5	0.9163	7.0	1.9459	45	3.8067
2.6	0.9555	7.1	1.9601	50	3.9120
2.7	0.9933	7.2	1.9741	55	4.0073
2.8	1.0296	7.3	1.9879	60	4.0943
2.9	1.0647	7.4	2.0015	65	4.1744
3.0	1.0986	7.5	2.0149	70	4.2485
3.1	1.1314	7.6	2.0281	75	4.3175
3.2	1.1632	7.7	2.0412	80	4.3820
3.3	1.1939	7.8	2.0541	85	4.4427
3.4	1.2238	7.9	2.0669	90	4.4998
3.5	1.2528	8.0	2.0794	95	4.5539
3.6	1.2809	8.1	2.0919	100	4.6052
3.7	1.3083	8.2	2.1041		
3.8	1.3350	8.3	2.1163		
3.9	1.3610	8.4	2.1282		
4.0	1.3863	8.5	2.1401		
4.1	1.4110	8.6	2.1518		
4.2	1.4351	8.7	2.1633		
4.3	1.4586	8.8	2.1748		
4.4	1.4816	8.9	2.1861		

Exponential Functions

x	e^x	e^{-x}	x	e^x	e^{-x}
0.00	1.0000	1.0000	2.5	12.182	0.0821
0.05	1.0513	0.9512	2.6	13.464	0.0743
0.10	1.1052	0.9048	2.7	14.880	0.0672
0.15	1.1618	0.8607	2.8	16.445	0.0608
0.20	1.2214	0.8187	2.9	18.174	0.0550
0.25	1.2840	0.7788	3.0	20.086	0.0498
0.30	1.3499	0.7408	3.1	22.198	0.0450
0.35	1.4191	0.7047	3.2	24.533	0.0408
0.40	1.4918	0.6703	3.3	27.113	0.0369
0.45	1.5683	0.6376	3.4	29.964	0.0334
0.50	1.6487	0.6065	3.5	33.115	0.0302
0.55	1.7333	0.5769	3.6	36.598	0.0273
0.60	1.8221	0.5488	3.7	40.447	0.0247
0.65	1.9155	0.5220	3.8	44.701	0.0224
0.70	2.0138	0.4966	3.9	49.402	0.0202
0.75	2.1170	0.4724	4.0	54.598	0.0183
0.80	2.2255	0.4493	4.1	60.340	0.0166
0.85	2.3396	0.4274	4.2	66.686	0.0150
0.90	2.4596	0.4066	4.3	73.700	0.0136
0.95	2.5857	0.3867	4.4	81.451	0.0123
1.0	2.7183	0.3679	4.5	90.017	0.0111
1.1	3.0042	0.3329	4.6	99.484	0.0101
1.2	3.3201	0.3012	4.7	109.95	0.0091
1.3	3.6693	0.2725	4.8	121.51	0.0082
1.4	4.0552	0.2466	4.9	134.29	0.0074
1.5	4.4817	0.2231	5	148.41	0.0067
1.6	4.9530	0.2019	6	403.43	0.0025
1.7	5.4739	0.1827	7	1096.6	0.0009
1.8	6.0496	0.1653	8	2981.0	0.0003
1.9	6.6859	0.1496	9	8103.1	0.0001
2.0	7.3891	0.1353	10	22026	0.00005
2.1	8.1662	0.1225			
2.2	9.0250	0.1108			
2.3	9.9742	0.1003			
2.4	11.023	0.0907			

Natural Trigonometric Functions

Angle					Angle				
Degree	Radian	Sine	Cosine	Tangent	Degree	Radian	Sine	Cosine	Tangent
0°	0.000	0.000	1.000	0.000					
1°	0.017	0.017	1.000	0.017	46°	0.803	0.719	0.695	1.036
2°	0.035	0.035	0.999	0.035	47°	0.820	0.731	0.682	1.072
3°	0.052	0.052	0.999	0.052	48°	0.838	0.743	0.669	1.111
4°	0.070	0.070	0.998	0.070	49°	0.855	0.755	0.656	1.150
5°	0.087	0.087	0.996	0.087	50°	0.873	0.766	0.643	1.192
6°	0.105	0.105	0.995	0.105	51°	0.890	0.777	0.629	1.235
7°	0.122	0.122	0.993	0.123	52°	0.908	0.788	0.616	1.280
8°	0.140	0.139	0.990	0.141	53°	0.925	0.799	0.602	1.327
9°	0.157	0.156	0.988	0.158	54°	0.942	0.809	0.588	1.376
10°	0.175	0.174	0.985	0.176	55°	0.960	0.819	0.574	1.428
11°	0.192	0.191	0.982	0.194	56°	0.977	0.829	0.559	1.483
12°	0.209	0.208	0.978	0.213	57°	0.995	0.839	0.545	1.540
13°	0.227	0.225	0.974	0.231	58°	1.012	0.848	0.530	1.600
14°	0.244	0.242	0.970	0.249	59°	1.030	0.857	0.515	1.664
15°	0.262	0.259	0.966	0.268	60°	1.047	0.866	0.500	1.732
16°	0.279	0.276	0.961	0.287	61°	1.065	0.875	0.485	1.804
17°	0.297	0.292	0.956	0.306	62°	1.082	0.883	0.469	1.881
18°	0.314	0.309	0.951	0.325	63°	1.100	0.891	0.454	1.963
19°	0.332	0.326	0.946	0.344	64°	1.117	0.899	0.438	2.050
20°	0.349	0.342	0.940	0.364	65°	1.134	0.906	0.423	2.145
21°	0.367	0.358	0.934	0.384	66°	1.152	0.914	0.407	2.246
22°	0.384	0.375	0.927	0.404	67°	1.169	0.921	0.391	2.356
23°	0.401	0.391	0.921	0.424	68°	1.187	0.927	0.375	2.475
24°	0.419	0.407	0.914	0.445	69°	1.204	0.934	0.358	2.605
25°	0.436	0.423	0.906	0.466	70°	1.222	0.940	0.342	2.748
26°	0.454	0.438	0.899	0.488	71°	1.239	0.946	0.326	2.904
27°	0.471	0.454	0.891	0.510	72°	1.257	0.951	0.309	3.078
28°	0.489	0.469	0.883	0.532	73°	1.274	0.956	0.292	3.271
29°	0.506	0.485	0.875	0.554	74°	1.292	0.961	0.276	3.487
30°	0.524	0.500	0.866	0.577	75°	1.309	0.966	0.259	3.732
31°	0.541	0.515	0.857	0.601	76°	1.326	0.970	0.242	4.011
32°	0.559	0.530	0.848	0.625	77°	1.344	0.974	0.225	4.332
33°	0.576	0.545	0.839	0.649	78°	1.361	0.978	0.208	4.705
34°	0.593	0.559	0.829	0.675	79°	1.379	0.982	0.191	5.145
35°	0.611	0.574	0.819	0.700	80°	1.396	0.985	0.174	5.671
36°	0.628	0.588	0.809	0.727	81°	1.414	0.988	0.156	6.314
37°	0.646	0.602	0.799	0.754	82°	1.431	0.990	0.139	7.115
38°	0.663	0.616	0.788	0.781	83°	1.449	0.993	0.122	8.144
39°	0.681	0.629	0.777	0.810	84°	1.466	0.995	0.105	9.514
40°	0.698	0.643	0.766	0.839	85°	1.484	0.996	0.087	11.43
41°	0.716	0.656	0.755	0.869	86°	1.501	0.998	0.070	14.30
42°	0.733	0.669	0.743	0.900	87°	1.518	0.999	0.052	19.08
43°	0.750	0.682	0.731	0.933	88°	1.536	0.999	0.035	28.64
44°	0.768	0.695	0.719	0.966	89°	1.553	1.000	0.017	57.29
45°	0.785	0.707	0.707	1.000	90°	1.571	1.000	0.000	

Powers and Roots

No.	Sq.	Sq. Root	Cube	Cube Root	No.	Sq.	Sq. Root	Cube	Cube Root
1	1	1.000	1	1.000	51	2,601	7.141	132,651	3.708
2	4	1.414	8	1.260	52	2,704	7.211	140,608	3.733
3	9	1.732	27	1.442	53	2,809	7.280	148,877	3.756
4	16	2.000	64	1.587	54	2,916	7.348	157,464	3.780
5	25	2.236	125	1.710	55	3,025	7.416	166,375	3.803
6	36	2.449	216	1.817	56	3,136	7.483	175,616	3.826
7	49	2.646	343	1.913	57	3,249	7.550	185,193	3.849
8	64	2.828	512	2.000	58	3,364	7.616	195,112	3.871
9	81	3.000	729	2.080	59	3,481	7.681	205,379	3.893
10	100	3.162	1,000	2.154	60	3,600	7.746	216,000	3.915
11	121	3.317	1,331	2.224	61	3,721	7.810	226,981	3.936
12	144	3.464	1,728	2.289	62	3,844	7.874	238,328	3.958
13	169	3.606	2,197	2.351	63	3,969	7.937	250,047	3.979
14	196	3.742	2,744	2.410	64	4,096	8.000	262,144	4.000
15	225	3.873	3,375	2.466	65	4,225	8.062	274,625	4.021
16	256	4.000	4,096	2.520	66	4,356	8.124	287,496	4.041
17	289	4.123	4,913	2.571	67	4,489	8.185	300,763	4.062
18	324	4.243	5,832	2.621	68	4,624	8.246	314,432	4.082
19	361	4.359	6,859	2.668	69	4,761	8.307	328,509	4.102
20	400	4.472	8,000	2.714	70	4,900	8.367	343,000	4.121
21	441	4.583	9,261	2.759	71	5,041	8.426	357,911	4.141
22	484	4.690	10,648	2.802	72	5,184	8.485	373,248	4.160
23	529	4.796	12,167	2.844	73	5,329	8.544	389,017	4.179
24	576	4.899	13,824	2.884	74	5,476	8.602	405,224	4.198
25	625	5.000	15,625	2.924	75	5,625	8.660	421,875	4.217
26	676	5.099	17,576	2.962	76	5,776	8.718	438,976	4.236
27	729	5.196	19,683	3.000	77	5,929	8.775	456,533	4.254
28	784	5.292	21,952	3.037	78	6,084	8.832	474,552	4.273
29	841	5.385	24,389	3.072	79	6,241	8.888	493,039	4.291
30	900	5.477	27,000	3.107	80	6,400	8.944	512,000	4.309
31	961	5.568	29,791	3.141	81	6,561	9.000	531,441	4.327
32	1,024	5.657	32,768	3.175	82	6,724	9.055	551,368	4.344
33	1,089	5.745	35,937	3.208	83	6,889	9.110	571,787	4.362
34	1,156	5.831	39,304	3.240	84	7,056	9.165	592,704	4.380
35	1,225	5.916	42,875	3.271	85	7,225	9.220	614,125	4.397
36	1,296	6.000	46,656	3.302	86	7,396	9.274	636,056	4.414
37	1,369	6.083	50,653	3.332	87	7,569	9.327	658,503	4.431
38	1,444	6.164	54,872	3.362	88	7,744	9.381	681,472	4.448
39	1,521	6.245	59,319	3.391	89	7,921	9.434	704,969	4.465
40	1,600	6.325	64,000	3.420	90	8,100	9.487	729,000	4.481
41	1,681	6.403	68,921	3.448	91	8,281	9.539	753,571	4.498
42	1,764	6.481	74,088	3.476	92	8,464	9.592	778,688	4.514
43	1,849	6.557	79,507	3.503	93	8,649	9.644	804,357	4.531
44	1,936	6.633	85,184	3.530	94	8,836	9.695	830,584	4.547
45	2,025	6.708	91,125	3.557	95	9,025	9.747	857,375	4.563
46	2,116	6.782	97,336	3.583	96	9,216	9.798	884,736	4.579
47	2,209	6.856	103,823	3.609	97	9,409	9.849	912,673	4.595
48	2,304	6.928	110,592	3.634	98	9,604	9.899	941,192	4.610
49	2,401	7.000	117,649	3.659	99	9,801	9.950	970,299	4.626
50	2,500	7.071	125,000	3.684	100	10,000	10.000	1,000,000	4.642

Table of Integrals

1. $\int u\,dv = uv - \int v\,du$

2. $\int a^u\,du = \dfrac{a^u}{\ln a} + C,\qquad a \neq 1,\qquad a > 0$

3. $\int \cos u\,du = \sin u + C$

4. $\int \sin u\,du = -\cos u + C$

5. $\int (ax + b)^n\,dx = \dfrac{(ax + b)^{n+1}}{a(n + 1)} + C,\qquad n \neq -1$

6. $\int (ax + b)^{-1}\,dx = \dfrac{1}{a}\ln|ax + b| + C$

7. $\int x(ax + b)^n\,dx = \dfrac{(ax + b)^{n+1}}{a^2}\left[\dfrac{ax + b}{n + 2} - \dfrac{b}{n + 1}\right] + C,\qquad n \neq -1, -2$

8. $\int x(ax + b)^{-1}\,dx = \dfrac{x}{a} - \dfrac{b}{a^2}\ln|ax + b| + C$

9. $\int x(ax + b)^{-2}\,dx = \dfrac{1}{a^2}\left[\ln|ax + b| + \dfrac{b}{ax + b}\right] + C$

10. $\int \dfrac{dx}{x(ax + b)} = \dfrac{1}{b}\ln\left|\dfrac{x}{ax + b}\right| + C$

11. $\int (\sqrt{ax + b})^n\,dx = \dfrac{2}{a}\dfrac{(\sqrt{ax + b})^{n+2}}{n + 2} + C,\qquad n \neq -2$

12. $\int \dfrac{\sqrt{ax + b}}{x}\,dx = 2\sqrt{ax + b} + b\int \dfrac{dx}{x\sqrt{ax + b}}$

13. (a) $\int \dfrac{dx}{x\sqrt{ax + b}} = \dfrac{2}{\sqrt{-b}}\tan^{-1}\sqrt{\dfrac{ax + b}{-b}} + C,\qquad$ if $\qquad b < 0$

    (b) $\int \dfrac{dx}{x\sqrt{ax + b}} = \dfrac{1}{\sqrt{b}}\ln\left|\dfrac{\sqrt{ax + b} - \sqrt{b}}{\sqrt{ax + b} + \sqrt{b}}\right| + C,\qquad$ if $\qquad b > 0$

14. $\int \dfrac{\sqrt{ax + b}}{x^2}\,dx = -\dfrac{\sqrt{ax + b}}{x} + \dfrac{a}{2}\int \dfrac{dx}{x\sqrt{ax + b}} + C$

15. $\int \dfrac{dx}{x^2\sqrt{ax + b}} = -\dfrac{\sqrt{ax + b}}{bx} - \dfrac{a}{2b}\int \dfrac{dx}{x\sqrt{ax + b}} + C$

16. $\int \dfrac{dx}{a^2 + x^2} = \dfrac{1}{a}\tan^{-1}\dfrac{x}{a} + C$

17. $\int \dfrac{dx}{(a^2 + x^2)^2} = \dfrac{x}{2a^2(a^2 + x^2)} + \dfrac{1}{2a^3}\tan^{-1}\dfrac{x}{a} + C$

18. $\int \dfrac{dx}{a^2 - x^2} = \dfrac{1}{2a}\ln\left|\dfrac{x + a}{x - a}\right| + C$

19. $\int \dfrac{dx}{(a^2 - x^2)^2} = \dfrac{x}{2a^2(a^2 - x^2)} + \dfrac{1}{2a^2}\int \dfrac{dx}{a^2 - x^2}$

20. $\int \dfrac{dx}{\sqrt{a^2 + x^2}} = \ln|x + \sqrt{a^2 + x^2}| + C$

21. $\int \sqrt{a^2 + x^2}\, dx = \dfrac{x}{2}\sqrt{a^2 + x^2} + \dfrac{a^2}{2}\log(x + \sqrt{a^2 + x^2}) + C$

22. $\int x^2\sqrt{a^2 + x^2}\, dx = \dfrac{x(a^2 + 2x^2)\sqrt{a^2 + x^2}}{8} - \dfrac{a^4}{8}\log(x + \sqrt{a^2 + x^2}) + C$

23. $\int \dfrac{\sqrt{a^2 + x^2}}{x}\, dx = \sqrt{a^2 + x^2} - a\log(x + \sqrt{a^2 + x^2}) + C$

24. $\int \dfrac{\sqrt{a^2 + x^2}}{x^2}\, dx = \log(x + \sqrt{a^2 + x^2}) - \dfrac{\sqrt{a^2 + x^2}}{x} + C$

25. $\int \dfrac{x^2}{\sqrt{a^2 + x^2}}\, dx = -\dfrac{a^2}{2}\log(x + \sqrt{a^2 + x^2}) + \dfrac{x\sqrt{a^2+x^2}}{2} + C$

26. $\int \dfrac{dx}{x\sqrt{a^2 + x^2}} = -\dfrac{1}{a}\ln\left|\dfrac{a + \sqrt{a^2 + x^2}}{x}\right| + C$

27. $\int \dfrac{dx}{x^2\sqrt{a^2 + x^2}} = -\dfrac{\sqrt{a^2 + x^2}}{a^2 x} + C$

28. $\int \dfrac{dx}{\sqrt{a^2 - x^2}} = \sin^{-1}\dfrac{x}{a} + C$

29. $\int \sqrt{a^2 - x^2}\, dx = \dfrac{x}{2}\sqrt{a^2 - x^2} + \dfrac{a^2}{2}\sin^{-1}\dfrac{x}{a} + C$

30. $\int x^2\sqrt{a^2 - x^2}\, dx = \dfrac{a^4}{8}\sin^{-1}\dfrac{x}{a} - \dfrac{1}{8}x\sqrt{a^2 - x^2}(a^2 - 2x^2) + C$

31. $\int \dfrac{\sqrt{a^2 - x^2}}{x}\, dx = \sqrt{a^2 - x^2} - a\ln\left|\dfrac{a + \sqrt{a^2 - x^2}}{x}\right| + C$

32. $\int \dfrac{\sqrt{a^2 - x^2}}{x^2}\, dx = -\sin^{-1}\dfrac{x}{a} - \dfrac{\sqrt{a^2 - x^2}}{x} + C$

33. $\int \dfrac{x^2}{\sqrt{a^2 - x^2}}\, dx = \dfrac{a^2}{2}\sin^{-1}\dfrac{x}{a} - \dfrac{1}{2}x\sqrt{a^2 - x^2} + C$

34. $\int \dfrac{dx}{x\sqrt{a^2 - x^2}} = -\dfrac{1}{a}\ln\left|\dfrac{a + \sqrt{a^2 - x^2}}{x}\right| + C$

35. $\int \dfrac{dx}{x^2\sqrt{a^2 - x^2}} = -\dfrac{\sqrt{a^2 - x^2}}{a^2 x} + C$

36. $\int \dfrac{dx}{\sqrt{x^2 - a^2}} = \ln\left|x + \sqrt{x^2 - a^2}\right| + C$

37. $\int \sqrt{x^2 - a^2}\, dx = \dfrac{x}{2}\sqrt{x^2 - a^2} - \dfrac{a^2}{2}\log\left|x + \sqrt{x^2 - a^2}\right| + C$

38. $\int (\sqrt{x^2 - a^2})^n\, dx = \dfrac{x(\sqrt{x^2 - a^2})^n}{n + 1} - \dfrac{na^2}{n + 1}\int (\sqrt{x^2 - a^2})^{n-2}\, dx, \qquad n \ne -1$

39. $\int \dfrac{dx}{(\sqrt{x^2 - a^2})^n} = \dfrac{x(\sqrt{x^2 - a^2})^{2-n}}{(2 - n)a^2} - \dfrac{n - 3}{(n - 2)a^2}\int \dfrac{dx}{(\sqrt{x^2 - a^2})^{n-2}}, \qquad n \ne 2$

40. $\int x(\sqrt{x^2 - a^2})^n\, dx = \dfrac{(\sqrt{x^2 - a^2})^{n+2}}{n + 2} + C, \qquad n \ne -2$

41. $\int x^2\sqrt{x^2 - a^2}\, dx = \dfrac{x}{8}(2x^2 - a^2)\sqrt{x^2 - a^2} - \dfrac{a^4}{8}\log\left|x + \sqrt{x^2 - a^2}\right| + C$

42. $\displaystyle\int \frac{\sqrt{x^2 - a^2}}{x}\, dx = \sqrt{x^2 - a^2} - a \sec^{-1}\left|\frac{x}{a}\right| + C$

43. $\displaystyle\int \frac{\sqrt{x^2 - a^2}}{x^2}\, dx = \log\left|x + \sqrt{x^2 - a^2}\right| - \frac{\sqrt{x^2 - a^2}}{x} + C$

44. $\displaystyle\int \frac{x^2}{\sqrt{x^2 - a^2}}\, dx = \frac{a^2}{2}\log\left|x + \sqrt{x^2 - a^2}\right| + \frac{x}{2}\sqrt{x^2 - a^2} + C$

45. $\displaystyle\int \frac{dx}{x\sqrt{x^2 - a^2}} = \frac{1}{a}\sec^{-1}\left|\frac{x}{a}\right| + C = \frac{1}{a}\cos^{-1}\left|\frac{a}{x}\right| + C$

46. $\displaystyle\int \frac{dx}{x^2\sqrt{x^2 - a^2}} = \frac{\sqrt{x^2 - a^2}}{a^2 x} + C$ 47. $\displaystyle\int \frac{dx}{\sqrt{2ax - x^2}} = \sin^{-1}\left(\frac{x - a}{a}\right) + C$

48. $\displaystyle\int \sqrt{2ax - x^2}\, dx = \frac{x - a}{2}\sqrt{2ax - x^2} + \frac{a^2}{2}\sin^{-1}\left(\frac{x - a}{a}\right) + C$

49. $\displaystyle\int (\sqrt{2ax - x^2})^n\, dx = \frac{(x - a)(\sqrt{2ax - x^2})^n}{n + 1} + \frac{na^2}{n + 1}\int (\sqrt{2ax - x^2})^{n-2}\, dx,$

50. $\displaystyle\int \frac{dx}{(\sqrt{2ax - x^2})^n} = \frac{(x - a)(\sqrt{2ax - x^2})^{2-n}}{(n - 2)a^2} + \frac{(n - 3)}{(n - 2)a^2}\int \frac{dx}{(\sqrt{2ax - x^2})^{n-2}}$

51. $\displaystyle\int x\sqrt{2ax - x^2}\, dx = \frac{(x + a)(2x - 3a)\sqrt{2ax - x^2}}{6} + \frac{a^3}{2}\sin^{-1}\frac{x - a}{a} + C$

52. $\displaystyle\int \frac{\sqrt{2ax - x^2}}{x}\, dx = \sqrt{2ax - x^2} + a\sin^{-1}\frac{x - a}{a} + C$

53. $\displaystyle\int \frac{\sqrt{2ax - x^2}}{x^2}\, dx = -2\sqrt{\frac{2a - x}{x}} - \sin^{-1}\left(\frac{x - a}{a}\right) + C$

54. $\displaystyle\int \frac{x\, dx}{\sqrt{2ax - x^2}} = a\sin^{-1}\frac{x - a}{a} - \sqrt{2ax - x^2} + C$

55. $\displaystyle\int \frac{dx}{x\sqrt{2ax - x^2}} = -\frac{1}{a}\sqrt{\frac{2a - x}{x}} + C$

56. $\displaystyle\int \sin ax\, dx = -\frac{1}{a}\cos ax + C$

57. $\displaystyle\int \cos ax\, dx = \frac{1}{a}\sin ax + C$

58. $\displaystyle\int \sin^2 ax\, dx = \frac{x}{2} - \frac{\sin 2ax}{4a} + C$

59. $\displaystyle\int \cos^2 ax\, dx = \frac{x}{2} + \frac{\sin 2ax}{4a} + C$

60. $\displaystyle\int \sin^n ax\, dx = \frac{-\sin^{n-1} ax \cos ax}{na} + \frac{n - 1}{n}\int \sin^{n-2} ax\, dx$

61. $\displaystyle\int \cos^n ax\, dx = \frac{\cos^{n-1} ax \sin ax}{na} + \frac{n - 1}{n}\int \cos^{n-2} ax\, dx$

62. (a) $\displaystyle\int \sin ax \cos bx\, dx = -\frac{\cos(a + b)x}{2(a + b)} - \frac{\cos(a - b)x}{2(a - b)} + C,\qquad a^2 \neq b^2$

(b) $\displaystyle\int \sin ax \sin bx\, dx = \frac{\sin(a - b)x}{2(a - b)} - \frac{\sin(a + b)x}{2(a + b)},\qquad a^2 \neq b^2$

(c) $\displaystyle\int \cos ax \cos bx\, dx = \frac{\sin(a - b)x}{2(a - b)} + \frac{\sin(a + b)x}{2(a + b)},\qquad a^2 \neq b^2$

63. $\displaystyle\int \sin ax \cos ax \, dx = -\frac{\cos 2ax}{4a} + C$

64. $\displaystyle\int \sin^n ax \cos ax \, dx = \frac{\sin^{n+1} ax}{(n+1)a} + C, \qquad n \neq -1$

65. $\displaystyle\int \frac{\cos ax}{\sin ax} \, dx = \frac{1}{a} \ln |\sin ax| + C$

66. $\displaystyle\int \cos^n ax \sin ax \, dx = -\frac{\cos^{n+1} ax}{(n+1)a} + C, \qquad n \neq -1$

67. $\displaystyle\int \frac{\sin ax}{\cos ax} \, dx = -\frac{1}{a} \ln |\cos ax| + C$

68. $\displaystyle\int \sin^n ax \cos^m ax \, dx = -\frac{\sin^{n-1} ax \cos^{m+1} ax}{a(m+n)} + \frac{n-1}{m+n} \int \sin^{n-2} ax \cos^m ax \, dx,$
$\qquad\qquad n \neq -m$ (If $n = -m$, use No. 86.)

69. $\displaystyle\int \sin^n ax \cos^m ax \, dx = \frac{\sin^{n+1} ax \cos^{m-1} ax}{a(m+n)} + \frac{m-1}{m+n} \int \sin^n ax \cos^{m-2} ax \, dx,$
$\qquad\qquad m \neq -n$ (If $n = -m$, use No. 86.)

70. $\displaystyle\int \frac{dx}{b + c \sin ax} = \frac{-2}{a\sqrt{b^2 - c^2}} \tan^{-1}\left[\sqrt{\frac{b-c}{b+c}} \tan\left(\frac{\pi}{4} - \frac{ax}{2}\right)\right] + C, \qquad b^2 > c^2$

71. $\displaystyle\int \frac{dx}{b + c \sin ax} = \frac{-1}{a\sqrt{c^2 - b^2}} \ln \left| \frac{c + b \sin ax + \sqrt{c^2 - b^2} \cos ax}{b + c \sin ax}\right| + C, \qquad b^2 < c^2$

72. $\displaystyle\int \frac{dx}{1 + \sin ax} = -\frac{1}{a} \tan\left(\frac{\pi}{4} - \frac{ax}{2}\right) + C$

73. $\displaystyle\int \frac{dx}{1 - \sin ax} = \frac{1}{a} \tan\left(\frac{\pi}{4} + \frac{ax}{2}\right) + C$

74. $\displaystyle\int \frac{dx}{b + c \cos ax} = \frac{2}{a\sqrt{b^2 - c^2}} \text{tar} \left[\sqrt{\frac{b-c}{b+c}} \tan\frac{ax}{2}\right] + C, \qquad b^2 > c^2$

75. $\displaystyle\int \frac{dx}{b + c \cos ax} = \frac{1}{a\sqrt{c^2 - b^2}} \ln \left| \frac{c + b \cos ax + \sqrt{c^2 - b^2} \sin ax}{b + c \cos ax}\right| + C, \qquad b^2 < c^2$

76. $\displaystyle\int \frac{dx}{1 + \cos ax} = \frac{1}{a} \tan\frac{ax}{2} + C$

77. $\displaystyle\int \frac{dx}{1 - \cos ax} = -\frac{1}{a} \cot\frac{ax}{2} + C$

78. $\displaystyle\int x \sin ax \, dx = \frac{1}{a^2} \sin ax - \frac{x}{a} \cos ax + C$

79. $\displaystyle\int x \cos ax \, dx = \frac{1}{a^2} \cos ax + \frac{x}{a} \sin ax + C$

80. $\displaystyle\int x^n \sin ax \, dx = -\frac{x^n}{a} \cos ax + \frac{n}{a} \int x^{n-1} \cos ax \, dx$

81. $\displaystyle\int x^n \cos ax \, dx = \frac{x^n}{a} \sin ax - \frac{n}{a} \int x^{n-1} \sin ax \, dx$

82. $\displaystyle\int \tan ax \, dx = -\frac{1}{a} \ln |\cos ax| + C$

83. $\displaystyle\int \cot ax \, dx = \frac{1}{a} \ln |\sin ax| + C$

84. $\displaystyle\int \tan^2 ax \, dx = \frac{1}{a} \tan ax - x + C$

85. $\displaystyle\int \cot^2 ax \, dx = -\frac{1}{a} \cot ax - x + C$

86. $\displaystyle\int \tan^n ax \, dx = \frac{\tan^{n-1} ax}{a(n-1)} - \int \tan^{n-2} ax \, dx, \qquad n \neq 1$

87. $\displaystyle\int \cot^n ax \, dx = -\frac{\cot^{n-1} ax}{a(n-1)} - \int \cot^{n-2} ax \, dx, \qquad n \neq 1$

88. $\displaystyle\int \sec ax \, dx = \frac{1}{a} \ln |\sec ax + \tan ax| + C$

89. $\displaystyle\int \csc ax \, dx = -\frac{1}{a} \ln |\csc ax + \cot ax| + C$

90. $\displaystyle\int \sec^2 ax \, dx = \frac{1}{a} \tan ax + C$

91. $\displaystyle\int \csc^2 ax \, dx = -\frac{1}{a} \cot ax + C$

92. $\displaystyle\int \sec^n ax \, dx = \frac{\sec^{n-2} ax \tan ax}{a(n-1)} + \frac{n-2}{n-1} \int \sec^{n-2} ax \, dx, \qquad n \neq 1$

93. $\displaystyle\int \csc^n ax \, dx = -\frac{\csc^{n-2} ax \cot ax}{a(n-1)} + \frac{n-2}{n-1} \int \csc^{n-2} ax \, dx, \qquad n \neq 1$

94. $\displaystyle\int \sec^n ax \tan ax \, dx = \frac{\sec^n ax}{na} + C, \qquad n \neq 0$

95. $\displaystyle\int \csc^n ax \cot ax \, dx = -\frac{\csc^n ax}{na} + C, \qquad n \neq 0$

96. $\displaystyle\int \sin^{-1} ax \, dx = x \sin^{-1} ax + \frac{1}{a} \sqrt{1 - a^2 x^2} + C$

97. $\displaystyle\int \cos^{-1} ax \, dx = x \cos^{-1} ax - \frac{1}{a} \sqrt{1 - a^2 x^2} + C$

98. $\displaystyle\int \tan^{-1} ax \, dx = x \tan^{-1} ax - \frac{1}{2a} \ln(1 + a^2 x^2) + C$

99. $\displaystyle\int x^n \sin^{-1} ax \, dx = \frac{x^{n+1}}{n+1} \sin^{-1} ax - \frac{a}{n+1} \int \frac{x^{n+1} \, dx}{\sqrt{1 - a^2 x^2}}, \qquad n \neq -1$

100. $\displaystyle\int x^n \cos^{-1} ax \, dx = \frac{x^{n+1}}{n+1} \cos^{-1} ax + \frac{a}{n+1} \int \frac{x^{n+1} \, dx}{\sqrt{1 - a^2 x^2}}, \qquad n \neq -1$

101. $\displaystyle\int x^n \tan^{-1} ax \, dx = \frac{x^{n+1}}{n+1} \tan^{-1} ax - \frac{a}{n+1} \int \frac{x^{n+1} \, dx}{1 + a^2 x^2}, \qquad n \neq -1$

102. $\displaystyle\int e^{ax} \, dx = \frac{1}{a} e^{ax} + C$

103. $\displaystyle\int b^{ax} \, dx = \frac{1}{a} \frac{b^{ax}}{\ln b} + C, \qquad b > 0, \qquad b \neq 1$

104. $\displaystyle\int xe^{ax} \, dx = \frac{e^{ax}}{a^2} (ax - 1) + C$

105. $\displaystyle\int x^n e^{ax} \, dx = \frac{1}{a} x^n e^{ax} - \frac{n}{a} \int x^{n-1} e^{ax} \, dx$

106. $\displaystyle\int x^n b^{ax} \, dx = \frac{x^n b^{ax}}{a \ln b} - \frac{n}{a \ln b} \int x^{n-1} b^{ax} \, dx, \qquad b > 0, \qquad b \neq 1$

107. $\displaystyle\int e^{ax} \sin bx \, dx = \frac{e^{ax}}{a^2 + b^2} (a \sin bx - b \cos bx) + C$

108. $\int e^{ax} \cos bx \, dx = \dfrac{e^{ax}}{a^2 + b^2} (a \cos bx + b \sin bx) + C$

109. $\int \ln ax \, dx = x \ln ax - x + C$

110. $\int x^n \ln ax \, dx = \dfrac{x^{n+1}}{n + 1} \ln ax - \dfrac{x^{n+1}}{(n + 1)^2} + C, \qquad n \neq -1$

111. $\int x^{-1} \ln ax \, dx = \dfrac{1}{2} (\ln ax)^2 + C$ \qquad 112. $\int \dfrac{dx}{x \ln ax} = \ln |\ln ax| + C$

113. $\int_0^\infty x^{n-1} e^{-x} \, dx = \Gamma(n) = (n - 1)!, \qquad n > 0.$

114. $\int_0^\infty e^{-ax^2} \, dx = \dfrac{1}{2} \sqrt{\dfrac{\pi}{a}}, \qquad a > 0$

115. $\displaystyle \int_0^{\pi/2} \sin^n x \, dx = \int_0^{\pi/2} \cos^n x \, dx = \begin{cases} \dfrac{1 \cdot 3 \cdot 5 \cdots (n - 1)}{2 \cdot 4 \cdot 6 \cdots n} \cdot \dfrac{\pi}{2}, & \text{if } n \text{ is an even integer} \geq 2, \\[2ex] \dfrac{2 \cdot 4 \cdot 6 \cdots (n - 1)}{3 \cdot 5 \cdot 7 \cdots n}, & \text{if } n \text{ is an odd integer} \geq 3 \end{cases}$

Greek Alphabet

Capital Letters	Name of Letter	Lower-case Letters
A	Alpha	α
B	Beta	β
Γ	Gamma	γ
Δ	Delta	δ
E	Epsilon	ϵ
Z	Zeta	ζ
H	Eta	η
Θ	Theta	θ, ϑ
I	Iota	ι
K	Kappa	κ
Λ	Lambda	λ
M	Mu	μ
N	Nu	ν
Ξ	Xi	ξ
O	Omicron	o
Π	Pi	π
P	Rho	ρ
Σ	Sigma	σ
T	Tau	τ
Υ	Upsilon	υ
Φ	Phi	ϕ, φ
X	Chi	χ
Ψ	Psi	ψ
Ω	Omega	ω

Answers to Selected Odd-numbered Problems

CHAPTER 1

Section 1

1.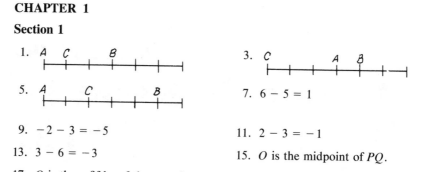

3.

5.

7. $6 - 5 = 1$

9. $-2 - 3 = -5$

11. $2 - 3 = -1$

13. $3 - 6 = -3$

15. O is the midpoint of PQ.

17. O is three-fifths of the way from Q to P.

19. If the points A and B have the coordinates a and b, respectively, then the signed distance from A to B is $b - a$. The sign of $b - a$ thus determines which direction is positive, and $|b - a|$, which is the distance between A and B, determines the unit of length. The origin is then the point whose signed distance from A is . . . ?

21. $(a + 2b)/3$

23.

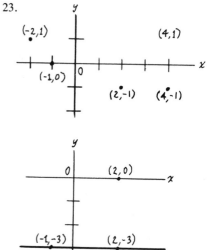

25. This is the straight line parallel to, and three units below, the x-axis.

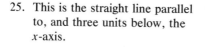

27.

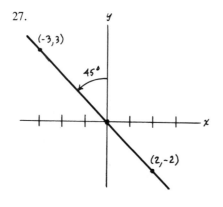

29.

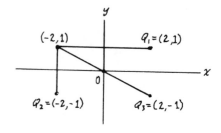

Section 2

1.

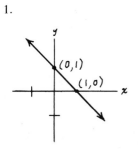

3.

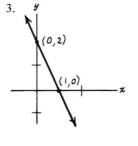

5.

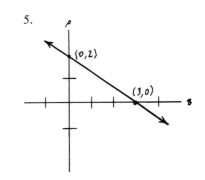

7.

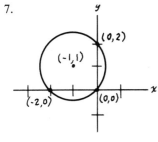

9.

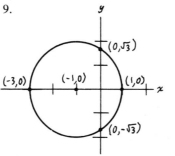

11.

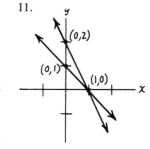

13.

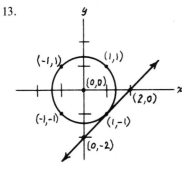

15.

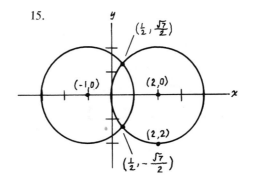

17. If $y = 3 - x$ and $x^2 + y^2 = 4$, then

$$x^2 + (3 - x)^2 = 4$$
$$\Rightarrow 2x^2 - 6x + 5 = 0$$
$$\Rightarrow x^2 - 3x + 5/2 = 0$$
$$\Rightarrow (x - 3/2)^2 + 5/2 - 9/4 = 0$$
$$\Rightarrow (x - 3/2)^2 = -1/4,$$

which is impossible.

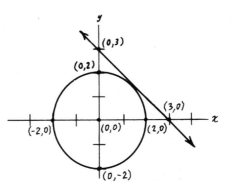

19.

x	y
±2	2
±3/2	1/4
±1	−1
±1/2	−7/4
0	−2

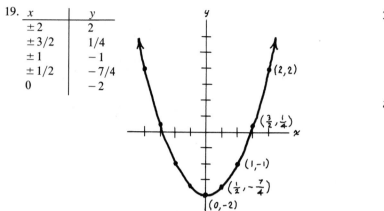

21.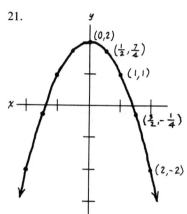

23.

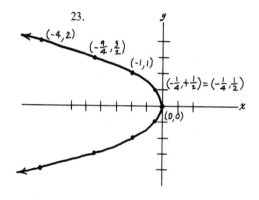

25.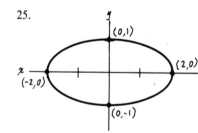

27.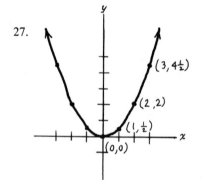

Section 3

1. $y = 3x - 6$; $(2,0)$, $(1,-3)$, $(7/3,1)$

3. $m = -1/2$, $b = 2$ 5. $m = -2/3$, $b = 2$ 7. $m = 1$, $b = 1$

9.

11.

13.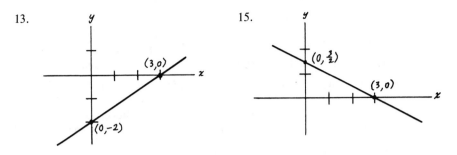

15.

17. $m = 1/2$, $y = x/2$ 19. No slope, $x = -2$ 21. $m = 0$, $y = 1$

23. Point–slope equation: $y - 2 = (-1/2)(x - 1)$, which is equivalent to $2y + x - 5 = 0$.

25. $3y - 2x - 1 = 0$ 27. $y = x + 1$ 29. $(-2,3)$

31. $y = -x + 1$ 33. $y = (-1/6)x + 4/3$

Section 4

1. $\sqrt{5}$ 3. 4 5. $2\sqrt{26}$ 7. $\sqrt{5}$ 9. $2\sqrt{2}$

11. $AB = 2\sqrt{2}$, $BC = 3\sqrt{5}$, $CA = \sqrt{41}$ 13. $AB = \sqrt{26}$, $BC = 5$, $CA = \sqrt{13}$

15. $P_1P_2 = P_2P_3 = \sqrt{41}$

17. $(2,1)$ and $(1,3)$ have slope -2; $(2,1)$ and $(8,4)$ have slope $1/2$.

19. $y = 3x - 6$

21. a) Show that the midpoint makes the slope $(b_2 - b_1)/(a_2 - a_1)$ with each end-point.

23. $3y - 4x + 1 = 0$ 25. $12y - 5x + 59 = 0$

27. P_1P_2 has midpoint $(7/2, -1/2)$. P_1P_3 has midpoint $(0, 3/2)$. The midpoints have slope $-4/7$. So does P_2P_3.

29. If the two points are (x_1, y_1) and (x_2, y_2), then the locus condition is
$$\sqrt{(x - x_1)^2 + (y - y_1)^2} = \sqrt{(x - x_2)^2 + (y - y_2)^2}.$$
Square; expand the squares, and simplify.

Section 5

1. $(x - 2)^2 + (y - 1)^2 = 9$; $x^2 + y^2 - 4x - 2y - 4 = 0$

3. $(x - 2)^2 + (y + 3)^2 = 1$; $x^2 + y^2 - 4x + 6y + 12 = 0$

5. $(x + 2)^2 + (y - 4)^2 = 25$; $x^2 + y^2 + 4x - 8y - 5 = 0$

7. $x^2 + y^2 = 16$; $x^2 + y^2 - 16 = 0$

9. Center $(-3,4)$; radius 5   11. Center $(-3/2, 5/2)$; radius 3

13. $c(-1,0)$; $r = 1$ 15. $C(0, -1/2)$; $r = \sqrt{5}/2$

17. No locus   19. $(x - 5)^2 + (y + 2)^2 = 85$

21. $(x - 5/2)^2 + y^2 = 25/4$ 23. $(x - 1/2)^2 + (y - 3/2)^2 = 10/4$

25. The circle $(x - 4)^2 + y^2 = 4$.

Section 6

1. Both axes and the origin   3. The origin   5. The y-axis

7. Both axes, the origin and the line $y = x$ 9. The origin and $y = x$

11. A line of symmetry l for a triangle must contain a vertex P. (Otherwise, there would be at least two vertices on one side of l, and hence, by symmetry, at least two on the other side also, making at least four vertices altogether.) Then l must be the perpendicular bisector of the side opposite P, by a similar argument, etc.

13. Put coordinate axes along the perpendicular lines of symmetry, and let $P = (x,y)$ by any point in the figure. Show that $Q = (-x,-y)$ must also be in the figure by proceeding in two steps, first to the point of symmetry across the y-axis, and from there across the x-axis.

Section 7

1.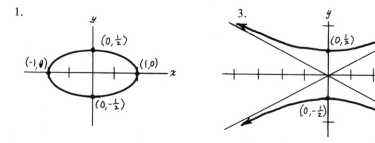

3.

5.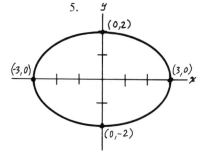

7.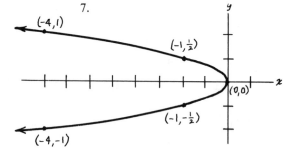

9.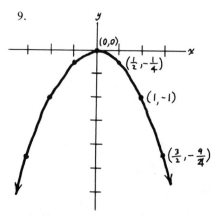

11.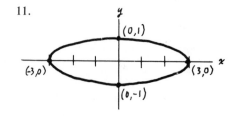

13.

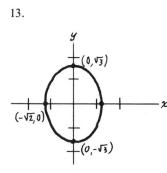

15.

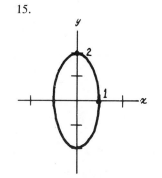

17.

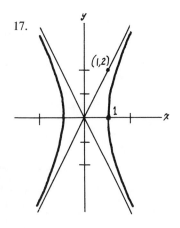

19.

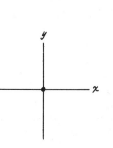

21.

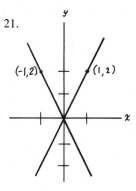

Section 8

1. a) $(2,-1)$ b) $X = x - 1;\ Y = y - 4$

3. a) $(3,9)$ b) $X = x - 2;\ Y = y + 4$

5. a) $(-2,-1)$ b) $X = x - 3;\ Y = y - 2$

7. a) $(2,1)$ b) $X = x + 3;\ Y = y + 2$

9. New origin at $(-2,3);\ X^2 + Y^2 = 18.$

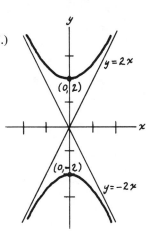

11. New origin at $(0,0)$
    (See answer to Problem 11 of Section 7.)

13. New origin at $(-2,1);\ \dfrac{X^2}{3^2} + \dfrac{Y^2}{2^2} = 1$
    (See answer to Problem 5 of Section 7.)

15. New origin at $(1,1);\ X^2 - \dfrac{Y^2}{2^2} = -1$

17. New origin at $(1,2)$; $\dfrac{X^2}{2^2} + Y^2 = 1$

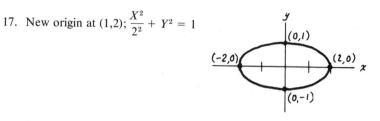

19. $\bar{X}^2 + \bar{Y}^2 + 4\bar{X} - 2\bar{Y} + 1 = 0$ 21. $\bar{X}\bar{Y} = 0$

Extra Problems for Chapter 1

1. Call the common origin of the two coordinate systems 0. We know that for some point, its signed distance from 0 is 2 in the y-coordinate system and 1 in the x-coordinate system. Thus the two coordinate systems have the same positive direction, but the unit of distance in the x-coordinate system is twice that in the y-coordinate system. Thus the y-coordinate of any point is twice its x-coordinate.

3. Signed distance in y-coordinate system $= a$ (signed distance in y-coordinate system). Thus $y = ax$.

5. The signed distance from the first point to the second point is 1 in the x-coordinate system and is a in the y-coordinate system. Then the signed distance between any 2 points in the y-coordinate system is a times that in the x-coordinate system. Then for any P, $\overline{O'P} = \overline{O'O} + \overline{OP} = b + ax$ in the y-coordinate system; thus $y = b + ax$.

7. $y = -1/4\,x - 1$; y-intercept $(0, -1)$, x-intercept $(-4,0)$

9. Slope is $-3/4$; $y - 2 = -3/4\,(x + 1/2)$.

11. Say two lines have the same slope m. $l_1 : y = mx + b$, $l_2 : y = mx + b_2$. If l_1 and l_2 intersect at a point (x_0, y_0), then $y_0 = mx_0 + b_1$, $y_0 = mx_0 + b_2$. Then subtracting: $b_1 - b_2 = 0 \Rightarrow b_1 = b_2 \Rightarrow l_1 = l_2$. Say two lines have different slope, $l_1 : y = m_1 x + b_1$; $l_2 : y = m_2 x + b_2$. Solve the equations simultaneously to get $x_0 = (b_2 - b_1)/(m_1 - m_2)$, $y_0 = (b_2 m_1 - b_1 m_2)/(m_1 - m_2)$. So l_1 and l_2 are not parallel. (Why?)

13. Yes; the three slopes are all $4/5$.   15. No; the slopes are $2/3$, $1/2$, $3/5$.

17. $(2,1)$

19. Is the given equation the equation of some line? If so, what are its intercepts?

21. Both points are on the side of the line opposite from the origin.

23.

25.

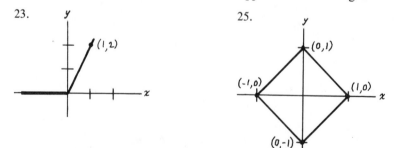

27. $y + 5 = -\dfrac{5}{6}\left(x - \dfrac{3}{2}\right)$

29. Choose the x-axis along one side of the triangle and the y-axis through the opposite vertex. The triangle then has vertices $(a,0)$, $(b,0)$, $(0,c)$. The perpendicular bisectors of the sides meet at the point

$$\left(\frac{a + b}{2}, \frac{c}{2} + \frac{ab}{2c}\right).$$

31. $4/\sqrt{5}$

33. $2x^2 + x + 2y^2 + 2y - 32 = 2(x^2 + x/2 + 1/16) + 2(y^2 + y + 1/4)$ so the standard form is that of a circle: $(x + 1/4)^2 + (y + 1/2)^2 + 16 + 5/16$; center $(-1/4, -1/2)$; radius $4 + \sqrt{5}/4$.

35. Let $Q_1 = (x_1,y_1)$, $Q_2 = (x_2,y_2)$. If distance from P to $Q_1 = k$ times distance from P to Q_2, then (distance from P to $Q_1)^2 = k^2$ (distance from P to $Q_2)^2$, and so $(x - x_1)^2 + (y - y_1)^2 = k^2(x - x_2)^2 + k^2(y - y_2)^2$, $(k^2 - 1)x^2 + (k^2 - 1)y^2 + 2(x_1 - k^2x_2)x + 2(y_1 - k^2y_2)y + k^2(x_2^2 + y_2^2) - (x_1^2 + y_1^2) = 0$. If $k = 1$, the terms in x^2 and y^2 drop out, leaving an equation of the form $Ax + By + C = 0$, so the locus is a straight line. If $k \neq 1$ we can divide by $k^2 - 1$, which changes the equation to the form $x^2 + y^2 + Ax + By + C = 0$. The locus is therefore either a circle, or a point, or empty, and since it is easy to see geometrically that the locus contains more than one point, it must be a circle.

In more detail, the new equation is

$$x^2 + y^2 + \frac{2(x_1 - k^2x_2)}{k^2 - 1}x + \frac{2(y_1 - k^2y_2)}{k^2 - 1}y + \frac{k^2(x_2^2 + y_2^2)}{k^2 - 1} - \frac{(x_1^2 + y_1^2)}{k^2 - 1} = 0.$$

37. In standard form, the circle has equation $(x + 3)^2 + (y + 4)^2 = 25$. Since the radius is 5, and since the distance from $(1,3)$ to the center $(-3, -4)$ is $\sqrt{65} > 5$, $(1,3)$ lies outside the circle.

39. For all very large x_1 and y_1, (x_1,y_1) lies outside the circle and $x_1^2 + y_1^2 + Ax_1 + By_1 + C > 0$. Since $x_0^2 + y_0^2 + Ax_0 + By_0 + C < 0$, we conclude that $x^2 + y^2 + Ax + By + C$ must cross through the value zero as (x,y) moves along the line segment joining (x_0,y_0) and (x_1,y_1). The segment thus intersects the circle $x^2 + y^2 + Ax + By + C = 0$. Now if (x_0,y_0) were outside the circle, we could choose very large x_1 and y_1 such that the line segment from (x_0,y_0) to (x_1,y_1) does *not* intersect the circle. This contradicts the above argument, so that (x_0,y_0) must lie inside the circle.

41. A regular pentagon has five lines of symmetry, each passing through a vertex and the midpoint of the opposite side. (A line of symmetry for any pentagon must contain a vertex.)

43. A line of symmetry through a vertex of a quadrilateral must also contain another vertex. (The remaining two vertices must be symmetric in this line for the same general reason.) Etc.

45. If a line of symmetry does not contain a vertex, then the four vertices must be two on each side of the line of symmetry, and each pair must be the symmetric image of the other pair. The figure is then an isosceles trapezoid. Now consider the effect of a second line of symmetry.

47. Let C be a bounded configuration with A and B points of symmetry. Call the distance from A to B, $d > 0$. Because C is bounded, there is a smallest circle of radius R about A which completely encloses C. Because R is least, there must be a point X in X at the distance R from A.

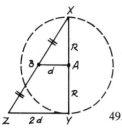

Let Y be the reflection of X through A.

Let Z be the reflection of X through B. We may assume that $\angle BAX$ is at least 90°; otherwise we just reverse the roles of X and Y. BA and ZY are parallel so $\angle ZYA$ is also at least 90°. Show, therefore, that $\overline{AZ} > R$. But this is a contradiction: Z is a point in the configuration C outside of the circle around A of radius R.

49. Substituting $X = x - 2$, $Y = y + 3/2$ gives a standard form equation for a point, the new origin: $X^2 + 4Y^2 = 0$.

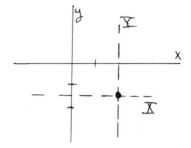

CHAPTER 2

Section 1

1. $x \ge 0$; $[0,\infty)$

3. $x \ne 2$; $(-\infty,2) \cup (2,\infty)$

5. $x \ne 2, x \ne -2$; $(-\infty,-2) \cup (-2,2) \cup (2,\infty)$

7. -1

9. $a^3 + 6a^2 + 11a + 6$

11. $u - 2$

13. $H(1/t) = \dfrac{t - 1}{t + 1} = -H(t)$

15. $H(H(t)) = t$

17. $f(x) = (1 - x^2)/4$; $(-\infty,\infty)$

19. $f(x) = (x - 2)/(1 - x)$; $x \ne 1$

21. $f(x) = -x^{2/3}$;   all x

23. $f(x) = -x$;   all x

25. $f(h) = \frac{1}{4}\sqrt{h}$

Section 2

1. Parabola

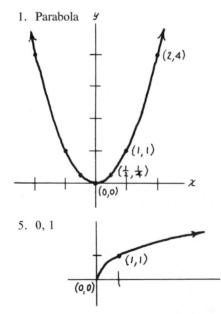

3. The graph is a semicircle, because $(f(x))^2 + x^2 = 1$ and because $f(x) \ge 0$.

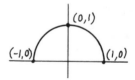

5. 0, 1

7.

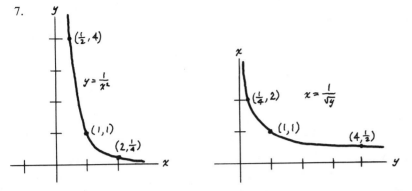

9. f is odd if and only if $f(-x) = -f(x)$, if and only if $(-x, -f(x))$ belongs to the graph of f whenever $(x, f(x))$ does, if and only if the graph of f is symmetric with respect to the origin.

Section 3

1. a) $\dfrac{f(3 + h) - f(3)}{h} = -4 - h;\ -0.41,\ -0.0401,\ -0.004001$

   b) $\dfrac{f(2 + h) - f(2)}{h} = -2 - h;\ -0.21,\ -0.0201,\ -0.002001$

3. 3, 5, and x are continuous by I, so by the sum and product rules of II, $3x^2 + 5x$ is continuous.

5. By I, $x^{1/2}$ is continuous $(x \geq 0)$; so by II, $1/\sqrt{x}$ is continuous for $x > 0$.

7. $f(x) = x^2$ is continuous by I and II; $g(x) = x^{1/2}$ is continuous by I, so $\sqrt{x^2} = g(f(x))$ is continuous by III.

9. $g(x) = \sqrt{x^2} = |x|$ is continuous (by Problem 7); $f(x) = x^{1/3}$ is continuous by I, so $g(f(x)) = |x^{1/3}|$ is continuous by III.

11. Here is one example. Draw a different one.

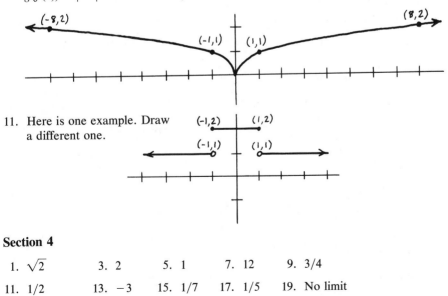

Section 4

1. $\sqrt{2}$ 3. 2   5. 1   7. 12   9. 3/4

11. 1/2   13. -3 15. 1/7   17. 1/5   19. No limit

21. No limit   23. -2 25. 1

27. $\lim\limits_{x\to0^-} \dfrac{|x|}{x} = -1;\ \lim\limits_{x\to0^+} \dfrac{|x|}{x} = +1$

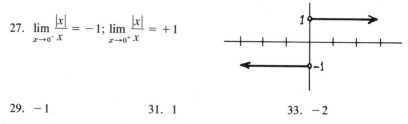

29. -1 　　　　　　　　 31. 1 　　　　　　　　 33. -2

35. $f(x) = [x]$ is continuous from the right everywhere. It is continuous from the left everywhere except at the integer points.

Section 5

1. $f(x) = x, g(x) = -4, h(x) = 2$ are all continuous. By the product rule gf is continuous, and so is f^2. Applying the rule again to f and f^2, f^3 is continuous. By the sum rule, $f^3 + gf$ is continuous. Then again by the sum rule $(f^3 + gf + h)$ is continuous, and this is p.

3. 　　　　　　　 $\sqrt{x}$ continuous by **L2**

　　　　　　 $1/\sqrt{x}$ continuous by **L3′** $(x > 0)$

　　　 $\sqrt{x} + 1/\sqrt{x}$ continuous by **L3′**

5. $1, \sqrt{x}, x$ continuous by **L2**

　　　　 x^2 continuous by **L3′**

　　 $1 + x^2$ continuous by **L3′**

　 $\dfrac{1}{1 + x^2}$ continuous by **L3′**

　 $\dfrac{\sqrt{x}}{1 + x^2}$ continuous by **L3′** (product rule)

7. If $\lim_{x\to a} f(x) = l$ and $\lim_{x\to a} g(x) = m$, then $\lim_{x\to a}(f(x) + g(x)) = l + m$, etc.

9. If $g(x) = f^3(x) + 4f(x) + 1 = y^3 - 4y + 1$, then

$$\lim_{x\to a^-} g(x) = \lim_{x\to a^-} (y^3 - 4y + 1) = (-2)^3 - 4(-2) + 1 = 1,$$

$$\lim_{x\to a^+} g(x) = \lim_{x\to a^+} (y^3 - 4y + 1) = 2^3 - 4(2) + 1 = 1,$$

$$g(a) = (y^3 - 4y + 1)_{y=0} = 1.$$

Thus $\lim_{x\to a^-} g(x) = g(a) = \lim_{x\to a^+} g(x) = 1$, and g is continuous at $x = a$.

11. By L3(a), $[f(x) - k] \to l - k > 0$ as $x \to a$. It then follows from **L6** that $f(x) - k > 0$ on some interval about a.

13. This is just like Problem 11.

15. Suppose $l < 0$. Then $f(x) - l \geq -l > 0$. But now **L5** says that
$$0 = \lim [f(x) - l] \geq -l \geq 0.$$
Therefore $-l = 0$, a contradiction. Therefore $l \geq 0$.

Section 6

1. -7 　　　 3. 1 　　　　　　 5. $-2/3$ 　　　 7. 1

9. $1/2$ 　　 11. $2/3$ 　　　　　 13. $-\infty$ 　　　 15. $+\infty$

17. $\lim\limits_{x\to 3}\dfrac{2x^2-5x-3}{x^2-x-6}=\lim\limits_{x\to 3}\dfrac{(2x+1)(x-3)}{(x+2)(x-3)}\underset{\text{L5}}{=}\lim\limits_{x\to 3}\dfrac{2x+1}{x+2}\underset{\text{L1}}{=}\dfrac{7}{5}$

This assumes that $(2x+1)/(x+2)$ is continuous at $x=3$, which in turn requires **L3'** and **L2**.

$$\lim_{x\to\infty}\frac{2x^2-5x-3}{x^2-x-6}=\lim_{x\to\infty}\frac{2-\dfrac{5}{x}-\dfrac{3}{x^2}}{1-\dfrac{1}{x}-\dfrac{6}{x^2}}=\frac{2-0-0}{1-0-0}=2,$$

by **L3** and Example 1.

Section 7

1. $[0,\infty)$

3. $[0,\infty)$

5. $(-\infty,0)\cup[\tfrac{1}{9},\infty)$

7. $(-\infty,36]$

9. $(-\infty,2]$

11. $(-\infty,-2]\cup[2,\infty)$

13. $f(1)=-1,f(2)=12$, and $-1<0<12$

15. $f(0)=0<2<2+4^{1/3}=f(4)$

Extra Problems for Chapter 2

1. $(-\infty,\infty)$

3. The union of $(-\infty,-3)$, $(-3,2)$ and $(2,\infty)$

5. $[-4,4]$

7. The union of $(-\infty,-4]$ and $[4,\infty)$

9. $1/x-1/y=(x-y)(-1/xy)$

11. $\dfrac{1}{x^2+1}-\dfrac{1}{y^2+1}=(x-y)(x+y)(-1/(x^2+1)(y^2+1))$

13. $D=-1\le x\le 1$. For $0\le x\le 1$, $y(y(x))=\sqrt{1-(\sqrt{1-x^2})^2}=\sqrt{x^2}=|x|=x$.

15. $g(y)=y^3$

17. We know that $f(x)=g(x)+h(x)$ and $f(-x)=g(-x)+h(-x)=g(x)-h(x)$. Adding these two equations together, we have $f(x)+f(-x)=2g(x)$, and thus $g(x)=(f(x)+f(-x))/2$. Subtracting the second equation from the first, we get $f(x)-f(-x)=2h(x)$, or $h(x)=(f(x)-f(-x))/2$. For any function f, the functions g and h, defined as above, are even and odd, respectively, and $f=g+h$.

19. The even component $=g(x)=(p(x)+p(-x))/2=a_0+a_2x^2+\cdots+a_mx^m$, where m is largest even integer $\le n$. Odd component $=(p(x)-p(-x))/2=a_1x+a_3x^3+a_5x^5+\cdots+a_lx^l$, where l is the largest odd integer $\le n$. $p(x)$ is even if only even powers of x appear in $p(x)$ with nonzero coefficients; $p(x)$ is odd if only odd powers appear in it.

21. $f(x)=f_1(x)f_2(x)=(g_1(x)+h_1(x))(g_2(x)+h_2(x))=(g_1g_2)(x)+(h_1h_2)(x)+(h_1g_2)(x)+g_1h_2)(x)$. By Problem 20, g_1h_2 is odd, etc. Then $g=g_1g_2+h_1h_2$ and $h=h_1g_2+g_1h_2$.

23. Even component $=g(x)$

$=(1/2)(f(x)+f(-x))$

$=(1/2)(|x|+x+|-x|-x)$

$=x$; odd component

$=h(x)=(1/2)(f(x)-f(-x))$

$=(1/2)(|x|+x-|-x|+x)$

$=x$.

$y=f(x)$

25.

x	0	1/2	1
x^3	0	1/8	1
x^5	0	1/32	1

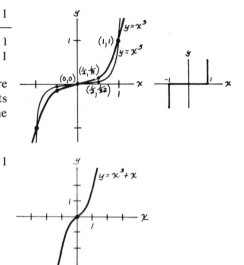

Since f and g are odd, the graphs are symmetric in the origin. As n gets very large, the graphs approach the configuration.

27. $(x^3 + x) = x(x^2 + 1)$. Since $x^2 + 1$ has no real roots, this is case (4).

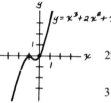

29. $x^3 + 2x^2 + x = x(x^2 + 2x + 1) = x(x + 1)^2$. Case (3).

31. If a fourth degree polynomial $p(x) = x^4 + \cdots$ has a root $x = a$ then it must have a second root $x = b$ (not necessarily distinct from a), because $p(x) = (x - a)q(x)$ where $q(x)$ is a cubic polynomial that has to have at least one root (Problem 26). The possibilities, with examples, are:

a) no roots; $p(x) = x^4 + 1$

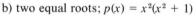

b) two equal roots; $p(x) = x^2(x^2 + 1)$

33. $y = (x + 1)^3$ so there is a triple root at $x = -1$. Graph looks just like x^3 moved one unit to the left.

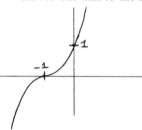

35. $(y - c)/(x - a) = (d - c)/(b - a)$. Solve for y.

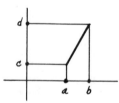

37. If $x = 2 + e$, then $x^2 - 4 = e(e + 4)$. If $|x - 2| = |e| < d$, then $|x^2 - 4| = |e(e + 4)| < d(d + 4)$.

39. If $|x - 2| < d < 1$, then $1 < x < 3$. Then $|x^3 - 8| = |x - 2|\,|x^2 + 2x + 4|$, where $|x^2 - 2x + 4| < 19$ and $|x - 2| < d$. Thus $|x^3 - 8| < 19d$.

41. Since $f(x) = (x^2 + x - 6)/(x - 2) = x + 3$ for $x \neq 2$, we let $f(2) = 5$, and then have $f(x) = x + 3$ everywhere, a continuous function.

43. For $x \neq 4, f(x) = (\sqrt{x} - 2)/(x - 4) = (\sqrt{x} - 2)/(\sqrt{x} - 2)(\sqrt{x} + 2) = 1/(\sqrt{x} + 2)$, so we let $f(4) = 1/4$, and then have $f(x) = 1/(\sqrt{x} + 2)$ for all $x \geq 0$, a continuous function.

47. If f is to be continuous, then
$$f(1/2) = c - 1/4 = \lim_{x \to (1/2)^-} f(x) =$$
$$\lim_{x \to (1/2)^-} x^2 = 1/4. \text{ Therefore } c = 1/2.$$

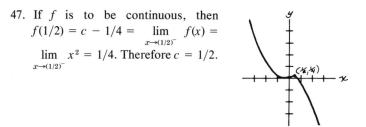

49. $\displaystyle \lim_{x \to 8} (\sqrt{x - 4} - 2)/(x - 8) = \lim_{x \to 8} (x - 4 - 4)/[(x - 8)(\sqrt{x - 4} + 2)]$
$$= \lim_{x \to 8} 1/(\sqrt{x - 4} + 2) = 1/4$$

51. $\displaystyle \lim_{x \to 2^-} \sqrt{4 - x^2}/\sqrt{2 - x} = \lim_{x \to 2^-} \sqrt{2 + x} = 2$

53. If $f(x) \to l$ as $x \to a^-$ and if $g(y)$ is continuous at $y = l$, then $g(f(x)) \to g(l)$ as $x \to a^-$.

55. $\displaystyle \lim_{x \to \infty} \sqrt{x - 1} - x = \lim_{x \to \infty} (\sqrt{x - 1} - x)[(\sqrt{x - 1} + x)/(\sqrt{x - 1} + x)]$
$$= \lim_{x \to \infty} (x - 1 - x^2)/(\sqrt{x - 1} + x)$$
$$= \lim_{x \to \infty} \frac{(1/x - 1/x^2 - 1)}{(\sqrt{x - 1}/x^2 + 1/x)} = -\infty$$

57. $\displaystyle \lim_{x \to -2^-} (x + 1)/|x + 2| = \lim_{x \to -2^-} (x + 1)/(-x - 2) = -\infty$

59. We must show that $\displaystyle \lim_{x \to \infty} \frac{r(x)}{(a_n/b_m)x^{n-m}} = 1$. If we divide the numerator and denominator of $\dfrac{r(x)}{(a_n/b_m)x^{n-m}}$ by x^{n-m} (which we may do for $x \neq 0$), we get

$$\frac{\left(\dfrac{a_n + a_{n-1}/x + \cdots + a_0/x^n}{b_m + b_{m-1}/x + \cdots + b_0/x^n} \right)}{a_n/b_m}.$$

As $x \to \infty$, this expression has limit $(a_n + 0)/(b_m + 0)/(a_n/b_m) = 1$, as we wished to prove.

61. $x^{-1/3} < 0.1 \Leftrightarrow$ (if and only if) $x^{1/3} > 10 \Leftrightarrow x > 10^3 = 1,000$. Similarly, $x^{-1/3} < 0.01 \Leftrightarrow x > (100)^3 = 10^6 = 1,000,000$, and $x^{-1/3} < 0.001 \Leftrightarrow x > 10^9$.

63. $x^{-1/3} < d$ if $x^{1/3} > 1/d$, $x > 1/d^3 = d^{-3}$.

65. $1/\sqrt{x}$ will be $> M$ if $\sqrt{x} < 1/M$, $x < 1/M^2$, so we let $d = 1/M^2$.

67. From Problem 63, we can take $M = d^{-3}$.

69. Domain: $x \geq 1/2$, range $y \geq 0$

CHAPTER 3

Section 1

1. $m = 2$

3. $m = -2$, $y = -2x - 1$

5. $m = -1$, $y = -x + 2$

7. $m = 1$, $y = x$

9. $m = 1 - 2a, y = (1 - 2a)x + a^2$. The tangent line is horizontal $(m = 0)$ when $a = 1/2$; its equation is then $y = 1/4$.

Section 2

1. $\dfrac{f(r) - f(x)}{r - x} = \dfrac{r^2 - x^2}{r - x}$

Use the factorization $r^2 - x^2 = (r - x)(r + x)$ to simplify. Then compute the limit.

3. $\dfrac{f(r) - f(x)}{r - x} = \dfrac{\dfrac{1}{r^2} - \dfrac{1}{x^2}}{r - x}$

Cross-multiply in the numerator and then use the factorization $r^2 - x^2 = (r - x)(r + x)$ to simplify. Then compute the limit.

5.   You will need to treat $r^{3/2} - t^{3/2}$ as a difference of cubes $(r^{1/2})^3 - (t^{1/2})^3$ and $r - x$ as a difference of squares $(r^{1/2})^2 - (t^{1/2})^2$, and use the formulas $b^3 - a^3 = (b - a)(b^2 + ab + a^2)$, $b^2 - a^2 = (b - a)(b + a)$. (Another method is to multiply the difference quotient top and bottom by $r^{3/2} + x^{3/2}$.)

7.   The difference quotient is
$$\frac{f(r) - f(s)}{r - s} = \frac{-3r - (-3s)}{r - s}.$$

9.   Your calculation should show that the difference quotient for the linear function $mx + b$ has the constant value m, regardless of the values of x and r.

11.   If $f(x) = x^a$, then $f'(x) = ax^{a-1}$.   13. $-\frac{1}{3}t^{-4/3}$

15.   Set $y = x^{1/4}$, $s = r^{1/4}$, and show that $(r^{1/4} - x^{1/4})/(r - x)$ simplifies to $1/(s^3 + s^2y + sy^2 + y^3)$ when $s - y$ is cancelled out. Now take the limit as $s \to y$ and then set $y = x^{1/4}$.

17. $(r^{2/3} - x^{2/3})/(r - x)$ simplifies to $(s + y)/(s^2 + sy + y^2)$, when $s - y$ is cancelled from top and bottom. Now take the limit as $s \to y$ and then set $y = x^{1/4}$.

19.   This should boil down to $2(a_1 + a_2)x + (b_1 + b_2) = (2a_1x + b_1) + (2a_2x + b_2)$.

Section 4

1. $7x^6 + \frac{1}{7}x^{-6/7}$ 3. $3x^2 + 6x$ 5. m 7. $16x^3 - 2x$

9. $2x - 1$ 11. $2x^{-1/2} + 4x - 3$ 13. $1 - x^{-2}$ 15. $1 - 2x^{-3}$

19. $dy/dx = 0$ at $x = 1$ 21. $y - 5 = \frac{1}{16}(x - \frac{1}{2})$

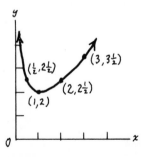

Section 5

1. $(x + 2)/\sqrt{x^2 + 4x}$ 3. $-(x + 2)^{-2}$ 5. $-2x/(x^2 + 1)^2$

7. $-x(1 - x^2)^{-1/2}$ 9. $\frac{8}{5}(2x + 1)^{-1/5}$ 11. $-x^2/(1 + x^3)^{4/3}$

13. $-\frac{1}{5}(x^{1/4} - 4x^3)^{-6/5}(\frac{1}{4}x^{-3/4} - 12x^2)$ 15. $-2(a - x)$

Section 6

1. $\dfrac{\Delta y}{\Delta x} = 2x + \Delta x$ 3. $\dfrac{\Delta y}{\Delta x} = \dfrac{1}{\sqrt{x + \Delta x} + \sqrt{x}}$

5. $\dfrac{\Delta y}{\Delta x} = 1 + 2x + \Delta x$ 7. $\dfrac{\Delta y}{\Delta x} = \dfrac{-2x - \Delta x}{[1 + (x + \Delta x)^2][1 + x^2]}$

9. $\Delta y = 3x^2\Delta x + 3x(\Delta x)^2 + (\Delta x)^3$, which can be thought of as a sum of seven terms. Your figure should show each of these seven terms as the volume of a rectangular solid.

Section 7

1. $v = 3t^2 - 4t + 1; t = \frac{1}{3}, 1$. Forward $(-\infty, \frac{1}{3})$, $(1, \infty)$; backward $(\frac{1}{3}, 1)$.

3. 144 ft   5. 3/4 sec; 16 ft/sec (up or down)

7. a) -11.73 b) 1.8   c) -10.73 9. -80 ft/sec; $t = 1\frac{1}{4}$ sec

11. Write $\Delta s = s - s_0$, $\Delta t = t - t_0$ in the increment equation, and solve for s.

Section 8

1. 5/2   3. $1/\sqrt{3}$ 5. $2/3^{1/3}$ 7. $\sqrt{6} - 1$

9. If $g(x) = x^3$, then $(f - g)'(x) = 0$ for all x. By Theorem 7, $(f - g)(x) = c \cdot f(x) = g(x) + c = x^3 + c$.

11. Set $g(x) = mx$ and follow the argument of Problem 9 ($b = c$).

Section 9

1. a) $4\pi r^2$unit^3/unit   b) $(36\pi)^{-1/3} V^{-2/3}$unit/unit^3 3. $-1\sqrt{15}$ ft/ft

5. $3r^2$cm^3/cm; volume must be increased by a factor of $2\sqrt{2}$.

7. $(75\pi/4)$ cubic inches per year; $(1875\pi/4)$ cubic inches per year.

9. $\dfrac{1}{l_0} \cdot \dfrac{dl_t}{dt} = \alpha + 2\beta t.$ Here β is small compared to α, so β can be ignored if t varies less than 20°. (See Example 2.)

Section 10

1. 0   3. $2/x^3$ 5. $6(x + 1/x^3)$ 7. $12x^{-5}$

9. 0   11. $-\frac{1}{4}x^{-3/2}$ 13. 0

15. $\dfrac{d^{m+1}}{dx^{m+1}} x^{m+1} = \dfrac{d^m}{dx^m}\left(\dfrac{d}{dx} x^{m+1}\right).$

Compute the inner derivative and then use the assumed formula. The formula holds for $n = 1$. Since it holds for $n = 1$ it holds for $n = 2$. Since it holds for $n = 2$ it holds for $n = 3$. So?

17. $f''(x) = \frac{3}{4}x^{-1/2}$ 19. $v_0 = 2$, $a_0 = -2$; $t = 1/3$

Extra Problems for Chapter 3

1. $f'(x) = \lim\limits_{r \to x} \dfrac{(r + 1) - (x + 1)}{r - x} = \lim\limits_{r \to x} 1 = 1$

3. $f'(x) = \lim\limits_{r \to x} \left(\dfrac{r^3/3 - x^3/3}{r - x} \right) = \lim\limits_{r \to x} \dfrac{(1/3)(r - x)(r^2 + rx + x^2)}{r - x}$

   $= \lim\limits_{r \to x} (1/3)(r^2 + rx + x^2) = x^2$

5. Difference quotient is $\dfrac{\sqrt{3r - 2} - \sqrt{3t - 2}}{r - t} = \dfrac{3}{(\sqrt{3r - 2} + \sqrt{3t - 2})}$;

   $g'(t) = \dfrac{3}{2\sqrt{3t - 2}}$

7. Difference quotient simplifies to $1 - \dfrac{1}{rx}$;

   $f'(x) = \lim\limits_{r \to x} (1 - 1/rx) = 1 - 1/x^2$.

9. Difference quotient simplifies to $\dfrac{-(s + r)}{(r^2 - 1)(s^2 - 1)}$, so $f'(s) = \dfrac{-2s}{(s^2 - 1)^2}$.

11. Difference quotient becomes $\dfrac{5}{(r + 1)(x + 1)}$; $f'(x) = \dfrac{5}{(x + 1)^2}$.

13. Difference quotient $= \dfrac{(r + 2)^{1/3} - (x + 2)^{1/3}}{r - x}$

    $= \dfrac{(r + 2)^{1/3} - (x + 2)^{1/3}}{(r + 2) - (x + 2)}$

    $= \dfrac{1}{(r + 2)^{2/3} + (r + 2)^{1/3}(x + 2)^{1/3} + (x + 2)^{2/3}}$;

    $f'(x) = \dfrac{1}{3(x + 2)^{2/3}} = (1/3)(x + 2)^{-2/3}$.

15. Difference quotient becomes $\dfrac{-1}{(1 + \sqrt{r})(1 + \sqrt{x})(\sqrt{x} + \sqrt{r})}$;

    $f'(x) = \dfrac{-1}{(1 + \sqrt{x})^2(2\sqrt{x})}$.

17. The tangent has slope 1 if $dy/dx = 3x^2 + 2x = 1$; this is true if $3x^2 + 2x - 1 = 0$; i.e., if $x = 1/3$ or $x = -1$.

19. Let (a, a^2) and (b, b^2) be the chord endpoints. Since they make equal slopes with $(0, 1/4)$ we have $a - 1/4a = b - 1/4b$. Thus $b - a = 1/4b - 1/4a = (a - b)/4ab$, so $4ab = -1$. Since the tangent slopes are $m_1 = 2a$ and $m_2 = 2b$, respectively, we have shown that $m_1 m_2 = -1$, and the tangents are perpendicular.

21. The tangent line at $(a, 1/a)$ has slope $f'(a) = -1/a^2$, so the equation of the line is $y - 1/a = -1/a^2(x - a)$. Its intercepts on the x and y axes are at $x = 2a$, $y = 2/a$, and the area of the triangle is $(1/2)bh = (1/2)(2a)(2/a) = 2$.

23. (Distance)^2 from (x_0, y_0) to $(0, 1/4) = x_0^2 - (y_0 - 1/4)^2 = y_0 + y_0^2 - 1/2\,y_0 + 1/16 = (y_0 + 1/4)^2 = $ (distance)^2 from $(0, -y_0)$ to $(0, 1/4)$.

25. The tangent line to $y = x^3$ at (x_0, x_0^3) has slope $3x_0^2$, and thus the equation of the line is $y - x_0^3 = 3x_0^2(x - x_0)$ or $y = 3x_0^2 x - 2x_0^3$. Substituting $x = -2x_0$ gives $y = (-2x_0)^3$, so the point $(-2x_0, -8x_0^3)$ lies on this line and on the graph of $y = x^3$.

27. As in Problem 26, the tangent line at (x_0, x_0^2) goes through (a, b) if and only if $2x_0 = (x_0^2 - b)/(x_0 - a)$; i.e., $x_0^2 - 2ax_0 + b = 0$. This equation has real solutions if and only if $(2a)^2 - 4b \geq 0$; i.e., if and only if $b \leq a^2$. Geometrically, this means that (a, b) must lie on or "outside" the parabola.

31. If we let $s = r^{1/n}$, $y = x^{1/n}$, then the difference quotient $\dfrac{r^{1/n} - x^{1/n}}{r - x}$ becomes

$$\frac{s - y}{s^n - y^n} = \frac{1}{s^{n-1} + s^{n-2} + \cdots + y^{n-1}}, \text{ so}$$

$$f'(x) = \lim_{s \to y} \frac{1}{s^{n-1} + s^{n-2}y + \cdots + y^{n-1}} = \frac{1}{ny^{n-1}} = \frac{1}{n(x^{1/n})^{n-1}} = \frac{1}{n} x^{(1/n - 1)}.$$

33. $(d/dx)(uv) = (d/dx)(x^m \cdot x^n) = (d/dx)(x^{m+n}) = (m + n)x^{m+n-1} = x^m(nx^{n-1}) + x^n(mx^{m-1}) = u \, dv/dx + v \, du/dx$.

35. Just use the fact that $|x| = x$ when $x > 0$ and $|x| = -x$ when $x < 0$. (Here is a tricky proof: $2x = (d/dx)(x^2) = (d/dx)(|x|^2) = 2|x|(d/dx)|x|$.)

37. Since $\dfrac{-r^2 + 0}{r - 0} \leq \dfrac{f(r) - 0}{r - 0} \leq \dfrac{r^2 - 0}{r - 0}$, and since now the left- and right-hand quantities approach 0 as $r \to 0$, we have $\dfrac{f(r) - 0}{r - 0} \to 0$, by the squeeze limit law (L5). But this limit is just $f'(0)$, by definition.

39. a) By Problem 11, Section 8, $v = ds/dt = -32t + b$ for some constant b. Let $g(t) = -16t^2 + bt$; then $\dfrac{d}{dt}(s(t) - g(t)) = v(t) - (-32t + b) = 0$. Then by Theorem 7, $s(t) = g(t) + c$ for some constant c. That is, $s = -16t^2 + bt + c$.   b) $s = -16t^2 + 50t$.

41. $X = a^{1/(1-a)}x$ 43. Use Theorem 7.

45. The normal line to the ellipse intersects the x-axis if and only if the tangent is not vertical; from Problem 44, this occurs if and only if $y_0 \neq 0$. If $y_0 \neq 0$, then the tangent has slope $-b^2 x_0/a^2 y_0$, and so the normal at (x_0, y_0) has slope $a^2 y_0/b^2 x_0$ (provided $x_0 \neq 0$). Then the equation of the normal is $y - y_0 = a^2 y_0/b^2 x_0(x - x_0)$. Setting $y = 0$, we find the x-intercept to be $x_0(a^2 - b^2)/a^2$. (This is correct also for the omitted case $x_0 = 0$.)

47. $f'(x) = \dfrac{d}{dx}(x + \sqrt{1 - x^2})^{1/2} = \dfrac{1}{2}(x + \sqrt{1 - x^2})^{-1/2} \dfrac{d}{dx}(x + \sqrt{1 - x^2})$

$= \dfrac{1}{2}(x + \sqrt{1 - x^2})^{-1/2}\left(1 + \dfrac{1}{2}(1 - x^2)^{-1/2}(-2x)\right) = \dfrac{1 - \dfrac{x}{\sqrt{1 - x^2}}}{2\sqrt{x + \sqrt{1 - x^2}}}$

49. $f'(x) = -(x\sqrt{x^2 - 1} + x^2 - 1)^{-1}$

51. $f'(x) = x^{a-1}(x^a + 1)^{1/a} - 1 = [x/f(x)]^{a-1}$

53. $f'(x) = -(x + g(x))^{-2}(1 + g'(x))$ 55. $f'(x) = \dfrac{-2(x + g(x)g'(x))}{(x^2 + (g(x))^2)^2}$

57. If $y = (c - x^2)^{1/3}$, $dy/dx = (-2x/3)(c - x^2)^{-2/3}$, and $3y^2 \dfrac{dy}{dx} = -2x$.

59. $h = 3V/\pi r^2$, $dh/dr = -6V/\pi r^3$

61. a) Since $m(0) = 10$, $c = 10$ if m has a minimum at $x_0 = 1000$, then 1000 is a critical value; i.e., $m'(1000) = 2a(1000) + b = 0$, $b = -2000a$. Since $m(1000) = 8$, $a(1000)^2 - 2000a(1000) + 10 = 8$, $-10^6 a = -2$, $a = 2 \cdot 10^{-6}$, $b = -4 \cdot 10^{-3}$, and $m(x) = (2 \cdot 10^{-6})x^2 - (4 \cdot 10^{-3})x + 10$.   b) $C(x) = (a/3)x^3 + bx^2/2 + cx + 500$.

CHAPTER 4

Section 1

1. $\frac{1}{7}x^7 + c$

3. $\frac{1}{2}x^2 + c$

5. $\frac{3}{4}x^{4/3} + c$

7. $\frac{3}{4}x^4 + \frac{1}{6}x^{-2} + c$

9. $\frac{2}{3}x^3 - \frac{1}{2}x^2 - 6x + c$

11. $2\sqrt{x} + c$

13. $ay^3 + c$

15. $2x^2 - 4\sqrt{x} + c$

17. $\frac{1}{3}(x^3 - 5)$

19. $\frac{1}{2}(x^2 - 1)$

Section 2

1. $f(x) = x^2/2 - 3x + 13$

3. $f(x) = (1/4)y^4 - (b^2/2)y^2 + 2b^2 - 4$

5. $f(t) = (2/3)t^{3/2} + 2t^{1/2} - 28/3$

7. $8p^2/3$

9. 80 ft/sec

11. $126\frac{2}{3}$ ft

Section 3

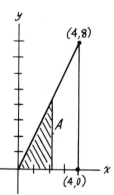

1. $\int 2x\, dx = x^2 + c$. When $x = 0$, $A = 0$, so $c = 0$; $A = 4^2 = 16$.

3. 64

5. 2

7. $15\frac{3}{4}$

9. 2

13. 16/3

15. 28/3

17. 8/3

Section 4

1. 1/6

3. $-4/15$

5. $y^3 - x^3$

7. $-1/2$

9. 64

11. 44/3

13. 63/4

15. 2

Section 5

1. The graph being rotated is $y = f(x) = \sqrt{r^2 - x^2}$, so $\pi f^2(x) = \pi(r^2 - x^2)$.

3. $\pi/7$

5. $\pi(6^4/5)$

7. 3π

9. $8/5\pi$

Section 6

1. 8/3

3. 18

5. 8

7. 32/3

9. 8

11. 3/4

13. 1/12

15. $2\sqrt{2}/3$

17. 8 (the sum of two pieces each having area = 4)

Section 7

1. $\frac{2}{3}\pi a^3$

3. $1024\pi/7 = 2^{10}\pi/7$

5. $8\pi/15$

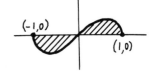

7. a) $-3x(4 - x^2)^{1/2}$ b) $28\pi\sqrt{21}$

Section 8

1. a) Show that $\rho(x) \geq 0$ at the critical points and endpoints.   b) $7/12$

3. a) $+1/6$ b) 0

5. $\sqrt{2}$

7. $\displaystyle\int_{-\infty}^{0} \rho(x)\,dx = \int_{0}^{\infty} \frac{2x\,dx}{(1 + x^2)^2} = \lim_{n\to\infty}\left[\frac{-1}{1 + x^2}\right]_{0}^{N} = \lim_{n\to\infty}\left(1 - \frac{1}{1 + N^2}\right) = 1$

Extra Problems for Chapter 4

1. $3t^3 + \frac{25}{2}t^2 + 14t + c$

3. $3\sqrt[3]{x} + c$

5. $(2/3)(x + 1)^{3/2} + c$

7. $(1/3)(2x + 1)^{3/2} + c$

9. $\dfrac{dV}{dP} = -\dfrac{2}{P^2}$, $V = 100$ at $P = \dfrac{1}{100}$, so $V = \displaystyle\int -2P^{-2}\,dP = 2P^{-1} + c$ and $100 = 200 + c \Rightarrow c = -100$. Thus $V = 2P^{-1} - 100$.

11. 2

13. The top line has the equation

$$y = \frac{b_2 - b_1}{h}x + b_1;$$

$$A = \frac{b_1 + b_2}{2}h.$$

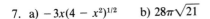

15. $V = \displaystyle\int_{0}^{1} \pi(x^3)^2\,dx = \pi x^7/7\Big]_{0}^{1} = \pi/7$ 17. $(1296/5)\pi$

19. 3π 21. $(8/5)\pi$ 23. $\dfrac{2\pi}{9}(2\sqrt{2} - 1)$ 25. $+\infty$

27. $(d/dx)\displaystyle\int_{a}^{x} f(t)\,dt = (d/dx)(F(x) - F(a))$, where F is an antiderivative of f.

Hence $(d/dx)\displaystyle\int_{a}^{x} f(t)\,dt = (d/dx)F(x) = f(x)$.

29. $f(x) = \int_0^{x^2} t^2\, dt = t^3/3 \Big]_0^{x^2} = x^6/3$, hence $f'(x) = 2x^5$.

31. $1/2x^2$ 37. $\pi h^2(R - (h/3))$ 39. $(\pi/4) - (1/3)$

41. If x and y both lie in the interval $[a,b]$, then $|x - y| \le b - a$. This is intuitively obvious and can be proved easily from the inequality laws. In the problem both ΔQ and $\rho(x)\Delta x$ lie between $l\ \Delta x$ and $u\ \Delta x$, so $|\Delta Q - \rho(x)\Delta x| \le (u - l)\Delta x$. That is, $\Delta Q \approx \rho(x)\Delta x$ with an error at most $(u - l)\ \Delta x$ in magnitude.

CHAPTER 5

Section 1

1. $(x - 1)^2(x + 2)^3(7x + 2)$

3. $3t^5(7t + 12)$

5. $5x^4 + 36x^3 + 33x^2 - 70x + 13$

7. $-4(1 + 2v)^3/v^5$

9. $(x + 1)^3(3x^2 - 1)^{-3/2}(9x^2 - 3x - 4)$

11. 1

13. $y\left[\dfrac{x - 4 - 1/2x^2}{x^2 - 8x + 1/x} + \dfrac{63x^{1/2}/8 - 9x^2}{7x^{3/2} - 4x^3 + 1}\right]$

15. $1/(1 + x)^2$

17. $4x/(x^2 + 1)^2$

19. $2(1 - 3t^2)/(1 + t^2)^3$

21. $(v^2 + 2v - 1)/(1 + v)^2$

23. $(1 + x)^{-1/2}(1 - x)^{3/2}$

25. $[xf'(x) - f(x)]/x^2$

27. $y\left[\dfrac{1}{x} + \dfrac{4}{(x - x^2)} + \dfrac{5}{1 - x} - \dfrac{6}{3x - 4}\right]$

Section 2

1. xe^x

3. x^2e^x

5. $e^x/2\sqrt{e^x - 1}$

7. $(\tfrac{x}{2}e^{x/2})/\sqrt{x - 1}$

9. ae^{ax}

11. $\tfrac{1}{3}(3e^{3x} + 1)(e^{3x} + x)^{-2/3}$

13. $e^x - e^{-x}$

15. $e^{-x}(1 - x - x^2 - x^3)/(1 + x^2)^2$

17. $1, -2$

19. $1, -3$

21. $\dfrac{dy}{dx} = (-2x + 1)e^{-2x};\qquad \dfrac{d^2y}{dx^2} = (4x - 4)e^{-2x}$

23. Show that $(d/dx)(ye^{-kx}) = 0$, by the product law and the assumed equation.

Section 3

1. $\cos t = 0,\ \sin t = -1$

3. $\cos t = \dfrac{\sqrt{2}}{2};\ \sin t = \dfrac{-\sqrt{2}}{2}$

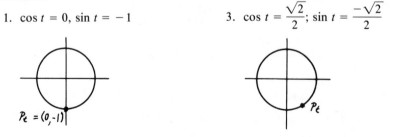

$P_t = (0, -1)$

5. Your estimates should be close to $\cos t = -0.415$, $\sin t = 0.91$, which come from tables.

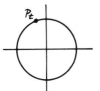

7. $\cos t = -1$; $\sin t = 0$

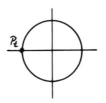

9. The correct values, to two decimal places, are $\cos t = -0.65$, $\sin t = -0.75$. Your estimates should be close.

11. $\cos t = \dfrac{\sqrt{2}}{2}$; $\sin t = \dfrac{\sqrt{2}}{2}$

13. $\frac{5}{4}\pi$

15. $\frac{2}{3}\pi$

17. About halfway between $\pi/6$ and $\pi/4$.

19. $\approx \dfrac{\pi}{2} + \dfrac{\pi}{7}$

21. $\sin(\pi + t) = -\sin(\pi - t)$. What is the corresponding cosine identity?

23. The 30°–60°–90° triangle is half an equilateral triangle. This gives $\sin \pi/6 = 1/2$, and then $\cos \pi/6$ is computed by the Pythagorean Theorem.

25. $\cos \theta = 1/\sqrt{5}$; $\sin \theta = 2/\sqrt{5}$

27. $\pi/4$

29. $\frac{3}{4}\pi$

31. If we guess the angle whose tangent is $1/2$ to be about $\pi/7$, then the answer here is $\approx 6\pi/7$.

33. $13\pi/7$, on the basis of the guess in Problem 31.

39. $\sin(x - y) = \cos((x - y) - \pi/2)$ (by 36)
$= \cos((x - \pi/2) - y) = \cdots$

41. $\sin(x + y) = \sin(x - (-y)) = \cdots$ (using Problem 39, etc.)

43. Divide the answer to Problem 51 by the answer to Problem 42, and then divide top and bottom by $\cos x \cos y$.

45. $\sin 2x = \sin(x + x) = \cdots$

47. $\tan 2x = \tan(x + x) = \cdots$

49. Set $y = 2x$ in one of the equations in Problem 46.

51. $\sin(x + y) - \sin(x - y) = 2 \sin y \cos x$

53. $A = \sqrt{a^2 + b^2}$; θ is the angle from the x-axis to the ray through (a,b).

Section 4

1. $\dfrac{d}{dx} \sec x = \dfrac{d}{dx} \left(\dfrac{1}{\cos x} \right) = \cdots$

3. $\dfrac{d}{dx} \cos x = \dfrac{d}{dx} \left(\dfrac{1}{\sin x} \right) = \cdots$

5. $e^x(\cos x - \sin x)$

7. $\sin 2x$

9. 0   **11.** $1/(1 + \cos x) = \csc x \tan \frac{x}{2} = \frac{1}{2} \sec^2 \frac{x}{2}$

13. $2 \tan x \sec^2 x$

15. $-\sin x/2\sqrt{1 + \cos x}$

17. $3 \sin^2 x \cos^2 x(\cos^2 x - \sin^2 x)$

19. 0   **21.** $e^x \cos x$

23. $\cos^2 x$

25. $(1 + x \cos x - \sin x)/(x + \cos x)^2$

Section 5

1. $f(x) = ax + b;\ \dfrac{d}{dx} \sin (f(x)) = \cos(f(x))f'(x) = a \cos(ax + b)$

5. $(\cos \sqrt{x})/2\sqrt{x}$ **7.** $-e^x \sin(e^x)$ **9.** $e^{(e^x + x)}$

11. $(3x^2 - 1) \cos (x^3 - x)$

13. If $e^{f(x)} = x$, then $\dfrac{d}{dx} e^{f(x)} = \dfrac{d}{dx} x = 1$. But $\dfrac{d}{dx} e^{f(x)} = ?$

15. Proceed as in Problems 13 and 14.

17. Let f be even, and set $h(x) = -x$. Then

$$\frac{d}{dx} f(x) = \frac{d}{dx} f(-x) = \frac{d}{dx} f(h(x))$$
$$= f'(h(x))h'(x) = f'(-x)(-1).$$

So $f'(x) = -f'(-x)$: f' is odd. Work out what happens when f is odd in a similar way.

21. $-xe^{\sqrt{1-x^2}}/\sqrt{1 - x^2}$

23. $(-\cos(1/x))/x^2$ $(x \neq 0)$.

25. $2x \sin(1/x) - \cos(1/x)$ $(x \neq 0)$

27. $(1 - 2x^2 - 2x)e^{-x^2}$

29. $-[\sec^2(\cos x)] \sin x$

31. $-[\cos(\cos(\sin x))][\sin(\sin x)]\cos x$.

Section 6

1. $-y/x = 5/2x^2$

3. $-x/y = -x/\sqrt{16 - x^2}$

5. $(-3x - 1)/y = (-3x - 1)/\sqrt{10 - 2x - 3x^2}$

7. $2x/y(x^2 + 1)^2 = 2x/(x^2 - 1)^{1/2}(x^2 + 1)^{3/2}$

9. $-\sqrt{x/y} = -x^{1/2}/(2 - x^{3/2})^{1/3}$

11. $(y^2 - 2xy - 2x)/(x^2 - 2xy + 2y)$

13. $-\csc y = -1/\sqrt{1 - x^2}$

15. $y/(\cos y - x)$

17. $\cos x/(3y^2 + 1)$

19. $y/(2 \sec^2 y \tan y - x)$

21. $-25/16y^3$

23. $y = (-1/2)x + 3/2$

Section 7

1. $3x^2\, dx + 2dx$

3. $2(2x + 1)^2(5x^2 + x + 6)dx$

5. $2xe^{x^2}\, dx$

7. $[(x^4 + 7x^2 - 2x + 6)/(x^2 + 3)^2]dx$

9. $(-1/x^2 + 2)dx$

11. $dy = 2 \times dx;\ dy/dx = 2$

13. $dy/dx = x(x + 2)^{-1/3} \times (x - 1)^{-2/3}$

15. $e^y \, dy = 2x \, dx$. Solve for dy/dx and manipulate, to get $dy/dx = 2/x$. Similarly, $dx/dy = \frac{1}{2}e^{y/2}$.

17. $(3y^2 + x)dy + (y - 5x^4)dx = 0$. Solve for dy/dx and for dx/dy.

Extra Problems for Chapter 5

1. $-1/(z - 1)(z^2 - 1)^{1/2}$

3. $-(1 - x)^{-1/3}(1 + x)^{-2/3}(x + 1/3)$

5. First apply the product law to w times (uv), etc.

7. $3x^2 - 7$

9. $dy/dx = -24x(x^2 - 1)/(x^2 + 1)^4$

11. $dy/dx = -2n(1 - x)^{n-1}/(1 + x)^{n+1}$

13. $dy/dx = a(f(x))^{a-1}f'(x)/(1 + f(x))^{a+1}$

15. $dy/dx = (-3x^2 + 6x + 7)/(1 - x)^2$

17. $dy/dx = -(3x^3 - 1)/(1 + 3x)^2(1 + x^2)^{3/2}$

19. If $m = 1$, then $(f^m)' = f' = 1 \cdot f^0 f'$. By Problem 17, the formula holds for $m = 2$, hence for $m = 3$, and so on. Since we can reach any positive integer p after p such steps, the formula holds for any p.

21. $0 = 1' = (f \cdot 1/f)' = f(1/f)' + (1/f)f'$; solve for $(1/f)'$.

23. $dy/dx = -x(1 - x^2)^{-1/2}$; $d^2y/dx^2 = -(1 - x^2)^{-1/2} - x[-(1/2)(1 - x^2)]^{-3/2}$
$(-2x) = -(1 - x^2)^{-3/2}(1 - x^2 + x^2) = (1 - x^2)^{-3/2}$

25. Let $f(x) = a_m x^m + \cdots + a_0$, $g(x) = b_n x^n + \cdots + b_0$. Then $r = f/g$, so that $r' = (gf' - fg')/(g)^2$. Both gf' and fg' have degree $m + n - 1$, so degree $(gf' - fg') \leq m + n - 1$. But the leading coefficient of $gf' - fg'$ is $a_m b_n (m - n)$, which is not zero. Thus degree $(gf' - fg') = m + n - 1$. Since degree $(g^2) = 2n$, degree $(f/g)' = m + n - 1 - 2n = (m - n) - 1 = p - 1$.

27. $(f/g)' = (fg^{-1})' = f(g^{-1})' + g^{-1}f' = f \cdot (-1(g)^{-2}g') + g^{-1}f' = (gf' - f'g)/g^2$

29. From Problem 6, Sec. 5–1, $\Delta y = u \cdot \Delta v + v \cdot \Delta u + \Delta u \cdot \Delta v$ and so $\Delta y/y = \Delta y/uv = \Delta v/v + \Delta u/u + \Delta u \cdot \Delta v/uv$

31. $\displaystyle\lim_{\Delta x \to 0} \frac{\Delta y}{\Delta x} = \lim_{\Delta x \to 0} \left(u\frac{\Delta v}{\Delta x} + v\frac{\Delta u}{\Delta x} + \Delta u \cdot \frac{\Delta v}{\Delta x} \right)$

$\displaystyle = \lim_{\Delta x \to 0} u\frac{\Delta v}{\Delta x} + \lim_{\Delta x \to 0} v\frac{\Delta u}{\Delta x} + \lim_{\Delta x \to 0} \Delta u \cdot \frac{\Delta v}{\Delta x}$ (by L3a)

$\displaystyle = u\lim_{\Delta x \to 0} \frac{\Delta v}{\Delta x} + v\lim_{\Delta x \to 0} \frac{\Delta u}{\Delta x} + \lim_{\Delta x \to 0} \Delta u \cdot \lim_{\Delta x \to 0} \frac{\Delta v}{\Delta x}$ (by L3b).

Since u is continuous by Theorem 5 of Chapter 3, $\displaystyle\lim_{\Delta x \to 0} \Delta u = 0$ by L1, so $dy/dx = $

$\displaystyle\lim_{\Delta x \to 0} \frac{\Delta y}{\Delta x} = u\frac{dv}{dx} + v\frac{du}{dx}.$

33. $f'(x) = (x - a)g'(x) + g(x)$, so $f'(a) = g(a)$. If g is only known to be continuous at a, it is still true that $f'(a)$ exists and equals $g(a)$, for the difference quotient $(f(a + \Delta x) - f(a))/\Delta x = g(a + \Delta x)$ has the limit $g(a)$ as $\Delta x \to 0$.

35. Let $g(t) = f(t)/t^n$. Then $g'(t) = (t^n f'(t) - nf(t)t^{n-1})/t^{2n} = (tf'(t) - nf(t))t^{n-1}/t^{2n} = (nf(t) - nf(t))t^{n-1}/t^{2n} = 0$. Therefore $g(t)$ is a constant function, $g(t) = c$, so $f(t) = ct^n$.

37. $dy/dx = 1/2 \left(\dfrac{x}{e^x + x} \right)^{-1/2} \left(\dfrac{(e^x + x) - x(e^x + 1)}{(e^x + x)^2} \right)$

39. $dy/dx = -1/2 \; e^{-x/2}$

41. $[f(x + h) - f(h)]/h = (a^{x+h} - a^x)/h = a^x(a^h - 1)/h$, so $f'(x) = \lim\limits_{h \to 0} a^x(a^h - 1)/h = a^x \lim\limits_{h \to 0} (a^h - a^0)/(h - 0) = a^x f'(0)$.

43. Let $y = f(x) = 2^x$. Then $f(-1) = 1/2$, $f(0) = 1$, and $f(1) = 2$. By the mean value property, there are numbers c and d with $-1 < c < 0 < d < 1$ such that $f'(c) = [f(0) - f(-1)]/(0 - (-1)) = 1/2$ and $f'(d) = [f(1) - f(0)]/(1 - 0) = 1$. Then, since $c < 0 < d$ and since f' is an increasing function of x, $f'(c) = 1/2 < f'(0) < f'(d) = 1$, and $1/2 < \alpha < 1$.

45. We know that $y = ce^x$ for some constant c. Since $f(1) = ce^1 = 1$, $c = e^{-1}$. Then $y = e^{-1}e^x = e^{x-1}$.

47. $dx/dt = 1 + e^{-t}$, acceleration $= d^2x/dt^2 = -e^{-t} = -(1 - x) = -(\text{distance from 1})$.

49. $\sin(x + \pi/2) = \sin x \cos \pi/2 + \cos x \sin \pi/2 = 0 \cdot \sin x + 1 \cdot \cos x = \cos x$

51. $\cos(x + \pi/2) = \cos x \cos \pi/2 - \sin x \sin \pi/2 = -\sin x$

53. $\sin(x + \pi) = \sin x \cos \pi + \cos x \sin \pi = -\sin x$

55. $\sin(x + \pi/4) = \sin x \cos \pi/4 + \cos x \sin \pi/4 = \sqrt{2}/2 \, (\sin x + \cos x)$

57. $\tan(x + \pi) = \sin(x + \pi)/\cos(x + \pi) = -\sin x/-\cos x = \tan x$

59. $f'(x) = Ag'(x) + Bh'(x)$, $f''(x) = Ag''(x) + Bh''(x)$. If $y = f(x)$, then $d^2y/dx^2 = f''(x) = Ag''(x) + Bh''(x) = -Ag(x) - Bh(x) = -f(x) = -y$.

61. Since $\cos x$ and $\sin x$ are solutions of (I), so is $\phi(x) = f(x) - f(0) \cos x - f'(0) \sin x$, by Problem 59. But $\phi(0) = f(0) - f(0) \cos 0 - f'(0) \sin 0 = f(0) - f(0) = 0$, and $\phi'(0) = f'(0) + f(0) \sin 0 - f'(0) \cos 0 = f'(0) - f'(0) = 0$. By Problem 60, we conclude that ϕ is identically zero, hence $f(x) = f(0) \cos x + f'(0) \sin x = A \cos x + B \sin x$, where $A = f(0)$, $B = f'(0)$.

63. $du/dx = e^{-x}(-y + dy/dx)$, $d^2u/dx^2 = e^{-x}(y - dy/dx - dy/dx + d^2y/dx^2) = e^{-x}(-y)$, since y is a solution. Then $d^2u/dx^2 + u = -e^{-x}y + e^{-x}y = 0$. Conversely, suppose $y = e^x u$, where $d^2u/dx^2 + u = 0$. Then a similar calculation shows that $d^2y/dx^2 - 2 \, dy/dx + 2y = e^x(d^2u/dx^2 + u) = 0$. And since the most general solution of $d^2u/dx + u = 0$ is $u = A \cos x + B \sin x$, the most general solution of the equation in Problem 62 is $y = e^x u = e^x(A \cos x + B \sin x)$.

65. a) $f'(x) = A(e^x \sin x + e^x \cos x) + B(e^x \cos x - e^x \sin x) = (A - B)e^x \sin x + (A+B)e^x \cos x$

    b) $f'(x)$ is of the form $A'e^x \sin x + B'e^x \cos x$, where $A' = A - B$, $B' = A + B$ are constants. Then the computation in (a) shows that $f''(x) = (A' - B')e^x \sin x + (A' + B')e^x \cos x = -2Be^x \sin x + 2Ae^x \cos x$. Similarly, $f'''(x) = -2(A + B)e^x \sin x + 2(A - B)e^x \cos x$; $f^{(4)}(x) = -4Ae^x \sin x - 4Be^x \cos x$; $f^{(5)}(x) = -4(A - B)e^x \sin x - 4(A + B)e^x \cos x$.

67. $1/(1 + \cos x)$

69. $(d/dx)(1 - \cos^2 x)^{1/2} = (1/2)(1 - \cos^2 x)^{-1/2} (2 \cos x \sin x)$

$$= \cos x \sin x(1 - \cos^2 x)^{-1/2}$$

$$= \cos x \sin x/|\sin x|$$

$$= \begin{cases} \cos x \text{ when } \sin x > 0 \\ -\cos x \text{ when } \sin x < 0. \end{cases}$$

71. $-4 \cos 2x/(1 + \sin 2x)^2$ 73. $\tan^4 x$
(using Problem 24 in Section 3)

75. $v = 3(2 \cos 2t) = 6 \cos 2t$ by Problem 24 in Section 3, $a = 6(-2 \sin 2t) = -12 \sin 2t$ by Problem 25 in Section 3. Hence $a = -4x$, and a is proportional to x but opposite in direction.

77. Since $\sin t = \sqrt{1 - \cos^2 t}$, $0 < \sqrt{1 - \cos^2 t} < t$, $0 < 1 - \cos^2 t < t^2$, and $1 - t^2 < \cos^2 t < 1$. Therefore $\cos^2 t \to 1$ as t approaches 0 through positive values. Since $\cos t > 0$ for small t, and since $\cos(-t) = \cos t$, we conclude that $\lim_{t \to 0} \cos t = \lim \sqrt{\cos^2 t}$ 1.

79. $(\sin r - \sin x)/(r - x) = [\sin((r - x)/2)/(r - x)/2] \cos ((r + x)/2) \to \cos x$ as $r \to x$

81. $(d/dx) \sinh x = (d/dx)((e^x - e^{-x})/2) = (e^x + e^{-x})/2 = \cosh x$

83. $(d/dx)\tanh x = (d/dx)(\sinh x/\cosh x) = (d/dx)[(e^x - e^{-x})/(e^x + e^{-x})] = 4/(e^x + e^{-x})^2 = \text{sech}^2 x$

85. $\cosh(-x) = [e^{(-x)} + e^{-(-x)}]/2 = (e^{-x} + e^x)/2 = \cosh x$

87. $\sinh x \cosh y + \cosh x \sinh y = (e^x - e^{-x})(e^y + e^{-y})/4 + (e^x + e^{-x})(e^y - e^{-y})/4 = (e^{x+y} - e^{-x-y})/2 = \sinh(x + y)$

89. $(1 - 1/x)e^{1/x}$ 91. $-b/2x^2\sqrt{a + b/x}$ 93. $\sin(2\sqrt{t})/4\sqrt{t}(1 + \sin^2(\sqrt{t}))$

95. $2x(\sin(x^2) + \cos(x^2) + 1)/(1 + \cos(x^2))^2$.

97. $2xf'(x^2)$ 99. $2f'(x)e^{2f(x)}$ 101. $f'(e^{g(x)})e^{g(x)}g'(x)$

103. Show that $f'(1/n\pi) = (-1)^{n+1}$, using the answer to Problem 25 in Section 5.

105. Slope at $(x_0,y_0) = m = (-x_0b^2)/(y_0a^2)$. Solve for k in $y_0 = mx_0 + k$, and recall that $(x_0^2/a^2) + (y_0^2/b^2) = 1$ to reduce $y = mx + k$ to
$$(xx_0/a^2) + (yy_0/b^2) = 1.$$

107. $\dfrac{dy}{dx} = \dfrac{e^{x+y}}{(1 - e^{x+y})} = \dfrac{y}{(1 - y)}; \dfrac{d^2y}{dx^2} = -\dfrac{y}{(1 - y)^3}$

109. $\dfrac{dy}{dx} = \pm 1; \dfrac{d^2y}{dx^2} = 0$

CHAPTER 6

Section 1

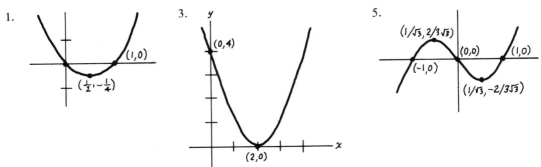

1.   (1,0) $(\frac{1}{2}, -\frac{1}{4})$

3. y (0,4)   (2,0)

5. $(1/\sqrt{3}, 2/3\sqrt{3})$ (0,0)   (1,0)   (-1,0) $(1/\sqrt{3}, -2/3\sqrt{3})$

7.

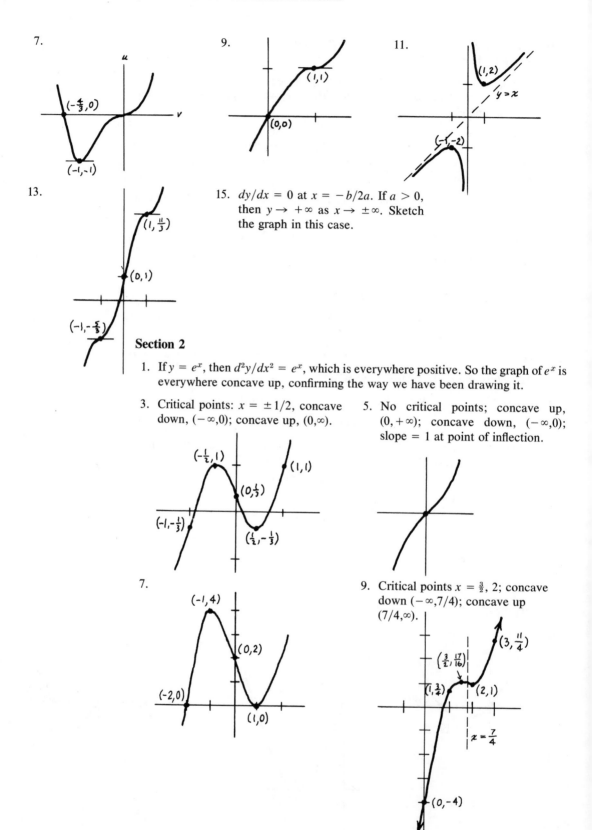

9.

11.

13.

15. $dy/dx = 0$ at $x = -b/2a$. If $a > 0$, then $y \to +\infty$ as $x \to \pm\infty$. Sketch the graph in this case.

Section 2

1. If $y = e^x$, then $d^2y/dx^2 = e^x$, which is everywhere positive. So the graph of e^x is everywhere concave up, confirming the way we have been drawing it.

3. Critical points: $x = \pm 1/2$, concave down, $(-\infty, 0)$; concave up, $(0, \infty)$.

5. No critical points; concave up, $(0, +\infty)$; concave down, $(-\infty, 0)$; slope $= 1$ at point of inflection.

7.

9. Critical points $x = \frac{3}{2}$, 2; concave down $(-\infty, 7/4)$; concave up $(7/4, \infty)$.

11. Critical points $x = -1, 2$; concave up $(-\infty, 0)$, $(2, \infty)$; concave down $(0, 2)$.

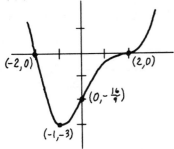

$(-2, 0)$ $(2, 0)$

$(0, -\frac{16}{9})$

$(-1, -3)$

15. Slope at $0 = 1$; slope at $1 = -2$.

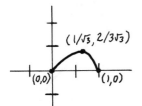

$(1/\sqrt{3}, 2/3\sqrt{3})$

$(0, 0)$ $(1, 0)$

17. Critical points $x = \pi/4, \pi, 7\pi/4$; concave down $(0, 7\pi/12)$, $(\pi, 17\pi/12)$, approx.; concave up $(7\pi/12, \pi)$, $(17\pi/12, 2\pi)$, approx.

13. Critical points $x = 2, 6$; concave down, $(-\infty, 4)$; concave up, $(4, \infty)$.

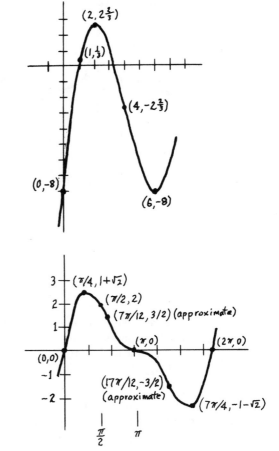

$(2, 2\frac{2}{3})$

$(1, \frac{1}{3})$

$(4, -2\frac{2}{3})$

$(0, -8)$

$(6, -8)$

3 $(\pi/4, 1+\sqrt{2})$

2 $(\pi/2, 2)$

$(7\pi/12, 3/2)$ (approximate)

1

$(\pi, 0)$ $(2\pi, 0)$

$(0, 0)$

-1

$(17\pi/12, -3/2)$

-2 (approximate)

$(7\pi/4, -1-\sqrt{2})$

$\frac{\pi}{2}$ π

Section 3

1.

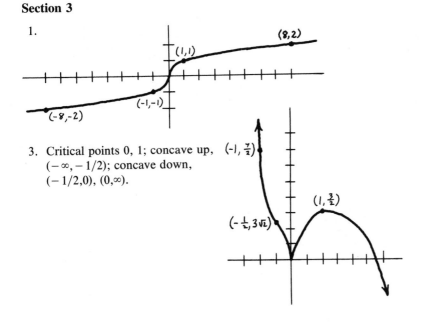

$(8, 2)$

$(1, 1)$

$(-1, -1)$

$(-8, -2)$

3. Critical points $0, 1$; concave up, $(-\infty, -1/2)$; concave down, $(-1/2, 0)$, $(0, \infty)$.

$(-1, \frac{7}{2})$

$(1, \frac{3}{2})$

$(-\frac{1}{2}, 3\sqrt{2})$

5. Singular points 0, 1; critical point 1/2; concave up $(-\infty,0)$, $(1,\infty)$; concave down $(0,1)$.

7. Critical points 0, $(2n + 1)\pi/2$; concave up, $((2n + 1)\pi, (2n + 2)\pi)$, $n \geq 0$; $(2n\pi, (2n + 1)\pi)$, $n < 0$; concave down, $(2n\pi,(2n + 1)\pi)$, $n \geq 0$; $((2n + 1)\pi, (2n + 2)\pi)$, $n < 0$.

9. Critical points -1, 1; concave up, $(-\infty, -1)$, $(-1,0)$, concave down, $(0,1)$, $(1, +\infty)$.

11. Critical point 0; concave up $(-\infty,4)$, $(4,\infty)$; concave down $(-4,4)$. Asymptotic to the line $x = -4$, $x = +4$, $y = 2$.

13. $x = 0$ is the critical point; $x = \pm 2/\sqrt{3}$ are the inflection points. The graph is concave down on $(-\infty, -2/\sqrt{3}]$ and $[2/\sqrt{3}, +\infty)$; concave up on $[-2/\sqrt{3},2/\sqrt{3}]$.

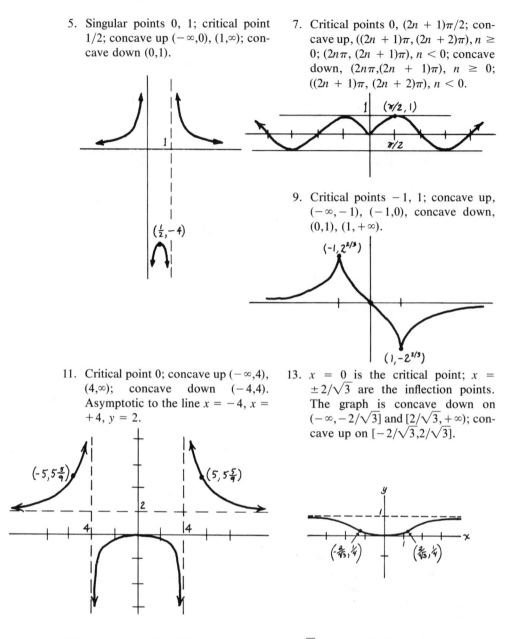

15. $x = \pm 1$ are the critical points; $x = 0$, $\pm\sqrt{3}$ are the inflection points. $y \to 0$ as $x \to \infty$. The graph is concave down on $(-\infty, -\sqrt{3}]$ and $[0,\sqrt{3}]$; concave up on $[-\sqrt{3},0)$ and $[\sqrt{3}, +\infty)$.

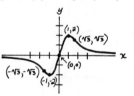

Section 4

1. $3\frac{1}{4}$, 1   3. $22\frac{22}{27}$, -28 5. 0, -6 7. $3\sqrt{3}/4$, 0

9. 3, $-3/8$ 11. $4/e^2$, 0   13. $5/4$, -1 15. max $= 3\frac{1}{4}$

17. max $= 2\frac{5}{27}$ 19. min $= -3/8$ 21. max $= 4/e^2$

Section 5

1. $\sqrt{2}$ 3. $1/2\sqrt{2}$ 5. $(2, -22)$ 7. -4

9. $r = \frac{2}{3}R$, $h = \frac{1}{3}H$

11. 6×8 yards (8 yard length borders the neighbor's lot)   13. $\sqrt{2A} \times \sqrt{A/2}$

15. Square piece $(96/(4 + \pi))$ in.; circle piece $24\pi/(4 + \pi)$

17. $\sqrt{3}$ 19. Width: depth $= 1: \sqrt{2}$

Section 6

1. By Theorem 4, $f'(x)$ changes sign from $-$ to $+$ as x crosses x_0. Thus f' is negative on a small interval (a, x_0), and positive on a small interval (y_0, b). Therefore, . . .

3. By Problem 1, f' has a relative minimum at x_0. Thus there is a small interval (a, b) about x_0 on which f' is positive, except at x_0 itself. So, . . .

6.

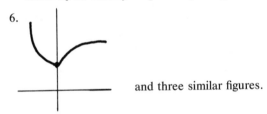

and three similar figures.

7. Relative minimum at $x = 2$; maximum at $x = 0$.

9. $x = -1/3$, relative maximum; $x = 1$, relative minimum.

11. $x = -2$, relative maximum; $x = 2$, relative minimum.

13. $x = -3$, relative maximum; $x = -9/5$, relative minimum; $x = 0$, case of Problem 3.

15. Relative minimum at $x = 1$; decreasing through the critical point $x = 0$.

Section 7

1. $-3/4$ 3. $25/12$ ft/sec   5. $9/5$ ft/sec

7. 65 mi/hr, 65 mi/hr   9. $20\sqrt{2}$ ft/min   11. 0 ft/sec

13. 10π cu.in./sec

Section 8

1. $\log 2/0.01 \approx 70$ hrs   3. $\log 2/24 \approx 2.9\%$ 5. 6.6 yrs

7. $y_0 = ce^{kt_0}$, $y_1 = ce^{kt_1}$; $\log y_0 = \log c + kt_0$; $\log y_1 = \log c + kt_1$. Subtracting: $k(t_1 - t_0)$ $= \log y_1 - \log y_0 = \log(y_1/y_0)$; $k = (\log(y_1/y_0))/(t_1 - t_0)$.

11. The temperature will reach $170°$ at time $t = 5 \log 1.3/\log(13/12) \approx 16.5$ min.

Section 9

1. $x = 2 \sin(3t - \pi/3)$

3. $x = 5\sqrt{2} \sin(t + 3\pi/4)$

5. $x = a \sin(2\pi f t + \alpha)$, so $dx/dt = $?

7. $x = \sqrt{53/2} \sin(\sqrt{2/5}\, t) + \arctan(2\sqrt{2}/3\sqrt{5})$

9. The motion $(d^2x/dt^2) = -kx$ has period $2\pi/\sqrt{k}$. Here $2 = 2\pi\sqrt{l/g}$.

11. 13 ft $\frac{1}{2}$in. (approx)

13. 13 in. (approx)

Section 10

1. If $f(x) = x^5$, then $(3.1416)^5 - \pi^5 = f'(X)(3.1416 - \pi)$ for some X between π and 3.1416. Show that the right side above is less than $5(10)^{-3}$, using the assumptions stated in the problem.

3. Here $f(x) = x^{1/5}$, and $(100,001)^{1/5} - 10 = f'(X)(100,001 - 100,000) = f'(X)$ where $100,000 < X < 100,001$. Show that $f'(X) < 10^{-4}/5$.

5. Given $|r - \pi| < 10^{-n}/2$ and $f(x) = x^{1/3}$. Then for some X between r and π,
$$|r^{1/3} - \pi^{1/3}| = |f'(X)(r - \pi)| < \tfrac{1}{3}X^{-2/3}\tfrac{1}{2}(10^{-n}).$$
Show that this is less than $10^{-(n+1)}$, since $X > 3$. (Note: $\sqrt{3} > 5/3$.)

7. $|\pi - r| < 9e$ (assuming $r \geq 3$)

9. $b^n > b^n - 1 > n(b - 1)$. So b^n will be larger than M if $n(b - 1) > M$, and hence if $n > M/(b - 1)$.

11. $\sqrt{1 + x} - 1 = 1/2\sqrt{1 + X} \cdot x$, etc. Here is a direct proof: $1 + x < 1 + x + x^2/4 = (1 + x/2)^2$. Taking the square root gives $\sqrt{1 + x} < 1 + x/2$.

Section 11

1. $8\frac{1}{16}$

3. $2\frac{107}{108}$

5. 1.02

7. $3 - \frac{1}{27} \approx 2.96$

9. $\frac{33}{128} \approx 0.258$

11. $0.08\pi r^3$

15. The tangent-line error is less than 0.001. Also, $1/16 \approx 0.06$, with an error of 0.0025. So, $\sqrt{65} \approx 8.06$, with an error less than $0.004 = 4(10)^{-3}$.

17. $|E| \leq 0.002$. Also, $1/108 \approx 0.01$, with an error less than 0.001. So $(80)^{1/4} \approx 2.99$, with an error less than 0.003.

19. $(0.98)^{-1} \approx 1.0200$ with an error less than $4(10)^{-5}$.

Section 12

1. $3.1\bar{6}$

3. 1, 3/4

7. Let $f(x) = \sin x = 2x/3$, $a = \pi/2$. Then $f'(x) = \cos x - 2/3$, and
$$x_1 = \frac{\pi}{2} - \frac{f(\pi/2)}{f'(\pi/2)} = \frac{3}{2}.$$
Between 3/2 and $\pi/2$, $\cos x = \sin(\pi/2 - x) < \pi/2 - x < (\pi - 3)/2 < 0.1$, and $|f'(x)| = |\cos x - 2/3| > 1/2 = L$. Also $|f''(x)| = \sin x \leq 1 = B$. So
$$|E| \leq \frac{B}{2L}(x_1 - a)^2 = (x_1 - a)^2 < 0.01.$$

9. $f(x) = x^2 - 10$, $x_0 = 3$, $x_1 = (19/6)$, $x_2 = (721/228) = 3.162280 \cdots$,
$|f''(x)| = 2 = B$, $|f'(x)| = 2x \geq 6 = L$,
$$|E| \leq \frac{B}{2L}(x_2 - x_1)^2 = \frac{2}{2 \cdot 6}\left(\frac{1}{19 \cdot 12}\right)^2 < \frac{1}{6(200)^2} = \frac{(0.005)^2}{6} < 5(10)^{-6}.$$

11. $x_1 = 10.050$, $|E| \leq 1.25(10)^{-4}$

13.
$$x_2 = \frac{17}{12} - \frac{1}{24 \cdot 17} = 1.414215 \cdots,$$
$$|E| < \frac{1}{2}\left(\frac{1}{24 \cdot 17}\right)^2 < 5 \cdot (10)^{-6}.$$

Thus $\sqrt{2} = 1.41421$ to the nearest 5 decimal places.

15.
$$x_2 = \frac{97}{56} - \frac{1}{2 \cdot 56 \cdot 97},$$
$$|E| < \frac{1}{2}\left(\frac{1}{2 \cdot 56 \cdot 97}\right)^2 < 5 \cdot (10)^{-9}.$$

Extra Problems for Chapter 6

1. $dy/dx = 3x^2 + 2x - 1 = (3x - 1)$
   $(x + 1) = 0$ if $x = 1/3$ or $x = -1$.

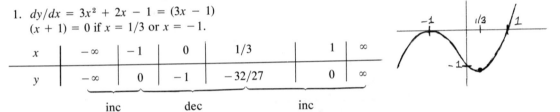

x	$-\infty$	-1	0	$1/3$	1	∞
y	$-\infty$	0	-1	$-32/27$	0	∞

         inc   dec   inc

3. $dy/dx = 1 + \cos x = 0$ if $x = \pi, 3\pi$; $d^2y/dx^2 = -\sin x = 0$ if $x = 0, \pi, 2\pi, 3\pi$

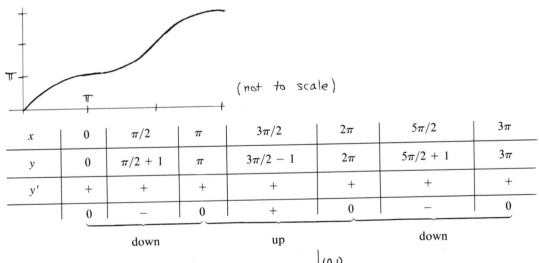

(not to scale)

x	0	$\pi/2$	π	$3\pi/2$	2π	$5\pi/2$	3π
y	0	$\pi/2 + 1$	π	$3\pi/2 - 1$	2π	$5\pi/2 + 1$	3π
y'	$+$	$+$	$+$	$+$	$+$	$+$	$+$
	0	$-$	0	$+$	0	$-$	0

       down   up   down

5. Critical point $x = 0$; concave down, $(-1,1)$; concave up, $(-\infty, -1)$, $(1,\infty)$.

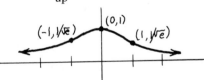

$(-1, 1/\sqrt{e})$ $(0,1)$ $(1, 1/\sqrt{e})$

7. Critical point, $x = 0$; concave up, $(-\infty,\sqrt{3})$, $(0,\sqrt{3})$; concave down, $(-\sqrt{3},0)$, $(\sqrt{3},\infty)$.

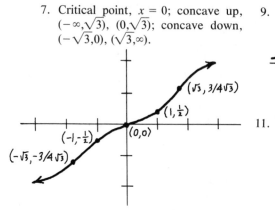

9.

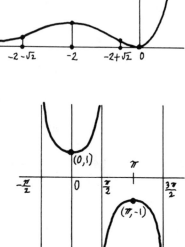

11.

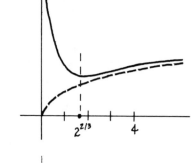

13. Concave up, $(0,\infty)$; concave down, $(-\infty,0)$.

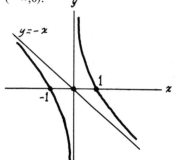

15. Critical point, $2^{2/3}$; concave up, $(0,4)$; concave down, $(4,+\infty)$.

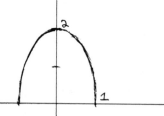

17. Singular points at $x = -1$, 1; zero derivative at $x = 0$; concave down for all x in domain.

19. $a^2 < 3b$ means no critical points;
$a^2 = 3b$ means one critical point;
$a^2 > 3b$ means two critical points.

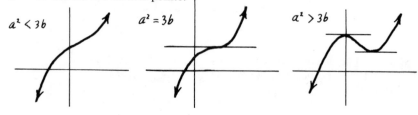

21. Take two nonoverlapping chords C_1 and C_2. By MVT there are tangents T_1 and T_2 with the same slope as C_1 and C_2 (see figure). Since the graph lies above T_1 and T_2, in particular, Y lies above T_1 and X lies above T_2 so $\sphericalangle XzY$ is less than π, in other words, slope $T_1 =$ slope $C_1 <$ slope $C_2 =$ slope T_2.

23. Say $x < y$. Let

$$s(u,v) = \frac{f(v) - f(u)}{v - u}.$$

Thus $s(u,v)$ is the slope of the chord from $(u,f(u))$ to $(v,f(v))$. The given is that Z is below XY so that slope $XZ <$ slope $XY <$ slope $ZY: s(x,z) < s(x,y) < s(y,z)$. Now fix z_0 as the midpoint of the interval $[x,y]$, and repeat the above argument for Z between X and z_0. Letting Z approach X, show that $f'(x) \leq s(x,z_0)$. In a similar manner, show that $f'(y) \geq s(z_0,y)$. Conclude that $f'(x) < f'(y)$.

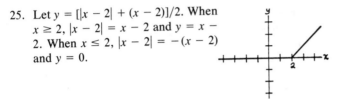

25. Let $y = [|x - 2| + (x - 2)]/2$. When $x \geq 2$, $|x - 2| = x - 2$ and $y = x - 2$. When $x \leq 2$, $|x - 2| = -(x - 2)$ and $y = 0$.

27. Let (x_0,y_0) be the vertex. By hypothesis, $y = f(x) = m_1(x - x_0) + y_0 = m_1|x - x_0| + y_0$ when $x \geq x_0$ and $y = f(x) = m_2(x - x_0) + y_0 = -m_2|x - x_0| + y_0$ when $x \leq x_0$. Then, for all $x, f(x) = [(m_1 - m_2)|x - x_0| + (m_1 + m_2)(x - x_0)]/2 + y_0 = a|x - x_0| + bx + c$, where $a = (m_1 - m_2)/2$, $b = (m_1 + m_2)/2$, and $c = y_0 - ((m_1 + m_2)/2)x_0$.

29. We may assume that $x_1 < x_2 < \cdots < x_n$. If x lies in $(-\infty,x_1]$, then $f(x) = -(a_1(x - x_1) + a_2(x - x_2) + \cdots + a_n(x - x_n)) + bx + c = [b - (a_1 + \cdots + a_n)]x + (c + a_1x_1 + \cdots + a_nx_n)$. Similarly for x in $[x_n, +\infty)$. If x lies in $[x_i,x_{i+1}]$, then $f(x) = a_1(x - x_1) + \cdots + a_i(x - x_i) - a_{i+1}(x - x_{i+1}) - \cdots - a_n(x - x_n) + bx + c$ which, again, is of the form $ax + b$. Thus f is linear on each of the intervals $(-\infty,x_1],[x_1,x_2], \ldots , [x_{n-1},x_n],[x_n,+\infty)$; hence f is polygonal.

31.

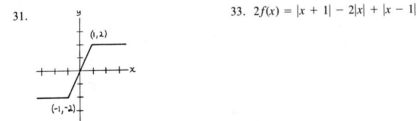

33. $2f(x) = |x + 1| - 2|x| + |x - 1|$

35. $f(x) = (1/2)|x + 2| - |x + 1| + |x - 1| - (1/2)|x - 2|$

37. D is maximum when D^2 is maximum, and $D^2 = (x - 1)^2 + (y - 4)^2 = (x - 1)^2 + (x^2/3 - 4)^2$. $dD^2/dx = (2/9)x^3 - (5/3)x - 1$, which equals 0 at $x = 3$. The minimum distance is $(2^2 + (-1)^2)^{1/2} = \sqrt{5}$.

39. Let $r =$ radius of cylinder, $h =$ height. Then

$$4\pi r + 4r + 4h = 16, \qquad h = 4 - (\pi + 1)r.$$

Volume $= V = r^2h = 4\pi r^2 - \pi(\pi + 1)r^3$, $dV/dr = 8\pi r - 3\pi(\pi + 1)r^2$. The maximum value of V occurs at the critical point $r = 8/3(\pi + 1)$.

41. L = length of wire = $a + b$, where $a^2 = 2000 - 80x + x^2$ and $b^2 = 900 + x^2$. Then $da/dx = (x - 40)/a$ and $db/dx = x/b$, and $dL/dx = (x - 40)/a + x/b$. Since this can be written $dL/dx = -\cos \theta + \cos \phi$, the critical point can most simply be determined by $\cos \theta = \cos \phi$, $\theta = \phi$.

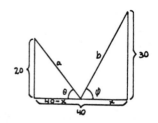

Then $20/(40 - x) = \tan \theta = \tan \phi = 30/x$, and so $x = 24$ ft, which should be the answer. (Then $a + b = 4\sqrt{41} + 6\sqrt{41} = 10\sqrt{41}$. The "endpoint" values for $a + b$ are $20 + 50 = 70$ and $30 + 20\sqrt{5} > 70$, each of which is larger than the critical point value $10\sqrt{41}$, which therefore is the minimum.)

43. Let r = radius of semicircle, h = height of rectangle. $2h + 2r + \pi r = 24$, $h = 12 - ((\pi + 2)/2)r$, and $A = 2rh + \pi r^2/2 = 24r - (\pi/2 + 2)r^2$. $dA/dr = 24 - (\pi + 4)r$. The max area occurs when $r = 24/(\pi + 4)$ft. $h = 12 - 12(\pi + 2)/(\pi + 4) = 24/(\pi + 4)$ ft.

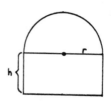

45. $D^2 = (10 - 5m)^2 + (5 - 10/m)^2 = (10 - 5m)^2(1 + 1/m^2)$. D is minimum when $m = -\sqrt[3]{2}$, and the path meets the two roads $10 + 5\sqrt[3]{2}$ and $5 + 10/\sqrt[3]{2}$ yds from the intersection, respectively.

47. L = light = $2rh + \pi r^2/4 = 24r - (3\pi/4 + 2)r^2$. $dL/dr = 24 - (3\pi/2 + 2)r$; L is max when $r = 48/(3\pi + 8)$ ft. $h = 12 - 24(\pi + 2)/(3\pi + 8) = (12\pi + 48)/(3\pi + 8) = ((\pi + 4)/4)r$.

49. Cost = $C = 5\pi r^2 + 2\pi rh$, so $0 = 10\pi r + 2\pi h + 2\pi r \, dh/dr$ and $dh/dr = -(5 + h/r)$. Then (as in above problem) $0 = dV/dr = \pi r[2r + 2h - r(5 + h/r)] = \pi r(h = 3r)$. Thus the maximum volume for a given total cost occurs when $h = 3r$.

51. Let A be the given surface area, r the radius of the cylinder, h its height. Then $A = 4\pi r^2 + 2\pi rh$ and $0 = 8\pi r + 2\pi h + 2\pi r \, dh/dr$; $dh/dr = -(4 + h/r)$. $V = (4/3)\pi r^3 + \pi r^2 h$, so $0 = dV/dr = 4\pi r^2 + 2\pi rh + \pi r^2 \, dh/dr = \pi r[4r + 2h - r(4 + h/r)] = \pi rh$. So $r = 0$ or $h = 0$. Since $r = 0 \Rightarrow V = 0$, max V occurs when $h = 0$ and the tank is spherical.

53. $P = 10x - (x^3 - 3x^2 + 4x + 1)$, $dP/dx = -3x^2 + 6x + 6$; $d^2P/dx^2 = 6x + 6$. Thus the max profit P is made when $x = 1 + \sqrt{3} \approx 2.73$. Thus he should produce 2,730 items/week to realize a maximum profit of about 1,739 dollars per week.

55. $L = k \sin \theta/r^2 = k \sin \theta \cos^2 \theta/(25)^2$. $dL/d\theta = (k/625)(-\sin^2 \theta \cos \theta + \cos^3 \theta)$. This is zero when $(\tan \theta)^2 = 1/2$ ($\cos \theta = 0$ being ruled out), so the proper height is $25 \tan \theta = 25/\sqrt{2}$.

57. Volume of water in trough = 8 $bh/2 = 8h^2/\sqrt{3}$, where h is depth of water and b = width of water surface. $16 = dV/dt = (16h/\sqrt{3})dh/dt$. Thus $dh/dt = \sqrt{3}/h$, and when $h = 1$, $dh/dt = \sqrt{3}$ ft/min.

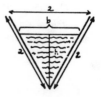

59. $u\, dV/dt + V\, du/dt = 0$, $du/dt = (-u/V)dV/dt = -2u/V = -32$ when $u = 4(V = 1/4)$.

61. $dy/dt = 2x\, dx/dt$, $dx/dt = (1/2x)dy/dt$. When $x = 2$, $dx/dt = (1/4)3 = 3/4$.

63. $dy/dt = 5 \cos \theta \cdot d\theta/dt = 5 \cos \theta = x$

65. $dV/dt = kA = k4\pi r^2$ where k is a positive constant. But also $dV/dt = (dV/dr)dr/dt = 4\pi r^2\, dr/dt$. Thus $4\pi k r^2 = 4\pi r^2\, dr/dt$ and $dr/dt = k$.

67. $dy/dx = 2(x^2 - 3x)(2x - 3)$, critical points are 0, 3, 3/2. $d^2y/dx^2 = 12x^2 - 36x + 18 = 18$ if $x = 0$, $= 18$ if $x = 3$, $= -9$ if $x = 3/2$; thus y takes on a maximum at $x = 3/2$ and local minima at $x = 0$ and $x = 3$.

69. $S \approx 4\pi r^2$; $\Delta S \approx (8\pi r)(\Delta r) = 8\pi$
    $S \approx 4\pi(3V/4\pi)^{2/3}$;
    $S \approx (8\pi/3)(3V/4\pi)^{-1/3}(3/4\pi)(\Delta V) = (-2\pi/3)(4/3)^{1/3}$.

71. Let $f(x) = x^{1/4}$, $f'(x) = (1/4)x^{-3/4}$, $f''(x) = (-3/16)x^{-7/4}$. Then $5^{1/4} = f(5) \approx f(5\ 1/16) + f'(5\ 1/16)(5 - 5\ 1/16) = 3/2 + (1/4)(2/3)^3(-1/16) = 3/2 - (1/64)8/27 = 3/2 - 1/216$. If $x \geq 5$, then $|f''(x)| \leq (3/16)1/5^{7/4} \leq (3/16)(1/5)(1/3) = 1/80$, hence $|E| \leq (1/2)(1/80)(1/16)^2 = 1/10 \cdot 2^{12} = 1/40{,}960 < 3 \times 10^{-5}$.

73. Let $f(x) = \cos x$. Then $\cos x = f(x) \approx f(0) + f'(0)(x - 0) = 1$, and since $|f''(x)| = |-\cos x| \leq 1$, $|E| \leq (1/2)(1)x^2 = x^2/2$.

75. If $f(h) = (1 + h)^a$, then $f'(h) = a(1 + h)^{a-1}$ and $f''(h) = a(a - 1)(1 + h)^{a-2}$. The tangent line approximation is $(1 + h)^a \approx 1 + ah$. If $h > 0$ and $a < 2$, then $|f''(h)| \leq |a(a - 1)|$, and $|E| \leq (1/2)|a(a - 1)|h^2$.

77.

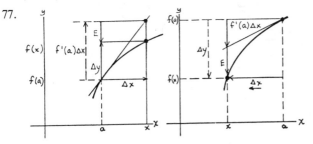

79. Let $f(x) = x^{1/2}$. $\sqrt{3} \approx f(25/9) + f'(25/9)(3 - 25/9) = 5/3 + (1/2)(3/5)(2/9) = 5/3 + 1/15 = 26/15 = 1.733 \cdots$. Since $f''(x)$ is negative for $x > 0$, E is also negative. Since $|f''(x)| \leq (1/4)(3/5)^3 = 27/500$ for $x \geq 5/3$, $|E| \leq (1/2)(27/500)(2/9)^2 = 2/15000 < 1.3(10)^{-3}$. Thus $1.732 < \sqrt{3} < 1.734$.

CHAPTER 7

Section 1

1. $\log (x^2 - 1)$ 3. $2 \log (x - 1)$

5. Already shown for $x > 0$. Let $h(x) = -x$. For $x < 0$,

$$\frac{d}{dx} \log |x| = \frac{d}{dx} \log (-x) = \frac{d}{dx} \log (h(x)) = \frac{1}{h(x)} h'(x) \qquad \text{(chain rule)}$$

$$= \frac{-1}{(-x)} = \frac{1}{x}.$$

7. $2x/(x^2 + 1)$ 9. $1 + \log x$ 11. $(1 - \log x)/x^2$

13. 2   15. $x^n \log x$

Section 2

1. $(\frac{1}{2} \log x + 1)x^{x^{1/2} - 1/2}$ 3. $9^x \log 9$

5. $(a \log x + a)x^{ax}$ 7. $(x \cot x + \log (\sin x))(\sin x)^x$

9. $(\log 10)(2ax + b)10^{ax^2 + bx + c}$ 11. $(\log x \sec^2 x + x^{-1} \tan x)x^{\tan x}$

13. See Problem 48 of Extra Problems   15. $x/(1 + x^2)^{1/2}$

17. $(m/n)x^{m/n - 1}$ 19. $(-1/x^3)4^{1/x}(x + \log 4)$

21. $(\sec^2 x \log x + x^{-1} \tan x)x^{\tan x}$ 23. $-(1 - x)^{-1/2}(1 + x)^{-3/2}$

Section 3

7. $f'(x) = 3(x - 1)^2$ 9. $f'(x) = 3(x - \frac{1}{3})^2 + \frac{2}{3}$

11. $f'(x) = 1/(1 + x^2)^{3/2}$ 19. They are inverses, where defined.

21. $x = (1/3)(y + 5)$ 23. $x = y^{5/3}$

25. $x = (1 - y)/(1 + y)$ 27. $x = y/\sqrt{1 - y^2}$

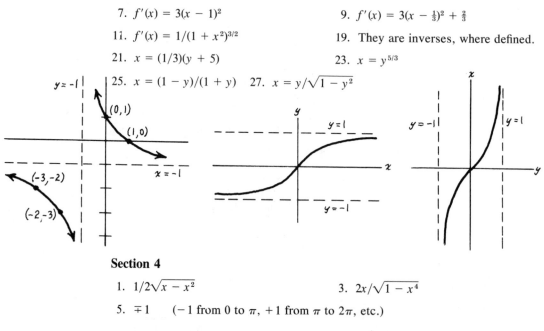

Section 4

1. $1/2\sqrt{x - x^2}$ 3. $2x/\sqrt{1 - x^4}$

5. ∓ 1 (-1 from 0 to π, $+1$ from π to 2π, etc.)

7. $\arcsin x$ 9. $2\sqrt{1 - x^2}$

11. $-1/\sqrt{2} \sin x + 2 \sin^2 x$ 13. 0

15. The problem is to furnish a value of y such that $\arcsin(\sin y) \neq y$.

17. Show that it must be of the form $2n\pi + \arcsin x$ or $(2n + 1)\pi - \arcsin x$ for some integer n.

19. The identity involved is $\cos 2\theta = 1 - 2 \sin^2 \theta$.

Section 5

1. $2x/(1 + x^4)$ 3. $a/(1 + a^2x^2)$

5. $1/x\sqrt{x^2 - 1}$ 7. $(1 + x)/(1 + x^2)$

11. Show that

$$\cot\left(\frac{\pi}{2} - \arctan x\right) = \frac{\cos\left(\dfrac{\pi}{2} - \arctan x\right)}{\sin\left(\dfrac{\pi}{2} - \arctan x\right)}$$

reduces to x (by the $\cos(s - t)$ identity, etc.).

13. Show that arcsec $x = \arccos(1/x)$, and then differentiate.

15. -1 17. $1/(1 + x^2)$ 19. $-1/(x(\log x)^2 + x)$

Extra Problems for Chapter 7

1. $\log x$ 3. $-\tan x$ 5. $\dfrac{1}{x \log x}$ 7. $x^{(1/x-2)}[1 - \log x]$

9. $x^{(x^x)}[x^x \log x(1 + \log x) + x^{x-1}]$

11. $\dfrac{\sec^2 x}{\sqrt{1 - \tan^2 x}}$ 13. $\dfrac{e^x}{1 + e^{2x}}$ 15. $\dfrac{2}{e^x + e^{-x}}$

17. Here $\log(e^u \cdot e^v) = \log e^u + \log e^v = u + v$. Then $e^u \cdot e^v = e^{\log(e^u \cdot e^v)} = e^{u+v}$.

19. $(d/dx) \log (uv) = (1/uv)(uv' + vu') = v'/v + u'/u$. $(d/dx) \log (uvw) = (1/uvw)(uvw' + uv'w + u'vw) = w'/w + v'/v + u'/u$.

21. Take x large enough that $x^{b/2} > 2/b$. Then by Problem 20, $(b/2) \log x < x^{b/2} - 1$, $\log x < (2/b)x^{b/2} - 2/b < (2/b)x^{b/2} < x^{b/2} \cdot x^{b/2} = x^b$.

23. Let $n = m + 1$. By Problem 22, $x^n = x^{m+1} < e^x$ for large enough x. Thus $0 < x^m e^{-x} = [x^{m+1}/e^x]1/x < 1/x$. By the squeeze limit law, $x^m e^{-x} \to 0$ as $x \to +\infty$.

25. $f'(x) = x + 2x \log x = 0$ when $\log x = -\frac{1}{2}$, $x = 1/\sqrt{e}$, minimum $= -\frac{1}{2}e$.

27. Show that the graph looks like this:

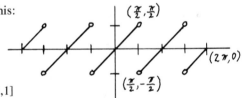

31. $1/\sqrt{2}$

35. $x = \sqrt{ab + b^2}$

39. Example: $f(x) = \sqrt{1 - x^2}$, on $[0,1]$

41. Since $f'(x) > 0$ for all x in $(-1,1)$, f is monotone, continuous, and invertible. $f(g(y)) = y$, $f'(g(y))g'(y) = 1$, $g'(y) = \sqrt{1 - (g(y))^2}$. Therefore $1 - (g(y))^2 = (g'(y))^2$ and $g^2 + (g')^2 = 1$.

43. $g(f(x)) = \sqrt{((x^2 + 1)^{1/2} + 1)((x^2 + 1)^{1/2} - 1)} = \sqrt{x^2 + 1 - 1} = \sqrt{x^2} = x$ if $x \geq 0$. $f(g(x)) = ((x^2 + 1)(x^2 - 1) + 1)^{1/4} = (x^4 - 1 + 1)^{1/4} = x$ if $x \geq 0$. f and g are mutually inverse if we restrict their domains to all nonnegative x.

45. a) If x is in domain f, then $g(y) = x$, where $y = f(x)$, hence x is in range g. Since $g(f(x))$ is defined for every x in domain f, range f is included in domain g.

b) Since $gf(x) = x$ for all x in domain f, domain $f \subseteq$ range g and range $f \subseteq$ domain g. Since $f(g(x)) = x$ for all x in domain g, domain $g \subseteq$ range f and range $g \subseteq$ domain f. Therefore domain $f =$ range g and range $f =$ domain g.

51. By Theorem 2, $E'(x) = \dfrac{1}{L'(E(x))} = \dfrac{1}{1/E(x)} = E(x)$.

CHAPTER 8

Section 1

1. $-\frac{1}{2}\cos 2x + C$

3. $\frac{1}{3}\sin^3 x + C$

5. $\frac{-1}{3}e^{-t^3} + C$

7. $-\sin(2 - x) + C$

9. $-\log(1 + \cos x) + C$

11. $\frac{1}{2}(\log x)^2 + C$

13. $\log|x^2 + x - 1| + C$

15. $\frac{3}{7}(a + y)^{7/3} - \frac{3a}{4}(a + y)^{4/3} + C$

17. $(4/\sqrt{3})\arctan\sqrt{\frac{1}{3}x - 1} + 2\sqrt{x - 3} + C$

19. $-\frac{1}{3}(x^3 + 2)^{-2} + C$

21. $2e^{\sqrt{x}} + C$

23. $\frac{1}{4}(e^x + 1)^4 + C$

25. $\frac{1}{3}(\log x)^3 + C$

Section 2

1. $\frac{1}{6}\log\left|\frac{3 - x}{3 + x}\right| + C$

3. $\frac{1}{9}\log\left|1 - \frac{9}{x}\right| + C$

5. $[1/(2\sqrt{3})]\log\left|\frac{x + 1 - \sqrt{3}}{x + 1 + \sqrt{3}}\right| + C$

7. $\log\left|\frac{x + 1}{x + 2}\right| + C$

9. $\frac{1}{6}\arctan\frac{2}{3}x + C$

11. $\frac{1}{6}\log|x - 1| + \frac{5}{6}\log|x + 5| + C$

13. $x - 2\log|x + 1| + (-1)/(x + 1) + C$

15. $\frac{2}{3}\log|x + 5| + \frac{1}{3}\log|x - 1| + C$

17. $(-5)/(x - 1) + 2\log|x| + C$

19. $\log|\sec\theta + \tan\theta| + C$

21. $\log\left(\frac{e^t + 1}{e^t + 2}\right) + C$

23. $\frac{1}{8}\log\left(1 - \frac{4}{x^2}\right) + C$

Section 3

1. $x/2 - \dfrac{\sin 2x}{4} + C = \frac{1}{2}(x - \sin x \cos x) + C$

3. $(-1/4)\sin^3 x \cos x - (3/8)\sin x \cos x + (3/8)x + C$

5. $(1/24)\sin^4 6x + C$

7. $(-1/3)\csc^3 x + \csc x + C$

9. $(1/6)\cos^5\theta \sin\theta + (5/24)\cos^3\theta \sin\theta + (5/16)\cos\theta \sin\theta + (5/16)x + C$

11. $(-3/4)(\cos 2x)^{2/3} + (3/16)(\cos 2x)^{8/3} + C$

13. $\log(\tan x) + C$

15. $(-1/3)\cot^3 x - (1/5)\cot^5 x + C$

17. $-4\cot\dfrac{x}{4} - (4/3)\cot^3\dfrac{x}{4} + C$

19. $+\frac{2}{9}\sec^{9/2} x - \frac{2}{5}\sec^{5/2} x + C$

21. $+\frac{1}{2}\sec^2 x + \log(\cos x) + C$

23. $\frac{1}{4}\tan^4 x - \frac{1}{2}\sec^2 x + \log(\sec x) + C$

Section 4

1. $\frac{1}{3}(4 - x^2)^{3/2} - 4(4 - x^2)^{1/2} + C$

3. $x/a^2\sqrt{a^2 + x^2} + C$

5. $-\frac{1}{3}(16 - x^2)^{3/2} + C$

7. $\log\left(\dfrac{\sqrt{1 + x^2} - 1}{|x|}\right) + C = -\log\left(\dfrac{\sqrt{1 + x^2} + 1}{|x|}\right) + C$

9. $\frac{1}{5} \log \left(\frac{5 - \sqrt{25 - x^2}}{|x|} \right) + C = -\frac{1}{5} \log \left(\frac{5 + \sqrt{25 - x^2}}{|x|} \right) + C$

11. $-(1 - x^2)^{1/2} + C$ 13. $(1 + x^2)^{1/2} + C$ 15. $(x^2 - a^2)^{1/2} + C$

17. $\arcsin \frac{y}{a} + C$ 19. $2\sqrt{x} - 2 \log (1 + \sqrt{x}) + C$

21. $\frac{6}{7} x^{7/6} - \frac{6}{5} x^{5/6} + \frac{3}{2} x^{2/3} + 2x^{1/2} - 3x^{1/3} - 6x^{1/6} + 3 \log (x^{1/3} + 1)$
    $+ 6 \arctan(x^{1/6}) + C$

Section 5

1. $\sin x - x \cos x + C$ 3. $\frac{2}{3}x(x + 1)^{3/2} - \frac{4}{15}(x + 1)^{5/2} + C$

5. $-y^2 \cos x + 2y \sin x + 2 \cos x + C$

7. $-x^2 e^{-x} - 2xe^{-x} - 2e^{-x} + C$ 9. $\frac{1}{2}(x - \sin x \cos x) + C$

11. $x \arcsin x + \sqrt{1 - x^2} + C$ 13. $\frac{1}{2}(\log x)^2 + C$

15. $2 \sin \sqrt{x} - 2\sqrt{x} \cos \sqrt{x} + C$ 17. $x \log (1 + x^2) - 2x + 2 \arctan x + C$

19. $\frac{1}{n - 1} \frac{\sin x}{\cos^{n-1} x} + \frac{n - 2}{n - 1} \int \sec^{n-2} x \, dx$

21. $x(\log x)^n - n \int (\log x)^{n-1} dx$

Section 6

1. $\log \left| \frac{x - 1}{x} \right| + \frac{1}{x} + \frac{1}{2x^2} + C$ 3. $x^{-1} + \frac{1}{2} \log \left| \frac{x - 1}{x + 1} \right| + C$

5. $-x^{-1} - \arctan x + C$ 7. $\frac{1}{4} \log \left| \frac{1 + x}{1 - x} \right| + \frac{x}{2(1 - x^2)} + C$

9. $\frac{1}{6} \log \left(\frac{x + 1}{x - 1} \right) + \frac{1}{12} \log \left(\frac{x - 2}{x + 2} \right) + C$

11. $\frac{1}{3} \arctan x - \frac{1}{6} \arctan \frac{x}{2} + C$

Extra Problems for Chapter 8

1. $\frac{1}{2} e^{2x} + C$ 3. $\frac{1}{2} \sec^2 x + C$

5. $e^{(e^x)} + C$ 7. $4\sqrt{x} (\log \sqrt{x}) - 4\sqrt{x} + C$

9. $\frac{1}{2} \arcsin x - \frac{x}{2} \sqrt{1 - x^2} + C$ 11. $\frac{2}{3bn} (a + bx^n)^{3/2} + C$

13. $-\frac{\sqrt{x^2 + 4}}{4x} + C$ 15. $-\frac{1}{2(\log x)^2} + C$

17. $\frac{1}{3} (x^2 + 1)^{3/2} - (x^2 + 1)^{1/2} + C$ 19. $2 \log |\sec \sqrt{x} + \tan \sqrt{x}| + C$

21. $-\cos x[\log (\cos x)] + \cos x + C$ 23. $\frac{1}{16} \log \left| \frac{(x - 4)^{17}}{x} \right| + \frac{1}{4x} + C$

25. $6x^{1/6} - 6 \arctan x^{1/6} + C$

27. $\dfrac{1}{5}\tan^5 x - \dfrac{1}{3}\tan^3 x + \tan x + C$

29. $\tfrac{1}{2}(\arctan t)^2 + C$

31. $-2\cos\sqrt{x} + C$

33. $-2\sqrt{1 - \sin t} + C$

35. $\tfrac{1}{3}(x^2 + 1)^{3/2} + C$ (if $x > 0$)

37. $\arctan(e^t) + C$

39. $\dfrac{1}{n+1}\tan^{n+1}\theta + C$

41. $\dfrac{1}{(n+1)}\sin^{n+1}\theta + C$

43. $\dfrac{1}{3}\log\left(\dfrac{\sqrt{9 + 4x^2} - 3}{|2x|}\right) + C$

45. $x/\sqrt{a^2 - x^2} - \arcsin(x/a) + C$

47. $-\dfrac{1}{4}(x - 1)(x^2 - 2x - 3)^{-1/2} + C$

49. $\displaystyle\int \dfrac{\cos t}{2 - \sin^2 t}\,dt = \dfrac{\sqrt{2}}{2}\int\dfrac{du}{1 - u^2}$ $\qquad (\sqrt{2}\,u = \sin t)$

$\qquad = \dfrac{\sqrt{2}}{4}\log\left|\dfrac{1 + u}{1 - u}\right| + C = \dfrac{\sqrt{2}}{4}\log\left|\dfrac{\sqrt{2} + \sin t}{\sqrt{2} - \sin t}\right| + C$

51. $\displaystyle\int\dfrac{\sqrt{1 - x}}{\sqrt{1 + x}}\,dx = \int\dfrac{\sqrt{1 - x}}{\sqrt{1 + x}}\dfrac{\sqrt{1 - x}}{\sqrt{1 - x}}\,dx = \int\dfrac{1 - x}{\sqrt{1 - x^2}}\,dx + \sqrt{1 - x^2} + C$

$\qquad = \arcsin x + \sqrt{1 - x^2} + C$

53. $\displaystyle\int\dfrac{x^3\,dx}{\sqrt{a^2 - x^2}} = \dfrac{-\sqrt{a^2 - x^2}}{3}(2a^2 + x^2) + C$ $\qquad (u = \sqrt{a^2 - x^2})$

55. $\displaystyle\int\dfrac{x^{1/3}\,dx}{1 + x} = 3\int\dfrac{u^3\,du}{1 + u^3} = 3\int du - 3\int\dfrac{du}{1 + u^3}$ $\qquad (x = u^3)$

$\qquad = 3u - 3\displaystyle\int\dfrac{du}{(u + 1)(u^2 - u + 1)}$

$\qquad = 3u - \displaystyle\int\dfrac{du}{u + 1} - \dfrac{1}{2}\int\dfrac{2u - 1}{u^2 - u + 1}\,du + \dfrac{3}{2}\int\dfrac{1}{u^2 - u + 1}\,du$

$\qquad = 3x^{1/3} + \log\dfrac{\sqrt{x} + 1}{\sqrt[3]{x^{2/3} - x^{1/3} + 1}}$

$\qquad = \sqrt{3}\arctan\dfrac{2}{\sqrt{3}}\left(x^{1/3} - \dfrac{1}{2}\right) + C$

57. $\displaystyle\int x\arcsin x\,dx = \dfrac{1}{4}(2x^2 - 1)\arcsin x + \dfrac{x}{4}\sqrt{1 - x^2} + C$

$\qquad (u = \arcsin x, \ dv = x\,dx)$

59. $\displaystyle\int x^3\sin x\,dx = -x^3\cos x + 3x^2\sin x + 6x\cos x - 6\sin x + C$

$\qquad (u_1 = x^3, u_2 = x^2, u_3 = x)$

61. $\displaystyle\int x^2\log(x + 1)\,dx = \dfrac{1}{3}(x^3\log(x + 1) - \dfrac{x^3}{3} + \dfrac{x^2}{2} - x + \log(x + 1)) + C$

$\qquad (u = \log(x + 1), \ dv = x^2\,dx)$

63. $\displaystyle\int(\log x)^2\,dx = x\log^2 x - 2x\log x + 2x + C$ $\qquad (u = \log^2 x, \text{ then } s = \log x)$

65. Equating the two expressions gives $a = A$, $b = B$, $c = C - A$ and $d = D - B$.

67. $\displaystyle\int \frac{dx}{(x^2 + a^2)^n}$ (let $x = a \tan \theta$, $dx = a \sec^2 \theta \, d\theta$)

$$= a^{-2n+1} \int \cos^{2n-2} \theta \, d\theta \qquad \text{(by Problem 20, Section 5)}$$

$$= \frac{a^{-2n+1}}{2n-2} \cos^{2n-3} \theta \sin \theta + a^{-2n} \frac{(2n-3)}{2n-2} \int \cos^{2n-2} \theta \, a \sec^2 \theta \, d\theta$$

$$= \frac{-1}{a^2(2n-2)} \frac{x}{(x^2 + a^2)^{n-1}} + \frac{1}{a^2} \frac{2n-3}{(2n-2)} \int \frac{dx}{(x^2 + a^2)^{n-1}}$$

69. $\pi^2/2$

71. $\displaystyle\int_0^1 \arctan x \, dx = \left(x \arctan x - \frac{1}{2} \log(x^2 + 1) \right) \Big]_0^1$ (by Problem 10, Section 5)

$$= \frac{\pi}{4} - \frac{1}{2} \log 2$$

73. $\displaystyle 4\pi \int_a^r x\sqrt{r^2 - x^2} \, dx = \frac{4\pi}{3} (r^2 - a^2)^{3/2}$

75. $\displaystyle 2\pi \int_1^2 x \log x \, dx = 4\pi \log 2 - \frac{5\pi}{2}$

77. $\displaystyle 4\pi \int_0^a x \, e^x \, dx = 4\pi(a - 1)e^a + 4\pi$

79. $\displaystyle 2\pi \int_1^2 (x^2 - x - x \log x) \, dx = \frac{19\pi}{6} + 4\pi \log \left(\frac{1}{2} \right)$

81. $\displaystyle 4\pi \int_0^a x\sqrt{a^2 - x^2} \, dx = \frac{4}{3} \pi a^3$

CHAPTER 9

Section 1

1. 2   3. 0   5. −5   7. 2   9. 1/2

11. $\displaystyle \lim n \sin \tfrac{1}{n} = \lim \frac{\sin \tfrac{1}{n}}{\tfrac{1}{n}} = \lim_{x \to 0} \frac{\sin x}{x} = 1$

The last limit was fundamental in Chapter 5, but it also follows from l'Hôpital's rule.

13. 1   15. −1

Section 2

1. $S_4 = 13/32$; $|E| = 1/32$ 3. $S_6 \approx 1.253$; $|E| \approx 0.03$

13. $S_4 = 21/32$; the integral is $5/8 = 20/32$, so the error is $1/32$. This is the same as given by the error estimate

$$\frac{1}{2} \Delta x [f(b) - f(a)] = \frac{1}{2} \cdot \frac{1}{8} \cdot \left[\frac{3}{2} - 1 \right] = \frac{1}{32}$$

17. $n = 40,000$ 21. $2 \cos x$ 23. 1

Section 3

1. $2/3$

3. $9/4$

5. $(18 - 10\sqrt{3})/27$

7. $\frac{1}{2} \log 3$

9. $\log(4/3)$

11. $\frac{3}{2} \arctan(1/\sqrt{2}) - 1/\sqrt{2}$

13. $(14\sqrt{3}/5)a^5$

15. $\dfrac{1}{a^2} \left(\dfrac{\sqrt{6} - 2}{\sqrt{3}} \right)$

17. $36/5$

19. 2

21. $\pi/2$

23. 3

25. $\pi/2$

27. Divergent

29. 6

Section 4

1. 0

3. $4/3$

5. 3

7. 30 mph

9. Average of instantaneous velocity $= 1/(t_1 - t_0) \int_{t_0}^{t_1} f'(t)\,dt = \cdots$.

Section 5

1. $2\pi/3$

3. $4\sqrt{3}/3$

5. $4/3$

7. $14/15$

9. $a^2h/3$

Section 6

1. The sum listed is the midpoint evaluation S_8 for the integral $\log 2 = \int_1^2 dx/x$. A more recognizable form is:

$$\frac{1}{8} \left[\frac{1}{17/16} + \frac{1}{19/16} + \cdots + \frac{1}{31/16} \right].$$

So compute the error estimate for S_8.

3. One must show that, in the accompanying figure, the shaded region R_1 with corners CDA is larger than the shaded region R_2 with corners DEB, because if that is the case, then the rectangle $ABZX$ misses more of the area under the curve than the extra amount it contains. To show this, use the fact that the graph lies above its tangent line at D, and that the tangent line cuts off congruent triangles.

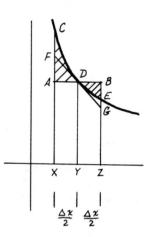

5. The sum is $\bar{S}_5$ in the integral $\pi/4 = \int_1^2 (dx/(1 + x^2))$. So work out the value of the error formula.

Section 7

1. $\log (\sqrt{2} + 1)$

3. $(1/a^2)((4/9 + a^2)^{3/2} - 8/27)$

5. $(8/27)(10\sqrt{10} - 1)$

7. $\sqrt{1 + e^2} - \sqrt{2} - \log[(\sqrt{1 + e^{-2}} + e^{-1})/(\sqrt{2} + 1)]$

9. $(b - a)\sqrt{1 + m^2}$

11. $2\sqrt{3} - \dfrac{2}{3}$

Section 8

1. $7/6$

3. 1

5. $105/62$

7. $\pi/2$

9. $(2 \log 2 - 3/4)/(2 \log 2 - 1)$

11. Since $\bar{x}_1 = \int_a^b x\rho(x)\,dx/m_1$, we have $\int_a^b x\rho(x)\,dx = m_1\bar{x}_1$, etc.

13. $M = \int_a^b (x - p)\rho(x)\,dx = \int_a^b x\rho(x)\,dx - p\int_a^b \rho(x)\,dx = m\bar{x} - pm$

Section 9

1. $(a/2,b/2)$

3. $(0,4/3)$

5. $(8/5,0)$

7. $(1/2,2/5)$

9. $(8\sqrt{2}/15,16\sqrt{2}/21)$

11. $([ae^a/(e^a - 1)] - 1,0)$

13. $(4a/3\pi,0)$

15. $(1,0)$

17. $(1/2,8/5)$

Section 10

1. $\dfrac{1}{10}\left[\dfrac{1 + 1/2}{2} + \displaystyle\sum_{k=1}^{9}\dfrac{10}{10 + k}\right] = \dfrac{3}{40} + \dfrac{1}{11} + \dfrac{1}{12} + \dfrac{1}{13} + \cdots + \dfrac{1}{19}$;

$K\dfrac{(b - a)(\Delta x)^2}{12} = \dfrac{1}{600}$

3. $S = \dfrac{1}{10}\left[\dfrac{3}{4} + \dfrac{100}{101} + \dfrac{25}{26} + \dfrac{100}{109} + \dfrac{25}{29} + \dfrac{4}{5} + \dfrac{25}{34} + \dfrac{100}{199} + \dfrac{25}{41} + \dfrac{100}{181}\right]$.

The main problem in the error estimate is finding a reasonable bound K for the second derivative of $f(x) = 1/(1 + x^2)$. Compute $f''(x)$ and show that $|f''(x)| \le 2$ (because $|1 - 3x^2| \le 1 + 3x^2$).

Extra Problems for Chapter 9

1. $\dfrac{1}{2}$

3. $\dfrac{\pi r^2}{4}$

5. $\sqrt{3} - \dfrac{1}{2}\log(2 + \sqrt{3})$

7. $\dfrac{\pi}{4}$

11. $\dfrac{2\sqrt{3}}{9}r^3$

13. $\dfrac{a^2 h}{3}$

15. $\dfrac{16}{3}r^3$

17. Let $0 = x_0 < x_1 < \cdots < x_n = a$ be a subdivision of the interval $[0,a]$ and let $y_k = f(x_k)$. Then $y_0 = b$, $y_n = 0$, and the polygon defined by sequentially joining the points (x_0,y_0), (x_1,y_1), $\ldots$, (x_n,y_n) determines an approximation to the graph of f. This polygon has length equal to

$$\sum_{k=0}^{n-1}\sqrt{(x_{k+1} - x_k)^2 + (y_k - y_{k+1})^2}.$$

If we replace $(0,1)$ and $(1,0)$ by $(0,b)$ and $(a,0)$ in the above problem, then L has length $\sqrt{a^2 + b^2}$ so length $P \ge \sqrt{a^2 + b^2}$. Also $\Sigma\Delta x_k = a$, $\Sigma|\Delta y_k| = -\Sigma\Delta y_k = -(0 - b) = b$, and length $P \le a + b$. Therefore, the length of any inscribed polygonal arc lies between $\sqrt{a^2 + b^2}$ and $a + b$. Hence, also, the length of the graph of f lies between $\sqrt{a^2 + b^2}$ and $a + b$.

If the graph of f increases from $(0,0)$ to (a,b), then the length of the graph necessarily lies between $\sqrt{a^2 + b^2}$ and $a + b$.

19. Consider the function $f(x) = 1 - x^n$. This is a smooth decreasing function whose graph runs from $(0,1)$ to $(1,0)$.

When $n = 1$, the graph is the straight line segment joining $(0,1)$ and $(1,0)$. Consequently, it has length, $\sqrt{2}$. Thus, the lower bound of $\sqrt{2}$ may not be improved.

To show that the upper bound of 2 may not be improved, it suffices to observe that $\lim_{n\to\infty} 2 - 2\sqrt{1/n} = 2$. Thus, by choosing a large enough n, the curve $1 - x^n$ may be chosen to have a length arbitrarily close to 2.

21. We may assume that the line is the y-axis. For each x, let l_x denote the height of the vertical cross section at x. Then $\bar{x} = \dfrac{\int x l_x \, dx}{\int l_x \, dx}$. But $\int l_x \, dx = $ area of the region $G = A$ and, by the shell formula, $\int 2\pi x l_x \, dx = $ volume of the solid generated $= V$, so $2\pi\bar{x} = \dfrac{V}{A}$, or $V = 2\pi\bar{x} A$.

23. The centroid of the given circle is at $(R,0)$ and the area of the circle is πr^2. Thus, the volume of the solid torus is $2\pi R \cdot \pi r^2 = 2\pi^2 r^2 R$.

25. Suppose the cylindrical shell has inner radius r, thickness Δr, and height h. The cylindrical shell may be obtained by rotating the rectangle with coordinates $(r,0)$, $(r + \Delta r,0)$, (r,h) and $(r + \Delta r,h)$, about the y-axis. This rectangle has $\bar{x} = r + \Delta r/2$ and area $A = (\Delta r)h$. Thus, the volume of the cylindrical shell is:

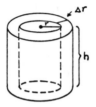

$$V = 2\pi\bar{x}A$$

$$= 2\pi\left(r + \frac{\Delta r}{2}\right)(\Delta r)h$$

$$= \pi(2r + \Delta r)(\Delta r)h.$$

27. Let G be the region lying between the graphs $y = f(x)$ and $y = g(x)$ from $x = a$ to $x = b$, where $a < b$ and $g(x) \leq f(x)$ over the interval $[a,b]$. Then the moment M, about the axis $x = p$, is given by:

$$M = \int_a^b (f(x) - g(x))(x - p)dx$$

$$= \int_a^b (f(x) - g(x))x \, dx - p \int_a^b (f(x) - g(x))dx.$$

$$\bar{x} = \frac{\displaystyle\int_a^b (f(x) - g(x))x \, dx}{\displaystyle\int_a^b (f(x) - g(x))dx} \quad \text{and} \quad A = \int_a^b (f(x) - g(x))dx.$$

Thus, $M = \bar{x}A - pA = A(\bar{x} - p)$.

29. Subdivide the interval $[a,b]$ into n equal subintervals. Let $x_0, \ldots, x_n$ be the subdividing points. The moment of G about the x-axis will be the sum of the moments of the incremental slices between x_k and $x_k + \Delta x = x_{k+1}$. Thus, the moment M_y of G about the x-axis satisfies

$$M_y \simeq \sum_{i=0}^{n-1} \frac{1}{2} [f^2(x_i) - g^2(x_i)]\Delta x.$$

Taking the limit as n goes to infinity, we obtain $M_y = \frac{1}{2} \int_a^b [f^2(x) - g^2(x)]dx$.

31. $\bar{x} = \dfrac{\displaystyle\int_0^2 x(x + 2 - 2x)dx}{\displaystyle\int_0^2 (x + 2 - 2x)dx} = \dfrac{\left(-\dfrac{1}{3}x^3 + x^2\right)\Big]_0^2}{\left(-\dfrac{1}{2}x^2 + 2x\right)\Big]_0^2} = \dfrac{4/3}{2} = \dfrac{2}{3}.$

$\bar{y} = \dfrac{\dfrac{1}{2}\displaystyle\int_0^2 (x + 2 - 2x)(x + 2 + 2x)dx}{\displaystyle\int_0^2 (x + 2 - 2x)dx}$

$= \dfrac{1}{2}\displaystyle\int_0^2 \dfrac{4 + 4x - 3x^2}{2}dx$

$= \dfrac{1}{4}\, 4x + 2x^2 - x^3 \Big]_0^2 = 2.$ The centroid is at $(2/3,2)$.

33. By symmetry $\bar{x} = 0$.

$\bar{y} = \dfrac{\dfrac{1}{2}\displaystyle\int_{-1}^1 [(x^2 + 1)^2 - (2x^2)^2]dx}{\displaystyle\int_{-1}^1 [x^2 + 1 - 2x^2]dx}$

$= \dfrac{\dfrac{1}{2}\displaystyle\int_{-1}^1 [-3x^4 + 2x^2 + 1]dx}{\displaystyle\int_{-1}^1 [1 - x^2]dx}$

$= \dfrac{\dfrac{1}{2}\left[-\dfrac{3}{5}x^5 + \dfrac{2}{3}x^3 + x\right]_{-1}^1}{\left(x - \dfrac{1}{3}x^3\right)\Big]_{-1}^1} = \dfrac{16/15}{4/3} = \dfrac{4}{5}.$

The centroid is at $(0,4/5)$.

35. Let G be the region described in Problem 34; then the volume obtained by rotating this region about the x-axis is $\int_a^b \pi f^2(x)dx$. On the other hand, by the Theorem of Pappus we know that $V = 2\pi\bar{y}A$, where A is the area of the region G. Also, $\bar{y} = M/A$, where M is the moment of G about the x-axis. Thus, $2\pi M = 2\pi\bar{y}A = V = \int_a^b \pi f^2(x)dx$, so $M = \frac{1}{2}\int_a^b f^2(x)dx$.

37. a) $\displaystyle\int_a^b (x - a)^2(b - x)^2 dx$

$= \displaystyle\int_a^b (x^2 - 2ax + a^2)(b^2 - 2bx + x^2)dx$

$= \displaystyle\int_a^b (x^4 - 2(b + a)x^3 + (b^2 + 4ab + a^2)x^2 - 2(ab^2 + a^2b)x + a^2b^2)dx$

$= \left[\dfrac{x^5}{5} - \dfrac{(b + a)}{2}x^4 + (b^2 + 4ab + a^2)\dfrac{x^3}{3} - (ab^2 + a^2b)x^2 + a^2b^2x\right]_a^b$

$= \dfrac{1}{30}b^5 - \dfrac{1}{6}ab^4 + \dfrac{1}{3}a^2b^3 - \dfrac{1}{3}a^3b^2 + \dfrac{1}{6}a^4b - \dfrac{1}{30}a^5 = \dfrac{1}{30}(b - a)^5.$

b) $\displaystyle\int_a^b (x - a)^2(b - x)^2 dx$

(let $x = y(b - a) + a$
$dx = (b - a)dy$)

$= \displaystyle\int_0^1 y^2(b - a)^2[b - a - y(b - a)]^2(b - a)dy$

$= (b - a)^5 \displaystyle\int_0^1 y^2(1 - y)^2 dy = \dfrac{1}{30}(b - a)^5.$

39. $\displaystyle\int_a^b (x-a)^2(x-b)^2 f^{(4)}\, dx$

$$= \underbrace{(x-a)^2(x-b)^2 f'''(x)\Big]_a^b}_{=\,0} - 2\int_a^b (x-a)(x-b)(2x-a-b)f'''(x)dx$$

$$= \underbrace{-2(x-a)(x-b)(2x-a-b)f''(x)\Big]_a^b}_{=\,0} + 2\int_a^b 6(x^2-(a+b)x+C)f''(x)dx$$

$$= \underbrace{12(x^2-(a+b)x+C)f'(x)\Big]_a^b}_{=\,0} - 12\int_a^b (2x-a-b)f'(x)dx$$

$$= \underbrace{-12(2x-a-b)f(x)\Big]_a^b}_{=\,0} + 24\int_a^b f(x)dx.$$

CHAPTER 10

Section 1

1. $y = x^2$; parabola $dy/dx = 2t$; critical point $t = 0$.

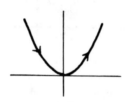

3. $y = (x+1)^3$; $dy/dt = 3t^2$, critical point at $t = 0$.

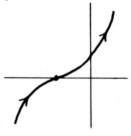

5. $y = 3x/2$; straight line;
$$\frac{dy}{dx} = \frac{dy/dt}{dx/dt} = \frac{3}{2}\frac{t^{-2/3}}{t^{-2/3}} = \frac{3}{2},$$
except at $t = 0$, where undefined.

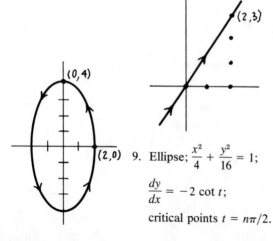

7. $\dfrac{x^2}{4} + \dfrac{y^2}{9} = 1$; ellipse;
$$\frac{dy}{dx} = -\frac{3}{2}\tan t;$$
critical points at $t = n\pi/2$.

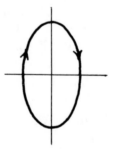

9. Ellipse; $\dfrac{x^2}{4} + \dfrac{y^2}{16} = 1$;
$$\frac{dy}{dx} = -2 \cot t;$$
critical points $t = n\pi/2$.

11. Circle; $x^2 + y^2 = 4$;

$\dfrac{dy}{dx} = -\cot t$;

critical points $t = n\pi/2$.

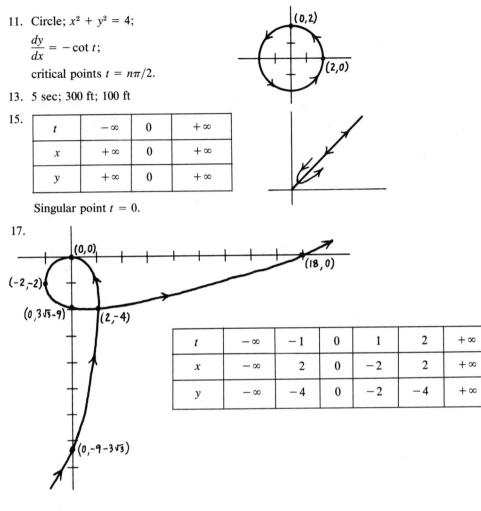

13. 5 sec; 300 ft; 100 ft

15.

t	$-\infty$	0	$+\infty$
x	$+\infty$	0	$+\infty$
y	$+\infty$	0	$+\infty$

Singular point $t = 0$.

17.

t	$-\infty$	-1	0	1	2	$+\infty$
x	$-\infty$	2	0	-2	2	$+\infty$
y	$-\infty$	-4	0	-2	-4	$+\infty$

Section 3

1. $(1 + 4t^2)^{1/2}$

3. $(9\cos^2 t + \sin^2 t)^{1/2} = \sqrt{1 + 8\cos^2 t}$

5. $a^2/2 + a$

7. a) $5\sqrt{10} - 8\sqrt{2}$ b) $32 - 8\sqrt{2}$

9. $\sqrt{2}(e - 1)$

11. 1/2

Section 4

1. [r,θ-plane]

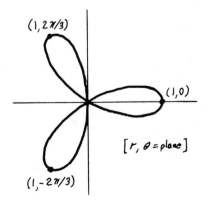

3.

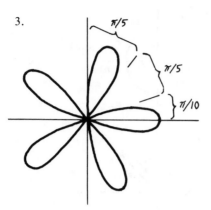

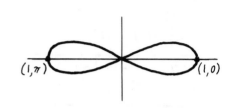

7. Circle of radius 1; center at $(0,1)$ (rectangular coordinates).

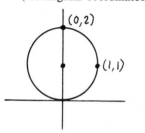

9.

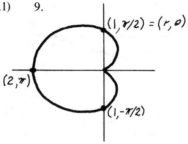

11.

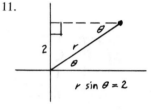

13. The equation in Cartesian coordinates is $y^2 = 1 + 2x$.

Section 5

1. $16\pi^5/5$

3. $\pi/8$

5. Area in the quadrant: $n\pi/2 \le \theta \le (n + 1)\pi/2 = \dfrac{e^{n\pi}}{4}(e^\pi - 1)$.

7. $3\pi/2$

9. 11π

11. a/n

Section 6

1. $(8/3)((\pi^2 + 1)^{3/2} - 1)$

3. 1

5. $\log\left(2[\sqrt{1 + \pi^2} + \pi]/[\sqrt{4 + \pi^2} + \pi]\right) + \dfrac{1}{\pi}[\sqrt{4 + \pi^2} - \sqrt{1 + \pi^2}]$

9. $\cot \psi = 1$. So ψ has the constant value $\pi/4$. The curve is turning counterclockwise at a constant rate equal to the turning rate of a ray from the origin. $\phi = \theta + \pi/4$.

11. $\cot \psi = -\tan \theta$, so $\psi = \theta + \pi/2$. Erect the perpendicular to the tangent line T at the point of tangency P, intersecting the axis $\theta = 0$ at C. The fact that $\psi = \theta + \pi/2$ means that $\triangle OCP$ is isosceles consistent with the fact that the graph is a circle centered at C.

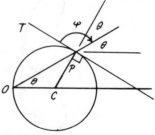

Section 7

1. $2/(1 + 4x^2)^{3/2}$ 3. $-x/(1 + x^2)^{3/2}$ 5. 0

7. The curvature takes on the value -1 at $x = 0$ and $+1$ at $x = \pi$. It is periodic, of period 2π, and is bounded by 1 in absolute value.

9. $2/3|a|$

11. If a function graph has 0 curvature, then $d^2y/dx^2 = 0$. Integrate this equation twice.

13. $K = ab/(a^2 \sin^2 t + b^2 \cos^2 t)^{3/2}$. Max and min: a/b^2, b/a^2 (supposing $0 < b < a$).

15. At $t = 0$ 17. None   19. $t = 1$

Section 8

1. 0 3. 1 5. $1/2$ 7. $1/2$

9. $-1/2$ 11. $-1/4$ 13. 1

15. 0 (but not by l'Hôpital; use the inequality in Problem 21 in the Extra Problems for Chapter 7)

17. 1 19. 0

Extra Problems for Chapter 10

1.

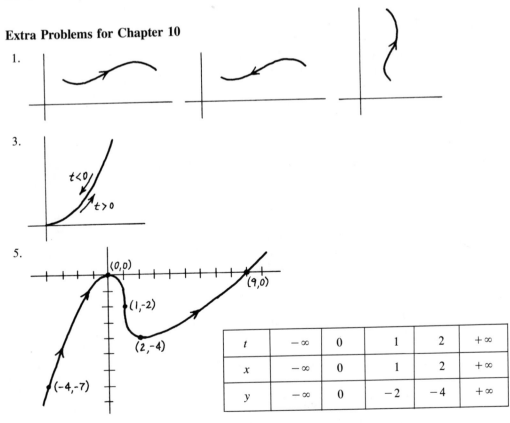

3.

5.

t	$-\infty$	0	1	2	$+\infty$
x	$-\infty$	0	1	2	$+\infty$
y	$-\infty$	0	-2	-4	$+\infty$

7. Singular point at $t = 0$.

t	$-\infty$	-1	0	$1/2$	1	$+\infty$
x	$-\infty$	-7	0	$-1/4$	1	$+\infty$
y	$+\infty$	-1	0	$-7/16$	-1	$+\infty$

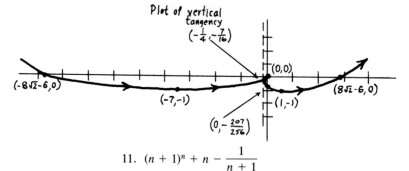

Plot of vertical tangency

11. $(n + 1)^n + n - \dfrac{1}{n + 1}$

15. $r = 1 + \epsilon r \cos \theta = 1 + \epsilon x$, $x^2 + y^2 = 1 + 2\epsilon x + \epsilon^2 x^2$, $(\epsilon^2 - 1)x^2 + 2\epsilon x - y^2 + 1 = 0$. $(\epsilon^2 - 1)[x + \epsilon/(\epsilon^2 - 1)]^2 - y^2 = \epsilon^2/(\epsilon^2 - 1) - 1 = 1/(\epsilon^2 - 1)$ which is the hyperbola $(x - \alpha)^2/a^2 - y^2/b^2 = 1$, $\alpha = -\epsilon/(\epsilon^2 - 1)$, $a = 1/(\epsilon^2 - 1)$, $b = 1/\sqrt{\epsilon^2 - 1}$.

17. The locus has the equation $r = \epsilon(x + k) = d + \epsilon x$, where $d = \epsilon k$. So it is a parabola, hyperbola, or ellipse depending on whether $\epsilon = 1$, $\epsilon > 1$, or $\epsilon < 1$.

19. $\dfrac{a^2 \pi}{4n}$ 　　21. $\pi - \dfrac{3\sqrt{3}}{2}$ 　　23. $\dfrac{\pi}{2(1 + \pi)}$ 　　25. $\dfrac{\pi}{2}$ 　　27. $\dfrac{1}{2}$

29. $(\overline{PF})^2 = (a \cos \theta - c)^2 + b^2 \sin^2 \theta$
$$= a^2 \cos^2 \theta - 2ac \cos \theta + c^2 + (a^2 - c^2)\sin^2 \theta$$
$$= a^2 - 2cx + c^2 \cos^2 \theta$$
$$= \left(a - \frac{c}{a} x\right)^2 = (a - ex)^2.$$

31. $(\overline{PF^2}) = (a \sec \theta - c)^2 + b^2 \tan^2 \theta$
$$= a^2 \sec^2 \theta - 2ac \sec \theta + c^2 + (c^2 - a^2)\tan^2 \theta$$
$$= a^2 - 2cx + c^2 \sec^2 \theta$$
$$= \left(a - \frac{c}{a} x\right)^2 = (a - ex)^2.$$

Thus $\overline{PF} = \pm(a - ex)$. On the right branch $ex > a$, since $e > 1$ and $x > a$, so then $\overline{PF} = ex - a$.

33. $x = r \cos \theta$, $r = f(\theta)$, $y = r \sin \theta$. $\dfrac{dx}{d\theta} = \dfrac{dr}{d\theta} \cos \theta - r \sin \theta$, $\dfrac{dy}{d\theta} = \dfrac{dr}{d\theta} \sin \theta + r \cos \theta$. $\left(\dfrac{dx}{d\theta}\right)^2 = \left(\dfrac{dr}{d\theta}\right)^2 \cos^2 \theta - 2r \dfrac{dr}{d\theta} \sin \theta \cos \theta + r^2 \sin^2 \theta$. $\left(\dfrac{dy}{d\theta}\right)^2 = \left(\dfrac{dr}{d\theta}\right)^2 \sin^2 \theta + 2r \dfrac{dr}{d\theta} \sin \theta \cos \theta + r^2 \sin^2 \theta$. $\left(\dfrac{dx}{d\theta}\right)^2 + \left(\dfrac{dy}{d\theta}\right)^2 = \left(\dfrac{dr}{d\theta}\right)^2 + r^2$.

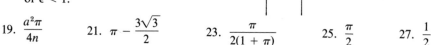

35. $\dot{x} = t \cos(t)$, $\ddot{x} = -t \sin(t) + \cos t$;   so $\dfrac{d\phi}{ds} = \dfrac{1}{t}$.

$\dot{y} = t \sin(t)$, $\ddot{y} = t \cos(t) + \sin t$;

37. $r = 2a \cos \theta$, $\dfrac{dr}{d\theta} = -2a \sin \theta$, $\dfrac{d^2 r}{d\theta^2} = -2a \cos \theta$,

$$k = \frac{4a^2 \cos^2 \theta + 8a^2 \sin^2 \theta + 4a^2 \cos^2 \theta}{(4a^2 \cos^2 \theta + 4a^2 \sin^2 \theta)^{3/2}} = \frac{1}{a}.$$

39. $\dfrac{1}{e^{a\theta}} \dfrac{1}{\sqrt{1+a^2}}$ 41. $-\dfrac{1}{3}$ 43. $\dfrac{1}{3}$

45. Let $F(x) = \int_a^x f(x)g(x)dx$ and $G(x) = \int_a^x g(x)dx$, where $a \le x \le b$, f and g are continuous on $[a,b]$ and $g(x) \neq 0$ on the open interval (a,b). Since $g(x) \neq 0$ on (a,b), $\int_a^x g(x)dx \neq 0$ when $a < x \le b$ and in particular, $G(b) \neq 0$. Since $G(a) = \int_a^a g(x)dx = 0$, the hypothesis $G(a) \neq G(b)$ is satisfied. We apply the parametric mean value theorem to obtain a point T, $a < T < b$, at which

$$\frac{F(b) - F(a)}{G(b) - G(a)} = \frac{f(T)g(T)}{g(T)}, \quad \frac{\int_a^b f(x)g(x)dx}{\int_a^b g(x)dx} = f(T), \quad \int_a^b f(x)g(x)dx = f(T)\int_a^b g(x)dx.$$

49. If we define $H(0) = 0$, then H becomes continuous at the origin, and $H'(y)/G'(y)$ turns out to be $h'(1/y)/g'(1/y)$.

CHAPTER 11

Section 1

1. $\dfrac{1}{7} - \dfrac{10}{71} = \dfrac{1}{497}$ 3. $\dfrac{385}{536} - \dfrac{334}{465} = \dfrac{1}{249,240}$

5. 3.8 (correct expansion; next place between 2 and 7)

7. 1.4 (closest; within 0.02)   9. 3.1 (closest; within 0.025)

11. 1.0 (correct); 1.1 (closest)   13. within 0.006

15. within 0.005; within 0.011

Section 2

1. 4/3   3. 3/5   5. 3   7. 1/6

9. 3/4   11. 27/16   13. 12   15. 1/6

19. Multiply by 2, and compare to $\Sigma 1/n$.

21. Multiply by 4 and compare to $\Sigma 1/n$.

23. $\displaystyle\sum_{n=1}^{\infty} \frac{23}{(100)^n} = \frac{23}{99}$ 25. $\displaystyle\sum_{n=1}^{\infty} \frac{315}{(1000)^n} = \frac{35}{111}$ 27. $2 + \displaystyle\sum_{n=1}^{\infty} \frac{1}{100^n} = 2\frac{1}{99}$

31. $2s_n = \dfrac{2}{1 \cdot 3} + \dfrac{2}{3 \cdot 5} + \dfrac{2}{5 \cdot 7} + \cdots + \dfrac{2}{(2n-1)(2n+1)}$

$= \left[\dfrac{1}{1} - \dfrac{1}{3}\right] + \left[\dfrac{1}{3} - \dfrac{1}{5}\right] + \left[\dfrac{1}{5} - \dfrac{1}{7}\right] + \cdots + \left[\dfrac{1}{2n-1} - \dfrac{1}{2n+1}\right]$

$= 1 - \dfrac{1}{2n+1} = \dfrac{2n}{2n+1}$.

Therefore $s = \frac{1}{2}$.

33. This series can be obtained as the sum of the series in Problem 31 and the series in Problem 32. Its sum is 3/4.

35. 390 ft

Section 3

1. Diverges   3. Diverges   5. Converges

7. Diverges   9. Converges (because log $n < n$)

11. Converges (sin $x < x$ for positive x)   13. Converges ($n!/n^n \leq 2/n^2$)

15. Converges   17. Diverges   19. Converges

Section 4

1. Converges   3. Converges

5. Test fails (but the series converges by comparison with $\Sigma 1/n^2$)

7. Test fails (but have divergence by comparison with harmonic series)

9. Converges   11. Diverges   13. Converges

15. Diverges   17. Diverges   19. Converges

Section 5

1. Converges   3. Converges   5. Diverges

7. $1/x^2$ is decreasing and continuous, so, by Theorem 7,

$$\sum_{n=2}^{\infty} \frac{1}{n^2} < \int_1^{\infty} \frac{dx}{x^2} = 1.$$

9. $\Sigma_2^{\infty} 1/n^3 < 1/2$, as in Problem 7

11. $\displaystyle\int_2^{\infty} \frac{x\,dx}{x^3 + 1} < \int_2^{\infty} \frac{dx}{x^2} = \frac{1}{2}$ 13. $\displaystyle\int_1^{\infty} e^{-x^2}\,dx < \int_1^{\infty} xe^{-x^2}\,dx = \frac{1}{2e}$

15. The trapezoidal approximation to $\int_a^b f$ is less than the integral if f is concave down. Therefore,

$$\frac{\log 1}{2} + \log 2 + \cdots + \log(n - 1) + \frac{\log n}{2} < \int_1^n \log x\,dx$$

or

$$\log[(n - 1)!\sqrt{n}] < n \log n - n + 1.$$

Section 6

1. Converges absolutely   3. Diverges

5. Diverges   7. Converges conditionally

9. Converges conditionally   11. Diverges

13. Converges conditionally   15. $\Sigma(-1)^n \dfrac{n + 1}{n}$

Section 7

1. $[-1,1)$ 3. $(-\infty,\infty)$ 5. $(-1,1)$ 7. $[-1,1]$

9. $(-1,1]$ 11. $[-1,1)$ 13. $(0,4]$ 15. $[-2,4)$

17. $(0,4)$ 19. $(\frac{1}{2},\frac{3}{2})$ 21. $[0,6]$ 23. 6

25. $1/(1 - x)^2$ 27. $(1 + x)/(1 - x)^3$

Section 8

1. $1 + x + x^2 + x^3 + \cdots$ 3. $0 + 0 + 0 + x^3 + 0 + \cdots$

5. $5 - 2x + 0 + x^3 + 0 + \cdots$ 7. $\displaystyle\sum_{n=0}^{\infty} \frac{(-1)^n x^{2n}}{n!}$

9. $\displaystyle\sum_{n=0}^{\infty} \frac{(-1)^n x^{6n+3}}{(2n + 1)}$ 11. $\displaystyle\sum_{n=0}^{\infty} \frac{(-1)^n x^{2n+1}}{2^n(2n + 1)n!}$

21. First two terms give $\sin 0.5 = 0.479 \ldots$

Extra Problems for Chapter 11

1. Converges absolutely   3. Converges absolutely

5. Converges absolutely   7. Converges absolutely

9. Converges conditionally   11. Diverges

13. Converges absolutely   15. Converges absolutely

17. Converge absolutely   19. Converges conditionally

21. Converges absolutely   23. Converges absolutely

25. Converges absolutely

27. Converges conditionally if x is not a multiple of π; converges absolutely if x is a
    multiple of π.

29. Converges conditionally

31. Converges absolutely if $|r| < 1$; diverges otherwise.

33. Diverges

35. Converges absolutely   37. Diverges

39. Converges absolutely if $|r| < 1$; diverges otherwise.

41. Converges absolutely   43. Converges conditionally

45. Converges conditionally   49. 0.40546510

CHAPTER 12

Section 1

3. Let
$$g(x) = f(x) - \left[\frac{f(b) - f(a)}{b - a}\right](x - a).$$

Assume there is no point X; then g' is not 0 in (a,b), so either $g' > 0$ on all (a,b)
or $g' < 0$ on all (a,b). In the former case, g is increasing. In the latter case, g is
decreasing. But $g(a) = g(b) = f(a)$, so g can be neither decreasing or increasing.
Contradiction. So the point X exists after all.

7. Say p has roots $x_1, \ldots, x_n$ with $x_1 < x_2 < \cdots < x_n$. By Rolle's theorem, p' is 0
   in each interval (x_k, x_{k+1}) $k = 1, \ldots, n - 1$. p' is of degree $n - 1$, so this accounts for all its roots and forces them to be distinct.

Section 2

1. $p_3(x) = x + x^3/3!$ 3. $p_3(x) = 1 + x^2/2$

5. $p_4(x) = (\sqrt{2}/2)(1 + x - (x^2/2!) - (x^3/3!) + (x^4/4!))$

7. $p_4(x) = x + (x^3/3)$

9. $f(x) = 1 + \dfrac{x^2}{2} + \dfrac{(5 \sec^3 X \tan X + \sec X \tan^2 X)}{6} x^3.$

   We can take $K = 1$. (Use the fact that $1/2 < \pi/6$.)

11. $f(x) = 1 - x + (x^2/2) - (x^3/3!) + (x^4/4!) - (e^{-X}/5!)x^5$ can take $K = 1/60$ (because $e^{1/2} < 2$).

13. $f(x) = 1 + (3/2)x + (3/8)x^2 - (1/16)x^3 + (\frac{9}{16}(1 + X)^{-5/2}/4!)x^4;\ K = 3\sqrt{2}/32.$

15. We proved earlier that the derivative of an odd function is even and vice versa, so if f is odd, then $f^{(2n)}$ is odd for all $n = 0, 1, \ldots$. So f has no even terms in its expansion. The argument for f even is similar.

17. $A = 1/2,\ B = -1/8,\ C = 1/16$ 19. $x + x^3/3$

21. For $-1 < x \le 0$, the remainder is less than $|x|^{n+1}$. For $0 < x < 1/2$, the remainder is less than $(2x)^{n+1}$.

23. Let $g(x) = f(x) - p_n(x)$. Then g has derivatives up through order $n - 1$ on an interval about the origin, $g^n(0)$ exists, and $g(0) = g'(0) = \cdots = g^n(0) = 0$.

25. Suppose f is even and suppose p_n is one of its Maclaurin polynomials.

$$\lim_{x \to 0} \frac{f(x) - p_n(x)}{x^n} = 0; \quad \lim_{x \to 0} \frac{f(-x) - p_n(-x)}{(-x)^n} = 0.$$

$$f(x) = f(-x), \text{ so } \lim_{x \to 0} \frac{f(x) - p_n(-x)}{x^n} = 0.$$

   Therefore, by Problem 24, $p_n(-x)$ is also the Maclaurin polynomial of f of degree n; that is, $p_n(-x) = p_n(x)$. So p_n is even. The proof for f odd is similar.

Section 3

1. $(x - 1) - ((x - 1)^2/2) + ((x - 1)^3/3)$

3. $1/2 + (\sqrt{3}/2)(x - \pi/6) - (1/4)(x - \pi/6)^2 - (\sqrt{3}/12)(x - \pi/6)^3 + (1/48)(x - \pi/6)^4$

5. $e^a + e^a(x - a) + (e^a/2)(x - a)^2 + (e^a/6)(x - a)^3$

7. $\dfrac{\sqrt{2}}{2} \left[1 - (x - \pi/4) - \dfrac{(x - \pi/4)^2}{2} + \dfrac{(x - \pi/4)^3}{3!} + \dfrac{(x - \pi/4)^4}{4!} \right]$

   $- \dfrac{\sin X}{5!}(x - \pi/4)^5$

9. p_4 as in 3. $r_4(x) = (\cos X/5!)(x - \pi/6)^5$

11. $1 + (1/3)(x - 1) - (1/9)(x - 1)^2 + (5/81)(x - 1)^3 - (10/243)X^{-11/3}(x - 1)^4$

13. $1 + r(x - 1) + \dfrac{r(r - 1)}{2}(x - 1)^2 + \dfrac{r(r - 1)(r - 2)}{3!}(x - 1)^3$

   $+ \dfrac{r(r - 1)(r - 2)(r - 3)}{4!} X^{r-4}(x - 1)^4$

15. $-\frac{1}{2}\log 2 - (x - \pi/4) - (x - \pi/4)^2 - \frac{2}{3}(x - \pi/4)^3$

$\qquad -\dfrac{4 \sec^2 X \tan^2 X + 2 \sec^4 X}{4!}(x - \pi/4)^4$

17. $1 + \frac{1}{2}(x - 1) - \frac{1}{8}(x - 1)^2 + \cdots$

$\qquad + \dfrac{\frac{1}{2}(-\frac{1}{2})(-\frac{3}{2}) \cdots ((3 - 2n)/2)}{n!}(x - 1)^n + \cdots$

19. $\displaystyle\sum_{n=0}^{\infty} \frac{(-1)^n(x - \pi/2)^{2n}}{(2n)!}$ \qquad 21. $\displaystyle\sum_{n=0}^{\infty} (-1)^{n+1}\frac{(x - \pi)^{2n+1}}{(2n + 1)!}$

25. $4 + 7(x - 1) + 5(x - 1)^2 + (x - 1)^3$

27. $|r_n(x)| = \left|\dfrac{(\overset{\sin}{\cos})(X)(x - \pi/2)^{n+1}}{(n + 1)!}\right| \le \dfrac{(x - \pi/2)^{n+1}}{(n + 1)!}.$

Since x is fixed, you must show that $(K^n/n!) \to 0$ as $n \to \infty$.

29. If $x > a$, then $|r_n(x)| \le e^x \left|\dfrac{(x - a)^{n+1}}{(n + 1)!}\right| = \dfrac{K_1(K_2)^{n+1}}{(n + 1)!}$

Section 4

1. $\frac{1}{3} \times 10^{-5}$; $\frac{1}{48} \times 10^{-5}$; $10^{-5}/3$; $10^{-5}/48$

5. Say we want $(p(a), p'(a), p(b), p'(b)) = (A, B, C, D)$. Let $p_1(x) = B(x - a) + A$; then $p_1(a) = A$ and $p_1'(a) = B$. Choose p_2 by (4) such that $p_2(a) = p_2'(a) = 0$ and $p_2(b) = C - p_1(b)$ and $p_2'(b) = D - p_1'(b)$. Then it is easy to check that $p = p_1 + p_2$ satisfies the correct equalities. For uniqueness, say there were another polynomial q that satisfied the same requirements. Show that $p - q = 0$ by the uniqueness in Problem 4.

7. If f is a third-degree polynomial, then $f'''' \equiv 0$, so one may set $K = 0$ in the error formula.

9. p a polynomial of degree at most 5;

$$p(b) - p(a) = \frac{p'(b) + p'(a)}{2}(b - a) - \frac{p''(b) - p''(a)}{12}(b - a)^2$$
$$+ \frac{p^{(5)}(0)(b - a)^5}{720}.$$

Section 5

1. A polynomial of degree three or less has fourth derivative 0, so one can set $K = 0$ in the error estimate.

CHAPTER 13

Section 1

1. Force along bow string ≈ 57 pounds

3. $\theta = 45°$ \qquad\qquad 5. speed ≈ 196 mph \qquad 7.

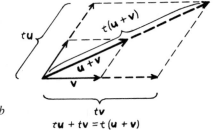

$\tau u + \tau v = \tau(u + v)$

9. $a + x = (a,b) + (x,y) = (a + x, b + y)$. Thus,

$\qquad\qquad a + x = a \leftrightarrow (a + x, b + y) = (a + b)$
$\qquad\qquad\qquad\quad \leftrightarrow a + x = a \quad$ and $\quad b + y = b$
$\qquad\qquad\qquad\quad \leftrightarrow x = 0 \quad$ and $\quad y = 0.$

11. $(\mathbf{u} + \mathbf{x}) + \mathbf{a} = ((u,v) + (x,y)) + (a + b)$
$= (u + x, v + y) + (a + b)$
$= ((u + x) + a,(v + y) + b)$
$= (u + (x + a),v + (y + b))$, etc.

13. $0\mathbf{a} = 0(a,b) = (0a,0b) = (0,0) = \mathbf{0}$

15. $\overrightarrow{a_1x_1} = \overrightarrow{a_2x_2} \Rightarrow x_1 - a_1 = x_2 - a_2$
$\Rightarrow a_2 - a_1 = x_2 - x_1$
$\Rightarrow \overrightarrow{a_1a_2} = \overrightarrow{x_1x_2}$

Section 2

1. $(x,y) = (1,2) + t(-3,0)$; $x = 1 - 3t$, $y = 2$

3. $(x,y) = t(3,4)$; $x = 3t$, $y = 4t$

5. $x = x_0 + t$, $y = y_0 + t$

7. $(-13/8, -15/4)$

9. $\mathbf{x} = (1,1) + t[(3,0) - (1,1)] = (1,1) + t(2,-1)$; $x = 1 + 2t$, $y = 1 - t$

11. $\mathbf{x} = (2,0) + t(0,1)$; $x = 2$, $y = t$

13. $(-11/3,1/3)$ 15. $(5/6,2/5)$ 17. $(-1/6,9/2)$

Section 3

1. $|\mathbf{v}| = 5$; $\dfrac{\mathbf{v}}{|\mathbf{v}|} = (3/5,4/5)$ 3. $|\mathbf{a}| = \sqrt{10}$; $\dfrac{\mathbf{a}}{|\mathbf{a}|} = \left(\dfrac{-3}{\sqrt{10}}, \dfrac{1}{\sqrt{10}}\right)$

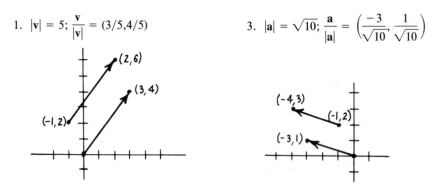

5. $|\mathbf{v}| = \sqrt{5}$; $\dfrac{\mathbf{v}}{|\mathbf{v}|} = \left(\dfrac{1}{\sqrt{5}}, \dfrac{-2}{\sqrt{5}}\right)$ 7. $\mathbf{a} \cdot \mathbf{x} = 5/2$; $\phi = \arccos(5/\sqrt{34})$ 9. $\mathbf{a} \cdot \mathbf{x} = 0$; $\phi = \pi/2$

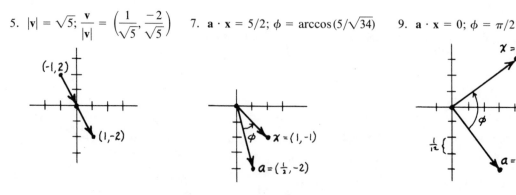

11. $\pm(2/\sqrt{5}, -1\sqrt{5})$

13. $\left(\dfrac{\sqrt{2} - \sqrt{6}}{4}, \dfrac{\sqrt{6} + \sqrt{2}}{4}\right)$ or $\left(\dfrac{\sqrt{6} + \sqrt{2}}{4}, \dfrac{\sqrt{2} - \sqrt{6}}{4}\right)$

15. $t(1,\sqrt{3}), s(1,-\sqrt{3})$ 17. $-2x + 3y = 0$

19. The equation of l is $\mathbf{a} \cdot (x - x_0) = 0$. The point x_p closest to the origin is on the line through the origin perpendicular to l, so $x_p = t\mathbf{a}$. So

$$\mathbf{a} \cdot (t\mathbf{a} - x_0) = 0 \Rightarrow t\mathbf{a} \cdot \mathbf{a} - \mathbf{a} \cdot x_0 = 0$$
$$\Rightarrow t = \mathbf{a} \cdot x_0 / \mathbf{a} \cdot \mathbf{a}.$$

21. $(2\sqrt{5},\sqrt{5})$ 23. $(1/\sqrt{5},2/\sqrt{5})$ 25. $(2/\sqrt{5},-1/5)$

27. 0   29. $10/\sqrt{101}$ 31. $1/\sqrt{101}$

Section 4

1.   3.   5.

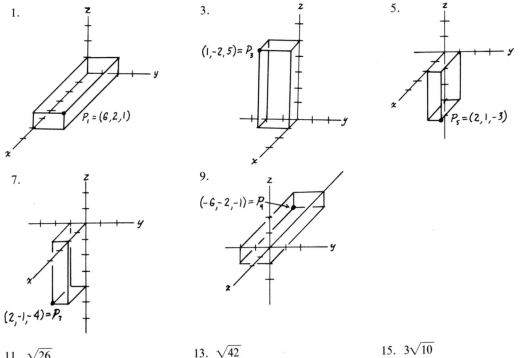

7.   9.

11. $\sqrt{26}$ 13. $\sqrt{42}$ 15. $3\sqrt{10}$

17. The plane perpendicular to the y-axis (parallel to the xz-coordinate plane) and cutting the y-axis at $y = 3$

Section 5

1. $(2/3,1/3,2/3)$ 3. $(7/11,-6/11,-6/11)$

5. $(2/3,2/3,1/3)$ 7. $0, \phi = \pi/2$

9. $0, \phi = \pi/2$ 11. $\mathbf{x} = (1 + t, -2 + 3t, 1 - 2t)$

13. $\mathbf{x} = (-2 + 3t, -1 + 4t, 2 + 2t)$ 15. $\mathbf{x} = (-1 + 3t, -2 + 2t, 1 + t)$

17. $x = 1, y = -1 + (3/2)t, z = 4 - 4t$ 19. $x = -2 + (5/2)t, y = -1, z = -2t$

21. $\mathbf{x} = (t, -2t, -t)$ 23. $\mathbf{x} = (1 + 2t, 1, 1 - t)$

25. $\mathbf{a} = (\cos \alpha, \cos \beta, \cos \gamma)$
$\mathbf{l} = (\cos \lambda, \cos \mu, \cos v)$
$\mathbf{a} \cdot \mathbf{l} = \cos \phi$

27. $(t, -t, -t)$

Section 6

1. $x + 2y + z = 0$

3. $x + 2y = 6$

5. $4x + 3y - 2z = 48$

7. $(1,0,1),\ (2,-1,0)$

9. $(1,5,0),\ (0,3,1)$

11. $t\mathbf{a}_1 + s\mathbf{a}_2$ is in the xy-plane if and only if $t + s = 0$. So $s = -t$, and points
$$t\mathbf{a}_1 + s\mathbf{a}_2 = t(\mathbf{a}_1 - \mathbf{a}_2) = t(1, -4, 0)$$
form the intersection.

13. $\mathbf{x} = t(3, -5, -7)$

15. $x + 9y + 7z = 12$

Extra Problems for Chapter 13

1. $x_1 = 3,\ x_2 = 1$

3. $x_1 = -1/5,\ x_2 = 3/5$

5. $x_1 = 8/5,\ x_2 = -9/5$

9. $(-6, -13)$

11. $(-5, -11)$

13. $(4,4)$

15. $(1,1)$

17. $(1,0)$ has components $(-1-2)$; $(0,1)$ has components $(1,1)$. So $(7,1)$ has components
$$7(-1, -2) + 1(1,1) = (-6, -13)$$
and $(-3,4)$ has components
$$-3(-1, -2) + 4(1,1) = (7,10),$$
and so on.

21. Note that reversing the roles of $\mathbf{a}_1$ and $\mathbf{a}_2$ reverses the sign of each expression in the formula for $\mathbf{a}_1 \times \mathbf{a}_2$.

23.
$$\begin{aligned}(\mathbf{a}_1 \times \mathbf{a}_2) \cdot \mathbf{a}_2 &= \mathbf{a}_1 \cdot (\mathbf{a}_2 \times \mathbf{a}_2) \\ &= \mathbf{a}_1 \cdot 0 \\ &= 0.\end{aligned}$$

(32)
(30)

25. Expand both sides in coordinate notation to verify the formula.

27. The parallelogram with base $|\mathbf{a}_2|$ has altitude $|\mathbf{a}_1|\sin \phi$. So its area is
$|\mathbf{a}_1||\mathbf{a}_2|\sin \phi = |\mathbf{a}_1 \times \mathbf{a}_2|$

33. $\{(2/\sqrt{6}, 1/\sqrt{6}, 1/\sqrt{6}),\ (1/\sqrt{3}, -1/\sqrt{3}, -1/\sqrt{3}),\ (0, 1/\sqrt{2}, -1/\sqrt{2})\}$

35. $\{(1/\sqrt{3}, 1/\sqrt{3}, 1/\sqrt{3}),\ (1/\sqrt{2}, -1/\sqrt{2}, 0),\ (1/\sqrt{6}, 1/\sqrt{6}, -2/\sqrt{6})\}$

37. $(2/3, 2/3, 1/3)$;
$$x_1B_1 + x_2B_2 + x_3B_3 = \frac{2}{9}(2,1,2) + \frac{2}{9}(2, -2 - 1) + \frac{1}{9}(1, 2, -2)$$
$$= \frac{1}{9}(4 + 4 + 1, 2 - 4 + 2, 4 - 2 - 2) = \frac{1}{9}(9,0,0) = (1,0,0).$$

39. $(2/3, -1/3, -2/3)$

41. $(4/3, 10/3, 8/3)$;

$$x_1 B_1 + x_2 B_2 + x_3 B_3 = \frac{4}{9}(2,1,2) + \frac{10}{9}(2,-2,-1) + \frac{8}{9}(1,2,-2)$$

$$= \frac{1}{9}(8 + 20 + 8, 4 - 20 + 16, 8 - 10 - 16)$$

$$= \frac{1}{9}(36, 0, -18) = (4, 0, -2)$$

43. $(-4/3, -1/3, -8/3)$

CHAPTER 14

Section 1

1. $\dfrac{d\mathbf{f}}{dt} = (1, -1/t^2); t = \pm 1 \qquad (\mathbf{f}(1) = \pm(1,1))$

3. $\dfrac{d\mathbf{f}}{dt} = (2t, 1); t = -1 \qquad (\mathbf{f}(-1) = (1,2))$

5. $\dfrac{d\mathbf{x}}{dt} = (-\sin t, \cos t)$; everywhere

7. $(d\mathbf{g}/dt) = (-3 \sin t, 2 \cos t); t = (n\pi/2) \qquad (\mathbf{g} = (0, \pm 2), (\pm 3, 0))$

9. $(40, -2); \quad \mathbf{x} = (40,14) + t(40,-2) = (40 + 40t, 14 - 2t), \quad$ or $\quad x = 40 + 40t,$
$y = 14 - 2t$

11. $(1,1); \mathbf{x} = (1,0) + u(1,1) = (1 + u, u)$, or $x = 1 + u, y = u$

13. $(0,2); \mathbf{x} = (2,0) + u(0,2) = (2,2u) \qquad$ 15. $(1,-2); x = u, y = 2 - 2u$

17. $x = x_0 - at \sin \theta_0, y = y_0 + bt \cos \theta_0$

19. The increment arrows are drawn to points on the other side of $\mathbf{x}_0$, but when divided by the *negative* Δt, the adjusted arrows approach the same limiting arrow (representing the tangent vector).

21. Let $h(t) = \int_0^t f(s) \, ds$; then $h'(t) = f(t)$. By hypothesis, $x'(t) = uf(t) \, y'(t) = vf(t)$, so $x(t) = uh(t) + k, y(t) = vh(t) + l$, and $\mathbf{x}(t) = \mathbf{k} + h(t)\mathbf{u}$. So $\mathbf{x}(t)$ is a straight line.

23. If d is the shortest distance between the two paths, then it is certainly the shortest distance from $\mathbf{x}_2$ to path $\mathbf{x} = \mathbf{g}(t)$. Therefore, by a slight generalization of Example 5, $\mathbf{x}_1 \mathbf{x}_2$ is perpendicular to $\mathbf{x} = \mathbf{g}(t)$. By a similar argument, $\mathbf{x}_1 \mathbf{x}_2$ is perpendicular to the other curve.

Section 2

1. By rule (5), $d\mathbf{x}/dt = \mathbf{a}$, a constant vector. Then by the example in this section $\mathbf{x} = \mathbf{a}t + \mathbf{c}$.

5. $\mathbf{x} \cdot d\mathbf{x}/dt = 0$ implies

$$0 = \mathbf{x} \cdot \frac{d\mathbf{x}}{dt} + \frac{d\mathbf{x}}{dt} \cdot \mathbf{x} = \frac{d}{dt}\mathbf{x} \cdot \mathbf{x} = \frac{d}{dt}|\mathbf{x}|^2.$$

So $|\mathbf{x}|^2 = c$, a constant. So $\mathbf{x} = \mathbf{f}(t)$ runs along the circle of radius $\sqrt{c}$ about O.

7. Assume that the path does not contain the origin (since the zero vector does not specify a direction). Solve for $|\mathbf{x}|\,d\mathbf{u}/dt$ in the identity

$$\frac{d\mathbf{x}}{dt} = \frac{d|\mathbf{x}|}{dt}\mathbf{u} + |\mathbf{x}|\frac{d\mathbf{u}}{dt}$$

and use the hypothesis of the problem, to conclude that $d\mathbf{u}/dt$ must be zero. Therefore, . . .

9. Your differentiated equation should be equivalent to $\cos\theta_1 + \cos\theta_2 = 0$, where θ_1 and θ_2 are the angles that $\mathbf{x} - \mathbf{f}$ and $\mathbf{x} - \mathbf{g}$ make with the tangent vector. Therefore $\theta_2 = \pi - \theta_1$, so the angle of incidence = angle of reflection = θ_1.

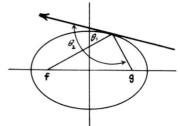

Section 3

1. a) $(x,y) = \left(\dfrac{s}{\sqrt{2}}\cos\left(\log\dfrac{s}{\sqrt{2}}\right), \dfrac{s}{\sqrt{2}}\sin\left(\log\dfrac{s}{\sqrt{2}}\right)\right)$

   b) $\mathbf{T} = \left(\dfrac{1}{\sqrt{2}}\left(\cos\left(\log\dfrac{s}{\sqrt{2}}\right) - \sin\left(\log\dfrac{s}{\sqrt{2}}\right)\right), \dfrac{1}{\sqrt{2}}\left(\sin\left(\log\dfrac{s}{\sqrt{2}}\right)\right.\right.$

   $\left.\left. + \cos\left(\log\dfrac{s}{\sqrt{2}}\right)\right)\right)$

   $\dfrac{d\mathbf{T}}{ds} = \dfrac{1}{s\sqrt{2}}\left(-\sin\left(\log\dfrac{s}{\sqrt{2}}\right) - \cos\left(\log\dfrac{s}{\sqrt{2}}\right), \cos\left(\log\dfrac{s}{\sqrt{2}}\right)\right.$

   $\left. - \sin\left(\log\dfrac{s}{\sqrt{2}}\right)\right)$

   c) $\kappa = 1/s$ (Note that in all the above, $0 < s < \infty$.)

3. Note: $0 \le t < \infty$, $0 \le s < \infty$.

   a) $(x,y) = \left(\dfrac{4}{3}\left((1 + 2s)^{1/2} - 1\right)^{3/2}, 2 - 2(1 + 2s)^{1/2} + s\right)$

   b) $\mathbf{T} = (2(1 + 2s)^{1/2} - 1)^{1/2}(1 + 2s)^{-1/2}, -2(1 + 2s)^{-1/2} + 1)$;
   $(d\mathbf{T}/ds) = (((1 + 2s)^{1/2} - 1)^{-1/2}(1 + 2s)^{-1} - 2((1 + 2s)^{1/2} - 1)^{1/2}(1 + 2s)^{-3/2},$
   $2(1 + 2s)^{-3/2})$

   c) $\kappa = 1/(1 + 2s)\sqrt{\sqrt{1 + 2s} - 1}$.

Section 4

1. $\mathbf{v} = (-3\sin 3t, 3\cos 3t)$; $\mathbf{a} = (-9\cos 3t, -9\sin 3t)$; $v = 3$, $dv/dt = 0$. We know that the unit circle has curvature $\kappa = 1$ (However, see the next answer); so $\kappa v^2 = 9$.

3. $\mathbf{v} = (-3\sin t, 2\cos t)$; $\mathbf{a} = (-3\cos t, -2\sin t)$; $v = (4 + 5\sin^2 t)^{1/2}$, $dv/dt = (5/2)\sin 2t/(4 + 5\sin^2 t)^{1/2}$. The absolute curvature κ can be computed from the formula $\kappa = |\dot{x}\ddot{y} - \ddot{x}\dot{y}|/v^3$ (Chapter 10, Section 7); so $\kappa v^2 = |\dot{x}\ddot{y} - \ddot{x}\dot{y}|/v$. Here, $\kappa v^2 = 6/(4 + 5\sin^2 t)^{1/2}$.

5. $\mathbf{v} = (1,2t)$; $\mathbf{a} = (0,2)$; $v = (1 + 4t^2)^{1/2}$, $dv/dt = 4t(1 + 4t^2)^{-1/2}$. $\kappa v^2 = |\dot{x}\ddot{y} - \ddot{x}\dot{y}|/v = 2(1 + 4t^2)^{-1/2}$.

7. $\mathbf{v} = (-6 \sin 3t, 6 \cos 2t)$; $\mathbf{a} = (-18 \cos 3t, -12 \sin 2t)$;
$v = 6(\sin^2 3t + \cos^2 2t)^{1/2}$, $dv/dt = (9 \sin 6t - 6 \sin 4t)(\sin^2 3t + \cos^2 2t)^{-1/2}$.
$\kappa v^2 = (18 \cos 3t \cos 2t + 12 \sin 3t \sin 2t)(\sin^2 3t + \cos^2 2t)^{-1/2}$.

9. $v \le 25\sqrt{2}$ ($\kappa = |d^2y/dx^2|/(1 + (dy/dx)^2)^{3/2} = 1/50$ and $\kappa v^2 = $ must
be ≤ 25.)

11. If the particle is moving ($v \ne 0$) and the path is curved, $\kappa \ne 0$, then the normal
component $\kappa v^2 \ne 0$.

13. $50\sqrt{2}$ mph ≈ 70 mph

Section 5

1. $(1, -2, 3)$ 3. $(3, 1, 1/2)$ 5. $(1/4, 1/4, -1/16)$

7. $\dfrac{d}{dt} w\mathbf{x} = \left(\dfrac{d}{dt} wx_1, \dfrac{d}{dt} wx_2, \dfrac{d}{dt} wx_3 \right)$

$= \left(w \dfrac{dx_1}{dt} + \dfrac{dw}{dt} x_1, w \dfrac{dx_2}{dt} + \dfrac{dw}{dt} x_2, w \dfrac{dx_3}{dt} + \dfrac{dw}{dt} x_3 \right)$

$= w \dfrac{d\mathbf{x}}{dt} + \dfrac{dw}{dt} \mathbf{x}$

9. Change to coordinate notation and use the ordinary chain rule, as in (21).

11. The function $|\mathbf{u} - \mathbf{x}_0|^2$ attains a minimum at $\mathbf{u}_0$, so

$$\frac{d}{dt} |\mathbf{u} - \mathbf{x}_0|^2 = 0$$

at this point. Apply Problem 8, and conclude that

$$2 \frac{d\mathbf{u}}{dt} \cdot [\mathbf{u}_0 - \mathbf{x}_0] = 0.$$

$(d\mathbf{u}/dt)$ is the vector tangent to the curve $\mathbf{u}$ at $\mathbf{u}_0$, so $\mathbf{u}_0 - \mathbf{x}_0$ is perpendicular to $\mathbf{u}$.
Similarly, $\mathbf{u}_0 - \mathbf{x}_0$ is perpendicular to $\mathbf{x}$ at $\mathbf{x}_0$.

13. Exactly similar to Section 2, Problems 1 and 3.

15. Prove that $|\mathbf{x}|^2 = \mathbf{x} \cdot \mathbf{x}$ is a constant.

$$0 = \frac{d\mathbf{x}}{dt} \cdot \mathbf{x}$$

implies

$$0 = \frac{d\mathbf{x}}{dt} \cdot \mathbf{x} + \mathbf{x} \cdot \frac{d\mathbf{x}}{dt} = \frac{d}{dt} [\mathbf{x} \cdot \mathbf{x}]$$

$$= \frac{d}{dt} (|\mathbf{x}|^2).$$

So $|\mathbf{x}|^2$ is constant.

17. $(d\mathbf{x}/dt) = \mathbf{x}f(t)$, where f is a scalar function. Let $\mathbf{u}|\mathbf{x}| = \mathbf{x}$; then follow the argument of Problem 7, Section 2.

21. $\kappa = (1 + 4t^2 + t^4)^{1/2}/(1 + t^2 + t^4)^{3/2}$

Extra Problems for Chapter 14

1. $\kappa = \dfrac{(t^4 + 4t^2 + 1)^{1/2}}{(t^4 + t^2 + 1)^{3/2}}$; $\tau = \dfrac{2}{t^4 + 4t^2 + 1}$

3. $\kappa = \dfrac{1}{b^2 + 1}$; $\tau = -\dfrac{b}{b^2 + 1}$

5. $\kappa = \dfrac{(a^4 b^2 c^2 + a^6 c^4)^{1/2}}{(a^2 c^2 + b^2)^{3/2}}; \tau = \dfrac{-ab}{b^2 + a^2 c^2}$

7. $\kappa = \dfrac{(4e^{6t} + 36e^{2t} + 4)^{1/2}}{(4e^{4t} + e^{2t} + e^{-2t})^{3/2}}; \tau = \dfrac{-12e^{2t}}{4e^{6t} + 36e^{2t} + 4}$

9. $\kappa = \dfrac{\sec t \sqrt{9 \sec^4 t - 8 \sec^2 t + 1}}{(2 \sec^4 t - \sec^2 t + 1)^{3/2}}; \tau = \dfrac{-2 \sec t(\sec^2 t + 1)}{9 \sec^4 t - 8 \sec^2 t + 1}$

CHAPTER 15

Section 1

1. Ellipsoid of revolution about the y-axis

3. Hyperboloid of two sheets; elliptical sections perpendicular to z-axis, and hyperbolic sections perpendicular to x- and y-axes.

5. Hyperboloid of one sheet, of revolution about the y-axis; elliptical sections perpendicular to y-axis, and hyperbolic sections perpendicular to x- and z-axes.

7. Elliptical cone about z-axis with vertex at the origin

9. Elliptic paraboloid; elliptic sections perpendicular to z-axis, and parabolic sections perpendicular to x- and y-axes.

11. Paraboloid of revolution about y-axis with vertex at origin; circular sections perpendicular to y-axis, and parabolic sections perpendicular to x- and z-axes.

13. Two planes intersecting in the y-axis, and both forming a 45° dihedral angle with the xy-plane.

15.
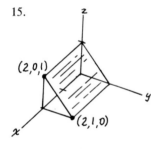

17.
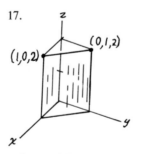

19.
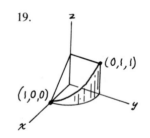

21. The intersection of the cylinders of radius 4 with axes the x- and z-axes. The curve lies in the plane $x = z$.

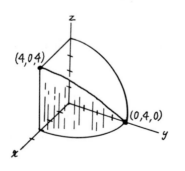

23. The intersection of the ellipsoid with semiaxes 4, 4, and 8, the first octant, and the region under the plane $y + z = 4$.

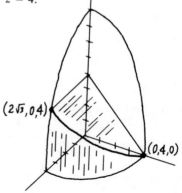

Section 2

1. $\dfrac{\partial z}{\partial y} = \lim\limits_{\Delta y \to 0} \dfrac{f(x,y + \Delta y) - f(x,y)}{\Delta y}$

3. $\dfrac{\partial z}{\partial x} = 6x - 2y;\ \dfrac{\partial z}{\partial y} = -2x + 1$

5. $\dfrac{\partial z}{\partial x} = ae^{ax+by};\ \dfrac{\partial z}{\partial y} = bc^{ax+by}$

7. $\dfrac{\partial z}{\partial x} = 2\cos(2x + 3y);\ \dfrac{\partial z}{\partial y} = 3\cos(2x + 3y)$

9. $\dfrac{\partial z}{\partial x} = ye^{-x^2/2} - x^2 y e^{-x^2/2};\ \dfrac{\partial z}{\partial y} = xe^{-x^2/2} = (1 - x^2)ye^{-x^2/2}$

11. $\dfrac{\partial z}{\partial x} = \dfrac{y}{x};\ \dfrac{\partial z}{\partial y} = \log(xy) + 1$

13. $\dfrac{\partial z}{\partial x} = a\cos ax \cos by;\ \dfrac{\partial z}{\partial y} = -b\sin ax \sin by$

15. $\dfrac{\partial z}{\partial x} = \dfrac{y}{x^2 + y_2};\ \dfrac{\partial z}{\partial y} = \dfrac{-x}{x^2 + y_2}$

17. $\dfrac{\partial z}{\partial x} = \dfrac{-2y}{(x - y)^2};\ \dfrac{\partial z}{\partial y} = \dfrac{2x}{(x - y)^2}$

19. $D_1 f(s,t) = (1 + st)e^{st}, D_2 f(s,t) = s^2 e^{st};\ D_1(2,3) = 7e^6, D_2(-2,1) = 4e^{-2}$

21. $D_1 f(x,y) = y^2 z^3, D_2 f(x,y) = 2xyz^3, D_3 f(x,y) = 3xy^2 z^2;\ -32;\ 0;\ 48.$

23. $2a \sin b;\ a^2$ 25. 4   27. -1

29. The boundary consists of just one point, the origin.

31. One example is
$$\log\left(\dfrac{1}{x^2 + y^2} - 1\right).$$
Find another (possibly by modifying this one).

Section 3

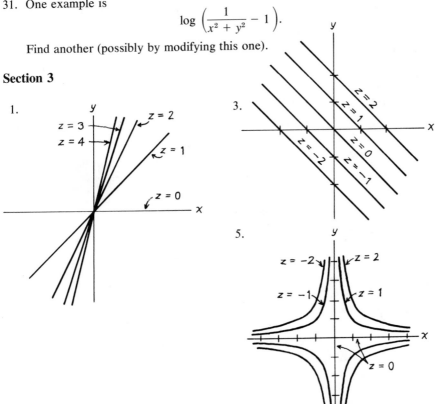

1.

3.

5.

7.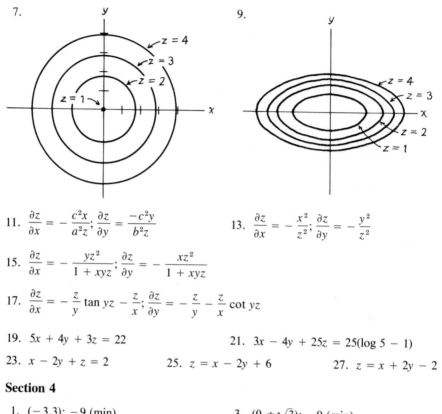

9.

11. $\dfrac{\partial z}{\partial x} = -\dfrac{c^2 x}{a^2 z}; \dfrac{\partial z}{\partial y} = \dfrac{-c^2 y}{b^2 z}$

13. $\dfrac{\partial z}{\partial x} = -\dfrac{x^2}{z^2}; \dfrac{\partial z}{\partial y} = -\dfrac{y^2}{z^2}$

15. $\dfrac{\partial z}{\partial x} = -\dfrac{yz^2}{1+xyz}; \dfrac{\partial z}{\partial y} = -\dfrac{xz^2}{1+xyz}$

17. $\dfrac{\partial z}{\partial x} = -\dfrac{z}{y}\tan yz - \dfrac{z}{x}; \dfrac{\partial z}{\partial y} = -\dfrac{z}{y} - \dfrac{z}{x}\cot yz$

19. $5x + 4y + 3z = 22$

21. $3x - 4y + 25z = 25(\log 5 - 1)$

23. $x - 2y + z = 2$

25. $z = x - 2y + 6$

27. $z = x + 2y - 2$

Section 4

1. $(-3,3); -9$ (min)

3. $(0, \pm\sqrt{3}); -9$ (min)

5. $\pm(\sqrt{\tfrac{2}{3}}, \sqrt{\tfrac{2}{3}}); -4/27$ (min)

7. $(\tfrac{1}{2}, \tfrac{1}{4}); -1/8$ (min)

9. $\tfrac{3}{8}\sqrt{2}$

11. $4, -\tfrac{1}{3}$

13. $2/\sqrt{5}$

15. $x = 6, y = 2, z = 3$

Extra Problems for Chapter 15

3. $x^2 + y^2 = -(z + 4)$. Paraboloid of revolution about z-axis, opening down, with vertex at $(0,0,-4)$.

5. $x^2 + (y + 1)^2 = z + 1$. Paraboloid of revolution about line $x = 0$, $y = -1$, opening up, with vertex at $(0,-1,-1)$.

7. $x^2 + 2(y + 1)^2 + (z + 1)^2 = 3$. Ellipsoid of revolution centered at $(0,-1,-1)$ about the axis parallel to the y-axis through $(0,-1,-1)$.

9. $(x + 2)^2 + 2(y - 1)^2 - z^2 = -1$. Hyperboloid of two sheets, centered at $(-2,1,0)$.

11. $-x^2 + 2(y - 1)^2 + (z + 2)^2 = 6$. Hyperboloid of one sheet, centered at $(0,1,-2)$.

13. $z = 0, y = \pm\dfrac{b}{a}x$

15. $(2,\sqrt{2},\sqrt{2}), (2,-\sqrt{2},-\sqrt{2})$

17. $32\sqrt{3}$

19. $x_i = a_i / \sqrt{\sum_1^n a_i^2}, i = 1, \cdots, n$

CHAPTER 16

Section 1

1. $32t$

3. $-\sec t$

5. $2(e^{2t} - e^{-2t})/(e^{2t} + e^{-2t})$

7. $\dfrac{\partial z}{\partial u} = 4u$; $\dfrac{\partial z}{\partial v} = 4v$

9. $\dfrac{\partial z}{\partial u} = 2[v^3 + 5v^2u + 3vu^2]e^{(u^2+2uv)(2uv+v^2)}$;

$\dfrac{\partial z}{\partial v} = 2[u^3 + 5u^2v + 3uv^2]e^{(u^2+2uv)(2uv+v^2)}$

11. $\dfrac{\partial z}{\partial x} = f'\left(\dfrac{y}{x}\right)\left(\dfrac{-y}{x^2}\right)$; $\dfrac{\partial z}{\partial y} = f'\left(\dfrac{y}{x}\right)\left(\dfrac{1}{x}\right)$

Section 2

1. If the functions $x = g(t)$, $y = h(t)$, and $z = k(t)$ are differentiable at $t = t_0$, and if $f(x,y,z)$ is smooth (continuously differentiable) inside a small sphere about $(x_0,y_0,z_0) = (g(t_0),h(t_0),k(t_0))$, then $F(t) = f(g(t),h(t),k(t))$ is a differentiable function of t at $t = t_0$, and
$$F'(t_0) = D_1f(x_0,y_0,z_0)g'(t_0) + D_2f(x_0,y_0,z_0)h'(t_0) + D_3f(x_0,y_0,z_0)k'(t_0).$$

3. If w is a smooth function of x_1, x_2, x_3, x_4 which in turn are differentiable functions of t, then w is a differentiable function of t and
$$\frac{dw}{dt} = \frac{\partial w}{\partial x_1}\frac{dx_1}{dt} + \frac{\partial w}{\partial x_2}\frac{dx_2}{dt} + \frac{\partial w}{\partial x_3}\frac{dx_3}{dt} + \frac{\partial w}{\partial x_4}\frac{dx^4}{dt}.$$

Section 3

1. Yes, everywhere

3. Everywhere but $(-3/4^{1/3}, 2^{1/3})$

5. Everywhere but $(0,0)$

Section 4

1. $(6,8)$

3. $(-\pi/12, -1/2)$

5. $(8,4)$

7. $(2,-1)$

9.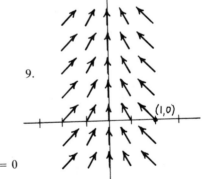

11. $(\sqrt{2}, -1)$

13. $(1,\sqrt{3})$

15. $2x + y = 4$

17. $3x + 4y = 25$

19. $3x + y = 5$

21. $-x + 2y = 4$

23. $(5/8 + \pi/4)(x - 2) + (5/2 + \pi/4)(y - 1/2) = 0$

25. $\operatorname{grad} f = (2x, 2y)$, $\operatorname{grad} g = (-y/x^2, 1/x)$. So $\operatorname{grad} f \cdot \operatorname{grad} g = 0$.

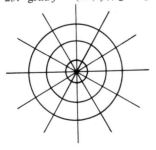

27. The vector $\mathbf{x}_0$ is perpendicular to the level curve at the point $\mathbf{x}_0$, by Example 5 of Section 1 of Chapter 14. So is grad $f(\mathbf{x}_0)$.

Section 5

1. Set $\mathbf{u} = \mathbf{a}/|\mathbf{a}|$; then apply Theorem 7.

3. $\sqrt{3}/2 + 1$ 5. $2\sqrt{10}/75$ 7. $e\sqrt{5}$

9. $54/169$ 11. 2 13. $(11\sqrt{2}; (1/\sqrt{2})(-1,-1)$

15. $32\sqrt{5}/15; (1/\sqrt{5})(1,-2)$ 17. $\sqrt{10}/2; (1/\sqrt{10})(3,-1)$

19. $2/\sqrt{5}; (1/\sqrt{5})(1,-2)$

21. Let $\mathbf{a} = \dfrac{d\mathbf{x}}{dt}$;

$$\frac{dz}{dt} = \text{grad } f \cdot \frac{d\mathbf{x}}{dt} = \frac{\text{grad } f \cdot \mathbf{a}}{|\mathbf{a}|} \, |\mathbf{a}| = (D_{a/|a|}f)|\mathbf{a}|.$$

23. grad $f(\mathbf{x}_0) = (4,9/2)$ 25. $u + v$ 27. $bu + av$

Section 6

1. $(-1,0,0)$ 3. $(2,3,4)$ 5. $(6/25,8/25,4)$

7. $19/3$ 9. $-5/18$ 11. $(3,1,-1)$

13. $\mathbf{u} = \left(\dfrac{-1}{\sqrt{117}}, \dfrac{-10}{\sqrt{117}}, \dfrac{4}{\sqrt{117}} \right); \sqrt{117}; -\mathbf{u}$

Section 7

1. $(-1,3,-3)$ 3. $(-1,0,1)$

5. a) $(2,3,1/6)$ if $(1/2,1/3,6)$ 7. $\mathbf{x} = (1 + et, \pi/2, e - t)$
   b) $(2,3,1/6)$

9. $\mathbf{x} = (2 + 2t, 3 + t, -1 + 5t)$ 11. $-x + 3y + 3z = 2$

13. $x = z$ 15. $x - y - z = 1$

17. Choose any positive r less than min $(1/|a|^3, 1/|b|^3)$. Show that if $x^2 + y^2 < r^2$, then, in particular, $x^2 < 1/a^{2/3}$ and $|x|^{1/3}|a| < 1$. Similarly, $|y|^{1/3}|b| < 1$. Show that, therefore, $|ax + by| < x^{2/3} + y^{2/3}$. Thus, within the circle of radius r, $ax + by > -(x^{2/3} + y^{2/3})$.

Section 8

1. $66/25$ 3. $9/4; 0$

5. The perpendicularity condition in Problem 23 of Chapter 14, Section 1, can be written

$$(x_1 - x_2, y_1 - y_2) = \lambda(2x_1/4, 2y_1) = \mu(-1,1).$$

The last equation requires

$$x_1 = -4y_1,$$

and the ellipse equation then gives

$$(x_1, y_1) = \pm \left(\frac{4}{\sqrt{5}}, \frac{-1}{\sqrt{5}} \right).$$

These are the points that maximize and minimize the distance to the line. A quick sketch shows the closer to be

$$\left(-\frac{4}{\sqrt{5}}, \frac{1}{\sqrt{5}}\right)$$

7. $(x^2/2) + (y^2/32) = 1$

9. Corner point $(2/\sqrt{6}, 1/\sqrt{6})$; perimeter $2\sqrt{6}$

Section 10

1. $\dfrac{\partial^2 z}{\partial x^2} = 6; \ \dfrac{\partial^2 z}{\partial x \partial y} = -2 = \dfrac{\partial^2 z}{\partial y \partial x}; \ \dfrac{\partial^2 z}{\partial y^2} = 0$

3. $\dfrac{\partial^2 z}{\partial x^2} = (4x^2 y + 2y)e^{x^2 y}; \ \dfrac{\partial^2 z}{\partial x \partial y} = (2x^3 y + 2x)e^{x^2 y} = \dfrac{\partial^2 z}{\partial y \partial x}; \ \dfrac{\partial^2 z}{\partial y^2} = x^4 e^{x^2 y}$

5. $\dfrac{\partial^2 z}{\partial x^2} = 2; \ \dfrac{\partial^2 z}{\partial x \partial y} = 3 = \dfrac{\partial^2 z}{\partial y \partial x}; \ \dfrac{\partial^2 z}{\partial y^2} = 2$ 7. $\dfrac{\partial^2 z}{\partial x^2} = 0 = \dfrac{\partial^2 z}{\partial y^2}; \ \dfrac{\partial^2 z}{\partial x \partial y} = \dfrac{\partial^2 z}{\partial y \partial x} = 1$

9. $\dfrac{\partial^2 z}{\partial x^2} = -\dfrac{6y}{x^4}; \ \dfrac{\partial^2 z}{\partial x \partial y} = \dfrac{2}{x^3} - \dfrac{2}{y^3} = \dfrac{\partial^2 z}{\partial y \partial x}; \ \dfrac{\partial^2 z}{\partial y^2} = \dfrac{6x}{y^4}$

11. Set $g(x) = x - ct$ (t fixed), $h(t) = x - ct$ (x fixed). Then:

$$\frac{\partial z}{\partial x} = f'(g(x))g'(x) = f'(g(x))$$

$$\frac{\partial^2 t}{\partial x^2} = f''(g(x))g'(x) = f''(g(x))$$

$$\frac{\partial z}{\partial t} = f'(h(t))h'(t) = -cf'(h(t))$$

$$\frac{\partial^2 z}{\partial t^2} = -cf''(h(t))h'(t) = c^2 f''(h(t))$$

$$c^2 \frac{\partial^2 z}{\partial x^2} = \frac{\partial^2 z}{\partial t^2}$$

13. $\dfrac{\partial^2 z}{\partial y^2} = \dfrac{2x^2 - 2y^2}{(x^2 + y^2)^2}; \ \dfrac{\partial^2 z}{\partial x^2} = \dfrac{2y^2 - 2x^2}{(x^2 + y^2)^2}$

15. $\dfrac{\partial z}{\partial y} = \dfrac{\partial^2 z}{\partial x^2} = (x^2 - 2y)e^{-x^2/4y}/4y^2\sqrt{y}$

Section 11

1. Rel. min. at $(-1, -1/4)$

3. Saddle point at $(0,0)$; rel. min. at $(1/6, 1/12)$

5. Saddle point at $(0,0)$ 7. Saddle point at $(1, -1/2)$

9. Saddle point at $(1/3, -1/3)$; rel. min. at $(3/4, 1/2)$

11. Rel. min. at $(-1,1)$ 13. Rel. min. at $(1,1)$ and $(-1, -1)$

15. Saddle point at $(0,0)$

Extra Problems for Chapter 16

29. $1/30$ 31. $16/\sqrt{3}$ 33. $\pm(98)^{3/4}/2$

CHAPTER 17

Section 1

1. 1/4   3. 1   5. 0   7. 0   9. 3/20

11. 1   13. 1

15. 5/12   17. 1/3   19. $\frac{4}{3}\pi abc$

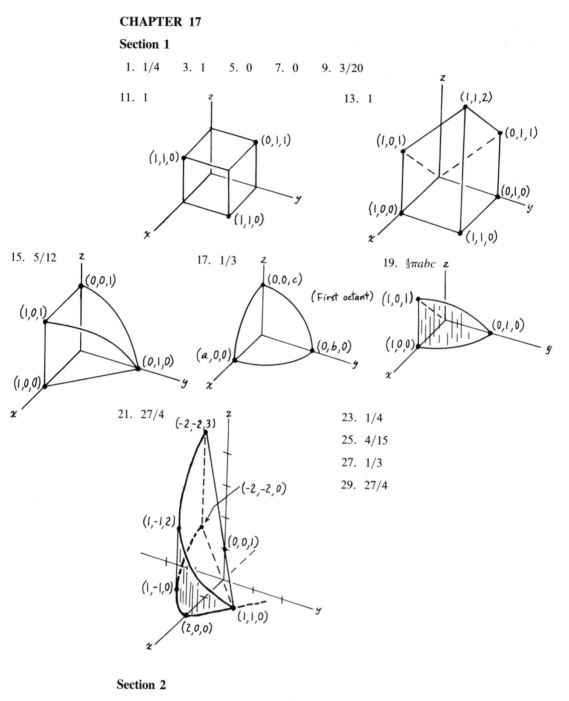

21. 27/4   23. 1/4

                                             25. 4/15

                                             27. 1/3

                                             29. 27/4

Section 2

1. 469/1800

7. $\int_0^1 \int_y^1 f(x,y)\,dx\,dy$ 9. $\int_0^1 \int_y^{\sqrt{y}} \cdots dx\,dy$

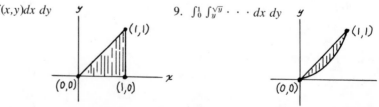

11. $\int_0^1 \int_{\sqrt{x}}^1 \cdots dy\, dx + \int_0^1 \int_{-1}^{\sqrt{x}} \cdots dy\, dx$ 13. $1/2 - 1/2e$

15. $1/3$

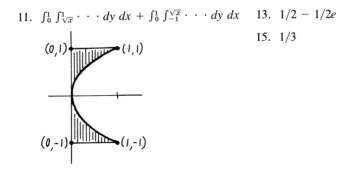

Section 3

1. Take a curve $H(s)$ from $(x_1 y_1)$ to (x_2, y_2) that stays in G. We may assume that H is parametrized by arc-length s, so that $H(0) = (x_1, y_1)$, $H(L) = (x_2, y_2)$, where L is the length of the path. Note $|dH/ds| = 1$. Use the dot-product form of the chain rule to show that

$$\left| \frac{d}{ds} f(H(s)) \right| = |\text{grad } f| \le K\sqrt{2}.$$

Show, therefore, that $|f(H(L)) - f(H(0))| \le K\sqrt{2}|L - 0|$, etc.

3. Let $M_i = f(a_i, b_i)$ be the maximum value of f in ΔG_i. Let $m_i = f(c_i, d_i)$ be the minimum value of f in ΔG_i. Now, ΔG_i has length less than e, so by (1),

$$|M_i - m_i| \le \sqrt{2}\, Ke.$$

Now reason exactly as in the text.

Section 4

1. $1/4$ 3. $4/35$ 5. $8k$ 7. $4k/3$

9. $8ka^3/3$ 11. πk 13. 1

Section 5

1. $(2/3, 2/3)$ 3. $(5/9, 5/11)$ 5. $(0, 4/3)$ 7. $(3/8, 0)$

9. $(0,0)$ 11. $((\pi^2 - 4)/\pi, \pi/8)$ 13. $(2, 1/8)$

Section 6

1. $122/3\pi$ 3. $\pi/3$ 5. $4\pi/3$ 7. $\pi/3 - 4/9$

9. $\int_0^{2\pi} \int_0^R (kr)r\, dr\, d\theta = 2\pi kR^3/3$

11. Mass $k\pi/12$; center of mass $(0, 6/5\pi)$ (rectangular coordinates)

13. $16a^3/3$

Section 7

1. $(\pi/6)(17\sqrt{17} - 1)$ 3. 20π 5. $\pi - 2$

7. $3\sqrt{14}$ 9. $\pi a\sqrt{a^2 + h^2}$

Section 8

1. 6 3. $abc/6$ 5. $18/35$ 7. $1/3$
9. $16/3$ 11. $64/15$ 13. $8/3$ 15. $(3/8)(a, b, c)$

Section 9

1. 8π

3. $\dfrac{16\pi}{3}\left(1 - \dfrac{\sqrt{3}}{2}\right) \approx \dfrac{4\pi}{5}$

5. $\frac{2}{3}\pi a^3(1 - \cos\alpha)$

7. $\frac{2}{3}\pi[6^{3/2} - 11]$

9. $2\pi a^2 k$, where k is the constant of proportionality.

11. $\frac{1}{2}\pi a^4 k$, where a is the radius, and k is the proportionality constant.

13. $\frac{1}{3}\pi a^3 b^2$

15. $\bar{x} = \bar{y} = 0; \bar{z} = 3/(16(1 - \sqrt{3}/2)) \approx 5/4$

17. $(3a/8)(1,1,1)$

Section 10

3. If Σ_1 and Σ_2 are Riemann sums for f over G_1 and G_2, respectively, then $\Sigma = \Sigma_1 + \Sigma_2$ is a Riemann sum for f over G. Each of these Riemann sums approaches the corresponding triple integral as the maximum subregion diameter approaches zero.

Extra Problems for Chapter 17

1. $\dfrac{2}{3}ab^3$

3. $\dfrac{72}{35}$

5. $4\pi\left(\dfrac{5\sqrt{2}}{12} - \dfrac{1}{3}\right)$

7. a) 1/4   b) 1/4   9. a) 2/3   b) 5/6   11. 3/4

13. a) $4 \cdot 1 + 8 \cdot 2 + 4 \cdot 4 = 36$
    b) $(1/36)(19/9)^2 = (19/18)^2(1/9)$
    c) 1/9

15. $(12/15,0)$

17. $(5a/6,0)$

27. $4\pi^2 ab$

29. $M_x = M_y = \dfrac{b^4}{8}\left(\dfrac{\pi}{2} - \arcsin\left(\dfrac{a}{b}\right)\right) + \dfrac{a}{4}\sqrt{b^2 - a^2}\left(\dfrac{b^2}{2} - a^2\right)$

$M_z = \dfrac{\pi}{16}(b^2 - a^2)^2$

31. Mass $= \dfrac{1}{6}\pi(b^3 - a^3)$

33. $\dfrac{49}{24} - \left[\dfrac{17^{3/2}}{24} - \dfrac{2^{3/2}}{3}\right]; \dfrac{49}{24} + \left[\dfrac{17^{3/2}}{24} - \dfrac{2^{3/2}}{3}\right]$ 35. $2\pi r^2$

CHAPTER 18

Section 1

1. 30

3. $2 \log 2$

5. $\dfrac{5}{2}A^2 = \dfrac{3}{2}V^2_{\max}$

7. $A = 5\sqrt{2}$

9. $16/(2 + m)^2$

11. 1,248,000 ft-lbs.

13. $62.4\ Vh - W$

17. $k \log(V_1/V_0)$ if $\alpha = 1$;

$\dfrac{k}{\alpha - 1}\ (V_1^{\alpha-1} - V_0^{\alpha-1})$ if $\alpha \neq 1$

Section 2

1. 2π 3. $-26/3$ 5. -10

7. a) 0   b) 18   11. 11

13. $1/r_0 - 1/r_1$, for all paths   15. $3/2$ 17. 18

Section 3

1. $\phi(x,y) = \frac{1}{2}(x^2 + y^2 - x_0^2 - y_0^2)$ 3. $\phi(x,y) = x^2y - x_0^2y_0$

5. $\phi(x,y) = x^2y + xy^2 - x_0y_0^2 - x_0^2y_0$ 7. $\phi(x,y) = \frac{1}{2}x^2y^2 + C$

9. $\phi(x,y) = x^2y + xy^2 + C$ 11. $\phi(x,y,z) = zx + y^2/2 + C$

13. $\phi(x,y,z) = 1/4r^4$

15. 7   17. 0   19. 0   21. $5^{3/2}$ 23. 2π

Section 4

1. 1   3. $-1/2$ 5. 0   7. πab

9. πab 11. $1/6$

13. -1; $-$(area of G), for any regular G 15. 2π; 2(area of G)

17. 1; area of G for any regular G not intersecting any line $x = \dfrac{\pi}{2} + k\pi$

19. 0 (for any G)   21. -2π; -2(area of G)

CHAPTER 19

Section 1

1. $y = Ce^{2x}$ 3. $y = Ce^{kx}$

5. $y = \log(2/(C - x^2))$ 7. $y = \sin(x + C)$

9. $y^2 - 2x^2 = C$ 11. $y = x^2/(Cx^2 + x - 2)$

15. $y = -\arctan(\log cx)$ or $y = \pi/2 + n\pi$ for fixed n

17. $P = \frac{1}{4}k^2t^2 + kP_0^{1/2}t + P_0$

19. $\log\left(\dfrac{c}{c - y}\right) = y + kt$

21. $T = 60 + 150e^{-0.6} \approx 142.35$

Section 2

1. $dy = dx$

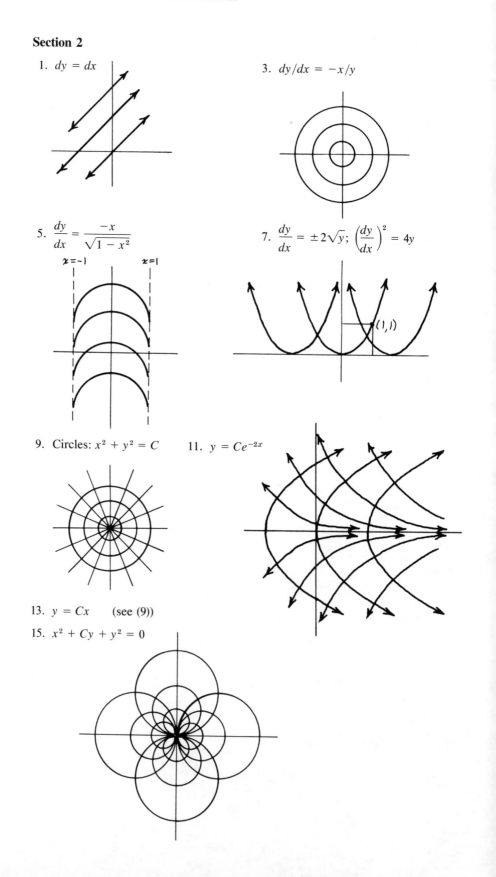

3. $dy/dx = -x/y$

5. $\dfrac{dy}{dx} = \dfrac{-x}{\sqrt{1-x^2}}$

$x = -1$ $x = 1$

7. $\dfrac{dy}{dx} = \pm 2\sqrt{y}$; $\left(\dfrac{dy}{dx}\right)^2 = 4y$

$(1,1)$

9. Circles: $x^2 + y^2 = C$

11. $y = Ce^{-2x}$

13. $y = Cx$ (see (9))

15. $x^2 + Cy + y^2 = 0$

Section 3

1.

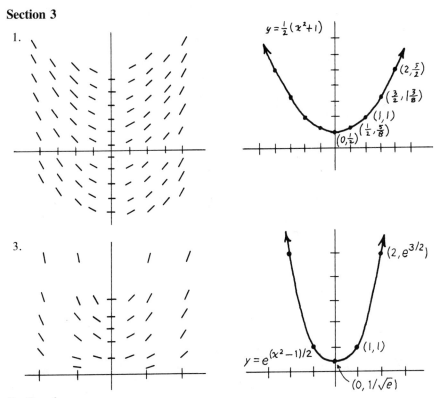

$$y = \tfrac{1}{2}(x^2+1)$$

$(2,\tfrac{5}{2})$

$(\tfrac{3}{2},\tfrac{13}{8})$

$(1,1)$

$(0,\tfrac{1}{2})$ $(\tfrac{1}{2},\tfrac{5}{8})$

3.

$(2,e^{3/2})$

$(1,1)$

$y = e^{(x^2-1)/2}$

$(0, 1/\sqrt{e})$

Section 4

1. $y = -e^x + Ce^{+2x}$

3. $y = e^x/(1 + a) + Ce^{-ax}$, for $a \neq -1$; $y = xe^x + Ce^x$ for $a = -1$

5. $y = x - 1 + Ce^{-x}$ 7. $y = x^2/3 + C/x$ 9. $y = 1 + C/x$

11. $y = (1/(1 + a^2))(a \sin x - \cos x) + Ce^{-ax}$

13. $y = (1/2)(x^2 - 1) + Ce^{-x^2}$ 15. $y = \tan x + C \sec x$

Section 5

1. $y = -e^x + Ce^{2x}$ 3. $y = x - 1 + Ce^{-x}$

5. $y = (1/a) + Ce^{-ax}(a \neq 0)$; $y = x + C(a = 0)$.

7. $y = (1/3)xe^x - (1/9)e^x + Ce^{-2x}$

9. $y = (1/2)(x \sin x = x \cos x + \cos x) + Ce^{-x}$

11. $\dfrac{d}{dx}(e^{ax}y) = e^{ax}r(x) \Leftrightarrow e^{ax}\dfrac{dy}{dx} + ae^{ax}y = e^{ax}r(x)$ (product formula)

13. Use the derivative product rule to expand $L(xy)$.

15. $y = xe^{-4x} + Ce^{-4x}$

Section 6

1. $y = x$ 3. $y = -\tfrac{1}{3}e^x$

5. $y = \tfrac{1}{4}\cos x$ 7. $y = \tfrac{1}{3}x^3 - x^2 + 2x$

9. $y = \frac{1}{3}e^x + \frac{2}{3}e^{-x}$

11. $y = \frac{1}{2}(xe^x - e^x)$

13. $y = -\frac{1}{2}e^{-x}$

15. $y = -\frac{3}{10}\sin x - \frac{1}{10}\cos x$

17. $y = \frac{1}{2}x^2 - x - \frac{1}{2}\sin x - \frac{1}{2}\cos x$

19. $y = \frac{1}{12}x^4 e^x$

21. $y = -\frac{1}{4}e^x(\sin x + \cos x)$

23. $y = e^x \sin x$

Section 7

1. $y = c_1 \sin \sqrt{2} + c_2 \cos \sqrt{2}x$

3. $y = \frac{1}{2}(x^2 - 1) + c_1 \sin \sqrt{2}x + c_2 \cos \sqrt{2}x$

5. $y = -\frac{1}{4}x \cos 2x + c_1 \sin 2x + c_2 \cos 2x = c_1 \sin 2x + (c_2 - (x/4))\cos 2x$

7. $y = \frac{1}{2}xe^x + c_1 e^x + c_2 e^{-x} = (c_1 + (x/2))e^x + c_2 e^{-x}$

9. $y = -\frac{1}{5}e^x \sin x - \frac{2}{5}e^x \cos x + c_1 e^x + c_2 e^{-x}$

11. $y = c_1 e^x + c_2 e^{-2x}$

13. $-\frac{3}{10}\sin x - \frac{1}{10}\cos x + c_1 e^x + c_2 e^{-2x}$

15. $y = +c_1 e^x + c_2 xe^x$

17. $y = \frac{1}{2}\cos x + c_1 e^x + c_2 xe^x$

19. $y = c_1 e^{-x} \sin x + c_2 e^{-x} \cos x$

21. $y = \frac{1}{5}e^x + c_1 e^{-x} \sin x + c_2 e^{-x} \cos x$

Section 8

1. For $m^2 > 4ms$, $m\ddot{x} + r\dot{x} + sx = 0$ has solutions $x = c_1 e^{-kt} + c_2 e^{-lt}$, where
$$k = \frac{r + \sqrt{r^2 - 4ms}}{2m} \quad \text{and} \quad l = \frac{r - \sqrt{r^2 - 4ms}}{2m}, \quad (0 < l < k).$$
$0 = \dot{x}(0) = -kc_1 e^{-k\cdot 0} - lc_2 e^{-l\cdot 0}$, so $c_2 = (-k/l)c_1$;
$0 < x(0) = c_1 e^{-k\cdot 0} - (k/l)c_1 e^{-l\cdot 0} = (1 - (k/l))c_1$; so $c_1 < 0$.
Let $c = -c_1/l$; then $c > 0$ and $x = c[ke^{-lt} - le^{-kt}]$.

3. Similarly, as in Problem 1, $x = c_1 e^{-kt} + c_2 e^{-lt}$ with $0 < l < k$.
$0 = x(0) = c_1 e^{-k\cdot 0} + c_2 e^{-l\cdot 0} = c_1 + c_2$, so $c_1 = -c_2$ and $x = c_2[e^{-lt} - e^{-kt}]$.
$0 < \dot{x}(0) = c_2[-le^{-l\cdot 0} + ke^{-k\cdot 0}] = c_2(k - l)$, so $c_2 = c > 0$.

5. As in (1) and (3), $x = c_1 e^{-kt} + c_2 e^{-lt}$, $0 < l < k$. $1 = x(0) = c_1 + c_2$, so $c_1 = 1 - c_2$. Let $a = c_2$; then $x = ae^{-lt} + (1 - a)e^{-kt}$.
$\dot{x}(0) = a(-l)e^{-l\cdot 0} + (1 - a)(-k)e^{-k\cdot 0} = a(k - l) - k$. So $a < 0$ implies
$\dot{x}(0) = v_0 = a(k - l) - k < -k$, and $v_0 = a(k - l) - k < -k$ implies $a < 0$.

7. $x = c_1 e^{-kt} + c_2 te^{-kt}$; $x_0 = 1 = c_1 + (c_2)(0) = c_1$, so $x = e^{-kt} + c_2 te^{-kt}$.
$0 = v_0 = \dot{x}(0) = -ke^{-k\cdot 0} + c_2(-k)(0)e^{-k(0)} + c_2 e^{-k\cdot 0} = -k + c_2$, so $c_2 = k$ and
$x = (1 + kt)e^{-kt}$.

9. This is a result of Theorem 9 (Section 7).

13. Differentiate (*) to get (**).

Section 9

1. $c_1 = 0$, $c_n = (n + 2)c_{n+2}$; converges for all x;
$$y = c_0 \left(1 + \frac{x^2}{2} + \frac{x^4}{2 \cdot 4} + \frac{x^6}{2 \cdot 4 \cdot 6} + \cdots + \frac{x^{2n}}{2^n \cdot n!} + \cdots\right)$$

3. $c_2 = 0$, $2c_2 = 1$, $(n + 3)c_{n+3} = c_n$;

$$y = c_0 \left(1 + \frac{x^3}{3} + \frac{x^6}{3 \cdot 6} + \cdots + \frac{x^{3n}}{3^n n!} + \cdots \right)$$

$$+ \left(\frac{x^2}{2} + \frac{x^5}{2 \cdot 5} + \frac{x^8}{2 \cdot 5 \cdot 8} + \cdots + \frac{x^{3n-1}}{2 \cdot 5 \cdot 8 \cdots (3n - 1)} + \cdots \right)$$

$$= c_0 f(x) = g(x).$$

Each series converges for all x.

5. c_0, $(n + 1)c_{n+1} = nc_{n-1}$;

$$y = c_1 \left(1 + \frac{x^2}{2} + \frac{3}{2 \cdot 4} x^4 + \frac{3 \cdot 5}{2 \cdot 4 \cdot 6} x^6 + \cdots \right);$$ converges on $(-1,1)$.

7. $c_3 = c_2 = 0$, $(n + 2)(n + 1)c_{n+2} = c_{n-2}$;

$$y = c_0 \left[1 + \frac{x^4}{4 \cdot 3} + \frac{x^8}{8 \cdot 7 \cdot 4 \cdot 3} + \cdots \right]$$

$$+ c_1 \left[x + \frac{x^5}{5 \cdot 4} + \frac{x^9}{9 \cdot 8 \cdot 5 \cdot 4} + \cdots \right].$$

Each series converges for all x.

9. $(n + 2)(n + 1)c_{n+2} = [n(n - 1) + 1]c_n$;

$$y = c_0 \left[1 + \frac{x^2}{2} + \frac{x^4}{8} + \cdots \right] + c_1 \left[x + \frac{x^3}{6} + \frac{7x^5}{120} + \cdots \right].$$

Converges on $(-1,1)$, by the ratio test.

Extra Problems for Chapter 19

1. $y = -2/(x^2 + C)$

3. $y = C\sqrt{1 - x^2} - \sqrt{1 - C^2}x$, or $y = \pm 1$

5. $y = x/(C - \log|x|)$, or $y = 0$

7. $y = \sqrt{x^2 - Cx}$

9. $y^2 = Cx$

11. $x = Cae^{at}/(b + bCe^{at})$

13. $50e^{-1.2} \approx 15.06$ lb.

15. $z = \dfrac{x_0 Ce^{(2y_0 - x_0)kt} - y_0}{\frac{2}{3}Ce^{(2y_0 - x_0)kt} - \frac{1}{3}}$, $\quad C > 0$

17. $t = 100(\sqrt{15} - \sqrt{10}) + \dfrac{400}{\pi} \log \dfrac{(\pi\sqrt{15}/4 - 1)}{(\pi\sqrt{10}/4 - 1)} \approx 112$ sec.

19. $xy = C$

21. $y = Cx$

23. $y = C|x|^{a^2/b^2}$

25. $y = \frac{1}{2}e^x + 1 + Ce^{-x}$

27. $y = -x - 1 + xe^x + Ce^x$

29. a) $3x^3$ b) x c) 0   d) xe^x

31. Use the derivative product rule to expand $L(u,v)$.

33. $y = \frac{1}{2}x^2 - x - \frac{1}{2}\sin x - \frac{1}{2}\cos x$

35. $y = \frac{1}{12}x^4 e^x$

37. $y = -\frac{1}{4}e^x(\sin x + \cos x)$

39. $y = e^x \sin x$

41. $(D - k_1)[(D - k_2)f] = (D - k_1)[f' - k_2 f]$, etc.

43. $y = c_1 e^{kx} + c_2 x e^{kx}$

45. $y = \frac{1}{3}xe^{2x} + c_1 e^{-x} + c_2 e^{2x}$

49. The coefficients c_0 and c_1 are arbitrary, and $c_2 = -c_0/2$, $c_3 = -(c_0 + c_1)/6$.
After that, the recursion formula

$$c_{n+2} = \frac{c_{n-1} + c_n}{(n + 2)(n + 1)}$$

determines the coefficients. There is no obvious pattern, but it can be shown that if k is the larger of $|c_0|$ and $|c_1|$, then

$$|c_n| < \frac{k}{\sqrt{n!}},$$

as follows. It is true by inspection for c_0, c_1, c_2 and c_3. Suppose that it is true up through an integer m. Then

$$|c_{m+1}| = \frac{|c_{m-2} + c_{m-1}|}{(m+1)m} \leq \frac{k}{(m+1)m} \left[\frac{1}{\sqrt{(m-2)!}} + \frac{1}{\sqrt{(m-1)!}} \right]$$

$$= \frac{k}{\sqrt{(m+1)!}} \left[\frac{\sqrt{m-1}+1}{\sqrt{(m+1)m}} \right] < \frac{k}{\sqrt{(m+1)!}}.$$

So it is true for $m+1$, too. Starting from our original list, the inequality has now been proved for $m+1 = 4$, then (taking $m = 4$) for $m+1 = 5$, and so on forever. This method of proof is called *mathematical induction*.

The solution series is thus dominated term by term by $k\Sigma |x|^n/\sqrt{n!}$ and hence converges for all x.

51. Let $y = c_0 + c_1 x + c_2 x^2 + c_3 x^3 + \cdots$.
Then

$$\begin{aligned} y' &= c_1 + 2c_2 x + 3c_3 x^2 + 4c_4 x^3 + 5c_5 x^4 + \cdots, \\ x^2 y' &= \qquad\quad c_1 x^2 + 2c_2 x^3 + 3c_3 x^4 + \cdots, \\ -2yx &= \quad -2c_0 x - 2c_1 x^2 - 2c_2 x^3 - 2c_3 x^4 + \cdots. \end{aligned}$$

So

$$(1 + x^2)y' - 2xy = 0$$
$$\begin{aligned} &= c_1 + (2c_2 - 2c_0)x + (3c_3 - c_1)x^2 \\ &\quad + (4c_4)x^3 + (5c_5 + c_3)x^4 \\ &\quad + (6c_6 + 2c_4)x^5 + (7c_7 + 3c_5)x^6 + \cdots. \end{aligned}$$

Solving: $c_1 = 0$, $c_2 = c_0$, $c_3 = c_1/3 = 0$, $c_4 = 0$, $c_5 = -c_3/5 = 0$, $c_6 = -2c_4/6 = 0$, etc. So $y = c + cx^2$. (This may be checked by solving the differential equation by separating variables.)

53. Let $y = c_0 + c_1 x + c_2 x^2 + c_3 x^3 + \cdots$.
Then

$$\begin{aligned} xy'' &= \qquad\quad 2c_2 x + 6c_3 x^2 + 12c_4 x^3 + 20c_5 x^4 + \cdots, \\ y' &= c_1 + 2c_2 x + 3c_3 x^2 + 4c_4 x^3 + 5c_5 x^4 + \cdots, \\ xy &= \qquad\quad c_0 x + c_1 x^2 + c_2 x^3 + c_3 x^4 + \cdots. \end{aligned}$$

So

$$xy'' + y' + xy = 0$$
$$\begin{aligned} &= c_1 + (4c_2 + c_0)x + (9c_3 + c_1)x^2 \\ &\quad + (16c_4 + c_2)x^3 + (25c_5 + c_3)x^4 \\ &\quad + (36c_6 + c_4)x^5 + \cdots. \end{aligned}$$

Solving: $c_1 = 0$, $c_2 = -c_0/4$, $c_3 = -c_1/9 = 0$, $c_4 = -c_2/16 = c_0/(16 \cdot 4)$, $c_5 = 0$, etc.

Choose $c_0 = 1$:

$$f(x) = 1 - \frac{x^2}{2^2} + \frac{x^4}{2^2 4^2} - \frac{x^6}{2^2 4^2 6^2} + \frac{x^8}{2^2 4^2 6^2 8^2} - \cdots.$$

General solution $= c \cdot f(x)$.

Interval of convergence:

ratio of consecutive terms $= \dfrac{x^{2n+2}}{2^2 \cdots (2n+2)^2} \cdot \dfrac{2^2 \cdots (2n)^2}{x^{2n}} = \dfrac{x^2}{(2n+2)^2} \to 0.$
So $f(x)$ converges for all x.